INTRODUCTION TO THE

THEORY OF CRITICAL PHENOMENA

Mean Field, Fluctuations and Renormalization

2nd Edition

INTRODUCTION TO THE

THEORY OF CRITICAL PHENOMENA

Mean Field, Fluctuations and Renormalization

2nd Edition

Dimo I. Uzunov

Bulgarian Academy of Sciences, Bulgaria

NEW JERSEY • LONDON • SINGAPORE • BEIJING • SHANGHAI • HONG KONG • TAIPEI • CHENNAI

Published by

World Scientific Publishing Co. Pte. Ltd.
5 Toh Tuck Link, Singapore 596224
USA office: 27 Warren Street, Suite 401-402, Hackensack, NJ 07601
UK office: 57 Shelton Street, Covent Garden, London WC2H 9HE

British Library Cataloguing-in-Publication Data
A catalogue record for this book is available from the British Library.

INTRODUCTION TO THE THEORY OF CRITICAL PHENOMENA (2nd Edition)
Mean Field, Fluctuations and Renormalization

ISBN-13 978-981-4299-48-0
ISBN-10 981-4299-48-0
ISBN-13 978-981-4299-49-7 (pbk)
ISBN-10 981-4299-49-9 (pbk)

Printed in Singapore by World Scientific Printers.

To my wife Diana and my children Ivan, Yulia, and Victor.

Foreword

It is a pleasure for me to present to the public this monograph on phase transitions and critical phenomena. In the past few years, along with our improving understanding of these phenomena, their theoretical description has made great progress evolving into more advanced forms. These also include, apart from new approximations elaborated in the framework of the mean-field theory, various novel approaches such as phenomenological and microscopic treatments of fluctuations, the scaling theory and the renormalization group theory. All of them are thoroughly explained in the present book. The author has managed to cover all these topics with a good balance, proceeding from the thermodynamics of phase transitions to the modern theory of the renormalization group. The way of presentation adopted by the author is very instructive. Detailed derivations of almost all the basic equations are included in the monograph. Thus, even readers who are too lazy or too busy to reproduce such calculations themselves can follow easily the discourse to its very end.

Even the introduction of the renormalization group theory does not diminish the importance of the mean-field theory, as we can see, for example, from the recent development of the coherent-anomaly method. Hence, the meticulous explanation of this theory in the present book seems to be justified even from the viewpoint of modern approaches to critical phenomena. Chapter 6 on fluctuations and Chapter 7 on the perturbation theory of fluctuation fields constitute a good introduction to Chapter 8 on the renormalization group theory. The last chapter contains a brief overview of some important applications of the latter method to various problems including the Bose gas, the classical-to-quantum crossover, the random-field problem, the anisotropy effect and multicritical behavior. This part of the book displays considerable expertise as Dr. Uzunov himself has made a series of interesting contributions in this field.

Masuo Suzuki
Professor of Statistical Physics
University of Tokyo

Preface to the Second Edition

In this second edition, Chapter 5 and Chapter 9 are considerably extended in order to ensure a more comprehensive presentation of the mean-field approximation and applications of the phase transition theory. A number of misprints and non-misleading mistakes in the first edition has been removed. Many new references are added, in particular, in new developing topics, as for instance, unconventional superconductivity, dilute Bose gases, and gauge effects in superconductors.

The rationale for a book on the theory of phase transitions and critical phenomena is still valid today — almost four decades after the advent of the ideas of scaling and universality and their integration in the Wilson–Fisher renormalization group approach. This is so, because the phase transition theory can be successfully applied to solve various problems of Physics and other natural sciences, even in economics and sociology.

Several new calculational variants of renormalization group appeared meanwhile aiming to ensure a more accurate calculation in real spatial dimensions. The oldest renormalization group method, known from the quantum electrodynamics of the fifties of 20th Century, received a vast development, too. But the Wilson-Fisher treatment has still remained most distinctly connected with the main ideas of homogeneity, scaling, and universality. It presents the basis for learning of renormalization group and performing calculations at an ample level of accuracy for the most part of problems of interest. For this reason, the second edition has not been widened with other than the Wilson-Fisher basic approach. For other renormalization schemes and applications of the phase transition theory the interested reader may use excellent books and review articles, cited in the text.

Since 1990–1991, when this book was written, until now many excellent books on phase transitions have been published, for example, by Goldenfeld (1992), Binney *et al.* (1992), Yeomans (1992), Ivanchenko ans Lisyanskii (1995), Zinn–Justin (1996), and Cardy (1996). A number of excellent and large collections of lectures and review articles have been also released (see the bibliography to this book), and the well-known book series of review articles *Phase Transitions and Critical Phenomena*, ed. by C. Domb and M. S. Green (after 1976, ed. by C. Domb and J. L. Lebowitz) was complemented by new volumes, containing both theory and applications.

Notwithstanding the exceptional collection of excellent text books, monographs and review articles, or, perhaps, because of its availability, the present book has its place in the literature on phase transitions and critical phenomena for several reasons. It covers the main stream of development of the theory for the last 100–120 years. As becomes apparent today, about forty years after the last revolutionary period in this theory - the advent of the renormalization group in 1971–1972, the mean field theories and some classic thermodynamic treatments, widely discussed in this book, have their considerable importance in the practical research, and cannot be substituted by the modern renormalization group methods. Further, the book ensures a gradual connection between the usual University courses and the advanced scientific literature. This might be convenient to readers among the talented undergraduates, doctorate students and non-experts. The close interrelationship between experimental and theoretical problems is discussed in details, in particular in Chapters 1 and 9. Outstanding problems are also reviewed, in particular, in Chapter 9. The numerous references, in particular, original papers, some of the review articles and books, are addressed to the attention of experts and doctorate students, involved in research on phase transitions.

The author thanks Professor Dr. Alvaro Ferraz for the hospitality at the International Institute of Physics (Natal, Brazil), where the second edition was accomplished. Besides, the author thanks Associate Professor Dr. Diana V. Shopova (Bulgarian Academy of Sciences) for the critical reading of new texts, included in the book.

D. I. Uzunov
March 2010, Natal, Brazil

Preface to the First Edition

This book is intended as an introduction to the theory of phase transitions and critical phenomena. The phase transitions are described by the methods of thermodynamics and statistical physics. On the contrary modern methods for the investigation of outstanding problems in statistical physics, thermodynamics, condensed matter physics and field theory may be partly ascribed to the development of the ideas and theoretical techniques intended to solve the questions of phase transitions. The sophistication of theoretical methods is often an obstacle to the quick understanding of important results currently reported in scientific literature. The main purpose of this book is to present a part of these methods in an intelligible form for students.

The physics of phase transitions is old and abundant in fascinating ideas, theories and experiments. Most of them are substantial for other fields in physics, chemical physics and even for other natural sciences. All the problems concerning phase transitions and even the main part of them cannot be comprised in only one book. I have chosen such a representation of the phase transition theory that goes along a general scheme for the description of phase transitions and critical phenomena. It includes the following topics: the thermodynamic stability theory (Chapter 2) and its connection with the classical and scaling picture of the phase transitions (Chapter 3); the Landau expansion as a further development of the thermodynamic approach and the notion of order and symmetry in condensed matter physics (Chapter 4); an introduction to statistical mechanics and the mean field approximation for homogeneous and inhomogeneous systems (Chapter 5); the fluctuation phenomena and the renormalization group method in treating the fluctuation fields (Chapters 6 – 8). I have made an attempt to establish a logical relationship between the different theories and their role in the interpretation of experiments. Special attention is paid to the problems which are less illuminated in books and review articles published in the last twenty years. On the contrary, I have only mentioned some important problems like the phenomenological scaling and the dynamic critical phenomena, for which several excellent books and reviews exist. The text contains a number of references to the literature intended for the reader who is interested in specific problems.

In addition to the presentation of basic ideas, a big part of Chapters 2–8 is devoted to the concrete methods of calculation, characterizing the different theories and approximations. This especially concerns Chapter 7, where the perturbation theory of fluctuations is discussed in detail. I suppose that in this way the renormalization group approach will be more easily comprehended both as a concept and as a technique of calculation.

In Chapters 2–6, I often follow the deductive way of presentation, which to my opinion reveals the wide application of the quasi-phenomenological theory to the description of phase transitions.

Chapter 7 and Chapter 8 are written in an opposite manner, since when the perturbation theory and diagrammatic technique are understood for simple models, their generalization to more complex cases can be made without difficulties.

Chapter 1 and Chapter 9 have in some sense a different role. In Chapter 1 the main stages in the development of theoretical and experimental studies on the phase transitions are outlined accompanied by corresponding references to the literature. The unexperienced reader may find this difficult. In this case I recommend reading it after an acquaintance with the contents of the remaining chapters. Chapter 9 presents the application of the mean field and fluctuation theories to concrete problems. In order to keep this survey within a manageable length it was necessary to avoid, as much as possible, problems which are widely discussed in other reviews. My intention has been to throw light on issues, which are not reviewed in the existing literature. Even within this limited scope it was impossible to present within a single chapter a detailed discussion of all developments; many interesting contributions and details of the technique of calculation are given in the numerous references to original papers. The predominant part of Chapter 9 can be thought as a sum of exercises together with a brief introduction to the problems and a short discussion of the main results. I hope that this part of the book may be interesting also to specialists working in the field of phase transitions.

In the Appendices, I give some mathematical details about the methods for calculation of the functional expansions and perturbation integrals.

This book would not have been completed without the substantial support of Prof. G. Scarpetta and the colleagues from the Department of Theoretical Physics at Salerno University and G. Nadjakov Institute of Solid State Physics of the Bulgarian Academy of Sciences in Sofia. I want to thank my colleagues, the stimulating discussions with them have improved my understanding of the problems of the phase transition theory. They are: Y. Brankov, M. Bushev, K. Chao, L. De Cesare, M. E. Fisher, R. Folk, V. L. Ginzburg, B. I. Halperin, J. A. Hertz, H. Iro, I. D. Lawrie, R. Micnas, L. P. Pitaevskii, V. L. Pokrovskii, D. I. Pushkarov, Kh. Pushkarov, J. Sznajd, N. S. Tonchev, M. Suzuki, and M. Zannetti.

The critical reading of parts of the manuscript by Ms. E. J. Blagoeva, Prof. L. De Cesare, Dr. Y. T. Millev, Dr. C. Noce, Dr. I. Rabuffo, and Dr. A. Romano as well as the technical assistance by Dr. A. Saggese and Dr. V. Scarano in the final preparation of the camera-ready version of the book are gratefully acknowledged.

I am grateful for the hospitality of the Salerno University where this book has been written.

D. I. Uzunov
Salerno, January 1992

Contents

Foreword vii

Preface to the Second Edition ix

Preface to the First Edition xi

1. Introduction 1
   - 1.1 Thermodynamic Description of Phase Transitions 1
   - 1.2 Statistical Description of Phase Transitions 3
   - 1.3 Scaling and Critical Exponents 8
   - 1.4 Dynamics 12
   - 1.5 Experiment 14
   - 1.6 Scaling and Homogeneity. Scaling Relations 17
   - 1.7 Length-Scale Invariance and Renormalization 20

2. Equilibrium States and Phases 25
   - 2.1 Phases 25
   - 2.2 Fundamental Relations 26
     - 2.2.1 Fundamental equations 27
     - 2.2.2 Properties of entropy and energy 28
     - 2.2.3 Notations 30
     - 2.2.4 Macroscopic scales 31
   - 2.3 Entropy Increase 33
   - 2.4 Conditions of Equilibrium and Stability 36
     - 2.4.1 Equilibrium conditions 36
     - 2.4.2 Stability conditions 39
   - 2.5 Classification of Equilibria 45

2.6 Thermodynamic Potentials 48
2.6.1 Energy minimum 48
2.6.2 Legendre transformations and extremum conditions 50
2.6.3 Legendre transformations in entropy representation 53
2.6.4 Application 55
2.7 Coexistence of Phases 57
2.7.1 Equilibrium conditions 58
2.7.2 Gibbs phase rule 60
2.7.3 Stability conditions 61
2.7.4 Limitation of the thermodynamic theory of stability 64

3. Thermodynamic Theory of Phase Transitions 67
3.1 Order Parameter 67
3.2 Coexistence of Two Phases 69
3.2.1 Phase diagram 70
3.2.2 Equilibrium isotherms 71
3.2.3 Relations on the coexistence line 73
3.3 Metastability 75
3.4 Critical Point 78
3.5 Classifications of Phase Transitions 82
3.6 Compressible Systems 85
3.6.1 Gas-liquid-solid diagram 86
3.6.2 The solid phase 87
3.6.3 The liquid state 91
3.7 Phase Separation in Mixtures 94
3.7.1 Binary mixtures 95
3.7.2 Stability conditions and critical states in binary mixtures 97
3.7.3 Coexistence surfaces, critical lines, and second Gibbs rule 99
3.7.4 Tricritical and critical-end points. Multicritical phenomena 100
3.8 Other Systems 103
3.8.1 General notations 103
3.8.2 Ferromagnets 106
3.8.3 Anisotropy in ferromagnets 110

3.8.4 Complex magnetic order 111
3.8.5 Structural and ferroelectric phase transitions . . . 115
3.8.6 Superfluidity and superconductivity 118
3.8.7 Liquid crystals 121
3.9 Advantages and Disadvantages of the Thermodynamic Theory 122

4. Landau Expansion 125

4.1 Introductory Remarks 125
4.2 Appropriate Variables 126
4.3 Generalized Gibbs Potential 129
4.3.1 Nonequilibrium potential F^t 129
4.3.2 Nonequilibrium potential Φ 130
4.3.3 Large reservoir and relation between F^t and Φ . . 134
4.4 Landau Potential 135
4.4.1 Taylor expansion 135
4.4.2 Equilibrium 136
4.4.3 Stability 137
4.4.4 Aims of the Landau theory 138
4.4.5 Examples 139
4.5 Phase Transition Line 140
4.6 Second Order Phase Transitions 143
4.7 First Order Transitions and the Isolated Critical Point . . 148
4.7.1 Phase diagram of the first order transition 148
4.7.2 Metastability. Specific heat 151
4.7.3 The isolated critical point 153
4.8 Extended Expansion: φ^6- and φ^8-theories 154
4.8.1 First order transition ($u < 0$) 155
4.8.2 The tricritical point ($u = 0$) 160
4.8.3 Irrelevance of the φ^6-term for $u > 0$ 162
4.8.4 φ^8-theory 162
4.9 Multicomponent Order Parameter and Symmetry Breaking 163
4.9.1 Thermodynamic functions 164
4.9.2 Symmetry breaking and symmetry conserving phase transitions. Landau criteria for second order phase transitions 164
4.10 Effect of External Field and Susceptibility 167
4.10.1 One-component order parameter 167

4.10.2 Two-component order parameter 172
4.11 Miscellaneous Topics 173
4.11.1 Summary of critical exponents within the Landau theory 173
4.11.2 Notion of crossover phenomena. Critical-to-tricritical crossover 174
4.11.3 Anisotropy 179
4.11.4 Coupling to strains 180
4.11.5 Validity and further development of the Landau theory 181

5. Statistical Mechanics and Mean-Field Approximation 185
5.1 Partition Function 185
5.1.1 Microscopic variables 186
5.1.2 Canonical distribution 187
5.1.3 Thermodynamic relations 189
5.2 Fixed External Field and Order Parameter Fluctuations 189
5.2.1 Continuous variable φ 190
5.2.2 Discrete variable φ 193
5.3 Inhomogeneous Systems and Translational Invariance . . 194
5.4 Coarse-Graining and Effective Hamiltonian 196
5.4.1 Coarse-graining 196
5.4.2 Transformation of the canonical partition function 198
5.4.3 Quasimacroscopic separability 200
5.4.4 Separability of the Hamiltonian 202
5.4.5 Effective Hamiltonian 204
5.4.6 Remarks 205
5.5 Generating Functionals and Correlation Functions 206
5.6 Ising Model 214
5.6.1 Definition of the Ising model 214
5.6.2 General exchange, general values of spin and angular momentum 217
5.6.3 Brief notes 220
5.7 Other Lattice Models 222
5.7.1 Heisenberg and XY model 222
5.7.2 Lattice gas model 224
5.7.3 n-component vector models 226
5.7.4 The s-state Potts model 226

5.7.5 The spherical model 226
5.7.6 Biquadratic exchange and single spin energy . . . 227
5.7.7 About the exact solutions of lattice models 228
5.8 Mean-Field Approximation 228
5.8.1 Mean field 229
5.8.2 Thermodynamic potential in mean-field approximation and symmetry breaking 231
5.8.3 Self-consistency of the mean-field approximation . 237
5.8.4 Free spins and Curie law 239
5.8.5 Uniform systems and equation of state 240
5.8.6 Expansion of the equation of state and the critical point 245
5.8.7 Helmholtz free energy and generalized potential . 248
5.9 Real Space Condensation and Statistical Mechanics 249
5.9.1 On the derivation of the Van der Waals equation . 249
5.9.2 Corresponding states and universality 252
5.9.3 Critical properties of Van der Waals fluids 253

6. Fluctuations and Fields 257

6.1 Description of Macroscopic Fluctuations 257
6.2 Spatially Independent Fluctuations and Anisotropy 260
6.2.1 Gaussian fluctuations 262
6.2.2 Gaussian-like and non-Gaussian fluctuations . . . 262
6.3 Partition Function of Spatially Independent Fluctuations 264
6.3.1 Treatment of Gaussian-like fluctuations 265
6.3.2 Beyond the Gaussian approximation. Weak coupling 267
6.3.3 Strong coupling and closed-form expression of the partition integral 272
6.4 Fluctuation Field 275
6.5 From Lattice to Field Models and Vice Versa 281
6.5.1 Continuum limit 281
6.5.2 The origin of fields in the generalized thermodynamics 284
6.5.3 Discretization of field models 285
6.5.4 Continuum limit in the momentum space 286
6.5.5 Functional integral 289
6.5.6 Truncated series 290

6.6 Integral Transformation of Microscopic Models 291
6.6.1 Integral transformation of the Ising model 293
6.6.2 Continuum limit for the Ising model and relation to the mean-field theory 296
6.6.3 Transformation of other models 303
6.7 Short-Range and Long-Range Translationally Invariant Interactions 308

7. Perturbation Theory of Fluctuation Fields 313

7.1 Basic Fluctuation Hamiltonian 313
7.2 Free Field 317
7.2.1 Equation of state 318
7.2.2 Green function 320
7.3 Thermodynamic Properties of Free Fields 324
7.4 Correlation Function of the Free Fields 332
7.5 Gaussian Approximation 340
7.5.1 Goldstone modes 342
7.5.2 Ginzburg–Levanyuk criterion and critical region . 343
7.6 Perturbation Expansion 346
7.6.1 Perturbation expansion for the partition function and the thermodynamic potential 346
7.6.2 Perturbation expansion for the averages 353
7.7 φ^4-theory. Coordinate Representation 354
7.7.1 First order perturbation contribution to the thermodynamic potential 355
7.7.2 Perturbation contribution to the correlation functions 357
7.7.3 Remarks and calculation of the combinatorial factors 363
7.8 φ^4-theory. Momentum Representation 368
7.8.1 First-order perturbation terms 368
7.8.2 Second-order perturbation terms 371
7.9 Dyson Equation and Self-energy Function 374
7.9.1 Dyson equation 375
7.9.2 Expansion of the self-energy function to order u_0^2 376
7.9.3 Remarks on the reducibility of the diagrams . . . 379
7.10 Notion of Renormalization 381
7.10.1 Mass renormalization from the perturbation series 383
7.10.2 Mass-renormalization counter-term 385

7.11 Properties of the Standard Perturbation Series 388
7.11.1 How to determine the critical region from the perturbation expansion 388
7.11.2 Effective expansion parameter 391
7.11.3 Dependence on the symmetry index n and the spatial dimension d. Universality classes . . . 393
7.11.4 Borderline dimensions 394
7.11.5 Breakdown of the standard perturbation expansion 396

8. Renormalization Group 399

8.1 Modified Perturbation Scheme 399
8.2 Perturbation Expansion for the Effective Hamiltonian . . 404
8.2.1 First order calculation 405
8.2.2 Vertex functions 407
8.2.3 Second order calculation 408
8.3 Loop Expansion 411
8.3.1 One-loop approximation 412
8.3.2 Two-loop approximation 413
8.3.3 Large-b limit and one-component field 415
8.3.4 Remarks on the renormalization 417
8.4 Large-n Limit and Hartree Approximation 418
8.4.1 Hartree diagrams 418
8.4.2 Large-n critical behaviour 420
8.4.3 Hartree approximation 423
8.5 Renormalization-group Recursion Relations 424
8.6 Renormalization Group in the One-loop Approximation 431
8.6.1 General features of the renormalization group equations 431
8.6.2 Renormalization group at the critical point 433
8.6.3 Critical Hamiltonian and fixed points 437
8.7 Linearized Renormalization Group in the One-loop Approximation 440
8.7.1 Recursion relations and fixed points 441
8.7.2 Linearization and exponents 443
8.7.3 Renormalization group flows and classification of the scaling fields 447
8.7.4 Remarks on the effect of the high-order interaction terms 449

8.7.5 Dimensional analysis 451
8.7.6 Differential renormalization group relations 452
8.8 Renormalization Group and Thermodynamics 454
8.8.1 Scaling form of the thermodynamic potential and classification of the scaling fields 456
8.8.2 Scaling relations 461
8.8.3 Remarks on the crossover phenomena in the critical region 464
8.8.4 Scaling and ϵ-expansion 466
8.9 Renormalization Group in the Two-loop Approximation 467
8.9.1 Recursion relations to order ϵ^2 468
8.9.2 Fixed points 469
8.9.3 Exponents and stability properties 473
8.10 Other Theoretical Schemes and Methods of Calculation 477
8.10.1 Direct calculation of the critical exponents 477
8.10.2 Renormalization group in three dimensions 478
8.10.3 $(1/n)$–expansion 480
8.10.4 Renormalization group in the x-space and references to other topics 481
8.11 Interrelations with other theories 481

9. Some Applications 485

9.1 Preliminary Remarks 485
9.2 Ideal Bose Gas 486
9.2.1 Thermodynamics 487
9.2.2 Bose–Einstein condensation at finite temperatures 490
9.2.3 Zero temperature condensation 491
9.3 Nonideal Bose Gas 491
9.3.1 Notion of universality of the critical behaviour at finite critical temperatures 491
9.3.2 Functional formulation 492
9.3.3 Renormalization group results 495
9.3.4 Notes 500
9.4 Quantum Critical Phenomena and Classical-to-Quantum Crossover 501
9.5 Disorder Effects 504
9.5.1 Random critical temperature 505

9.5.2 Extended impurities and long-range random correlations 508
9.5.3 Random field 510
9.5.4 Technical remarks 512
9.5.5 Ginzburg criterion 512
9.6 Anisotropy, Coupled Order Parameters and Multicritical Behaviour 513
9.6.1 Anisotropic systems 513
9.6.2 Mean-field analysis of anisotropic systems 515
9.6.3 Cubic anisotropy in mean-field approximation . . 517
9.6.4 Renormalization group studies of anisotropic systems 520
9.6.5 Tricritical points and renormalization group . . . 521
9.6.6 Bicritical and tetracritical points 522
9.6.7 Lifshitz's and tricritical Lifshitz's points 525
9.7 Critical Properties of Superconductors and Liquid Crystals 526
9.7.1 Conventional superconductors 526
9.7.2 Unconventional superconductors 533
9.7.3 Layered superconductors 536
9.7.4 Fluctuations of the superconducting order parameter 537
9.7.5 Magnetic fluctuations in superconductors 538
9.7.6 Weakly-first-order smectic A–nematic phase transition 541
9.7.7 Renormalization group investigations 542
9.8 Magnetic Fluctuations in Thin Superconducting Films . . 553
9.8.1 Preliminary notes 553
9.8.2 General form of the effective free energy 556
9.8.3 Dimensional crossover 559
9.8.4 Free energy for particular spatial dimensions . . . 562
9.8.5 Limitations of the theory 565
9.8.6 Bulk superconductors 566
9.8.7 Thermodynamics of quasi-2D films 572
9.9 Phase Transitions in Unconventional Ferromagnetic Superconductors 579
9.9.1 Coexistence of unconventional superconductivity and itinerant ferromagnetism 579

9.9.2 Free energy of ferromagnetic superconductors with spin-triplet electron pairing 581
9.9.3 Phases 583
9.9.4 Phase diagram, quantum phase transitions, and classification of ferromagnetic superconductors . . 588
9.9.5 Peculiar renormalization group analysis and unusual universality class of critical behaviour . . 594

Appendix A Useful Mathematical Formulae 603
A.1 Homogeneous Functions and Euler Theorem 603
A.2 Useful Formulae for the Gamma Function 603
A.3 Logarithm of Matrices 604

Appendix B Gaussian Integrals and Transformations 605
B.1 Gaussian Integrals 605
B.2 Multiple Gaussian Integrals 606

Appendix C Some Functionals 609
C.1 Definition 609
C.2 Functional Power Series 609
C.3 Functional Differentiation 611
C.4 Functional Integration 612
C.5 Functional Gaussian Integral 612

Appendix D D-dimensional Integration and Some Fourier Transformations 615
D.1 D-dimensional Angular Integration 615
D.1.1 Angular integration 615
D.1.2 Relationship of Cartesian and spherical coordinates in d-dimensional space 617
D.2 Some Fourier Transformations 620

Appendix E Perturbation Integrals and Sums 623
E.1 Simple Perturbation Integrals 623
E.2 Treatment of Perturbation Integrals 624
E.3 ϵ-Expansion of Perturbation Integrals 628
E.4 Summation over Matsubara Frequencies 632

Appendix F Fourier Amplitudes of Lattice Exchange Interactions 635

Bibliography 639

Index 663

Chapter 1

Introduction

Under certain conditions a number of systems in nature undergo phase transitions. Let us remember the transformations, for example, of a gas to a liquid or a solid, and of a paramagnet to ordered magnetic states; the transition of a normal metal to a superconducting phase; the normal to superfluid transition in helium, and so on. These remarkable phenomena occur as a consequence of the inter-particle correlations. Their description is made with the help of the thermodynamics and the statistical physics.

1.1 Thermodynamic Description of Phase Transitions

The thermodynamics provides a general framework for the description of phases and phase transitions. Within the thermodynamic approach the phase transition is considered as an abrupt change of one phase into another caused by the variation of the parameters of the state like the temperature T, the pressure P, the magnetic field $\boldsymbol{H}$, or, the magnetic field magnitude $H = |\boldsymbol{H}|$, etc. As usual in thermodynamics, throughout this book we use the hydrodynamic pressure $P = |\boldsymbol{P}|$.

Remember that the phases are bodies of the same substance that can be either stable, metastable or unstable thermodynamic states. Usually the thermodynamic systems are defined for fixed external conditions — temperature, volume V, etc.

For a given set of fixed parameters there always exists a function called thermodynamic potential having a minimum in the states, for which the system is in a thermodynamic equilibrium. As an example we would like to mention the Gibbs potential Φ; it depends on intensive parameters like T, P, and $\boldsymbol{H}$. In order to make an appropriate use of the introduced thermodynamic potentials we should apply to them the Gibbs stability conditions. This gives the opportunity to outline the main properties of the phase transitions and the phase diagrams (for details see Chapters 2 and 3).

Depending on whether the phases can coexist in an equilibrium contact or their distinctions vanish as the transition point is approached, two different types of thermodynamic behaviour may exist. The first one is referred to the *first-order* phase transitions while the latter usually characterizes the *second-order* phase transitions. The *order* and the main properties of the phase transition depend on the way, in which the exchange of the stability between the possible phases participating in it takes place. There is another important point for discussion — the so-called *critical state.* This is a homogeneous thermodynamic state occurring at the second-order transition point, i.e., the critical point T_c where the phases are indistinguishable (Gibbs, 1876, 1878); see also Gibbs (1948). On approaching the critical state the distinctive features of the phases gradually disappear and the phase transition is continuous; a special type of continuous transitions is the second-order phase transition.

The phenomena taking place in a close vicinity of the critical points (or lines) are called *critical phenomena.* The critical behaviour at (and near to) the critical point is a special subject of the "critical" thermodynamics. The "phases" in the critical state and in the near-to-critical states have almost equal stability. For example, let us consider para-to-ferromagnetic transition (Section 3.8.2). Below the Curie point T_c the low-temperature (ferromagnetic) phase is stable. It possesses a nonzero spontaneous magnetization $\boldsymbol{M}$ in a zero external magnetic field ($\boldsymbol{H} = 0$)[1]. When the temperature approaches T_c from the low-temperature side ($T \to T_c^- = T_c - \delta$, $\delta \to 0$) the stability of the ferromagnetic phase is reduced and at T_c the high-temperature ($T > T_c$) paramagnetic ($M = 0$) phase becomes stable. The ordered (ferromagnetic) phase cannot exist above T_c either as stable or metastable state and this is a characteristic feature of the second-order phase transition.

Within the framework of the rational thermodynamics the possibility of the critical points to appear is related to the existence of nonanalytic

[1]Henceforth the vectors $\boldsymbol{H}$ and $\boldsymbol{M}$, and the scalar product $\boldsymbol{H}.\boldsymbol{M}$ will be often written by H, M, and $H.M$ (except for cases when this may introduce a confusion). The bold face will be avoided also in the notation of other non-scalar quantities, for example, the spatial vector notations $\boldsymbol{x}$, $\boldsymbol{R}$, or, $\boldsymbol{r}$, and the wave vector symbols $\boldsymbol{k}$, $\boldsymbol{p}, \dots$ will be often written as x, R, r, k, p, ... ; the scalar products will be denoted, for example, by $k.p$, or, even by kp, the vector product — by $k \times p$, and the absolute values, for example, $|\boldsymbol{k}|$ will be denoted either by the same symbol, k, or, by $|k|$.

singularities of the thermodynamic functions. For instance, the thermodynamic susceptibilities, i.e., the second derivatives of the thermodynamic potential Φ may exhibit at the transition point either simple discontinuities or power and logarithmic singularities. The thermodynamics itself allows all these possibilities but it does not answer to the question concerning which of them actually occur in real systems. The study of the mechanism of the phase transitions and the structure of the ordered phases requires additional information about the interactions and the symmetry properties of the system. These problems are beyond the scope of the pure thermodynamic treatment. That is why the thermodynamics of phase transitions is closely connected with the statistical mechanics and other fields in physics, for example, condensed matter theory.

1.2 Statistical Description of Phase Transitions

The statistical physics treats the phase transitions and critical phenomena as a result from the collective behaviour of macroscopic ensembles of particles (or, more precisely, statistical degrees of freedom). According to the theorems proven by van Hove (1949) and Yang and Lee (1952) the phase transitions can be described in the thermodynamic limit $[V, N \rightarrow \infty, \rho = (N/V) \neq 0, \infty]$, where N is the number of particles (or the degrees of freedom) and ρ is the (number) density. In this limit the results obtained from different statistical ensembles coincide; see books on statistical physics, e.g., Isihara (1971), Landau and Lifshitz (1980), and Ma (1985). On one side, the results of van Hove, Yang and Lee demonstrate that for systems having a finite size (volume) an abrupt phase transition cannot take place. This is because the (Gibbs) canonical distribution $\exp(-\mathcal{H}/T)$, ($\mathcal{H}$ is the Hamiltonian, $k_\mathrm{B} = 1$ — the Boltzmann constant) of the finite system is bounded analytic function of T, V, N and the shape of the domain where the system is situated. So the partition function is an integral (for quantum systems – *trace*) over a finite number of degrees of freedom in restricted boundaries of integration (or summation). Consequently, the partition function of finite systems must be an absolutely convergent sum of simple exponents and no singularities typical of phase transitions can appear.

On the other hand the sharp phase transition, i.e., that at a given point of the phase diagram $(T, P, H, \dots)$ is obtained from the statistical calculation of the thermodynamic potential density (Φ/V) of infinite interacting systems (V, $N \to \infty$). The results can be extrapolated for finite (i.e., real) macroscopic systems of interacting particles to an accuracy of the order $N^{-1/2}$, where $N \geq 10^{22} \div 10^{23}$. No doubt that this accuracy is sufficient and, hence, we can apply the statistical calculations in the thermodynamic limit in order to predict the behaviour of finite systems. Note, that for temperatures T and pressures P typical for our Planet imply typical mean interatomic (intermolecular) distances $l \sim 1 \div 10$ Å in condensed matter systems and some dense gases. As the methods of statistical physics work approximately but fairly well up to particle numbers $N \sim 10^3 \div 10^5$, one may investigate small systems of characteristic size $L \sim V^{1/3} \sim 10 \div 100$ Å, i.e., nano-size systems (note, that 1 Å= 0.1 nm $= 10^{-10}$ m).

Except for the theorems of the existence of sharp phase transitions and critical points, the microscopic van Hove–Yang–Lee approach is not easy to be applied to an explicit calculation of the thermodynamic functions and, in particular, to complex microscopic interactions in real systems. For an excellent introduction in this field see Uhlenbeck and Ford (1963); for rigorous proofs of the existence theorems for classical and quantum fluid and spin systems see Ruelle (1963a,b, 1969), Fisher (1964a,b,c), (Griffiths, 1964, 1965a; Suzuki, 1968); for a detailed discussion, see also the book by Münster (1969).

Despite the overwhelming amount of specific features and concrete mechanisms the main properties of the phase transitions and critical phenomena can be unified by several global concepts. The first of them is the notion of ordering which arises from the inter-particle interactions or/and from quantum correlations (pseudo-interactions). For example, owing to the intermolecular forces a classical gas turns into a liquid and this is the usual condensation while the quantum correlations in the ideal Bose gas produce a completely different cooperative phenomenon — the Bose–Einstein condensation (BEC) in the space of the particle momenta (see Section 9.2).

The quantum statistical correlations and the "usual" interactions act simultaneously in real systems. Often the notion of quasiparticles (excitations) is introduced for the description of the collective quantum effects in interacting systems. An example is the transition from normal to superconducting state in metals where the electrons form Cooper pairs — quasiparticles of Bose type (see Section 3.8).

The phases are distinguished by a physical quantity called *an order parameter*, denoted by φ. Within the thermodynamic approach this is one of the extensive thermodynamic variables or its density. Let us consider a phase transition between two phases: 1 and 2. The order parameter described by the extensive variable X is usually chosen so, $\varphi = X - X_1$, that it is zero in one of the phases (phase 1; $X_1 - X_1 = 0$) and it is positive ($X_2 - X_1 > 0$) in the other phase ($X_2 > X_1$). The phase 2, where $\varphi > 0$ will be then called "ordered" phase and the phase 1 will be the disordered phase.

For example, for the gas–liquid transition the order parameter φ is the difference $\Delta V = (V_G - V)$ between the volume V of the system and the specific volume V_G of the gas (N is fixed). In terms of the inverse density $1/\rho = v = V/N$ we may denote the order parameter by $\varphi = \Delta v$, where $\Delta v = (V_G - V)/N$. As you see, according to our general notations, here we have chosen $\varphi = (X_1 - X)$ instead of $\varphi = (X - X_1)$ and in this way we keep the quantity φ positive. This choice of positive φ has some advantages and is often made in particular studies of phase transitions. In the liquid phase $\varphi_L = (V_G - V_L) > 0$ and in the gas phase, $\varphi_G = 0$. When V is fixed but N varies it is convenient to choose $\varphi = \Delta\rho = \rho - \rho_G$ — the difference between the density $\rho = 1/v$ and the specific density of the gas $\rho_G = 1/v_G$. Then in the "ordered" (liquid) phase we have $\varphi = (\rho_L - \rho_G) > 0$, which exactly corresponds to our definition $\varphi = (X_2 - X_1)$. This order parameter, $\varphi \sim \Delta v$, or, alternatively, $\varphi \sim \Delta\rho$, has a purely quantitative meaning and it does not describe any symmetry difference between the "ordering" in the phases 1 and 2 (vapour and liquid, respectively); see also Section 3.6.

In other cases, the order parameter describes a breaking of the invariance of the system with respect to certain symmetry transformations (symmetry breaking; see Section 5.8.2). An example is the paramagnetic-to-ferromagnetic transition. The appearance of a nonzero magnetization M below the Curie point T_c of the ferromagnet in a zero external magnetic field ($H = 0$) is related with the spontaneous breaking of the rotational symmetry in the ferromagnetic phase (for more details, see Sections 3.8.2 and 4.9.2). As we see, in this case the order parameter is not a simple real scalar but a vector — a mathematical object with a structure.

Another order parameter which may describe a spontaneous symmetry breaking and difference between the symmetries of the "disordered" and the "ordered" phases is the complex number $\varphi = \varphi' + i\varphi''$, i.e., the two-component real vector $\varphi = (\varphi', \varphi'')$. Such type of ordering occurs, for example, in certain superfluids and supersonductors. This type of ordering is similar to the ordering in certain ferromagnets with a plane of easy magnetization (bi-axial ferromagnets), where the magnetization lies in the XY plane, and $M = (M_x, M_y, 0)$.

Depending on their origin, the microscopic interactions may cause symmetry-breaking or symmetry-conserving phase transitions. The successful description of the phase transitions and the symmetry of the ordered phases depends to a great extent on the appropriate choice of the order parameter φ. It is not always obvious how to define the order parameter of a concrete system and both experimental and theoretical difficulties can arise. The order parameter φ is either a scalar as, for example, in fluids or a vector with $n = 1, 2, 3, \ldots$ components – $\varphi = (\varphi_1, \varphi_2, \ldots)$. In some cases φ has a more complex tensor representation (liquid crystals, superfluid ^{3}He; see Sections 3.8.6 and 3.8.7).

We should bear in mind that a successful thermodynamic description of a phase transition requires the system to be open towards such an exchange with its surroundings which allows the variation of the order parameter φ. Otherwise the order parameter is fixed and no phase transition can occur. Usually the other parameters of state, like T, P, etc., may also vary but if we want to understand the essential features of the phase transition we should keep them fixed. Thus we define the equilibrium thermodynamic states (and phases) at given (fixed) parameters $T, P, \ldots$. Then we may produce the phase change by varying the same thermodynamic parameters $(T, P, \ldots)$ within certain quasi-static thermodynamic processes. In this situation the equilibrium state (phase) of the system will be characterized only by some equilibrium value $\bar{\varphi}(T, P, \ldots)$ of the order parameter φ.

Let us denote the fixed parameters by T, h and Y, where h is an intensive thermodynamic variable ("field") conjugate to φ and Y stands for a set of additional variables which will be neglected in our further discussion (such variables Y often exist but they are irrelevant to the main properties of the system). Obviously, for a ferromagnet, h is equal to the external magnetic field H.[2] The important point is to choose the appropriate thermodynamic potential, for which the natural (fixed) variables are T and h. Such is the equilibrium Gibbs potential $\Phi(T, h)$ which plays a fundamental role in the description of phase transitions, a topic widely discussed in Chapters 2 and 3 and, in particular, in Chapter 4. Φ is a function also of the equilibrium order parameter $\bar{\varphi}(T, h)$ that gives the minimum $\Phi(T, h, \bar{\varphi})$ of the nonequilibrium potential $\widetilde{\Phi}(T, h, \varphi)$ which depends on an extra variable — the nonequilibrium value φ of the order parameter; recall, for a gas–liquid system, $\varphi = (V - V_L)$, $\bar{\varphi} = (V_G - V_L)$.

[2] According to the footnote in page 2 we shall often avoid the boldface for vectors like $\boldsymbol{h}$, and the thermodynamically conjugate order parameter vectors $\boldsymbol{\varphi}$ and $\bar{\boldsymbol{\varphi}}$.

The quantity φ can be represented by the sum $\varphi = \bar{\varphi} + \delta\varphi$ of the equilibrium value $\bar{\varphi}$ and the fluctuation part $\delta\varphi$ of φ. The fluctuations are very important, in particular near critical points, where $\bar{\varphi}$ tends to zero and the stabilities of the phases 1 and 2 are almost equal. Thus the description of the critical and near-to-critical states comprises both the equilibrium states and the fluctuations around them (generalized thermodynamic approach).

From the view of statistical mechanics, $\bar{\varphi}$ is equal to the statistical averages $\langle\varphi\rangle$ of φ. By definition the average $\langle\delta\varphi\rangle$ of the fluctuations is zero but the fluctuations can exhibit strong correlation properties so as their correlation functions $\langle\delta\varphi^m\rangle, m = 2, 3, \ldots$ to be different from zero and, under certain circumstances to reach large values. The phase transition properties are described with the help of the thermodynamic and correlation functions which essentially depend on the properties of the equilibrium order parameter $\bar{\varphi}$ and its fluctuations $\delta\varphi$. The equilibrium order parameter $\bar{\varphi}$ is adequately determined by the classical mean field (MF) theories of phase transitions (Chapter 5). These theories are based on the notion of a *molecular* (*internal, mean*) field. It reflects fairly well the real inter-particle interaction for those values of the thermodynamic parameters, for which fluctuations $\delta\varphi$ of the order parameter can be neglected. The MF theories describe the phases themselves rather than the specific properties near phase transition points.

The fact that the order parameter is relatively small near the phase transition point has been used by Landau (1937a) to develop an approach to the description of phase transition properties based on quite general thermodynamic and symmetry arguments (for more details, see Chapter 4). The Landau conditions imposed on the analytic properties of the parameters of the theory lead to results similar to those from the MF approximation in the statistical mechanics of many-body systems (Chapter 5).

The reduced stability of the phases near the phase transition point emerges from the appearance of large fluctuation correlations. In particular, this is the case of the continuous transitions, where the point of the stability exchange between the phases coincides with the point of the equilibrium phase transition.

The fluctuation effects have been known for a long time from the phenomenon of the critical opalescence taking place at the gas–liquid critical point; for an excellent discussion see the book of Stanley (1971). An approach for taking into account the influence of the fluctuations has been developed by Ornstein and Zernicke (see also Section 7.2) for the gas–liquid critical points (Ornstein and Zernicke, 1914). This approximation of free, uncorrelated or Gaussian fluctuations describes qualitatively well the fluctuation effects but in a close vicinity of the critical point (the critical region) it is insufficient, because the thermodynamic properties there essentially depend on the fluctuation interactions.

The treatment of interacting fluctuations requires other methods created to interpret the precise critical experiments performed in the 50's–60's. In the remaining part of this Chapter we shall briefly outline the recent period in the study of the critical phenomena. The theory of phase transitions and critical phenomena is based on the methods of statistical mechanics and concepts of condensed matter physics but it has a substantial stimulus on the advance of a number of other fields of physics and natural sciences (astrophysics, quantum field theory and elementary particle physics, hydrodynamics, biophysics, some fields in chemistry, and some new inter-disciplinary sciences as, for example, economic physics (econophysics) and sociological physics (socio-physics). The literature on these problems extends over thousands of articles, and several hundreds excellent review articles and books. A comprehensive review of the experimental situation in this field comprising the period 1950–1970 is presented by Heller (1967) and de Jongh and Miedema (1974). Both experimental and theoretical problems can be found in the reviews of Fisher (1965, 1967, 1970), and in the book of Stanley (1971); see also Stanley (1973).

1.3 Scaling and Critical Exponents

The experiments on critical behaviour give very useful information about the main properties of the critical phenomena. The thermodynamic and correlation functions follow a power law dependence on $(T - T_c)$ over a temperature interval close to the critical point $(T_c, h = 0)$. Similar power laws express the dependence of the thermodynamic functions on other parameters as, for example, the field h conjugate to φ, which stands for the difference between the fixed thermodynamic parameter (h) and its critical value $(h_c = 0)$. We shall denote by x all quantities like $t = (T - T_c)/T_c, h$, etc.

In order to describe the behaviour of a thermodynamic (or correlation) function $f(x)$ of a thermodynamic parameter x in a close vicinity of the critical point $(x_c = 0)$ we shall introduce the power law

$$f(x) = \begin{cases} f_0^{(+)} x^{-\lambda_f^{(+)}} & \text{if } x > 0, \\ f_0^{(-)} (-x)^{-\lambda_f^{(-)}} & \text{if } x < 0, \end{cases} \tag{1.1}$$

where $\lambda_f^{(+)}$ is the critical exponent of the function f with respect to the variable x for $x > 0$, $\lambda_f^{(-)}$ is the same for $x < 0$ and $f_0^{(\pm)}$ are the critical amplitudes; note that usually, $0 < |\lambda_f^{(\pm)}| < \infty$. In Eq. (1.1) x could be $t = (T - T_c)/T_c$, h or another thermodynamic quantity which tends to zero at the critical point. When negative values of x are impossible, as it is often the case for the external field h, one may work only with the first-line expression on the r.h.s. of Eq. (1.1).

In the current literature $\lambda_f^{(+)}$ and $\lambda_f^{(-)}$ are usually denoted by λ_f and λ'_f, respectively. In experiments, the critical exponents are usually given by the ($\ln f$, $\ln x$) diagram (*alias log–log* diagram) of the system and then the following definition of $\lambda_f^{(\pm)} \neq 0$ is applied

$$\lim_{x \to 0\pm} \frac{\ln |f(x)|}{\ln |x|} = -\lambda_f^{(\pm)} \tag{1.2}$$

(provided $\lambda_f^{(\pm)} \neq 0$!). The limiting cases $\lambda_f = \pm\infty$ indicate that the divergence ($\lambda_f = \infty$) of f or the convergence ($\lambda_f = -\infty$) are exponential: $f(x) \sim \exp\left[\text{const}^{(\pm)}/|x|\right]$, where $\text{const}^{(\pm)} > 0$. Note that the function $f(x)$ defined by Eq. (1.1) is one of a series of functions $f(x)$ which satisfy the Eq. (1.2). This series of functions is given by the variation of the amplitudes values $f_0^{(\pm)} \neq 0$. When $\lambda_f^{(\pm)} = 0$ the function $f(x)$ coincides with its (scaling) amplitudes $f_0^{(\pm)}$ and the Eq. (1.2) is redundant.

The Eq. (1.1) gives the leading dependence of f on x for small x. The asymptotic limit ($x \sim 0$) of the function $f(x)$ in Eq. (1.2) is described by the power law (1.1); provided the exponents $\lambda_f^{(\pm)}$ are finite (for short, we shall often denote $\lambda_f^{(\pm)}$, $|f(x)|$, $f_0^{(\pm)}$, and $|x|$ by λ_f, $f(x)$, f_0, and x, respectively). The only thing which the existence of the limit (1.2) confirms is that $f(x)$ is asymptotically proportional to $x^{-\lambda_f}$ and, of course, this does not mean $f(x)$ to have such a simple form as given by Eq. (1.1).

Following Fisher (1971) we shall discuss the formal aspects of Eqs. (1.1) and (1.2) and after that we shall apply them to particular physical quantities. The more general form of the function $f(x)$ will be

$$f(x) = x^{-\lambda} f_0(x) \tag{1.3}$$

(here the amplitude f_0 depends itself on x). If the function $f_0(x)$ from Eq. (1.3) is analytic at $x = 0$ it can be represented by the series

$$f_0(x) = f_0 + f_1 x + f_2 x^2 + \dots \tag{1.4}$$

but if it is nonanalytic, we can consider two cases:

(i) *coincident weaker singularities*

$$\lim_{x\to 0} |\ln f_0(x)| < \infty, \tag{1.5}$$

for example

$$f_0(x) = f_0 + f_1 x^{\mu} + \dots, \quad 0 < \mu < 1 \tag{1.6}$$

and

(ii) *divergent coincident singularities*

$$|\ln f_0(x)| \to \infty \quad \text{as} x \to 0, \tag{1.7}$$

for example

$$f_0(x) = [\ln(1/x)]^{\mu} f_{00}(x), \quad \mu < 1. \tag{1.8}$$

Eqs. (1.1) and (1.3)–(1.4) determine the simple branch points ($x = 0$) while the nonanalytic function $f_0(x)$ represents less tractable singularities of $f(x)$.

When the critical exponents $\lambda_f \equiv \lambda_f^{(\pm)}$ are small ($\lambda_f \sim 0$) the power law (1.1) is hardly distinguished from some weaker dependence. The logarithmic singularity can serve as an example of such weak singularity

$$f(x) = f_0 \ln(1/x) + f_1 \ln\ln(1/x) + const. \tag{1.9}$$

Another example is the simple discontinuity $\Delta f = f^{(+)} - f^{(-)} \neq 0$:

$$f(x) = f_0^{(\pm)} + f_1 x^{\mu}, \quad 0 < \mu < 1, \tag{1.10}$$

without ($f_1 = 0$) or with *a cusp* — $f_1 \neq 0$, $(\partial f/\partial x) \to \infty$ for $x \to 0$. Certainly, the asymptotic power law (1.2) with $\lambda_f = 0$ describes both the weak logarithmic singularity (1.9) and the simple discontinuity (1.10).

It is readily seen from the above considerations that the power laws are characterized by a leading dependence on x expressed by the main (or

Table 1.1 Definition of critical exponents of the thermodynamic quantities.

quantity	definition	scaling law and validity domain
order parameter φ	$\varphi = -(\partial\Phi/\partial h)_T$	$\varphi \sim \|t\|^{\beta}$, if $t \sim 0^-$, $h = 0$ $\|\varphi\| \sim \|h\|^{1/\delta}$, if $t = 0$, $h \sim 0$ $(\delta = \delta')$
specific heat C_φ at fixed φ	$C_\varphi = -T\left(\partial^2\Phi/\partial T^2\right)_\varphi$	$C_\varphi \sim \begin{cases} t^{-\alpha} & \text{if } t \sim 0^+ \\ \|t\|^{-\alpha'} & \text{if } t \sim 0^- \end{cases}$
isothermal susceptibility χ_T in a zero field h	$\chi_T = (\partial\varphi/\partial h)_{T,h=0}$	$\chi_T \sim \begin{cases} t^{-\gamma} & \text{if } t \sim 0^+ \\ \|t\|^{-\gamma'} & \text{if } t \sim 0^- \end{cases}$

asymptotic) critical exponent λ_f. This is the asymptotic power law. There can be one or more correction terms to the asymptotic law like the terms $f_1, \ldots$ in Eqs. (1.4), (1.6) and (1.9)–(1.10).

Experimentally the order parameter φ behaves according to the scaling law

$$\varphi \sim |t|^{\beta}, \quad t = \frac{T - T_c}{T_c} \to 0^-, \tag{1.11}$$

where the scaling exponent $\beta = -\lambda_\varphi$ is defined for $t < 0$ (for $t > 0$, $\varphi \equiv 0$). The power laws and critical exponents for the specific heat at fixed φ and the isothermal susceptibility in a zero external field h conjugate to φ are shown in Table 1.1 (the power amplitudes and corrections to the asymptotic behaviour are omitted).

Another quantity, observable by light, X-rays and neutron scattering experiments is the pair (or two-point) correlation function $G(\boldsymbol{R}) = \langle \delta\varphi(\boldsymbol{x})\delta\varphi(\boldsymbol{x}+\boldsymbol{R})\rangle$, where $\boldsymbol{x}$ and $\boldsymbol{R}$ are spatial vectors. For relatively large $|t| < 1$,

$$G(\boldsymbol{R}, t) \equiv G(R, t) \sim \frac{e^{-R/\xi}}{R^{(d-1)/2}}, \quad R \equiv |\boldsymbol{R}| > \xi, \tag{1.12}$$

where $d > 2$ denotes the spatial dimensionality, usually $d = 3$, and ξ is the characteristic length of the fluctuation correlations (correlation length or correlation radius). The correlation length ξ itself obeys the law

$$\xi \sim \begin{cases} t^{-\nu} & \text{if } t \sim 0^+, \\ |t|^{-\nu'} & \text{if } t \sim 0^-. \end{cases} \tag{1.13}$$

When the critical point ($t = 0$) is approached ξ tends to infinity. For $R \ll \xi$ and these are the distances in a finite body near its critical point, the exponential law (1.12) changes to

$$G(R) \sim \frac{1}{R^{d-2+\eta}}. \tag{1.14}$$

This relation with $\eta = 0$ describes the critical fluctuations within *the Ornstein–Zernicke approximation* of noninteracting fluctuations (Ornstein and Zernicke, 1914); see also Section 7.2.2.

The exponent η has been introduced by Fisher (1964b) in order to explain the experiments on critical ($t \sim 0$) scattering. Since $G(R)$ is expressed by the fluctuations $\delta\varphi(\boldsymbol{x})$ the exponent η is obviously connected with the behaviour of the x-dependent (field) variable $\delta\varphi(x)$ (for the use of x instead of $\boldsymbol{x}$, see the footnote in page 2). This exponent is often called the Fisher exponent or the *anomalous dimension* of the spatially dependent fluctuations $\delta\varphi(x)$; in fact, the anomalous dimension of $\delta\varphi(x)$ differs from η by a number factor $-1/2$ and is equal to $-\eta/2$ (see Section 8.6).

The power law (1.14) for $R < \xi$ corresponds to an anomalous energy spectrum of the critical fluctuations ($t = 0$). The Fourier transform of $G(R)$ at $T = T_c$ is of the form

$$G(k) \sim k^{-2+\eta}, \qquad k \sim 0, \tag{1.15}$$

where k ($\equiv |\boldsymbol{k}|$) is the momentum modulus (or wave number)[3] describing fluctuations; as mentioned in the footnote in page 2, the notation k will be used either for the wave vector $\boldsymbol{k}$, or, for the modulus $|\boldsymbol{k}|$ (provided this does not introduce any confusion). The departure ($\eta \neq 0$) from the quadratic spectrum $G_0^{-1}(k, t = 0) = \varepsilon_0(k) \sim k^2$ of the order parameter fluctuations $\delta\varphi(k)$ is experimentally observed in a close vicinity of T_c.

1.4 Dynamics

In a thermal equilibrium the order parameter does not depend on time but if the system is taken out of the equilibrium φ becomes time-dependent. The quantity which characterizes the relaxation of the system to the equilibrium is the relaxation time τ. Near the critical point it tends to infinity through the law

[3]Sometimes, we shall use the terms "momentum" and "momenta" as synonymous of wave vector(s), although the respective physical quantities are not identical.

$$\tau \sim |t|^{-\nu z} \sim \xi^z, \tag{1.16}$$

where z is called the dynamic critical exponent. Let us denote the time, as usual, by t; the reader should distinguish this variable from $t = (T - T_c)/T_c$.

The relaxation time τ and the dynamic exponent z are determined by an appropriate time-dependent differential equation for $\varphi(t) = \bar{\varphi} + \delta\varphi(t)$. When the analysis is made at a quasi-macroscopic level we may think of $\varphi(x,t) = \bar{\varphi}(x) + \delta\varphi(x,t)$ as corresponding to a quasi-equilibrium (quasi-static) state; see Landau and Lifshitz (1980). Within this (macroscopic) approach the dynamic critical phenomena are determined by the time evolution of the quasistatic fluctuations $\delta\varphi(x,t)$ and their time correlations $G(t-t') = \langle\delta\varphi(t)\delta\varphi(t')\rangle$ in a way analogous to that for the static critical phenomena. If the static critical phenomena of a system are described by the potential $\Phi[\varphi(x)]$ — functional of the order parameter field $\varphi(x)$, the dynamic critical behaviour of the same system can be treated, and this is the simplest way of investigation, in terms of the Landau–Khalatnikov equation

$$\frac{\partial\varphi}{\partial t} = -\Gamma\frac{\delta\Phi}{\delta\varphi(x,t)}, \tag{1.17}$$

where Γ is a time-independent parameter; see, e.g., Luban (1976), Lifshitz and Pitaevskii (1979), and Landau and Lifshitz (1980). From such equations one obtains the relation between the characteristic frequency $\omega_c^{(\pm)}$ above and below the critical point which depends on $(T - T_c)$ and the momentum k of the fluctuations; see Stanley (1971). The power law $\omega_c^{(\pm)} \sim k^z$ is realized for $T = T_c$.

The dynamic critical exponent z usually takes on values between 1 and 5 and they depend on the type of the conservation laws, which the system obeys. Except for Sections 9.3 and 9.4, where the dynamic critical behaviour of Bose system is deduced within the Matsubara–Abrikosov–Gor'kov–Dzyaloshinskii temperature diagrammatic technique we shall not be involved in the dynamic critical phenomena; for a comprehensive presentation of these phenomena see Halperin and Hohenberg (1967, 1969), Stanley (1971), Ma (1976a), Hohenberg and Halperin (1977), and Patashinskii and Pokrovskii (1979). The knowledge of the dynamic exponent z and the static ones $\alpha, \beta, \ldots$ gives the opportunity to investigate the transport properties near T_c (heat flow, sound propagation modes, thermal and electrical conductivity, etc).

1.5 Experiment

The experimental quantities for the static critical exponents are shown in Table 1.2. It is obvious that they are quite different from the values obtained in the classical theories. It should be mentioned that the experimental values vary substantially when the temperature interval t (or h) about the critical point $(T_c, h = 0)$ changes. Far from $(T_c, h = 0)$ the exponents take effective values which confirm the results from the classical theories. A departure from these predictions is found in a close vicinity of T_c (Section 7.5.2).

Moreover, the precise experiments demonstrate that the exponents and, hence, the main features of the critical behaviour gradually change from their classical values observed in early experiments far from T_c to the non-classical (anomalous) values when T_c is approached enough. This is clearly shown in the experiments of Voronel (1976), Ahlers (1976, 1980), Lipa and Chui (1983), Singsaas and Ahlers (1984), and Belanger and Yoshizawa (1987). So, there are two different asymptotic types of critical behaviour: 1) that corresponding to the classical theories and to experiments which do not measure the behaviour in the narrow critical region around T_c, and 2) that corresponding to the non-classical behaviour in the critical region.

The critical region has usually the size $t \lesssim 10^{-2}$; in some cases it extends to $t \gtrsim 10^{-2}$. For usual superconductors the critical region is very small ($t \lesssim 10^{-10} \div 10^{-14}$) and cannot be observed in experiments but in oxide superconductors with $T_c \sim 100$ K the critical region is of the order $t \sim 10^{-3} \div 10^{-1}$ (see Section 9.7). The usual critical experiments penetrate in a temperature interval $10^{-4} \div 10^{-5}$ about T_c but in exceptional cases as, for example, the experiments on the λ-line in ^{4}He the accuracy of the measurements is raised up to $10^{-7} \div 10^{-8}$ K from T_c (Lipa and Chui, 1983).

The more detailed experimental studies of the critical behaviour reveal that it depends on the spatial dimensionality d and the symmetry index n of the system. The positive quantity n is the number of the order parameter components. For the gas–liquid transition $n = 1$ — the order parameter is the scalar $\varphi = \Delta\rho$, for ferromagnets, it varies from 1 to 3 depending on the type of magnetic ordering but in systems with complex structure it can take values $n > 3$; see, e.g., Section 3.8, or, Pfeuty and Toulouse (1975), and Tolédano and Tolédano (1987).

Certain experiments in particular, in magnets (de Jongh and Miedema, 1974) display that there is a substantial difference between the critical properties of $(d = 3)$–three dimensional (usual) systems (in short, "3D systems"), quasi-one ($d \approx 1$) and quasi-two ($d \approx 2$) dimensional systems (in short, q1D and q2D systems)[4]. It will be convenient to consider n and d as positive integers or, even as non-integer nonnegative parameters ($n, d \geq 0$). There are extreme cases: ($d = 0$)–zero dimensional systems (0D systems), where the spatial dependence of the fluctuations can be neglected (an approximation for systems with a relatively small size), and ($n = 0$)–systems, describing the analogy between the critical phenomena in ferromagnets and the excluded volume problem for polymer chains; see de Gennes (1972, 1979) and Pfeuty and Toulouse (1975).

While the important thermodynamic quantities like the critical exponents λ_f and the amplitude ratios, $f_0^{(+)}/f_0^{(-)}$, depend on d and n the same quantities do not depend on the physical nature of the system considered; for example, they are the same for the gas–liquid transition (in noble gases, carbon dioxide, etc) and in ($n = 1$, 3D)-ferromagnets as FeF_2 (within the accuracy of the measurement). We must bear in mind that the experimental results are approximate, in the sense that they are influenced by the specific experimental conditions. Near the critical point, the system is very sensitive to small perturbations as, for example, the effect of the gravitational field on the gas–liquid system or the superfluid ^{4}He, impurities in solids, incomplete equilibrium due to the divergence of the relaxation time ($\tau \to \infty$ at T_c); for details see Levelt–Sengers (1974), Ahlers (1976, 1980), Voronel (1976), Moldover *et al.* (1979), and Wilks and Betts (1987).

The deviation of the critical exponents from their classical values is known from the late thirties; see, e.g.,, Fisher (1965). The interest in the precise study of critical phenomena arose when Guggenheim in 1945 compared the coexistence curves of a number of gases (Guggenheim, 1945). He has shown that near the critical point (T_c, P_c) they obey the (1/3)-law, $\Delta\rho \sim |t|^{1/3}$ in contrast to the prediction $\Delta\rho \sim |t|^{1/2}$ of the Van der Waals theory of condensation. In fact, the "1/3–law" is a new law of the corresponding states (Section 5.9.2).

The accuracy of the critical experiments performed in the fifties and the sixties was usually of order $\Delta T \sim (T - T_c) \sim 10^{-2} \div 10^{-3}$ K which was sufficient to indicate the departure of the critical exponents from their classical values but absolutely insufficient to indicate the dependence on n and d. For example, the first measurements of the specific heat of the λ-transition in ^{4}He were made with precision $|t| \lesssim 10^{-2}$ on both sides of T_c (Fairbank, Buckingham and Kellers, 1958; Buckingham and Fairbank, 1961; Fairbank,

[4]We denote a d-dimensional system by dD, for example, 1D, 2D, 3D,..., and quasi-dD systems — by qdD

Table 1.2 Values of static critical exponents $[\lambda_f^{(+)} = \lambda_f^{(+)}]$.

exponent	classical theory	2D Ising model	experimental values
α	0 (discont.)	0 (log.)	$-0.10 \div 0.14$
β	$\frac{1}{2}$	$\frac{1}{8}$	$0.32 \div 0.39$
γ	1	$\frac{7}{4}$	$1.20 \div 1.50$
δ	3	15	$4.00 \div 5.00$
ν	$\frac{1}{2}$	1	$0.63 \div 0.72$
η	0	$\frac{1}{4}$	$0.03 \div 0.08$

1963). The result was interpreted as the appearance of a logarithmic singularity ($\alpha = 0$); see Table 1.1 which was predicted by Onsager (1944) for the 2D Ising model (Section 5.6). The more precise experiments, see Ahlers (1976, 1980), with an accuracy $|t| \lesssim 10^{-4} \div 10^{-5}$ yield $\alpha = -0.026 \pm 0.004$ rather than $\alpha = 0$. Moreover, in the interval $10^{-5} \lesssim |t| \lesssim 10^{-8}$ the experimental value of the exponent α is $\alpha = -0.0127 \pm 0.0026$ (Lipa and Chui, 1983) which is in agreement with the contemporary theoretical predictions on the basis of the renormalization group (RG) (see, e.g., Le Guillou and Zinn–Justin (1980) and the discussion in the remainder of this Chapter).

In Table 1.2 the typical experimental values of the critical exponents are compared with the exponents from the Onsager (1944) exact solution of the 2D Ising model; see also Baxter (1982). The theoretical results for this exact solution of a 2D problem are obviously far from the experimental values for 3D systems. The effective (experimental) exponents as $\alpha = 0$ concerning the above discussed "ln-divergence" of the specific heat in ^{4}He should not be interpreted as coinciding with the results from one or another exact solution of models intended to describe a completely different situation; for the experimental study of 2D magnets see the review by de Jongh and Miedema (1974). Despite the essential advance which the exact solutions ((Stanley, 1971; Baxter, 1982) provide for the theoretical studies they often investigate models which have no wide application in the interpretation of experiments.

1.6 Scaling and Homogeneity. Scaling Relations

The discrepancy between the precise critical measurements and the predictions of the classical theories has stimulated the development of essentially new ideas for the theoretical approach to the critical phenomena. The further progress of theoretical study has been connected with the proper understanding what actually the thermodynamic laws tell us about phase transitions. In fact, they do not give the definite values of the critical exponents; rather the thermodynamic stability theory of equilibrium states stipulates inequalities and under certain conditions equalities, from which the relations between the critical exponents can be found. One of these relations, namely $\delta = 1 + \gamma'/\beta$ — the Widom equality, was introduced by Widom (1964). At about the same time other inequalities and equalities between the critical exponents were introduced by Fisher; see Essam and Fisher (1963); Fisher (1967, 1971); Rushbrooke (1963); Griffiths (1965b,c); Josephson (1967) and other authors. The simplest and the most straightforward idea for the general thermodynamic approach to the scaling description of the critical state was proposed by Widom (1965); effectively equivalent idea was that of Domb and Hunter (1965).

It is important to realize that the inequalities which relate the critical exponents can be obtained from the thermodynamic stability conditions and appropriate thermodynamic equations without the introduction of additional assumptions. In this sense the mentioned inequalities are exact and quite general thermodynamic relations between the exponents. The experimental test of the inequalities introduced in the sixties demonstrates that they are "near to" equalities. The more precise experiments performed afterwards confirm this hypothesis. Moreover the precise experiments show that the symmetry relation $\lambda_f = \lambda'_f$ between the exponents for $T < T_c$ and $T > T_c$ is satisfied. These experimental results have played an important role in the further development of the theory.

The Widom (1965) homogeneity (or scaling) hypothesis implies that the thermodynamic functions like the Gibbs (or other) potentials and the equation of state are *generalized homogeneous functions.* The mathematical expression for a generalized homogeneous function of m variables $(y_1, \ldots, y_m)$ is

$$f(\lambda^{a_1} y_1, \lambda^{a_2} y_2, \ldots, \lambda^{a_m} y_m) = \lambda^{a_f} f(y_1, \ldots, y_m), \tag{1.18a}$$

for any real positive parameter $\lambda > 0$. In Eq. (1.18a) a_i are exponents describing the order of homogeneity with respect to each variable y_i. Such functions possess an invariance with respect to *scale transformations.* Let us change the parameter λ to $\lambda = \eta^{1/a_f}$. Then Eq. (1.18a) becomes

$$f(\eta^{a_1/a_f} y_1, \ldots, \eta^{a_m/a_f} y_m) = \eta f(y_1, \ldots, y_m), \tag{1.18b}$$

where (a_1/a_f) is the scaling exponent for y_1, (a_2/a_f) is the exponent of y_2, etc. The number of the independent exponents $a_1, \ldots, a_m, a_f$ is not $(m+1)$ but m. Remember that an (ordinary) homogeneous function of order p satisfies the relation

$$f(\lambda y_1, \lambda y_2, \ldots, \lambda y_m) = \lambda^p f(y_1, \ldots, y_m) \tag{1.18c}$$

(cf. Eq. (A.1) in Appendix A; henceforth referred to as Section A). Clearly, if $a_1 = a_2 = \cdots = a_m$ the function 1.18a) will be an ordinary homogeneous function of order a_f.

For two variables, x and y, the most frequently investigated case of critical phenomena, we can write

$$f(\lambda^a x, \lambda^b y) = \lambda^p f(x, y), \tag{1.19a}$$

or, equivalently,

$$f(\lambda^{a/p} x, \lambda^{b/p} y) = \lambda f(x, y), \tag{1.19b}$$

where we have used the same notation (λ) for the scaling variable $\eta = \lambda^{1/p}$. The homogeneity hypothesis for the mathematical form of the thermodynamic quantities turns out to be the necessary framework of the phenomenological description of the power law dependencies (1.1) observed in experiments.

Unlike the classical theories which seem to impose strong restrictions on the possible thermodynamic behaviour, the , given by Eq. (1.18a) and, for two variables, by Eq. (1.19a), is quite flexible and, therefore, applicable to any experimental result on the asymptotic power laws. In order to show how the homogeneity relations work let us analyze Eq. (1.19a) and Eq. (1.19b) setting $y = 0$. Then $f(x, 0)$ is a homogeneous function of x of order (p/a).

Obviously, the thermodynamic quantities from Eqs. (1.11), (1.13), (1.16), and Table 1.1 are also simple homogeneous functions of $x = t^{-1}$ whose order is given by the corresponding critical exponent $(\alpha, \beta, \ldots)$. Of course, there are more complicated situations. Consider a ferromagnet where t and h are simultaneously different from zero. In this case the thermodynamic quantities will not be ordinary but generalized homogeneous functions of t and h. This is so because the critical exponents with respect to t and h are different.

The homogeneity hypothesis describes the possibility to obtain the power laws for the singularities of the thermodynamic and correlation quantities near T_c. In addition a general form of the equation of state can be found. We shall briefly illustrate this on the example of an isotropic ferromagnet, where the magnetization vector M is parallel to the external magnetic field H. Let Eqs. (1.19a) and (1.19b) with $f(x,y) = H$, $x = t$, and $y = M$ be the equation of state of this ferromagnet. It can be written in the following way

$$\frac{H}{M^{\delta}} = g\left(\frac{t}{M^{1/\beta}}\right). \tag{1.20a}$$

In order to derive Eq. (1.20a) we have set $\lambda = M^{-1/b}$, the scaling function

$$g\left(\frac{t}{M^{1/\beta}}\right) \equiv f\left(\frac{t}{M^{a/b}}, 1\right), \tag{1.20b}$$

and the choice $\beta = (b/a)$ and $\delta = (p/b)$ for the critical exponents. In the same way one can represent the Gibbs potential $\Phi(t,h)$ or the correlation function $G(\xi, R)$. We shall not enter into a detailed discussion of the phenomenological scaling because we shall come back to the scaling features of the critical state in Chapter 8 where the further development of the scaling theory will be presented by means of the renormalization group (RG) approach to critical phenomena.

The power dependencies like (1.11) and (1.13) and the relations between the critical exponents which follow from the homogeneity (scaling) hypothesis are often called scaling laws. Certainly the scaling laws and scaling equations of state, like Eq. (1.20a) will be valid sufficiently near T_c where the leading power dependencies of the thermodynamic quantities on the free parameters t and h are much larger than corrections to the scaling (power) laws; see, e.g., Eqs. (1.3), (1.4), and (1.9). Assuming the Widom scaling hypothesis, we can derive the equalities relating the six static critical exponents α, β, γ, δ, ν, and η. Under the symmetry requirement ($\lambda_f = \lambda'_f$) for all these exponents, four scaling relations can be obtained

$$\begin{aligned} \text{Widom's relation:} \quad & \gamma = \beta(\delta - 1) \\ \text{Fisher's relation:} \quad & \gamma = (2 - \eta)\nu \\ \text{Rushbrooke's relation:} \quad & \alpha + 2\beta + \gamma = 2 \\ \text{Josephson's relation:} \quad & \nu d = 2 - \alpha. \end{aligned} \tag{1.21}$$

These scaling laws are valid to the same extent as the Widom scaling hypothesis. If we know two of the exponents from the experiments, the remaining four can be found with the help of the relations Eq. (1.21). For considering both the static and the dynamic critical behaviour, one should know three quantities: two static exponents and the dynamic critical exponent z. The dynamic critical exponent z essentially depends on certain thermodynamic constraints on the system. In many relevant cases the dynamic critical exponent z is independent on the static critical exponents which means that this exponent does not enter in algebraic relations of type Eq. (1.21) with other (static) critical exponents; for details, see, Stanley (1971) and Hohenberg and Halperin (1977).

1.7 Length-Scale Invariance and Renormalization

The scaling approach has been also introduced on a quasi-macroscopic level of description by Kadanoff (1966) and Patashinskii and Pokrovskii (1964, 1966). Kadanoff (1966) derived the scaling for the static correlation functions using the idea of the length-scale invariance which later became the basis of the RG approach to the critical phenomena. Dividing the system in cells (blocks) ν of size L_ν much less than the correlation length ξ, he has shown that the parameters t and h are renormalized by scaling factors depending on the size L_ν of the cells. In fact, the thermodynamic functions, the correlation function and the parameters of the Hamiltonian of the system will satisfy scaling relations like Eqs. (1.19) where λ is a function of L_ν.

The results not only confirm the phenomenological homogeneity hypothesis but the block construction introduced by Kadanoff gives new theoretical opportunities. It has been shown for the first time that a system consisting of the original particles can be mapped to an effective system built up of blocks, for whose description a reduced number of variables is necessary. The mapping is carried out by recursion (scaling) relations connecting the parameters of the original (microscopic) system to the block variables. This construction is quite similar to the static variant of the coarse-graining procedure in statistical mechanics, e.g., the Bogoliubov–Born–Green–Kirkwood–Yvon hierarchy applied to the Liouville equation (see Uhlenbeck and Ford (1963), Münster (1969) and the references therein).

The Kadanoff block construction is based on the notion that the short (finite)-range effects may turn out to be unessential to the asymptotic critical behaviour characterized by extremely long-range (inter-cell) correlations ($\xi \to \infty$) of the order parameter fluctuations $\delta\varphi(x)$. The enlargement of the block size from one value L to another L', $L < L' \ll \xi$ will lead to the removal of the relatively short-range ($\lesssim L'$) effects of the inter-particle correlations in the new lattice of blocks of size L'. In fact, the effects having a range $\sim L \div L'$ are taken into account by the effective change of the parameters of the lattice (lattice renormalization).

Similar ideas, i.e., that the finite-scale effects will lead to a simple renormalization of the parameters of the system and the asymptotic ($T \to T_c$) critical behaviour should be governed by the fluctuation correlations at long distances are used as a basis of the microscopic scaling theory of the nonideal Bose gas developed by Patashinskii and Pokrovskii (1964, 1979). These authors have shown also how the theory of interacting fluctuations can be applied to the study of critical fluctuations with the help of the nontrivial parquet (skeleton) diagrammatic technique; for an introduction see Roulet, Gavoret and Nozières (1969). In a subsequent paper Patashinskii and Pokrovskii (1966) have introduced the method of the correlation functions making a generalization of the Ornstein–Zernicke theory for interacting fluctuations. After these authors several attempts involving additional ideas have been made to derive the scaling properties of the critical state from effective Hamiltonians of interacting fluctuations (Polyakov, 1968; Migdal, 1968; De Pasquale, Di Castro and Jona–Lasinio, 1971).

These theories are based on the re-summation (parquet) method for the perturbation diagrams first introduced for the description of the ultraviolet asymptotes of the propagator in the quantum electrodynamics (Landau, Abrikosov and Khalatnikov, 1954; Fradkin, 1955; Landau and Pomeranchuk, 1955). The studies in quantum electrodynamics have stimulated the development of the field–theoretical RG method (Stueckelberg and Petermann, 1953; Gell–Mann and Low, 1954; Bogoliubov and Shirkov, 1955); see also Bogoliubov and Shirkov (1959). Abrikosov (1965) has re-summed the most divergent perturbation terms for the Kondo problem which concerns the singular behaviour of magnetic impurities in metals at low temperatures. The obtained logarithmic divergences are similar to those in the quantum field theory (a result, reproduced by Fowler and Zawadowsky (1971) with the help of the Gell–Mann–Low renormalization scheme).

Later on, Larkin and Khmel'nitskii (1969) have discovered the logarithmic divergences of the correlation function (fluctuation propagator) in uniaxial ferroelectrics, which can be effectively considered in the 4D space. In this way they have opened the road for the study of interacting fluctuations for the so-called critical (borderline) dimension $d = 4$ and the investigation of the logarithmic corrections to the Ornstein–Zernicke correlation function of free (noninteracting) fluctuations. A similar perturbation techniques for the self-consistent treatment of interacting fluctuations in a close vicinity of T_c have been developed by Vaks and Larkin (1965), Vaks, Larkin and Pikin (1966), Larkin and Pikin (1969), and Larkin, Mel'nikov and Khmel'nitskii (1971).

All above mentioned studies are qualitatively correct and give the opportunity for successful predictions about a number of fluctuation and scaling properties near phase transition points. In the sixties the development of sophisticated resummation perturbation methods and the Gell–Mann–Low RG was directed in three main fields — the quantum field theory, the Kondo impurities in metals, and the critical phenomena — all of them concerning the problem of how to treat strongly interacting fields.

Returning to the problem of phase transitions and critical phenomena, it becomes obvious that the quantitative description cannot be carried out from the first principles of the statistical mechanics, i.e., from the microscopic Hamiltonian of the system. Since the phenomena in this field are quasi-macroscopic (macroscopic or quasi-macroscopic ordering and long-scale correlation properties) the microscopic phenomena can safely be ignored or taken into account by the coarse graining procedure. This procedure can be thought of as a formal logical tool because it cannot be explicitly applied to real systems. In this sense we can always imagine that the system is described by a (quasi)macroscopic fluctuation Hamiltonian (the free energy or the Gibbs potential) of fluctuation fields-quantities, like the order parameter $\varphi(x)$, that depend on the spatial position x. This effective Hamiltonian $\mathcal{H}[\varphi(x)]$ — functional of $\varphi(x)$, can be suggested by adequate phenomenological and symmetry arguments, or, as it is in some cases confirmed by calculations of microscopic models (see, e.g., the Hubbard–Stratonovich transformation in Section 6.6). Thus the phase transition theory becomes a theory of a fluctuating classical field $\varphi(x)$.

The quantum effects can be ignored in a close vicinity of the critical points ($T_c > 0$), where the correlation length ξ of the thermal fluctuations is much larger than the (de Broglie) thermal length $\lambda_T = (2\pi\hbar^2/mk_{\rm B}T)^{1/2}$ of the quantum correlations between the particles (or, quasiparticles) of (effective) mass (Sections $9.2 - 9.4$). Instead of a coarse-graining procedure from microscopic to macroscopic variables the Kadanoff block picture can be used to perform an effective coarse-graining from a quasi-macroscopic characteristic length-scale (L) to another $L' > L$.

In 1971 K. Wilson published his "shell" analysis, in which it was demonstrated how the Kadanoff transformation could be represented by an approximate recursion formula for the parameters of the block variables. This recursion formula is a renormalization transformation for the parameters of the theory and it makes possible to calculate for the first time the nonclassical values of the critical exponents (Wilson, 1971). Soon after (Wilson and Fisher, 1972) have shown how the critical exponents can be calculated from new (exact) recursion relations and the newly introduced $\epsilon = (4-d)$-expansion. In this way a new Kadanoff–Wilson–Fisher RG scheme has been introduced.

The new RG theory received in the seventies an impressive application to a variety of problems in phase transitions and other fields in Physics. This period in the theory of phase transitions is discussed in Chapter 8. Some applications are considered in Chapter 9.

Chapter 2

Equilibrium States and Phases

2.1 Phases

The thermodynamics of phase transitions originates from the general thermodynamic concepts of equilibrium. The thermodynamic state, at which a phase changes to another phase, namely, the phase transition point, is described to a great extent in terms of equilibrium thermodynamics. The study of the relation between the concepts of thermodynamic equilibrium and the phenomenon of a phase change is the way to retrace the foundations of the theory of phase transitions. Obviously, the notion of *phase* is the one to start with. In classical thermodynamics the phases are homogeneous bodies of the same substance that exist in equilibrium and are distinguished by suitably chosen thermodynamic parameters. There are two essential points in this definition: (i) the phases are *equilibrium states*, and (ii) they *can be distinguished* from each other by specific thermodynamic features. These can be the particular values of certain *extensive thermodynamic variables* such as the volume, magnetization, electric polarization and so on. Besides, let us emphasize, that the different phases have different energies (at fixed thermodynamic parameters), except for certain isolated states of coexistence of two or more phases, where their energies are equal (Section 2.7).

In this Chapter we focus our attention on the equilibrium properties of phases. As it is thermodynamics itself that provides the description of thermal equilibrium, our consideration will be inevitably related to basic thermodynamic principles and equations. The general aspects of the second problem concerning the distinction between phases are also established below. However, this second topic is elucidated in greater detail in the next Chapter where we apply the general conditions of equilibrium and stability to problems of the theory of phase transitions.

In classical thermodynamics phases do not arise — they rather exist. This is an *ad hoc* assumption related to the general concept of thermal equilibrium. Remember that the notion of equilibrium, namely, that a system will always reach an equilibrium state when kept under fixed conditions, may be considered as a postulate of the classical thermodynamics. The different phases usually appear as manifolds of equilibrium states in the space of the thermodynamic degrees of freedom and in this aspect we make difference between phases and states.

Since the very concept of a phase refers to equilibrium, it obeys *the principle of increase of entropy*. The latter can be used to select and classify the equilibrium states and, consequently, the possible phases. This is carried out by subjecting the entropy of the system to a variational procedure. As a result equations for the thermodynamic degrees of freedom are obtained. From these equations one may deduce the general conditions for equilibrium and stability of thermodynamic systems (Gibbs, 1876, 1878); see also Gibbs (1948). In this way, questions of extreme importance in the phase transition theory can be answered to, for instance, whether a substance can exist in equilibrium as a homogeneous (single-phase) state or it splits into several phases so that a heterogeneous state sets in. This problem is directly related to the description of phase coexistence and to the construction of the phase diagrams for various thermodynamic systems. Furthermore, the variational approach provides the link between the thermodynamic and the statistical theories of phase transitions.

In the next sections we review the general conditions of equilibrium and stability of thermodynamic systems. Besides, we outline the derivation of these conditions and some basic thermodynamic relations. Details concerning the explanation of thermodynamic equations are supposed known.

The reader may recall the general background and the details from advanced university courses such as those written by Pippard (1957), Callen (1960), Landau and Lifshitz (1980), and Reichl (1980).

2.2 Fundamental Relations

A thermodynamic system is defined by a set of extensive parameters, the most universal of which are the entropy S and the internal energy U. The choice of additional extensive parameters depends on the problem under consideration. For example, we call *simple fluids* those thermodynamic systems, in which the volume V and the number N of particles are taken as relevant parameters together with S and U. In this simple case, one can successfully treat condensation processes which manifest themselves as jumps of the volume V from one value to another. By varying the volume one can describe the peculiarities of the vapour, liquid, and solid phases, and their mutual transformations. If a substance consists of several chemical components, it can no longer be a simple fluid. Now, in addition to condensation, the phenomena of phase separation can be investigated. In magnetic or dielectric bodies, one needs new extensive variables: the total magnetization or the total electric polarization, respectively. In systems where surface effects are important the surface area has to be included into consideration. In the studies of more complex problems, a number of *additional extensive parameters* have to be introduced. Fortunately, the usual situation is such that the essential features of the phenomena can be described in terms of a few variables.

2.2.1 *Fundamental equations*

The fundamental relation for a thermodynamic system is the equation involving the extensive parameters S, U and the chosen set $a = \{a_i\}$ of additional extensive variables. The fundamental equation of the system then can be written in the form

$$S = S(U, a). \tag{2.1}$$

If, as in Eq. (2.1), the entropy is considered to depend on U and a, we shall refer to the analysis that follows as being carried out in the *the entropy representation.* In this case, the extensive variables U and a_i are the natural variables of the entropy function. Once the fundamental relation (2.1) is known, all conceivable thermodynamic information about the system is ascertainable from it.

One can consider as a postulate of the equilibrium thermodynamics that the entropy S is a single-valued, piecewise continuous, differentiable and monotonically increasing function of U at constant a_i. The monotonic property of S yields the temperature T as a positive quantity:

$$\left(\frac{\partial S}{\partial U}\right)_a = \frac{1}{T} > 0. \tag{2.2}$$

Besides, the enumerated properties imply that the entropy function $S(U, a)$ can be inverted with respect to U,

$$U = U(S, a). \tag{2.3}$$

so the energy can be thought of as a function of state with properties similar to those of the entropy S. Then Eq. (2.3) is the fundamental relation in *the energy representation.* Both representations (2.1) and (2.3) of the fundamental relation (equation) are equivalent and can be used on equal grounds.

2.2.2 *Properties of entropy and energy*

The extensivity of S allows the introduction of the scaling transformation

$$S(U, a_1, \ldots, a_{m-1}, a_m) = a_m S\left(\frac{U}{a_m}, \frac{a_1}{a_m}, \ldots, \frac{a_{m-1}}{a_m}\right), \tag{2.4}$$

where a_m is usually chosen among the volume V, the number N_ν of particles of one of the chemical components or the total number N of particles, $N = \sum_\nu N_\nu$. With the choice $a_m = N$, the fundamental relation (2.4) is written in the form

$$s = s(u, a_{1,d}, \ldots, a_{m-1,d}, 1), \tag{2.5}$$

in terms of the *densities*: $s = S/N$, $u = U/N$, and $a_{i,d} = a_i/N$; the d-subscript merely indicates that $a_{i,d}$ is a density. Now one of the extensive parameters is redundant and the number of the parameters of state in terms of densities is reduced from $(m+1)$ to m. The same procedure can be performed for the energy: $u = u(s, a_{id}, 1)$; hereafter $a_{i,d} \equiv a_{id}$.

The extensivity of all quantities in the relations (2.1) and (2.3) implies that S and U are first-order homogeneous functions of their variables. For example,

$$S(\lambda U, \lambda a) = \lambda S(U, a), \tag{2.6}$$

for any λ. This expression is a generalization of Eq. (2.4). With the help of the Euler theorem for homogeneous functions (see, e.g., Callen (1960), Münster (1970), and the Section A), we obtain

$$S = \left(\frac{1}{T}\right) U + \sum_i \left(\frac{\partial S}{\partial a_i}\right) a_i\,; \tag{2.7}$$

cf. Eq. (A.2). In Eq. (2.7), the suffixes of the partial derivatives $(\partial S/\partial a_i)_{a_j \neq a_i}$ which are in current use in thermodynamics have been omitted as being clear enough.

The same theorem can be applied to the energy function (2.3), so that

$$U = TS + \sum_i \left(\frac{\partial U}{\partial a_i}\right) a_i. \tag{2.8}$$

To cast the relations (2.7) and (2.8) in a more usual form, we choose variables which are suitable for fluid and isotropic magnetic systems. Then, the fundamental relations (2.1) and (2.3) become

$$S = S(U, V, \boldsymbol{M}, N_\nu), \tag{2.9}$$

and

$$U = U(S, V, \boldsymbol{M}, N_\nu), \tag{2.10}$$

and, respectively. Here $\boldsymbol{M}$ is the total magnetization. The Euler equations (2.7) and (2.8) take the form

$$S = \left(\frac{1}{T}\right) U + \left(\frac{P}{T}\right) V - \left(\frac{\boldsymbol{H}}{T}\right) \cdot \boldsymbol{M} - \sum_\nu \left(\frac{\mu_\nu}{T}\right) N_\nu, \tag{2.11}$$

and

$$U = TS - PV + \boldsymbol{H} \cdot \boldsymbol{M} + \sum_\nu \mu_\nu N_\nu, \tag{2.12}$$

where

$$P = -\frac{\partial U}{\partial V}, \tag{2.13}$$

where is the pressure,

$$\boldsymbol{H} = \frac{\partial U}{\partial \boldsymbol{M}}, \tag{2.14}$$

is the magnetic field, and

$$\mu_\nu = \frac{\partial U}{\partial N_\nu}, \tag{2.15}$$

with $\nu = 1, 2, \ldots, l$ are the chemical potentials of the $l \geq 1$ chemical components contained in the substance. In thermodynamic relations like (2.11), (2.12) and (2.14) we shall often assume that the external magnetic field $\boldsymbol{H}$ and the magnetization $\boldsymbol{M}$ are parallel so that the respective thermodynamic equalities could be written by the magnitudes H and M of the vectors $\boldsymbol{H}$ and $\boldsymbol{M}$. According to our stipulation (see the footnote in page 2) we may avoid the bold face for the vectors like $\boldsymbol{H}$ and $\boldsymbol{H}$ even in the general case when these vectors are not parallel provided this does not lead to any confusion.

If the thermodynamic investigation is performed in the energy representation the derivatives $(\partial U/\partial a_i)$ are accepted as a definition of the *intensive parameters*, $h_i^U = (\partial U/\partial a_i)$. For example, when the fundamental equation is written in the form (2.10), we define the temperature as $T = (\partial U/\partial S)_a$ and the other intensive parameters P, H, and μ_ν, with the help of Eqs. (2.13)–(2.15). The negative sign in (2.13) defines the pressure in accordance with the mechanical notion of pressure. When the entropy representation is applied the intensive parameters are defined by the first derivatives of the entropy $h_i^S = (\partial S/\partial a_i)$. For example, a system with a fundamental equation (2.9) has *entropic intensive parameters* $1/T$, P/T, $-H/T$, and $-\mu_\nu/T$ as this is readily seen from Eq. (2.11).

The densities defined by Eq. (2.5) are also intensive parameters in the sense that they do not depend on the size of the system. However, the densities $(s, u, a_{i,d})$ and the intensive variables, such as T, H, and P, have quite different properties (see, e.g., Section 3.2.1). Variables like T, P, etc, which are the first derivatives of the extensive parameters are often called *thermodynamic fields* or the intensive parameters of the system. One should distinguish between the two types of intensive parameters: the densities and the thermodynamic fields.

2.2.3 *Notations*

It should be clear from the foregoing discussion that the general notations are useful in many aspects. In certain cases, it will be unnecessary and cumbersome to stick to them throughout. To keep close to the usual notations, we shall illustrate the results in terms of fluid variables (V, N_ν) when necessary. In these variables Eq. (2.1) reads

$$S = S(U, V, N_\nu). \tag{2.16}$$

Many derivations of thermodynamic formulae can be presented in these notations. The results can readily be extended to other systems. This is a way of presentation which will be often followed in our review of thermodynamic results. In the more general cases, one needs only to replace the volume V with a set of variables $\{X_i\}$ and to interpret X_i according to the physical features of the system considered. For example, in the case of ferromagnets one has to replace $V \to -M$ in Eq. (2.16). In studies of purely magnetic phenomena, the numbers N_ν are not of importance and can safely be excluded from consideration. In more complicated cases the variations of the fluid variables are considered simultaneously with other extensive parameters. Under such circumstances, the description for fluid systems is generalized by a fundamental relation of the type

$$S = S(U, V, X, N_\nu). \tag{2.17}$$

In what follows one should become convinced that the essence of the problem is neither in the notations used, nor in the possibility for generalization. Utmost generality is never needed in the description of real systems. A moderate generality is however desirable in at least two aspects and these are: to make compact derivations of theoretical relations and to clarify the underlying universality of physical phenomena which might seem to have very little in common at first glance. We shall use the general symbols a_i whenever this does not lead to obscure mathematical expressions.

2.2.4 *Macroscopic scales*

The quantities S, U and a_i characterize the thermodynamic system as a whole. In this sense they are *macroscopic parameters.* A problem arises when the thermodynamic system is divided into subsystems (cells) or, which is the same, when it represents only a part of a bigger system. In this case, the questions to be answered are whether the total system possesses *the property of additivity* and under what conditions this property is fulfilled. Now we shall dwell upon this important point.

Let us denote our macroscopic system by $\sigma(S, U, a)$. Everything which is not included in σ comprises its *surroundings* $\sigma^0(S^0, U^0, a^0)$. The total system will be denoted by $\sigma^t (= \sigma + \sigma^0)$ with parameters S^t, U^t and a^t. This simple division is quite common in thermodynamics. It is schematically illustrated in Fig. 2.1.

The total parameters S^t, U^t and a^t can be written as sums

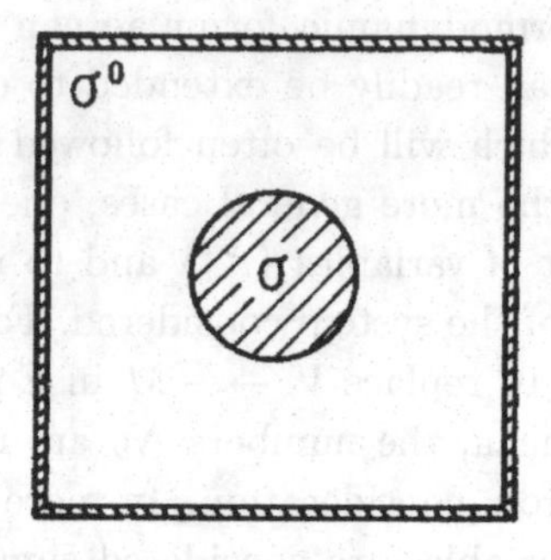

Fig. 2.1 A composite system $(\sigma^0 + \sigma)$.

$$S^t = S^0 + S, \tag{2.18}$$

$$U^t = U^0 + U, \tag{2.19}$$

and

$$a^t = a^0 + a. \tag{2.20}$$

It is important to realize that the representation of S^t, U^t and a^t as sums of the corresponding quantities of the cells does not represent the additivity property yet. This merely indicates, and sometimes it is quite conventional, that a portion of each extensive quantity is suitably attributed to σ and the remaining portion is referred to σ^0. Additivity implies the fulfilment of stronger condition: each of the parameters S, U, and a_i considered as a function of state obeys Eqs. (2.18) to (2.20) and, when related to a given cell σ, it is a function of the extensive parameters of that cell *alone*. For example,

$$S^0 = S^0(U^0, a_i^0). \tag{2.21}$$

does not depend on U and a_i and $S(U, a)$ from Eq. (2.1) does not depend on U^0 and a_i^0 provided additivity is available.

Clearly, macroscopic thermodynamics relies on this property and it is implicitly involved in the fundamental relations (2.1) and (2.3). Additivity makes possible to define the macroscopic parameters of a thermodynamic system and is closely related to the homogeneity property (2.6). It should be accepted in all cases of validity of the thermodynamic theory, i.e., within the thermodynamic treatment in terms of macroscopic (mean) parameters S, U, and a_i.

There is a simple but rather trivial example of a system in which the additivity holds exactly. This is when σ^0 and σ are isolated from each other by boundaries which are absolutely impermeable for energy or particles exchange. Obviously σ^0 and σ can be separated away without any physical change. This is an example of mental and physical additivity of macroscopic cells. Surely, the exact additivity excludes correlations between the cells and, hence, between their extensive parameters. Physical systems are however correlated and in most of the cases, these correlations are rather strong. It is left to our choice to think of them as divided into cells as a possible way of investigation.

The additivity assumption should certainly be correct in the case of macroscopic cells σ^0 and σ of sizes larger than the characteristic length, over which the correlations die off. Then one might neglect the correlations as being microscopic or quasi-microscopic phenomena of no importance to macroscopic studies. It is in the style of thermodynamics to deal with macroscopic effects only and to ignore phenomena of a relatively smaller scale. The mathematical expression of this statement takes the form of the additivity relations (2.18)–(2.21). The sums in Eqs. (2.18)–(2.20) can always be constructed because of the extensivity of S, U, and a_i; however in cases where relations such as Eq. (2.1) and Eq. (2.21) for σ and σ^0 are not correct any more, it is said that σ and σ^0 are no longer macroscopic systems. Then, for example, S will depend on both (U, a) and (U^0, a^0): $S = S(U, a; U^0, a^0)$. The same can be written for S^0, U, and U^0 as well as for any other function of state. Now the problem of description extends beyond the framework of classical thermodynamics and requires the methods of statistical thermodynamics.

2.3 Entropy Increase

Let us discuss our composite system σ^t in terms of the fluid variables S, U, V, and N_ν. Now it is convenient to denote $a \equiv (U, V, N_\nu)$. The cell σ interacts with its surroundings σ^0. Our task is to deduce the equations for the thermodynamic degrees of freedom a_i and the behaviour of the entropy S of the open (interacting) system σ which is allowed to exchange heat, work, and matter with its surroundings σ^0. Now the parameters a_i may undergo a redistribution between σ and σ^0. When some of these processes of exchange are forbidden, this will be settled by agreement. With this statement of the problem there is no loss of generality in imagining that the total system $\sigma^t (= \sigma + \sigma^0)$ is isolated.

In order to obtain information about σ, we turn to the second law of thermodynamics in one of its most useful versions — the principle of the increase of entropy. In the case of an isolated system σ^t it affirms that the entropy change ΔS^t is positive and tends to zero for any thermodynamic processes which come close to *reversibility*, i.e.,

$$\Delta S^t = \sum_i \left[\left(\frac{\partial S}{\partial a_i} \right) \Delta a_i + (\partial S^0 \partial a_i^0) \, \Delta a_i^0 \right] \geq 0, \tag{2.22}$$

where the equality is attained for reversible processes (see below). The change ΔS^t is produced by the redistribution of the quantities a_i between the subsystems σ and σ^0. These exchanges are denoted by Δa_i in Eq. (2.22).

The principle (2.22) determines the direction of the evolution of an isolated system, in which $\Delta a_i^0 = -\Delta a_i$. It is obvious that the entropy function $S^t(a)$ tends to its maximal value which will be inevitably reached after an adequate number of *irreversible processes* have taken place ($\Delta S^t > 0$). Thus the maximum of S^t is a particular value which describes a special state — the state of equilibrium (or just: *the equilibrium*) of the system σ^t. All other states are called *nonequilibrium states.*

Having reached equilibrium, an isolated system may only change by reversible processes, in which $\Delta S^t = 0$. On the ground of inequality (2.22), the irreversible processes which tend to equilibrium are often called *natural processes* in the sense that they actually happen in σ^t. An unnatural process is one proceeding in a direction away from equilibrium with

$$\Delta S^t < 0 \qquad \textit{(unnatural processes)} \tag{2.23}$$

and, hence, it never occurs (in macroscopic thermodynamics). A borderline case between these two processes is the *reversible process,* for which Eq. (2.22) holds.

It should be noted that the principle (2.22) itself does not ensure a maximum of the entropy function $S^t(a)$ in the well known mathematical sense. $S^t(a)$ would rather possess a maximum provided that appropriate analytic properties are assumed (see Section 2.2.1). In what follows, requirements of this type will often be taken for granted without further comments.

If the composite system σ^t is in an equilibrium, $S^t(a)$ takes its maximum value. It is common to think of S^t and consequently of S^0 and S as possessing more than one maximum. The local properties of σ^t in a close vicinity of the supposed equilibria are established by the behaviour of the neighbouring nonequilibrium states in the same way, in which the local extremal properties of any function are determined by its behaviour in the infinitesimal vicinity of the extrema. The neighbouring nonequilibrium states cannot be reached by natural or reversible infinitesimal processes as they correspond to a loss of entropy.

Infinitesimal changes in σ^t that lower the entropy ought to be related to unnatural processes (2.23). Both the hypothetical infinitesimal processes which bring the system out of equilibrium and the natural infinitesimal processes will be denoted by Δa_i in contrast to the finite variations Δa_i. The infinitesimal reversible changes which always leave the system in equilibrium will be denoted by da_i.

As a matter of fact the total system σ^t is isolated so the parameters a_i^t are fixed and $S^t(a^t)$ can hardly be accepted as a function which reaches extrema at some values of its variables a_i^t. We are interested, however, in the cell σ, in which the variables a_i may vary to produce a maximal value of the entropy function $S^t(a) = S^t(a+a^0)$. Within this scheme the entropy $S^t(a + a^0)$ is supposed to be a well-defined function for a wide range of values of the extensive parameters a_i which cover both equilibrium and nonequilibrium states of the system σ^t.

Using the principle of increase of entropy, the task of finding the equilibrium properties of our open system σ can be formulated as a variational problem. We can imagine that infinitesimal unnatural processes from equilibrium to nonequilibrium pseudo-states are initially admitted to occur in σ^t and that they are subject to the constraints of fixed a^t,

$$\delta a^t = 0, \qquad a^t = (U^t, V^t, N^t). \tag{2.24}$$

The variational procedure together with the criterion of entropy increase will select the natural (real) processes and will rule out the unnatural (impossible) ones.

Perhaps it is convenient to remember that a usual and fruitful approach in theoretical physics is to obtain the basic equations by a *variational principle.* For example, in classical mechanics and electrodynamics the equations of motion for the degrees of freedom are derived by a similar variational procedure using a special function called *the action* of the system. In the action of mechanical or electrodynamic systems all "motions" of the degrees of freedom subject to certain constraints are permitted, until a certain law, the principle of *least action* selects the real motions and rules out the impossible ones. As a result of the variational procedure one obtains the equations for the real motions. In a thermodynamic description, we assume that a generalized entropy function $S(a)$ exists and gives both the equilibrium and nonequilibrium states.

The variational procedure applied to $S^t(a)$ will enable us to rederive the basic thermodynamic relations and, which is our main purpose, to deduce the conditions of equilibrium and stability of thermodynamic systems; see, e.g., Prigogine and Defay (1954), Glansdorf and Prigogine (1971), and Reichl (1980).

2.4 Conditions of Equilibrium and Stability

Suppose that owing to some spontaneous process, infinitesimal variations (δa_i) occur in the energy U, volume V, and particle numbers N_ν in each of subsystems σ and σ_0, subject to the constraints of fixed $a^t = \{a_i^t\}$; see Eqs. (2.24). By Eqs. (2.18)–(2.20) and (2.24) the constraints take the form

$$\delta a_i^0 = -\delta a_i. \tag{2.25}$$

Because of this connection, we can choose for our purposes δa_i as the independent variations. The change of the entropy ΔS^t of the total system produced by δa_i can be taken into account to the first, second or higher-order terms in δa_i, that is

$$\Delta S^t = \delta S^t + \delta^2 S^t + \dots. \tag{2.26}$$

2.4.1 *Equilibrium conditions*

The first variation δS^t is obtained by Eq. (2.18) and Eq. (2.25). In terms of the fluid variables it reads

$$\delta S^t = \left(\frac{\partial S}{\partial U} - \frac{\partial S^0}{\partial U^0}\right)\delta U + \left(\frac{\partial S}{\partial V} - \frac{\partial S^0}{\partial V^0}\right)\delta V + \sum_\nu \left(\frac{\partial S}{\partial N_\nu} - \frac{\partial S^0}{\partial N_\nu^0}\right)\delta N_\nu. \tag{2.27}$$

Note that in deriving Eq. (2.27), we have supposed that the additivity property (2.21) is fulfilled. All derivatives in Eq. (2.27) are taken in the equilibrium so we can use Eq. (2.2) and the thermodynamic relations,

$$\frac{\partial S}{\partial V} = \frac{P}{T}, \tag{2.28}$$

and

$$\frac{\partial S}{\partial N_\nu} = -\frac{\mu_\nu}{T}. \tag{2.29}$$

Then δS^t will be

$$\delta S^t = \left(\frac{1}{T} - \frac{1}{T^0}\right)\delta U + \left(\frac{P}{T} - \frac{P^0}{T^0}\right)\delta V - \sum_\nu \left(\frac{\mu_\nu}{T} - \frac{\mu_\nu^0}{T^0}\right)\delta N_\nu. \tag{2.30}$$

The *stationary* values of S^t are given by the equation

$$\delta S^t = 0. \tag{2.31}$$

This equation will be satisfied for any variations $\delta a = (\delta U, \delta V, \delta N_\nu)$, if only

$$T = T^0, \tag{2.32}$$

$$P = P^0, \tag{2.33}$$

and

$$\mu_\nu = \mu_\nu^0. \tag{2.34}$$

In equilibrium, the temperatures, pressures and chemical potentials of the subsystems σ and σ^0 ought to be equal. Eq. (2.32) is called *thermal condition* for equilibrium, Eq. (2.33) is the *mechanical condition* and Eq. (2.34) represents a condition of equilibrium with respect to the matter flows (*material condition*). The external (with respect to σ) "forces" T^0, P^0, and μ^0 are equal to the internal ones (T, P, μ_ν).

In deriving relations (2.32)–(2.34), we have not used arguments specifying any similarities or differences between the cells σ and σ^0. Therefore, these results are certainly valid both for homogeneous and inhomogeneous states of the total system σ^t. The Eqs. (2.32)–(2.34) define the *extrema* of the entropy of a fluid and are known as the *equilibrium conditions.* For a general system, one immediately writes these relations in the form

$$\frac{\partial S}{\partial a_i} = \frac{\partial S^0}{\partial a_i^0}, \qquad i = 1, \dots \tag{2.35a}$$

The derivatives are identified according to the physical properties of the particular systems, namely, the physical nature of a_i. For example, if $a_i = M$, $\partial S/\partial a_i = -H/T$. The equilibrium relations (2.35a) are the equations of the (equilibrium) state in the entropy representation. Using the notations h_i^S for the entropy intensive parameters (Section 2.2.2) we can write the relations (2.35a) in the form

$$h_i^S = h_{0i}^S, \tag{2.35b}$$

i.e., the internal fields ("forces") h_i^S are equal to the external "forces". For each number i, the equation $h_i^S = h_{0i}^S$ is obtained for fixed values of the parameters $a_j \neq a_i$.

The extensive variable a_i conjugate to h_i^S in entropy representation is not fixed, rather it varies ($\delta a_i \neq 0$) and receives a particular (equilibrium) value $a_i = \bar{a}_i$ that "maximizes" the entropy S^t (at fixed $a_j \neq a_i$): generally, $\bar{a}_i$ is a stationary point of S^t. It is the value $\bar{a}_i$ of the parameter a_i which takes part in the equilibrium equation of state $h_i^S = h_i^S(\bar{a}_i, a_j)$. Solving this equation, we can find $\bar{a}_i$ as a function of h_i^S and the fixed parameters $a_j \neq a_i$. If all $a_j \neq a_i$ are fixed (the system σ_i is open only with respect to the parameter a_i) we have to solve only one equation.

In general, a subset of parameters can vary freely subject to the constraint $a_i = (a_i^t - a_i^0)$ so we must solve several equations of state. For example, the equations of state for a fluid system are $T = T(a)$, $P = P(a)$, and $\mu_\nu = \mu_\nu(a)$, where $a = (U, V, N_\nu)$; see Eqs. (2.32)–(2.34). At fixed V and N_ν, say $V = V'$, $N_\nu = N'_\nu$ we have to solve the single equation $T = T(\bar{U}, V', N'_\nu)$ and from it, $\bar{U} = \bar{U}(T, V', N'_\nu)$; here $\bar{U}$ is the equilibrium internal energy (the bar of $\bar{U}$ will be often omitted).

In our derivation of equilibrium conditions (2.32)–(2.34), we have also obtained the fundamental equation for the open system σ in a differential form

$$dS = \frac{dU}{T} + \frac{P}{T}dV - \sum_\nu \frac{\mu_\nu}{T} dN_\nu. \tag{2.36}$$

Now dS is related to $dS^t = 0$ (reversible processes: $U \to U + dU$, $V \to V + dV, \dots$) and we write da_i instead of δa_i (see Section 2.3). In general notations it is

$$dS = \frac{dU}{T} + \sum_i \left(\frac{\partial S}{\partial a_i}\right) da_i, \qquad (U \neq a_i), \tag{2.37}$$

where U has been excluded from the set of a-parameters. The case $dS = 0$ corresponds to reversible *adiabatic processes* ($U \to U + dU$, $a_i \to a_i + da_i$).

Comparing Eq. (2.37) with the differential of the Euler equation (2.7),

$$dS = \frac{dU}{T} + Ud\left(\frac{1}{T}\right) + \sum_i \left[\left(\frac{\partial S}{\partial a_i}\right) da_i + a_i d\left(\frac{\partial S}{\partial a_i}\right)\right], \tag{2.38}$$

we obtain the equation

$$Ud\left(\frac{1}{T}\right) + \sum_i a_i d\left(\frac{\partial S}{\partial a_i}\right) = 0 \tag{2.39}$$

(*Gibbs–Duhem relation*). This means that the variations of the intensive parameters T and $\partial S/\partial a_i$ away from their equilibrium values are related and one of these variations can always be represented as a linear combination of the others.

In fluid variables the Gibbs–Duhem relation (2.39) takes the usual form:

$$Ud\left(\frac{1}{T}\right) + Vd\left(\frac{P}{T}\right) - \sum_\nu N_\nu d\left(\frac{\mu_\nu}{T}\right) = 0. \tag{2.40}$$

The differentials in Eq. (2.40) can be expressed with the help of dT, dP, and $d\mu_\nu$. A subsequent application of Eq. (2.12) with $M = 0$ yields

$$SdT - VdP + \sum_\nu N_\nu d\mu_\nu = 0, \tag{2.41}$$

which is the most used form of Gibbs–Duhem relation. All these relations are true at the stationary points (extrema) of the entropy.

2.4.2 *Stability conditions*

To determine the type of the entropy extrema, we need to know the sign of the second-order variation $\delta^2 S$. The maxima of S^t are looked for among the stationary points ($\delta S = 0$), satisfying the requirement

$$\Delta S^t \approx \delta^2 S^t \leq 0. \tag{2.42}$$

In terms of δa_i, $\delta^2 S^t$ becomes

$$\delta^2 S^t = \frac{1}{2} \sum_{ij} \left(\frac{\partial^2 S}{\partial a_i \partial a_j} + \frac{\partial^2 S^0}{\partial a_i^0 \partial a_j^0} \right) \delta a_i \delta a_j, \tag{2.43}$$

where $a = (U, V, \dots)$ and the constraint (2.25) for the total (isolated) system has been applied again: $\delta a_i \delta a_j = \delta a_i^0 \delta a_j^0$, for any ij.

The variation $\delta^2 S^0$ of the entropy S^0 takes part in Eq. (2.43) through the second derivative of S^0 and cannot be eliminated or connected to $\delta^2 S$ without making additional assumptions about the structure of the total system σ^t. The stability analysis can be carried out without such assumptions (see, e.g., Section 2.7.3).

Here our main purpose is to determine the properties of $\delta^2 S$ and, hence, the stability conditions for the system σ. Besides, our intention is to present these stability conditions as relations between the parameters of the system σ, in which the parameters (a^0) of the surroundings σ^0 do not appear.

To do this we can proceed in two ways. The first is to suppose that the supplementary system σ^0 is much larger than σ. This is equivalent to the statement that an extensive variable a_m exists such that $a_m \ll a_m^0 \sim a_m^t$. In fact, any extensive parameter (a_m^0 and S^0) of σ^0 will be much greater than the corresponding parameter (a_i and S) of σ. For such a structure of the total system σ^t, the variation $\delta^2 S^0$ can be neglected as small compared to $\delta^2 S$. This point is explained below in more details.

The second assumption does not concern the relative size of the subsystems σ and σ^0 but implies that the composite system σ^t is homogeneous, i.e., the cells σ and σ^0 consist of identical phases (*a single-phase state*). It will be more clear if we present this condition of homogeneity as a mathematical equality. Let the extensive parameter $a_m \neq S$ have a fixed value (a'_m) in σ. Then the constraint (2.20) of fixed a_m^t means that $a_m^0 = a_m^t - a'_m$ cannot vary too. Usually, a_m is chosen among the particle number N and the volume V.

Defining all densities of the extensive parameters S and $a_i \neq a_m$ with respect to a_m (for example, $s = S/a_m$, $s^0 = S^0/a_m^0, \dots$) and using the notations from Eq. (2.5) we can write the condition of homogeneity for σ^t in the form

$$\frac{\partial^2 s}{\partial a_{id} \partial a_{jd}} = \frac{\partial^2 s^0}{\partial a^0_{id} \partial a^0_{jd}}. \tag{2.44}$$

[$a_{id} = a_i/a_m$; $i \neq m$; cf. notations in Eq.(2.5)]. This equality can be presented in simplified notations, i.e., $s^{(2)}_{ij} = s^{0(2)}_{ij}$. It means that if the phases in cells σ and σ^0 are identical the density derivatives (2.44) will be equal.

Simple calculations show that Eq. (2.44) gives the relation $(a_m/a^0_m) = S^{0(2)}_{ij}/S^{(2)}_{ij}$ between the derivatives S_{ij} and S^0_{ij} in Eq. (2.43). Now we can express the second derivative of S^0 in Eq. (2.43) by the second derivative of S and to obtain the relation between $\delta^2 S^t$ and $\delta^2 S$ with a factor $(a^t_m/a^0_m) > 1$. Here we use the condition that the system σ^0 ("reservoir") is much larger than σ so that $a_m \ll a^0_m$ and $\delta^2 S \gg \delta^2 S^0$; see, e.g., Callen (1960). As a result,

$$\delta^2 S^t \approx \delta^2 S, \tag{2.45}$$

and

$$\delta^2 S \leq 0 \tag{2.46}$$

instead of Eq. (2.42), which means that if the total entropy S^t has a maximum in equilibrium, the entropy of any subsystem (σ) will have a maximum in equilibrium, too.

Note that the condition $a_m \ll a^0_m$ cannot be applied to the first variation δS^t because $\partial S/\partial a_i$ and $\partial S^0/\partial a^0_i$ are intensive variables and do not depend on the size of σ and σ^0. In the remaining part of this subsection we shall consider that the system σ is much smaller than its surroundings σ^0.

The variation $\delta^2 S$ in Eq. (2.43) is then

$$\delta^2 S = \frac{1}{2} \sum_{ij} \left(\frac{\partial^2 S}{\partial a_i \partial a_j} \right) \delta a_i \delta a_j. \tag{2.47}$$

Remember that $\partial^2 S/\partial a_i \partial a_j$ is a symmetric matrix which leads to the well known Maxwell relations (see any book on thermodynamics).

The variation $\delta^2 S$ can be calculated directly from Eq. (2.47). However, the simplest way to do this is to apply the variational operator

$$\delta = \delta U \frac{\partial}{\partial U} + \delta V \frac{\partial}{\partial V} + \sum_\nu \delta N_\nu \frac{\partial}{\partial N_\nu} \tag{2.48}$$

to δS, given by Eq. (2.36) with the change of $d \to \delta$. So

$$\delta^2 S = \frac{1}{2}\left[\delta U \delta\left(\frac{1}{T}\right) + \delta V \delta\left(\frac{P}{T}\right) - \sum_\nu \delta N_\nu \delta\left(\frac{\mu_\nu}{T}\right)\right], \tag{2.49}$$

and after simple differentiations and taking into account the Eq. (2.36), the result will be

$$\delta^2 S = -\frac{1}{2T}\left[\delta T \delta S - \delta P \delta V + \sum_\nu \delta\mu_\nu \delta N_\nu\right]. \tag{2.50}$$

In our example, we have taken δU, δV, and δN_ν as initial variations. As a result, in Eq. (2.50) the six types of variations are equally presented and certainly three of them should be taken as dependent on the others. These variations δU, δV, and δN_ν yield δS and variations of the intensive parameters T, P, and μ_ν. However, any set of three possible variations can be chosen as independent one. One can persuade oneself of this statement by an alternative derivation of Eq. (2.50). It flows directly from Eq. (2.30) after setting $\delta T = (T - T^0)$, $\delta P = (P - P^0)$, and $\delta\mu_\nu = (\mu_\nu - \mu_\nu^0)$, and making simple calculations. In this calculation the terms of order higher than second in powers of variations ought to be neglected.

Let now choose δT, δP, and δN_ν as independent variations. Then

$$\delta S = \frac{\partial S}{\partial T}\delta T + \frac{\partial S}{\partial P}\delta P + \sum_\nu \frac{\partial S}{\partial N_\nu}\delta N_\nu, \tag{2.51}$$

$$\delta V = \frac{\partial V}{\partial T}\delta T + \frac{\partial V}{\partial P}\delta P + \sum_\nu \frac{\partial V}{\partial N_\nu}\delta N_\nu, \tag{2.52}$$

and

$$\delta\mu_\nu = \frac{\partial\mu_\nu}{\partial T}\delta T + \frac{\partial\mu_\nu}{\partial P}\delta P + \sum_{\nu'} \frac{\partial\mu_\nu}{\partial N_{\nu'}}\delta N_{\nu'}. \tag{2.53}$$

Making use of the appropriate Maxwell relations,

$$\left(\frac{\partial S}{\partial P}\right)_{T,N_\nu} = -\left(\frac{\partial V}{\partial T}\right)_{P,N_\nu},$$

$$\left(\frac{\partial S}{\partial N_\nu}\right)_{T,P} = -\left(\frac{\partial \mu}{\partial T}\right)_{P,N_\nu},$$

$$\left(\frac{\partial V}{\partial N_\nu}\right)_{T,P} = \left(\frac{\partial \mu}{\partial P}\right)_{T,N_\nu},$$

the variation $\delta^2 S$ from Eq. (2.50) is obtained in the form

$$\delta^2 S = -\frac{1}{2T}\left[\frac{C_P}{T}(\delta T)^2 + VK_T(\delta P)^2 - 2V\alpha_P \delta T \delta P \right.$$
$$\left. + \sum_{\nu\nu'}\left(\frac{\partial \mu_\nu}{\partial N_{\nu'}}\right)_{T,P} \delta N_\nu \delta N_{\nu'}\right], \tag{2.54}$$

where

$$C_P = T\left(\frac{\partial S}{\partial T}\right)_{P,N_\nu} \tag{2.55}$$

is the specific heat capacity at constant pressure,

$$K_T = -\frac{1}{V}\left(\frac{\partial V}{\partial P}\right)_{T,N_\nu} \tag{2.56}$$

is the isothermal compressibility, and

$$\alpha_P = \frac{1}{V}\left(\frac{\partial V}{\partial T}\right)_{P,N_\nu} \tag{2.57}$$

is the coefficient of thermal expansion. The index N_ν of the derivatives in Eqs. (2.55)–(2.57) is usually omitted.

The form (2.54) which is bilinear in δT and δP can be diagonalized. Straightforward calculations show that it is convenient to introduce the displacement $(\delta V)_{N_\nu}$ of the volume at fixed particle numbers N_ν,

$$(\delta V)_{N_\nu} = \left(\frac{\partial V}{\partial T}\right)_{P,N_\nu} \delta T + \left(\frac{\partial V}{\partial P}\right)_{T,N_\nu} \delta P,$$

and the relation

$$C_P - C_V = \frac{TV\alpha_P^2}{K_T}, \tag{2.58}$$

where $C_V = T(\partial S/\partial T)_V$ is the specific heat at a fixed volume. The final result for $\delta^2 S$ in terms of δT, δV, and δN_ν is

$$\delta^2 S = -\frac{1}{2T}\left[\frac{C_V}{T}(\delta T)^2 + \frac{1}{VK_T}(\delta V)^2_{N_\nu} + \sum_{\nu\nu'}\left(\frac{\partial \mu_\nu}{\partial N_{\nu'}}\right)_{T,P} \delta N_\nu \delta N_{\nu'}\right]. \tag{2.59}$$

The maxima of S^t are given by inequality (2.46) which must be true for any infinitesimal variation δa. Let us consider a variation of the kind $\delta N_\nu = 0$, $\delta T \neq 0$, and $\delta V \neq 0$. Then the condition (2.46) reads

$$C_V \geq 0, \tag{2.60}$$

and

$$K_T \geq 0. \tag{2.61}$$

In some discussions it is more convenient to represent Eq. (2.61) in the form

$$\left(\frac{\partial P}{\partial V}\right)_{T,N_\nu} \leq 0. \tag{2.62}$$

If the variations δN_ν are taken into account we receive a third condition

$$\sum_{\nu\nu'}\left(\frac{\partial \mu_\nu}{\partial N_{\nu'}}\right)_{T,P} \delta N_\nu \delta N_{\nu'} \geq 0, \tag{2.63}$$

which is satisfied when the eigenvalues of the symmetric matrix $(\partial\mu_\nu/\partial N_{\nu'})$ are all non-negative. This condition becomes very simple in the case of one-component systems ($\nu = 1$, $N_1 = N$):

$$\left(\frac{\partial \mu}{\partial N}\right)_{T,P} \geq 0. \tag{2.64}$$

The extrema of the entropy S^t will be maxima, if all inequalities (2.60)–(2.64) are fulfilled. When some of them reduce to equalities, the maximum is no longer guaranteed and variations of S of the order higher than second are to be investigated. The problem is mathematically clear and for completeness we shall write down the conditions for maxima of the entropy:

$$\delta S^t = 0, \qquad \delta^{k+1} S = 0, \tag{2.65}$$

and

$$\delta^{2m} S < 0, \tag{2.66}$$

where $0 < k < (2m - 1)$; k and m – natural numbers. The value $m = 1$ describes "usual" maxima ($\delta^2 S < 0$) whereas $m > 1$ corresponds to "nontrivial" cases, in which the high order analysis is required. Note that such sophisticated higher-order calculations are often necessary in dealing with real systems.

If

$$\delta^{2m} S > 0 \tag{2.67}$$

is obtained for some stationary state (U, V, N_ν), then there is a series of unnatural states with increasing entropy in the neighbourhood of this state. So it is not an allowed state of the system σ, and it corresponds to a (local or absolute) minimum of the entropy.

Note that the conditions (2.65)–(2.67) for $m > 1$ can be derived for a system σ in a large reservoir σ^0 as supposed in the beginning of this subsection. If the sizes of σ and σ^0 are comparable, one should use the variations of the total entropy S^t, i.e., to substitute S in Eqs. (2.65)–(2.67) by S^t.

2.5 Classification of Equilibria

We started with the notion that the maximal value of entropy determines the possible equilibrium states of the system. The foregoing analysis, however, implies a wider understanding of the concept of equilibrium. All stationary points of the entropy defined by Eq. (2.31) describe equilibria. There are several types of stationary points.

The maxima can be of two kinds: an absolute maximum which always exists in a stable system, and local (relative) maxima which may also occur. If only a single entropy maximum is present in the system, it is the absolute maximum. Although the maximal value of the entropy is unique it can be typically achieved over a whole domain of states (U, V, N_ν). An absolute maximum of the entropy or, which is the same, an equilibrium state satisfies not only the relations (2.65) and (2.66) for infinitesimal variations but also the general inequality (2.23), written for both infinitesimal (δa) and finite-size variations (Δa). For any displacement Δa of the extremal state out of equilibrium the system will retrieve this state thus obeying the principle of entropy increase.

The absolute maximum of the entropy describes a *stable equilibrium* or what is normally accepted as *equilibrium* in thermodynamics. The corresponding states and phases are called *stable states* and *stable phases.*

If the largest value of entropy is given by more than one maxima at different points of the phase diagram, the system will be in the so-called *neutral equilibrium* (two or more phases coexist). If local maxima exist, they will satisfy the requirements (2.65) and (2.66) for infinitesimal variations δa from equilibrium, but the inequality (2.23) is broken. In this situation the system could move away to a higher local maximum (or to the absolute one). According to this mechanism a thermodynamic system relaxes to the (stable) equilibrium after some time. The local maxima of the entropy describe *metastable equilibria* and, respectively, *metastable states* and *phases.*

We have established quite a formal mathematical explanation of the metastability which means, in particular, that if a system is in a metastable equilibrium, it will never be strictly in an equilibrium. The thermodynamic and statistical interpretations of metastable phenomena, which we suppose the reader is familiar with, make possible to relate the above *mini–max* picture to the physical processes in real substances. Being in a metastable phase the substance changes with time to the corresponding (stable) equilibrium phase as a result of internal quasi-macroscopic (or lower scale) processes. Some of these processes may turn out to be unnatural and this does not contradict the principle of maximal entropy which, strictly speaking, holds for stable equilibrium only. If the relaxation time of such processes is considerably longer (in some systems it is extremely long) in comparison to the experimental time of observation and if a rapid change is not artificially stimulated, a system in the metastable equilibrium can be looked upon as being in an equilibrium. Under this condition the second thermodynamic law will be valid (at least, approximately) for metastable states, too.

These arguments concern qualitative aspects of the description of metastability and do not change the mathematical scheme outlined above. To consider a metastable state as a truly stable one, we have to neglect the fact that at least one higher-lying maximum of the entropy exists. In order to pass from one extremum to the other, the system goes through a minimum of the entropy. The minima of the entropy are given by Eq. (2.65) and Eq. (2.67). They correspond to the so-called *unstable states.* The absolute and local minima of S do not present qualitatively different physical situations. Note that the introduction of the idea of metastability in the above scheme is bounded to the existence of a higher maximum of entropy. This inevitably leads to the appearance of unstable (minimal) states. Having claimed that our metastable state can be taken as a true equilibrium state because of extremely long relaxation times in the system, we have actually excluded the availability of other maxima of S and within this restricted picture the phenomena of metastability and instability have been ignored. In this way the metastable state enters in the rights of a stable equilibrium.

We have seen how the principle of entropy increase makes possible to introduce several types of equilibria and to establish their principal properties. The unstable states can never occur, the metastable states can occur under some special conditions but, in fact, these states also do not obey the entropy maximum principle and, finally, the stable states do satisfy completely the requirements of the equilibrium thermodynamics. These circumstances are to be accepted as important rules in thermodynamic studies of states and phases and, in particular, that the existence of metastable and unstable states within equilibrium thermodynamics is, strictly speaking, inconsistent with the principle of the entropy increase.

We shall also focus our attention on the special cases of extrema, for which $\delta^2 S^t$ (or $\delta^2 S$ in the case of homogeneous equilibrium) turns to zero. This may happen either for all variations or, as it occurs in common cases, for some of the variations. The last case is illustrated by Eq. (2.54) or Eq. (2.59) when $\delta N_\nu = 0$. So from Eq. (2.60) and Eq. (2.61) it follows that $\delta^2 S = 0$. Such an extremum (maximum or minimum) is called *critical equilibrium*; see Gibbs (1876), and Tisza (1961).

The corresponding states and phases are *critical states* or, more uncommonly, *critical phases.* The critical states, in case of stability can be classified according to the number $m \geq 2$ which is defined by inequality (2.66). It seems possible that in particular cases the number m of a critical equilibrium state may happen to be very big or even infinity. This can occur only in extraordinary cases or in special theoretical models; see, e.g., Zimm (1951). In fact, at $m = 2$ the parameters of state U, V, and N_ν of a fluid system should satisfy the three equations (2.65): $\delta S^t = \delta^2 S = \delta^3 S = 0$.

For simple fluids, the parameters of state are three (U, V, N) and the above three equations define a unique point (provided the system of equations is not degenerate). In the case of $m = 3$, the equations $\delta S^t = \delta^{k+1} S = 0$ are five and they give one or more critical equilibria only in two cases: (i) a high level of degeneration of the system of critical equations, and (ii) a very complex system with many components or work coordinates $(V, M, \ldots)$.

The term marginal stability is also used in cases of nontrivial extrema ($\delta^2 S = 0$) and then it means the same as *critical stability* (or instability). Strictly speaking, this term should be used in the case of $m = \infty$.

It will be tutorial to trace the connection between the classification of critical equilibria (plus neutral equilibrium) made on the basis of the entropy extrema and the classifications of phase transitions presented in Section 3.5 of the next Chapter. The use of the general Gibbs criteria of stability (2.60)–(2.62) in developing a theory of phase transitions as proposed by Tisza (1951) is quite attractive but this approach meets difficulties in studies of phases with a stability of order $m > 2$. Even the case $m = 2$ involves "multivariant effects" and one remains in uncertainty about the correct predictions. They should be derived from the experimental results for real systems rather than from a self-consistent theory.

We recommend to the reader who is interested in the stability theory of phase transitions the original papers by Tisza (1951, 1961). The same theory is presented in detail by Callen (1960) and Münster (1970).

2.6 Thermodynamic Potentials

We have discussed the criteria of equilibrium and stability in terms of the entropy function. It is more convenient in most of the theoretical studies to formulate these criteria in terms of other functions of state. Here we outline the derivation of the extremum conditions for the thermodynamic functions (potentials) which we shall use in our further studies.

2.6.1 *Energy minimum*

The relation (2.22) can be represented in the form

$$\Delta S \geq \left(\frac{\partial S^0}{\partial a_i^0}\right) \Delta a_i, \tag{2.68}$$

where we have used the constraints $\Delta a_i^0 = -\Delta a_i$ for the isolated system σ^t, and $\Delta S^t = \Delta S^0 + \Delta S$. Further we shall extract the energy U from the set $a = \{a_i\}$. Using Eq. (2.68) and $(\partial S^0/\partial U^0) = 1/T^0$, that is, Eq. (2.2) for the surroundings σ^0, we obtain the relation

$$\Delta U \leq T^0 \Delta S - T^0 \left(\frac{\partial S^0}{\partial a_i^0}\right) \Delta a_i, \qquad (U \neq a_i), \tag{2.69a}$$

which represents the law (2.22) in another form. It can be written in fluid variables:

$$\Delta U \leq T^0 \Delta S - P^0 \Delta V + \sum_\nu \mu_\nu^0 \Delta N_\nu. \tag{2.69b}$$

In the relations (2.69a) and (2.69b) the variations ΔS, ΔU, and Δa_i of all extensive parameters of the open system σ are related to the intensive parameters of the surroundings σ^0.

It is obvious from Eq. (2.69a) that, when natural processes at fixed S and a_i take place, the energy change ΔU in the system σ is negative and approaches zero for reversible processes, i.e.,

$$(\Delta U)_{S,a} \leq 0. \tag{2.70a}$$

For fluid systems,

$$(\Delta U)_{S,V,N_\nu} \leq 0. \tag{2.70b}$$

In equilibrium, the internal energy U of our system σ has a minimum at fixed S and a_i. Thus the maximization of entropy at fixed U and a_i is now replaced by minimization of energy at fixed S and a_i.

For reversible processes in equilibrium ($\Delta S^t = 0$), Eqs. (2.68)–(2.70b) are fulfilled. It follows from Eq. (2.69a), Eq. (2.35a), Eq. (2.35b) and Eq. (2.2) that

$$dU = TdS - T\left(\frac{\partial S}{\partial a_i}\right) da_i, \qquad (U \neq a_i). \tag{2.71}$$

Comparing this relation with the differential

$$dU = TdS + \sum_i \left(\frac{\partial U}{\partial a_i}\right) da_i. \tag{2.72a}$$

of the fundamental equation (2.3) we obtain the relation

$$\left(\frac{\partial U}{\partial a_i}\right)_{S,a_j} = -T\left(\frac{\partial S}{\partial a_i}\right)_{U,a_j} \tag{2.72b}$$

between the intensive parameters $(\partial U/\partial a_i)$ in the energy representation and $(\partial S/\partial a_i)$ in the entropy representation. In this relation, $a_i \neq (U, S)$ and $a_j \neq a_i$.

For fluid systems, the differential dU is

$$dU = TdS - PdV + \sum_{\nu} \mu_\nu dN_\nu. \tag{2.72c}$$

Thus we have re-derived Eq. (2.36).

In deriving Eqs. (2.72a) and (2.72b), we have used the equilibrium conditions (2.32)–(2.34), and (2.35a). But the same equations as well as the equilibrium conditions can be obtained directly from the criterion (2.70a) in a way, similar to that used in Section 2.4.1.

2.6.2 *Legendre transformations and extremum conditions*

Suppose that in the system σ the temperature is fixed. This means that σ is interacting with its surroundings σ^0 and they serve as a heat reservoir which is in a diathermal contact with σ. As the temperature T is fixed $(T = T^0)$, we find from Eq. (2.69b) that

$$\Delta(U - TS) \leq -P^0 \Delta V + \sum_{\nu} \mu_\nu^0 \Delta N_\nu. \tag{2.73}$$

Identifying

$$F = U - TS \tag{2.74}$$

as the known *Helmholtz function,* we can write the criterion

$$\Delta F)_{T,V,N_\nu} \leq 0, \tag{2.75}$$

which directly follows from the relation (2.73). The system is in an equilibrium with a "reservoir of temperature" at any fixed V and N_ν, for which the Helmholtz function has a minimum. Comparing the criteria (2.70b) and (2.75), we see that the condition of fixed S is now replaced by a condition of fixed T. We have replaced an extensive parameter of state – the entropy S, by an intensive one – the temperature T, and this has become possible with the help of the relation (2.74). The differential form of Eq. (2.74) reads

$$dF = dU - TdS - SdT. \tag{2.76}$$

Combining it with Eq. (2.72c) we obtain

$$dF = -SdT - PdV + \sum_{\nu} \mu_{\nu} dN_{\nu}. \tag{2.77}$$

The natural variables of the Helmholtz function are T, V, and N_{ν} and it has a minimum with respect to these variables when σ is in an equilibrium.

If the energy of a fluid system is a known function, one may solve the fundamental relation

$$U = U(S, V, N_{\nu}) \tag{2.78}$$

together with the relation

$$T = \frac{\partial U}{\partial S}, \tag{2.79}$$

to obtain U and S as functions of T, V, and N_{ν}. The subsequent replacement of the solutions for U and S in Eq. (2.74) yields the function $F(T, V, N_{\nu})$, that is, the Helmholtz free energy in terms of its natural variables. The procedure of such a change of natural variables based on equations like (2.74) and (2.79) is known as *Legendre transformation.*

If the system σ is in both temperature and pressure reservoirs, that is, the surroundings σ^0 act in a way to keep the temperature and the pressure constant ($T = T^0$ and $P = P^0$) the relation (2.69b) yields

$$\Delta(U - TS + PV) \leq \sum_{\nu} \mu_{\nu}^{0} \Delta N_{\nu}. \tag{2.80}$$

Now we define the *Gibbs function*

$$\Phi = U - TS + PV \tag{2.81}$$

and the corresponding criterion

$$(\Delta\Phi)_{T,P,N_{\nu}} \leq 0 \tag{2.82}$$

follows immediately from Eq. (2.80). Here we have used another Legendre transformation: from independent extensive variables S and V to the intensive variables T and P. To express the Gibbs function $\Phi(T, P, N_\nu)$ in its natural variables, the Eqs. (2.78) and (2.79) together with the relation

$$P = -\frac{\partial U}{\partial V} \tag{2.83}$$

must be solved with respect to U, S, and V, and the result must be put back in Eq. (2.81). The differential $d\Phi$ is obtained from Eq. (2.71) and Eq. (2.81) in the form

$$d\Phi = -SdT + VdP + \sum_\nu \mu_\nu dN_\nu. \tag{2.84}$$

Using the Euler equation (2.12) for fluid systems ($M = 0$), we get

$$\Phi = \sum_\nu \mu_\nu N_\nu. \tag{2.85}$$

For a one-component system ($\nu = 1$, $N_1 = N$) the chemical potential μ is identical with the Gibbs function per particle, $\mu = \Phi/N$.

With the help of Eq. (2.69b) one may demonstrate in an analogous way that the *grand canonical potential,* defined as

$$\Omega = U - TS - \sum_\nu \mu_\nu N_\nu, \tag{2.86}$$

reaches a minimal value in equilibrium. Accordingly, one may explain the corresponding Legendre transformation. The criterion of stability in terms of Ω is

$$(\Delta\Omega)_{T,V,\mu_\nu} \leq 0. \tag{2.87}$$

The differential $d\Omega$ is given by

$$d\Omega = -SdT - PdV - \sum_\nu N_\nu d\mu_\nu. \tag{2.88}$$

The Eqs. (2.74), (2.81), (2.85) and (2.86) yield

$$\Omega = F - \Phi = -PV. \tag{2.89}$$

Clearly all relations (2.75), (2.82) and (2.87) are an application of the principle of entropy increase to the characteristic functions F, Φ, and Ω which are often called *thermodynamic potentials.* F and Φ are also called free energies. Here they will be referred to as the *Helmholtz free energy* and the *Gibbs thermodynamic potential* or, shortly, the free energy and the thermodynamic potential.

Another example of a thermodynamic potential is the *enthalpy function* of fluid systems $W = U + PV$, the natural variables of which are S, P and N_ν. For magnetic systems, $W(S, H, a) = U(S, M, a) - MH$, where a are supplementary extensive variables; cf. Eq. (5.73), where $W(S, h, a)$ is represented by the field h conjugate to the equilibrium order parameter $\bar{\varphi}$. The potential W is treated in the same way as the potentials F, Φ, and Ω.

2.6.3 *Legendre transformations in entropy representation*

It seems redundant to repeat the derivation of the thermodynamic potentials F, Φ, and Ω from the entropy representation of the fundamental relation, but this will give a direct link to statistical thermodynamics. So we shall briefly review this point.

Let us take again expressions (2.22), (2.30), and (2.36) where the infinitesimal variations ("δ" or "d") are replaced by finite displacements (Δ). Then ΔS^t is written in the form

$$\Delta S^t = -\frac{1}{T^0}\left(\Delta U - T^0\Delta S + P^0\Delta V - \sum_\nu \mu_\nu^0 \Delta N_\nu\right) \geq 0. \qquad (2.90)$$

By the way, the expression in the brackets of this equation is called *availability* or *available energy* of the system and is used in the formulation of the criterion of the minimal work, which is used here quite implicitly; for the explicit use of the minimal work, see, e.g., Pippard (1957), and Landau and Lifshitz (1980). Now the contact of our system σ with the reservoir σ^0, which has a temperature $T = T^0$ leads to

$$\Delta S^t = -\frac{1}{T}\left(\Delta F + P^0\Delta V - \sum_\nu \mu_\nu^0 \Delta N_\nu\right). \qquad (2.91)$$

At fixed $T = T^0$ and $P = P^0$,

$$\Delta S^t = -\frac{1}{T}\left(\Delta\Phi - \sum_\nu \mu_\nu^0 \Delta N_\nu\right), \tag{2.92}$$

and, finally, at fixed $T = T^0$ and $\mu_\nu = \mu_\nu^0$,

$$\Delta S^t = -\frac{1}{T}\left(\Delta\Omega + P^0 \Delta V\right). \tag{2.93}$$

Furthermore,

$$-T\Delta S^t = (\Delta F)_{T,V,N_\nu} = (\Delta\Phi)_{T,P,N_\nu} = (\Delta\Omega)_{T,V,\mu_\nu}. \tag{2.94}$$

Besides, it is not difficult to prove the following relations:

$$-T\Delta S^t = (\Delta F^t)_T = (\Delta\Phi^t)_{T,P} = (\Delta\Omega^t)_{T,\mu_\nu}, \tag{2.95}$$

where F^t, Φ^t, and Ω^t are the potentials of the composite system σ^t: $\tilde{\Psi}^t = (\tilde{\Psi}^0 + \tilde{\Psi})$, $\tilde{\Psi} = (F, \Phi, \Omega)$.

At this point, we want to draw the attention to the subtle difference between Eqs. (2.94) and (2.95). For this purpose we shall derive the first equation in Eq. (2.95). From Eq. (2.18), Eq. (2.74), and $S_0 = (U^0 - F^0)/T^0$, the production of entropy ΔS^t in the composite system σ^t can be written in the form

$$\Delta S^t = \Delta\left(\frac{U}{T}\right) - \Delta\left(\frac{F}{T}\right) + \Delta\left(\frac{U^0}{T^0}\right) - \Delta\left(\frac{F^0}{T^0}\right).$$

Having in mind that $\Delta U^0 = -\Delta U$ and under the condition of constant temperature ($T = T^0$), it is readily seen that $\Delta S^t = -\Delta(F + F^0)/T$. The other relations in Eq. (2.95) are derived in the same way using Eqs. (2.18), (2.81), and (2.86), and the constraints $\Delta V = -\Delta V^0$ and $\Delta N_\nu = -\Delta N_\nu^0$.

The extremum conditions (2.65) and (2.66) can be written in terms of the potentials in the form

$$d^{k+1}\tilde{\Psi} = 0, \qquad d^{2m}\tilde{\Psi} > 0. \tag{2.96}$$

The classification of equilibria, presented in Section 2.5, can be straightforwardly applied to the extrema of the potentials $\tilde{\Psi}$. One should keep in mind that the conditions like (2.96) are derived for a system σ in a large reservoir σ^0 (see Section 2.4.2 and Section 2.6.4).

The maximum of S^t at fixed T is given by the maximum of the function $(-F/T)$ and, for fixed T and P — by the maximum of $(-\Phi/T)$. In fact, functions like $(-F/T)$ and $(-\Phi/T)$ which are Legendre transforms of S have been first introduced by Massieu and are often called *Massieu functions* (potentials); see, e.g., Callen (1960), and Münster (1970). They are directly connected with the fluctuation and the statistical properties of macroscopic systems (see Chapters 5 and 6).

2.6.4 *Application*

The relations (2.95) are used in studies of equilibrium states under conditions of fixed T, P, or μ. For instance, in equilibrium and at fixed T, $dF^t)_T = 0$ or $(dF)_T + (dF^0)_T = 0$. The constraints (2.25) yield

$$\left[\left(\frac{\partial F}{\partial V}\right) - \left(\frac{\partial F}{\partial V}\right)_0\right] \delta V + \sum_\nu \left(\frac{\partial F}{\partial N_\nu} - \frac{\partial F^0}{\partial N_\nu^0}\right) \delta N_\nu = 0, \qquad (2.97)$$

which, using $(\partial F/\partial V)_{T,N_\nu} = -P$ and $(\partial F/\partial N_\nu)_{T,V} = \mu_\nu$, leads to $P = P^0$ and $\mu_\nu = \mu_\nu^0$ as in Eq. (2.33) and Eq. (2.34), respectively. However the condition (2.32) cannot be obtained as it has already been taken into account. In the same way, one can obtain from $(d\Phi^t)_{T,P} = 0$ that $\mu_\nu = \mu_\nu^0$. From $(d\Omega^t)_{T,\mu_\nu} = 0$, we get $P = P^0$.

Let all N_ν in σ be fixed by an impermeable wall separating σ and σ^0 (an additional constraint). Then $\delta N_\nu = 0$ and $F(T, V, N_\nu)$ can vary only because of the exchange of the volume V between σ and σ^0. The first variation δF^t is thus reduced to the first term in Eq. (2.97) which gives the equation of state $P = P(T, \bar{V}, N_\nu)$; $\bar{V}$ is the volume that minimizes F^t. Suppose that the equation of state can be solved with respect to $\bar{V} : \bar{V} = \bar{V}(T, P, N_\nu)$. The second variation $\delta^2 F^t$ can be expressed by the second derivatives of F and F^0 with respect to V and V^0. They can be written in the forms $(\partial P/\partial V)_{T,N_\nu}$ and $(\partial P^0/\partial V^0)_{T,N_\nu^0}$ or, equivalently, $N^{-1}(\partial P/\partial v)_{T,N_\nu}$ and ${N^0}^{-1}(\partial P^0/\partial v^0)_{T,N_\nu^0}$; v and v^0 are the volumes per particle in σ and σ^0, respectively. Obviously, $|\partial^2 F/\partial V^2|_{\bar{V}} \gg |\partial^2 F^0/\partial V_0^2|_{\bar{V}_0}$ if $(\partial P/\partial v) \sim (\partial P^0/\partial v^0)$ and the reservoir σ^0 is much larger than the system σ ($N^0 \gg N$). In this case, we can use that $\delta^{(l)} F^t \approx \delta^{(l)} F$, $l \geq 2$, which makes possible to apply the relation (2.96) to the Helmholtz free energy F of the system σ.

The second variation $\delta^2 F \approx \delta^2 F^t$ of F about the equilibrium state (T, V, N_ν) becomes

$$\delta^2 F = A_\nu(T, \bar{V}, N_\nu)(V - \bar{V})^2,$$

where the coefficient A_ν, i.e., the second derivative $\frac{1}{2}(\partial^2 F/\partial V^2)_{\bar{V}}$ will be positive provided the state $(T, \bar{V}, N_\nu)$ is stable.

Let us consider the potential $\Phi(T, P, N_\nu)$ and its relation with the volume V. We shall assume that N_ν are fixed numbers. Then the equality $(d\Phi^t)_{T,P} = 0$ is trivially fulfilled because T and P are kept fixed. For the same reason, we can write $d\Phi = (d\Phi)_{T,P} = 0$ for the system σ. In the equilibrium the potential Φ has a stationary point (eventually, minimum) at fixed T and P. Now the main question is: With respect to which variables Φ exhibits this extremum?

Using the formal dependence $\Phi(V) = F(V) + PV$ in the Legendre transformation (2.89), where V is the nonequilibrium volume ($V \neq \bar{V}$) it is easy to see that $(\partial\Phi/\partial V) = 0$ for $V = \bar{V}(T, P)$ (here P is the equilibrium pressure, $P = -(\partial F/\partial V)_T$). The nonequilibrium Gibbs potential $\Phi(T, P, V)$ of the system has a minimum for the equilibrium states at fixed T and P. So, two equivalent ways can be used to obtain the equilibrium value $\bar{V}$ of the system σ at fixed T and N_ν. The first is to consider the minimum of the total free energy F^t which gives the equation of state $P = P(T, \bar{V}, N_\nu)$ as an equilibrium condition. The second way is to minimize the nonequilibrium potential $\Phi(T, P, V)$ of the system σ.

In the last case the parameters T and P are fixed and the volume V is the nonequilibrium parameter. Minimizing Φ with respect to V, we obtain the equilibrium value $\bar{V}$. Then $\bar{\Phi}(T, P, \bar{V}(T, P))$ is the equilibrium Gibbs potential of the system σ. These remarks are important for our discussion in Chapter 4.

The advantage of the thermodynamic analysis in terms of potentials as F and Φ is that one can consider systems, in which one or more intensive parameters are fixed. The conjugate extensive parameters vary so as to reach the equilibrium values, which are extrema of the potential Φ. For example, we can introduce a generalized Helmholtz free energy $F(S)$ with the help of Eq. (2.74), where F is considered as a function of U, T, and S subject to the condition $\partial F(S)/\partial S = 0$. The last equation, or, equivalently, $(\partial U/\partial S) = T$, gives the caloric equation of state. The solution of the caloric equation with respect to S yields the equilibrium value $\bar{S} = \bar{S}(T, a_i)$ where a_i are the fixed thermodynamic parameters.

There are limitations in finding the equilibrium conditions from the potentials $\tilde{\Psi}$, because in the formulation of the extremum principle for them it is imposed from the very beginning that the total system σ^t is in thermal equilibrium with respect to some of the intensive parameters. For example, for F, T is assumed at its equilibrium value, for G both T and P are at their equilibrium values and for Ω, T, and μ_ν are at their equilibrium values. This is demonstrated by Eqs. (2.94). So it is the entropy or the energy extrema which define the equilibrium conditions for any type of interaction between the system considered (σ) and its surroundings (σ^0). The extrema of the thermodynamic potentials define the equilibrium conditions for the same system at certain (but not arbitrary) conditions of its interaction with the surroundings.

2.7 Coexistence of Phases

We shall now demonstrate that the former results are generalized for heterogeneous states, that is, when the composite system σ^t consists of different phases. Let us divide σ^t in p cells: $\sigma^\alpha(S^\alpha, a^\alpha); \alpha = 1, \ldots, p$. This division is shown in Fig. 2.2. In this mental division some of the cells may consist of one and the same phase and, if the system is homogeneous, all of them will contain the same phase. In the last case, the previous consideration for two cells is generalized straightforwardly for p cells. Studying the potentials, one has to take one of the cells as representing the appropriate reservoir (of T, P, or μ_ν). We have already demonstrated in Section 2.4.1 that the equilibrium conditions (2.32)–(2.34) are realized for both homogeneous and inhomogeneous systems. The application of this result is directly extended for p cells, while the stability conditions in case of a multi-phase coexistence need a special attention.

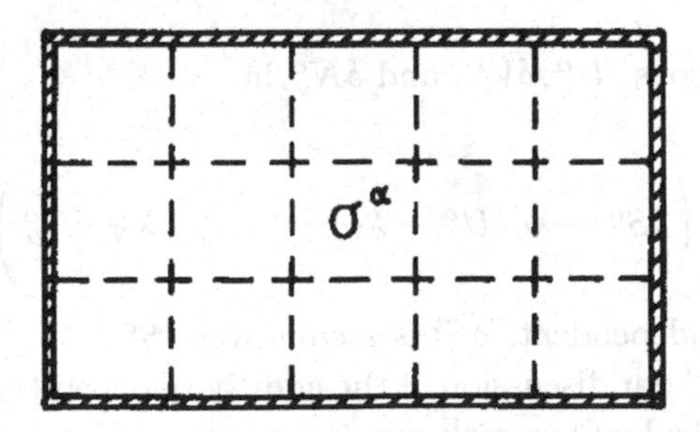

Fig. 2.2 An isolated composite system divided into cells.

2.7.1 *Equilibrium conditions*

It is convenient to account for the constraints of fixed a^t with the help of Lagrange undetermined multipliers λ_a (see books on mechanics or thermodynamics). In this subsection we denote $a^\alpha = (U^\alpha, V^\alpha, N_\nu^\alpha)$. We write the sums

$$a^t = \sum_\alpha a^\alpha, \tag{2.98}$$

and

$$S^t = \sum_\alpha S^\alpha, \tag{2.99}$$

over the cells and having the constraint $\delta a^t = 0$, i.e.,

$$\sum_\alpha \delta a^\alpha = 0 \tag{2.100}$$

we subtract the identity

$$\sum_\alpha \left(\lambda_U \delta U^\alpha + \lambda_V \delta V^\alpha + \sum_\nu \lambda_{N_\nu} \delta N_\nu^\alpha \right) = 0 \tag{2.101}$$

from

$$\delta S^t = \sum_\alpha \delta S^\alpha, \tag{2.102}$$

so that the variations $\delta U^\alpha, \delta V^\alpha$, and δN_ν^α in

$$\delta S^t = \sum_\alpha \left(\delta S^\alpha - \lambda_U \delta U^\alpha - \lambda_V \delta V^\alpha - \sum_\nu \lambda_{N_\nu} \delta N_\nu^\alpha \right) \tag{2.103}$$

can be taken as independent. δS^t is a sum over δS^α.

Having in mind our discussion of the additivity property in Section 2.2.3 and taking the cells large enough, we can write

$$S^\alpha = S^\alpha(U^\alpha, V^\alpha, N_\nu^\alpha), \tag{2.104}$$

i.e., the extensive parameters of a cell α depend only on the extensive parameters of the same cell. The first variation δS^α is obtained from Eq. (2.104):

$$\delta S^\alpha = \frac{1}{T^\alpha}\delta U^\alpha + \frac{P^\alpha}{T^\alpha}\delta V^\alpha + \sum_\nu \frac{\mu_\nu^\alpha}{T^\alpha}\delta N_\nu^\alpha. \tag{2.105}$$

It can be written also in terms of differentials ($\delta \to d$). From Eqs. (2.103) and (2.105) the equilibrium conditions take the form

$$T^\alpha = T, \tag{2.106}$$

$$P^\alpha = P, \tag{2.107}$$

and

$$\mu_\nu^\alpha = \mu_\nu. \tag{2.108}$$

We have expressed the Lagrangian multipliers by T, P, and μ_ν : $\lambda_U = T^{-1}$, $\lambda_V = P/T$, and $\lambda_{N_\nu} = -\mu_\nu/T$. All temperatures T^α, pressures P^α, and chemical potentials μ_ν^α of the cells are equal to the equilibrium temperature T, pressure P and chemical potential μ_ν, respectively. We have merely confirmed Eqs. (2.32)–(2.34) in a more general form. Under the conditions (2.106)–(2.108), p phases of a substance may coexist in an equilibrium.

Consider briefly the derivation of the same conditions with the help of the thermodynamic potentials. For example, let our total system σ^t be attached to reservoirs of T and P so these quantities are fixed for any part of σ^t. The total Gibbs potential Φ^t of σ^t is a sum over the cells,

$$\Phi^t = \sum_\alpha \Phi^\alpha. \tag{2.109}$$

The equilibrium requires $d\Phi^t = 0$. From the assumed total additivity

$$\Phi^\alpha = \Phi^\alpha(T, P, N_\nu^\alpha), \tag{2.110}$$

and Eqs. (2.85) and (2.109), it follows that

$$d\Phi^t = \sum_{\nu\alpha} \mu_\nu^\alpha dN_\nu^\alpha. \tag{2.111}$$

In equilibrium, $d\Phi^t = 0$ and, hence, in terms of μ_ν^α and N_ν^α, the equilibrium condition is

$$\sum_{\nu\alpha} \mu_\nu^\alpha dN_\nu^\alpha = 0. \tag{2.112}$$

σ^t is a closed system which means that its boundaries are impermeable for particles, i.e.,

$$\sum_{\alpha} dN_\nu^\alpha = 0. \tag{2.113}$$

Taking into account this constraint in Eq. (2.112), with the help of Lagrange multipliers we find $\mu_\nu^\alpha = \mu_\nu$ as in Eq. (2.108).

2.7.2 *Gibbs phase rule*

Let all cells α consist of different phases. The equilibrium condition (2.108) for l chemical components and p phases gives $(p-1)l$ equations, which are to be solved in order to find the independent parameters of state in equilibrium. The chemical potentials μ_ν^α of the phase α depend on the concentrations of the components ν rather than on the particle numbers N_ν^α. This follows from the extensivity property (2.5) of the entropy and, hence, of the Gibbs potential Φ, and the fact that μ_ν^α are derivatives of S (and Φ). In the case of l chemical components, $(l-1)$ independent concentrations can be defined so that the independent variables of any function μ_ν^α are T, P, and $(l-1)$ concentrations: $n_1^\alpha, \dots, n_{l-1}^\alpha$; $n_\nu^\alpha = N_\nu^\alpha/N_l$. The total number of independent parameters for p phases is $2+(l-1)p$. We have therefore a set of $(p-1)l$ equations ($\mu_\nu^{\alpha_1} = \mu_\nu^{\alpha_2}$) for $2+p(l-1)$ unknown variables. The number f of variables, which may take arbitrary values, is

$$f = 2 + l - p. \tag{2.114}$$

This equation is the *Gibbs phase rule*; see Gibbs (1876, 1948). This rule does not depend on the type of chemical components or on other specific physical properties of the system. For instance, it can be deduced in the same way for magnetic ($P \to H$) or other systems.

Furthermore the external variables $(T, P, H, \dots)$ for different systems are different and their number may be smaller or greater than two. Denoting this number by k, the rule (2.114) becomes

$$f = k + l - p. \tag{2.115}$$

To see how the rule works we take a one-component system ($l = 1$). It is obvious from Eq. (2.114) that two phases ($p = 2$) will coexist in an equilibrium on a line ($f = 1$) and three phases ($p = 3$) will coexist at a single point in the phase diagram, because the number of free parameters of state is zero ($f = 0$).

Note that the rules (2.114) and (2.115) are valid in the space of intensive parameters of state. The quantity f is then referred to as the *number of thermodynamic degrees of freedom* and is to be interpreted precisely as the number of *intensive* parameters capable of independent variation (Callen, 1960). The Gibbs rule does not work for phase diagrams where extensive parameters like volume V are included into consideration (see, e.g., the discussion in Section 3.2. Another Gibbs rule can be derived for the case of critical equilibrium (see Section 3.7.3).

2.7.3 *Stability conditions*

Since the entropy S^α of a cell α does not depend on the parameters of the other cells, the second variation $\delta^2 S^t$ of σ^t is a sum of $\delta^2 S^\alpha$ and the stability condition (2.42) for the possible maxima of S^t can be written as

$$\delta^2 S^t = \sum_\alpha \delta^2 S^\alpha \leq 0. \tag{2.116}$$

Now one may trace out again our derivation of Eq. (2.50). The generalization for p cells is straightforward and yields

$$\delta^2 S^\alpha = -\frac{1}{2T}\left(\delta T^\alpha \delta S^\alpha - \delta P^\alpha \delta V^\alpha + \sum_\nu \delta\mu_\nu^\alpha \delta N_\nu^\alpha\right). \tag{2.117}$$

This equation is easy to obtain by applying the operator $\frac{1}{2}\delta_\alpha$,

$$\delta_\alpha = \delta U^\alpha \frac{\partial}{\partial U^\alpha} + \delta V^\alpha \frac{\partial}{\partial V^\alpha} + \sum_\nu \delta N_\nu^\alpha \frac{\partial}{\partial N_\nu^\alpha}, \tag{2.118}$$

to Eq. (2.105). In Eq. (2.117) the variations $\delta T^\alpha, \delta P^\alpha$, and $\delta\mu_\nu^\alpha$ are a result of the initial displacements $\delta U^\alpha, \delta V^\alpha$, and δN_ν^α; so they can be written in the form $\delta_\alpha T^\alpha, \delta_\alpha P^\alpha$, and $\delta_\alpha \mu_\nu^\alpha$. For example, $\delta T^\alpha \equiv \delta_\alpha T^\alpha$, where

$$\delta_\alpha T^\alpha = \delta U^\alpha \frac{\partial T^\alpha}{\partial U^\alpha} + \delta V^\alpha \frac{\partial T^\alpha}{\partial V^\alpha} + \sum_\nu \delta N_\nu^\alpha \frac{\partial T^\alpha}{\partial N_\nu^\alpha}. \tag{2.119}$$

All derivatives are taken at the equilibrium values of parameters and, hence, the equilibrium conditions (2.106)–(2.108) can be taken into account in the subsequent calculations.

As in Section 2.4.2, we can choose the variations $\delta T^\alpha, \delta V^\alpha$, and δN_ν^α as independent ones and then the result (2.59) is generalized for the α-th cell

$$\delta^2 S^\alpha = -\frac{1}{2T}\left[\frac{C_V^\alpha}{T}(\delta T^\alpha)^2 + \frac{1}{V^\alpha K_T^\alpha}(\delta V^\alpha)^2_{N_\nu^\alpha} + \sum_{\nu\nu'}\left(\frac{\partial \mu_\nu^\alpha}{\partial N_{\nu'}^\alpha}\right)_{T,P}\delta N_\nu^\alpha \delta N_{\nu'}^\alpha\right]. \tag{2.120}$$

Note that here we have not used the homogeneity relation (2.44) and our present study is true for multi-phase systems as well.

The variation $\delta^2 S^t$ is obtained by Eqs. (2.116) and (2.120). Because of the complicated form of $\delta^2 S^\alpha$, it is now difficult to deduce simple criteria like the relations (2.60)–(2.64). For this reason and in order to establish the difference between the stability conditions in homogeneous and heterogeneous systems we consider again the case of two cells ($\alpha = 1, 2$) and make equal the notations to those in Section 2.4; we omit the superscript $\alpha = 1$ of the "first" cell and replace the superscript $\alpha = 2$ of the "surroundings" by a superscript "0". Then the simple constraints (2.25) are valid once more.

We choose $\delta V = \delta N_\nu = 0$, and examine the stability towards δT-variations. From Eqs. (2.119) and (2.120), we find $\delta T = (1/C_V)\delta U$ and $\delta T^0 = -(1/C_V^0)\delta U$; $\delta U = -\delta U^0$. Then

$$\delta^2 S^t = -\frac{1}{2T^2}\left(\frac{1}{C_V} + \frac{1}{C_V^0}\right)(\delta U)^2, \qquad (\delta V = \delta N_\nu = 0). \tag{2.121}$$

Clearly, the stability of the phases which coexist in equilibrium with one another is guaranteed by their intrinsic stability ($C_V > 0, C_V^0 > 0$). This sufficient condition is however rather strong. The actual requirement,

$$\frac{1}{C_V} + \frac{1}{C_V^0} \geq 0, \tag{2.122}$$

as it is readily seen from Eq. (2.121), is weaker, and allows for one of the specific heat values to become zero or negative. In general, an inhomogeneous system may have regions with negative specific heat (C_V) provided the relation (2.122) is fulfilled. For example, $C_V < 0$ means that the cell σ is unstable towards energy (δU-) perturbations.

Now let a variation be of the type $\delta T = \delta N_\nu = 0$, $\delta V \neq 0$. One finds in a similar way, that

$$\frac{1}{VK_T} + \frac{1}{V^0K_T^0} \geq 0, \qquad (\delta T = \delta N_\nu = 0), \tag{2.123}$$

which is the stability condition towards volume fluctuations $(\delta V)_{N,T}$. In the relation (2.123), we denote $K_T^0 = -(1/V^0)(\partial V^0/\partial P)_T$. The relation (2.123) can be written in a more simple form

$$\left(\frac{\partial P}{\partial V}\right)_{T,N_\nu} + \left(\frac{\partial P^0}{\partial V^0}\right)_{T,N_\nu^0} \leq 0, \qquad (\delta T = \delta N_\nu = 0). \tag{2.124}$$

Again one of the phases, say σ, may be unstable $(\partial P/\partial V > 0)$ so the isothermal compressibility will be negative $(K_T < 0)$ within the validity of inequality (2.124).

Finally, one may consider variations of the particle numbers δN_ν, $(\delta T = \delta V = 0)$. The essential features in this case are well illustrated by the example of a simple fluid. For two phases the relation (2.64) is replaced by

$$\left(\frac{\partial \mu}{\partial N}\right)_{T,P} + \left(\frac{\partial \mu^0}{\partial N^0}\right)_{T,P} \geq 0. \tag{2.125}$$

The case of l-component fluids requires an investigation of the matrix sum

$$\tilde{\mu} = \sum_\alpha \mu^\alpha_{\nu\nu'}, \qquad \mu^\alpha_{\nu\nu'} = \left(\frac{\partial \mu^\alpha_\nu}{\partial N^\alpha_{\nu'}}\right)_{T,P}. \tag{2.126}$$

The stability depends on the definiteness of $\tilde{\mu}$, and $\mu^\alpha_{\nu\nu'}$ (for an example, see Section 3.7.2).

The constraints (2.25) indicate that $\partial/\partial U^0 = -\partial/\partial U, \partial/\partial V^0 = -\partial/\partial V$, and $\partial/\partial N_\nu^0 = -\partial/\partial N_\nu$. This fact does not necessarily reduce the relations (2.122)–(2.125) to those for a homogeneous substance. Despite the equilibrium conditions (2.32)–(2.34) for two cells, or the conditions (2.106)–(2.108) for p cells, the derivatives of the temperatures, pressures, and chemical potentials of the cells may be different. For instance, using the definition $C_V^{-1} = (\partial T/\partial U)_{V,N_\nu}$, it is easy to check that C_V and C_V^0 in Eq. (2.122) will be different if the derivatives $\partial T/\partial U$ and $\partial T^0/\partial U$ do not coincide at the equilibrium point $(T = T^0)$. The same conclusion can be made for the criterions (2.123)–(2.125). In other cases, the conditions of stability for homogeneous and inhomogeneous systems coincide.

We have not used in our present analysis the condition for the large reservoir. The dimensions of all cells σ^{α} are supposed to be of the same order ($a_i^{\alpha} \sim a_i^{\beta}$) and this is so in the case of two subsystems ($a_i^0 \sim a_i$) discussed with the help of the relations (2.121)–(2.125). This treatment corresponds to that in Section 2.4.2 when the contribution to $\delta^2 S^t$ from the subsystem σ^0 can be neglected. For example, if the specific heat $C_V^0 = V_0 \tilde{C}_V^0$, where $\tilde{C}_V^0$ is the specific heat per unit volume is much larger than the specific heat $C_V = V\tilde{C}_V$ of the system σ we can neglect the $\tilde{C}_V^0$ — terms in the Eq. (2.121) and the relation (2.122); cf. Eq. (2.60). We can treat in the same way the relations (2.123)–(2.125). If the Eq. (2.116) is taken in nonequilibrium states, S^t will increase for any variation δa_i so that we must use the inequality $\delta^2 S^t > 0$ instead of inequality (2.116). When there are nonequilibrium states in the total isolated system σ^t the inequalities (2.121)–(2.125) must be inverted. It is now clear that for an isolated system in a nonequilibrium state a negative total specific heat can exist but the specific heat of its parts may remain positive. It is obvious from Eqs. (2.116) and (2.120) that at fixed T ($T^{\alpha} = T$), the specific heat of the total system σ^t is $C_V^t = \sum_{\alpha} C_V^{\alpha}$; therefore, C_V^t obeys the additivity property as any other extensive thermodynamic quantity.

2.7.4 *Limitation of the thermodynamic theory of stability*

We have derived the general conditions of equilibrium and stability of thermodynamic systems which are, as it must have become evident from the above analysis, a direct application of the principle of the entropy increase. The equilibrium conditions for multiphase states are not the same as those for single-phase (homogeneous) states. The intrinsic stability of each cell in a composite system is a sufficient condition for its overall stability. However a composite system may be stable as a whole under weaker conditions. This means that under the variations of the thermodynamic parameters a homogeneous state may be decomposed into different phases.

Many of the thermodynamic relations derived above have been written in terms of fluid variables but when the expressions are used in general notations it becomes clear that there are no restrictions of the above results with respect to their applicability to any thermodynamic system. It must however be underlined, in accordance with the discussion in Section 2.2.4, that the above theory is limited to macroscopic systems.

We have essentially used the additivity property of the cells α, which means that these cells are taken as macroscopic bodies in a sense that their typical size is larger than all characteristic lengths of inter-particle interactions or, more generally, of the intrinsic correlations in the cells. Because the individual particles and the interactions between them are not a matter of interest to macroscopic thermodynamics, within the frame of this theory, the physical requirements of the macroscopic approach are expressed by the additivity properties of the cells α. This is the reason why it becomes possible to treat both infinitesimal (δa^{α}) and finite amplitude (Δa^{α}) variations which describe changes in a cell α as a whole, but we do not take into account variations of its parts, for example. Besides, the variations of the parameters in any cell α have been considered as independent on the variations of the parameters in the other cells (apart from the secondary conditions: a^t =constant). It should become clear in the following studies that this limitation is an obstacle to the development of a successful theory of phase transitions in the framework of macroscopic thermodynamics. The variations δa, or Δa, that have been studied here are nothing else but the *spontaneous fluctuations* of the parameters in a cell (σ) of a thermodynamic system (σ^t). Thus the stability conditions for the phases α in any cell of σ^t mean the stability of the phase α with respect to macroscopic fluctuations.

In the completed analysis we have not been able to describe nonequilibrium phenomena. Under the assumption that the entropy or the other thermodynamic functions are given by initial (hypothetical) functions of state and performing a subsequent variational procedure, we have obtained the conditions of stability in the equilibrium, but nothing can be said about the properties of the nonequilibrium states. More details about the last topic are given, for example, by de Groot and Mazur (1969), Glansdorf and Prigogine (1971), Reichl (1980), and Kreuzer (1981). Thus the present study is closely related to the fluctuation theory within the equilibrium thermodynamics.

Chapter 3

Thermodynamic Theory of Phase Transitions

3.1 Order Parameter

In this Chapter we apply the general requirements for equilibrium and stability to the problem of phase transitions. We show how the basic equations and inequalities derived in Chapter 2 can be used to distinguish between phases in equilibrium. Besides, we make an outline of the phase diagrams and describe phase transitions in terms of classical thermodynamics.

The distinction between the phases is expressed by one or more thermodynamic quantities, *the order parameter(s)*. The theories change, contradict or reconcile with each other but the general concept of order parameter as a tool to distinguish between different phases retains in all studies of phase transitions. A variety of theoretical and experimental investigations provides for a corresponding variety of ways to get insight into the mechanism of the phase change or, which is the same, into the physical properties of the order parameter. The different theories are sometimes based on different approaches but all of them are united by the idea of the order parameter. Within the framework of classical thermodynamics such concepts as order and order parameter, which imply the existence of a kind of structure are unusual. So here the term order parameter is supposed to describe a simple quantitative difference. This terminology historically comes from the parallel investigations of phase transitions by methods of the solid state physics and statistical mechanics.

The order parameters are chosen among the extensive variables a_i, or their densities: a_i/N. For each phase there is a characteristic value of the order parameter. If the latter is identical for the phases considered, which may happen for some definite values of the thermodynamic parameters, it is said that the phases are identical too (a single-phase state). For example in compressible (fluid and solid) systems the order parameter is the difference in the specific volumes of the gas, liquid, and solid phases, such is the total magnetization in ferromagnets or both the total and staggered magnetizations in antiferromagnets; see, e.g., Kittel (1971), the macroscopic wave function of a superfluid liquid or of the electron pairs in superconductors, respectively; see, e.g., Tilley and Tilley (1974), and Tinkham (2004).

As a quantity which is introduced to distinguish between the phases, the order parameter represents a number of physical properties which are revealed in the course of theoretical and experimental studies. Usually it is chosen to be identically zero in one of the phases and to take some nonzero value when this phase changes into another. In some cases this choice is not necessarily implied by physical requirements; rather it is a matter of convenience as, for example, when we choose the difference between the specific volumes (or densities) in a gas-liquid system as an order parameter. To the same accuracy the condensation can be studied by considering the volume (or density) of the fluid as an order parameter in which case the former will be nonzero for each value of the external parameters (T and P).

So the choice of the order parameter depends on the physical properties of the particular system (or the class of systems) considered but the main thermodynamic properties of the phase change in a lot of systems are rather more similar than different because some general thermodynamic rules have to be satisfied at the phase transition points. Generally speaking, the behaviour of the order parameter near such points can be of two types. There is either a discontinuous jump or a continuous change of the order parameter when the system passes through the phase transition point. Thus one can classify the possible types of phase transitions and, as we shall see, this classification is necessarily somewhat more complicated.

Here we shall study relatively simple systems, starting with the example of a two-phase coexistence in one-component fluids. This is an adequate example illustrating the thermodynamic methods involved in the treatment of phase transitions. As it has been shown in Chapter 2, the matrix $(\partial^2 S/\partial a_i \partial a_j)$ which describes the stability properties of a thermodynamic system is often a diagonal matrix or can be diagonalized because of its symmetry. This means that, in most of the cases, fluctuations of the extensive parameters do not interfere with each other and can be examined separately. So, no essentially new ideas or mathematical methods are needed for treating of more complex thermodynamic systems.

3.2 Coexistence of Two Phases

Consider a classical fluid of N identical particles. The Gibbs potential is $\Phi = \mu N$ and Eq. (2.108) can now be written either for Φ or μ. The thermodynamic approach implies that the phases are already present and what remains is to test their properties of coexistence in equilibrium. Note that one does not need to specify the physical properties of the possible phases; one may still find it convenient to think of them as of vapour, liquid or solid phases of a simple (one-component) fluid. Let us take two of these phases which from now on will be referred to as phases α; $\alpha = 1, 2$. From Eqs. (2.106)–(2.108) follows that these phases have equal equilibrium temperatures T^{α}, pressures P^{α}, and chemical potentials μ^{α} and, hence, all equilibrium conditions are given by the equality

$$\mu_1(T, P) = \mu_2(T, P). \tag{3.1}$$

From Eq. (3.1) one may obtain either the *equilibrium* pressure P as a function of the *equilibrium* temperature T, $P(T)$ or, vice versa, $T(P)$. In accordance with the Gibbs rule (2.114), the condition for coexistence of phases 1 and 2 may hold place on a line, $P(T)$ or, which is the same $T(P)$, on the (P, T)-diagram of the system; see Fig. 3.1(a).

It is called *a coexistence line,* or *a phase transition line.* From a theoretical point of view the relations (2.106)–(2.108) as well as the intensive parameters T, P, and μ can be used on equal grounds so one may choose each of the equations: (3.1), $T_1(\mu, P) = T_2(\mu, P)$ and $P_1(T, \mu) = P_2(T, \mu)$ as a starting point in making the phase coexistence analysis. However, the experimentalists often look upon the representation (3.1) as more convenient because in experiments T and P can be directly measured and controlled. Therefore, it is more suitable to think of T and P as independent parameters of state.

The functions $\mu_{\alpha}(T, P)$ define two surfaces in the space of the parameters T, P, and μ; see Fig. 3.1(b). On these surfaces the phases α are stable (or metastable) depending on whether the corresponding $\mu_{\alpha}(T, P)$ represent an absolute (or local) minimum of the hypothetical chemical potential $\mu(T, P)$ of the fluid. Remember that, as explained in Chapter 2, the functions $\mu_{\alpha}(T, P)$ are the extrema of a hypothetical (nonequilibrium) potential $\mu(T, P)$. If the latter has more than one minimum at given T and P it must also have a maximum. We shall see below that the presence of a maximum follows as an inevitable consequence of the assumption that two minima can exist in a certain domain of parameters T and P.

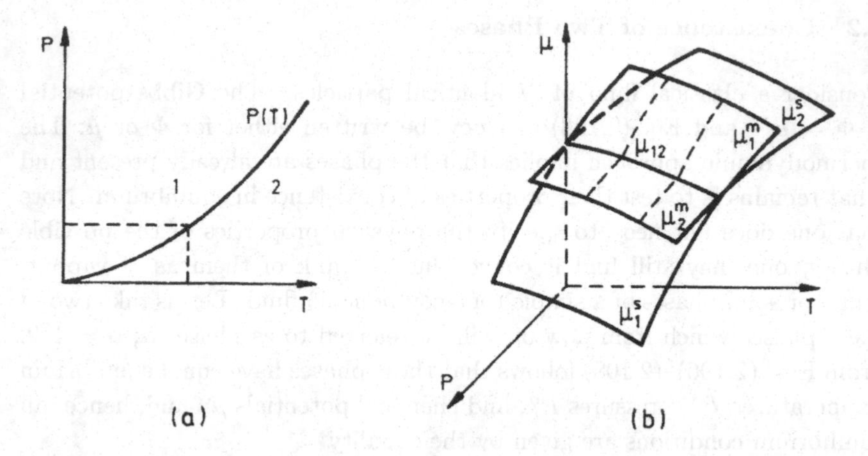

Fig. 3.1 (a) Coexistence line, (b) Intersection of $\mu_\alpha(T,P)$-surfaces; μ_α^s stability branches of μ_α, μ_α^m-metastability branches of μ_α.

3.2.1 *Phase diagram*

As our aim is to study a phase transition between phases 1 and 2, it is appropriate to suppose that the μ_α-surfaces intersect as shown in Fig. 3.1(b). Note that Eq. (3.1) can also be satisfied in other cases but these are of no importance to our intentions. The projection of the intersection line μ_{12} onto the (P,T) plane is just the coexistence line in Fig. 3.1(a). The supposed behaviour of the functions $\mu_\alpha(T)$ and $\mu_\alpha(P)$ at fixed P and T is illustrated in Fig. 3.2(a) and Fig. 3.2(b), respectively. The dotted branches of these functions indicate the presence of metastable states (see Section 2.5 and Section 3.3). Their existence is a mere consequence of our assumption that the surfaces $\mu_\alpha(T,P)$ continue beyond the intersection line. According to this picture phase 1 is stable ($\mu_1 < \mu_2$) on the low-temperature, high-pressure side of the transition line $P(T)$. It is called the *low-temperature phase* or, more uncommonly, the high-pressure phase. Correspondingly, phase 2 is the *high-temperature* (or low-pressure) phase. When crossing the line $P(T)$, the phases exchange their stability.

Clearly there must be a third variable which will depend on T and P and which will make possible to distinguish between the phases. In our treatment such a variable is the volume V of the fluid: $V = N(\partial\mu/\partial P)_T$. As T and P are the free parameters of state, the volume V is not independent, i.e., $V(T,P)$, and could be used to mark the difference between phases 1 and 2, so it is supposed to play the role of the order parameter. In fact, the specific volumes v_α of the phases per particle, $v_\alpha = V_\alpha/N$, are

$$v_\alpha = \left(\frac{\partial \mu_\alpha}{\partial P}\right)_T, \tag{3.2}$$

and the difference $\Delta v(T, P) = (v_2 - v_1)$ is a suitably chosen order parameter which is zero in phase 2 and takes a positive value at T and P corresponding to phase 1. Alternatively, one may choose $\Delta v = (v_2 - v) \geq 0$,where $v = V/N$. The entropies $s_\alpha = S_\alpha/N$ per particle, which correspond to the potentials μ_α are

$$s_\alpha = \left(\frac{\partial \mu_\alpha}{\partial P}\right)_T, \tag{3.3}$$

Eqs. (3.2) and (3.3) directly follow from the relation (2.41) which can be written for each of the phases 1 and 2.

In the theory of phase transitions the intensive variables are often classified in two categories (Fisher, 1968, 1971; Griffiths and Wheeler, 1970). Those which are equal in the coexisting phases such as T, P, μ_ν or H are called *thermodynamic fields* and those which are not equal — *thermodynamic densities* such as s, v, $\rho = N/V$ and so on; briefly, one may use the terms "fields" and "densities". Each density is a ratio of two extensive variables and in this way it is independent on the size of the system. Despite of this similar property the densities and the fields behave in essentially different way. The fields take equal values in each of the coexisting phases, whereas the values of the densities vary from one phase to another. *The order parameter is chosen among the densities.*

If, as shown in Figs. 3.2(a, b), the slopes of $\mu_\alpha(T, P)$ are different on the transition line $P(T)$, the volume V and the entropy S of the system undergo jumps at the transition point: $v_1 \neq v_2$ and $s_1 \neq s_2$. In this case the specific volumes v_α on the line $P(T)$ are represented in the (P, v) and (T, v) planes by two lines, 1 and 2, as depicted in Figs. 3.2(c, d). These *coexistence* lines separate the regions of stability of phases 1 and 2. They are often called *binodal lines.*

3.2.2 *Equilibrium isotherms*

Since at any fixed temperature (T_1) on the line $P(T)$ the pressure P is fixed, $P_1 = P(T_1)$ and by the stability condition (2.62), each equilibrium isotherm will look as shown in Fig. 3.2(c). The isotherm T_1 has unavoidably

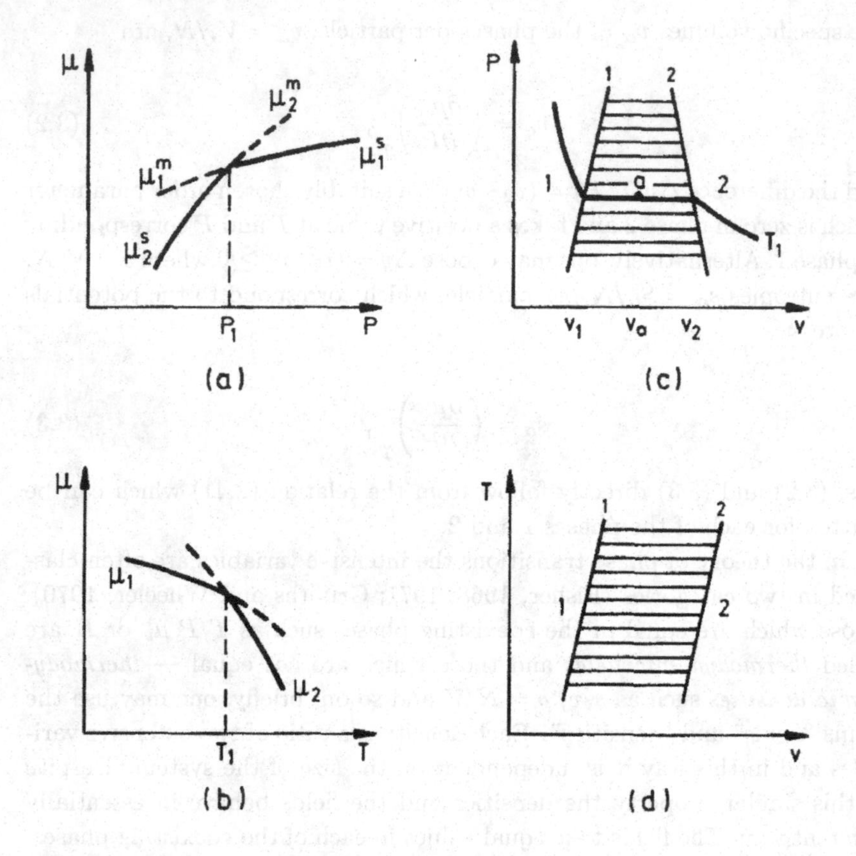

Fig. 3.2 (a) The functions $\mu_\alpha(P)$ at a fixed temperature $T_1 = T(P_1)$. (b) The functions $\mu_\alpha(T)$ at fixed $P_1 = P(T_1)$. The dotted lines represent the metastable branches μ_α^m of μ_α. The superscript s denotes the stability branches μ_α^s of the μ_α-surfaces. (c) and (d) represent (P, v) and (v, T) diagrams with binodal lines 1 and 2. The thick line T_1 in the (P, v) diagram is an isotherm.

a horizontal part confined between the binodal curves 1 and 2, which is sometimes called a *connodal.*

A subtle point is connected with the structure of states on these connodals, where Eq. (2.62) holds place. A homogeneous state could be in equilibrium if

$$\left(\frac{\partial^2 P}{\partial v^2}\right)_T = 0, \tag{3.4}$$

and

$$\left(\frac{\partial^3 P}{\partial v^3}\right)_T < 0, \tag{3.5}$$

as it is easy to see from the general stability analysis with the help of relations (2.96), replacing $\tilde{\Psi}$ by F or μN and using $(\partial F/\partial V)_T = -P$ or $(\partial\mu/\partial P)_T = v$. Eqs. (2.62) and (3.4) will rather define a unique state in the (P, v) plane than a line of states and so we are forced to conclude that on the flat parts of the isotherms the system splits into an inhomogeneous mixture, consisting of phases 1 and 2.

The shaded regions confined within the binodals 1 and 2 in Figs. 3.2(c, d) are the *the domains of coexistence* of phases 1 and 2 in the (T, v) and (P, v) planes of the parameter space (T, P, v). In these domains the phases coexist in equilibrium and can be distinguished as different bodies of the same substance. The treatment of the phase coexistence in the parameter space (T, P, v) apparently differs from that in terms of the intensive parameters (T, P, μ), but these different descriptions do not contradict to each other.

Consider the (P, v) diagram in Fig. 3.2(c). On the connodals in the shaded region the volume v runs over the values $v_1 < v < v_2$ with the exception of the boundary points. The values $v_1 < v < v_2$ of v describe the inhomogeneous states, in which definite amounts of phases 1 and 2 coexist. The volume v will depend on the fractions x_1 and x_2 of the phases: $v = x_1 v_1 + x_2 v_2$, where $x_1 + x_2 = 1$. The ratio of the fractions,

$$\frac{x_1}{x_2} = \frac{v_2 - v}{v - v_1}, \tag{3.6}$$

is given by the simple *lever rule*, namely, the fractions x_1 and x_2 of phases 1 and 2 represented by a point a, lying on a connodal are in inverse ratio to the distances from this point to the endpoints of the connodal ($v = v_a$; see Fig. 3.2(c). On each connodal the phases 1 and 2 are in a neutral equilibrium (which is explained in Section 2.5).

3.2.3 *Relations on the coexistence line*

Simple relations can be derived for the variations of the temperature, the pressure and the chemical potential for points lying on the coexistence curve $P(T)$. Let us consider two neighbouring points on this line: (T, P) and $(T + dT, P + dP)$. Expanding $\mu_\alpha(T + dT, P + dP)$ in Eq. (3.1) to first order in powers of dT and dP and taking into account the relations (3.2) and (3.3), we get

$$d\mu_1 = d\mu_2 \tag{3.7}$$

and the Gibbs–Duhem relations

$$d\mu = -s_\alpha dT + v_\alpha dP, \tag{3.8}$$

where we have set $d\mu_\alpha = d\mu$ in accord with Eq. (3.7). Eq. (3.8) connect the variations $d\mu$, dT, and dP for reversible processes on the coexistence curve. Obviously, only one of these variations is independent. Choosing, for example, dT as independent, we obtain from Eq. (3.8) two relations

$$\frac{dP}{dT} = \frac{s_2 - s_1}{v_2 - v_1}, \tag{3.9}$$

and

$$\frac{d\mu}{dT} = \frac{s_2 v_1 - s_1 v_2}{v_2 - v_1}. \tag{3.10}$$

Here dP and $d\mu$ are expressed by dT and the parameters of the system. The other ratios, as $(d\mu/dP)$, are directly given by Eqs. (3.9) and (3.10): $(d\mu/dP) = (d\mu/dT)\ /\ (dP/dT)$.

The pressure and the temperature are usually controlled in experiments and the relation (3.9) is often used in discussion of experimental results. This is the well-known *Clapeyron–Clausius equation*, or the vapour-pressure equation; for a generalization to l-component fluids, see Münster (1970). It expresses the variation of vapour pressure with temperature in terms of other measurable quantities. This equation can be written in terms of *the latent heat* per particle

$$q = \int_{s_1}^{s_2} T ds. \tag{3.11}$$

On a connodal the integral (3.11) yields $T(s_2 - s_1)$, and

$$\frac{dP}{dT} = \frac{q}{T\Delta v}. \tag{3.12}$$

For a particle to pass from the low-temperature phase 1 to the high-temperature phase 2, an amount of heat $q = T\Delta s$ is required, where $\Delta s = s_2 - s_1$.

In the case of condensation $q > 0$, $\Delta v > 0$ and, therefore, $(dP/dT) > 0$. As the temperature is raised, the vapour pressure of the liquid increases too. Then the pressure coefficient $(q/T\Delta v)$ at the transition is positive. Usually, in the liquid–solid transitions q is also positive. In this case a solid which expands on melting $(v_2 > v_1)$ has a positive pressure coefficient at the melting temperature while a solid, such as the ice for instance, which contracts, has a negative pressure coefficient. The melting curve of ^{3}He is an exception. As predicted by Pomeranchuk (1950) the relatively high-spin entropy of the solid ^{3}He at low temperature $(T \leq 0.3$ K) produces $\Delta s = (q/T) < 0$. In this case the negative pressure coefficient, $(dP/dT) < 0$, results from the behaviour of the magnetic entropy (London, 1954; Trikey, Kirk and Adams, 1972; Wilks and Betts, 1987).

3.3 Metastability

The isotherm depicted in Fig. 3.2(c) describes stable states. It is therefore related to the stability branches μ_α^s of μ_α-surfaces. The metastability effects are excluded from consideration in Section 3.2.2 because we have adopted the stability requirements (2.62). As a result the equilibrium isotherms with flat parts (connodals) describe systems where the μ_α-surfaces rather terminate at the line μ_{12} in Fig. 3.1(b) than intersect. It has been already explained in Section 2.5 that the metastability branches of the thermodynamic functions unavoidably go along with unstable states, so in this picture the requirement (2.62) breaks down in the instability domains of the phase diagram.

To illustrate the topics discussed here we consider the Gibbs–Duhem relation (2.41) in the form

$$d\mu = -sdT + vdP. \tag{3.13}$$

A formal integration of this equation yields

$$\mu(T,P) = \mu(T,P_0) + \int_{P_0}^{P} v(P)dP, \tag{3.14}$$

where the pressure P_0 is fixed. The integrand $v(P)$ is unknown as we do not have in mind some explicit form of the equation of state. We now use the isotherm in Fig. 3.2(c) to outline the shape of the function $\mu(P)$ at fixed T; see Fig. 3.3(a). Let us take P_0 at point A in the phase 1. The other points of importance along the equilibrium isotherm are denoted by B, C, and D. Replacing v by v_1 in Eq. (3.14), the integration from A to B will give the length of the curve AB in Fig. 3.3(b). The formal integration from C to D with the integrand v_2 gives the other stability branch (μ_2) of the chemical potential. The different slopes of the isotherm $P(v)$ in phases 1 and 2 result in different slopes of $\mu_1(P)$ and $\mu_2(P)$ in Fig. 3.3(b). Clearly the equilibrium isotherm $ABCD$ in Fig. 3.3(a) is a good fitting to the solid curves in Fig. 3.3(b). In this picture, the metastability branches μ_α^m of μ_α remain redundant.

Since the initial functions $\mu_\alpha(T, P)$ have branches of metastability, each phase may occur on both sides of the transition line $P(T)$: on one side as a stable phase and on the other as a metastable one. For example, phase 1 is stable above the line $P(T)$ but under certain conditions it can appear below this line as a metastable phase ($\mu_1 > \mu_2$). This phenomenon is called *superheating* of the phase 1. It may occur because a region below the curve $P(T)$ exists, in which phase 1 corresponds to a local minimum (μ_1) of the chemical potential. When decreasing the pressure from P to P_1, the system reaches the branch point B; see Fig. 3.3(a). Beyond this point it either follows the stability branch μ_2^s (Fig. 3.2a) or remains on the line μ_1 and follows the metastability branch μ_1^m. Here we focus our attention on the last case.

At pressures $P < P_1$ the volume v_1 changes according to the Eq. (3.2) and is represented by the segment BE on the isotherm (Fig. 3.3a). The curve BE extends to lower pressures as long as the metastability branch μ_1^m exists. It may continue even to negative pressures which is permitted for unstable or metastable states. Here we assume that the metastability branch μ_1^m and, consequently, the curve BE terminates at the pressure P_1'' as shown in Figs. 3.3(a, b). As the point E is approached along the line μ_1^m, the inequality (2.62) turns into equality and the isothermal compressibility K_T from Eq. (2.56) becomes infinite. At the point E the system is no longer a homogeneous substance in equilibrium.

One can explain the *supercooling* of the phase 2 in an analogous way. The supercooled states of the this phase lie on the segment CF of the isotherm in Fig. 3.3(a). The special boundary points E and F can be connected by an isotherm segment under the natural assumption that the isothermal process is continuous. Moreover it may be speculated that, if the isotherms are analytic functions in the whole domain of states, the points like E and F represent the minima and maxima of these functions. The minima and the maxima of the isotherms lie on lines which are often called *spinodal lines* (curves $1'$ and $2'$ in Fig. 3.3(a). These lines bound a region in the (P, v) plane, where the homogeneous states of the system are

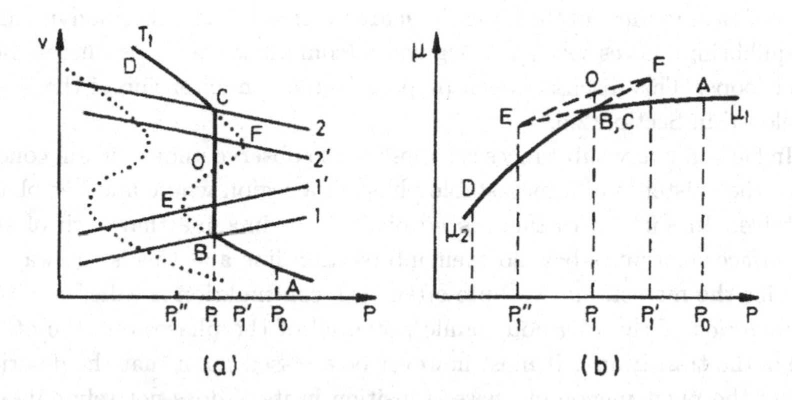

Fig. 3.3 (a) An isotherm with loops (dotted); (b) The chemical potential obtained by integrating the isotherm T_1 (cf. Callen, 1960).

unstable since there $(\partial P/\partial v)_T \geq 0$. In this region the system splits into a mixture of phases 1 and 2.

The integration along the isotherm $ABEOFCD$ in Eq. (3.14) gives the line $\mu(P)$ shown in Fig. 3.3(b). The segment EOF corresponds to an instability branch, namely, to a maximum of the chemical potential. The maximum is connected with the unstable states in the region between the spinodal curves. This fact has not been explicitly supposed in our initial assumptions but it naturally follows from the availability of local minima and the assumed analytic properties of the chemical potential (see Section 2.5). The two areas confined by the closed loops $BEOB$ and $OFCO$ in Fig. 3.3(a) are equal (*the Maxwell rule*). To see this, we use Eq. (3.14) and the fact that the chemical potentials at points B and C are equal. The pressure P_0 in Eq. (3.14) can now be chosen to coincide to P_1 and then the integral becomes zero. The same integral can be taken along the line $BEOFC$ and then a simple account of the areas beneath the segments BE, OE, OF, and FC leads to the Maxwell rule.

As a result of our assumption of the existence of local minima of the chemical potential we have succeeded to describe the isotherms $P(v)$ with instability loops. Such isotherms pass from one phase to another through metastable and unstable states. In this treatment, we have rather used heuristic arguments than an actual thermodynamic analysis. The reason is, that along the loops the system is not in a (stable) equilibrium and the relations of stability within equilibrium thermodynamics are not valid. In view of equilibrium thermodynamics our basic hypotheses about the possibility of local minima of the potential μ are untenable. The thermodynamics of equilibrium states yields flat segments (connodals) of the isotherms but never loops. This discussion can be perceived as an extension of the ideas developed in Section 2.5.

In fact, in many substances it is possible to observe under certain conditions, the existence of a metastable phase in a region where another phase is stable. In such cases there is no obstacle to imagine that each of the μ_α-surfaces continues beyond their intersection line and this is the way to describe the metastability. Quite often such continuation is called *analytic continuation* of the thermodynamic potential of the phases on "the other side of the transition". It must however be stressed upon that the description of the phenomenon of phase transition in itself does not rely only on the possibility of analytic continuation of the thermodynamic functions. There are cases in which the μ_α-surfaces terminate at the phase transition line $P(T)$ instead of intersecting. Then the phenomena of metastability do not occur. One of the cases in which the metastability branches vanish at a point and where the transition line terminates is discussed in the next section.

3.4 Critical Point

The order parameter $\Delta v = (v_2 - v_1)$ varies along the transition line $P(T)$, and if the temperature is raised enough, it may tend to zero. This occurs in the gas–liquid transitions. Let us take this example and consider the case when the order parameter gradually vanishes as the point (P_c, T_c) is approached, and is equal to zero for any $T \geq T_c$. At this point and above it the phases 1 and 2 cannot be distinguished and, consequently, no phase transition can occur. Then the transition line $P(T)$ terminates at the *critical point* (T_c, P_c) as shown in Fig. 3.4(a). We shall call this point shortly, although incorrectly, *the critical temperature* T_c.

Historically, the terms critical point and critical parameters (P_c, T_c) have been used for establishing that the gas condensation cannot occur for $T > T_c$ or $P > P_c$. There are however other important physical phenomena which take place at such points and this justifies terms like critical points, *critical phenomena* and so on.

In contemporary physics, critical points of several types are known in a number of systems. They may appear as endpoints of phase transition lines as in our present example of condensation or they may form *a line*

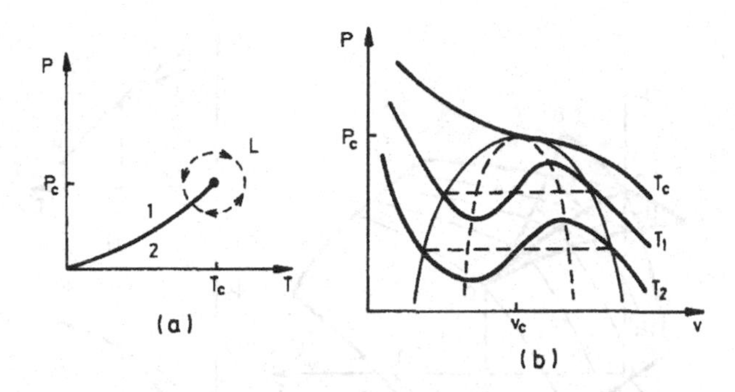

Fig. 3.4 (a) (P, T) diagram with a critical point (P_c, T_c), (b) (P, v) diagram with a critical point.

of critical points. This question will be discussed in details throughout the following text.

Here we focus on the consequences of our assumption, that the transition line $P(T)$ terminates at a critical point (Fig. 3.4a). In this case the phases 1 and 2 *transform* into each other in a continuous way without crossing the transition line. Such a reversible process is shown schematically by the dotted curve L in Fig. 3.4(a). The continuous decrease of the order parameter Δv to zero as the temperature is raised to T_c means that the (P, v) diagram will look like the one in Fig. 3.4(b).

The form of the μ_α-surfaces, which is appropriate to the present case, is given in Fig. 3.5(a); see Pippard (1957).

Obviously the metastable branches AOC of μ_2 and BOC of μ_1 vanish at the critical point C; see also Figs. 3.5(b, c). Above it the μ-surface has a simple shape which describes a single (gaseous) phase. The coexistence region is limited from above and the binodal and spinodal lines terminate at the critical point (Fig. 3.4b). That is why $\Delta v \to 0$ as $T \to T_c$ from below ($T \to T_c^-$), at the same time the connodals become shorter with the increase of T and as a result, the critical point is the only point on the equilibrium critical isotherm (T_c) where $(\partial P/\partial V)_T = 0$. This equation follows from the fact that $\Delta v = 0$ at T_c and the assumption that $P(V, T)$ is an analytic function of V. So the pressure can be expanded in terms of the variations δv of the volume about the critical value $v_c = v(T_c, P_c)$. To perform this procedure we write the equation

$$P(v, T) = P(v + \delta v, T), \tag{3.15}$$

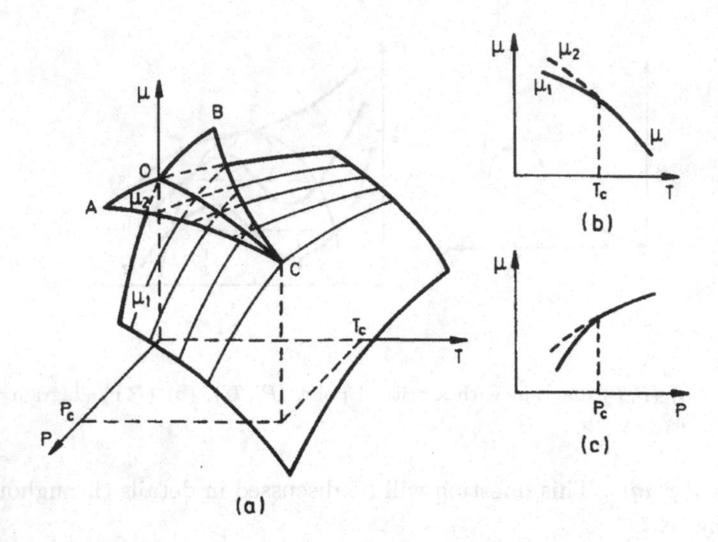

Fig. 3.5 (a) μ_i-surfaces for systems with a critical point C; (b) and (c) the functions $\mu(T)$ and $\mu(P)$.

which holds for connodals when $T < T_c$. We assume $\delta v < \Delta v$. In the nontrivial case of $\delta v \neq 0$, expanding the r.h.s. of Eq. (3.15) in power series of δv we obtain

$$\left(\frac{\partial P}{\partial v}\right)_T + \frac{1}{2}\left(\frac{\partial^2 P}{\partial v^2}\right)_T \delta v + O(\delta^2 v) = 0, \tag{3.16}$$

where $O(\delta^n)$ denotes terms of order equal or higher than δ^n. As $T \to T_c^-$, $\Delta v \to 0$ and, hence, δv tends to zero as well. Therefore,

$$\left(\frac{\partial P}{\partial v}\right)_{T_c} = 0. \tag{3.17}$$

Eq. (3.17) together with the relations (3.4) and (3.5) define only one critical point at which the fluid is in a stable homogeneous state. The critical parameters T_c and P_c are obtained as solutions of Eqs. (3.4) and (3.17).

At this stage another subtle problem exists. It is related to the thermal stability condition $C_V > 0$ (the number N of particles is supposed to be constant; see Section 2.4.2, inequality (2.60). If one assumes that C_V has a finite value (C_V^c) at the critical point, the overall stability of the system is doubtless. The results in this version of the theory are summarized in the course written by Landau and Lifshitz (1980).

The assumption that $C_V(T)$ may increase to infinity when $T \to T_c$ ($C_V^c = \infty$), requires additional investigation. Actually, replacing $C_V^0 = C_V$ in Eq. (2.122) (because of homogeneity), it can be seen that the stability of the substance towards δU-perturbations remains undeterminable within the second-order analysis in δU. To establish the stability properties the derivatives $(\partial^3 S/\partial U^3)$ and $(\partial^4 S/\partial U^4)$ must be investigated. The requirements for stability of the critical state include the analysis of the equation $(\partial^3 S/\partial U^3)_{V,N} = 0$. It is easy to see from Eq. (2.122) that $(\partial^2 S/\partial U^2) = (-1/T^2 C_V)$, and after a subsequent differentiation using the relation $C_V = (\partial U/\partial T)_V$, we get:

$$\left(\frac{\partial^3 S}{\partial U^3}\right)_{V,N} = \frac{2}{C_V^2 T^3}\left[1 + \frac{T}{2C_V}\left(\frac{\partial C_V}{\partial T}\right)_V\right]. \tag{3.18}$$

If $C_V \to \infty$ as $T \to T_c$, the third-order derivative of S in Eq. (3.18) will tend to zero when

$$\frac{1}{C_V^3}\left(\frac{\partial C_V}{\partial T}\right)_V \longrightarrow 0. \tag{3.19}$$

Assuming that C_V behaves as

$$C_V \sim \frac{1}{(T - T_c)^\alpha} \tag{3.20}$$

in a close vicinity of T_c, with α being a positive number, obviously the limit (3.19) is satisfied only if

$$\alpha > \alpha_2 = \frac{1}{2}. \tag{3.21}$$

Then it is easy to see that $(\partial^m S/\partial U^m)_{V,N}$ will tend to zero (with C_V given by Eq. (3.20)), if

$$\alpha > \alpha_m = \frac{m-1}{m}; \tag{3.22}$$

and α_m will tend to unity, if $m \to \infty$. Therefore, the critical point will have a marginal stability towards U-perturbations, if the exponent α in the relation (3.20) is equal to unity. The limit (3.19) becomes zero (for $\alpha > 1/2$) or infinity (for $\alpha < 1/2$). The exponents $\alpha = 1$ and $\alpha = 1/2$ are of particular significance because for these values one type of critical behaviour is changed to another.

The expression (3.20) does not follow from general thermodynamic principles, it is predicted by experimental data on the specific heat behaviour in a number of real systems. The experimental results are in favour of values $\alpha < 1/2$ so that it may be supposed that the limit of the expression (3.19), as far as real substances are concerned, is rather infinity than zero. Besides, for $\alpha < 1/2$, $(\partial^m S/\partial U^m)$ will tend to infinity as the critical point is approached for any $m \geq 2$. Therefore, in this version of the theory the critical state is unstable towards U-perturbations of the system, and from a mathematical point of view it corresponds to an inflection point of the entropy function (provided $\alpha < 1/2$).

3.5 Classifications of Phase Transitions

The system of Eq. (3.8) is degenerate at the critical point, where $\Delta v = 0$. It is obvious from these equations, that $\Delta s \to 0$ as $\Delta v \to 0$. Then the ratios on the r.h.s. of Eqs. (3.9) and (3.10) are indeterminable. One way to solve this problem was proposed by Ehrenfest (1933).

Consider the equations $s_1 = s_2$ and $v_1 = v_2$, or

$$ds_1 = ds_2, \qquad dv_1 = dv_2 \tag{3.23}$$

in the same way as we have done for Eq. (3.7). Using

$$\left(\frac{\partial \Delta s}{\partial T}\right)_P dT + \left(\frac{\partial \Delta s}{\partial P}\right)_T dP = 0 \tag{3.24}$$

and

$$\left(\frac{\partial \Delta v}{\partial T}\right)_P dT + \left(\frac{\partial \Delta v}{\partial P}\right)_T dP = 0 \tag{3.25}$$

with $\Delta s = s_2 - s_1$ and $\Delta v = v_2 - v_1$, and the Maxwell relations $(\partial v/\partial T)_P = -(\partial s/\partial P)_T$, one obtains from Eqs. (3.23):

$$\frac{dP}{dT} = \frac{1}{vT}\frac{C_P^{(2)} - C_P^{(1)}}{\alpha_P^{(2)} - \alpha_P^{(1)}} = \frac{\alpha_P^{(2)} - \alpha_P^{(1)}}{K_T^{(2)} - K_T^{(1)}}, \tag{3.26}$$

where the specific heat at constant pressure C_P, the isothermal compressibility K_T and the expansion coefficient α_P are given by Eqs. (2.55)–(2.57).

Note that the above equations can be derived from the equality of the second differentials of μ_α: $d^2\mu_1 = d^2\mu_2$; provided $d\mu_1 = d\mu_2$. In case of $s_1 = s_2$ and $v_1 = v_2$, the system of Eq. (3.7)–Eq. (3.8) is degenerate and this is the reason to use the second differentials of μ_α.

The Ehrenfest equation (3.26) hold place on the transition line $P(T)$ provided the differences $\Delta C_P (= C_P^{(2)} - C_P^{(1)})$, $\Delta \alpha_P$ and ΔK_T between the susceptibilities of phases 1 and 2 are finite and nonzero. Thereforc the first derivatives of the chemical potential are continuous, but the second derivatives ("susceptibilities") undergo finite jumps at the critical point.

So, the behaviour of the first and second derivatives serves as a basis, on which the classification of phase transitions is made (*Ehrenfest classification*). Ehrenfest (1933) has proposed, that the phase transition in which one or more first derivatives of the Gibbs potential Φ are discontinuous to be called *first order* phase transitions. If the first derivatives of the thermodynamic potential are continuous at the transition point, but one or more second derivatives are discontinuous, the phase transition is of *second order*. This criterion can be used to define a transition of nth order, when all derivatives of the thermodynamic potential up to the $(n-1)$th order are continuous, but one or more nth order derivatives are discontinuous.

For example, let us consider an Ehrenfest transition of *third order*. In our fluid notations all quantities s, v, C_P, K_T and α_P are continuous at the transition point, while the derivatives of the last three quantities are discontinuous. Investigating $d(\Delta C_P) = 0$, $d(\Delta K_T) = 0$ and $d(\Delta \alpha_P) = 0$ in the previously outlined way we obtain the resultant Ehrenfest equations

$$\frac{dP}{dT} = \frac{1}{vT}\frac{(\partial \Delta C_P/\partial T)_P}{(\partial \Delta \alpha_P/\partial T)_P} = \frac{(\partial \Delta \alpha_P/\partial T)_P}{(\partial \Delta K_T/\partial T)_P} = \frac{(\partial \Delta \alpha_P/\partial P)_T}{(\partial \Delta K_T/\partial P)_T}. \tag{3.27}$$

In deriving Eqs. (3.9) and (3.10) for first-order transitions and Eqs. (3.26) and (3.27) for higher-order transitions, we have assumed the existence of finite jumps of the corresponding quantities, namely, we have presumed *a priori* the type of transition we are going to study. However, whether the behaviour at the transition points is described by finite jumps or by another peculiarity of the thermodynamic quantities is a problem which cannot be deduced from general thermodynamic arguments. So, the main question is how the Ehrenfest theory is related to the behaviour of real systems.

We have already seen that the isothermal compressibility K_T tends to infinity at the critical point. The experimental results on the specific heats C_P and C_V at critical points in gas-liquid and other systems usually exhibit

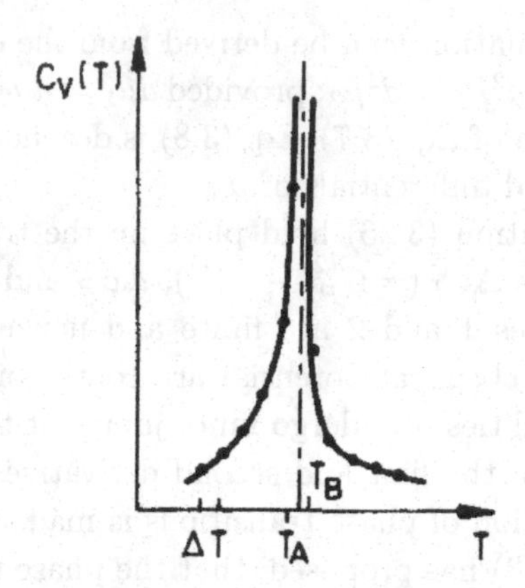

Fig. 3.6 Typical behaviour of C_V near a second-order transition point.

a divergency of these quantities. Therefore, it is clear that the Ehrenfest equations and the corresponding classification are not valid for a lot of transitions in real systems. However, this theory has been used for a long time and, even now, we can employ it in discussions of experimental results. Why this can be done we shall explain with the help of Fig. 3.6.

An experiment can be carried out within a finite accuracy. Let us suppose that the accuracy in temperature is ΔT. Then if the real curve $C_V(T)$ is the continuous line in Fig. 3.6 and the dots represent the observed experimental quantities of accuracy ΔT, the result of the experiment will be seen as a finite jump $(C_V^A - C_V^B)$ at T_c and the actual singularity of C_V will be obscure. As far as the experiments are not precise, as it stands for early experimental studies, the divergence of C_V cannot be observed. Within this precision the results correspond qualitatively to the predictions of the Ehrenfest theory. There are however modern high resolution experimental studies of second order transitions where divergences of the susceptibilities rather than finite jumps are observed. Consequently, the advanced studies require another methods of approach in describing the divergences of the susceptibilities and their derivatives. We shall see later that the main problems of both experimental and theoretical investigations in the modern theory of phase transitions are connected with attempts to surmount the shortcomings of the Ehrenfest theory.

An attempt made in this direction is the classification presented by Pippard (1957). The same classification is discussed in details by Münster (1969). It is based on a prediction of the possible types of the specific heat

(C_P) behaviour using the results either from experiments in real systems or the analysis of the particular theoretical models. Thus several types of behaviour of C_P are predicted, one of which is the finite jump within the Ehrenfest classification. Here, in contrast to the Ehrenfest theory, the disadvantage is in the lack of a unifying theoretical approach. A natural interrelationship of the Pippard classification with the stability theory of phase transitions can be established in principle (Tisza, 1951), but this approach could hardly be related to definite theoretical predictions. Besides, during the last thirty years, the studies of phase transitions followed another way of development in which new ideas of classification had been set up.

It seems that a third thermodynamic classification scheme presented by Fisher (1967), is the most comprehensive one and meets the requirements of the present status in the field. It is simple, since the first order transitions are defined as previously suggested by Ehrenfest and all others are called *continuous transitions* in the sense that the first derivatives of the thermodynamic functions change continuously at the transition point. All possible singularities as finite jumps and divergences of various types are likely to be included in the Fisher classification and this fits well the pure thermodynamic approach, within which no particular predictions about the singularities can be made. This classification offers a stable basis for further analysis of continuous transitions, founded on the modern concept of universality (see Sections 5.9.2 and 7.11.3).

Brout (1965) has proposed that the phase transitions can be classified by their symmetry characteristics. This idea binds together the problem of classification and the original Landau ideas about the close relation between the phase transition properties and the type of the symmetry change at the phase transition point (Section 4.9.2). Note that within the Brout classification the order parameter is the "response coordinate" of the system and this "coordinate" exhibits the loss of the symmetry below the transition point T_c; for details, see Brout (1965).

3.6 Compressible Systems

So far our reasoning has been restricted in the limits of the simple case of phase transition of two phases in fluid variables. Using this example we have presented the main aspects of the thermodynamic theory of phase transitions. Now it can be applied to any substance where two phases change one into another or to be directly generalized to complex systems where three or more phases may coexist. To do this one needs certain knowledge of the relationship between thermodynamics and other fields of theoretical physics, because some special forms of the fundamental equations have to be studied. Here we shall briefly discuss several examples of one-component compressible systems.

3.6.1 *Gas-liquid-solid diagram*

The two-phase coexistence, discussed in Sections 3.2 and 3.3, is straightforwardly applied to gas–liquid, liquid–solid and gas–solid transitions , provided the distinctions between the gas (G), the liquid (L) and the solid (S) phases are depicted by simple volume or, which is the same, density differences. So the order parameter can be represented as the density difference $\Delta\rho = (\rho_1 - \rho_2)$, where $\rho_\alpha = 1/v_\alpha$ ($\alpha = 1, 2$), and, hence, $\Delta v = (v_2 - v_1)$ is proportional to $\Delta\rho$: $\Delta v = \Delta\rho/\rho_1\rho_2$. Furthermore, imposing the condition of coexistence in equilibrium

$$\mu_G = \mu_L = \mu_S, \tag{3.28}$$

i.e., Eq. (2.108), we directly obtain that the gas, liquid, and solid phases in a simple substance coexist at a single point of the (P, T) diagram (Fig. 3.7). It is called the *triple point.* At this point the three phases which are separated by three first-order transition lines coexist in equilibrium.

We show in Fig. 3.7 that the condensation line (1) terminates at the critical point T_c and the melting line (2) continues indefinitely with the increase of T and P but this conclusion is mainly due to our experience in interpreting experimental behaviour of these systems rather than to thermodynamic arguments. Just on the basis of thermodynamic arguments we cannot decide whether a first order transition line will continue indefinitely or will terminate at a critical point. Another possibility, which is realized in a number of systems is that the first order transition may change its properties, say, to become a second-order transition for some values of T and P.

The pure thermodynamics does not contradict to any of these possibilities and provides the general framework for their treatment. The experimental results, however, can be explained involving some knowledge about the intrinsic nature of the phases considered. So at this stage it is impos-

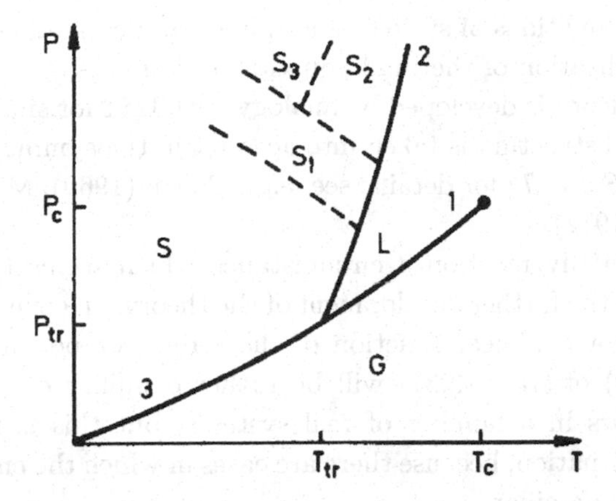

Fig. 3.7 (P, T) diagram of a gas–liquid–solid system. S_1, S_2 and S_3 denote domains of existence of different crystal modifications.

sible to continue our analysis without the knowledge of relationship of the thermodynamics to the other fields of theoretical physics. In the remaining part of this section we shall remind some features of the solid and liquid states.

3.6.2 *The solid phase*

A comprehensive thermodynamic study of the solid state is done with the help of the theory of elasticity. The fundamental relation taken, for example, in energy representation will be

$$U = U(S, V, N). \tag{3.29}$$

The volume V has to be replaced by the quantity $(V_0\hat{S})$, where $\hat{S} = \{S_{ij}; i, j = 1, 2, 3\}$ is the symmetric strain tensor and V_0 is the unstrained volume of the solid. Therefore the basis for developing of the thermodynamics will be the fundamental relation in the form:

$$U = U(S, V_0\hat{S}, N). \tag{3.30}$$

The isotropic pressure P of the fluid is replaced by components

$$T_{ij} = \left(\frac{\partial U}{\partial S_{ij}}\right)_{S,N}. \tag{3.31}$$

These are the equations of state of amorphous and crystalline solids which are the generalization of the fluid equation $P = P(V, T)$. The subsequent macroscopic theory is developed by analogy with that for simple fluids, but now the crystal structure is taken into account in the symmetry properties of the tensors $\hat{S}$ and $\hat{T}$; for details, see, e.g., Callen (1960), Münster (1970), and Wallace (1972).

We must briefly mention a circumstance which seems to be of great importance for the further development of the theory. In general, the energy (3.30) may be a nonlinear function of the stress components S_{ij} and so the equation(s) of state (3.31) will be rather complicated. This feature obviously occurs in a number of real systems, but this is not the most unfavourable situation, because there are cases in which the energy function is completely unfamiliar.

In both cases the *Hooke approximation*, known from mechanics, is useful in making predictions of the thermodynamic properties of solid phases. Its basic assumption is that the derivative in Eq. (3.31) is a linear function of S_{ij}, namely,

$$T_{ij} = \sum_{kl} C^{(S)}_{ijkl} S_{kl}. \tag{3.32}$$

This equation corresponds to a quadratic dependence of U on S_{ij}, which is seen from Eqs. (3.30) and (3.31); see below the derivation of $F(\hat{S})$, Eq. (3.36), within the same approximation. The linear dependence expressed by Eq. (3.32) is always valid for small ("harmonic") strains (S_{ij}).

The response parameters $C^{(S)}_{ijkl}$ are called *adiabatic* (or *isoentropic*) *stiffness coefficients.* In fact, $C^{(S)}_{ijkl}$ are the derivatives of T_{ij} with respect to S_{kl} at fixed entropy S, number N, and $S_{k'l'} \neq S_{kl}$. As $\hat{S}$ and $\hat{T}$ are symmetric tensors, each of them has six independent components. They are denoted by S_μ and T_μ where $\mu = (1, \ldots, 6)$ so Eq. (3.31) becomes $T_\mu = (\partial U/\partial S_\mu)_{S,N}$ and, hence, Eq. (3.32) takes the form

$$T_\mu = \sum_{\nu=1}^{6} C^{(S)}_{\mu\nu} S_\nu. \tag{3.33}$$

The structure of the (6×6) matrix $\hat{C}^{(S)} = \{C^{(S)}_{\mu\nu}\}$ depends on the symmetry of the crystal. This makes possible to investigate how the liquid–crystal transition from a pure isotropic (liquid or amorphous) to a crystalline phase or *structural phase transitions* from one crystal structure to another (see the subdomains bound by the dotted lines in Fig. 3.7) take place.

By Eqs. (2.74), (3.30), and (3.31) the Helmholtz free energy F is obtained in the form

$$dF = -SdT + V_0 \sum_{\mu} T_\mu dS_\mu + \mu dN. \tag{3.34}$$

Suppose the entropy S is independent on the strain variables S_μ. The approximation (3.32)–(3.33) can now be written in terms of the *isothermal stiffness coefficients*

$$C^{(T)}_{\mu\nu} = \left(\frac{\partial T_\mu}{\partial S_\nu}\right)_{T,N,S_{\nu'}\neq S_\nu}. \tag{3.35}$$

By integrating the Eq. (3.34) F can be obtained as a function of S_μ:

$$F = F_0(T,N) + \frac{V_0}{2} \sum_{\mu\nu} C^{(T)}_{\mu\nu} S_\mu S_\nu, \tag{3.36}$$

where F_0 is the free energy of the unstrained state; see, e.g., Callen (1960).

We have used the Hooke law to demonstrate that it produces a bilinear dependence of F on S_μ. Within this approximation one can describe the distinction between the liquid, amorphous bodies and the variety of crystal modifications in solids but it is not sufficient in determining in what way the phases change into one another. The study of the liquid–solid and structural phase transitions requires an expansion to order higher than the second in S_μ, which means to go over to the Landau MF theory (see Chapter 4).

Thus we see the way in which Hooke's ideas of describing elasticity are included in the basis of the modern theory of phase transitions. In our further studies the reader will find the correct relation between the Hooke approximation in the thermodynamic theory of solids and the Gaussian approximation in the statistical theory of phase transitions. These two approximations are not equivalent but they arise from one and the same general idea of an expansion of the thermodynamic potential in terms of small order parameters (in the present case, S_μ).

The stability of the solid phase at fixed temperature T depends on the definiteness of the matrix $\hat{C}^{(T)}$ in Eq. (3.36). If $\hat{C}^{(T)}$ is a positive definite matrix, the Helmholtz free energy (3.36) will have a minimum at zero dilatations, which corresponds to a stable solid or liquid phase — the phase for which the stiffness constants are calculated. Remember that in an isotropic liquid, the "stiffness coefficient" is only one — the inverse isothermal compressibility $K_T^{-1} = -(\partial P/\partial V)_T$; in an isotropic (amorphous) solid the stiffness matrix has two independent components (the *Lamé constants*) and possesses quite a simple structure. The same structure of the stiffness matrix is displayed in the most symmetric (cubic) crystals but here, three independent stiffness coefficients are available (Callen, 1960; Münster, 1970).

In other crystal structures one has to investigate stiffness matrices of lower symmetry with higher number of independent coefficients. The investigation of the stability of a phase towards dilatations, which correspond to lower symmetries, can be done by calculating the stiffness matrix of the phase with the higher symmetry and a subsequent analysis of its definiteness. Different symmetries of the crystal phase may correspond to different minima of the thermodynamic potential and only in the absolute minimum the stable phase exists. When varying the free parameters of state an absolute minimum may change to another which describes a new crystal symmetry. In this case, a structural phase transition takes place. In such cases the quadratic form in Eq. (3.36) changes sign and the stability analysis requires an investigation of derivatives of F of order higher than second in the dilatations S_μ and, obviously, this is beyond the simple Hooke approximation.

The strain tensor $\hat{S}$ is initially set at a spatial point $\boldsymbol{x}$ in a small volume $dV = (dxdydz)$. The size of this domain (cell) is small compared to the size of the total system and large enough compared to the mean inter-particle distance. It then becomes evident that the elasticity theory can be developed rather as a quasi-macroscopic than as a macroscopic description of the solid state. The dilatations S_μ can be treated as quasi-macroscopic field variables, which depend on the space position $\boldsymbol{x}$ as in the wave theory of elasticity. Then the free energy F will be a functional of the (quasi-macroscopic) fields $S_\mu(\boldsymbol{x})$. The wave theory of elasticity incorporates both macroscopic and quasi-macroscopic (large-scale) phenomena (Wallace, 1972; Khachaturyan, 1983).

As we focus our attention on the macroscopic description we have supposed in the previous section that the tensor $\hat{S}$ is one and the same for the total solid body of unstrained volume V_0. Therefore we have failed to reach what is aimed in a quasi-macroscopic (long-wave length) theory of the solid order. We shall see later why the pure macroscopic thermodynamics is deficient in describing the essential features of phase transitions and critical effects and why the large-scale (quasi-macroscopic) theories provide the appropriate basis of description of phase transitions.

3.6.3 *The liquid state*

Another disadvantage in our attempts to describe gas–liquid–solid transitions in terms of macroscopic thermodynamics is,that within this approach, nothing can be said about the structure of the liquid state. This state has no macroscopic order and, hence, within macroscopic (or quasi-macroscopic) treatment one obtains a simple fluid picture in which the liquids are distinguished only by their specific volumes. However these substances exhibit a special form of ordering which is called the *short-range order.* The short-range order and hence, the structure of liquids is not easy to describe. A number of theories on this topic exist, but the solutions of the problem about the liquid structure are quite controversial; see, e.g., Temperley and Travena (1978), and Ziman (1979).

To settle the problem, it is sufficient to remember some qualitative differences between liquids and solids. When a solid is melting the long-range crystalline order is destroyed because the particles go away from their equilibrium positions in the crystalline lattice. However a short range, *local* order remains at typical distances of about $20 \div 30$ Åaround any particle and it looks very much like the corresponding crystalline order. This fact is confirmed by X-ray and neutron diffraction experiments. Clearly the theories of macroscopic or large-scale phenomena can hardly be used in treating the short-range order. We have to turn our look to more powerful theoretical methods.

The structure of any substance is described by the density distribution function $\rho(\boldsymbol{x})$, where $\boldsymbol{x}$ is the spatial coordinate. It must be emphasized that $\rho(\boldsymbol{x})$ represents a mean (statistically averaged) density at any site $\boldsymbol{x}$; fluctuations from the mean value $\rho(\boldsymbol{x})$ will always occur due to the thermal movement of the particles but this point may safely be ignored as unessential in our present discussion. The density function (or operator) is well defined in both classical and quantum mechanics, so the problem now is how this quantity can be calculated for dense phases like liquids and solids. As the thorough discussion of this central problem of statistical physics will lead us away from our initial aims we shall restrain ourselves to simple qualitative arguments.

Consider a substance consisting of spherically symmetric molecules as those of neon and argon liquids and imagine that, quite usually, the two-particle interactions are sufficient to describe the density distribution. The structure is given by the functional dependence of $\rho(\boldsymbol{x})$ on the pairwise potential $u(\boldsymbol{x}_1, \boldsymbol{x}_2)$ which, in our approximation, will depend only on the

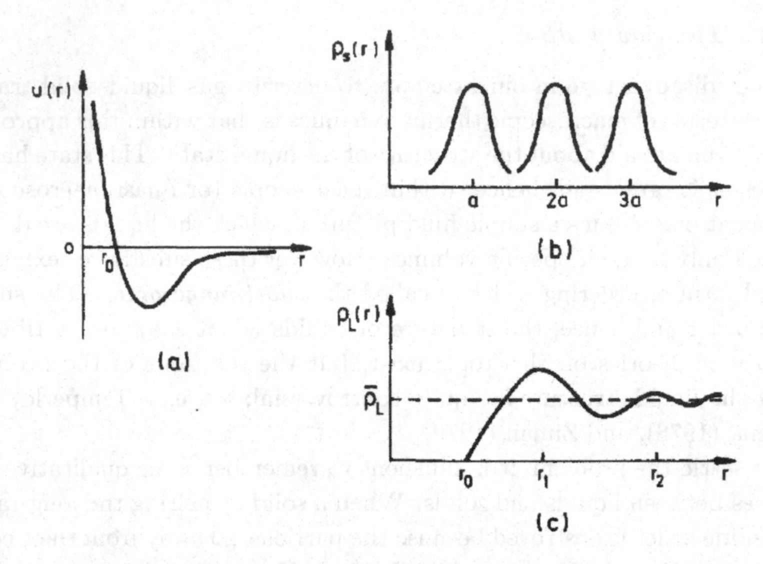

Fig. 3.8 (a) Interaction potential $u(r)$ with a hard–core radius r_0; (b) density distribution function $\rho_S(r)$ of a solid; (c) density distribution function $\rho_L(r)$ of a liquid; $\overline{\rho}_L$ is the mean density N/V.

inter-particle distance: $u = u(r), r = |\boldsymbol{x}_1 - \boldsymbol{x}_2|$. A typical pairwise potential $u(r)$ is shown in Fig. 3.8(a). If a particle is located at point $\boldsymbol{x} = 0$, the quantity which describes the variation of local density at a distance r from the center of this particle is obviously the density function $\rho(r)$. Now we can compare the behaviour of $\rho(r)$ for solids and liquids with the help of the pictures in Figs. 3.8(b, c).

The deviation of the distribution function $\rho_L(r)$ of the liquid from the mean value $\overline{\rho}_L(= N/V)$ can be accepted as a measure of the short-range order. Taking a particle at position $r = 0$, the first peak in the distribution function is at a distance r_1, which is twice the radius of the hard-core potential ($r_1 \approx 2r_0$) and hence, this peak determines the position of the nearest neighbour particles to the given one. The next peak is at $r_2(\approx 3r_0)$ and determines the position of the next nearest neighbours, etc., but at large distances ($\geq 20 - 30$ Å) $\rho_L(\boldsymbol{x})$ takes its mean value $\overline{\rho}_L$. This means that at small distances from the "initial" particle the liquid preserves a crystalline-like structure but at larger distances this picture is violated, i.e., there is no translational invariance as in crystals. So one can treat gas–liquid and liquid–solid phase transitions with the help of the distribution function $\rho(r)$. In macroscopic thermodynamics one takes the mean values $\overline{\rho}_L$ and

$\overline{\rho}_S$ of ρ of the liquid and solid phases and the description reduces merely to an account of volume (density) differences. Note that a gas, at least the ideal one, is regarded as a phase without structure: $\rho_G(r) = \overline{\rho}_G = N/V$. For the real gases of low density, $\rho_G(r)$ is proportional to the Boltzmann factor $exp[-u(r)/k_B T]$. This dependence ought to be more complicated for dense substances and such studies will lead us to the Kirkwood (1935) integral equation which connects $\rho(r)$ and $u(r)$ or, to a more simple integral equation considered by Percus and Yévick (1958). With regard to dense substances (liquids and solids) explicit mathematical results about the dependence of $\rho(\boldsymbol{x})$ on the relevant inter-particle interactions cannot be obtained and we remain with the alternative to develop phenomenological or quasi-phenomenological theories. For some problems, general theoretical tools such as symmetry arguments can be used to deduce the proper answer.

We have outlined one of the liquid models which seems to be correct, especially, near the melting curve where the short-range order of the liquid state resembles crystalline lattice. However this picture may become rather crude near the condensation line and, in particular, near the critical point. It may be expected that in this domain of the phase diagram the liquid density $\rho_L(r)$ will be similar to that of a dense interacting gas: $\rho_L(r) \approx \rho_G(r)$. The distinction both in specific volume and structure of the vapour and the liquid vanishes as the critical point is approached and for this reason, one of the phases can be continuously transformed to the other.

However *the distinction* between the solid and liquid phases always exists and these phases cannot be transformed one into another without crossing the transition point. This conclusion implies that a macroscopic symmetry property, in our particular case the solid order, is either present or absent and no intermediate states are possible (Landau, 1937a); see also Landau (1967). Now it is widely accepted that the change in the macroscopic symmetry takes place abruptly, at the point of the phase transition. A rigorous basis of the statement, that a liquid–solid transition cannot possess a critical point, can be given by investigations of statistical models; see, e.g., Euch, Knops, and Verboven (1970). One has however to keep in mind that the model proofs are never general to the extent we need for making reliable predictions about real systems.

Another question is whether the crystallization of a liquid is always a phase transition of first order. As it seems, the order of this transition is closely related to the specific volume difference between the liquid and the solid phases. According to the Ehrenfest theory (Section 3.5) a necessary condition for a melting transition of second order is the gradual equalization of the specific volumes of these phases. Whether such a condition can be realized at extremely high pressures is an open question. The available high-pressure experiments are in favour of the widely accepted opinion, that the melting transition is always of first order, but they are limited in pressure and cannot be conclusive. The theory of melting is not advanced enough to give a reliable answer to such questions. In any way, general symmetry arguments do not contradict to the possibility of a second order melting transition at high pressure; for discussion on this topic see also Pippard (1957) and Anderson (1984a).

3.7 Phase Separation in Mixtures

Let us return to the thermodynamics of multi-component systems. It is outlined in Chapter 2 where we have defined a l-component system consisting of the chemical components ν with particle numbers N_ν; $\nu = 1, \ldots, l$. Such l-component systems are called *mixtures* or *solutions*. In thermodynamics we speak of "mixtures" when the components of the system are treated on equal grounds and we use the term "solution" when one of the components, usually that in the biggest proportion, is chosen as a *a solvent* then the others are regarded as *soluble substances* (*solutes*). Therefore there is no fundamental difference between a mixture and a solution except for the manner of description; see, e.g., Guggenheim (1967).

In order to clarify this point we should remember that the thermodynamic densities (see Section 3.2.1), appropriate for the treatment of the mixtures are *the number fractions:*

$$x_\nu = \frac{N_\nu}{N}, \tag{3.37}$$

where $N = \sum_\nu N_\nu$ is the total number of particles. The densities which correspond to the extensive parameters N_ν in solutions are the *concentrations* (or *solute-solvent ratios*):

$$c_\nu = \frac{N_\nu}{N_l}, \qquad \nu \neq l, \tag{3.38}$$

provided the l-component is taken as the solvent. In both cases there are $(l-1)$ independent densities: c_ν concentrations $(1 \leq \nu \leq l-1)$ or l fractions x_ν which obey the relation

$$x_1 + \cdots + x_l = 1. \tag{3.39}$$

It is easy to obtain from Eqs. (3.37) and (3.38) that $c_\nu = x_\nu/x_l$ for $\nu \neq l$.

Now the analysis in Section 3.2 and 3.6 for one-component compressible systems can be generalized for multi-component systems as gaseous, liquid and solid mixtures, called *alloys*. For this purpose we have to add to the independent densities x_ν the entropy density $s = S/N$ and the volume $v = V/N$ per particle; in this way the number of the independent variables becomes $(l+1)$. Alternatively, the description can be made in terms of the conjugate $(l+2)$ intensive variables T, P, and μ_ν, the variations of which obey the Gibbs–Duhem relation

$$sdT - vdP + \sum_\nu x_\nu d\mu_\nu = 0, \tag{3.40}$$

where $s = -N^{-1}(\partial\Phi/\partial T)_{P,N_\nu}$ and $v = N^{-1}(\partial\Phi/\partial P)_{T,N_\nu}$; cf. Eq. (2.41) and Eq. (2.84).

The thermodynamic investigation of multi-component systems includes the possibility of chemical reactions between the substances ν but in our short discussion we shall not consider this case; for details, see, e.g., Münster (1970), Landau and Lifshitz (1980). Besides the chemical reactions and the phase transitions related to changes of the specific volume (discussed in Section 3.6), in the mixtures the so-called *phase* (or *chemical*) *separation* may occur under the variation of the parameters of state. This means that phases α with different chemical composition can be formed in a homogeneous mixture. The thermodynamic description of this phenomenon is made by applying the general equilibrium conditions (2.106)–(2.108) and the stability criteria (2.60)–(2.63).

3.7.1 *Binary mixtures*

There is no difference in principle between the phase separation in two-component ($l = 2$) systems (*binary systems*) and that in *ternary* ($l = 3$), quaternary ($l = 4$) etc. systems. For the sake of simplicity we shall mainly discuss *binary mixtures*. Examples of binary systems where the phase separation occurs are gaseous mixtures as He-Xe, classic fluid mixtures as methanol cyclohexane, the liquid quantum mixture ^{3}He–^{4}He, and alloys as β-brass (CuZn).

For binary systems ($\nu = 1, 2$) the phase separation can be described in terms of the concentration, say, $c_1 = N_1/N_2$ of the component 1 dissolved in the component 2 or, equivalently, in terms of the corresponding fraction $x_1 = N_1/N$; see also Section 2.7.2. In this section we denote c_1 by c, and x_1 by x. From $N = N_1 + N_2$ one gets $c = x/(1-x)$, where $0 \leq x < 1$.

We choose x as a variable; then the Eq. (2.108) for the two phases ($\alpha = 1, 2$), which have number fractions $x^{(\alpha)}$ take the form

$$\mu_\nu^{(1)}(T, P, x^{(1)}) = \mu_\nu^{(2)}(T, P, x^{(2)}), \tag{3.41}$$

where the equilibrium temperature T and the equilibrium pressure P are given by the Eqs. (2.106) and (2.107).

The number fractions $x^{(\alpha)}$ of the component 1 in phases α are related to x by

$$x = x^{(1)} + x^{(2)}. \tag{3.42}$$

So for binary systems one can use as independent variables either $x^{(\alpha)}$ or, equivalently, x from Eq. (3.42) and

$$\Delta x = x^{(2)} - x^{(1)} \tag{3.43}$$

(it is assumed that $x^{(2)} > x^{(1)}$, therefore $\Delta x > 0$). Here we suppose that no other work coordinates except those given by Eq. (2.41) are taken into consideration. If in this case the specific volumes $v_\nu^{(\alpha)}$ of the phases α are equal ($v_\nu^{(\alpha)} = v_\nu$), where

$$v_\nu^{(\alpha)} = \left(\frac{\partial \mu_\nu^{(\alpha)}}{\partial P} \right)_{T, x^{(\alpha)}}, \tag{3.44}$$

[cf. Eq. (??), the phases α can differ from one another only by the quantity

$$\Delta x = \frac{N_1^{(2)} - N_1^{(1)}}{N} = \frac{\Delta N_1}{N}. \tag{3.45}$$

If $\Delta x = 0$, the phases 1 and 2 are identical and the mixture is homogeneous; $\Delta x \neq 0$ means that a phase separation takes place.

Now we are prepared to discuss the coexistence states, which because of the relatively large number of independent variables lie on *coexistence surfaces.* What approach we shall use to study the equilibrium is connected with our choice of the independent variables. If T, P and $x^{(\alpha)}$ are taken as independent, we do the analysis with the help of Eqs. (3.41). The results from Section 3.2 can be generalized by expanding $\mu_\nu^{(\alpha)}$ to the first order in dT, dP, and $dx^{(\alpha)}$. For weak solutions the investigation is more adequate in terms of concentration in order to make use of the fact that $c^{(\alpha)} \ll 1$; a treatment of this problem can be found in the books of Guggenheim (1967), and Landau and Lifshitz (1980).

An alternative study can be done in terms of the intensive variables T, P, and $\mu^{(\alpha)}$. For the states of coexistence $\mu_\nu^{(\alpha)} = \mu_\nu$, so the Eq. (3.40) for two phases can be written in the form:

$$s^{(\alpha)} dT - v^{(\alpha)} dP + \sum_\nu x_\nu^{(\alpha)} d\mu_\nu = 0. \tag{3.46}$$

This approach is comprehensively discussed by Münster (1970) for l-component mixtures. The solutions of equations such as Eq. (3.46) are given by matrices of the densities $s^{(\alpha)}, v^{(\alpha)}$ and $x_\nu^{(\alpha)}$; in this way a generalization of the Clapeyron–Clausius equation (3.9) can be obtained.

These two ways of description are equivalent and it depends on the purposes of the particular investigation which one we shall choose. Comparing them one has to be careful because the densities $s^{(\alpha)}$ and $v^{(\alpha)}$ from Eq. (3.46), see also Eqs. (2.41) and (2.85), and those given by the derivative

$$s_\nu^{(\alpha)} = -\left(\frac{\partial \mu_\nu^{(\alpha)}}{\partial T}\right)_{T, x^{(\alpha)}} \tag{3.47}$$

and Eq. (3.44) are different.

3.7.2 *Stability conditions and critical states in binary mixtures*

The relation (2.63) ensures the stability of the homogeneous state ($\Delta x = 0$) with respect to diffusion. The homogeneous phase is stable when the symmetric matrix $\hat{\mu}$ whose elements $\mu_{\nu\nu'}$ are given by the derivatives $(\partial \mu_\nu / \partial N_\nu{}')_{T,P}$ is non-negatively definite, i.e., when the determinants of all principal minors of $\hat{\mu}$ are nonnegative. For binary systems ($\nu = 1, 2$), the condition (2.63) reads

$$s_\nu^{(\alpha)} = -\left(\frac{\partial \mu_\nu^{(\alpha)}}{\partial T}\right)_{T, x^{(\alpha)}} \tag{3.48a}$$

and

$$\det \begin{pmatrix} \mu_{11} & \mu_{12} \\ \mu_{21} & \mu_{22} \end{pmatrix} \geq 0. \tag{3.48b}$$

These relations however are not independent. Because of Eq. (2.41), at fixed T and P:

$$\sum_{\nu} N_{\nu} d\mu_{\nu} = 0, \tag{3.49}$$

and we obtain

$$\sum_{\nu\nu'} N_{\nu} \mu_{\nu\nu'} dN_{\nu}{}' = 0. \tag{3.50}$$

The former equation must be satisfied for each dN_{ν}; so

$$\sum_{\nu} \mu_{\nu'\nu} N_{\nu} = 0, \tag{3.51}$$

which, combined with Eqs. (3.37) and (3.39) gives:

$$c\mu_{11} + \mu_{12} = 0, \tag{3.52a}$$

and

$$\mu_{12} + \mu_{22} = 0, \tag{3.52b}$$

where we have used that $c = N_1/N_2$, or, as a final result,

$$\mu_{22} = c^2 \mu_{11}, \qquad \mu_{12} = -c\mu_{11}. \tag{3.53}$$

Therefore the relations (3.48a) are equivalent ($c \neq 0$) and when the homogeneous state is stable, $\mu_{12} \leq 0$.

The relation (3.48b) always gives an equality: $\det(\hat{\mu}) = 0$, which follows from Eqs. (3.53) and the symmetry of the $\hat{\mu}$-matrix ($\mu_{12} = \mu_{21}$). This means that variations δN_{ν} of special type exist corresponding to a marginal stability of the system. To establish the type of the variations δN_{ν} which give rise to the marginal stability, we replace μ_{22} and μ_{12} from Eq. (3.53) in Eq. (2.63) and receive the following relation

$$\mu_{11}(\delta N_1 - c\delta N_2)^2 \geq 0. \tag{3.54}$$

For $\mu_{11} > 0$ the equality is possible, if only

$$\frac{\delta N_1}{\delta N_2} = c. \tag{3.55}$$

It can be shown that the variations δN_ν, for which Eq. (3.54) is fulfilled and $\mu_{11} > 0$, do not change the concentration c. In fact,

$$\delta c = \delta\left(\frac{N_1}{N_2}\right) = \frac{\delta N_1 - c\delta N_2}{N_2} \tag{3.56}$$

and when δN_ν satisfy Eq. (3.54), $\delta c = 0$. All other variations δN_ν obey the inequality (3.54).

When $\mu_{11} = 0$, the states become critical. Usually the critical equation is written as

$$\left(\frac{\partial \mu_1}{\partial c}\right)_{T,P} = 0. \tag{3.57}$$

This equality is similar to Eq. (3.17) which describes the critical condensation point in simple fluids. The *critical line* in (T, P, c) coordinates is determined by Eq. (3.57) and

$$\left(\frac{\partial^2 \mu_1}{\partial c^2}\right)_{T,P} = 0. \tag{3.58}$$

The critical states are stable, if $(\partial^3 \mu_1/\partial c^3)_{T,P} \geq 0$.

3.7.3 *Coexistence surfaces, critical lines, and second Gibbs rule*

It is easy to outline the phase diagrams topology of binary mixtures in terms of independent "fields": T, P, μ_1, and μ_2. Taking into account that when the phases 1 and 2 are in equilibrium the variations of these thermodynamic fields are connected by Gibbs–Duhem relations (3.46), we can choose two of them as the independent variables. Excluding, for example, $d\mu_2$ from the Eq. (3.46) we obtain that the domains of stability of phases 1 and 2 are three-dimensional (T, P, μ_1) and are separated by the *coexistence surface*.

So the phases 1 and 2 coexist on a surface. The limit of this surface is marked by the *critical line*, defined by Eqs. (3.57) and (3.58). We have considered the coexistence only of two phases but in binary systems one may also investigate the coexistence of three phases (on the line of coexistence) or of four phases (in a single point) on the (T, P, μ_1)-phase diagram.

In a l-component mixture there are $(l+1)$ independent fields, for example, $T, P, \mu_1, \ldots, \mu_{l-1}$. For two phases the $(l+1)$-dimensional stability domains of the phases $(\alpha = 1, 2)$ intersect along a *coexistence hypersurface* (that is, on a first-order phase transition surface) of dimension l, which may be bounded by a *critical hypersurface* of dimension $(l-1)$. In general, when p phases exist the dimension of their coexistence hypersurface is $(l+2-p)$ in accordance with the Gibbs rule (2.114). The coexistence surfaces of dimension $f = (l+2-p)$, or $f = (l+k-p)$ as given by Eq. (2.115) are bounded by critical hypersurfaces of dimension $f_c = (f-1)$. If we denote the number of independent thermodynamic fields by n, i.e., $n = (l+k-1)$, the dimensionality f_c of the critical hypersurface is

$$f_c = n - p, \tag{3.59a}$$

or, for two phases $(p = 2)$ in the case of ordinary critical points

$$f_c = n - 2. \tag{3.59b}$$

The last equation, or its generalization (3.59a), is sometimes called *the second Gibbs rule* (Gibbs, 1948); see also Tisza (1961). This rule is valid under the conditions discussed above; see Section 3.8.4 for an exception. On the critical manifold of dimensionality f_c, p phases are in a critical "coexistence", which means that they are indistinguishable because their characteristic densities (order parameters) are equal. The phase rule (3.59a) as well as the first phase rule given by (2.114) and (2.115) can be applied to any complex system where several types of critical behaviour may occur (Griffiths and Wheeler, 1970); for gas–liquid critical points in binary mixtures, see Saam (1970). The books by Rowlinson (1959) and Anisimov (1987) present a lot of information for critical phenomena in mixtures.

3.7.4 *Tricritical and critical-end points. Multicritical phenomena*

In binary alloys as FeAl and, especially, in ternary and quaternary fluid mixtures critical points which have no analogue in simple fluids can be observed (Griffiths, 1974; Lang and Widom, 1975; Fisher, 1984).

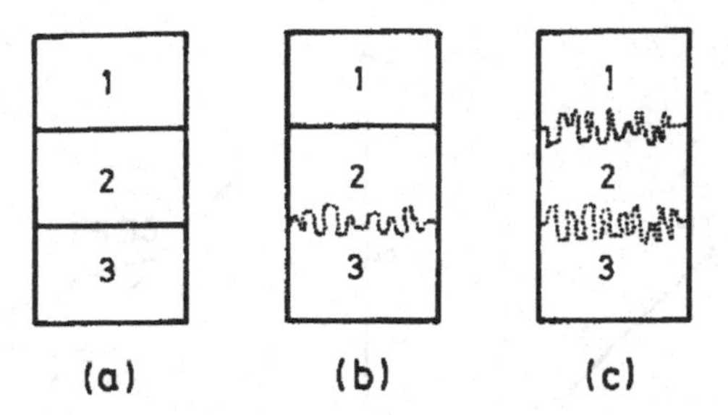

Fig. 3.9 (a) Three–phase coexistence in mixtures; (b) a state of coexistence at a critical end point; (c) coexistence at a tricritical point.

These are *the tricritical points* — critical points, where three phases become identical simultaneously. Obviously for $p = 3$, tricritical points can exist when $f_c = 0$, and lines of tricritical points when $f_c = 1$; see Eq. (3.59a). The possibility of such phenomena in mixtures was first pointed out by Kohnstamm (1926). Another type of critical behaviour in mixtures is described by the so-called *critical end points* — critical states, where two of the three coexisting phases become identical (critical).

Following Griffiths (1974) we show quite schematically in Fig. 3.9 the states of a mixture, corresponding to the tricritical and critical end points. Imagine that there are three (coexisting) phases (1, 2, and 3) of a mixture enclosed in a vessel as shown in Fig. 3.9(a). In Fig. 3.9(b) phases 2 and 3 become identical at the critical end point. At this point (or near it) the separation surface (meniscus) between phases 2 and 3 vanishes (or is not well established) and this is illustrated by the dotted curve. So, the critical end point describes a situation in which a critical (2 − 3) phase coexists with another phase (1). When the distinctive features between all the three coexisting phases vanish, as shown in Fig. 3.9(c), the mixture is in a tricritical state, i.e., at a tricritical point or *a critical point of third order*; see, e.g., Aharony (1983).

The experiments on systems which possess tricritical points show that at such points the first order phase transition line in the space of thermodynamic fields (T, P, μ) changes to a line of critical points (second order transition) as shown in Fig. 3.10(a). Because of this property, the tricritical points are sometimes referred to as *critical points of second order transitions* Landau and Lifshitz (1980). At the critical end point, two lines of the first order transitions and a line of the second order transition terminate (Fig. 3.10b).

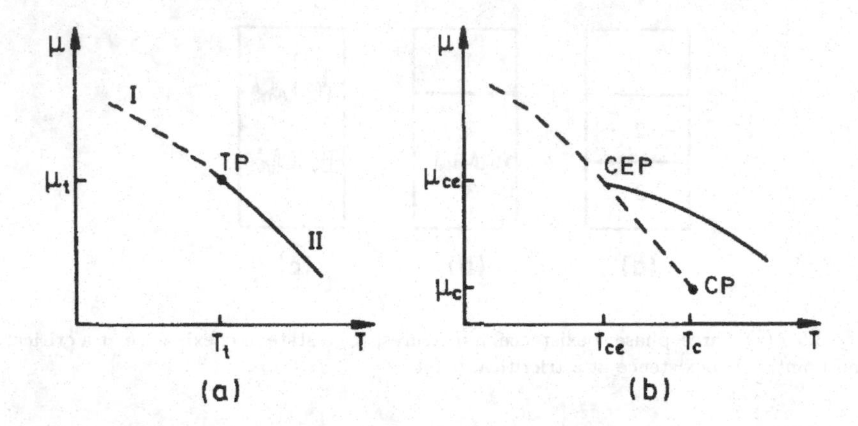

Fig. 3.10 (a) Tricritical point (TP) where a first order (I) transition changes to second order (II); (b) a critical end point (CEP); in comparison, an ordinary critical point (CP) at the end of the first order transition line is shown.

Up to now we have used general thermodynamics rules to outline the topology of the phase diagrams in complex systems. The same can be done by using particular theoretical models; see Griffiths (1973, 1974), Aharony (1983), Fisher (1984), and references therein. Complex critical points as critical-end and tetracritical often appear in various systems and are called *multicritical points.* The phenomena at, and in the close vicinity of these special thermodynamic states are called *multicritical phenomena.*

The above thermodynamic analysis gives a convincing argument that multicritical phenomena occur in systems described by a sufficiently large number of thermodynamic fields $(T, P, \mu, \dots)$. The various multicritical points can be classified either by the number of the existing neighbouring phases or by the number and the type of the intersecting lines. An interesting systematics based on the phase rules (2.115) and (3.59a) is given by Griffiths and Wheeler (1970), and Chang, Hankey and Stanley (1973); see also Hankey, Chang and Stanley (1973), and Stanley *et al.* (1976).

We shall not enter into details of the pure thermodynamic theory of multicritical phenomena because, after making it, we shall not gain anything else but persuade ourselves that the general thermodynamic arguments give a number of new complex phase diagrams and μ-surfaces. In the next Chapters we shall be often concerned with multicritical phenomena in complex systems.

3.8 Other Systems

Let us again consider for simplicity one-component ($l = 1$) systems. Using the notations from Eq. (2.5) and Eq. (2.17) the fundamental relation in the energy density representation can be written in the form

$$u = u(s, v, x) \tag{3.60}$$

where $x = X/N$. The phase transitions connected with the changes of the specific volume v have been discussed in Section 3.6. Here we focus on phase transitions related to an abrupt change of the density x.

3.8.1 *General notations*

We define the order parameter $\varphi = x_2 - x_1$, where x_α are relevant to phases α; $\alpha = 1, 2$. If we choose that $x_2 > x_1$, φ will be positively definite. Alternatively one may choose as an order parameter $\varphi = x - x_1$ and to introduce the density $x' = x - x_1$; then φ coincides with x'. The last definition is more general and allows to consider also nonequilibrium values of φ (i.e., of the density x). Moreover it gives a more clear picture, in which the order parameter $\varphi = (x - x_1)$ describes a transition from the disordered phase 1 ($\varphi = 0$) to the ordered phase 2 ($\varphi > 0$). For the states of a coexistence at equilibrium $x = x_2$, and $\varphi = \varphi_{coex} = x_2 - x_1$; cf. the definitions of Δv and $\Delta \rho$ in Sections 3.2.1 and 3.6.1.

Now one can repeat the thermodynamic treatment from Sections 3.2–3.5 of the first- and second-order transitions in terms of the general order parameter φ. For example, we shall briefly outline the generalization of the Clapeyron–Clausius equation (3.9). The fundamental relation (3.60) corresponds to the Gibbs potential $\Phi = N\mu$ given by the Legendre transformation

$$\mu = u - sT + vP - \varphi h, \tag{3.61}$$

where the thermodynamic field

$$h = \left(\frac{\partial u}{\partial \varphi} \right)_{s,v} \tag{3.62}$$

is conjugate to the order parameter φ.[1] At this stage we shall preserve the explicit dependence of the energy density u on v. In this way the order parameter φ and the other relevant densities are described in terms of three intensive variables: T, P, and h. Excluding v, or Δv, means neglecting the P-dependence of the densities. The chemical potentials $\mu_\alpha = \Phi_\alpha/N$ of phases α are equal on the coexistence (phase transition) surface; see Eq. (3.1). Following the derivation of Eq. (3.9) we obtain

$$\Delta s dT = \Delta v dP + \varphi dh, \tag{3.63}$$

where $\varphi = \varphi_{\text{coex}} = \varphi_2 - \varphi_1$. The Eq. (3.63) gives a connection between the variations dT, dP, and dh at a given point on the coexistence surface of the (T, P, h) space.

The first term on the r.h.s. of Eq. (3.63) is zero on the coexistence isobars ($P = const$) or in the case of $\Delta v = v_2 - v_1 = 0$, when the phases have no difference in the specific volumes or this difference is small and may be neglected. Then

$$\frac{\partial h}{\partial T} = \frac{\Delta s}{\varphi}, \tag{3.64}$$

which is Eq. (3.9) in terms of φ and h. Quite often the phase transitions are accompanied by striction effects ($\Delta v \neq 0$), and in such cases the relation (3.63) can be used.

We may write the relation (3.63) in the form (3.64) with the help of the two-component order parameter $\tilde{\varphi} = (\varphi, \Delta v)$ and the conjugate field $\tilde{h} = (h, P)$; Δv and P are then a secondary order parameter and a secondary conjugate field, respectively. In the above "linear algebra", the order parameters φ and Δv are independent. If the second order differentials of μ_α must be taken into account ($d^2\mu_1 = d^2\mu_2$), as it is for the second order transitions within the Ehrenfest theory (Section 3.5), the secondary order parameter will be coupled to the primary one φ by a term of the type $\Delta v \varphi$.

In case of an "adiabatic" transition ($\Delta s = 0$) or on the coexistence isotherm Eq. (3.63) reduces to

[1] Note, that according to our convention (see the footnote in page 2), φ and h in Eq. (3.61) and Eq. (3.62) may be scalars, vectors, or, even more complex tensor quantities. Then one should define the multiplication φh accordingly. For vectors, φh in Eq. (3.61) means the scalar product $\boldsymbol{\varphi}.\boldsymbol{h} = \sum_i \varphi_i h_i$, the vectorial derivative in Eq. (3.62) is given by the components $h_i = (\partial u/\partial \varphi_i)$ of the vector $\boldsymbol{h}$, and the susceptibility $\hat{\chi}$ is a 2nd rank symmetric tensor with elements $\chi_{ij} = (\partial \varphi_i/\partial h_j)$.

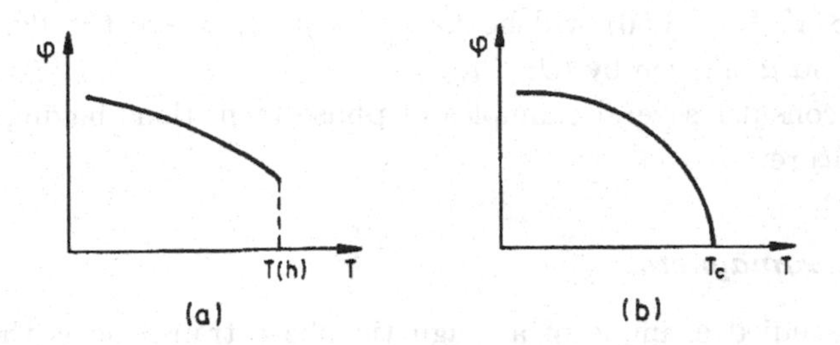

Fig. 3.11 The behaviour of $\varphi(T)$ at fixed h: (a) at a first order coexistence line $T(h)$, (b) at the critical point (CP).

$$\frac{\partial h}{\partial P} = -\frac{\Delta v}{\varphi}. \tag{3.65}$$

Keeping in mind the effect of Δv as a secondary order parameter we shall exclude the specific volume v from our discussion below. Then the fundamental equation (3.60) will be

$$u = u(s, \varphi). \tag{3.66}$$

These densities describe either a line of first-order transitions, on which the order parameter undergoes a jump, as shown in Fig. 3.11(a), or a second-order transition (critical point) where the order parameter φ is continuous (Fig. 3.11b). The location of the critical point (or line) of a second order transition is given by

$$\chi^{-1} = \left(\frac{\partial h}{\partial \varphi}\right)_T = 0 \tag{3.67a}$$

and

$$\left(\frac{\partial^2 h}{\partial \varphi^2}\right)_T = 0; \tag{3.67b}$$

cf. Eqs. (3.4) and (3.17). In Eq. (3.67a), χ is *the isothermal susceptibility* or, shortly, *the susceptibility* of the system with respect to a field, conjugate to the order parameter.

In many discussions it is convenient to define the order parameter per unit volume by $\varphi = X/V$. Then the thermodynamic treatment is performed in the same way as before by defining the densities in Eq. (3.66) $u = U/V$

and $s = S/V$. Eq. (3.60) will be $u = u(s, \rho, x)$, where the density ρ is $\rho = N/V$, and μ is given by $(\partial u/\partial \rho)_{s,x}$.

Let us consider several examples of phase transitions having different physical nature.

3.8.2 *Ferromagnets*

The most studied example of a magnetic phase transition is the change from a paramagnetic to a ferromagnetic state in *ferromagnets* like Fe, Ni, EuO etc. The ferromagnetic (ordered) phase is spontaneously magnetized, which means that a nonzero macroscopic magnetization (the overall magnetic moment) $\boldsymbol{M}$ exists in the body at a zero external magnetic field $\boldsymbol{H}$. In the paramagnetic phase $\boldsymbol{M} = 0$, so we can define a *vector* order parameter in ferromagnets by $\boldsymbol{M}$ or, equivalently by $\varphi \equiv \boldsymbol{m}$, where $\boldsymbol{m} = \boldsymbol{M}/N$ is the magnetization per particle (alternatively, one may use the density defined by $\boldsymbol{m}' = \boldsymbol{M}/V = \rho\boldsymbol{m}$, where $\rho = N/V \neq 0$ is the number density). The quantity $\boldsymbol{m}$ is a thermodynamic density, it is *not* the magnetic moment of a *free* particle from the same substance. We shall not consider the pressure effects[2], so fixing the pressure P at an appropriate value and neglecting the v-dependence in Eq. (3.60) we can begin the analysis with Eq. (3.66), where $\varphi = \boldsymbol{m}$, i.e.,

$$u = u(s, \boldsymbol{m}). \tag{3.68}$$

Because u is a scalar and $\boldsymbol{m}$ is a vector, in an isotropic ferromagnet u will depend only on $\boldsymbol{m}^2$: $u = u(s, m^2)$. The equations of state are then obtained in the form

$$T = T(s, \boldsymbol{m}), \tag{3.69a}$$

and

$$\boldsymbol{h} = \boldsymbol{h}(s, \boldsymbol{m}), \tag{3.69b}$$

where $\boldsymbol{h} = (\partial u/\partial \boldsymbol{m})_s$ is the magnetic field conjugate to $\boldsymbol{m}$. We shall assume that $\boldsymbol{h}$ is proportional to (or coincides with) the external magnetic field $\boldsymbol{H}$, ($\boldsymbol{h} \equiv \boldsymbol{H}$; note, that in the remainder of this Chapter we shall use the

[2] While pressure effects on the magnetic properties and, in particular, on the magnetization M are possible in a number of systems, for the sake of simplicity, here we avoid their consideration.

notation $\boldsymbol{H}$) ; for the difference between a local field $\boldsymbol{H}'$ and an external (applied) field $\boldsymbol{H}$ see, e.g., Münster (1970); for an example, where $\boldsymbol{h} \sim \boldsymbol{H}$ see Sec. 5.8.6, Eq. (5.153d), where $\bar{h}$, which plays the role of $\boldsymbol{h}$ in Eq. (3.69b) is given by $\bar{\boldsymbol{h}} = g_J\mu_B(J+1)\boldsymbol{H}/3T$.

Following our convention of avoiding bold face for the vector notations (see the footnote in pages 2) we shall use Eq. (3.69b) in the form

$$H = H(s, m). \tag{3.69c}$$

Excluding the entropy s from Eqs. (3.69a) and (3.69b) we have the most usable equation of state

$$H = H(m, T), \tag{3.70a}$$

or, which is the same,

$$m = m(H, T). \tag{3.70b}$$

To describe the ferromagnetic transition we need an explicit form of the function $u(s, m)$ or, equivalently, of the equation of state but as in all previous cases this information is not available from rational thermodynamics. Thus we are forced to make a general analysis based on Eq. (3.1). Because there is no presence of different chemical components the equilibrium condition (3.1) can be expressed by the Gibbs potentials of the phases: $\Phi_1 = \Phi_2$ or, using densities ϕ_α,

$$\phi_f(T, H) = \phi_p(T, H), \tag{3.71}$$

where the subscripts f and p stand for the ferromagnetic and paramagnetic phases, respectively. One may consider $\phi_\alpha = \Phi_\alpha/N$, then $\phi_\alpha = \mu_\alpha$, or $\phi_\alpha = \Phi_\alpha/V$, which implies that the order parameter $m = M/V$ is taken per unit volume. Solving Eq. (3.71) we find the coexistence line:

$$H = H(T). \tag{3.72}$$

To continue the analysis one needs general arguments and basic experimental facts. Suppose that $m \neq 0$ when $H \neq 0$. Therefore no paramagnetic state ($m = 0$) exists for $H \neq 0$ and the paramagnetic-to-ferromagnetic phase transition is impossible at any point $(T, H \neq 0)$ of the (T, H) diagram. What remains is to focus our attention on the line $H = 0$ of the same diagram. Now we shall make use of two experimental facts: the ferromagnetic phase is usually the low-temperature phase and the transition is usually of a second order. Therefore a temperature $T_c \neq 0$ exists on the coexistence line $H(T) = 0$, below which the ferromagnetic phase is stable; see Fig. 3.12(a). Because the transitions is of second order the equilibrium value of m at T_c equals zero $m(T_c) = 0$. The paramagnetic and the ferromagnetic phases are indistinguishable at T_c. The function $m(T)$ gradually increases with the decrease of T below T_c and reaches its maximal value $m_0 = m(0)$ at $T = 0$; see Fig. 3.12(b). At $T < T_c$ and $H = 0$ one expects a first order transition to occur because of the change $H \to -H$ in Fig. 3.12(a) will give rise to a jump of $m(T, H)$ at $H = 0$:

$$\Delta m(T, 0) = \lim_{H \to 0} [m(H) - m(-H)]. \tag{3.73}$$

In our idealized system $m(H) = m(-H)$ for any T, so Δm in Eq. (3.73) is equal to $2m(T, 0)$. The function $m(T, 0)$ is shown by the solid line in Fig. 3.12(b). The broken line in the same figure describes $m(T, H)$ for $H \neq 0$. Note that a "paramagnetic effect" is expected for $H \neq 0 : \Delta m(H) = m(T, H) - m(T, 0) > 0$. For $H \neq 0$, one can think of a *field-induced* phase transition from a "phase" m to a another one, $-m$, but these phases cannot be distinguished for $H = 0$. Accepting such an interpretation an attempt to outline the (H, m) diagram can be made; see Fig. 3.12(c). For $T < T_c$, $m(H)$ will decrease to $m(0)$ when H decrease to zero. At $H = 0$, m will take the values described by the solid line in Fig. 3.12(b). On the flat solid line AB the behaviour of the ferromagnet will be similar to that of liquids; see Fig. 3.12(c) or Fig. 3.3(a). Here the phases 1 and 2, which follow the lever rule (3.6) are the above mentioned "phases" m and $-m$. The solid line AB and the broken line AB in Fig. 3.12(c) have the same meaning as those for the liquid–vapour transition. They correspond to the AB-line in Fig. 3.12(b).

The picture of the ferromagnetic transition outlined above will be obtained from a particular model in Section 4.10; see also Section 5.8. It is obvious that the thermodynamic description is the same as that of the liquid–vapour transition, apart from some differences in the symmetry requirements.

Expanding, for small m, $u(m)$ from Eq. (3.68) to the second order in m we find

$$u_f = u_p + \frac{1}{2} C^s m^2, \tag{3.74}$$

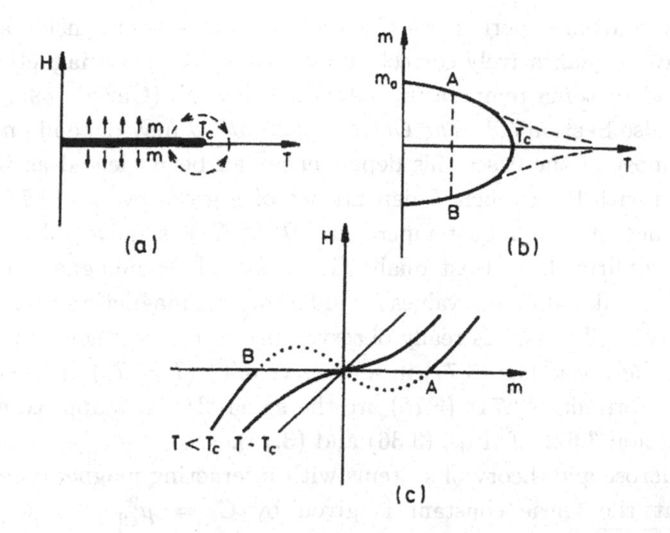

Fig. 3.12 (a) (H, T) diagram of a usual ferromagnet. The broken line describes *a* process without phase transition. (b) (m, T) diagram. The solid line $m(T)$ describes the case $H = 0$, and the broken line stands for $H \neq 0$. (c) (H, m) diagram. The lines 1, 2 and 3 correspond to $T < T_c, T = T_c$ and $T > T_c$. The broken line describes metastable and unstable states.

where C^s is the (*magnetic*) stiffness coefficient at constant entropy s. The corresponding expression for the free energy density $f = F/N$ is

$$f_f = f_p + \frac{1}{2}C^T m^2, \tag{3.75}$$

where C^T is the respective stiffness coefficient at constant temperature.

The Gibbs potential density ϕ_f is $\phi_f = f_f - mH$. The isothermal magnetic susceptibility $\chi_T = (\partial m/\partial H)_T$ can be written in the form

$$\chi_T = \frac{1}{C^T} \tag{3.76}$$

and is also taken at fixed P, so $\chi_T = \chi_{T,P} = \chi$, in accordance with the above assumption; cf. Eq. (3.67a).

Accepting that χ depends on the temperature according to the *the Curie–Weiss law*:

$$\chi = \frac{C}{|T - T_c|}, \tag{3.77}$$

which follows from experimental observations and model considerations,[3] one can give a qualitatively correct description of the paramagnetic phase $T > T_c$ and in some temperature interval below T_c (Curie, 1895; Weiss, 1907); see also Section 5.8. *The Curie "constant"* C may depend on T and P but in most of the cases this dependence can be neglected as weak in comparison with the explicit T-dependence of χ given by Eq. (3.77). This T-dependence of χ at high temperature ($T \gg T_c$) is universal, in sense, that it is confirmed, at least qualitatively for all ferromagnets, but the "constant" C takes different values for different ferromagnetic systems. The divergence of $\chi(T)$ at T_c is really observed in the real systems but it does not follow the linear law (3.77) in a close vicinity ($T \sim T_c$) of the critical point. The formulae (3.74)–(3.76) are the same "Hooke's approximation" used in Section 3.6.2; cf. Eqs. (3.36) and (3.75).

The microscopic theory of systems with interacting magnetic moments shows that the Curie constant is given by $C = \mu_{\text{eff}}^2$, where $\mu_{\text{eff}} = g_J\mu_{\text{B}}\sqrt{J(J+1)}$ is the effective magnetic moment of the atom (J is the angular momentum quantum number); see Sections 5.8.4–5.8.6. One may claim that the angular momentum J varies with the variation of the pressure and the temperature (the dependence on the chemical composition is obvious).

In isotropic ferromagnets in a zero external field $\boldsymbol{H}$ the magnetization vector $\boldsymbol{m} = (m_x, m_y, m_z)$ takes an arbitrary direction in space. In nonzero field $\boldsymbol{H}$ this rotational invariance of $\boldsymbol{m}$ vanishes because the thermodynamic functions depend on $\boldsymbol{H}.\boldsymbol{m}$, i.e., on the field orientation; see Eq. (3.61) with $\varphi \to \boldsymbol{m}, h \to \boldsymbol{H}$ (broken continuous symmetry; see Section 5.8.2).

3.8.3 *Anisotropy in ferromagnets*

We have neglected *the crystal anisotropy* but when it is of importance the m-dependent terms in Eqs. (3.74) and (3.75) are replaced by more general expressions. For example, the m^2-term in Eq. (3.75) can be replaced by

$$\frac{1}{2}C_{ij}^T \boldsymbol{m}_i.\boldsymbol{m}_j, \tag{3.78a}$$

[cf. Eq. (3.36)], where $i, j = (x, y, z)$.

The symmetric matrix $C^T = \{C_{ij}^T\}$ can be diagonalized, so this term becomes

[3] The usual form of the known Curie law and the Curie constant C are discussed for both noninteracting magnetic moments and Ising model in Sections 5.8.4–5.8.6

$$\frac{1}{2}C_i^T m_i^2, \tag{3.78b}$$

where C_i^T are the eigenvalues of $\hat{C}^T$. The susceptibility changes too:

$$\chi_i = \frac{1}{C_i^T} \tag{3.79}$$

and is different in the different spatial directions ($i = x, y, z$).

When all C_i^T in Eq. (3.78b) are positive, $f_f > f_p$ and the paraphase ($m_i = 0$) is stable at fixed temperature T. If $C_z^T < 0$, but C_x^T and C_y^T remain positive at some temperature, a nonzero magnetization $m_z = m$ along the z-axis is energetically favourable ($f_f < f_p$). When all C_i^T are negative but $|C_z^T| \gg |C_x^T| \sim |C_y^T|$, the energy contribution from the m_z component of $\boldsymbol{m}$ will be maximal. Again, the preferred direction of $\boldsymbol{m}$ will be the z-axis because the ferromagnetic states with $m_z \gg m_x, m_y$ correspond to lower free energy $f_f(m)$. In such cases we can accept the approximation of the *the extreme uniaxial anisotropy*: $m_x = m_y = 0$ and $m_z \neq 0$. Systems possessing this property are called *uniaxial ferromagnets* (or ferromagnets with *an axis of easy magnetization*); see Section 5.6. for the treatment of these systems in the frame of the Ising model.

In the same way one can introduce the notion of *a plane of easy magnetization.* If $|C_x^T| \sim |C_y^T| \gg |C_z^T|$ the magnetic vector $\boldsymbol{m} \approx (m_x, m_y, 0)$ lies in the X–Y plane of easy magnetization. This is an *XY system* in which the two-component vector $\boldsymbol{m} = (m_x, m_y)$ or, which is the same, the complex scalar $m = |m|e^{i\vartheta}$, can rotate in the X–Y plane (or in the complex plane) giving equivalent ferromagnetic states (for $H = 0$ or $\boldsymbol{H} \parallel \hat{z}$); see also Section 5.7.1.

3.8.4 *Complex magnetic order*

We have sketched out the thermodynamic description of a usual ferromagnetic transition using the notion of a macroscopic (mean) magnetic moment $\boldsymbol{m}$. Below T_c, $\boldsymbol{m}$ spontaneously arises out of the ferromagnetic interaction between the particle magnetic moments (see also Section 5.6); to be more clear we accept that the magnetic moments of the particles (atoms) are localized at the vertices of the crystalline lattice. In *antiferromagnets* (such as Cr, Mn, FeF_2, etc.) the magnetic interactions tend to an antiparallel orientation of the neighbouring magnetic moments. In the simplest case the lattice consists of *two* equivalent interpenetrating sublattices ($i = 1, 2$)

with oppositely oriented magnetic moments, as shown in Fig. 3.13(a). Each sublattice has its own spontaneous (sub)magnetization m_i. For zero external field the total (*alternating*) magnetization $m = \sum_i m_i$, generally, $i \geq 2$ is equal to zero; see Fig. 3.13(b), where $m_{1,2} \equiv |\boldsymbol{m}_{1,2}|$. The field H gives rise to asymmetry, and $\boldsymbol{m} \neq 0$. For isotropic antiferromagnets the energy density $u(\boldsymbol{m}_1, \boldsymbol{m}_2)$ will depend on the quadratic form

$$\frac{1}{2}C_1^s m_1^2 + C_{12}^s \boldsymbol{m}_1 \cdot \boldsymbol{m}_2 + \frac{1}{2}C_2^s m_2^2, \tag{3.80}$$

by analogy with Eq. (3.74). Because the sublattices are equivalent $C_2^s = C_1^s$; and because the ordering is antiferromagnetic we have $\boldsymbol{m}_1 \cdot \boldsymbol{m}_2 = -m_1^2$. The antiferromagnetic ground state energy will be lower than the paramagnetic one if the quadratic form (3.80) is negatively definite below some temperature, that is if $C_{12}^s > C_1^s$.

Quite often another form of the order parameters is used: the alternating magnetization $\boldsymbol{m} = \boldsymbol{m}_1 + \boldsymbol{m}_2$ which is nonzero for $H \neq 0$, and *the staggered magnetization* $\boldsymbol{m}_s = \boldsymbol{m}_1 - \boldsymbol{m}_2$. The field $\boldsymbol{H}$ is conjugate to $\boldsymbol{m}$ as in ferromagnets, whereas the *staggered field* $\boldsymbol{H}_s$ is conjugate to $\boldsymbol{m}_s$. The staggered field is somewhat strange because it enhances both $\boldsymbol{m}_1$ and $\boldsymbol{m}_2$. In other words it acts as a magnetic field in $\boldsymbol{m}_1$-direction and, simultaneously in $\boldsymbol{m}_2$-direction. It cannot be caused by external effects and is rather a result from the competing magnetic interactions in some (anti)ferromagnets.

In zero fields ($H = H_s = 0$) the antiferromagnetic order appears at the *Neél temperature* T_N (Neél, 1932); see Fig. 3.13(b) and Fig. 3.14. The magnetic susceptibility χ above T_N is given by

$$\chi = \frac{C}{T + T_N}, \tag{3.81}$$

so the divergence like that in Eq. (3.77) occurs now for $T = -T_N$.

At T_N the susceptibility of antiferromagnets exhibits rather a cusp than a divergence and below T_N the susceptibilities along ($\chi_{\|}$) and perpendicular ($\chi_{\perp}$) to the field H are different; see, e.g., Mattis (1965, 1985), Kittel (1971), or Barbara, Gignoux and Vettier (1988).

We have seen in Section 3.8.2 that a finite external field H destroys the ferromagnetic transition because the overall disorder ($m = 0$) of the magnetic moment directions of the particles is thermodynamically improbable for $H \neq 0$. Equivalently an antiferromagnet in a staggered field H_s will not undergo a paramagnetic-to-antiferromagnetic transition because m_s will be

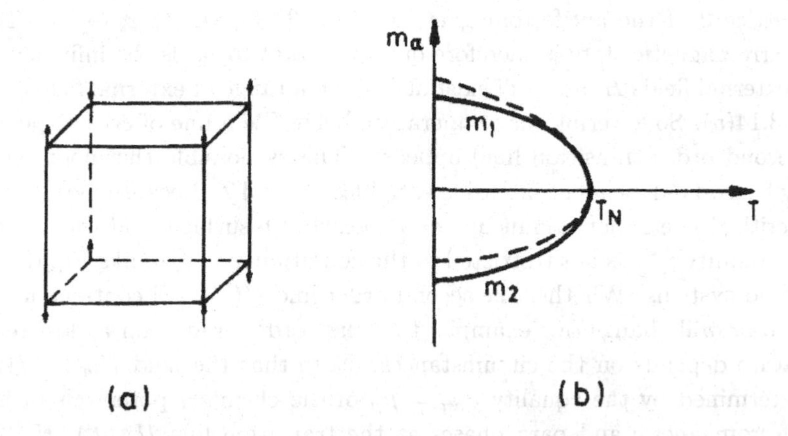

Fig. 3.13 (a) Antiferromagnetic order for a simple cubic magnetic unit cell. (b) (m, T) phase diagram of an antiferromagnet. The solid line represents $m_\alpha(T)$ for $H = 0$, and the broken line stands for $H \neq 0$.

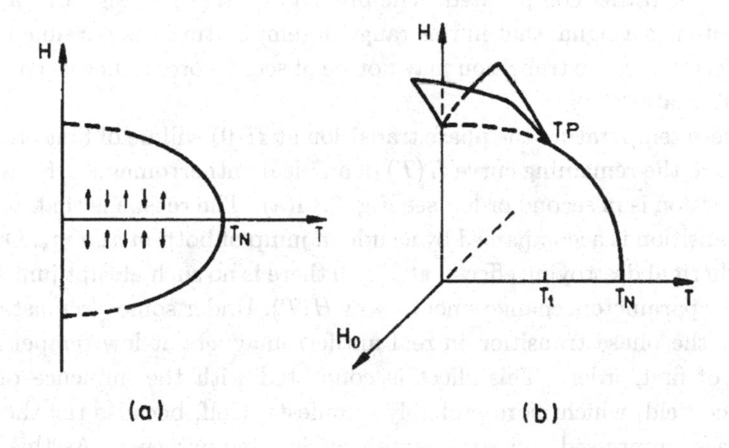

Fig. 3.14 Antiferromagnets: (a) (T, H) diagram. (b) (H_s, T, H) diagram. The solid lines are second order transitions. The meaning of the broken lines is given in the text.

nonzero for any $T \geq 0$. However in most of the systems it is difficult to reproduce experimentally staggered field H_s and in the usual cases the effect of the external field H can only be observed; see Fisher (1967) and Blume *et al.* (1974).

Let $\boldsymbol{H}$ be along the $\boldsymbol{m}_1$-direction. Then the energy of order $2m_2H$ will be necessary to alter the direction of the mean magnetic moment $\boldsymbol{m}_2$ in the sublattice 2 of the antiferromagnet; see Eq. (3.61), where $\varphi \to \boldsymbol{m}$. The antiferromagnetic state is therefore quite resistant towards the influence of the external fields H, and will exist at least in a range of external fields; see Fig. 3.14(a). So lowering the temperature below T_N a line of critical points (a second order transition line) appears. This is possible thermodynamically for two independent intensive variables (H and T) because in this case the critical line is not a limit line of a coexistence surface, and the inverse susceptibility χ^{-1} is not restricted to the condition $\chi^{-1} = 0$; like Eq. (3.17) for fluid systems. Whether the second order line $H(T)$ will continue up to $T = 0$ or will change, for example, to a first order transition at low temperature depends on the circumstances. Note that the field $H_m \equiv H(T)$ is determined by the equality $\mu_{an} = \mu_p$ of the chemical potentials of the antiferromagnetic and para phases at the transition line $H_m(T)$. $H_m(T)$ is sometimes called *the mean thermodynamic magnetic field* of the phase transition. In proper ferromagnets $H_m(T) = 0$; see Fig. 3.12(a). In general, the phase transition behaviour of the antiferromagnets at low temperature ($T \ll T_N$) is rather complicated. The broken line $H(T)$ in Fig. 3.14(a) can be accepted as a signal that in this range of temperature the paramagnetic-to-antiferromagnetic transition may not be of second order. Let us consider some alternatives.

At zero temperature the phase transition at $H(0)$ will be of first order in contrast to the remaining curve $H(T)$ of an ideal antiferromagnet, for which the transition is of second order; see Fig. 3.14(a). The reason is that at $T = 0$ the transition is accompanied by a sudden jump of both m and m_s. Owing to the thermal destroying effects, at $T > 0$ there is no such abrupt jump and the order parameters change smoothly at $H(T)$. Under some circumstances however the phase transition in real antiferromagnets at low temperature can be of first order. This effect is connected with the influence of the staggered field, which more probably manifests itself, because the thermal effects are suppressed to a great extent at low temperature. As this field enhances the antiferromagnetic order the phase diagram in the (H_s, T, H) space will be as shown in Fig. 3.14(b). The staggered field H_s breaks the symmetry of the antiferromagnetic order, that is it acts in the same way as the external field H in a ferromagnet. So it is easy to understand why the broken line in Fig. 3.14(b) and the corresponding lines in Fig. 3.14(a) are first order transition lines. The tricritical point in Fig. 3.14 naturally follows from the change of the order of the antiferromagnetic transition.

A phase diagram with a first order transition at low temperature and a second order transition at higher temperatures connected by a tricritical point is observed in *metamagnets* like $FeCl_2$; see Birgeneau *el al.* (1974). In layered structures with strong ferromagnetic interactions within the layers and weaker antiferromagnetic interactions between the layers the antiferromagnetic state at $T < T_N$ and low fields H is stable but at higher fields the magnetic moments of some of the layers can alter their directions and hence, states with net magnetization $\boldsymbol{m}$ can occur (metamagnetic states).

In uniaxial antiferromagnets with strong anisotropy when the magnetic field $\boldsymbol{H}$ is along the magnetization axis the antiferromagnetic state very often changes to the so-called *spin-flop states* in which the magnetic moments (or spins) lie in planes perpendicular to the field $\boldsymbol{H}$.

One may describe thermodynamically *the ferrimagnetic state* assuming that the parameters C_1^s and C_2^s in Eq. (3.80) are not equal. Then m_1 will be different from m_2 ($m_\alpha = |\boldsymbol{m}_\alpha|$). Despite of some similarity between the description of this type of order and the antiferromagnetism by sublattices, the ferrimagnetic ground state is quite different and is characterized by a relatively weak magnetization ($\boldsymbol{m} \neq 0$).

The *weak ferromagnetism* is another "modification" of antiferromagnetism in which the sublattice magnetizations are equal in magnitude ($m_1 = m_2$) but their orientations are not completely antiparallel; so the total magnetization is weak, as in ferrimagnet, but the reason for it is quite different.

We have recalled some of the most frequently discussed systems with magnetic order; for details related to the above discussion as well as for other magnetic phase transitions (Mattis, 1965, 1985; Kittel, 1971; Aharony, 1983; Barbara, Gignoux and Vettier, 1988). The similarities between the critical behaviour in metamagnets and mixtures is discussed by Liu and Fisher (1973), Lawrie and Sarbach (1984), and Knobler and Scott (1984).

3.8.5 *Structural and ferroelectric phase transitions*

Some notion of the thermodynamic description of structural phase transitions has been already given in Section 3.6.2. The phase transition properties depend on what is the type of the crystal symmetry change, that is on the symmetry of the $C_{\mu\nu}^{(T)}$-matrix in Eq. (3.36). The microscopic description can be done with the help of the methods of solid state physics. In general the structural transitions in crystals can be either *order-disorder* or *displacive* type; see Scott (1974), Shirane (1974), and Cowley (1980).

The order-disorder type of structural transitions is easy to demonstrate on the example of ordering in alloys such as CuZn (β-brass). Above the critical temperature ($T_c = 741$ K) the Cu and Zn atoms are randomly distributed in the body-centered cubic lattice and below T_c they arrange in a regular way: the nearest neighbours of Cu are Zn and vice versa.

The order-disorder transitions can occur simultaneously with the appearance of another variable (order parameter) which describes another property of the system different from the crystal symmetry change. For example, in *ferroelectrics*, this is the *spontaneous electric polarization* $\boldsymbol{P}$ which appears in the low temperature phase at zero applied electric field, owing to the reconstruction of the crystal lattice (Landau and Lifshitz, 1960); an example is the ferroelectrics $NaNO_2$. The thermodynamic treatment of the ferroelectric transition is similar to that of ferromagnetic transition, but instead of the magnetization $\boldsymbol{M}$ we use the electric polarization $\boldsymbol{P}$. In antiferroelectrics the sublattice polarizations ($\boldsymbol{P}_\nu$) play the role of the sublattice magnetizations $\boldsymbol{M}_\nu$ in antiferromagnets (see Section 3.8.4). The Landau theory (see Chapter 4) of the ferroelectric transitions was developed by Devonshire (1949, 1951, 1954) and Ginzburg (1945, 1949); see also Fatuzzo and Merz (1967), Blinc and Ẑekš (1974).

The cause of the displacive transitions is the instability of the crystal towards a particular vibrational normal mode ("soft mode") in the high-temperature phase, so they are described in the best way by the models of interacting phonons
indexphonon. For example, the displacive transition in $SrTiO_3$ where the order parameter is the distortion coordinate corresponding to the $\boldsymbol{q} = (\pi/2a)(1, 1, 1)$ soft phonon mode, is responsible for the doubling of the unit cell; a is the lattice constant. The displacive transition can occur simultaneously with the appearance of a ferroelectric phase, as the ferroelectric displacive transition in $BaTiO_3$.

We have mentioned that for structural transitions the displacement variables like S_ν in Eq. (3.36) can be considered as order parameter. A formal description of the symmetry change can be done with the help of the group theory; see Landau (1937a, 1967), Landau and Lifshitz (1980), Gufan (1982), and Tolédano and Tolédano (1987). The crystal structure is described by the density distribution $\rho(\boldsymbol{x})$; see Section 3.6.3. The function $\rho(\boldsymbol{x})$ can be represented by a linear combination of normalized functions $f(\boldsymbol{x})$ which transform into one another under the symmetry operations of the group G of the higher-symmetry phase. Then the group G_1 of the distorted phase will be a subgroup of G, $G_1 \subset G$. If the density distribution $\rho_0(\boldsymbol{x})$ corresponds to the high-symmetry phase and $\rho(\boldsymbol{x})$ to the distorted phase, $\delta\rho = \rho - \rho_0$ can be represented as the following linear combination:

$$\delta\rho(\boldsymbol{x}) = \sum_{i\nu} C_i^{(\nu)} f_i^{(\nu)}(\boldsymbol{x}). \tag{3.82}$$

The complete sets of functions $f_i^{(\nu)}$ form the bases of the irreducible representations $G^{(\nu)}$ of G; the superscript ν of f labels the irreducible representations and the subscript $i = (1, \cdots, n_\nu)$ labels the functions of a given representation ν. Then the free energy $F = F(T, h, \rho)$ where h denotes a set $(h_1, \dots)$ of intensive variables, will depend on the amplitudes $C_i^{(\nu)}$ of the expansion (3.82), i.e., $F(T, h, C_i^{(\nu)})$ or, shortly, $F(C_i^{(\nu)})$. So the amplitudes $C_i^{(\nu)}$ can be considered as a (multi-component) order parameter. The free energy must be invariant under each operation of the irreducible representation $G^{(\nu)}$ and, therefore, it depends on an invariant expression like $\sum_\nu (C_i^{(\nu)})^2$. The distorted states $(C_i^{(\nu)} \neq 0)$ will be stable, if $F(0) > F(C_i^{(\nu)})$ for some T and h. In the expansion

$$F(C_i^{(\nu)}) = F(0) + \sum_\nu r_\nu \left(\sum_i C_i^{(\nu)} \right)^2, \tag{3.83}$$

similar to that given by Eq. (3.36), the problem of the stability is connected with the sign of the (stiffness) coefficients r_ν. The high-symmetry phase is usually the high-temperature one. Let us assume that starting to lower the temperature of the high-symmetry phase $(r_\nu > 0)$, one of these coefficients, say r_1, becomes negative for the first time below some temperature T_1. Then the phase change at T_1 is described by the corresponding irreducible representation $G^{(1)}$ or equivalently by an n_1-component order parameter $\varphi = (C_1^{(1)}, \cdots, C_{n_1}^{(1)})$. It may happen that the transition is described by two (or more) irreducible representations of the group G. Let those be $G^{(1)}$ and $G^{(2)}$. So the equations are two, instead of one, and are given by

$$r_\nu = 0, \qquad (\nu = 1, 2). \tag{3.84}$$

The Eqs. (3.84) describe two lines (or surfaces) of two different structural transitions; the intersection point (or line) of these transition gives the initial transition described by the two irreducible representations.

In structural phase transitions the order parameter usually couples to elastic strain parameters or other secondary order parameters. The *improper ferroelectrics* are materials where the secondary order parameter is the ferroelectric distortion; see Cochran (1971), Dvorak (1974), and Cowley (1980).

3.8.6 *Superfluidity and superconductivity*

These phenomena occur at low temperatures as a result of the simultaneous action of strong quantum correlations and rather specific inter-particle interactions. A classical example of superfluidity is the liquid ^{4}He. At $\sim$ 2.2 K the liquid ^{4}He at normal pressure P undergoes a second order phase transition to the low temperature ($T < 2.2$ K) phase, the superfluid ^{4}He (it is often referred to as He II; and the normal liquid ^{4}He–He I). The transition at the saturated vapour pressure is signalized by a specific heat $C_P(T)$ change, whose "λ-looking" shape is shown in Fig. 3.6; this has provoked the name λ-point. The term "λ-point" is often used as a synonym of a critical point. Moreover, the superfluid transition in ^{4}He is one of the most studied transitions in the physics of critical phenomena. Because the superfluid critical temperature T_λ varies with the pressure P, there is rather a λ-line, $T_\lambda(P)$ of critical points in the phase diagram of ^{4}He; see, e.g., London (1954), Pippard (1957), and Wilks and Betts (1987).

The superfluidity means an absence of viscosity and this fact can be understood from the special energy spectrum of the Bose liquid (^{4}He). We shall not enter into this extremely interesting physics (London, 1954; Khalatnikov, 1965; Tilley and Tilley, 1974; Tinkham, 2004). Rather we shall recall that the profound phenomenological description is performed within the Landau theory of Bose liquids (Landau, 1941, 1947, 1949). In these liquids a condensation in the momentum ($\boldsymbol{k}$-) space occurs which differs (although the formal resemblance is evident) from the BEC of the ideal Bose gas; see, e.g., Lifshitz and Pitaevskii (1980). The most essential difference is that the Bose liquids are able to become superfluids below the condensation (critical) point. The superfluidity of interacting Bose systems and the form of the energy spectrum of the quasiparticles $\varepsilon(\boldsymbol{k})$ has been postulated by Landau and theoretically derived from the model of the degenerate nonideal Bose gas for small momenta $\boldsymbol{k}$ by Bogoliubov (1947).

The order parameter of the superfluid transition in ^{4}He is the macroscopic wave function $\psi(\boldsymbol{x}, t)$ of the superfluid phase (condensate), and $|\psi|^2 = \rho_s$ is the density of the superfluid "portion" of the liquid. At $T \ll T_\lambda$, $|\psi|^2 \to 1$ whereas at T_λ, $|\psi|^2 = 0$. When the superfluid flows are absent in equilibrium, ψ is $(\boldsymbol{x}, t)$-independent as every variable in macroscopic thermodynamics. The complex number $\psi = \psi_1 + i\psi_2$ can be treated either as a two-component real vector (ψ_1, ψ_2) or, equivalently, as a complex scalar rotating in the complex plane, $\psi = |\psi_0|e^{i\theta}$. While $|\psi_0|$ is fixed by the superfluid density ($|\psi_0| = \rho_s^{1/2}$) for given T and P, θ can take arbitrary values.

This degeneracy of the superfluid state with respect to θ determines the similarity to the phase transition in XY-ferromagnets (see Section 3.8.3).

The superconductivity is a superflow of charged particles. Numerous metals and metal compounds are superconductors at temperatures $T \leq 20$ K. The reason for it is the occurrence of the Bose–Einstein-like condensation of composite bosons, which are the famous Cooper pairs of electrons (Cooper, 1957). These quasiparticles have spin $S = 0$ and orbital angular momentum $L = 0$. The mechanism of pairing is the effective attraction between the electrons on the Fermi surface due to exchange of virtual phonons. A successful microscopic description of this phenomenon is given by the BCS theory of superconductivity (Bardeen, Cooper and Scrieffer , 1957).

At macroscopic level of description the internal energy $U(S, V, \boldsymbol{M}, \psi)$ is considered, where ψ is the macroscopic wave function of the charged superfluid. Due to the electric charge $(2e)$ of the effective electron pairs we have to take into account their interaction with the external magnetic field $\boldsymbol{H}$ and therefore, the intensive variable $\boldsymbol{M}$ conjugate to $\boldsymbol{H}$ enters in U. The thermodynamic properties and the phase diagram (T, P, H) can be studied with the help of the thermodynamic potential

$$\Phi(T, P, \boldsymbol{H}) = U - TS + PV - \boldsymbol{H} \cdot \boldsymbol{M}. \tag{3.85}$$

The thermodynamic theory of the superconducting transition is comprehensively presented by London (1950), Pippard (1957), and Landau and Lifshitz (1960). One has to keep in mind that, because their origin, the order parameters in superfluids and superconductors do not possess thermodynamically conjugate fields. The latter may be introduced as "fictitious" complex fields h and, when Legendre transformations have to be made a term like $(h^*\psi + h\psi^*)$ is used, where the asterisk $(*)$ means complex conjugation.

A fundamental property of the superconductors is their *superdiamagnetism* (Meissner–Ochsenfeld effect), i.e., there is a stable superconducting phase (Meissner phase) in nonzero applied magnetic field, in which the magnetic induction $\boldsymbol{B} = \boldsymbol{H} + 4\pi\boldsymbol{M}$ is equal to zero ($\boldsymbol{M} = -\boldsymbol{H}/4\pi$). Taking into account in our present discussion only the bulk properties (any surface effects are ignored) we point out the possibility of existence of another superconducting phase (*the mixed phase*) in which the magnetic field $\boldsymbol{H}$ penetrates, under certain conditions, in the superconductor in the form of magnetic vortex lines. The cores of these vortices are in a normal (non-superconductive) state. If $\boldsymbol{H}$ is in the $\hat{z}$-direction of space the vortex lines along $\boldsymbol{H}$ form a 2D XY lattice (Abrikosov's (1957) lattice). At strong enough external field $\boldsymbol{H}$ the normal phase ($\psi = 0$) is stable.

One may write the thermodynamic potential Eq. (3.85) in the form

$$\Phi = \Phi(T, P, \psi, \boldsymbol{B}), \tag{3.86}$$

where

$$\boldsymbol{B} = \text{rot}\boldsymbol{A}, \tag{3.87}$$

and $\boldsymbol{A}$ is the vector potential of the magnetic induction $\boldsymbol{B}$, and to try to describe the stable bulk phases outlined above. This is done within the Ginzburg–Landau theory of superconductivity (Ginzburg and Landau, 1950); see also Section 9.7. Vortex phases can exist also in the rotating superfluid ^{4}He; see, e.g., Tilley and Tilley (1974).

The superfluidity exists also in the liquid ^{3}He (a phenomenon theoretically predicted by Pitaevskii (1959) and experimentally discovered in the range of mK temperature by Osheroff, Richardson and Lee (1972). The magnetic fluctuations induced pairing of the spin-1/2 ^{3}He atoms leads to the formation of composite Cooper-like bosons with spin $S = 1$ and orbital angular momentum $L = 1$ (triplet- , or, p-pairing). As a result, the order parameter is no longer a simple scalar but a multi-component quantity, so the possible superfluid phases of the $(T, P, \boldsymbol{H})$ diagram are three, and their description is more complicated compared to that of the superfluid ^{4}He; for details, see Leggett (1975), and Vollhardt and Wölffe (1990).

The usual superconductivity mentioned above is described by Cooper s-pairs of electrons. When, as in ^{3}He, p-pairs, or d-pairs of fermions are formed we call them Cooper-like, or "unconventional" pairs. The corresponding multi-component order parameter ψ describes *unconventional superconductors* (or *unconventional superfluids* such as ^{3}He). The existence of unconventional superconductivity is not yet experimentally confirmed but it is quite probable from a theoretical point of view that this superconductivity may appear in real systems. For example, whether some of the heavy-electron superconductors, like UBe_{13} and UPt_3 and high-temperature superconductors like YBaCuO–systems are of an unconventional type is still an open question (Anderson, 1984b; Volovik and Gor'kov, 1985); for a review, see Pines (1990), Annett (1990), Sigrist and Ueda (1991), and Uzunov (1991).

3.8.7 *Liquid crystals*

In Section 3.6.3 we have discussed the liquid state of spherically symmetric molecules. If the molecules cannot be approximated by spheres, the pair potential $u(\boldsymbol{x}_1, \boldsymbol{x}_2)$ of two molecules at sites $\boldsymbol{x}_1$ and $\boldsymbol{x}_2$ depends no longer only on the intermolecular distance $r = |\boldsymbol{x}_1 - \boldsymbol{x}_2|$ but also on their mutual orientation. This gives rise to a new type of ordering phenomena which usually occur in *liquid crystals.*

The intermolecular potential of complex molecules is a difficult problem and is solved by the quantum chemistry. From a phenomenological point of view it is clear that in the case of oblong molecules, for example, like ellipsoids or dumbbells, the potential energy with parallel molecule orientation may turn out lower in comparison with the energy corresponding to the nonzero angle θ between their long axes. Then lowering the temperature of a macroscopically isotropic liquid consisting of long molecules (approximately "linear objects") it may undergo a phase transition to the *nematic state,* in which the "lines" have a *preferred orientation.* This spontaneous preferred orientation can be described introducing the *director vector* $\boldsymbol{n}$, which plays the role of the order parameter in nematics; see Fig. 3.15(a).

Other orderings, all allowed by the underlying uniaxial symmetry of the "lines", are also possible in this system. In the nematic phase the orientational ordering is not connected with a position ordering in space of the molecules but this, at least partially, is observed in the *smectic phase.* In a three dimensional system the position (the centers of masses) of the long molecules lie, quite approximately, in planes perpendicular to the director vector $\boldsymbol{n}$ (smectic A phase); see Fig. 3.15(b).

The isotropic liquid-to-nematic state transition is of first order but the nematic-to-smectic A transition may be of the second order or to exhibit tricritical points. We can also mention the smectic C order where the director vector $\boldsymbol{n}$ is not perpendicular to the parallel planes of the molecular mass centers; see Fig. 3.15(c).

The liquid crystals present a large field for investigation in the theory of phase transitions. More information about the liquid crystal orderings and the phase transitions in liquid crystals can be found in the specialized books, e.g., by de Gennes (1974), and Anisimov (1987). A treatment of the possible liquid crystal orderings in the frame of the group theory is presented by Tolédano and Tolédano (1987). More specialized problems of the fluctuation theory of the nematic - smectic A phase transition, in

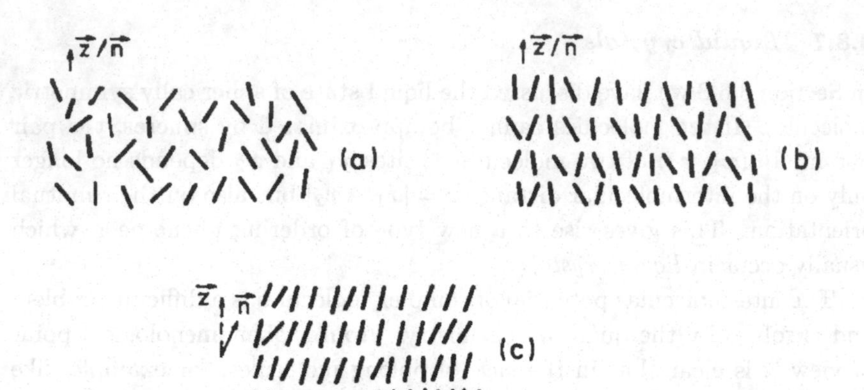

Fig. 3.15 Two-dimensional representation of: (a) nematic, (b) smectic A, and (c) smectic C orders in liquid crystals.

some respects similar to the superconducting transition are discussed in Chapter 9 (Section 9.7).

3.9 Advantages and Disadvantages of the Thermodynamic Theory

We have enumerated some of the most often studied orderings and phase transitions in condensed matter physics. In the course of our discussion we have convinced ourselves that the Gibbs–Duhem stability theory presented in Chapter 2 gives completely general thermodynamic framework for the description of the phase changes. In fact all phase transitions arising from thermal effects can be included in the thermodynamic scheme of classification and description of phase diagrams. To do that one has to make an appropriate choice of the primary and, when necessary, secondary order parameter(s). In most of the examples given in this Chapter we have found convenient to use terminology from solid state physics and statistical mechanics but in a formal thermodynamic treatment this can easily be avoided.

The topology of the phase diagram for various systems is obtained usually from experimental observations but always, even in the extremely complex systems the possible phase transitions and multicritical points can, in principle, be included into the thermodynamic scheme. This is because the Gibbs–Duhem stability theory is based on general laws — the first and second laws of thermodynamics. Each theory which describes phase transitions as a result of interplay of the thermal effects and the inter-particle interactions in many body systems has to obey the requirements of the thermodynamic description.

Despite of the advantages of the thermodynamic theory it does not answer to such important questions as what is the precise form of the equations of state, or which is the same — of the thermodynamic potential. For the description of the phase transitions in particular systems one needs their equations of state at least in a vicinity of the phase transition points. Without this information nothing can be said about the properties of the thermodynamic quantities near phase transition points.

We have seen that it is possible to simplify the thermodynamic description by expanding the thermodynamic potential for small values of the order parameter φ; cf.,e.g., Eqs. (3.36), (3.75), and (3.83). Such procedure, which we have already performed to the second order in φ may be further developed in the Landau theory of phase transitions (Chapter 4). It gives also a rather general approach to the main task of the phenomenological theory of phase transitions, which is to gain more information about the explicit dependence of the thermodynamic quantities on thermodynamic parameters as $T, P, \boldsymbol{H}$, etc. Since such expansions do not follow from any law in Nature they should be additionally justified and their limitations determined.

A more general thermodynamic approach is based on *the scaling hypothesis*, which supposes that the thermodynamic potential is a generalized homogeneous function of the thermodynamic parameters (see Chapter 1 and Chapter 8). Certainly the main scaling (homogeneity) assumption that the thermodynamic potential is a generalized homogeneous function (see, e.g., Eq. (1.18a)) rather than a usual (ordinary) homogeneous function, see Eqs. (1.18c) and (A.1), does not contradict to the Eq. (2.6) which expresses the fundamental thermodynamic equation and, consequently, the potentials $(F, \Phi, \Omega, \ldots)$ as first-order homogeneous functions; see also Eqs. (2.12), (2.74), (2.81), and (2.89).

The simple first-order homogeneity is a general thermodynamic rule which means that any thermodynamic potential, $\tilde{\Psi}$, is a linear homogeneous function of its natural variables X_i, where the coefficients are the parameters $Y_i = \left(\partial\tilde{\Psi}/\partial X_i\right)$ thermodynamically conjugate to X_i. The generalized homogeneity is concerned with the representation of any potential $\tilde{\Psi}$ in the terms of its natural variables (X_i) only, i.e., when the corresponding conjugate parameters Y_i in $\tilde{\Psi}(X_i, Y_i)$ are substituted by X_i. For this reason one has to solve the equations of state $f(X_i, Y_i) = 0$ with respect to Y_i and to substitute the result in $\tilde{\Psi}(X_i, Y_i)$ so that to obtain the generalized homogeneous function $\tilde{\Psi}(X_i) \equiv \tilde{\Psi}\left[X_i, Y_i\left(X_i\right)\right]$.

The phenomenological scaling theory which results from the scaling (homogeneity) hypothesis gives a general description of the singularities at critical and multicritical points; see, e.g. Eq. (3.20). The scaling approach is inspired by the analysis of critical experiments. In comparison with the Ehrenfest theory or the Landau expansion, the scaling theory gives weaker restrictions on the critical behaviour.

The phenomenological theories, even in the cases when the explicit form of the equation of state is known do not give the relation between the macroscopic properties and the underlying microscopic mechanism of phase transitions. Obviously, the critical temperature T_c, in energy units ($k_{\mathrm{B}} = 1$) is connected with the microscopic interaction which is responsible for the ordering. To describe such problems is possible only by the methods of statistical physics. Besides there are statistical quantities as, for example, the correlation functions, which give essential information about the phenomena occurring near phase transition points (see Chapters 5 – 8).

As the most interesting statistical models of interacting systems cannot be treated exactly, one is faced with two alternatives: to examine simple models, solving them exactly or to use approximate methods for study of more realistic systems. In both cases the results are compared with the thermodynamic approach and are used to test assumptions as those in the Landau expansion and the scaling theory.

Chapter 4

Landau Expansion

4.1 Introductory Remarks

Up to now we have tried to maintain our discussion close to the general thermodynamic requirements for equilibrium and stability of phases and states of phase coexistence (phase transition points). Thus we have explained the phase transition as a change in the stability properties of the phases that may occur simultaneously with the appearance of a new macroscopic mode (the order parameter) that describes new (ordered) phases. Following the general thermodynamic laws it has become possible to deduce several thermodynamic relations for the order parameter as a function of state.

Throughout our detailed discussion in Chapter 2 and Chapter 3 we have had several occasions to involve some comprehensive assumptions about the behaviour of the thermodynamic quantities. For example the notion of a finite jump of the susceptibilities at transition points of second order has led us to the results of the Ehrenfest theory (Section 3.5). However, the general thermodynamic approach is not restricted to the mentioned finite jump and allows different types of singularities of the susceptibilities at the critical points; see, e.g., Section 3.4, where the behaviour of the specific heat is discussed. The thermodynamic laws are too general and additional suppositions have to be made in order to gain more definite information about the phase transition properties. Thc most notable assumptions, as the Ehrenfest one, are usually inspired by experimental results and are applicable to a number of systems irrespectively of the particular nature of the phase transition considered. Certainly, every successful thermodynamic theory of phase transitions is built on additional restrictive suppositions, which do not contradict the thermodynamic laws and make possible to explain important features of the phase change.

In this Chapter we shall consider the Landau (1937a) theory of second order transitions. This theory is based on the Taylor expansion of the thermodynamic potential in terms of the order parameter. In order to obtain a particular form of the thermodynamic potential Landau has applied quite general assumptions about the behaviour of the expansion coefficients. A part of these assumptions are concerned with the general requirements for stability of phases and states of coexistence of phases. The rest, are the famous *Landau symmetry criteria* for second order phase transitions. The symmetry criteria indicate whether a phase transition is of the second or of the first order. This theory is in the roots of the contemporary theory of phase transitions and is verified by subsequent studies of statistical models (Chapter 5). The theory can be used to describe both first and second order transitions of several types.

In order to establish the interrelationship between the Landau theory and the general thermodynamic stability theory we shall at first summarize some of the results previously presented in Chapters 2 and 3. This summary is revealed in the next two sections. It will be useful to those readers who are interested particularly in the Landau expansion and have omitted to read the preceding Chapters.

4.2 Appropriate Variables

Suppose that the thermodynamic system $\sigma(a_i)$ is determined by a set of extensive parameters a_i. One of these parameters, for example, $a_j = X$ is related to the order parameter $\varphi = X - X_1$, where X_1 is the equilibrium value ($\bar{X} = X_1$) of X in the disordered phase 1. The order parameter is chosen such that its equilibrium value $\bar{\varphi} = \bar{X} - X_1$ is equal to zero in the disordered phase and is different from zero ($\bar{\varphi} = X_2 - X_1$) in the ordered phase ($\bar{X} = X_2 \neq X_1$). We denote the system σ by $\sigma(\varphi, a_i)$, $a_i \neq a_j$, $i = 1, 2\ldots$. The thermodynamic treatment of this system is performed considering the isolated system $\sigma^t(\varphi^t, a_i^t)$, which consists of σ and its surrounding $\sigma^0(\varphi^0, a_i^0)$, i.e., $\sigma^t = \sigma + \sigma^0$. Having the relations $\varphi^t = \varphi + \varphi^0$ and $a_i^t = a_i + a_i^0$ one must keep in mind the requirements for a complete isolation of σ^t : $\varphi^t = const$ and $a_i^t = const$. The notion of a composite system σ^t is basic in thermodynamics and has been widely used in Chapter 2. Here we shall investigate the equilibrium coexistence of the system σ with its surroundings in the particular case when the parameters $a_i \neq \varphi$ are *fixed*. The interaction between σ and σ^0 is then possible only through the exchange of quantities $\delta\varphi$.

If we intend to carry out a description in entropy representation we must exclude the entropy S of σ from the set $a = \{a_i\}$ of the fixed parameters. Within the above scheme the equilibrium value $\bar{\varphi}$ of φ will maximize the total entropy S^t of the composite system σ^t. The equilibrium order parameter $\bar{\varphi}$ is given by the solutions of the equation of state

$$h^S = \frac{\partial S}{\partial \varphi}, \tag{4.1}$$

where h^S is the equilibrium field conjugate to $\bar{\varphi}$ in entropy representation; from now on, the superscript "S", denoting the adopted thermodynamic representation will be omitted. Remember that the Eq. (4.1) is a direct result from the equilibrium condition $\delta^{(1)}S^t = 0$, where $\delta^{(1)}S^t$ is the first variation of S^t. Further one may obtain the stability conditions for the $\bar{\varphi}$-phases by investigating the second and, possibly, the higher order variations $\delta^{(l)}S^t$ of S^t, as discussed in Chapter 2.

The description of a phase transition in entropy representation is possible but inconvenient and one of the reasons is that the entropy $S(\varphi, a_i)$ has a maximum in equilibrium only for fixed φ, that is, when σ is totally isolated from σ^0. In our construction the system σ is open towards the variable φ and hence, $S(\varphi, a_i)$ has no maximum for states of equilibrium contact between σ and σ^0. This fact forces us to consider the composite system σ^t and the extrema of S^t. Without wrong physical consequences, the surroundings σ^0 can always be treated as much larger than the system σ ($\varphi^t \gg \varphi, a_i^t \gg a_i$) but this does not eliminate the clumsy procedure of investigating a composite auxiliary system σ^t instead of a single system σ.

This disadvantage of the entropy approach can be avoided by using another thermodynamic potential (Massieu's function) $\hat{\Psi}_S(h, a_i)$ obtained by the Legendre transformation $S = \Psi_S + h\varphi$. Both functions S and Ψ_S, however, have another inconvenient property and this is unavoidable. One of the fixed variables a_i in S- and Ψ_S-representations is the internal energy U. The energy U cannot be chosen as an order parameter because it rather determines a general quantitative property than a specific feature of the phases. Therefore the energy U is one of the complementary fixed parameters a_i. However it can neither be fixed nor measured and this disadvantage of the S- and Ψ_S-representations cannot be avoided anyway.

If the energy fundamental equation $U(\varphi, a_i)$ is chosen, the entropy S will be amongst the fixed variables a_i. It is also not directly measurable in experiments. The problem is easy to solve, if our treatment is carried out in the terms of a thermodynamic potential $\hat{\Psi}(T)$, for which the temperature T is a natural variable.

Thermodynamic potentials like $\hat{\Psi}(T)$ can be easily constructed by an appropriate Legendre transformation of the energy function $U(S, a_i, \varphi)$ with respect to S and, when necessary, with respect to φ and some of the fixed variables a_i. Such potentials are the (Helmholtz) free energy F, the (Gibbs) thermodynamic potential Φ, the grand canonical potential Ω etc. (see Section 2.6). For all these potentials, generally denoted by $\hat{\Psi}(T)$, the analogue of an open (with respect to S) system σ is the same system to be at constant (fixed) temperature T (equal to the temperature T_0 of the reservoir σ^0).

Now let the system σ be open with respect to S and to the part $(a_1, \dots, a_k)$ of the extensive variables a_i. All unfixed variables S, φ, $a_1, \dots, a_k$ will get their equilibrium values $\bar{S}$, $\bar{\varphi}$, $\bar{a}_1, \dots, \bar{a}_k$ when the total system σ^t is in equilibrium. These equilibrium values will be solutions of the corresponding equations of state and will be expressed by their equilibrium conjugate fields T, h, and $h_1, \dots, h_k$. The analogue of the open system is now a system with extensive parameters $\bar{S}, \bar{\varphi}$ and $(\bar{a}_1, \dots, \bar{a}_k)$ at constant temperature and fields $h, h_1, \dots, h_k$.

Generally, one can write a $\hat{\Psi}(T)$ potential in the form

$$\hat{\Psi} = \hat{\Psi}(T, h_1, \dots, h_k, a_{k+1}, \dots, \varphi, \dots) \tag{4.2}$$

when φ is among its natural variables, and

$$\hat{\Psi}' = \hat{\Psi}(T) - h\varphi, \tag{4.3}$$

with

$$\hat{\Psi}' = \hat{\Psi}(T, h_1, \dots, h_k, a_{k+1}, \dots, h, \dots), \tag{4.4}$$

when the field h conjugate to φ is among the natural variables of the potential. Whatever the number $k \geq 1$ is, we can denote all the variables $(h_1, \dots, h_k, a_{k+1}, \dots)$ except φ by the general symbol Y and to treat them as a single thermodynamic parameter. The reason is that we are mainly interested in the description of the order parameter φ and its conjugate field. The parameters Y then are important only to indicate that the phase transition $(\bar{\varphi} = 0) \rightarrow (\bar{\varphi} \neq 0)$ may be driven by both T- and Y-variations. Thus the variable Y traces the Y-dependence of the equilibrium order parameter

$\bar{\varphi}(T, Y)$ and makes possible, when this dependence is known, to outline the shape of the phase transition "line" on the (T, Y) diagram of the system. In this aspect T and Y play an equivalent role in our description of the phase transition $0 \to \bar{\varphi}$. For the sake of simplicity of the foregoing formulae we can absorb the symbol Y into T, i.e, to replace the couple (T, Y) by T. In many cases we shall omit to write explicitly T- and Y- dependencies of the thermodynamic functions keeping in mind that all the thermodynamic potentials of interest actually depend on T and Y and their derivatives are taken at fixed T and Y: for example, $(\partial\hat{\Psi}/\partial\varphi) = (\partial\hat{\Psi}/\partial\varphi)_{T,Y}$ as suggested by Eq. (4.2).

We have therefore an appropriate scheme of description in terms of the variables T, Y, and φ and the thermodynamic function $\hat{\Psi}(T, Y, \varphi)$ from Eq. (4.2). Concerning its dependence on the order parameter φ, the potential $\hat{\Psi}(T, Y, \varphi)$ or shortly $\hat{\Psi}(\varphi)$ will be referred to as the (Helmholtz) free energy, denoted by $F(\varphi)$. This free energy does not depend on S or U and in this respect the thermodynamic treatment in F-representation is an advantage. However it has the order parameter φ as a natural variable and therefore, $F(\varphi)$ has no minimum for the open system σ. Once again we have to construct the composite isolated system σ^t and to deduce the most appropriate potential for studies of the phase transitions.

4.3 Generalized Gibbs Potential

4.3.1 *Nonequilibrium potential* F^t

We continue our discussion of a system σ at fixed parameters T and Y, which is a part of the total system $\sigma^t = \sigma + \sigma^0$ and exchanges the extensive variable φ with its surroundings σ^0 subject to the constraint $\varphi^t = const.$ A thermodynamic function that has extrema at the (stable or unstable) equilibrium coexistence of σ with its surroundings σ^0 is the total free energy F^t of the composite system σ^t. In fact, $F^t(\varphi^t)$ can be considered as a function of φ, $F^t(\varphi, \varphi^t - \varphi) = F(\varphi) + F^0(\varphi^t - \varphi); \varphi + \varphi^0 = \varphi^t$. We have already discussed in Section 2.6.4 the extrema of the function $F^t(V, V^t - V)$ of a fluid system at fixed temperature T and number of particles N and here our brief summary is just a repetition having another purpose. We shall relate the extremum properties of F^t to those of another, more suitable thermodynamic function and this aim can be achieved either in terms of the order parameter φ or the corresponding extensive variable $X = \varphi + X_1$.

The function $F^t(\varphi) = F^t(\varphi, \varphi^t - \varphi)$ at constant φ^t (and T, Y) can be expanded in the variations $\delta\varphi = \varphi - \varphi'$ around a given state φ'. The relevant Taylor series are

$$F^t(\varphi) = F(\varphi') + F^0(\varphi^{0\prime}) + (h' - h_0')\delta\varphi + \sum_{l>1} \delta^{(l)}(F + F^0), \tag{4.5}$$

where

$$h' = \left(\frac{\partial F}{\partial \varphi}\right)_{\varphi'} \tag{4.6a}$$

and

$$h_0' = \left(\frac{\partial F^0}{\partial \varphi^{0\prime}}\right)_{\varphi^{0\prime}} \tag{4.6b}$$

are the fields conjugate to φ' and $\varphi^{0\prime}$, respectively. If φ' is an extremal state ($\varphi' = \bar{\varphi}$) of σ^t, the equilibrium internal (with respect to σ) field $\bar{h}$ is equal to the equilibrium external (applied) field $\bar{h}_0$.

The fields $\bar{h}$ and $\bar{h}_0$ are conjugate to $\bar{\varphi}$ and $\bar{\varphi}^0 = \bar{\varphi}^t - \bar{\varphi}$ and are given by the derivatives (4.6a) and (4.6b) taken at $\bar{\varphi}, T$, and Y. The fields $\bar{h}$ and $\bar{h}_0$ are then indistinguishable and will be further denoted by $\bar{h}$ and determined by the equilibrium equation of state

$$\bar{h} = \left(\frac{\partial F}{\partial \varphi}\right)_{\bar{\varphi}}. \tag{4.7}$$

From this equation one obtains the equilibrium order parameter $\bar{\varphi} = \bar{\varphi}(T, Y, \bar{h})$.

4.3.2 *Nonequilibrium potential* Φ

The problem now is how to find another potential, which like F^t has extrema towards φ at equilibrium, but unlike F^t is written only in terms of quantities describing the system σ. The potential $F^t(\varphi)$ from Eq. (4.5) can formally be represented in the following way:

$$F^t(\varphi) = \tilde{F}'(\varphi) + \tilde{F}^{o\prime}(\varphi^0) + (h' - h_0')(\varphi - \bar{\varphi}), \tag{4.8}$$

where

$$\tilde{F}'(\varphi) = F(\varphi) - h'(\varphi - \bar{\varphi}) \tag{4.9a}$$

and

$$\tilde{F}^{0\prime}(\varphi^0) = F^0(\varphi^0) - h'_0(\varphi^0 - \bar{\varphi}^0). \tag{4.9b}$$

The free energy $\tilde{F}'(\varphi)$ from Eq. (4.9a) has extrema for each $\varphi'(h')$ given by Eq. (4.6a). The extremal points $\varphi'(h')$ of $\tilde{F}'(\varphi)$ coincide with those $(\bar{\varphi})$ of F^t only for $h' = \bar{h}$. At the equilibrium field $h' = h'_0 = \bar{h}$ we shall denote $\tilde{F}'$ with

$$\tilde{F}(\varphi) = F(\varphi) - \varphi\bar{h} + \bar{\varphi}\bar{h}. \tag{4.10}$$

This free energy has extrema $\tilde{F}_{\text{ext}}$ at $\bar{\varphi}$ defined by Eq. (4.7). The function $\tilde{F}(\varphi)$ of the variable φ can therefore be considered as the nonequilibrium potential of the system σ. When σ is not in equilibrium, φ can be different from $\bar{\varphi}$. The extremal value $\tilde{F}_{\text{ext}}$ coincides with the values of $F(\varphi)$ at $\bar{\varphi}$: $\tilde{F}_{\text{ext}}(\bar{\varphi}) = F[\bar{\varphi}(\bar{h})]$. So $\tilde{F}_{\text{ext}}(\bar{\varphi})$ is the equilibrium free energy of the system. The function $\tilde{F}(\varphi)$may happen to be inconvenient for some studies because it is expressed by both the equilibrium $(\bar{\varphi})$ and the nonequilibrium values of the order parameter.

Another thermodynamic function that depends only on the parameters of the system σ is

$$\Phi(\varphi, \bar{h}) = F(\varphi) - \varphi\bar{h}. \tag{4.11}$$

Its extrema $\Phi_{\text{ext}} = \bar{\Phi}$,

$$\bar{\Phi}(\bar{h}) = F(\bar{\varphi}) - \bar{\varphi}\bar{h} \tag{4.12}$$

are the same as the extremal points $\bar{\varphi}$ of F^t and $\tilde{F}$. Substituting $\bar{\varphi}$ from Eq. (4.7) in Eq. (4.12) we have

$$\bar{\Phi}(T, Y, \bar{h}) = F[T, Y, \bar{\varphi}(\bar{h})] - \bar{h}\bar{\varphi}(T, Y, \bar{h}). \tag{4.13}$$

The function $\bar{\Phi}(\bar{h})$, described in this way by Eq. (4.12) or Eq. (4.13), is the equilibrium Gibbs potential, $\Phi_{\text{ext}} = \bar{\Phi}(\bar{h})$.

More generally, the function Φ can be defined for an arbitrary (nonequilibrium) field h, i.e.,

$$\Phi(\varphi, h) = F(\varphi) - \varphi h, \tag{4.14}$$

where

$$h = \frac{\partial F}{\partial \varphi} \tag{4.15}$$

is the equation that determines the field h conjugate to the variable φ. The derivative (4.15) is taken at fixed T and Y and is supposed to exist for any (equilibrium and nonequilibrium) state φ, for which the energy $F(\varphi)$ can be defined (the possibility to define thermodynamic potentials for nonequilibrium, but near-to-equilibrium states has been discussed in Chapter 2).

One may connect h and φ from Eq. (4.14) with h' and φ' from Eq. (4.5). In result the total free energy $F^t(\varphi)$ is expressed by $\Phi(\varphi, h)$:

$$F^t(\varphi) = \Phi(\varphi, h') + \Phi^0(\varphi^0, h'_0) + (h' - h'_0)\varphi + h'_0\varphi^t, \tag{4.16}$$

or, in equilibrium

$$F^t(\varphi) = \Phi(\varphi, \bar{h}) + \varphi\bar{h} + \Phi^0(\varphi^0, \bar{h}) + \varphi^0\bar{h}. \tag{4.17}$$

When φ in the last equation is replaced by $\bar{\varphi}(\bar{h})$ from Eq. (4.7), F^t reaches its extremal value $F^t(\bar{\varphi})$.

Now we can substitute our original conditions for the open system σ with their analogue — the constant external field $\bar{h}$. For any given set of variables T, Y, and $\bar{h}$, the nonequilibrium potential $\Phi(\varphi, h)$ will reach its extremal value $\bar{\Phi}(\bar{h})$ when the system is in equilibrium at fixed T, Y and external field $\bar{h}$. The nonequilibrium potential $\Phi(\varphi, h)$ can be regarded as a function of both φ and h. The corresponding partial derivatives (fixed T and Y) are

$$\left(\frac{\partial \Phi}{\partial h}\right)_\varphi = -\varphi, \tag{4.18}$$

and

$$\left(\frac{\partial \Phi}{\partial \varphi}\right)_h = 0. \tag{4.19}$$

The last equation is another form of the Eq. (4.15). When φ is one of the extrema $\bar{\varphi}$, of $F^t(\varphi)$, Eq. (4.15) is identical to the equilibrium equation of state (4.7). Within our present treatment in terms of the potential $\Phi(\varphi, h)$, the surroundings σ^0 of the system σ are replaced by an external field h conjugate to the order parameter φ.

Eqs. (4.14)–(4.15) on the one hand and Eqs. (4.7) and (4.11) on the other hand are identical, provided h is considered as the external equilibrium field. In order to make difference between h in Eqs. (4.14) and (4.15) and some equilibrium field $\bar{h}$ we should accept by convention that a nonzero difference exists

$$\Delta h = (h - \bar{h}) \tag{4.20}$$

between the equilibrium and the nonequilibrium fields. Obviously $\Delta h = \Delta h'$, where $\Delta h' = (h' - h'_0)$ is the field difference in Eq. (4.16). An important case for the difference $\Delta h \neq 0$ is when the equilibrium value $\bar{\varphi}$ of φ corresponds to $\bar{h}$ and the field $h(0)$ corresponds to $\varphi = 0$. Then $\Delta h = [\bar{h} - h(0)]$. This particular difference is considered in the next section. The thermodynamic potentials $\Phi(\varphi, h)$ and $\Phi(\bar{\varphi}, \bar{h})$ describing the states $\varphi(h)$ and $\bar{\varphi}(\bar{h})$ will be different. Subtracting Eq. (4.12) from Eq. (4.14) and taking the first differential of $F(\varphi)$ at $\varphi = \bar{\varphi}$ we have:

$$\Phi(h) - \Phi(\bar{h}) = \Delta h \varphi(h). \tag{4.21}$$

Whether $\Phi(h) > \Phi(\bar{h})$ depends on the sign of $\Delta h \varphi$.

The main assumption is that whatever the external field is, the system will arrive at equilibrium contact with it, i.e., the extrema $\bar{\varphi}$ of $\Phi(\varphi, h)$ will be given by Eq. (4.7) or Eq. (4.19), where $h = \bar{h}$ and $\varphi = \bar{\varphi}$. In our previous example of a composite system σ^t, the external field h is nothing else but the derivative $(\partial F^0/\partial \varphi^0)_{T,Y}$ of the free energy F^0 of the surroundings. In most of the cases we are interested in the equilibrium states $(T, Y, \bar{h})$. That is why we can use Eqs. (4.14), (4.15), (4.18), and (4.19), where h is interpreted as the equilibrium field ($h = \bar{h}$); further the use of nonequilibrium fields will be specially stipulated.

Very often the thermodynamic potential Φ is calculated for a zero external field ($h = 0$). In such calculations Φ coincides with the free energy F. Then the equation of state (4.19) is replaced with

$$\frac{\partial F}{\partial \varphi} = 0. \tag{4.22}$$

The function $\Phi(T, Y; h, \varphi)$ or, shortly $\Phi(\varphi)$ of two (or more) fixed variables (T, Y) and the two conjugate to each other non-equilibrium parameters (φ, h), the extrema of which describe equilibrium or near-to-equilibrium states is often called the *generalized* or *non-equilibrium Gibbs potential.* The potential $\Phi(\varphi)$ is the appropriate thermodynamic function to describe the phase transition $0 \to \bar{\varphi}$. Landau (1937a) expanded this function, in fact its free energy part $F(\varphi)$ in Taylor series at $\bar{\varphi} = 0$ to investigate phase transitions. The above described construction of the potential $\Phi(\varphi)$ fits well to the results from statistical calculations of microscopic models (Chapter 5). The macroscopic approach based on the non-equilibrium potential $\Phi(\varphi)$ gives the opportunity to examine the fluctuations $\delta\varphi = (\varphi - \bar{\varphi})$ near the equilibrium states and to account the fluctuation contributions to the thermodynamic quantities by methods of statistical mechanics (Chapter 6).

The relation between the total free energy F^t and the potential Φ of the system σ can be further clarified for a large reservoir σ^0.

4.3.3 *Large reservoir and relation between F^t and Φ*

For a large reservoir $(\delta^{(l)}F \gg \delta^{(l)}F^0; l > 1)$ the potential $F^t(\varphi)$ from Eq. (4.3) at $\varphi' = \bar{\varphi}$ and $h (= \bar{h})$ takes the form

$$F^t(\varphi) = F^0(\bar{\varphi}^0) + F(\varphi) - h\delta\varphi. \tag{4.23}$$

Using

$$-h\delta\varphi = \delta(\Phi - F) \tag{4.24}$$

as presented by Eq. (4.11), it is easy to obtain the elucidating formula:

$$\delta\Phi = \delta F^t, \tag{4.25}$$

where

$$\delta F^t = F^t(\varphi) - F^t_{\text{ext}}, \tag{4.26}$$

is expressed by $F^t_{\text{ext}} = F^t(\bar{\varphi})$, and the variation $\delta\Phi$ is given by

$$\delta\Phi = \Phi(\varphi, h) - \Phi_{\text{ext}}; \tag{4.27}$$

Φ_{ext} is from Eq. (4.12). Using Eqs. (4.25)–(4.27) one obtains

$$F^t(\varphi) = (F^t_{\text{ext}} - \Phi_{\text{ext}}) + \Phi(\varphi, h). \tag{4.28}$$

Finally we should note that in all formulae of this subsection the variational symbol δ means both infinitesimal and finite variations ($\Delta = \delta$). For any thermodynamic potential Ψ^t of the composite system σ^t and its Legendre transform Φ_Ψ with respect to the free extensive variables of the open subsystem σ, one can write the relation

$$\Delta\Psi^t = \Delta\Phi_\Psi. \tag{4.29}$$

4.4 Landau Potential

4.4.1 *Taylor expansion*

Let us expand the nonequilibrium thermodynamic potential (4.14) in fluctuations $\delta\varphi = (\varphi - \bar{\varphi})$ around the equilibrium state $\bar{\varphi}(\bar{h})$. Using Eq. (4.7) we have

$$\Phi(\varphi, h) = \bar{\Phi}(\bar{h}) - (h - \bar{h})\bar{\varphi} - (h - \bar{h})\delta\varphi + u_2(\bar{\varphi})(\delta\varphi)^2 + u_3(\bar{\varphi})(\delta\varphi)^3 + \dots, \tag{4.30}$$

where $\bar{\Phi}(\bar{h})$ is given by Eq. (4.12) and the coefficients $u_l(T, Y, \bar{\varphi})$ are expressed by the derivatives of the free energy $F(\varphi)$ at fixed T, Y and $\bar{\varphi}$:

$$u_l(\bar{\varphi}) = \frac{1}{l!}\left(\frac{\partial^l F}{\partial \varphi^l}\right)_{\varphi = \bar{\varphi}}, \qquad l > 1. \tag{4.31}$$

The Taylor series of $\Phi(\varphi, h)$ at $\varphi = 0$,

$$\Phi(\varphi, h) = F(0) - [h - h(0)]\varphi + u_2\varphi^2 + u_3\varphi^3 + \dots \tag{4.32}$$

with $u_l = u_l(0)$,

$$F(0) = \Phi[0, h(0)], \tag{4.33}$$

and

$$h(0) = \left(\frac{\partial F}{\partial \varphi}\right)_{\varphi = 0}. \tag{4.34}$$

are obtained by a direct expansion of $\Phi(\varphi, h)$ from Eq. (4.14) or by setting $\bar{\varphi} = 0$ and $\delta\varphi = \varphi$ in Eq. (4.30). Note that the coefficient $u_2(\bar{\varphi})$ is proportional to the inverse equilibrium susceptibility $\chi^{-1}(\bar{h})$ defined as

$$\chi(\bar{h}) = \left(\frac{\partial\varphi}{\partial h}\right)_{\bar{h}} = \left(\frac{\partial^2 F}{\partial\varphi^2}\right)^{-1} . \tag{4.35}$$

The coefficient $u_2(0)$ is then proportional to the inverse equilibrium susceptibility when $\bar{h} = h(0)$, namely, when the equilibrium external field is equal to the field $h(0)$ corresponding to the equilibrium value $\bar{\varphi} = 0$ of φ. The thermodynamic potential that has extrema at $\varphi = 0$ is

$$\Phi_0[\varphi, h(0)] = \Phi(\varphi, h) + [h - h(0)]\varphi \tag{4.36}$$

or, equivalently,

$$\Phi_0[\varphi, h(0)] = \Phi(\varphi, h) + [h - h(0)]\varphi \tag{4.37}$$

Φ_0 coincides with Φ for $h = h(0)$.

If $h = \bar{h}$, the phases $\bar{\varphi}$ will be in stable (metastable, unstable or neutral) equilibria. In particular, the state $\bar{\varphi} = 0$ is the equilibrium one for $h = \bar{h} = h(0)$. Further we shall examine the case $h = \bar{h}$, for which the terms linear in φ in the expansion (4.30) of Φ vanish because the nonequilibrium field difference $\Delta h = (\bar{h} - h)$ is not present. From now on the bar of $\bar{h}$ will be omitted.

4.4.2 *Equilibrium*

The equilibrium transition $(0 \to \bar{\varphi})$ occurs at the points of the (T, Y, h) diagram, where the thermodynamic potentials $\bar{\Phi}(\bar{\varphi}, h)$ and $\bar{\Phi}(0, h)$ have one and the same value. Note that the equality of these potentials,

$$\bar{\Phi}(\bar{\varphi}, h) = \bar{\Phi}[0, h(0)] \tag{4.38}$$

is equivalent to the equality

$$h = h(0). \tag{4.39}$$

Actually inserting $\bar{\varphi}(h)$ in Eq. (4.37) gives $\bar{\Phi}(h) = \bar{\Phi}(h(0))$ which is fulfilled only if $h = h(0)$.One may therefore describe the phase coexistence making use of Eq. (4.39) in the way explained in Section 3.2, where the equality of the chemical potentials has been applied. On the other hand we can look upon Eq. (4.38) as a requirement for finding out $\bar{\varphi}$ on the coexistence (transition) line defined by Eq. (4.39).

From Eqs. (4.37), (4.39), (4.33), and (Eq. (4.32) we obtain the equation for $\bar{\varphi}$ on the phase transition line:

$$u_2\varphi^2 + u_3\varphi^3 + u_4\varphi^4 + \cdots = 0, \tag{4.40}$$

where, as often below, the bar of $\bar{\varphi}$ has been omitted. While the Eq. (4.40) for $\bar{\varphi}$ is valid only on the phase transition line, the equation of state (4.7) for the expanded potential (4.32) when $(h = \bar{h})$ determines the equilibrium states in the whole phase diagram (T, Y, h). It can be written in the form

$$\Delta h = 2u_2\varphi + 3u_3\varphi^2 + 4u_4\varphi^3 + \ldots, \tag{4.41}$$

with

$$\Delta h = h - h(0); \qquad h \equiv \bar{h}. \tag{4.42}$$

The Eq. (4.41) can also be established by setting $(h = \bar{h})$ in Eq. (4.30) and expanding Eq. (4.7) for $\bar{\varphi} = 0$. When $\Delta h = 0$, as it is on the transition line Eq. (4.39), Eq. (4.41) reads

$$2u_2\varphi + 3u_3\varphi^2 + 4u_4\varphi^3 + \cdots = 0. \tag{4.43}$$

The common solutions of Eqs. (4.40) and (4.43), which give the equilibrium order parameter at the phase transition points are discussed in the next section.

4.4.3 *Stability*

Because we have already identified the nonequilibrium field h, it is readily seen from the expansion (4.31) that the stability conditions for the solutions $\bar{\varphi}(\Delta h)$ of the equation of state (4.41) are

$$u_{2m}(\varphi) > 0, \tag{4.44}$$

where u_{2m} is the first nonvanishing even coefficient for a given phase $\bar{\varphi}$, then $u_k = 0$, for $k = 2, \ldots, 2m - 1$; cf. Eqs. (2.65), (2.66), and (2.96). The solution $\bar{\varphi} = 0$ of Eq. (4.41) corresponding to the disordered phase exists only for $h = h(0)$. Its stability is given by the inequality $u_{2m}(0) \equiv u_{2m} > 0$ which follows from the general condition (4.44). It is easy to establish the relation between the coefficients $u_l(\bar{\varphi})$ and u_l for two different solutions ($\bar{\varphi}$ and $\bar{\varphi} = 0$) of the equation of state (4.41). Replacing φ with $\bar{\varphi} + \delta\varphi$ in Eq. (4.32) and comparing with Eq. (4.30) for $h = \bar{h}$, we have

$$u_2(\varphi) = 2u_2 + 6u_3\varphi + 12u_4\varphi^2 + \cdots + m(m-1)u_m\varphi^{m-2} + \ldots, \quad (4.45a)$$

with $\bar{\varphi} = \varphi$, and the general formula

$$u_l(\varphi) = \sum_{m=l}^{\infty} \frac{m!}{(m-l)!} u_m \varphi^{(m-l)}. \quad (4.45b)$$

By this procedure one is able to re-derive the Eqs. (4.32), (4.41) and (4.42) for the equilibrium value $\bar{\varphi}$ of φ.

4.4.4 *Aims of the Landau theory*

In determining the stability of the possible phases $\bar{\varphi}$ we need to check only the sign of the susceptibility ($\chi^{-1} \sim u_2(\bar{\varphi})$) or, in special cases, of some higher coefficients $u_l(\bar{\varphi})$; $l > 2$. Confining ourselves to this simple aim we remain within the framework of the thermodynamic stability theory. The problem stated by Landau in 1937 however is to receive more detailed results, namely, to describe the thermodynamic quantities by making use somehow of the expansion (4.32). The idea is to take into account the φ^l-terms in Eq. (4.32) as fluctuations $\delta\varphi = \varphi$ in the case of a stable disordered phase $\bar{\varphi} = 0$ or, in the case of instability of this phase to investigate the solutions $\bar{\varphi} \neq 0$ of the equation of state (4.41). The fluctuations $\delta\varphi = (\varphi - \bar{\varphi})$ about the stable states ($\bar{\varphi} = 0$ or $\bar{\varphi} \neq 0$) can at least to the lowest order approximation be neglected. This programme can certainly be carried out, if we truncate in some way the infinite Taylor series. If some suitable cutoff of the infinite series (4.32) can be done the problem will be reduced to the investigation of simple algebraic polynomials. Having a definite form of the thermodynamic potential Φ as a result of an abbreviated expansion of the free energy F we can derive the explicit form for all thermodynamic quantities. As we shall see the expansions in a Taylor series and their adequate truncation is just the first step to the description of the phase transitions.

We shall not dwell upon the mathematical requirements for the validity of infinite expansions as Eq. (4.30) or Eq. (4.32) or their truncated variants. These requirements are clear enough and are not among the most essential problems to be discussed here.

Let all φ-terms in Eq. (4.32) of order higher than fourth, denoted by $O(\varphi^5)$ be small enough so that the difference between the infinite and the abbreviated series, i.e., the sum $O(\varphi^5)$ can be ignored. In this case we can replace the infinite series for Φ with the abbreviated variant

$$\Phi = -\Delta h\varphi + u_2\varphi^2 + u_3\varphi^3 + u_4\varphi^4, \tag{4.46}$$

where the potential $\Phi(0)$ of the disordered phase is subtracted from Φ; $\Phi \to \Phi - \Phi(0)$.

It will become clear from our investigation of the potential (4.46) that for the study of a number of first and second order phase transitions the terms $O(\varphi^5)$ are irrelevant indeed. The model approximation (4.46) of the full potential $\Phi(\varphi)$ describes fairly well the ordinary phase transitions of second order (those without multicritical points) and some types of first order phase transitions that come about from the nonzero φ^3-term.

The potential (4.46), especially when the φ^3-term is absent, often is referred to as the φ^4-model of the second order phase transitions (or, φ^4-theory). When multicritical points are of interest the expansion of Φ includes terms up to φ^6 or φ^8, namely, we have to deal with φ^6- or φ^8-theories.

As we shall see in our further discussions the reliability of this approach rather depends on the methods involved in the study than on the order of truncation of the infinite series (4.32). In the next sections we shall analyze the thermodynamic function (4.46), which is often called *the Landau potential.* This term is used also for the general series expansion (4.32). To be precise, we should mention that the thermodynamic potential introduced by Landau in 1937 is actually of the type of Eq. (4.46) but the coefficient functions $u_l(T, Y)$ are chosen in a particular way to describe the equilibrium first and second order transitions.

4.4.5 *Examples*

Let us apply the above scheme to two simple examples.

In ferromagnets the free energy $F(\boldsymbol{M})$ in a zero external magnetic field ($\boldsymbol{h} = 0$) depends on M^2: $F = F(M^2)$, $M = |\boldsymbol{M}|$. Consequently, the field $h(0) = (\partial F/\partial M)_0$ in ferromagnets is equal to zero. So $\Delta \boldsymbol{h} = \boldsymbol{h}$, which enters in Eqs. (4.41), (4.42), and (4.46), coincides with the external magnetic field $\boldsymbol{h}$; $\Delta h\varphi = \boldsymbol{h}\cdot\boldsymbol{M}$. When $\boldsymbol{M}$ is parallel to $\boldsymbol{h}$, $\boldsymbol{h}\cdot\boldsymbol{M} = HM$. The paraphase ($M = 0$) does not exist at $H \neq 0$.

The nonequilibrium potential Φ of the liquid–vapour system is

$$\Phi(V) = F(V) + PV. \tag{4.47}$$

Denoting by $\varphi = V - V_l$, where V_l is the usual volume of N particles in the liquid state, the expansion of $\Phi(V)$ at $V = V_l$ takes the form:

$$\Phi(V) \approx F(V) + PV_l + (P - P_l)\varphi + O(\varphi^2), \tag{4.48}$$

where $P_l = -(\partial F/\partial V)_{V_l}$ is the field (pressure) that corresponds to $h(0)$ in Eq. (4.42). The field difference $\Delta h = h - h(0)$ with $h(=\bar{h})$ — the equilibrium field (pressure, $-P$), is equal to $-(P - P_l)$. The equilibrium solution $\varphi = 0$ ($V = V_l$) of the equation $(\partial\Phi/\partial\varphi) = 0$ does not exist for pressure $P \neq P_l$. The term PV_l in Eq. (4.48) can be represented as a sum of $(P - P_l)V_l$ and P_lV_l. The difference $(P - P_l)$ is of order $O(\varphi)$ as it can be seen expanding the function $P(V,T)$ at $V = V_l$. The field P_l is a function of T and V_l. The equilibrium potential of the liquid state is $\Phi(T, P_l) = F(T, V_l) + P_lV_l$. Usually the critical point of the liquid–vapour transition is studied by an expansion as that in Eq. (4.48), in which V_l and P_l are replaced by V_c and P_c, the critical volume and the critical pressure.

4.5 Phase Transition Line

The equilibrium phase transition line of systems, described by the model (4.46), is defined by the Eq. (4.39) or, equivalently, by the condition for the existence of solutions common to Eqs. (4.40) and (4.43). Let us make use of the last condition. Eqs. (4.40) and (4.43 always have a trivial common solution $\varphi = 0$, which means that the disordered phase is always present on the transition (coexistence) line.

The nonzero solutions

$$\varphi_{\pm} = \frac{-3u_3 \pm (9u_3^2 - 32u_2u_4)^{1/2}}{8u_4}, \qquad u_4 \neq 0, \tag{4.49}$$

of Eq. (4.43) exist, provided

$$9u_0 > 8u_2, \qquad u_0 = \frac{u_3^2}{4u_4}. \tag{4.50}$$

Comparing $\varphi_\pm$ from Eq. (4.49) with the corresponding values $\varphi'_\pm$ from Eq. (4.41) we obtain that a common nonzero solution

$$\bar{\varphi} = -\frac{u_3}{2u_4} \tag{4.51}$$

of Eqs. (4.41) and (4.43) exists when

$$u_0 = u_2, \qquad u_3 \neq 0. \tag{4.52}$$

This result is valid irrespectively of the value of the field difference Δh or, in other words Δh is always equal to zero on the coexistence line of the transition $0 \to \bar{\varphi}$.

It is easy to check from the stability condition

$$u_2 + 3u_3\varphi + 6u_4\varphi^2 \geq 0 \tag{4.53}$$

that the para-phase ($\varphi = 0$) is stable in the domain of the phase diagram $(T, Y, h = h(0))$, where

$$u_2(T, Y) > 0 \tag{4.54}$$

or

$$u_2 = u_3 = 0, \qquad u_4(T, Y) > 0. \tag{4.55}$$

The ordered phase (4.51) is stable for

$$u_0 \geq 0, \tag{4.56}$$

which gives a domain of stability defined by

$$u_4(T, Y) > 0, \tag{4.57}$$

or, equivalently, by the inequality (4.54); cf. Eq. (4.52).

Obviously the stability of both phases on the transition line is ensured by the condition (4.57). The same condition is easily derived from the φ^4-thermodynamic potential (4.46). In fact, if $u_4 < 0$, the potential $\Phi(\varphi)$ will reach a minimal value $\Phi = -\infty$ for $\varphi \to \pm\infty$ and stable phases with finite values $\bar{\varphi}$ will not occur at all. The case $u_4 = 0$ refers to another model (see our discussion at the end of this section and Section 4.7). In the framework of the φ^4-theory the only one sensible choice is $u_4 > 0$.

Since $\Delta h = 0$ on the phase transition line and obviously the potentials of the coexisting phases are equal $\Phi(0) = \Phi(-u_3/2u_4)$, the equilibrium properties of the system on this line will be the same as those of the disordered phase. Our prior study of the thermodynamic system with the help of potential (4.46) has indicated that the cases $u_3 \neq 0$ and $u_3 = 0$ are different and may serve as examples of qualitatively different types of phase transitions.

Certainly, for $u_3 = 0$, coexistence of different phases cannot take place because the phases become identical ($\bar{\varphi} = 0$). Such coincidence of phases is observed for continuous phase transitions and we may expect that the phase transition described by the model (4.46) without φ^3-term is of second order. The possibility of equilibrium coexistence between the disordered phase and the phase $\bar{\varphi}$ given by Eq. (4.51) makes us believe that the φ^3-term in Eq. (4.46) will give rise to first order transitions. In the next two sections we shall examine the model (4.46) with and without the φ^3-term separately.

Until now we have studied the phase transition line where $\Delta h = 0$. In order to investigate the domains of the phase diagram $(T, Y, \Delta h)$ near the transition line we have to consider states with $\Delta h \neq 0$. The equation of state (4.41) has no solutions $\bar{\varphi} \neq 0$ for nonzero field difference Δh. Consequently, no phase transitions of the type $0 \to \bar{\varphi}$ can occur for $\Delta h \neq 0$. Mainly we shall focus our attention on the case of zero field. The special transition $\varphi \to -\varphi$ driven by the field variation will be discussed in Section 4.10.

In the rest of this section we shall briefly mention the auxiliary example of the φ^3-theory ($u_4 = 0$). This theory does not give a $\bar{\varphi} \neq 0$ solution on the phase transition line. Outside the line and for $\Delta h = 0$, the disordered phase ($\varphi = 0$) is stable for $u_2 > 0$ whereas the phase $\bar{\varphi} = (-2u_2/u_3)$ is stable for $u_2 < 0$. Certainly the transition is located on the line $T_c(Y)$ defined by the equation $u_2(T_c, Y) = 0$. This model can be used to describe a second order transition with the not-worth-having result that the critical state $\bar{\varphi}(T_c, Y)$ is unstable. This state is an inflection point of the function $\Phi(\varphi)$. The effect of the field difference Δh can be accounted by simple calculations.

4.6 Second Order Phase Transitions

In systems like ferromagnets the free energy F is an even function of φ, $F(\varphi^2)$ (see also Section 3.8.2). This means that the odd powers in the expansion of $F(\varphi)$ are not present and, hence, the potential (4.46) for $\Delta h = 0$ can be written in a simple form:

$$\Phi = u_2\varphi^2 + u_4\varphi^4. \tag{4.58}$$

The equation of state

$$\varphi(u_2 + 2u_4\varphi^2) = 0 \tag{4.59}$$

has solutions

$$\varphi = 0, \tag{4.60}$$

and

$$\varphi_\pm = \pm\left(\frac{-u_2}{2u_4}\right)^{1/2}, \qquad u_2 < 0. \tag{4.61}$$

The stability condition

$$u_2 + 6u_4\varphi^2 \geq 0 \tag{4.62}$$

shows that the disordered phase ($\varphi = 0$) is stable in the domain

$$u_2(T, Y) > 0 \tag{4.63}$$

of the (T, Y) plane and on the line

$$u_2(T, Y) = 0. \tag{4.64}$$

Inserting $\varphi_\pm$ from Eq. (4.61) in Eq. (4.62) we obtain that the solutions $\varphi_\pm$ correspond to stability when the opposite to (4.63) inequality is fulfilled. Therefore the line $T_c(Y)$, defined by the solution of Eq. (4.64) with respect to T gives the phase transition line; equivalently, one can get the solution $Y_c(T)$. The line $T_c(Y)$ separates the domains of stability of the ordered and the disordered phase. On this line only the "para" phase solution ($\varphi = 0$) exists. This phase is stable on the critical line $T_c(Y)$ because in our model

Eq. (4.58) u_3 is identically zero and $u_4 > 0$. The phase transition line $T_c(Y)$ can be obtained from the equilibrium condition $\Phi(\varphi) = 0$ putting $u_3 = 0$; cf. Eq. (4.40).

Note that the disordered phase exists for any u_2 and this simply means that it is always a solution of the equation of state. However it gives a minimum of Φ if only $u_2 \geq 0$. The solutions $\varphi_\pm$ behave in a different way. They do not exist for $u_2 > 0$, that is, in the region of stability of the disordered phase they do not present any type of equilibrium (either stable, unstable or metastable). It is important to realize that in the domain $u_2 \geq 0$ only the disordered phase may appear and it really does (minimizing the potential Φ). In the region $u_2 < 0$ all the phases ($\varphi = 0$) and $\varphi_\pm$ exist but the disordered phase ($\varphi = 0$) corresponds to a maximum of Φ and this phase practically does not occur (absolute instability).

The solutions $\varphi_\pm$ describe one and the same ordered phase $\bar{\varphi} = \varphi_\pm$ because they cannot be physically distinguished, of course, in the absence of an external field. The equilibrium Gibbs potential $\Phi(\varphi) = \Phi(T, Y)$ for $\bar{\varphi} = \varphi_\pm$ is

$$\Phi(T, Y) = \frac{u_2}{2}\bar{\varphi}^2 = -\frac{u_2^2}{4u_4} < 0. \tag{4.65}$$

The energy $\Phi(\bar{\varphi})$ of the system at fixed T and Y is lower for the ordered phase $\bar{\varphi}$ in comparison with the corresponding energy $\Phi(0) = 0$ of the disordered phase. Some of the described effects are illustrated by Fig. 4.1.

Now our task is to understand the dependence of Φ on T and Y in Eq. (4.65). Certainly this dependence cannot be obtained before some additional considerations to be involved. The criterion selecting the good suggestions is only one — the successful description of the second order transition. We have already shown that such a transition can be described by the model (4.58) and we have determined the stability properties of the phases. The thermodynamic properties of the transition will depend on the form of the function $\Phi(T, Y)$ in Eq. (4.65), i.e., on the functions u_2 and $\bar{\varphi}$ (or u_4). The function $u_2(T, Y)$ should change sign at $T_c(Y)$ and the order parameter $\bar{\varphi}$ should gradually decrease to zero when T increases to T_c.

The Landau proposal presents the simplest and elegant solution of the problem. The coefficient $u_2(T, Y)$ in the first equality (4.65) is an analytic function of T for any Y. The only way to represent the analytic function $u_2(T)$ having the property to change sign at T_c, taking zero value at the same T_c, is

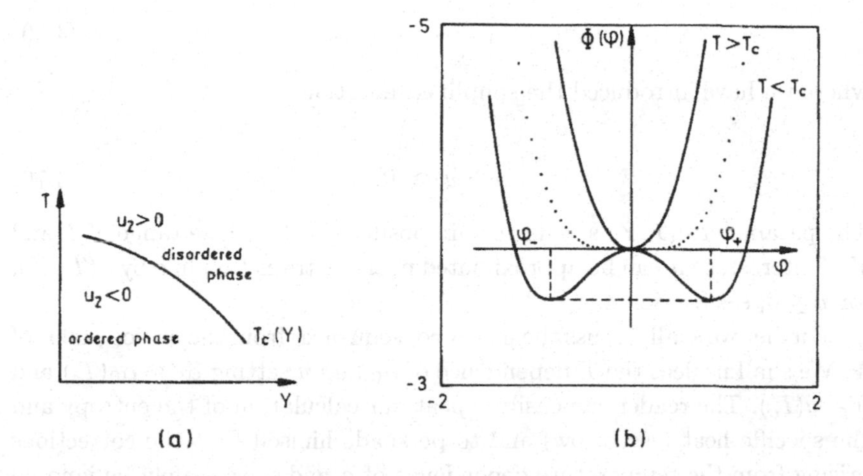

Fig. 4.1 (a) (T, Y) phase diagram corresponding to the potential (4.58). (b) The function $\Phi(\varphi)$ from Eq. (4.58). The thick line illustrates the stable state $\bar{\varphi} = \varphi_{\pm}$ for $T < T_c$. The dashed line gives an idea of the function $\Phi \sim \varphi^4$ at $T = T_c$.

$$u_2(T, Y) = u_2'(T, Y)(T - T_c), \tag{4.66}$$

with $u_2'(T, Y)$ an even function, which does not change sign (and remains nonzero) at any T and Y of interest. One may think of u_2' as the first derivative of u_2 in a Taylor expansion around T_c with $u_2(T_c, Y) = 0$. Of course this is the case when the difference

$$\Delta T_c = (T - T_c) \tag{4.67a}$$

is small. In dimensionless units it is:

$$t = \frac{T - T_c}{T_c}. \tag{4.67b}$$

Taking into account the terms of order $O(t^2)$ one simply finds the expansion of the function $u_2'(T, Y)$. Usually the ordered phase is the low temperature ($T < T_c$) one and as a consequence $u_2' > 0$. In the following discussions we shall denote u_2' by α_0' and u_2 by r:

$$r = \alpha_0 t = \alpha_0' \Delta T_c, \tag{4.68}$$

where $\alpha_0 = \alpha_0' T_c$. With this choice of u_2 we can write the potential (4.65) in the form:

$$\Phi = -\frac{r^2}{4u}, \tag{4.69}$$

where we have introduced the simplified notation

$$u = u_4 > 0. \tag{4.70}$$

The parameter $u(T, Y)$ should remain positive in the whole range of T and Y of interest, so it can be approximated near the transition line by $u(T_c, Y)$; for $u \leq 0$; see Section 4.8.

Further we shall discuss the major consequences from the choice (4.69) of Φ. We shall neglect the T-dependence of α_0 and u setting $\alpha_0 \approx \alpha_0(T_c)$ and $u \approx u(T_c)$. The reader can easily repeat our calculation of the entropy and the specific heat (see below) and to persuade himself that the corrections arising from the temperature dependence of α and u can safely be ignored near T_c.

To calculate the entropy S we must recall that Φ in Eq. (4.58) is the difference $\Delta\Phi = \Phi(\bar{\varphi}) - \Phi(0)$ between the thermodynamic potentials of the ordered and disordered phases; which means that we in fact calculate the difference ΔS between the entropies of the phases: $\Delta S = (S_{\text{ord}} - S_{\text{dis}})$. Using the thermodynamic equality

$$\Delta S = -\left(\frac{\partial \Phi}{\partial T}\right)_Y \tag{4.71}$$

and Eq. (4.69) for $\Phi (= \Delta\Phi)$ we obtain

$$\Delta S = \frac{\alpha_0^2 t}{2uT_c} < 0. \tag{4.72}$$

ΔS tends to zero when t approaches zero from the low temperature side ($t \to 0; T \to T_c^-$). The special choice of the coefficients u_2 and u_4 is the reason for ΔS to be a linear function of T for small $t < 0$.

The difference $\Delta C = (C_{\text{ord}} - C_{\text{dis}})$ between the specific heats C_{ord} and C_{dis} of the phases at fixed Y and $\Delta h = 0$ is given by $\Delta C = T(\partial \Delta S/\partial T)_Y$. With the help of Eq. (4.72) we have

$$\Delta C(T) = \frac{\alpha_0^2 T}{2uT_c^2}, \tag{4.73}$$

or,

$$\Delta C(T_c) = \frac{\alpha_0^2}{2uT_c} \tag{4.74}$$

on the critical line $T_c(Y)$. The entropy ΔS and the specific heat jump $\Delta C(T_c)$ at the transition point have been calculated from the function $\Phi(T)$ given by Eqs. (4.68) and (4.69). Both quantities correspond to a fixed external field ($\Delta h = 0$); see also Section 4.7.2.

The above theory results in a continuity of the entropy and a finite jump of the specific heat at the critical point T_c. In this aspect it confirms the original Ehrenfest suggestion (Section 3.5) about the properties of second order phase transitions. Within the Landau theory however, other susceptibilities (second derivatives of Φ or F) may undergo divergences rather than finite jumps. For example the susceptibility (4.35) at fixed T, Y and $\bar{h} = \Delta h = 0$ is divergent at T_c (see Section 4.10).

Owing to the special choice of the coefficients u_2 and u_4 in the expansion (4.58), we have obtained Φ as a simple analytic function of T; see Eq. (4.69). On the contrary, from the requirement of the analyticity of Φ near T_c the behaviour of the coefficients u_2 and u_4 can be derived. If $\Phi(T, Y)$ is an analytic function of T, we can write

$$\Phi = a(Y)t + \frac{1}{2}b(Y)t^2 + \ldots, \tag{4.75}$$

where the coefficients a and b are taken at T_c.

Because the entropy S has no jump at the critical points $T_c(Y)$, the coefficient $a(Y)$ is identically zero. Then ΔC is

$$\Delta C = -\frac{bT}{T_c^2}. \tag{4.76}$$

Using $a = 0$, Eq. (4.65), Eq. (4.68) and having in mind that the ordering appears for $t < 0$ we obtain

$$u_2 = \pm t\sqrt{-2u_4 b}, \tag{4.77}$$

or, equivalently,

$$u_2 = \alpha_0 t, \tag{4.78}$$

where

$$\alpha_0 = \sqrt{-2u_4 b}, \qquad b < 0. \tag{4.79}$$

From Eqs. (4.76) and (4.79), we rederive Eqs. (4.73) and (4.74). The temperature dependence of α_0 and $u = u_4$ remains unspecified but, surely, $(-\alpha_0^2/u_4) = b$ is t-independent.

4.7 First Order Transitions and the Isolated Critical Point

Our discussion of the Landau paper (1937*a*) continues with the study of first order transitions caused by the presence of the φ^3-term in Eq. (4.46). Here we shall investigate the model (4.46) with $\Delta h = 0$.

4.7.1 *Phase diagram of the first order transition*

The equation of state will be

$$\varphi\left(2r + 3u_3\varphi + 4u\varphi^2\right) = 0, \tag{4.80}$$

where $u = u_4 > 0$ and $r = u_2$, as given by Eqs. (4.68) and (4.70). The stability condition (4.53) now reads

$$r + 3u_3\varphi + 6u\varphi^2 \geq 0. \tag{4.81}$$

The trivial solution $\varphi_0 = 0$ of Eq. (4.80) minimizes $\Phi(\varphi)$ for $r > 0$ or $r = u_3 = 0$ $(u_4 > 0)$. The nonzero real solutions $\varphi_\pm$ of Eq. (4.80) are given by Eq. (4.49); see also the conditions (4.50). The potential $\Phi_\pm = \Phi(\varphi_\pm)$ can be written in the for

$$\Phi_\pm = \frac{\varphi_\pm^2}{32u}\left[16ru - 3u_3^2 \pm u_3\left(9u_3^2 - 32ru\right)^{1/2}\right]. \tag{4.82}$$

The stability regions of the phases $\varphi_\pm$ are obtained with the help of the condition (4.81). Here we face a situation which usually occurs for first order transitions, namely, the thermodynamic potential Φ has more than one minimum for equal values of the parameters r, u_3 and u. As a result there is an overlap of the domains of stability of the phases in the (T, Y) diagram of the system. In such a situation the stable states correspond to the absolute minimum of Φ whereas the relative minima of Φ give metastable states.

Owing to the symmetry $(u_3, \varphi) \to (-u_3, -\varphi)$ of the potential (4.46) with $\Delta h = 0$, we have $\varphi_{\pm}(u_3) = \varphi_{\mp}(-u_3)$; see Eq. (4.49). It is then sufficient to analyze only $u_3 > 0$ or $u_3 < 0$. The case $u_3 = 0$ is discussed in the next subsection.

Consider $u_3 < 0$. For $r < 0$, $\Phi(0)$ is the maximum of $\Phi(\varphi)$ and the paraphase ($\varphi_0 = 0$) cannot exist. Because both $\Phi(\varphi_{\pm})$ are minima of Φ and $\Phi_+ < \Phi_- < 0$, the phase φ_+ will be stable; so the phase φ_- may appear as a metastable one. For $r = 0$, $\varphi_- = \varphi_0 = 0$ and $\Phi(\varphi)$ for $\varphi = 0$ has a double inflection point. Under the same condition $r = 0$, $\varphi_+ = (-3u_3/4u)$ determines the minimum of Φ :

$$\Phi_+ = -\frac{27u_3^4}{256u^3}. \tag{4.83}$$

In the domain

$$0 < r < \frac{u_3^2}{4u}, \tag{4.84}$$

the phase φ_+ is stable, the paraphase is metastable and φ_- gives a maximum of Φ (instability). When

$$\frac{u_3^2}{4u} < r < \frac{9u_3^2}{32u}, \tag{4.85}$$

$\Phi(\varphi_-)$ is again a maximum but the phases φ_+ and φ_0 exchange their roles: the paraphase φ_0 is stable and the phase φ_+ is metastable. For $r > (9u_3^2/32u)$ the paraphase gives a single minimum of Φ and, hence, this is the stable phase. The line $T_c(Y)$ on the phase diagram (Fig. 4.2a) is determined by the equation $r = 0$. The equations $r = u_3^2/4u$ and $r = 9u_3^2/32u$ define two lines: $T_1(Y)$ and $T_2(Y)$, respectively. It is obvious that $T_c < T_1 < T_2$; see Fig. 4.2(a).

Up to now we have not been forced to specify the form of the function $u_l(T, Y)$ in the model (4.46); we use the notation $u_2 = r$ but the explicit dependence of r on T has not been applied. If we set, as usual $u_2 = r = \alpha_0 t$, see Eqs. (4.67a)–(4.68), the characteristic temperatures T_1 and T_2 following from the potential (4.46) can be written in the form

$$T_1 = T_c\left(1 + \frac{u_3^2}{4\alpha_0 u}\right) \tag{4.86}$$

and

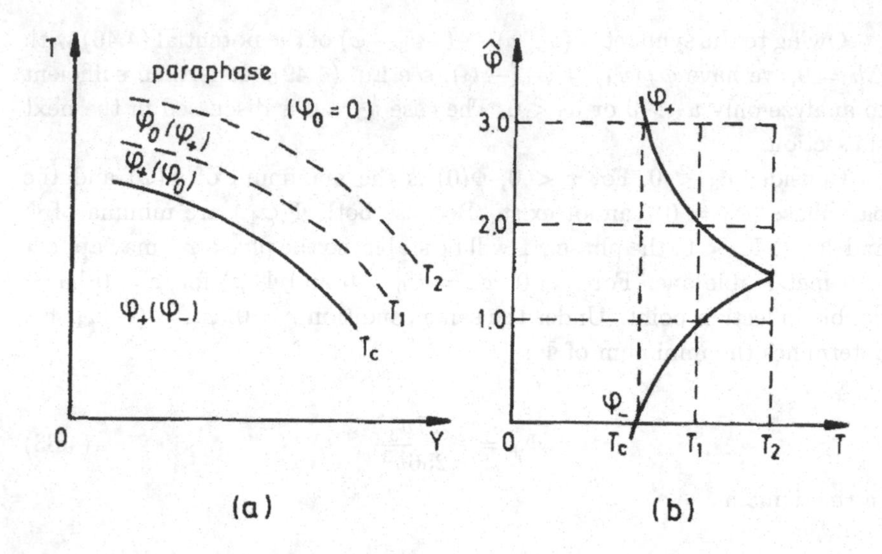

Fig. 4.2 (a) Borderlines: T_c, T_1 and T_2. The metastable phases are given in brackets. (b) The behaviour of $\hat{\varphi}_\pm = (-4u/u_3)\varphi_\pm$. Both figures correspond to $u_3 < 0$.

$$T_2 = T_c\left(1 + \frac{9u_3^2}{32\alpha_0 u}\right). \tag{4.87}$$

The behaviour of the potential $\Phi(\varphi)$ with $u_3 < 0$ is depicted in Fig. 4.3 for different temperatures. If the temperature T is lower than T_c the phase φ_+ is stable but the phase φ_- may also occur as a metastable state. The phase φ_0 is absolutely unstable. When T is increased up to T_c the phases φ_0 and φ_- change their roles. At T_c the paraphase $\varphi_0 = \varphi_- = 0$ is unstable; for $T > T_c$ φ_0 can appear as a stable ($T > T_1$) or metastable ($T_c < T < T_1$) phase whereas the phase φ_- cannot occur because it is either unstable (see the maxima of Φ for $T_c < T < T_2$) or does not exist ($T > T_2$) as a solution of Eq. (4.80).

Between the lines $T_c(Y)$ and $T_1(Y)$ in Fig. 4.2(a) the phase φ_+ is stable while φ_0 may appear only as a metastable phase. At $T_1(Y)$, $\Phi_+(T_1) = 0$ and the phases φ_0 and $\varphi_+ = (-u_3/2u)$ are equally stable — on this line they coexist in equilibrium. The energy $\Phi(\varphi_-)$ at $T_1(Y)$, corresponding to $\varphi_- = -u_3/4u$ is $\Phi_-(T_1) = (r^2/16u_4) > 0$. Between the lines $T_1(Y)$ and $T_2(Y)$ the phase φ_0 is stable; the phase φ_+ may exist as a metastable one. $T_2(Y)$, is the maximal temperature at which the phases $\varphi_\pm$ may appear. The equal values $\varphi_\pm(T_2) = (-3u_3/8u)$ correspond to a double

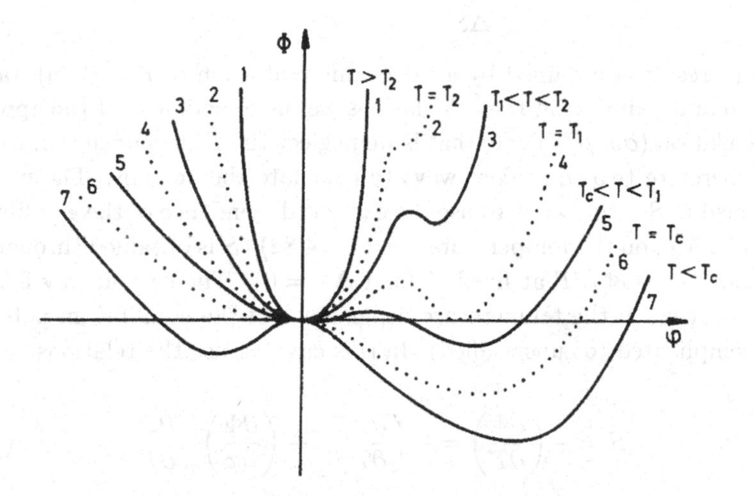

Fig. 4.3 The shape of $\Phi(\varphi)$ for $u_3 < 0$. The case $u_3 > 0$ is obtained by the reflection $\varphi \to -\varphi$.

inflection point of Φ: $\Phi_{\pm}(T_2) = (r^2/12u_4) > 0$. The dimensionless functions $\hat{\varphi}_{\pm} = (-4u/u_3)\varphi_{\pm}$ are depicted in Fig. 4.2(b).

4.7.2 *Metastability. Specific heat*

We have shown that the equilibrium phase transition takes place at T_1; T_c and T_2 determine domains of metastability (overcooling and overheating). Increasing the temperature from the low-temperature ($T < T_1$) side, the equilibrium order parameter φ_+ abruptly changes from $\varphi_+ \neq 0$ at temperatures $T \leq T_1$ to the value $\varphi_0 = 0$ at $T \geq T_1$. When there is overheating this change of φ_+ to zero may occur at some temperature T in the interval $T_1 < T \leq T_2$. In this case the system is in a metastable (overheated) ordered state $\varphi_+ \neq 0$. The overcooling can be described in an analogous way.

The entropy ($S = \Delta S$) and the specific heat jump ΔC for this transition are calculated as explained in Section 4.6. Instead of u_2 we shall use r from Eq. (4.68), the temperature dependence of α_0 and u will be neglected. Within this scheme the difference $\Delta S_{\pm}$ between the entropy of the $\varphi_{\pm}$ phases and that (S_0) of the paraphase $\varphi_0 = 0$ is

$$\Delta S_{\pm} = -\alpha_0 \varphi_{\pm}^2(T). \tag{4.88}$$

This result is obtained by a direct differentiation of Eq. (4.46), taking $\Delta h = 0$ and fixing φ; in the calculations we have made use of the approximate relation $(\partial u_3/\partial T) \approx 0$, that is we neglect the T-dependence of u_3. We have therefore two equivalent ways to calculate the entropy. The first, as explained in Section 4.6 is to use the explicit dependence of the equilibrium potential $\Phi_{\pm}$, on the temperature, see Eq. (4.82). S is obtained through the relation $S = -d\Phi/dT$ at fixed Y (and $\Delta h = 0$). The second way is more convenient when the temperature dependence of the equilibrium potential $\bar{\Phi}$ is complicated (or unspecified). In this case we use the relations

$$S = -\left(\frac{\partial \Phi}{\partial T}\right) = -\left(\frac{\partial \Phi}{\partial T}\right)_{\varphi} - \left(\frac{\partial \Phi}{\partial \varphi}\right)_T \frac{\partial \varphi}{\partial T}, \tag{4.89}$$

Φ is the nonequilibrium potential and all derivatives are taken at fixed Y and $\Delta h = 0$. As $(\partial\Phi/\partial\varphi) = 0$ at $\varphi = \bar{\varphi}$, the equilibrium entropy is given by $S = -(\partial\Phi/\partial T)_{\bar{\varphi}}$. Now using Eq. (4.46) with $\Delta h = 0$ and $(\partial u_3/\partial T) = (\partial u/\partial T) = 0$ we obtain Eq. (4.88).

The formula (4.88) is a general expression for the entropy jump at the first order phase transition and can be successfully used to distinguish between the different cases ($T = T_c, T_1$, or T_2). At T_c, $\varphi_+ = 0$ so that $\Delta S_+ = (S_+ - S_0) = 0$ but $\varphi_- \neq 0$ and $\Delta S_-(T_c) = -(\alpha_0 u_3^2/16u^2)$. The latent heat

$$q = -T_c \Delta S \tag{4.90}$$

of the transition $0 \to \varphi_-$ is then $q = -T_c\Delta S(T_c)$, or

$$q = +\frac{\alpha_0 T_c u_3^2}{16u^2}. \tag{4.91}$$

The jump of the specific heat $\Delta C = -T(\partial^2\Phi/\partial T^2)$ corresponding to $\Delta S_{\pm}$ from Eq. (4.88) is

$$\Delta C_{\pm} = -2\alpha_0 T_c \left(\frac{\partial \varphi_{\pm}}{\partial T}\right)_{T_c}. \tag{4.92}$$

Using u_3 and u_4 as temperature independent parameters, it is easy to see that the derivative of φ_+ in the last equation is negative and the derivative of φ_- is positive; so $\Delta C_+ > 0$ and $\Delta C_- < 0$; see, e.g., Mróz *et al.* (1971).

4.7.3 *The isolated critical point*

Unlike the case from Section 4.6 where the parameter u_3 is identically zero, this parameter may become zero on the line $T_3(Y)$, defined by

$$u_3(T_3, Y) = 0. \tag{4.93}$$

The interesting point is when the line $T_3(Y)$ crosses $T_c(Y)$ as shown in Fig. 4.4. If u_3 gradually tends to zero as T approaches T_3, the lines $T_1(Y)$ and $T_2(Y)$ will look as depicted in Fig. 4.4; cf., for example, the model equations (4.86) and (4.87) for $u_3(T) \to 0$. Except for the intersection point O in Fig. 4.4 all the other transition points (T_c, T_1, and T_2) have the same properties as described in Sections 4.7.1 and 4.7.2. The phase transition at the point O is of second order. This is the so-called *isolated critical* (or *Landau*) *point,* lying on the first order transition line (T_1).

We shall briefly discuss the second order transition at the point O and the behaviour of the system on the line T_3O, where the order parameter of the ordered phase changes from φ_+ to φ_-.

On T_3O, $\varphi_+ = -\varphi_- = (|u_2|/2u_4)^{1/2}$ and the jump of φ is $\Delta\varphi_3 = (\varphi_+ - \varphi_-) = 2(|u_2|/2u_4)^{1/2}$; here we do not make use of the explicit dependence $u_2 = r = \alpha_0 t$. Because of $\varphi_+ = -\varphi_-$ only the φ^3-term in $\Phi(\varphi)$ will change when the line T_3O is crossed. So the difference $\Delta S = S(\varphi_+) - S(\varphi_-)$ will be

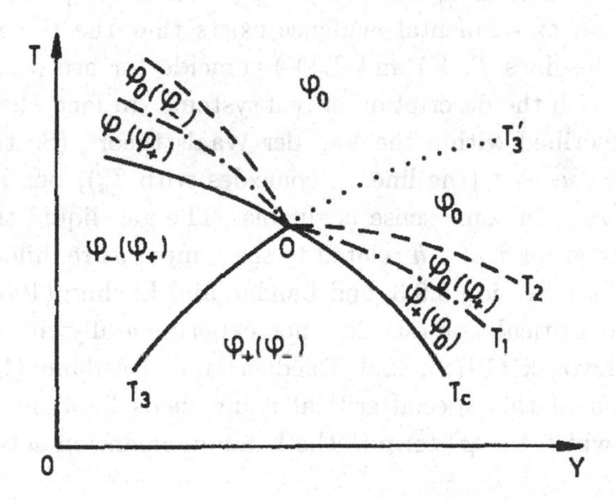

Fig. 4.4 The phase diagram in the case of intersection of the line $T_c(Y)$, defined by $u_2 = 0$, and the line T_3OT_3', defined by $u_3 = 0$. The phases in brackets are metastable.

$$\Delta S = -2\left(\frac{\partial u_3}{\partial T}\right)\left(\frac{|u_2|}{2u_4}\right)^{3/2} . \tag{4.94}$$

This result is obtained without any assumptions about the temperature dependence of u_2, u_3 and u_4; except for the change of the sign of u_2 at T_c. As u_3 changes sign on T_3O it is expected that the $u_3(T)$-dependence cannot be neglected and, hence, $(\partial u_3/\partial T) \neq 0$. If we assume $u_3 \sim (T - T_3)$ and $u_2 = r = \alpha_0 t$, $\Delta S \sim (-t)^{3/2}$ and, consequently, the corresponding jump $\Delta C = (C_+ - C_-)$ of the specific heat on the line T_3O will be $\Delta C \sim (-t)^{1/2}$; $t < 0$. Both ΔS and ΔC turn zero in the limit $t \to 0$; i.e., at the point O.

Surely the discontinuities of the entropy (4.88) and the specific heat (4.92) created by the first order transition $0 \to \varphi_\pm$ on the equilibrium transition line $T_1(Y)$, also vanish at the point O. Near the isolated critical point O, the equilibrium transition along the line $T_1(Y)$ becomes quite similar to a second order transition because of the weak discontinuities of φ and S; such a transition is called *weakly first order transition* or a *first order transition near to the second.*

It seems quite reasonable to conclude that the line T_3O is an example of a phase change between two ordered phases (φ_+ and φ_-). However the thermodynamic processes, described by crossing of the line T_3O can hardly be called phase transitions because the change $\varphi_+ \to \varphi_-$ unavoidably results from the change $u_3 \to -u_3$ and, hence, φ_+ and φ_- give one and the same phase; a case similar to that for $\bar{\varphi} = \varphi_\pm$ within the φ^4-model (4.58).

Up to now no experimental evidence exists that the theoretical predictions when the lines $T_3(Y)$ and $T_c(Y)$ coincide (or are parallel), have anything to do with the description of real systems. In fact, the gas–liquid transition is described within the Van der Waals theory (Section 5.9) by coefficients $u_2 \sim u_3 \sim t$ (the line T_3 coincides with T_c), but in this case the coefficient u_3 is in some sense negligible. The gas–liquid transition is driven by the external field Δh related to the temperature difference t and the pressure P; see Section 5.9.3, and Landau and Lifshitz (1980).

The isolated critical point is also not experimentally discovered; see Ishibashi and Dvorak (1978), and Tolédano and Tolédano (1987). But the investigation of this special critical point sheds light on the precise mechanism, in which the φ^3-term in the Landau expansion acts.

4.8 Extended Expansion: φ^6- and φ^8-theories

In systems as, for example, ferromagnets the thermodynamic potential $\Phi(\varphi)$ in a zero external field is invariant with respect to the change $\varphi \to -\varphi$, i.e.,

$\Phi(\varphi) = \Phi(-\varphi)$. So $\Phi(\varphi)$ depends on φ^2 and the expansion up to sixth order in φ is

$$\Phi = r\varphi^2 + u\varphi^4 + u_6\varphi^6, \qquad u_6 > 0, \tag{4.95}$$

where the parameter r is given by Eqs. (4.68). We shall show that the term u_6 does not introduce any new effects for $u > 0$. For $u > 0$ the description of the second order phase transition within the φ^4-theory (Section 4.6) remains valid and in this case the φ^6-term in the theory (4.95) is irrelevant. When $u \leq 0$ the φ^4-theory is however not well defined and the φ^6-term should be included into consideration. It describes essentially new phenomena — a type of first order transitions ($u < 0$) and the tricritical point ($u = 0$), discussed in Section 3.7. In order to avoid the appearance of instability within the φ^6-theory ($\Phi \to \infty$ as $\varphi \to \pm\infty$), we assume that u_6 is a positive parameter.

4.8.1 *First order transition (u < 0)*

The equation of state and the susceptibility at zero field $\chi_0(\varphi)$ for the potential (4.95) are

$$\varphi(r + 2u\varphi^2 + 3u_6\varphi^4) = 0 \tag{4.96}$$

and

$$\chi_0^{-1} = 2r + 12u\varphi^2 + 30u_6\varphi^4 \geq 0. \tag{4.97}$$

The last relation gives the stability condition, which is the requirement for the nonnegative definiteness of the susceptibility χ_0 at zero external field; cf. Eq. (4.35) with $\bar{h} = 0$.

The trivial solution $\varphi_0 = 0$ of Eq. (4.96) exists for any r, u, and u_6, i.e., in the whole plane (T, Y). The paraphase φ_0 is stable for $r > 0$. For $r < 0$, i.e., $T < T_c$ the paraphase is unstable ($\Phi(0)$ is a maximum of Φ).

The other solutions of Eq. (4.96) are

$$\varphi_\pm = \pm\bar{\varphi}, \qquad \bar{\varphi} = \left(\frac{-u + D}{3u_6}\right)^{1/2} \tag{4.98}$$

and

$$\varphi'_{\pm} = \pm \left(\frac{-u - D}{3u_6} \right)^{1/2}, \tag{4.99}$$

where

$$D = (u^2 - 3ru_6)^{1/2}. \tag{4.100}$$

The solutions (4.98) exist for

$$r \leq \frac{u^2}{3u_6}, \tag{4.101}$$

whereas the solutions (4.99) exist for

$$0 < r \leq \frac{u^2}{3u_6}. \tag{4.102}$$

In order to study the phases $\varphi_{\pm}$ and $\varphi'_{\pm}$, it is convenient to write the stability condition (4.97) in the form

$$D^2 \geq \pm uD, \tag{4.103}$$

where the sign $(+)$ corresponds to $\varphi_{\pm}$ and the sign $(-)$ corresponds to $\varphi'_{\pm}$. As $u < 0$ and $D \geq 0$, it is obvious that the solutions $\varphi'_{\pm}$ always give maxima of Φ. According to the condition (4.103) the phases $\varphi_{\pm}$ are always stable or metastable ($\Phi(\varphi_{\pm})$ are minima of Φ). The case $D = 0$ will be considered separately.

We focus our attention on the properties of the stable phases φ_0 and $\varphi_{\pm}$ and the minima $\Phi(0) = 0$ and $\Phi(\varphi_{\pm})$ of Φ. From Eqs. (4.95) and (4.98) we have $\Phi(\bar{\varphi}) = \Phi(\varphi_{\pm})$, and

$$\Phi(\bar{\varphi}) = \left(\frac{\bar{\varphi}^2}{9u_6} \right) \left(6u_6 r - u^2 + uD \right). \tag{4.104}$$

Our task is to understand what are the domains of stability of the phases φ_0 and $\bar{\varphi} = \varphi_{\pm}$ for given values of r, u, and u_6 in the parameter space (r, u, u_6) or, which is the same, for given T and Y in the (T, Y) diagram of the system. Therefore we should compare the magnitude of the possible minima of Φ. From Eq. (4.104) it follows that $\Phi(\bar{\varphi}) < 0$ for

$$r < \frac{u^2}{4u_6} \tag{4.105}$$

and $\Phi(\bar{\varphi}) > 0$ for

$$\frac{u^2}{4u_6} < r < \frac{u^2}{3u_6}. \tag{4.106}$$

The equality $r = u^2/4u_6$ defines the temperature

$$T_1 = T_c\left(1 + \frac{u^2}{4\alpha_0 u_6}\right) > T_c. \tag{4.107}$$

At $T = T_1$, $\Phi(\bar{\varphi}) = 0$ and consequently, the phases φ_0 and $\varphi_\pm$ coexist in equilibrium. It is clear that the phase $\bar{\varphi} = \varphi_\pm$ is stable for $T < T_1$. When $T > T_1$ the absolute minima of Φ is $\Phi(0)$ and, hence, the disordered phase is stable. The equilibrium transition takes place at T_1.

Another characteristic temperature is given by the Eq. (4.101), which can be written in the form

$$T_2 = T_c\left(1 + \frac{u^2}{3\alpha_0 u_6}\right). \tag{4.108}$$

When $T > T_2$ the solutions $\varphi_\pm$ of the Eq. (4.96) do not exist. The only possible (and stable) phase in this temperature region is $\varphi_0 = 0$. For $T_c < T < T_2$, $\bar{\varphi} = \varphi_\pm$ is either stable or metastable phase.

The phase changes, when the temperature increases from the low-temperature side ($T < T_c$) seems quite interesting. Raising the temperature up to T_c, the phase φ_0 transforms from an unstable to a metastable state and can exist for $T_c < T < T_1$; see Fig. 4.5(a). At T_1 the stability of the phases changes again and for $T_1 < T < T_2$ the ordered phase $\bar{\varphi}$ can appear only as a metastable state. The order parameter $\bar{\varphi}$ behaves in a way shown in Fig. 4.5(a). The values $\bar{\varphi}_c = \bar{\varphi}(T_c)$, $\bar{\varphi}_i = \bar{\varphi}(T_i)$ are given by

$$\bar{\varphi}_c = \left(\frac{-2u}{3u_6}\right)^{1/2}, \tag{4.109a}$$

$$\bar{\varphi}_1 = \left(\frac{-u}{2u_6}\right)^{1/2}, \tag{4.109b}$$

and

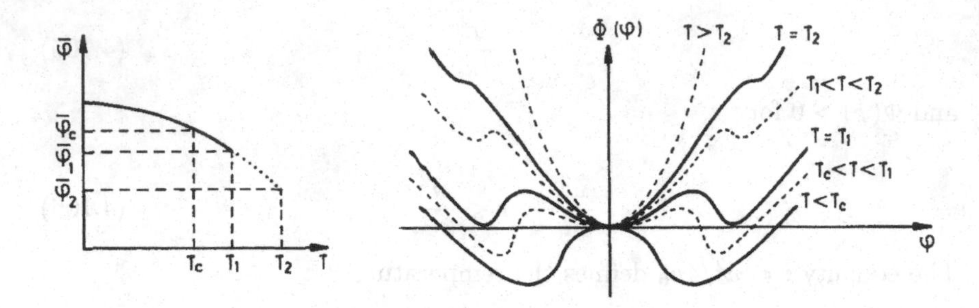

Fig. 4.5 (a) The order parameter $\bar{\varphi}(T)$ for $u < 0$. The dotted lines indicate the metastability of the phase $\bar{\varphi} \neq 0$ for $T_1 < T < T_2$, and the metastability of the phase $\varphi = 0$ for $T_c < T < T_1$. (b) The shape $\Phi(\varphi)$ for fixed $u_4 < 0$ and different temperatures.

$$\bar{\varphi}_2 = \left(\frac{-u}{3u_6}\right)^{1/2} .$$

The function $\Phi(\varphi)$ is schematically shown in Fig. 4.5(b). For $T < T_c$, $\Phi(\varphi)$ has two equal minima (at φ_+ and φ_-). These minima describe two physically indistinguishable ordered states (phase $\bar{\varphi} = \varphi_\pm$). With the increasing of T, the unstable phases $\varphi'_\pm$ first appear at T_c; $\varphi'_\pm(T_c) = 0$, see Eq. (4.99). At T_c, $\Phi(\varphi)$ has a triple maximum ($\varphi'_\pm = \varphi_0 = 0$). At T_2, $\bar{\varphi}(T_2)$ is a saddle (inflection) point of $\Phi(\varphi)$.

In summary we have found that this first order transition may occur, under special circumstances, at any temperature $\tilde{T}$ between T_c and T_2. The equilibrium transition however is at T_1. No matter how the transition occurs: from a metastable to a stable state at $\tilde{T} \neq T_1$ or precisely at T_1, the order parameter φ undergoes a finite jump. We have described two important features of the first order transitions within the model (4.95): the discontinuity of the order parameter and the metastability phenomena.

The latent heat $q(T_1)$ of the transition is obtained by the relation

$$q(T) = T(S_0 - S), \tag{4.110}$$

where S_0 and S are the entropies of the disordered (φ_0) and the ordered ($\bar{\varphi}$) phases, respectively; cf. Eq. (3.12). In our notation $S_0 = 0$. Using $S = -(\partial\Phi/\partial T)_\varphi$ and neglecting the T-dependence of u and u_6 (see subsection 4.7.2) we obtain the simple formula

$$q(T) = \alpha_0 T T_c \bar{\varphi}^2(T). \tag{4.111}$$

At T_1,

$$q(T_1) = -\alpha_0 \frac{T_1}{T_c} (u2u_6) . \tag{4.112}$$

The specific heat jump $\Delta C = C$ at T_1 is calculated from the relation $C = T(\partial S/\partial T)_Y$, where the entropy $S = -q(T)/T$. Using Eq. (4.111) we have $S = -\alpha_0 \bar{\varphi}^2/T_c$. The temperature dependence of $\bar{\varphi}^2(T)$ is determined by the function $r(T)$ in Eq. (4.98). In this way we obtain

$$\frac{\partial \bar{\varphi}^2}{\partial T} = -\frac{\alpha_0}{|u|T_c} \tag{4.113}$$

and

$$\Delta C = \frac{\alpha_0^2 T_1}{|u|T_c^2} \tag{4.114}$$

for $T = T_1$. This discontinuity is quite similar to that for second order transitions; see Eq. (4.74). But in our present case the entropy S exhibits a discontinuity at the transition point T_1, too.

The susceptibility (4.97) is calculated for the stable phases. For $T > T_1$, χ_0 of the paraphase is

$$\chi_0^{>}(T) = \frac{1}{2r} . \tag{4.115}$$

In the limit $T \to T_1$, from Eq. (4.107) we obtain

$$\chi_0^{>}(T_1) = \frac{2u_6}{u^2}, \qquad (T = T_1^+). \tag{4.116}$$

For $T < T_1$ the susceptibility χ_0 of the stable phase $\bar{\varphi}$ is calculated with the help of Eqs. (4.97) and (4.98). The result is

$$\chi_0^{<}(T) = \left(\frac{3u_6}{8D}\right) \frac{1}{|u| + D} . \tag{4.117}$$

As $T \to T_1^-$,

$$\chi_0^{<}(T_1) = \frac{u_6}{2u^2} . \tag{4.118}$$

The jump $\Delta\chi = (\chi_0^{>} - \chi_0^{<})$ is

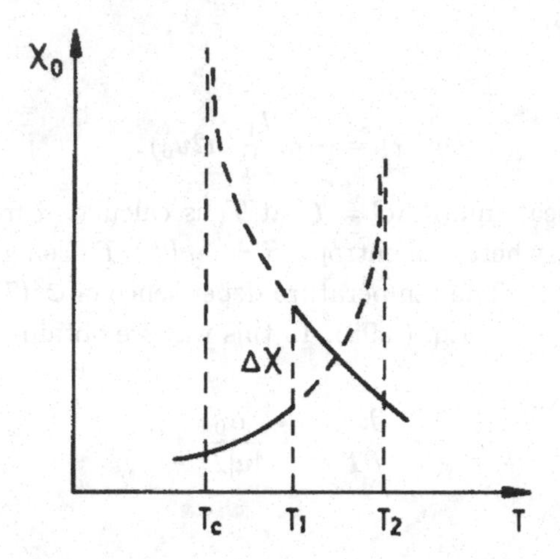

Fig. 4.6 Susceptibility near T_1.

$$\Delta\chi = \frac{3u_6}{2u^2}. \tag{4.119}$$

The behaviour of χ_0,described by Eqs. (4.115)–(4.119) is depicted in Fig. 4.6. The susceptibilities $\chi_0^>$ and $\chi_0^<$ are divergent at T_c and T_2, respectively; see Eqs. (4.115) and (4.117) as well as the dashed lines in Fig. 4.6. These divergences, however, can be observed only when there are extreme overcooling and overheating.

4.8.2 *The tricritical point* ($u = 0$)

The equation $u(T, Y) = 0$ defines a line $T_t(Y)$, which crosses the critical line $T_c(Y)$ determined by $r = 0$. This is shown in Fig. 4.7. The intersection point $T_t(Y_t)$ of $T_c(Y)$ and $T_t(Y)$ is the tricritical point (see our discussion in Section 3.7.4). Here we shall show how this multicritical point is described within the Landau expansion.

On the line $T_t(Y)$ the disordered phase φ_0 is stable for $r \geq 0$, i.e., at the tricritical point too. On line 3, t below T_c, the ordered phase $\bar{\varphi} = \varphi_\pm$ is given by Eqs. (4.92), where

$$\bar{\varphi} = \left[\frac{-r(T_t)}{3u_6}\right]^{1/4} \tag{4.120}$$

and $r(T_t) = \alpha_0(T - T_t)$, $T_t = T_c(Y_t)$. This ordered state is stable for $T < T_t$. From Eq. (4.120) and Eq. (4.104) we receive

$$\Phi(\bar{\varphi}) = -\frac{2}{3}\left(\frac{|r|^3}{3u_6}\right)^{1/2} < 0. \tag{4.121}$$

The entropy S is a continuous function of T at T_t but the specific heat $C = C(h = 0)$ is divergent in the limit $T \to T_t^-$. For $T < T_t$, using Eq. (4.121) and the relation $C = -T(\partial^2\Phi/\partial T^2)$ we get

$$C(T) = \left(\frac{\alpha_0^2 T}{2T_t^2}\right)\frac{1}{[3u_6 r(T_t)]^{1/2}}. \tag{4.122}$$

The susceptibility χ_0 is calculated from the Eq. (4.97). It is

$$\chi_0 = \frac{1}{2r}, \quad \text{for} \quad T > T_t, \tag{4.123a}$$

and

$$\chi_0 = \frac{1}{8|r|}, \quad \text{for} \quad T < T_t. \tag{4.123b}$$

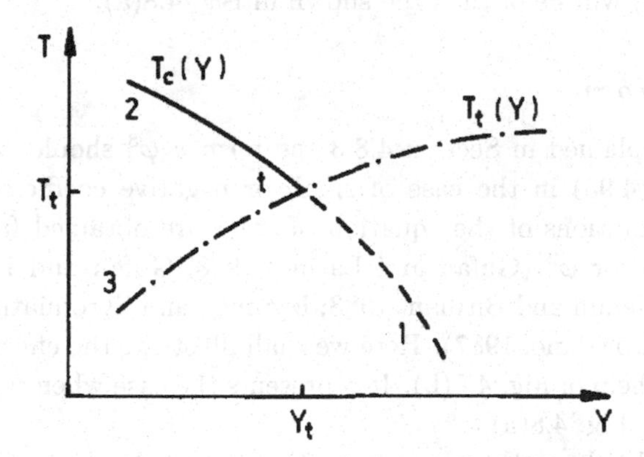

Fig. 4.7 Tricritical point (T_t, Y_t). The order of the transition at $T_c(Y)$ changes from second (the solid line) to first (the dashed line) at T_t.

4.8.3 *Irrelevance of the φ^6-term for $u > 0$*

The present remark is important for the understanding of the Landau theory. It gives insight into conditions for the truncation of the expansion (4.32).

Let us try to describe a second order transition with the help of the potential (4.95), when $u > 0$. The solutions $\varphi'_{\pm}$ (4.99) of Eq. (4.96) do not exist for $u > 0$ and $\bar{\varphi} = \varphi_{\pm}$ in Eq. (4.98) can be represented by the series expansion

$$\bar{\varphi}^2 = \bar{\varphi}_s^2 \left[1 + \frac{1}{4x^2} + O(x^{-4}) \right], \tag{4.124}$$

provided

$$|x| \gg 1, \qquad x = \frac{u}{(3|r|u_6)^{1/2}}. \tag{4.125}$$

In Eq. (4.124), $\bar{\varphi}_s^2 = (-r/2u)$ is the square of the order parameter within the φ^4-theory; cf. Eq. (4.61). The effect of the coefficient u_6 is insignificant in the vicinity of T_c, where $\bar{\varphi}_s^2 = (-r/2u)$ is very small. Clearly, the criterion (4.125) cannot be fulfilled for $u \to 0$. Within the same approximation (4.125), the potential $\Phi(\bar{\varphi})$ from Eq. (4.104) becomes equal to $\Phi(\bar{\varphi}_s)$ given by Eq. (4.69).

Thus for any fixed $u_6 > 0$ the phase diagram (r, u) resulting from the potential (4.95) will be of the type shown in Fig. 4.8(a).

4.8.4 *φ^8-theory*

For reasons explained in Section 4.8.3 the term $u_8\varphi^8$ should be added to the potential (4.95) in the case of small or negative coefficients u_4 and u_6. Now the solutions of the equation of state are obtained from a third order equation for φ^2 (Gufan and Larin , 1978; Gufan and Larin, 1979; Gufan, 1982; Galam and Birman, 1983; Izyumov and Syromiatnikov, 1984; Tolédano and Tolédano, 1987). Here we shall illustrate the effect of the φ^8-term with the help of Fig. 4.8(b). It represents the case when $u > 0, u_6 < 0$ and $u_8 > 0$; cf. Fig. 4.8(a).

Figure 4.8(b) shows the phase transitions at fixed values of $u_6 < 0$ and $u_8 > 0$. The effect of the coefficient u_8 is rather drastic for $u_6 < 0$. The second order transition at $r = 0$ is shifted to the right with respect to the same transition in Fig. 4.8(a). The tricritical point does not appear, but

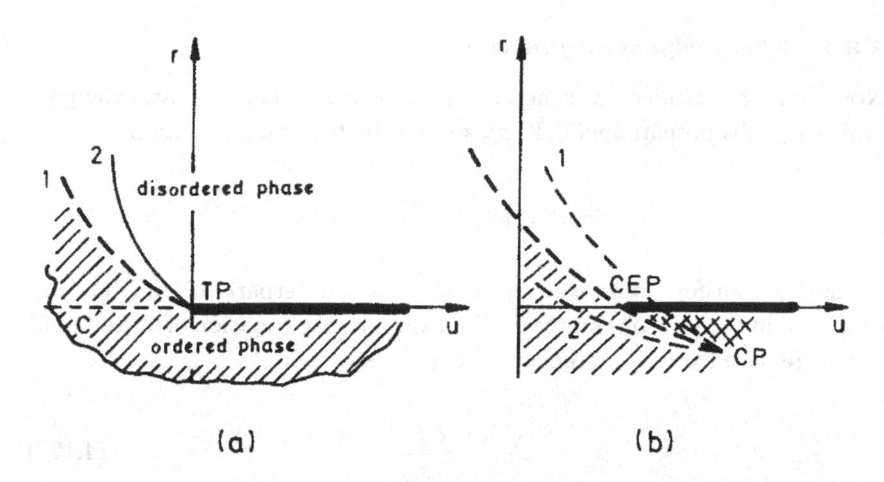

Fig. 4.8 (a) (r, u) diagram for fixed $u_6 > 0$. At the tricritical point (TP) the second order transition (the thick line $r = 0$) changes to a first order transition (the thick dashed line $1 - TP$). The lines $C - TP$, $1 - TP$ and $2 - TP$ correspond to $T_c(Y)$, $T_1(Y)$ and $T_2(Y)$ given by Eqs. (4.101) and (4.102). (b) (r, u) diagram from the φ^8-theory ($u > 0, u_6 < 0, u_8 > 0$). The shaded regions indicate two different ordered phases.

instead of it there are two other special critical points: the critical point (CP) similar to that for the gas–liquid transition and the critical-end point (CEP); see Section 3.7.4. The ordered phases are two. The metastability regions near the first order transition are marked by the spinodal lines $1 - CP$ and $2 - CP$.

4.9 Multicomponent Order Parameter and Symmetry Breaking

The Landau theory can be generalized for two, three and more real components φ_α of the order parameter φ, which form a real vector, $\varphi = \{\varphi_\alpha, \alpha = 1, \dots, n; n \geq 1\}$. When the order parameter has complex components ($\varphi_\alpha = \varphi'_\alpha + i\varphi''_\alpha$) this will be specially stipulated. A $(n/2)$-component complex vector (or scalar, $n = 1$) can always be represented as a n-component real vector. For example, the complex scalar order parameter, usually denoted by $\psi = \psi' + i\psi''$ in ordinary superconductors (and the superfluid ^{4}He) can be represented as a two component real vector $\varphi = (\psi', \psi'') \equiv (\varphi_1, \varphi_2)$.

4.9.1 *Thermodynamic functions*

Now we shall consider the generalized free energy $F(T, Y, \varphi)$ and the generalized Gibbs potential $\Phi(T, Y, \varphi)$, related by the transformation

$$\Phi(\varphi) = F(\varphi) - \sum_{\alpha} h_\alpha \varphi_\alpha. \tag{4.126}$$

From this transformation we can deduce the counterparts of all equations discussed in the previous sections. For example, the equation of state (4.15) is now replaced with a system of n equations

$$h_\alpha = \frac{\partial F}{\partial \varphi_\alpha}. \tag{4.127}$$

The equilibrium equations of state are obtained from Eq. (4.127) when $\varphi_\alpha = \bar{\varphi}_\alpha$ (and $h_\alpha = \bar{h}_\alpha$).

The susceptibility χ from Eq. (4.35) becomes a susceptibility matrix $\hat{\chi} = \{\chi_{\alpha\alpha'}\}$,

$$\chi_{\alpha\alpha'} = \left(\frac{\partial \varphi_\alpha}{\partial h_{\alpha'}}\right)_{\bar{h}} = \left(\frac{\partial^2 F}{\partial \varphi_\alpha \partial \varphi_{\alpha'}}\right)^{-1}_{\varphi=\bar{\varphi}}. \tag{4.128}$$

Using the equilibrium Gibbs potential $\bar{\Phi}[\bar{\varphi}_\alpha(h_\alpha)]$, the components $\chi_{\alpha\alpha'}$ of χ will be

$$\chi_{\alpha\alpha'} = -\frac{\partial^2 \Phi}{\partial h_\alpha \partial h_{\alpha'}}, \tag{4.129}$$

where h is the equilibrium external field.

4.9.2 *Symmetry breaking and symmetry conserving phase transitions. Landau criteria for second order phase transitions*

The difficulties in the treatment of a multi-component order parameter first arise when one tries to perform the Taylor expansion (4.32). The linear term is simple and can always be written in the form

$$-\sum_{\alpha} \Delta h_\alpha \varphi_\alpha. \tag{4.130}$$

The higher order differentials of $\Phi(\varphi)$ are however complicated and, surely, the pure mathematical treatment may lead us to difficulties and even in the wrong direction. In Sections 4.6 and 4.8 we have investigated systems possessing the symmetry $F(\varphi) = F(-\varphi)$ when the external field is zero ($\Delta h = h = 0$). In these systems the first derivative of the even function $F(\varphi^2)$ at $\bar{\varphi} = 0$ is always equal to zero,

$$\left(\frac{\partial F}{\partial \varphi}\right)_0 = 0 \tag{4.131}$$

and, hence, the equilibrium external field $h(0)$ at $\bar{\varphi} = 0$ is equal to zero,

$$h(0) = 0, \qquad \Delta h = h. \tag{4.132}$$

This result was received in the original paper by Landau (1937a) with the help of symmetry arguments.

Certainly odd powers of φ cannot appear in the expansion of an even function $F(\varphi)$. The even powers of φ give the abbreviated potentials (4.46) and (4.95) with $u_3 = 0$, where

$$\varphi^2 = \sum_\alpha \varphi_\alpha^2. \tag{4.133}$$

In the case of one-component system ($n = 1$) the thermodynamic potential is invariant under the change of the sign of the order parameter ($\varphi \to -\varphi$). For many-component systems ($n > 1$) the thermodynamic potential is invariant with respect to rotations of the vector φ in the space formed by its components ($\varphi_1, \dots, \varphi_n$). In the disordered phase ($\bar{\varphi} = 0$) all directions in the order parameter space $\{\varphi_\alpha\}$ are equivalent whereas in the ordered phase the equilibrium vector $\bar{\varphi} \neq 0$ fixes one of these directions. As the ordering $\varphi \neq 0$ spontaneously arises (without the effect of an external field conjugate to φ) the appearance of ordered phases of this type is accompanied with the *spontaneous symmetry breaking.* The corresponding phase transitions are called *symmetry breaking transitions.* To avoid misunderstanding we shall emphasize that the phenomenon of symmetry breaking is the appearance of a stable phase with a symmetry lower than that of the thermodynamic potential $\Phi(\varphi)$ of the system.

Within the framework of the one-component $\Phi(\varphi^2)$-model the simple "inversion" symmetry ($\varphi \to -\varphi$) is broken while within many-component $\Phi(\varphi^2)$-models the continuous symmetry towards rotations in the n-dimensional space $\{\varphi_\alpha\}$ is broken. These two different kinds of broken symmetry bring about quite different phase transition properties. The breaking of the continuous symmetry is always accompanied by the appearance of the so-called *Goldstone modes* (Section 7.5.1).

The phenomenon of symmetry breaking is more complicated in systems with anisotropy (see Sections 3.8.3 and 4.11.2). The anisotropy restricts the possible rotations in the order parameter space. Some of these rotations become forbidden (strong anisotropy) or less probable (weak anisotropy). In an isotropic ferromagnet the ordered state ($M \neq 0$) is invariant with respect to rotations of the magnetization vector M. However once the direction of the vector M is fixed, say, along the $\hat{z}$-axis, the magnetic properties will be different along the various spatial directions. For example, the response of the system to an external field h is given by the term $-h.M$ and, hence, will depend on the angle θ between h and the $\hat{z}$-axis of magnetization. When the anisotropy of type "plane of easy magnetization" is present (see Section 3.8.3), the magnetization vector either prefers the directions along the plane of magnetization (weak anisotropy) or is confined to lay in this plane. In an external field h no spontaneous breaking of symmetry is possible because the same symmetry has been already broken by the field h. The field h breaks the isotropy of the order parameter space $\{\varphi_\alpha\}$.

We have seen from our treatment of φ^4-, φ^6- and φ^8- theories that the symmetry breaking transition can be of first or second order. Besides, it may occur under special circumstances at special multicritical points.

The Landau criteria for second order transitions are referred to the expansion of $\Phi(\varphi)$ up to fourth order in φ. The first of these criteria is expressed by the Eq. (4.132), that is the first coefficient in the expansion of the free energy is identically equal to zero (for any T and Y). The second criterion requires the third order coefficient u_3 to be identically zero.

When the terms $h(0)\varphi$ and $u_3\varphi^3$ are not identically zero they may become zero at some point of the (T, Y, h)diagram of the system. But these zeros are not a consequence of a symmetry of the ordering (remember the critical point of the gas–liquid transition; see Sections 3.6 and 5.9.

The phase transitions, which are not followed (or caused) by a symmetry breaking (as the gas–liquid transition) are often called *symmetry conserving transitions.* From a macroscopic point of view these transitions drive the system to a new ("ordered") phase, which has the same symmetry as the previous ("disordered") one. For a symmetry conserving phase transition all odd (and even) terms can be present in the Landau expansion.

Landau is perhaps the first who applied symmetry arguments in studying general properties of thermodynamic systems and, in particular, in the phase transition theory. The reader who is interested more about the symmetry properties of ordered states and phase transitions is referred to the specialized literature (Landau and Lifshitz, 1958, 1980; Gufan, 1982; Tolédano and Tolédano, 1987; Izyumov and Syromiatnikov, 1984).

4.10 Effect of External Field and Susceptibility

Now we shall take into account the effect of the external field h conjugate to the order parameter φ for a symmetry breaking ordering.

4.10.1 *One-component order parameter*

The thermodynamic potential within the φ^4-theory is written in the form

$$\Phi = -h\varphi + r\varphi^2 + u\varphi^4. \tag{4.134}$$

The equation of state is

$$h = 2r\varphi + 4u\varphi^3. \tag{4.135}$$

It will be more convenient to analyze the function $h(\varphi)$ instead of solving the third order equation (4.135) with respect to $\varphi = \varphi(h)$.

The shape of the function $h(\varphi)$ given by Eq. (4.135) is shown in Fig. 4.9(a) for different temperatures. For $T \geq T_c$ this function has only one zero ($h(0) = 0$). For $T < T_c$ the picture drawn in Fig. 4.9(a) is obtained by investigating the extrema of $h(\varphi)$. These extrema are located at

$$\varphi^e_\pm = \pm\left(\frac{-r}{6u}\right)^{1/2} = \frac{1}{\sqrt{3}}\varphi_\pm(0), \tag{4.136}$$

where $\varphi_\pm(0)$ are the solutions of Eq. (4.135) for $h = 0$; cf. Eq. (4.61), where $u_2 = r$. The sign of the derivative $\partial h/\partial\varphi$ indicates that φ^e_+ corresponds to a minimum of $h(\varphi)$ and φ^e_- corresponds to a maximum. The extrema $h^e_\pm = h(\varphi^e_\pm)$ are

$$h_\pm = \mp\frac{4}{3}\left(\frac{|r|^3}{6u}\right)^{1/2}, \qquad T < T_c. \tag{4.137}$$

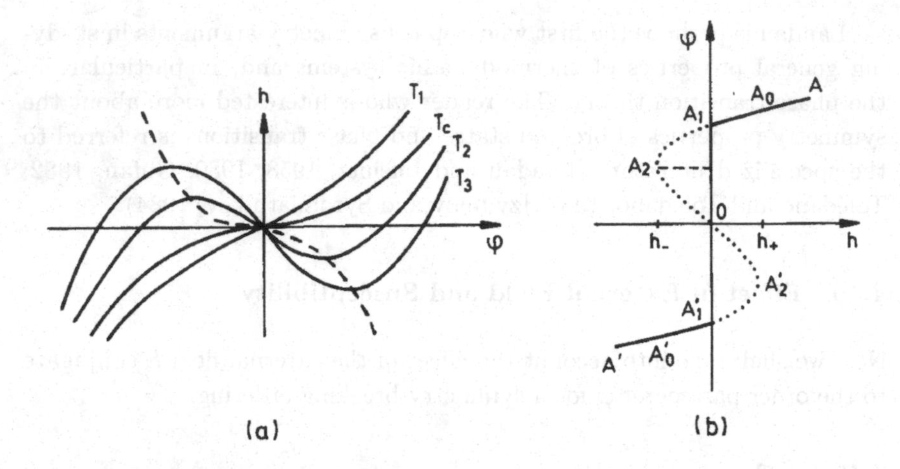

Fig. 4.9 (a) The function $h(\varphi)$ for temperatures $T_1 > T_c > T_2 > T_3$. The dashed line is the line of the extrema $h_{\pm}(\varphi)$. (b) The function $\varphi(h)$ for temperatures $T < T_c$.

The function $\bar{\varphi} = \varphi(h)$ is depicted in Fig. 4.9(b) for a temperature $T < T_c$; cf. Fig. 3.12(c).

The points A_1 and A_1' of this (φ, h) diagram correspond to the values φ_+ and φ_- of the order parameter, respectively. The coordinate of points A_2 and A_2' are (φ_+^e, h_-), and (φ_-^e, h_+). Obviously for each field h in the interval (h_-, h_+) the function $\varphi(h)$ is three-valued. One of the values $(|\varphi| < |\varphi_{\pm}^e|)$ lies on the line A_2A_2' and corresponds to a maximum of $\Phi(\varphi)$. The other values of $\varphi(h)$ correspond to minima of $\Phi(\varphi)$. One of these minima (on the line A_1A_0 for $h > 0$ or on the line $A_0'A_1'$ for $h < 0$) is the absolute minimum of $\Phi(\varphi)$ and describes the stable phase whereas the other is a relative minimum of $\Phi(\varphi)$ and describes metastable states.

The equilibrium function $\varphi(h)$ has a jump equal to $2\varphi_+$ at $h = 0$. The system undergoes a first order phase transition driven by the change of h (from $h < 0$ to $h > 0$ and vice versa). The same type of symmetry conserving ($h \neq 0$) phase transition has been discussed in Section 3.8.2. There is a similarity between some of the properties of this field-induced transition and some features of the gas–liquid transition.

The entropy $S = -(\partial\Phi/\partial T)_h$ is calculated from Eq. (4.134) at fixed φ and h because $(\partial\Phi/\partial\varphi) = 0$; see also Eq. (4.89). Thus the entropy

$$S = -\frac{\alpha_0}{T_c}\varphi^2(h) \tag{4.138}$$

corresponds to a stable phase $\varphi(h)$. The specific heat is then

$$C_h(T) = -\frac{\alpha_0 T}{T_c}\frac{\partial}{\partial T}\varphi^2(h). \tag{4.139}$$

The susceptibility

$$\chi(h) = \left(\frac{\partial \varphi}{\partial h}\right)_T \tag{4.140}$$

is obtained from Eq. (4.135). It is

$$\chi(h) = \frac{1}{2r + 12u\varphi^2(h)}. \tag{4.141}$$

Obviously, the entropy, the specific heat C_h and the susceptibility χ do not depend on the sign of h. The calculation of S, C, and χ from Eqs. (4.138)–(4.141) can be performed explicitly, if the function $\varphi^2(T,h)$ is known.

The function $\varphi(T,h)$ is very simple for the paraphase ($\varphi = 0$) but this phase exists only for $h = 0$. We can calculate S, C_h, and χ at zero field, where the order parameter has the discontinuity $2\varphi_+$; see Fig. 4.9(b). Having in mind the quadratic dependence of S on φ in Eq. (4.138), it is easy to conclude that the entropies at points A_1 and A_1' are equal. At the same points C_h from Eq. (4.139) is equal to the specific heat jump given by Eq. (4.73). For the critical point $(T_c, h = 0)$ S and $C_{h=0}$ are obtained from Eqs. (4.72) and (4.74).

The susceptibility χ at zero field $\chi_0 = \chi(0)$ for the paraphase $\varphi = 0$ is

$$\chi_0 = \frac{1}{2r}, \qquad T > T_c, \tag{4.142}$$

and for the ordered phase $\bar{\varphi} = \varphi_\pm$ is given by

$$\chi_0 = -\frac{1}{4r}, \qquad T < T_c. \tag{4.143}$$

Note that χ_0 from Eq. (4.143) is the tangent to the equilibrium line AA' in Fig. 4.9(b) at the points A_1 and A_1', whereas χ_0 from Eq. (4.142) is the tangent to the line T_1 in Fig. 4.9(a). Eq. (4.142) shows that the susceptibility at the point O of the line AA' in Fig. 4.9(b) is negative ($T < T_c$) and this indicates the instability of the paraphase solution in this temperature interval. The susceptibility at zero field χ_0 rapidly increases when $T \to T_c$ and $h \to 0$ and diverges at T_c. This type of divergence is a consequence of the choice $u_2 = r = \alpha_0 t$. The formulae (4.142) and (4.143) show that our model describes the Curie–Weiss law Eq. (3.77).

For weak fields h, when h in Eq. (4.135) can be considered as a small perturbation we can write the order parameter $\varphi(h)$ in the following way:

$$\varphi(h) = \varphi(0) + \Delta\varphi(h), \tag{4.144}$$

where

$$\Delta\varphi(h) \ll \varphi(0) \tag{4.145}$$

is a small correction to $\varphi(0)$. Substituting $\varphi(h)$ from Eq. (4.144) in Eq. (4.135) and neglecting the terms of order $O(\Delta\varphi^2)$ we have

$$\Delta\varphi\chi_0 h. \tag{4.146}$$

Now one can check that the condition (4.145) a field to be weak, holds true if only the system is in the ordered state $\varphi(0) \neq 0, r < 0$. Using $\chi_0 \sim 1/|r|$ and $\varphi(0) \sim (|r|/u)^{1/2}$, we can rewrite the inequality (4.145) in the form

$$h \ll h^e = |h_\pm| \sim \left(\frac{|r|^3}{u}\right)^{1/2}. \tag{4.147}$$

For $T > T_c$, $\varphi(0) = 0$, so the condition (4.145) cannot be fulfilled for any h, which means any field $h \neq 0$ to be considered strong. In this case the formula (4.146) stands for $\varphi = \chi_0 h$ — a field-induced order parameter. One may calculate the contribution of the field-induced term $\Delta\varphi(h)$ to the entropy S and the specific heat C_h making use of Eqs. (4.138), (4.139), (4.144), and (4.146).

The region of very strong fields is defined by $|h| \gg h^e$. It is easy to see that in this limit the linear term in φ in Eq. (4.135) can be neglected, and

$$h \approx 4u\varphi^3. \tag{4.148}$$

For $T = T_c$ the Eq. (4.148) is exact. When $\varphi = 0$, $h(\varphi)$ has an inflection point. This function behaves very much like the curve $h(\varphi)$ for $T = T_c$ in Fig. 4.9(a). For strong fields the entropy S will be

$$S = -\frac{\alpha_0}{T_c}\left(\frac{h}{4u}\right)^{2/3}. \tag{4.149}$$

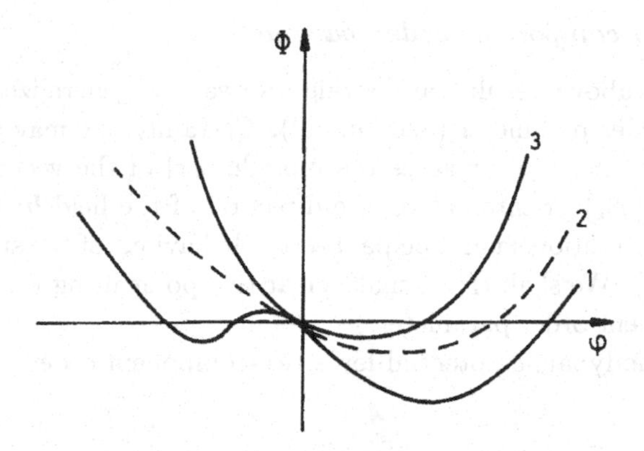

Fig. 4.10 The function $\Phi(\varphi)$ for $T < T_c$ (curve 1), $T = T_c$ (curve 2) and $T > T_c$ (curve 3).

In fact, this is the entropy difference of the states in the presence and without the external field h. As this entropy difference is temperature independent, the specific heat will not depend on the field h.

The shape of the thermodynamic potential $\Phi(\varphi)$ is outlined in Fig. 4.10; cf. Fig. 4.1(b). The weak field limit is presented by the curve 1 for $T < T_c$. It can be shown, making use of Eqs. (4.144) and (4.146) with χ_0 from Eq. (4.143), that $\Phi(\varphi)$ has three extrema for $T < T_c$, provided the condition (4.147) is fulfilled. The absolute minimum on the curve 1 corresponds to the solution of Eq. (4.135) $\varphi(h)$, which obeys the inequality $\varphi h > 0$; see also Eq. (4.134). The maximum on the curve 1 is given by Eqs. (4.146) and (4.143); $\varphi = \Delta\varphi$, $\varphi(0) = 0$. The curves 2 and 3 correspond to the strong field limit. From Eqs. (4.134) and (4.148) follows that in this approximation the potential Φ has a minimum

$$\Phi = -3u|\varphi(h)|^4, \tag{4.150}$$

for

$$\varphi(h) = \left(\frac{h}{4u}\right)^{1/3}. \tag{4.151}$$

The tangent to the curves $\Phi(\varphi)$ at $\varphi = 0$ is always equal to $-h$.

4.10.2 *Two-component order parameter*

Most of the above results are straightforwardly generalized for a n-component order parameter (Section 4.9). Certainly one may always consider φ in the above formulae as the modulus $|\varphi|$ of the vector order parameter $\varphi = \{\varphi_\alpha\}$ oriented along the direction of the field $\boldsymbol{h}$; $\boldsymbol{h} \cdot \boldsymbol{\varphi} \equiv h\varphi$. More special attention should be paid to the behaviour of the susceptibility matrix (4.128). We shall try to make clear this point using the example of a two-component order parameter.

The thermodynamic potential for a two-component order parameter is

$$\Phi = -h_1\varphi_1 - h_2\varphi_2 + r\varphi^2 + u\varphi^4. \tag{4.152}$$

Without loss of generality one can choose for example, the vector $h = (h_1, h_2)$ to be along the 2–nd axis in the order parameter space and, hence, $h = (0, h_2)$. For any $h_2 \neq 0$, the equations of state $(\partial\Phi/\partial\varphi_\alpha) = 0, (\alpha = 1, 2)$, have no solution of the type (φ_1, φ_2) with $\varphi_1 \neq 0$. The result of substituting $h_1 = 0, h_2 = h$ and $\varphi_2 = \varphi$ in Eq. (4.152) is equivalent to the model (4.134). The special choice $h = (0, h_2)$ of the field does not influence the final results, because the general case $h = (h_1, h_2)$ can always be presented in this form through an appropriate rotation in the order parameter space:

$$\varphi_1 = a\varphi'_1 + b\varphi'_2,$$

$$\varphi_2 = -b\varphi'_1 + a\varphi'_2, \tag{4.153}$$

where

$$a^2 + b^2 = 1. \tag{4.154}$$

The quadratic and the quartic terms in Eq. (4.152) are invariant under the transformation (4.153)–(4.154). The field h is transformed to $h' = (h'_1, h'_2)$. The components h'_α are obtained from the relation $h'\varphi' = h\varphi$. Imposing the condition $h'_1 = 0$ we get $b = (h_1/h_2)a$ and after simple calculations the potential (4.152) takes the form

$$\Phi = -h'_2\varphi_2 + r\varphi^2 + u\varphi^4, \tag{4.155}$$

where $h_2' = (h_1^2 + h_2^2)^{1/2}$ and the "primes" of φ_2 and φ have been omitted. The potential (4.155) describes a system in an external field $h = (0, h_2')$.

The reciprocal susceptibility χ_0^{-1} at zero external field is given by the matrices

$$\chi_0^{-1} = 2r \begin{pmatrix} 1 \\ 0 \end{pmatrix}, \qquad T > T_c, \tag{4.156a}$$

and

$$\chi_0^{-1} = 8u \begin{pmatrix} \varphi_1^2 & \varphi_1\varphi_2 \\ \varphi_1\varphi_2 & \varphi_2^2 \end{pmatrix}, \qquad T < T_c, \tag{4.156b}$$

where φ_α $(= \bar{\varphi}_\alpha)$ stands for the stable phase. The expression (4.156a) is a direct generalization of Eq. (4.142). The eigenvalues of χ_0^{-1} for $T < T_c$ are $\lambda_1 = 0$ and $\lambda_2 = -4r$. The eigenvalue $\lambda_1 = 0$ corresponds to the susceptibility of the system toward an external field perpendicular to the vector (φ_1, φ_2). This is *the transverse* (or *perpendicular*) *susceptibility* $(\chi_\perp)$. In our case this susceptibility $(\chi_\perp^0 = 1/\lambda_1)$ is infinite (any infinitesimal transverse field will produce a finite variation of the order parameter direction). This effect is related to our previous conclusion that the ordering of an isotropic system is always along the external field h. The eigenvalue $\lambda_2 = -4r$ corresponds to the susceptibility of the system with respect to fields parallel to the vector $\bar{\varphi}$ — *the longitudinal* (or *parallel*) *susceptibility* $(\chi_\parallel)$; cf. Eq. (4.143). The different response of the system to the external field along and perpendicular to the order parameter orientation is a consequence of the symmetry breaking below T_c; see Sections 4.9 and 7.5.

The effect of external field can also be studied within the framework of the φ^6- or φ^8- theories. For example, in the presence of an external field the φ^6- model gives a phase diagram quite similar to that described in Fig. 4.9; see, e.g., Western *et al.* (1978), Gufan and Larin (1979), Gufan (1982), Izyumov and Syromiatnikov (1984), where additional extrema of the thermodynamic potential are investigated.

4.11 Miscellaneous Topics

4.11.1 *Summary of critical exponents within the Landau theory*

The Landau theory gives definite predictions about the thermodynamic behaviour near the critical points of the ordinary phase transitions of second

Table 4.1 Thermodynamic behaviour near critical and tricritical points

Critical point; $t = \dfrac{T - T_c}{T_c}$	Tricritical point; $t = \dfrac{T - T_t}{T_t}$
$\lvert\varphi\rvert \sim (-t)^{1/2}, \quad \beta = \dfrac{1}{2}$	$\lvert\varphi\rvert \sim (-t)^{1/4}, \quad \beta = \dfrac{1}{4}$
C_h - finite jump ($\alpha = 0$)	$C_h \sim (-t)^{1/2}$
$\chi_0 \sim \lvert t\rvert^{-1}, \quad \gamma = 1$	$\chi_0 \sim \lvert t\rvert^{-1}, \quad \gamma = 1$
$\varphi_c \sim h^{1/3}, \quad \delta = 3$	$\varphi_t \sim h^{1/5}, \quad \delta = 5$

order. We have seen that predictions on the basis of this theory can be made for certain multicritical points by extending the Landau expansion to the order in φ higher than fourth. So within the standard Landau theory of uniform order parameter field φ one can always obtain the values of the static critical exponents α, β, γ, and δ. The static critical exponents ν and η are related to the correlation length and correlation function scaling laws and for this reason the calculation of this exponents requires a generalization $[\varphi \to \varphi(x)]$ of the theory by including spatial (x-) variations of the order parameter field (see Chapters 6 and 7). The calculation of the dynamical critical exponent $\hat{z}$ given by Eq. (1.16) could be performed by the extension of the theory to a time-dependent order parameter, $\varphi(t)$. For this aim one may use, for example, the Landau- Khalatnikov Eq. (1.17).

In Table 4.1 we summarize the results obtained in the preceding sections for the thermodynamic behaviour near critical and tricritical points. In this table we show the temperature dependence of the modulus $|\varphi|$ of the order parameter, the specific heat C_h (at $h = 0$), the susceptibility χ_0 at zero external field and the h-dependence of the order parameter (φ_c and φ_t) on the critical ($T = T_c$) and tricritical ($T = T_t$) isotherms. It is clear that the critical and tricritical behaviour of φ and C_h are quite different.

4.11.2 *Notion of crossover phenomena. Critical-to-tricritical crossover*

We have demonstrated that the system may be driven from a critical to tricritical behaviour by such variations of the thermodynamic parameters T and Y, that lead to the decrease of the positive parameter $u = u_4$ to zero. These changes of the type of critical behaviour and more generally, of the type of phase transition are often called *crossover phenomena*.

In order to get some notion about the mathematical description of the crossover phenomena we shall briefly outline how the critical-to-tricritical crossover is treated in the framework of the Landau expansion.

The modulus $\bar{\varphi}$ of the order parameter (4.98) within the φ^6-theory can be written in the form

$$\bar{\varphi} = \left(\frac{|r|}{3u_6}\right)^{1/4} \tilde{\varphi}(x), \tag{4.157}$$

where

$$\tilde{\varphi}(x) = \left[-x + \sqrt{x^2 - \text{sign}(r)}\right]^{1/2} \tag{4.158}$$

and $x = u/(3|r|u_6)^{1/2}$ is the dimensionless parameter given by Eq. (4.125). The function $\tilde{\varphi}(x)$ is real-valued for $r < 0$. It is often called *the crossover function*, which describes the equilibrium order parameter in two limiting cases: tricritical ($x \to 0$) and critical ($x \to \infty$). Besides, it is possible with the help of this function to consider the gradual crossover between these limiting cases. Clearly, $\tilde{\varphi}(0) = 1$ for $r < 0$ and $x = 0$ (i.e., $u = 0$). Then $\bar{\varphi}$ from Eq. (4.157) coincides with the order parameter below the tricritical point ($T < T_t$); cf. Eq. (4.120). When $u \neq 0$ and $|u| < (3|r|u_6)^{1/2}$, which corresponds to large r, the variable x is small; so we can expand $\tilde{\varphi}(x)$ in powers of x :

$$\tilde{\varphi}(x) = 1 - \frac{1}{2}x + \frac{1}{8}x^2 + O(x^3). \tag{4.159}$$

The x-terms give the corrections to the tricritical behaviour. The asymptotic tricritical behaviour can be observed when the function $\tilde{\varphi}(x) = \tilde{\varphi}(0)$.

When $u \neq 0$ and r tends to zero, x is very large and we can do the following expansion:

$$\tilde{\varphi}(x) = \frac{1}{\sqrt{2x}}\left[1 + O\left(\frac{1}{x^2}\right)\right]. \tag{4.160}$$

Neglecting the correction terms we obtain from Eq. (4.160) and Eq. (4.157) that $\bar{\varphi} = (-r/2u)^{1/2}$ as given by the Eq. (4.61) for ordinary critical points.

Instead of the expression (4.157) one may equivalently use

$$\tilde{\varphi}(x) = \frac{1}{\sqrt{2x}}\left[1 + O\left(\frac{1}{x^2}\right)\right]. \tag{4.161}$$

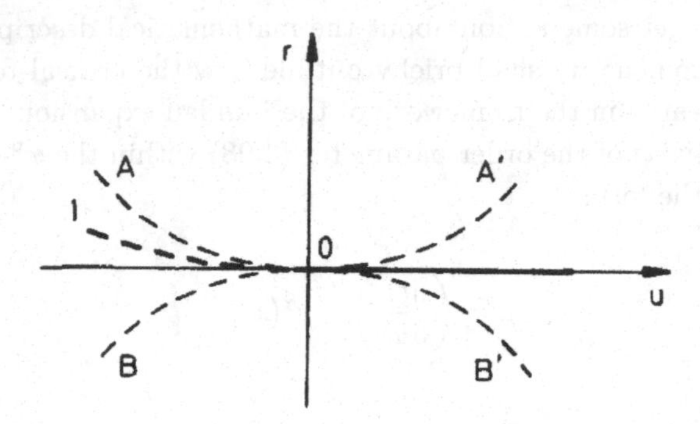

Fig. 4.11 Crossover regions in the (u, r) diagram of the φ^6-model.

The expansion of the "new" crossover function $(2x)^{1/2}\tilde{\varphi}(x)$ is given by Eq. (4.124).

The borderline between the critical and tricritical regimes can be defined by $x^2 = 1$ or

$$u^2 = 3|r|u_6. \tag{4.162}$$

This equation determines two parabolas in the (u, r) diagram of the system, one for $r > 0$ and another for $r < 0$ (Fig. 4.11); cf. Fig. 4.8(a), where the line $2 - OP$ coincides with the branch AO of the parabola AOA' $(r > 0)$.

The tricritical regime, for example the dependence Eq. (4.120) should be experimentally observed along trajectories in the (u, r) diagram that approach the tricritical point O and lie below the parabola BOB' in the ordered phase $(\bar{\varphi} \neq 0)$. The trajectories (thermodynamic processes) that approach the point O by ordered thermodynamic states lying above the parabola BOB', that is tangentially to the transition lines $1 - O$ and $r = 0$, obey the condition $x \gg 1$ and, hence, will not exhibit tricritical behaviour. Surely, the parabola BOB' traces out quite approximately the two limiting regimes. In the intermediate region $u_2^2 \sim 3|r|u_6$, near the line BOB' the system does not follow well established critical or tricritical behaviour (intermediate behaviour). The states above the parabola AOA' and in the regions between the lines $1 - O - A$ and $u - O - A'$ are disordered $(\bar{\varphi} = 0)$. The present theory cannot tell us anything about the critical behaviour in the domain of the disordered phase. The parabola AOA' also plays a role of a bordering line between the critical and tricritical regimes, which can appear after taking into account the fluctuations (for the effect of fluctuations (see Chapters 6 and 7).

The thermodynamic potential in a zero external field ($h = 0$) from Eq. (4.104) after simple calculation can be rewritten in the form:

$$\Phi(\bar{\varphi}) = \frac{|r|\bar{\varphi}^2}{3}\left[2(\text{sign}(r) + x\tilde{\varphi}^2(x)\right]. \tag{4.163}$$

Substituting $\bar{\varphi}^2$ from Eq. (4.157),

$$\Phi(\bar{\varphi}) = -\frac{2|r|^{3/2}}{3(3u_6)^{1/2}}\widetilde{\Phi}(x) \tag{4.164}$$

the crossover function $\widetilde{\Phi}$ is obtained

$$\widetilde{\Phi}(x) = -\left[(\text{sign}(r)\tilde{\varphi}^2(x) + \frac{1}{2}x\tilde{\varphi}^4(x)\right]. \tag{4.165}$$

When $x = 0$ and $\tilde{\varphi}(0) = 1$, the thermodynamic potential below the tricritical point($r = \alpha_0(T - T_t) < 0$) is

$$\Phi_t = -\frac{2}{3}\left(\frac{|r|^3}{3u_6}\right)^{1/2}. \tag{4.166}$$

When $x \gg 1$, the potential (4.69) for the ordinary critical point is received, provided $r < 0$.

The effect of the external field h can be taken into consideration, if we know the dependence of the equilibrium order parameter $\bar{\varphi}$ on h: $\bar{\varphi} \equiv \varphi(h)$. It is not easy to calculate this dependence in the φ^6-theory but in the limits of weak and strong fields (Section 4.10) the problem can be treated analytically. Let us denote the equilibrium thermodynamic potential Φ for a zero field by $\Phi[\varphi(0)]$. For nonzero h, the equilibrium potential will be:

$$\Phi_h = -h\varphi(h) + \Phi[\varphi(h)]. \tag{4.167}$$

In the weak field limit the above expression can be simplified using the Eqs. (4.144) and (4.146) for $\varphi(h)$. After neglecting terms of order $O(\Delta\varphi^2)$ and $O(h^2)$, Eq. (4.167) becomes:

$$\Phi_h = -h\varphi(0) + \Phi[\varphi(0)]. \tag{4.168}$$

In Eq. (4.168) we put $\Phi[\varphi(0)]$ given by Eqs. (4.164), (4.165) and $\varphi(0) = \pm\varphi$ given by Eqs. (4.157), and (4.158). The result is

$$\Phi_h = -\frac{2}{3}\left(\frac{|r|^3}{3u_6}\right)^{1/2} \widetilde{\Phi}_h(x, y), \tag{4.169}$$

where the crossover function $\widetilde{\Phi}_h$ depends on two variables:

$$\widetilde{\Phi}_h = \widetilde{\Phi}(x) \pm \frac{3}{2} y\tilde{\varphi}(x); \tag{4.170}$$

y is connected with the (weak) external field h through:

$$y = h\left(\frac{3u_6}{|r|^5}\right)^{1/4}. \tag{4.171}$$

From the above three equations we can get the potential $\Phi_h^c = \Phi_h(x \to 0)$ in the critical regime

$$\Phi_h^c \approx \mp\left(\frac{|r|}{2u}\right)^{1/2} h - \frac{r^2}{4u} \tag{4.172}$$

and the potential $\Phi_h^t = \Phi_h(x \gg 1)$ in the tricritical regime

$$\Phi_h^t \approx \mp\left(\frac{|r|}{3u_6}\right)^{1/4} h - \frac{2}{3}\left(\frac{|r|^3}{3u_6}\right)^{1/2}. \tag{4.173}$$

It is obvious from Eqs. (4.172) and (4.173) with $h = 0$ or Eqs. (4.166) and (4.69) that

$$\frac{\Phi^c(t)}{\Phi^t(t)} \sim |t|^{\phi}, \qquad \phi = \frac{1}{2}. \tag{4.174}$$

The exponent ϕ is called *the crossover exponent.* Within the present theory it is equal to $1/2$ but this value may change owing to the fluctuation effects. One can check that the change of the order parameter $\varphi(0)$ in a zero external field from the critical to tricritical behaviour is given by a relation like (4.174) and this transformation is described by the so-called *shift exponent* ψ. The parameter y from Eq. (4.171) is another example, for which an equation like Eq. (4.174) can be written. This variable is very suitable for the description of the field effect near the tricritical point and in this sense it plays the same role as the parameter h^e in Eqs. (4.147) and (4.137) for the description of ordinary critical points. The ratio of these two parameters can also be represented by a relation like Eq. (4.174) and the crossover exponent $\phi = 1/2$. Certainly, in most of the cases the crossover exponent ϕ is equal to the shift exponent ψ, which gives the shift of the transition line $T_c(y)$ to $T_1(y)$ for $u < 0$; see Eq. (4.107). It is obvious from Eqs. (4.107) and (4.108) that

$$(T_1 - T_c) \sim (T_2 - T_c) \sim |u|^{1/\psi}, \qquad \psi = \frac{1}{2}. \tag{4.175a}$$

The difference $(T_1 - T_t)$ can also be represented by this exponent:

$$(T_1 - T_t) \sim |u|^{1/\psi}. \tag{4.175b}$$

4.11.3 *Anisotropy*

The symmetry of the ordered phase is identical to the symmetry of the equilibrium value $\bar{\varphi}$ of φ in the n-dimensional space of the order parameter components $\{\varphi_\alpha\}$ (Section 4.9). The φ^4-, φ^6- and φ^8- theories, which possess a rotational invariance are not always adequate to the description of phase transitions in systems with a multi-component ($n \geq 2$) order parameter $\varphi = \{\varphi_1, \ldots, \varphi_n\}$; as in Section 4.9 we assume that the order parameter has real components φ_α. For example, the potential

$$\Phi = r\varphi^2 + u\varphi^4 + v\sum_\alpha \varphi_\alpha^4 \tag{4.176}$$

includes a term, which is anisotropic in the order parameter space. The anisotropy energy given by the third term in Eq. (4.176) describes a type of anisotropy often called *cubic anisotropy.* In ferromagnets $\varphi = M$ is a vector and such anisotropy term takes into account the cubic symmetry of the crystal. Depending on the values of the coefficients u and v, the magnetization vector $M = (M_x, M_y, M_z)$ may preferably orientate itself either along the edges or along the body diagonals of the elementary cubic cell; see, e.g., Landau and Lifshitz (1960). When the number n of the order parameter components φ_α is greater than three, this anisotropy is often called *hyper-cubic anisotropy.* The different types of anisotropy in the order parameter space admit different φ^4-terms in the thermodynamic potential Φ. All these terms are invariant under these transformations (translations, rotations, reflections) that belong to the symmetry group of the ordered phase ($\bar{\varphi} \neq 0$). They are often called *independent invariants* (or *Landau invariants*). The group symmetry treatment of the quartic and higher order invariants is discussed by Landau and Lifshitz (1980) and in specialized books as, e.g., Tolédano and Tolédano (1987).

Another example is the case with ($n = 4$)-component order parameters. For them the general form of $\Phi(\varphi)$ is

$$\Phi = r\varphi^2 + u\varphi^4 + v\sum_{\alpha=1}^{4}\varphi_\alpha^4 + v_1(\varphi_1^2\varphi_2^2 + \varphi_3^2\varphi_4^2) \\ + v_2(\varphi_1^2\varphi_3^2 + \varphi_2^2\varphi_4^2) + v_3\varphi_1\varphi_2\varphi_3\varphi_4. \tag{4.177}$$

In real systems not all the invariants included in Eq. (4.177) are allowed from symmetry considerations. Landau expansions like (4.177) are widely used in the studies of structural phase transitions; see, e.g., Cowley (1980), Gufan (1982), Izyumov and Syromiatnikov (1984). Models with anisotropy quartic invariants will be discussed in Sections 9.6.1–9.6.4. It is often necessary to apply sophisticated group theoretical investigations in order to obtain the appropriate invariants of order higher than φ^4; see, e.g., Deonarine and Birman (1983). The inclusion of higher-order terms in the Landau expansion may lead to essential effects in the phase diagram of the system (Galam and Birman, 1983; Galam, 1985).

4.11.4 *Coupling to strains*

To take account of strain effects and secondary order parameters (Section 3.8) new terms must be added to the thermodynamic potential Φ. The elastic strains are often essential to the description of structural phase transitions. Within the Landau theory the potential Φ will depend on the strain variable η and the corresponding expansion will be

$$\Phi = r\varphi^2 + u\varphi^4 + A\varphi^2\eta + B\eta^2 + \dots, \tag{4.178}$$

where A is the lowest order coupling between the order parameter φ and η. Minimizing Φ with respect to both φ and η one obtains what the strain effect is on the equilibrium value of φ. Let us first apply the condition $(\partial\Phi/\partial\eta) = 0$ for "a stress-free" material. It yields:

$$\bar{\eta} = -\frac{A}{2B}\varphi^2. \tag{4.179}$$

Substitution of the equilibrium value $\bar{\eta}$ of η in Eq. (4.178) gives the modified φ^4-potential

$$\Phi = r\varphi^2 + \left(u - \frac{A^2}{4B}\right)\varphi^4. \tag{4.180}$$

Now, results from the usual φ^4-theory can be applied but instead of u there will be a new parameter

$$u_{\text{eff}} = u - \frac{A^2}{4B}. \tag{4.181}$$

The strain produces a shift of the coefficient u and as a consequence a change of the order of the phase transition may occur; for example, from second ($u > 0$) to first ($u_{\text{eff}} < 0$) order. These effects will be more complicated in systems with multi-component order parameters; for details, see Devonshire (1954), Slonczewski and Thomas (1970), and Cowley (1980).

4.11.5 *Validity and further development of the Landau theory*

The original Landau theory was created for the second order phase transitions within the expansion of the thermodynamic potential to the fourth order in φ. We have seen how this theory is extended to describe different types of first order transitions and multicritical points. It is difficult to present all applications of this approach to the description of phase transitions because they are numerous and affect almost every system, which undergoes a phase change.

There are systems as, for example some types of superconductors where the Landau theory gives excellent results. There are however cases, for which the theory loses its validity in a close vicinity of the critical point or far from it. Note that the expansion in powers of φ implies that this quantity is small in some sense. The comparison of the sixth order term $u_6\varphi^6$ with the fourth order term $u\varphi^4$ near the ordinary critical point, where $\varphi^2 = (-r/2u)$, shows that the former is irrelevant in the temperature interval

$$\frac{T_c - T}{T_c} \ll \frac{2u^2}{\alpha_0 u_6} \tag{4.182}$$

below T_c. So the theory is well-based in some region near T_c. The validity of the theory can be extended to lower temperatures, provided $2u^2 \gg \alpha_0 u_6$. For $T = 0$, the Landau theory gives the order parameter $\varphi_0^2 = (\alpha_0/2u_0)$, with $u_0 = u(T = 0)$; see Eq. (4.61) and Eq. (4.68). Then the ratio (φ^2/φ_0^2) below T_c is

$$t = \frac{T_c - T}{T_c} \ll 1, \tag{4.183}$$

if $u_0 \sim u$. However, very close to the critical point the theory may not be correct. This is due to the instability of the system ($r \sim 0$) with respect to fluctuations of the order parameter. It should be generalized in a way to include the fluctuation effects (Chapter 6).

Until now we have discussed the original variant of the theory where the ordering φ is spatially uniform, i.e., the order parameter φ does not depend on the spatial point x. The stability of such uniform ordering with respect to spatial variations of the order parameter has been investigated by Lifshitz (1941, 1944), whose original arguments are summarized in the book of Landau and Lifshitz (1980); see also Tolédano and Tolédano (1987). Here we shall restrict our discussion to the origin of the spatial dependence. It may arise in the equilibrium order parameter $\bar{\varphi}$, if only this parameter is coupled to other spatially dependent macroscopic variables (modes) like nonuniform strains or nonuniform external magnetic field. The systems, we describe in the phase transition theory are represented by the order parameter φ or, in other words, each system is identified with φ. It is then natural to assume that the thermodynamic potential $\Phi(\varphi)$ may describe nonuniform stable states, if the system (φ) is placed in a nonuniform medium (external fields). Although the stable state $\bar{\varphi}$ of a uniform system is uniform, the nonequilibrium order parameter $\varphi = \bar{\varphi} + \delta\varphi$ may exhibit x-dependence due to the fluctuations $\delta\varphi = \delta\varphi(x)$. We shall come back to the requirements of stability for nonuniform states (Lifshitz criterion) in Section 9.7.

Another problem is connected with the possible modification of the Landau assumptions about the form of the coefficients u_l, $l \geq 2$. In order to extend the applicability of the theory near the critical point T_c some authors have proposed a general form of the coefficients u_2 and u_4. Then the treatment of the phase transition is carried out within a generalized Landau expansion; for details, see Luban (1976), and Ginzburg and Sobyanin (1976, 1988). The more general form of the expansion coefficients is obtained either from phenomenological considerations or by successive account for the fluctuation effects.

Chapter 5

Statistical Mechanics and Mean-Field Approximation

5.1 Partition Function

Our task now is to develop the phase transition theory through the methods of statistical mechanics and to relate the phase transition properties to the inter-particle interactions. This problem has been briefly discussed in Chapter 1 where we have mentioned some general aspects of the statistical foundations of the phase transition theory. In this Chapter we shall consider classical models of phase transitions. The treatment of quantum correlations and their influence on the phase transition properties is postponed to Chapter 9.

We assume that the reader is well acquainted with the connection of the thermodynamic stability theory with the properties of the Gibbs distribution in statistical physics. In both cases the stability properties of an isolated macroscopic system in equilibrium result from the requirement of the maximal entropy S. The entropy approach would lead us to the known microcanonical distribution in statistical mechanics; see, e.g., Landau and Lifshitz (1980), Ma (1985), Münster (1969) or Isihara (1971). However this statistical distribution is inconvenient for practical calculations. This reason as well as the explanations given in Section 4.2 direct our considerations to statistical ensembles at fixed (T, Y, φ) or (T, Y, h).

According to our convention in Section 4.2, the parameters T, Y, and φ are the natural variables of the Helmholtz free energy F and, correspondingly, the statistical ensemble at fixed (T, Y, φ) is given by the Gibbs canonical distribution, which relates the Hamiltonian $\mathcal{H}$ of the system to the distribution function $w(\mathcal{H})$. Within this ensemble one has to apply the constraint of a fixed order parameter φ. In order to avoid this we shall introduce the statistical distribution at fixed T, Y and external field h conjugate to φ (Section 5.2).

In each of these cases we are faced with the problem to connect the microscopic properties of the system, expressed by the energy of the microscopic degrees of freedom, with the thermodynamic behaviour at several fixed macroscopic parameters. Microscopically, the energy of the system is expressed by the Hamiltonian function of the microscopic degrees of freedom — *the Hamiltonian.* The statistical mechanics relates the Hamiltonian to the thermodynamic functions.

5.1.1 *Microscopic variables*

The first task is to pick out among all microscopic degrees of freedom those variables of the Hamiltonian that are directly related with the formation of the macroscopic ordering φ. In other words we have to find out the microscopic variables, which being averaged over the appropriate statistical distribution, give the density of the order parameter φ of the system. Thus we can write the Hamiltonian of the system only in terms of these microscopic degrees of freedom, denoted by $\sigma_1, \ldots, \sigma_N$; $N \gtrsim 10^{23}$, but in various small systems (including nano-size systems) one may consider $N \gtrsim (10^2)^d$, where d is the spatial dimensionality (see, e.g., Uzunov and Suzuki (1994)). All other microscopic degrees of freedom can be neglected. One has to accept at the beginning this assumption and thereby to take easy the numerous approximations of statistical mechanics.

In ferromagnets the quantity σ_i is relevant to the spin of an atom i (in classical approximation). The order parameter φ, that is the magnetization $\boldsymbol{M}$, may result from the electron motion and from the presence of electron spins (neglecting the small contribution from the nuclear spin). If the electrons can be considered as localized near the atoms of the crystal lattice, it is convenient to define the microscopic variable σ_i as proportional to the localized spins (or magnetic moments) of the atoms. The microscopic degrees of freedom of a classical gas are the generalized coordinates q_i and the generalized momenta p_i of the particles.

The generalized coordinates q_i are in some way connected with the nonequilibrium volume $V = V(q_i)$ of the system and, hence, one may expect $V(q_i)$ to be related to the equilibrium volume V and thus the difference $(V_g - V_l) \sim \varphi$ between the specific volumes of the gas and the liquid near the gas–liquid transition point. So we shall consider the nonequilibrium order parameter φ as a function of $\sigma = \{\sigma_i\}$: $\varphi = \varphi(\sigma)$. In the bigger part of this Chapter, $\varphi(\sigma)$ and the conjugate external field h are scalar quantities but all results can be straightforwardly generalized for n-component order parameters: $\varphi = \{\varphi_\alpha, \alpha = 1, \ldots, n\}$ with conjugate vector fields $h = \{h_\alpha, \alpha = 1, \ldots, n\}$.

At this stage of consideration we do not need to know the specific nature of the set $\sigma = \{\sigma_i\}$ of microscopic variables. It is however convenient to assume that the suffix i indicates a given site ($i \equiv \boldsymbol{r}_i$) in the volume of the system. The microscopic degrees of freedom σ_i and their relation with the order parameter $\varphi(\sigma)$ will be further discussed in Sections 5.3 and 5.4.

5.1.2 *Canonical distribution*

The next task is to express the energy of the system through the microscopic variables σ_i, that is the Hamiltonian $\mathcal{H}(\sigma)$. If we know the Hamiltonian, we can write down the Gibbs distribution function

$$w(\sigma) = \frac{1}{\mathcal{Z}} e^{-\mathcal{H}(\sigma)/T}, \tag{5.1}$$

where we use energy units, in which the Boltzmann constant k_B is equal to unity ($k_\mathrm{B}T \to T$). In Eq. (5.1)

$$\mathcal{Z} = \mathrm{Tr} e^{-\mathcal{H}(\sigma)/T} \tag{5.2}$$

is *the canonical partition function.* The symbol Trace (Tr) means the summation (or integration) over all possible values of the degrees of freedom $\sigma = \{\sigma_i\}$. Obviously

$$\mathrm{Tr}[w(\sigma)] = 1. \tag{5.3}$$

Now we shall recall some useful relations. *The statistical average* $\langle A \rangle$ of any function $A(\sigma)$ of the microscopic variables σ_i is defined by

$$\langle A \rangle = \mathrm{Tr}(Aw) = \frac{1}{\mathcal{Z}} \mathrm{Tr}\left[A(\sigma) e^{-\mathcal{H}(\sigma)/T}\right]. \tag{5.4}$$

$\langle A \rangle$ is also called *the mean value* of $A(\sigma)$. As we can see the mean value $\langle A \rangle$ of any function $A(\sigma)$ of σ is a functional of the distribution function $w(\sigma)$, that is of the Hamiltonian $\mathcal{H}(\sigma)$. For example the mean value of the Hamiltonian $\mathcal{H}(\sigma)$ is the internal energy U,

$$U = \langle \mathcal{H}(\sigma) \rangle = \frac{1}{\mathcal{Z}} \mathrm{Tr}\left(\mathcal{H} e^{-\mathcal{H}/T}\right). \tag{5.5}$$

The equilibrium entropy S is

$$S = -\langle \ln w \rangle = -\mathrm{Tr}(w \ln w). \tag{5.6}$$

As $0 \leq w \leq 1$, S is always nonnegative. Taking the logarithm of Eq. (5.1) we obtain

$$S = \frac{U - F}{T}, \tag{5.7}$$

where

$$F = -T \ln \mathcal{Z} \tag{5.8}$$

is the equilibrium Helmholtz free energy.

The origin of the temperature dependence of the equilibrium potential $F = F(T, Y, \varphi)$ is clear enough. The additional variable(s) Y could represent the number of particles (often equal to the number N of the degrees of freedom), the volume V or parameters describing the microscopic interparticle interactions (the last are included in $\mathcal{H}(\sigma)$).

The dependence of F on the order parameter φ remains ambiguous because we have not established explicitly what the condition of fixed φ means. The nonequilibrium parameter φ is a function of the microscopic state (configuration of values $\{\sigma_i\}$ of σ_i) like the Hamiltonian $\mathcal{H}(\sigma)$ and it is often necessary to know the actual dependence $\varphi(\sigma)$.

In the present case, however, the constraint $\varphi(\sigma) = const$ is imposed on the system. For a statistical ensemble at fixed (T, Y, φ) the allowed values of the microscopic degrees of freedom lie on the manifold (hypersurface), defined by $\varphi(\sigma) = \bar{\varphi}$, where we have denoted the constant value of φ by $\bar{\varphi}$. The summation (or integration) over the possible microscopic states σ is carried out on the "surface" $\varphi = \bar{\varphi}$, which can be mathematically written as

$$\mathrm{Tr}\,(Aw) \to \mathrm{Tr}\left\{\delta\left[\bar{\varphi} - \varphi\left(\sigma\right)\right] Aw\right\}, \tag{5.9}$$

where δ is Kronecker's symbol when φ is a discrete variable and Dirac's δ-function when the variable φ is continuous. Now the relation (5.8) becomes

$$F = -T \ln \mathcal{Z}\left(T, Y, \bar{\varphi}\right) \tag{5.10}$$

with

$$\mathcal{Z}(\bar{\varphi}) = \mathrm{Tr}\left\{\delta\left[\bar{\varphi} - \varphi(\sigma)\right] e^{-\mathcal{H}(\sigma)/T}\right\}. \tag{5.11}$$

5.1.3 *Thermodynamic relations*

If the function $F(T, Y, \bar{\varphi})$ is known, the entire thermodynamics of the system is available. For example the entropy S can be obtained from Eqs. (5.5), (5.7), and (5.8) or directly from the relation $S = -(\partial F/\partial T)$ and Eq. (5.8). The specific heat C_φ at fixed $\varphi = \bar{\varphi}$ is $C_\varphi = -T(\partial^2 F/\partial T^2)$; equivalently, it can be obtained from $C_\varphi = (\partial U/\partial T)_\varphi$ and Eq. (5.5). The result is:

$$C_\varphi = \frac{1}{T^2}\left[\langle \mathcal{H}^2\rangle - \langle \mathcal{H}\rangle^2\right] = \frac{1}{T^2}\langle (\mathcal{H} - \langle \mathcal{H}\rangle)^2\rangle. \tag{5.12}$$

As we shall see later (Section 5.3) this expression gives the connection between C_φ and the irreducible energy–energy correlation function, provided $\mathcal{H}(\sigma)$ can be written as a sum (or integral) over the Hamiltonian density $\mathcal{H}_i(\sigma)$. In fact, if the Hamiltonian $\mathcal{H}$ can be represented as a sum

$$\mathcal{H}(\sigma) = \sum_i \mathcal{H}_i(\sigma), \tag{5.13}$$

C_φ becomes

$$C_\varphi = \frac{1}{T^2}\sum_i \langle (\mathcal{H}_i - \langle \mathcal{H}_i\rangle)(\mathcal{H}_j - \langle \mathcal{H}_j\rangle)\rangle. \tag{5.14}$$

Another useful formula for C_φ is derived by the formal replacement $\mathcal{H} \to \lambda\mathcal{H}$ so that

$$C_\varphi = \left[\frac{\partial^2}{\partial \lambda^2}\ln \mathcal{Z}\right]_{\lambda=1}. \tag{5.15}$$

The relation $h = (\partial F/\partial \bar{\varphi})_T$ yields the field h conjugate to $\bar{\varphi}$. It can alter with the variations of the "fixed" surface ($\varphi = \bar{\varphi}$), that is when the external (boundary) conditions are changed.

5.2 Fixed External Field and Order Parameter Fluctuations

The δ-constraint in Eqs. (5.9) and (5.11) involves calculational difficulties and the statistical studies are frequently performed for ensembles at fixed (T, Y, h). We shall at first consider the usual case of a homogeneous system, when φ is a continuous variable.

5.2.1 *Continuous variable* φ

The main step is to modify the partition function (5.11), making the substitution

$$\delta\left[\bar{\varphi}-\varphi(\sigma)\right] \longrightarrow e^{-\lambda[\bar{\varphi}-\varphi(\sigma)]}, \tag{5.16}$$

where $\lambda > 0$. The replacement (5.16) gives rise to a new distribution function $w'(\sigma)$ and, hence, to another statistical ensemble. Our aim is to present avoiding the rigorous mathematical treatment the relation between the canonical ensemble at fixed $\bar{\varphi}$ and the ensemble at a fixed homogeneous field h. This procedure is quite similar to the transformation $U \rightarrow T$ from the microcanonical ensemble $(U, Y, \bar{\varphi})$ to the canonical one $(T, Y, \bar{\varphi})$; for details, see, e.g., Münster (1969).

The new partition function will be

$$\mathcal{Z}'(\bar{\varphi}) = \mathrm{Tr}\left\{e^{-\mathcal{H}(\sigma)/T-\lambda[\bar{\varphi}-\varphi(\sigma)]}\right\}. \tag{5.17}$$

Using the identity

$$e^{-\lambda[\bar{\varphi}-\varphi(\sigma)]} = \int_{-\infty}^{+\infty} d\varphi' \delta\left[\varphi'-\varphi(\sigma)\right] e^{-\lambda\left[\bar{\varphi}(\sigma)-\varphi'\right]} \tag{5.18}$$

and bearing in mind that $\bar{\varphi}$ and φ' do not depend on σ, we obtain

$$\begin{aligned}\mathcal{Z}'(\bar{\varphi}) &= \int_{-\infty}^{+\infty} d\varphi' e^{-\lambda(\bar{\varphi}-\varphi')}\mathcal{Z}(\varphi') \\ &= \int_{-\infty}^{+\infty} d\varphi' e^{-\lambda(\bar{\varphi}-\varphi')-F(\varphi')/T}. \end{aligned} \tag{5.19}$$

The function

$$F'(\varphi') = F(\varphi') + \lambda T(\bar{\varphi}-\varphi') \tag{5.20}$$

looks like the free energy (4.10). Choosing the undetermined multiplier λ to be equal to h/T, where h is the equilibrium field, the minima of $F'(\varphi')$ are found from the equation of state $(\partial F'/\partial\varphi')_{T,Y} = 0$. Expanding $F'(\varphi')$ in terms of the fluctuations $\delta\varphi' = (\varphi' - \bar{\varphi}')$ we have

$$\mathcal{Z}'(\bar{\varphi}) = e^{-F(\bar{\varphi})/T} \int_{-\infty}^{+\infty} d\delta\varphi \exp\left\{-\frac{u_2(\bar{\varphi})}{T}(\delta\varphi)^2 + O\left[\delta\varphi)^3\right]\right\}, \tag{5.21}$$

where the second-order expansion coefficient $u_2(\varphi)$ is consistent with the formula Eq.(4.31). In Eq. (5.21) the "prime" of φ' and $\delta\varphi'$ has been omitted and the integration variable has been changed ($\varphi \rightarrow \delta\varphi, d\varphi = d\delta\varphi$).

Note, that here a single spatially uniform thermodynamic variable φ is considered and within this restriction of our consideration we deal with a single fluctuation $\delta\varphi$ for the system as a whole - a single (spatially uniform) fluctuation; alternatively, one may use the term *spatially independent fluctuation.* But having in mind that $\delta\varphi$ may take various values, we may focus on the latter and use the term "fluctuations."

The exponent in the integrand of Eq. (5.21) is precisely equal to the expansion (4.30) applied for the difference $-\Delta\Phi/T$, where

$$\Delta\Phi(\bar{\varphi}, \delta\varphi) = \Phi(\varphi, h) - \Phi[\bar{\varphi}(h)]. \tag{5.22}$$

Here $\Phi(\bar{\varphi}) \equiv \overline{\Phi}(h)$ is given by the Eq. (4.12) and $h = \bar{h}$. So the terms of order $O(\delta\varphi^3)$ can be taken from the expansion (4.30). From Eqs. (5.21) and (5.8) we obtain the modified free energy

$$\widetilde{F}'(\bar{\varphi}) = F(\bar{\varphi}) + F_f(\bar{\varphi}), \tag{5.23}$$

where

$$F_f(\bar{\varphi}) = -T \ln \int_{-\infty}^{+\infty} d\delta\varphi e^{-\Delta\Phi/T} \tag{5.24}$$

is the fluctuation part of $\widetilde{F}'$.

The free energy $\widetilde{F}'$ may coincide with F provided the fluctuation contribution F_f is relatively small ($F_f \ll F$) and can be neglected. Assuming that the main contribution to the integral in Eq. (5.24) comes from the small fluctuations ($\delta\varphi \sim 0$) we can neglect the terms $O(\delta\varphi^3)$; see Eq. (5.21). Then $F_f(\bar{\varphi})$ is approximately given by the simple Gaussian integral:

$$F_f(\bar{\varphi}) \approx -T \ln \int_{-\infty}^{+\infty} dx \exp\left[-\frac{u_2(\bar{\varphi})}{T} x^2\right]; \tag{5.25}$$

see also Eq. (B.1). After the integration we get the relation between F_f and the susceptibility $\chi = 1/2u_2$:

$$F_f(\bar{\varphi}) = -T \ln \left[2\pi T \chi(\bar{\varphi})\right]^{1/2}. \tag{5.26}$$

The extensive quantities F and φ are proportional to the number of particles N and therefore, $\chi^{-1} = \partial^2 F/\partial\varphi^2$ is proportional to N too. In Eq. (5.23), $\widetilde{F}' \sim N$ and $F \sim N$ so that $F_f \approx \ln N$ and can be neglected in the asymptotic limit $N \gg 1$. When $N \to \infty$, $\widetilde{F}'(\bar{\varphi}) = F'(\bar{\varphi})$, and

$$\widetilde{F}'(\bar{\varphi}) = F(\bar{\varphi}). \tag{5.27}$$

As $N \to \infty$ the effect of the fluctuations $\delta\varphi$ vanishes and the system at fixed (T, Y, h) behaves just like that at fixed $(T, Y, \bar{\varphi})$. This result will hold true, if the susceptibility per particle χ/N is not divergent for large N.

For the ensemble at fixed h we have determined $\bar{\varphi}$ as the stationary (saddle) point of the function $F'(\varphi')$, see Eq. (5.21), at $\lambda = h/T$ or, which is the same, as the stationary point of $\Phi(\varphi') = F(\varphi') - h\varphi'$. Now we shall define the statistical distribution and the partition function, which correspond to the Gibbs potential. The Eq. (5.17) can be written in the form

$$\mathcal{Z}(h) = \mathcal{Z}'(\varphi)e^{h\bar{\varphi}/T} = \text{Tr}\left(\exp\left\{-\frac{1}{T}\left[\mathcal{H}(\sigma) - h\varphi(\sigma)\right]\right\}\right). \tag{5.28}$$

This partition function conforms to the Gibbs potential

$$\widetilde{\Phi}(h) = -T\ln\mathcal{Z}(h) = \Phi(\bar{\varphi}) + \Phi_f(\bar{\varphi}). \tag{5.29}$$

Here we have taken into account that $\Phi(\bar{\varphi}) = \overline{\Phi}(h)$ in accordance with Eq. (4.12) for $h = \bar{h}$; $\Phi_f(\bar{\varphi})$ is in fact equal to the fluctuation part $F_f(\bar{\varphi})$ given by Eq. (5.24).

The saddle point potential $\Phi(\bar{\varphi})$ and the total potential $\widetilde{\Phi}(h)$ are not equivalent. Their difference $\Phi_f(\bar{\varphi})$ depends on the fluctuation contribution to the Gibbs potential describing an open towards the exchange of φ system. The saddle point approximation $\widetilde{\Phi}(h) \approx \Phi(h)$ is often identical to the MF approximation (Section 5.8) which comes from another conceptions.

The statistical distribution $w_h(\sigma)$ for the ensemble at fixed h can be found from the relation $\text{Tr}(w_h) = 1$ and Eq. (5.28):

$$w_h(\sigma) = \mathcal{Z}^{-1}(h)\exp\left\{-\frac{1}{T}\left[\mathcal{H}(\sigma) - h\varphi(\sigma)\right]\right\}. \tag{5.30}$$

We shall illustrate our discussion with the example of *the isothermal–isobaric ensemble*; see, e.g., Hill (1956), Kubo (1965), Münster (1969) and Salsburg (1971). For this ensemble the order parameter $\varphi(\sigma)$ will be the nonequilibrium volume $V(\sigma)$ and the field h is equal to the pressure $(-P)$. The distribution function $w(P)$ is

$$w_P(\sigma) = \exp\left\{\frac{1}{T}\left[\widetilde{\Phi}(P) - \mathcal{H}(\sigma) - PV(\sigma)\right]\right\}, \qquad (5.31)$$

where

$$\begin{aligned}\widetilde{\Phi}(P) &= -T\ln \mathcal{Z}(P)\\ &= -T\ln \mathrm{Tr}\left(\exp\left\{-\frac{1}{T}\left[\mathcal{H}(\sigma) + PV(\sigma)\right]\right\}\right).\end{aligned} \qquad (5.32)$$

Now we can write the thermodynamic functions. The potential $\widetilde{\Phi}(P)$ will coincide with the saddle point potential $\Phi[V(P)]$ when the fluctuations $\delta V = V(\sigma) - V$ are neglected; $V = \bar{V}$.

5.2.2 *Discrete variable* φ

When φ is a discrete variable δ in Eq. (5.16) and Eq. (5.18) is the Kronecker symbol and the integration in Eq. (5.18) is replaced by the sum over φ'. The equations for $\mathcal{Z}'(\bar{\varphi})$ are derived absolutely in the same way as Eq. (5.19). Therefore, the partition function for a discrete φ will be:

$$\mathcal{Z}'(\bar{\varphi}) = e^{-h\bar{\varphi}/T}\sum_{\varphi'}\exp\left\{-\frac{1}{T}\left[F(\varphi') - h\varphi'\right]\right\}, \qquad (5.33)$$

where we have set $\lambda = h/T$. If we denote the discrete values, which φ takes by $\varphi_1, \ldots, \varphi_\nu$, Eq. (5.33) can be rewritten in the form

$$\mathcal{Z}(h) = \mathcal{Z}'(\bar{\varphi})e^{h\bar{\varphi}/T} = \sum_{\nu}\exp\left\{\frac{1}{T}\left[h\varphi_\nu - F(\varphi_\nu)\right]\right\}. \qquad (5.34)$$

Here we define a new partition function $\mathcal{Z}(h)$.

If we set $\varphi_\nu = N$ and $h = \mu$ — the chemical potential, we shall obtain the partition function $\mathcal{Z}(T, Y, \mu)$ of *the grand canonical ensemble:*

$$\mathcal{Z}(\mu) = \sum_{N} e^{[\mu N - F(N)]/T} = \mathrm{Tr}\left(\exp\left\{\left[\mu N(\sigma) - \mathcal{H}(\sigma)\right]/T\right\}\right). \qquad (5.35)$$

The corresponding grand canonical distribution will be

$$w_\mu(\sigma) = \exp\left\{\left[\widetilde{\Omega}(\mu) - \mathcal{H}(\sigma) + \mu N(\sigma)\right]/T\right\}, \tag{5.36}$$

where $\widetilde{\Omega}(\mu)$ is the grand canonical potential

$$\widetilde{\Omega}(\mu) = -T \ln \mathcal{Z}(\mu). \tag{5.37}$$

$\widetilde{\Omega}(\mu)$ is the sum of the stationary potential $\Omega[\mu, \bar{N}(\mu)]$ and the fluctuation part $\Omega_f(\mu, \bar{N})$; here $\bar{N}$ is the mean value of $N = N(\sigma)$. The connection between $\widetilde{\Omega}(\mu)$ and the Helmholtz function $F(N)$ can be elucidated with the help of the saddle point method; see, e.g., Isihara (1971) and Pathria (1972). In our notations Ω is a function of T, Y, and μ. For usual fluids there are no additional work coordinates, so $Y = V$.

Let us come back to Eq. (5.34). The sum over ν can be evaluated by heuristic arguments. We shall present them because they are very useful for our further considerations. The macroscopic quantity φ_ν can be considered as a discrete variable but the difference $(\varphi_{\nu+1} - \varphi_\nu)/\varphi_\nu$ can always be considered negligible. In fact each φ_ν depends on the set of microscopic variables σ_i, the values of which are of the order φ_ν/N. For example, if σ_i represents the magnetic moment m_i of an atom at site i, φ_ν will be the nonequilibrium magnetization $M_\nu(\sigma) = \sum_i m_i \sim Nm_i \sim 10^{22} m_i$.

The variation of the discrete magnetic moments m_i at the sites i of the macroscopic system leads to a quasicontinuous set of values $\mathcal{M}_\nu$ for the instantaneous magnetization $\mathcal{M}(m_i)$. If $\sigma_i = v_i$ represents the specific volume of a particle at the site i in the volume of a gas, $\varphi_\nu = V_\nu(v_i)$, where $V_\nu(v_i)$ is one possible value of the nonequilibrium volume $V(v_i)$. Then $V(v_i) = \sum_i v_i$ will also be a quasicontinuous variable provided $V \gg v_i$.

Since our discussion is intended to describe macroscopic systems we can always consider the macroscopic functions $A(\sigma)$ like $\mathcal{M}(m_i)$ and $V(v_i)$ as continuous variables. So we can approximate the summation over φ' in Eq. (5.34) by integration, which will give the same results for the fluctuation part $F_f(\bar{\varphi})$ as those presented in Section 5.2.1. The rigorous treatment by the saddle point method (Isihara, 1971) confirms our simple heuristic arguments.

5.3 Inhomogeneous Systems and Translational Invariance

In the major part of our thermodynamic studies we have considered spatially uniform systems. At the microscopic level all real systems are nonuniform. But their macroscopic properties are given by the mean values of functions $A(\sigma)$ of the degrees of freedom $\sigma = \{\sigma_i\}$ and these statistically averaged quantities are very often independent on the particular site i. Then we say that the thermodynamic system is spatially uniform and the mean values of the thermodynamic quantities have the property of *translational invariance*. For example, if $\langle \mathcal{H}_i(\sigma) \rangle$ in Eq. (5.14) is independent on the site i, the energy density $\langle \mathcal{H}_i \rangle$ will be translationally invariant: $\langle \mathcal{H}_i(\sigma) \rangle = \langle \mathcal{H}_{i+j}(\sigma) \rangle; j \neq 0$. It may happen in certain theoretical models and approximations that $\mathcal{H}_i$ itself is translationally invariant (see, e.g., Section 5.6).

In a homogeneous system the equilibrium order parameter $\langle \varphi \rangle$ is uniform, that is its densities per unit volume or per particle are spatially independent. The nonequilibrium order parameter $\varphi(\sigma)$ depends on the configurations $\sigma = \{\sigma_i\}$ of the microscopic variables or, that is the same, on the microscopic states of the system. One can define this quantity as a sum

$$\varphi(\sigma) = \sum_i \varphi_i(\sigma) \tag{5.38}$$

of the densities φ_i at the sites i of a discrete system. In this case, the macroscopic translational invariance means that there is no dependence of $\langle \varphi_i \rangle$ on the site i but the same cannot be said about the quantity φ_i itself.

A cause for breaking of the uniformity of the system is the presence of spatially dependent external forces (fields); for example, the gravitational field acting on a fluid. Owing to the interaction with the nonuniform surroundings, the system can be driven to inhomogeneous equilibrium states. Thus the system becomes a constituent of a total inhomogeneous "mixture" consisting of our system itself and its surroundings.

It is important to realize that at the macroscopic level the system of interest to our studies is characterized by the order parameter (s) $\varphi(\sigma)$. So we identify the considered system with its order parameter(s) φ. Then the influence of the nonuniform fields (surroundings) is represented by the field h conjugate to φ.

In complex systems the order parameter $\varphi(\sigma)$ may interact with other internal macroscopic modes, $\eta = \eta(\tau)$, where $\tau = \{\tau_i\}$ is a supplementary set of microscopic degrees of freedom. The interaction between the order parameter and the supplementary mode(s) η results from the microscopic interactions between the degrees of freedom σ and τ. If such $(\sigma - \tau)$ interactions are present in the Hamiltonian $\mathcal{H}(\sigma, \tau)$, the macroscopic variable φ will depend on η and *vice versa*. Then our macroscopic system will consist of two macroscopic modes φ and η. The translational invariance of the primary order parameter φ will depend on the translational invariance of the secondary order parameter η.

In this Chapter we shall avoid the consideration of supplementary modes as this will obscure the main picture of description. The supplementary modes can always be taken into account for complex systems and this procedure does not involve methodical difficulties (see Section 6.6.3 and models discussed in Chapter 9).

For the canonical ensemble at fixed $\bar{\varphi}$ the mean value $\langle\varphi\rangle$ of φ is equal to $\bar{\varphi}$ and therefore, inhomogeneous systems can be described supposing nonuniform boundary conditions $\varphi_i(\sigma) = \bar{\varphi}_i = const$. It is however more convenient for theoretical studies to make the statistical treatment of the total composite system (the initial system and its surroundings). Instead of dealing with the composite isolated system at fixed overall order parameter we can equivalently consider the initial system in a nonuniform external field h_i, which corresponds to an ensemble at fixed (T, Y, h_i). In the spatially uniform case the external field h_i will be constant $(h = h_i)$. This topic will be discussed in the following section.

5.4 Coarse-Graining and Effective Hamiltonian

In a homogeneous system the mean order parameter

$$\langle\varphi(\sigma)\rangle = \sum_i \langle\varphi_i(\sigma)\rangle \tag{5.39}$$

is a sum over the densities $\langle\varphi_i\rangle$. It will be also a restriction to suppose that all $\langle\varphi_i\rangle$ are different: $\langle\varphi_i\rangle \neq \langle\varphi_j\rangle$ for each couple of sites ij. The general case is given by the following picture.

5.4.1 *Coarse-graining*

Let us assume that our discrete system of points i is represented by a lattice of microscopic cells i of size a_i (Fig. 5.1). The microscopic variable σ_i is defined at a point in the cell i or, equivalently, at the vertex i of the lattice of microscopic cells i. The quasi-macroscopic variable $\varphi_\nu(\sigma)$ is defined for every quasi-macroscopic cell ν of size $L_\nu : a_i \ll L_\nu \ll L$, where L is the characteristic size of the system. Now $\varphi_\nu(\sigma)$ will tend to σ_i as $L_\nu \to a_i$ and $\varphi_\nu(\sigma)$ will tend to $\varphi(\sigma)$ as $L_\nu \to L$.

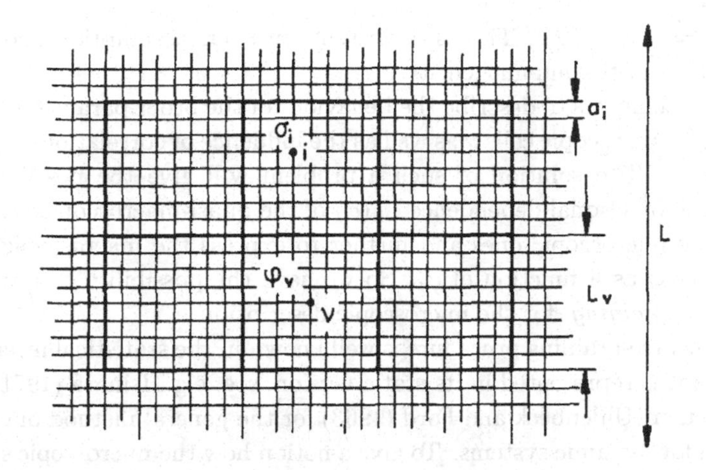

Fig. 5.1 A square lattice consisting of microscopic (i) and quasi-macroscopic (ν) cells: $a_i \ll L_\nu \ll L$.

Within this picture we can write Eq. (5.38) in the form

$$\varphi(\sigma) = \sum_\nu \varphi_\nu(\sigma) = \sum_\nu \Big(\sum_{i_\nu} \varphi_{i_\nu}(\sigma)\Big) \tag{5.40}$$

where the densities $\varphi_{i_\nu}(\sigma)$ belong to the cell ν. The mean order parameter $\langle\varphi\rangle$ will be

$$\langle\varphi(\sigma)\rangle = \sum_\nu \langle\varphi_\nu(\sigma)\rangle = \sum_\nu \sum_{i_\nu} \langle\varphi_{i_\nu}(\sigma)\rangle. \tag{5.41}$$

Certainly, the boundary conditions of the inhomogeneous system can be written as $\bar{\varphi}_\nu = const$ and, consequently we denote the nonuniform external field h conjugate to $\bar{\varphi}_\nu$ by h_ν. Quite generally, we can consider the characteristic size of the inhomogeneity as given by the size L_ν of the cells ν. Varying L_ν from a_i to L we can study short-range ($L_\nu \sim a_i$) and long-range ($L_\nu \le L$) inhomogeneities. This is very convenient in the particular case of external fields h which change over distances L_ν.

In Fig. 5.1 we show two interpenetrating quadratic lattices of cells i and ν. A d-dimensional system can be divided, for example, in hypercubic cells i and ν. The form of the cells and the regularity of the lattices are not important in our present discussion. The only important quantity is the number N_ν of cells i included in the cell ν. For a hypercubic lattice N_ν will be given by $(L_\nu/a_i)^d$. If all $\langle\varphi_{i_\nu}(\sigma)\rangle$ within a cell ν are equal,

$\langle\varphi_\nu(\sigma)\rangle = N_\nu\langle\varphi_{i_\nu}(\sigma)\rangle$. This is consistent with our assumption about the length-scale of the inhomogeneity.

Our task now is to describe the behaviour of the order parameter within the quasi-macroscopic cells ν as well as the influence of correlations between these cells. The solution of such a problem will give the possibility to obtain the functional dependence $\varphi_\nu(\sigma)$ of the quasi-macroscopic variables φ_ν on the microscopic ones and further to express the macroscopic order parameter φ as a function of φ_ν. So we have the possibility to introduce the *coarse-graining* for the microscopic description.

The coarse-graining procedure is well known in the statistical mechanics and here it is represented in its static version; see, e.g., Isihara (1971) and, in particular, Uhlenbeck and Ford (1963) for the general method of coarse-graining for dynamic systems. To give a notion how the microscopic system $\sigma = \{\sigma_i\}$ presented by the cells i can be coarse-grained we shall generalize the calculations from Section 5.2.1.

5.4.2 *Transformation of the canonical partition function*

For the canonical ensemble at fixed $\bar{\varphi} = \{\bar{\varphi}_\nu\}$ the partition function (5.11) can be written in the form

$$\mathcal{Z}(\bar{\varphi}) = \mathrm{Tr}\left\{\prod_\nu \delta[\bar{\varphi}_\nu - \varphi_\nu(\sigma)]e^{-\mathcal{H}(\sigma)/T}\right\}. \tag{5.42}$$

As the numbers N_ν are very large ($N_\nu \gg 1$) we can consider φ_ν as continuous variables. The δ-functions in Eq. (5.42) are replaced according to Eq. (5.16), where we substitute $\bar{\varphi}$ by $\bar{\varphi}_\nu$, $\varphi(\sigma)$ by $\varphi_\nu(\sigma)$ and λ by λ_ν. The modified partition function is

$$\mathcal{Z}'(\bar{\varphi}) = \mathrm{Tr}\left(\exp\left\{-\mathcal{H}(\sigma)/T - \sum_\nu \lambda_\nu\left[\bar{\varphi}_\nu - \varphi_\nu(\sigma)\right]\right\}\right). \tag{5.43}$$

We can derive the relation between $\mathcal{Z}'(\bar{\varphi})$ and $\mathcal{Z}(\bar{\varphi})$ with the help of the identity (5.18), which must be now written for each ν. Thus, we obtain a generalization of Eq. (5.19):

$$
\begin{aligned}
\mathcal{Z}'(\bar{\varphi}) &= \prod_{\nu} \int_{-\infty}^{+\infty} d\varphi'_{\nu} \exp\left[-\frac{1}{T}\sum_{\nu} h_{\nu}\left(\bar{\varphi}_{\nu} - \varphi'_{\nu}\right)\right] \mathcal{Z}(\varphi') \\
&= \prod_{\nu} \int_{-\infty}^{+\infty} d\varphi'_{\nu} \exp\left\{-\frac{1}{T}\left[\sum_{\nu} h_{\nu}\left(\bar{\varphi}_{\nu} - \varphi'_{\nu}\right) + F(\varphi')\right]\right\},
\end{aligned}
\tag{5.44}
$$

where $\mathcal{Z}(\varphi')$ with $\varphi' = \{\varphi'_{\nu}\}$ is the canonical partition function given by Eq. (5.42) and $\bar{\varphi} \to \varphi'$; $h_{\nu} = T\lambda_{\nu}$. The function

$$
F'(\varphi') = F(\varphi') + \sum_{\nu} h_{\nu}(\bar{\varphi}_{\nu} - \varphi'_{\nu}) \tag{5.45}
$$

has extrema for those configurations $\{\varphi'_{\nu}\}$ of the order parameter, which are solutions of the equations of state

$$
h_{\nu} = \left[\frac{\partial F(\varphi')}{\partial \varphi'_{\nu}}\right]_{\bar{\varphi}}. \tag{5.46}
$$

As $h = \{h_{\nu}\}$ is the equilibrium external field configuration, the solutions $\bar{\varphi}_{\nu}(h)$ of Eq. (5.46) will coincide with the boundary values $\bar{\varphi}_{\nu}$ of the nonuniform order parameter $\varphi(\sigma) = \{\varphi_{\nu}(\sigma)\}$. Following the calculations in Section 5.2.1 we shall expand $F'(\varphi')$ at the saddle point solution $\bar{\varphi} = \{\bar{\varphi}_{\nu}\}$ in terms of the fluctuations $\delta\varphi'_{\nu} = (\varphi'_{\nu} - \bar{\varphi}_{\nu})$. Introducing the notation

$$
\mathcal{Z}(h) = \mathcal{Z}'(\bar{\varphi}) \exp\left(\frac{1}{T}\sum_{\nu} h_{\nu}\bar{\varphi}_{\nu}\right), \tag{5.47}
$$

we obtain the Gibbs potential $\widetilde{\Phi}(h)$ in the form

$$
\widetilde{\Phi}(h) = \Phi(\bar{\varphi}) + \Phi_f(\bar{\varphi}) \tag{5.48}
$$

with

$$
\Phi(\bar{\varphi}) = F(\bar{\varphi}) - \sum_{\nu} h_{\nu}\bar{\varphi}_{\nu}, \tag{5.49}
$$

$$
\Phi_f(\bar{\varphi}) = -T \ln\left[\prod_{\nu} \int_{-\infty}^{+\infty} d\delta\varphi_{\nu} e^{-\Delta\Phi(\bar{\varphi},\delta\varphi)/T}\right], \tag{5.50}
$$

and

$$\Delta\Phi = \frac{1}{2}\sum_{\mu\nu} \chi_{\mu\nu}^{-1}(\bar{\varphi})\delta\varphi_\mu\delta\varphi_\nu + O[(\delta\varphi_\nu)^3], \tag{5.51}$$

(the 'prime' of $\delta\varphi'_\nu$ has been omitted). In Eq. (5.51),

$$\chi_{\mu\nu}^{-1} = \left[\frac{\partial^2 F(\varphi)}{\partial\varphi_\nu\partial\varphi_\mu}\right]_{\bar{\varphi}} \tag{5.52a}$$

is the second derivative of $F(\varphi)$. For a n-component order parameter $\varphi = \{\varphi^\alpha\}$ we obtain

$$\chi_{\mu\nu;\alpha\beta}^{-1} = \left[\frac{\partial^2 F(\varphi)}{\partial\varphi_\nu^\alpha\partial\varphi_\mu^\beta}\right]_{\bar{\varphi}} \tag{5.52b}$$

for fixed $\bar{\varphi} = \{\bar{\varphi}_\nu^\alpha\}$.

5.4.3 *Quasimacroscopic separability*

The Gibbs potential (5.48) again is a sum of the saddle point potential $\Phi(\bar{\varphi})$ and the fluctuation part Φ_f. Let us consider the saddle point potential (5.49). The first question is whether $\Phi(\bar{\varphi})$ can be represented as a sum over the potentials $\Phi^\nu(\bar{\varphi})$ of the cells ν. This depends on the properties of the free energy $F(\bar{\varphi})$ in Eq. (5.49) and, hence, we shall focus our attention on Eqs. (5.10) and (5.42). If the Hamiltonian $\mathcal{H}(\sigma)$ is additive over the cells ν, it can be represented by the sum

$$\mathcal{H}(\sigma) = \sum_\nu \mathcal{H}_\nu(\sigma_\nu), \tag{5.53}$$

where $\sigma_\nu = \{\sigma_i\}_\nu$ is the subset of variables σ_i that belong to the cell i of the quasi-macroscopic cell ν; if the cell i does not belong to the cell ν, σ_i will not enter in the subset σ_ν. Then the total Hamiltonian $\mathcal{H}(\sigma)$ is additive at the quasi-macroscopic level for characteristic lengths $L_\nu \gg a_i$; sometimes this property is called *separability*. In certain theoretical models $\mathcal{H}(\sigma)$ is separable at the microscopic level of lengths a_i. Our division of the system in microscopic and quasi-macroscopic cells makes possible to study two types of microscopic and quasi-macroscopic separabilities. Systems with additive (or separable) Hamiltonians are called *separable systems.*

Taking into account the Eq. (5.53) in Eq. (5.42), $\mathcal{Z}(\bar{\varphi})$ is written as a sum over the cells ν :

$$\mathcal{Z}(\bar{\varphi}) = \prod_{\nu} \mathcal{Z}^{\nu}(\bar{\varphi}) = \prod_{\nu} \mathrm{Tr}\left\{\delta[\bar{\varphi}_{\nu} - \varphi_{\nu}(\sigma_{\nu})]e^{-\mathcal{H}_{\nu}(\sigma)/T}\right\}. \tag{5.54}$$

In result, the free energy $F(\bar{\varphi}) = -T \ln \mathcal{Z}(\bar{\varphi})$ is

$$F(\bar{\varphi}) = \sum_{\nu} F_{\nu}(\bar{\varphi}_{\nu}) \tag{5.55}$$

with

$$F_{\nu}(\bar{\varphi}_{\nu}) = -T \ln \mathcal{Z}^{\nu}(\bar{\varphi}). \tag{5.56}$$

Now $\Phi(\bar{\varphi})$ becomes

$$\Phi(\bar{\varphi}) = \sum_{\nu} \Phi_{\nu}(\bar{\varphi}_{\nu}) = \sum_{\nu} [F_{\nu}(\bar{\varphi}_{\nu}) - h_{\nu}\bar{\varphi}_{\nu}]. \tag{5.57}$$

Eq. (5.55) implies that $\chi_{\mu\nu}^{-1}$ is a diagonal matrix: $\chi_{\mu\nu}^{-1} = \chi_{\nu}^{-1}\delta_{\mu\nu}$, where χ_{ν}^{-1} are the respective eigenvalues. As $\chi_{\mu\nu}^{-1}$ is positively definite, $\chi_{\nu}^{-1} > 0$. If we assume that the terms $O(\delta\varphi_{\nu}^{3})$ in $\Delta\Phi$ are negligible, $\Phi_{f}(\varphi)$ from Eq. (5.50) can be calculated with the help of a multiple Gaussian integral (see Section B):

$$\Phi_{f} = -T \sum_{\nu} \ln(2\pi T \chi_{\nu})^{1/2}. \tag{5.58a}$$

When φ is a multi-component vector $\varphi = \{\varphi^{\alpha}\}$, one should substitute φ_{ν} and $\bar{\varphi}_{\nu}$ by φ_{ν}^{α} and $\bar{\varphi}_{\nu}^{\alpha}$ in all the above formulae. The fact that $F_{\nu}(\bar{\varphi}_{\nu}^{\alpha})$ does not depend on the components φ_{μ}^{α} with $\mu \neq \nu$ yields a matrix $\chi_{\mu\nu;\alpha\beta}^{-1} = \chi_{\nu;\alpha\beta}^{-1}\delta_{\mu\nu}$. $\chi_{\nu;\alpha\beta}^{-1}$ can be diagonalized with respect to the indices $\alpha\beta$ by a rotation in the n-dimensional order parameter space. Performing such a rotation (Landau and Lifshitz, 1980) we change the initial vectors $\{\varphi_{\alpha}\}$ to a new ones $\{\varphi_{\alpha}'\}$ and the quadratic form $\delta\varphi_{\nu}^{\alpha}\delta\varphi_{\nu}^{\beta}$ becomes a sum of the squares $(\delta\varphi_{\nu}^{\alpha})^{2}$ with coefficients $\chi_{\alpha}^{-1}(\nu) \equiv \chi_{\alpha;\nu}^{-1}$ — the eigenvalues of the matrix $\chi_{\nu;\alpha\beta}^{-1}$. Now the calculation of the corresponding Gaussian integral over $\delta\varphi_{\nu}^{\alpha}$ gives

$$\Phi_{f} = -T \sum_{\nu\alpha} \ln\left[2\pi\chi_{\alpha}(\nu)\right]^{1/2}. \tag{5.58b}$$

The reader is advised to repeat all calculations in this section using the multi-component vectors $\varphi = \{\varphi^\alpha\}$ and $h = \{h^\alpha\}$; see also Section B.2 and Section A.3.

So the system under consideration can be represented as a sum of quasi-macroscopic systems (cells) ν of the number $\widetilde{N} \sim (L/L_\nu)^d$. The cells ν are not correlated and behave as an ideal classic gas of coarse-grained objects described by the coarse-grained effective variables $\varphi_\nu(\sigma_\nu)$. When the system is homogeneous, $\bar{\varphi}_\nu = \bar{\varphi}/\widetilde{N}$, where $\bar{\varphi}$ is the total order parameter and $\chi_\nu = \chi/\widetilde{N}$, χ is the susceptibility of the total system ($\chi^{-1} = \partial^2 F/\partial\varphi^2$). The coarse-grained variables φ_ν are indistinguishable.

The present treatment with the help of quasi-macroscopic cells ν looks like our consideration of macroscopic cells α performed in Section 2.7. It is accepted as a postulate in the macroscopic thermodynamics that each thermodynamic system is separable in macroscopic cells and the extensive thermodynamic quantities obey the additivity property. Here the thermodynamic additivity is a result of the statistical independence of the cells ν on each other. If the inter-cell statistical correlations are much weaker than those in the cells, we can neglect the inter-cell correlations. Certainly this can be done for sufficiently large cells ν when L_ν are macroscopic lengths.

5.4.4 *Separability of the Hamiltonian*

Let us consider $\widetilde{\Phi}(h)$ for fixed h_ν. From Eqs. (5.43), (5.47) and $\widetilde{\Phi} = -T \ln \mathcal{Z}(h)$ we have

$$\widetilde{\Phi}(h) = -T \ln \mathrm{Tr}\left[e^{-\mathcal{H}_h(\sigma)/T}\right], \tag{5.59}$$

where

$$\mathcal{H}_h(\sigma) = \mathcal{H}(\sigma) - \sum_\nu h_\nu \varphi_\nu(\sigma) \tag{5.60}$$

is the Hamiltonian of the system at fixed $h = \{h_\nu\}$. The Hamiltonian $\mathcal{H}(\sigma)$ in Eq. (5.60) can be always represented as a sum over the Hamiltonian densities $\mathcal{H}_\nu(\sigma)$ corresponding to the cells ν. Then

$$\mathcal{H}_h(\sigma) = \sum_\nu \mathcal{H}_h^{(\nu)}(\sigma) = \sum_\nu \left[\mathcal{H}_\nu(\sigma) - h_\nu \varphi_\nu(\sigma)\right]. \tag{5.61}$$

One cannot be certain that $\mathcal{H}_h(\sigma)$ will be additive, unless $\mathcal{H}_\nu(\sigma)$ and $\varphi_\nu(\sigma)$ depend only on $\sigma_\nu = \{\sigma_i\}_\nu : \mathcal{H}_\nu(\sigma_\nu)$ and $\varphi_\nu(\sigma_\nu)$.

If the system is not separable, we can write

$$\mathcal{H}_h(\sigma) = \sum_\nu [\mathcal{H}_\nu(\sigma_\nu) - h_\nu \varphi_\nu(\sigma_\nu)] + \Delta\mathcal{H}_h^{\text{corr}}(\sigma), \tag{5.62}$$

where

$$\Delta\mathcal{H}_h^{\text{corr}}(\sigma) = \Delta\mathcal{H}^{\text{corr}}(\sigma) - \sum_\nu h_\nu \Delta\varphi_\nu^{\text{corr}}(\sigma). \tag{5.63}$$

Neglecting the correlation terms

$$\Delta\mathcal{H}_h^{\text{corr}}(\sigma) = \mathcal{H}_h(\sigma) - \sum_\nu \mathcal{H}_\nu(\sigma_\nu) \tag{5.64}$$

and

$$\Delta\varphi_\nu(\sigma) = \varphi_\nu(\sigma) - \varphi_\nu(\sigma_\nu) \tag{5.65}$$

in Eq. (5.59) we obtain $\widetilde{\Phi}(h)$ as a sum over the logarithms of the grand canonical partition function $\mathcal{Z}_\nu(h)$ of the cells ν

$$\begin{aligned}\widetilde{\Phi}(h) &= -T\sum_\nu \ln \mathcal{Z}_\nu(h) \\ &= -T\sum_\nu \ln \mathrm{Tr}\left[e^{-\mathcal{H}_h^\nu(\sigma_\nu)/T}\right].\end{aligned} \tag{5.66}$$

Here

$$\mathcal{H}_h^{(\nu)} = \mathcal{H}_h^{(\nu)}(\sigma_\nu) - h_\nu \varphi_\nu(\sigma_\nu). \tag{5.67}$$

If the Hamiltonian $\mathcal{H}_h(\sigma)$ is known, Eq. (5.59) should lead us to the macroscopic representation (5.48) of $\widetilde{\Phi}(h)$ with $\Phi(\bar{\varphi})$ and $\Phi_f(\bar{\varphi})$ given by Eqs. (5.49) and (5.50). In the particular case of separability of the system, Eq. (5.56) should be equivalent to Eq. (5.48) with $\Phi(\bar{\varphi})$ and $\Phi_f(\bar{\varphi})$ from Eqs. (5.57), (5.58a), and (5.58b).

5.4.5 *Effective Hamiltonian*

There is an important difference between the expression (5.59) for $\widetilde{\Phi}(h)$ and the expression

$$\widetilde{\Phi}(h) = -T \ln \left[\prod_{\nu} \int_{-\infty}^{+\infty} d\varphi_\nu e^{-\Phi(\varphi)/T} \right], \tag{5.68}$$

which is obtained from Eqs. (5.44)–(5.48) and

$$\Phi(\varphi) = F(\varphi) - T \sum_{\nu} h_\nu \varphi_\nu . \tag{5.69}$$

$\Phi(\varphi)$ is the generalized Gibbs potential corresponding to a given configuration $\varphi = \{\varphi_i\}$ for the lattice of cells i and $\varphi = \{\varphi_\nu\}$ for the lattice of cells ν. While Eq. (5.59) gives $\widetilde{\Phi}(h)$ as a sum (or an integral for continuous variables σ_i) over the microscopic degrees of freedom of the number $(L/a_i)^d \sim \widetilde{N} \cdot N_\nu \sim 10^{22}$, the multiple integral in Eq. (5.68) is over the quasi-macroscopic degrees of freedom φ_ν of the number $\widetilde{N} \ll \widetilde{N} \cdot N_\nu$. Because $\Phi(\varphi)$ in Eq. (5.68) plays a role quite analogous to that of $\mathcal{H}_h(\sigma)$ in Eq. (5.59), the nonequilibrium Gibbs potential $\Phi(\varphi)$ is often called *the effective Hamiltonian* and is denoted by $\mathcal{H}_{\text{eff}}(\varphi)$, as proposed first by L. D. Landau in 1958; see, e.g., Landau and Lifshitz (1980). The effective Hamiltonian is also referred to as the generalized free energy or the (Gibbs) free energy functional.

There is an essential difference between our definition of the nonequilibrium Gibbs potential $\Phi(\varphi)$ in Section 4.3.2 and the effective Hamiltonian introduced in our present discussion. The nonequilibrium potential $\Phi(\varphi)$ is a function of one degree of freedom — the nonequilibrium order parameter φ. The effective Hamiltonian $\Phi(\varphi)$ in Eq. (5.68) has $(L/L_\nu)^d$ degrees of freedom $\varphi = \{\varphi_\nu\}$ — the nonequilibrium order parameters of the cells ν. As the cells ν are not of a macroscopic size ($\sim L$) the potential $\Phi_\nu(\varphi)$ of each cell ν can depend on the variables φ_ν attached to the other cells ($\mu \neq \nu$). The nondiagonal elements of the matrices $\chi^{-1}_{\mu\nu}$ and $\chi^{-1}_{\mu\nu;\alpha\beta}$ given by Eqs. (5.52a) and (5.52b) are a result of the inter-cell correlations represented by dependencies of the type $\Phi_\nu(\varphi_\mu)$; $(\mu \neq \nu)$; nonseparability of $\Phi(\varphi)$. Enlarging the size L_ν to the macroscopic scale L, $\Phi(\varphi)$ is expected to become separable: $\Phi_\nu = \Phi_\nu(\varphi_\nu)$ as it is in the macroscopic theory of fluctuations (Section 2.7).

All (quasi)macroscopic effects related with the behaviour of the order parameter $\varphi = \{\varphi_\nu\}$ can be described with the help of $\Phi(\varphi)$. The coarse-graining procedure should give the relation between $\mathcal{H}_h(\sigma)$ and $\Phi(\varphi)$. It can be mathematically derived for certain Hamiltonians $\mathcal{H}_h(\sigma)$ with the help of the integral transformations (Section 6.6). For relatively complex Hamiltonians, the correspondence $\mathcal{H}_h(\sigma) \leftrightarrow \Phi(\varphi)$ cannot be derived in an explicit form.

Very often the mathematical form of the effective Hamiltonian is suggested on the basis of general symmetry arguments and/or requirements of the order and the properties of the phase transition. This phenomenological approach is very similar to the Landau theory for a single macroscopic order parameter φ (see Chapter 4). In Section 5.8 we shall demonstrate how the effective Hamiltonian can be derived within the MF approximation, where the fluctuations $\delta\varphi$ are neglected.

5.4.6 *Remarks*

The distribution function $w(h)$ for fixed $h = \{h_\nu\}$ can be written in the form

$$w(h) = \exp\left[\frac{\widetilde{\Phi}(h) - \mathcal{H}_h(\sigma)}{T}\right]; \tag{5.70}$$

cf. Eqs. (5.30), (5.59), and (5.60). The mean value of any function $A(\sigma)$ of $\sigma = \{\sigma_i\}$ is obtained from Eq. (5.4) with the change $\mathcal{Z} \to \mathcal{Z}(h)$ and $\mathcal{H} \to \mathcal{H}_h$. The mean value $\langle\varphi_i\rangle$ of the order parameter density per particle $\varphi_i(\sigma)$ is

$$\langle\varphi_i\rangle = \frac{1}{\mathcal{Z}}(h)\mathrm{Tr}\left[\varphi_i(\sigma)e^{-\mathcal{H}_h(\sigma)/T}\right]. \tag{5.71}$$

The mean energy $\langle\mathcal{H}_h\rangle$ is equal to the enthalpy W of the system

$$W = \langle\mathcal{H}_h\rangle = U - \sum_\nu h_\nu\langle\varphi_\nu\rangle, \tag{5.72}$$

where $\langle\varphi_\nu\rangle$ is the sum of all $\langle\varphi_i\rangle$ belonging to the cell ν. For uniform systems $(h = h_\nu)$,

$$W = U - h\bar{\varphi}. \tag{5.73}$$

In fact the natural variables of the enthalpy W are the entropy S and the field h while Eqs. (5.72) and (5.73) give W as a function of T and h. However in principle the equation $S = -\partial\Phi/\partial T$ can be solved with respect to $T, T = T(S, h)$ and the result can be substituted in Eq. (5.72) or Eq. (5.73). Then both W and U will be functions of the entropy S. From the same Eq. (5.73) the internal energy U can be obtained as a function of $h = \{h_\nu\}$. In order to express U as a function of its natural variable $\bar{\varphi} = \{\bar{\varphi}_\nu\}$, the equations $\bar{\varphi}_\nu = \bar{\varphi}_\nu(h_\nu)$, that is Eq. (5.71) with $\varphi_i \to \varphi_\nu$ should be inverted and the function $h_\nu(\bar{\varphi}_\nu)$ should be substituted in U : $U \to U[h_\nu(\bar{\varphi}_\nu)]$.

These calculations are made only in special theoretical studies because in the most of the cases one wants to known the thermodynamic quantities as a function of directly measurable parameters as T and h. For the ensemble at fixed $h = \{h_\nu\}$, equations analogous to Eqs. (5.12) and (5.14) can be derived for the specific heat C_h. The result is obtained by replacing $\mathcal{H}(\sigma)$ with $\mathcal{H}_h(\sigma)$ in Eq. (5.12) and $\mathcal{H}_i(\sigma)$ with $\mathcal{H}_h^{(\nu)}(\sigma)$ in Eq. (5.14); $i = \nu$.

5.5 Generating Functionals and Correlation Functions

Within the framework of statistical mechanics one can study both thermodynamic and correlation properties. The notion of correlations between the various physical quantities is a new point in our study of phase transitions. *The spatial correlations* between the states $\bar{\varphi}_i(\sigma)$ and $\bar{\varphi}_j(\sigma)$ at two spatial points (i and j) can serve as an appropriate example. The correlations are described by *the correlation functions* and here we shall discuss them from a quite general point of view. In Section 5.4 we have used indices ν for the quasi-macroscopic cells ν. The correlation functions can be defined for the cells ν but it seems more general to discuss this functions for the order parameter density $\varphi_i(\sigma)$ per particle.

Let us denote h_i/T by $h(i)$ and consider the expansion of the partition function

$$\mathcal{Z}(h) = \mathrm{Tr}\left[e^{-\mathcal{H}_h(\sigma)/T}\right], \tag{5.74}$$

with

$$\mathcal{H}_h(\sigma) = \mathcal{H}(\sigma) - \sum_i h_i\varphi_i(\sigma) \tag{5.75}$$

in powers of $\delta h(i) = [h(i) - \bar{h}(i)]$, where $\bar{h} = \{\bar{h}_i = T\bar{h}(i)\}$ is a fixed external field; it may happen that in a quasi-macroscopic cell ν, all h_i are equal ($h_i = h_\nu$). In result we obtain

$$\mathcal{Z}(h) = \mathcal{Z}(\bar{h}) + \sum_{l=1}^{\infty} \frac{1}{l!} \sum_{(i_1,\ldots,i_l)} \left[\frac{\delta^{(l)} \mathcal{Z}(h)}{\delta h(i_1) \ldots \delta h(i_l)} \right]_{\bar{h}} \delta h(i_1) \ldots \delta h(i_l). \tag{5.76}$$

This expansion can be written with the help of the correlation functions for a given external field configuration $\bar{h} = \{\bar{h}_i\}$:

$$G^{(l)}(i_1, \ldots, i_l/\bar{h}) = \langle \varphi_{i_1}(\sigma) \ldots \varphi_{i_l}(\sigma) \rangle_{\bar{h}} \tag{5.77}$$

where $\langle \ldots \rangle_{\bar{h}}$ is the statistical average at $h = \bar{h}$. In fact, from Eq. (5.77), the first equality (5.4), Eqs. (5.70) and (5.74) for w and $\mathcal{Z}(h)$ we obtain that the *l-point* (or *l-th order*) correlation function $G^{(l)}$,

$$\begin{aligned} G^{(l)}(i_1, \ldots, i_l/\bar{h}) &= \frac{1}{\mathcal{Z}(\bar{h})} \left[\frac{\delta^{(l)} \mathcal{Z}(h)}{\delta h(i_1) \ldots \delta h(i_l)} \right]_{\bar{h}} \\ &= \frac{T^l}{\mathcal{Z}(\bar{h})} \left[\frac{\delta^{(l)} \mathcal{Z}(h)}{\delta h_{i_1} \ldots \delta h_{i_l}} \right]_{\bar{h}}, \end{aligned} \tag{5.78}$$

is equal to the l-th order derivative of the functional

$$\widetilde{\mathcal{Z}}(h) = \frac{\mathcal{Z}(h)}{\mathcal{Z}(\bar{h})} = \langle \exp \left[\frac{1}{T} \sum_i (h_i - \bar{h}_i) \varphi_i(\sigma) \right] \rangle_{\bar{h}}. \tag{5.79}$$

So, $G^{(l)}$ are coefficients in the expansion

$$\widetilde{\mathcal{Z}}(h) = 1 + \sum_{l=1}^{\infty} \frac{1}{l!} \sum_{(i_1,\ldots,i_l)} G^{(l)}(i_1, \ldots, i_l/\bar{h}) \delta h(i_1) \cdots \delta h(i_l). \tag{5.80}$$

The functional $\widetilde{\mathcal{Z}}$ and the partition function $\mathcal{Z}(h)$ are very often called *the generating functionals* for *the correlation functions of the order parameter.* If we know $\widetilde{\mathcal{Z}}$ or $\mathcal{Z}(h)$ we can obtain the correlation functions $G^{(l)}$ and *vice versa.* In many cases it is more convenient to find some of the correlation functions instead of the partition function.

The correlation functions $G^{(l)}$ and their relation with $\mathcal{Z}(h)$ provide a scheme of description that gives both the thermodynamic and correlation properties of the system. Within the statistical mechanics the precise form of the order parameter $\varphi = \{\varphi_i(\sigma)\}$ cannot be determined, but it is possible to investigate its mean value

$$\langle\varphi\rangle_{\bar{h}} = \sum_i G^{(1)}(i) = \sum_i \langle\varphi_i(\sigma)\rangle_{\bar{h}} \tag{5.81}$$

and the l-order correlation between the states $\langle\varphi_{i_k}(\sigma)\rangle$ at sites $i_1, \ldots, i_l$ ($l \geq 2$). A quantitative measure of the correlation is given by the functions $G^{(l)}$. If the order parameters at positions i and j are not statistically correlated:

$$\begin{aligned} G^{(2)}(i, j/\bar{h}) &= \langle\varphi_i(\sigma)\varphi_j(\sigma)\rangle_{\bar{h}} \\ &= \langle\varphi_i(\sigma)\rangle_{\bar{h}}\langle\varphi_j(\sigma)\rangle_{\bar{h}} \\ &= G^{(1)}(i)G^{(1)}(j). \end{aligned} \tag{5.82}$$

The net correlation effect is hidden in the irreducible correlation function of second order ($l = 2$):

$$\begin{aligned} \chi_{ij}^{(2)}(\bar{h}) &= \langle\varphi_i\varphi_j\rangle - \langle\varphi_i\rangle\langle\varphi_j\rangle \\ &= \langle(\varphi_i - \langle\varphi_i\rangle)(\varphi_j - \langle\varphi_j\rangle)\rangle \\ &\equiv \langle\langle\varphi_i\varphi_j\rangle\rangle. \end{aligned} \tag{5.83}$$

The irreducible correlation functions like $\chi^{(2)}$ are functional derivatives of the generating functional $\ln \widetilde{\mathcal{Z}}(h)$; see Eq. (5.79). From Eq. (5.79) we find that

$$\chi_i^{(1)} = G_i^{(1)} = \frac{\delta \ln \widetilde{\mathcal{Z}}(h)}{\delta h(i)} = \langle\varphi_i\rangle_{\bar{h}} \tag{5.84}$$

and

$$\chi_{ij}^{(2)} = \left[\frac{\delta^2 \ln \widetilde{\mathcal{Z}}(h)}{\delta h(i)\delta h(j)}\right]_{\bar{h}}. \tag{5.85}$$

In general the l-th order irreducible correlation function $\chi^{(l)}$ is

$$\chi_{i_1,\ldots,i_l}^{(l)} = \left[\frac{\delta^l \ln \widetilde{\mathcal{Z}}(h)}{\delta h(i_1)\cdots\delta h(i_l)}\right]_{\bar{h}}. \tag{5.86}$$

The functional $\widetilde{\mathcal{Z}}(h)$ is expressed by the Gibbs potential (5.29) in the following way:

$$\begin{aligned}\widetilde{\Phi}(h) &= -T \ln \mathcal{Z}(h) \\ &= \Phi(\bar{h}) - T \ln \widetilde{\mathcal{Z}}(h).\end{aligned} \tag{5.87}$$

Quite often the functions $\chi^{(l)}$ are calculated as derivatives of $\widetilde{\Phi}(h)$:

$$\begin{aligned}\chi^{(l)}_{i_1,\ldots,i_l} &= -T\left[\frac{\delta^l \Phi}{\delta h(i_1)\cdots\delta h(i_l)}\right]_{\bar{h}} \\ &= -T^{l-1}\left[\frac{\delta^l \Phi}{\delta h_{i_1}\cdots\delta h_{i_l}}\right]_{\bar{h}}.\end{aligned} \tag{5.88}$$

The irreducible correlation functions $\chi^{(l)}$ are coefficients in the expansion of $\widetilde{\Phi}(h)$ at $h(i) = \bar{h}(i)$:

$$\widetilde{\Phi}(h) = \Phi(\bar{h}) - T\sum_{l=1}^{\infty}\frac{1}{l!}\sum_{(i_1,\ldots,i_l)}\chi^{(l)}(i_1,\ldots,i_l/\bar{h})\delta h(i_1)\cdots\delta h(i_l). \tag{5.89}$$

The connection between the two types of correlation functions, $G^{(l)}$ and $\chi^{(l)}$ can be obtained by expanding the logarithm in the equation

$$\begin{aligned}&\ln\left[1+\sum_{l=1}^{\infty}\frac{1}{l!}\sum_{(i_1,\ldots,i_l)}G^{(l)}\delta h(i_1)\cdots\delta h(i_l)\right] \\ &\quad=\sum_{l=1}^{\infty}\frac{1}{l!}\sum_{(i_1,\ldots,i_l)}\chi^{(l)}\delta h(i_1)\cdots\delta h(i_l).\end{aligned} \tag{5.90}$$

This equation follows from Eq. (5.80), Eq. (5.87), and Eq. (5.89). The same relations can be derived by differentiation of the functions $\chi^{(l)}$ with respect to $h(i)$. For example, the direct differentiation of $\chi^{(2)}$ from Eq. (5.83) yields an expression for $\chi^{(3)}_{ijk}$ through the G-functions:

$$\begin{aligned}\chi^{(3)}_{ijk} &= \frac{\delta\chi^{(2)}}{\delta h(k)} \\ &= G^{(3)}_{ijk} - G^{(2)}_{ij}G^{(1)}_{k} - G^{(2)}_{ik}G^{(1)}_{j} - G^{(2)}_{kj}G^{(1)}_{i} + 2G^{(1)}_{i}G^{(1)}_{j}G^{(1)}_{k}.\end{aligned} \tag{5.91}$$

Using the first equation (5.84) and Eq. (5.83) in the form

$$\chi_{ij} = G_{ij} - G_i G_j. \tag{5.92a}$$

or, equivalently

$$G_{ij} = \chi_{ij} + \chi_i \chi_j, \tag{5.92b}$$

where the superscripts $(l) = (2)$ and (1) have been omitted, the relation (5.91) can be inverted and we obtain:

$$G_{ijk} = \chi_{ijk} + \chi_{ij}\chi_k + \chi_{ik}\chi_j + \chi_{kj}\chi_i + \chi_i\chi_j\chi_k. \tag{5.93}$$

Now it is easy to derive the connection between the functions $\chi^{(l)}$ and $G^{(l)}$ when $l > 3$.

Very often we shall use the correlation functions at zero field $\bar{h}$, $\chi^{(l)}(\bar{h} = 0)$ and $G^{(l)}(\bar{h} = 0)$, denoting them shortly by the same symbols $\chi^{(l)}$ and $G^{(l)}$. All formulae discussed above can be rewritten for the particular case $\bar{h} = 0$. Then $\chi^{(l)}$ and $G^{(l)}$ describe the response of the system to the field perturbations $\delta h(i) = h(i) = h_i/T$.

One of the most frequently used correlation function is $\chi_{ij}^{(2)}$ for $\bar{h} = 0$ as given by Eq. (5.83). When $\langle\varphi_i\rangle = 0$ as it is in the disordered phase, $\chi_{ij}^{(2)} = G_{ij}^{(2)}$. It is connected with the susceptibility $\chi = \chi(\bar{h} = 0)$ of the system at zero field by

$$\chi = \sum_i \left[\frac{\delta\langle\varphi\rangle}{\delta h_i}\right]_{\bar{h}=0} = \frac{1}{T}\sum_{ij}\chi_{ij}^{(2)}, \qquad \langle\varphi\rangle = \sum_j \langle\varphi_j\rangle; \tag{5.94}$$

cf. Eqs. (5.46), (5.52a), and (5.52b).

For uniform external fields ($h = h_i$) the above formulae become much more simple. In this case the Hamiltonian (5.60) will be

$$\mathcal{H}_h = \mathcal{H}(\sigma) - h\sum_i \varphi_i(\sigma), \tag{5.95}$$

and one could find the correlation function, that is the derivatives of the corresponding partition function $\mathcal{Z}(h)$ and the Gibbs potential $\widetilde{\Phi}(h)$ with respect to the uniform field h, or the ratio $\tilde{h} = h/T$. The field h is now conjugate to the total order parameter $\varphi(\sigma)$. In result the simplified version of the analysis presented above will give statistical averages of sums like (5.39) or, more generally,

$$G^{(n)} = \langle \varphi^n(\sigma) \rangle = \sum_{i_1,\ldots,i_n} \langle \varphi_{i_1}(\sigma), \cdots, \varphi_{i_n}(\sigma) \rangle$$
$$= \sum_{i_1,\ldots,i_n} G^{(n)}(i_1,\ldots,i_n), \qquad n \geq 1, \tag{5.96}$$

where $G^{(n)}(i_1,\ldots,i_n)$ is the correlation function defined by Eq. (5.78). For example, the second derivative of $\widetilde{\mathcal{Z}}(h)$ gives the function

$$G^{(2)} = \left[\frac{\partial^2 \widetilde{\mathcal{Z}}(h)}{\partial \tilde{h}^2}\right]_{\bar{h}} = \sum_{ij} \langle \varphi_i \varphi_j \rangle_{\bar{h}} \tag{5.97}$$

or

$$G^{(2)} = \frac{1}{\mathcal{Z}(0)} \left[\frac{\partial^2 \mathcal{Z}(h)}{\partial \tilde{h}^2}\right]_{h=0} = \sum_{ij} \langle \varphi_i \varphi_j \rangle_{\bar{h}=0}, \tag{5.98}$$

for $\bar{h} = 0$. The susceptibility χ will be equal to $G^{(2)}$ when $\langle \varphi \rangle = 0$, but if $\langle \varphi \rangle \neq 0$,

$$\chi = \frac{\partial \langle \varphi \rangle}{\partial h} = \frac{1}{T} \frac{\partial \langle \varphi \rangle}{\partial \tilde{h}} = \frac{1}{T} \sum_{ij} \langle\langle \varphi_i \varphi_j \rangle\rangle \tag{5.99}$$

cf. Eqs. (5.83) and (5.94).

Very often we are interested only in the effect of an uniform external field h on the behaviour of homogeneous systems. It is clear from Eqs. (5.97)–(5.99) that this case is also treated in terms of l-point correlation functions ($G^{(l)}$ and $\chi^{(l)}$).

To construct the generating functionals for the correlation functions of a homogeneous system in an uniform external field h, we should use the formal generalization of the potential $\Phi(h)$ and of the partition function $\mathcal{Z}(h)$ by adding the term $\sum_i h_i \varphi_i$ to the Hamiltonian (5.85). After that we can perform the analysis of the corresponding inhomogeneous system ($h_i \neq 0$) and to set the nonuniform component h_i of the joint field $(h + h_i)$ equal to zero as a final step of all mathematical procedures. The same can be done by replacing of the initial homogeneous field h in Eq. (5.85) by h_i. The spatial uniformity of the system is restored when all the calculations of interest to the particular study are made.

There are systems as, for example, superconductors and superfluids, for which the field conjugate to the order parameter does not really exist. In this case the generating functionals are constructed by adding to the

Hamiltonian of the system fictitious terms of the type $\varphi_i h_i^f$, where h_i^f is a fictitious field serving as a convenient mathematical tool.

No matter whether h_i are real or fictitious fields they are often called "sources" of the field (order parameter) φ_i. This terminology is taken from the quantum field theory and is largely applied; see, e.g., Vassilev (1976), Amit (1978), Popov (1987) as far as correlation functions and their investigation are concerned (see Chapter 6 and Chapter 7, where the correlation functions are investigated by a standard perturbation expansion). For example, the correlation functions $G^{(l)}$ are also called Green's functions, a term introduced in the field theory from the theory of the linear differential equations.

As we shall see in Section 7.2, the function $G^{(2)}$ in its continuous limit $G_{ij}^{(2)} \to G^{(2)}(x, y)$ obeys a nonlinear differential equation. In this sense it is not the Green function itself but a more complex mathematical object represented in terms of the Green function of the corresponding linear differential equation (Abrikosov, Gor'kov, and Dzyaloshinskii, 1962; Fetter and Walecka, 1971). Having in mind the present terminology the correlation functions $G^{(2)}$ and, generally, $G^{(l)}$ will be frequently referred to as the two-point and the l-point Green functions. The functions $G^{(l)}$ are called *cumulants* because they enter in the expression for the cumulant expansion in statistical physics; see, e.g., Ma (1985). For reasons that shall become clear in Chapter 6, the irreducible correlation functions $\chi^{(l)}$ are called *the connected Green functions.* In the continuum limit ($i \to \boldsymbol{x}$), see Section 6.5, we shall denote the functions $\chi^{(l)}$, that is the connected Green functions by $G_c^{(l)}$.

We have considered the generating functionals for the order parameter $\varphi(\sigma)$. The method of generating functionals can however be used for every physical quantity $A(\sigma)$ or, when necessary, for several quantities: $A(\sigma), B(\sigma)$, $C(\sigma)$, etc. Now we shall introduce (real or fictitious) fields $h_A, h_B, h_C, \dots$ conjugate to $A(\sigma), B(\sigma), C\sigma), \dots$, respectively, or, in case of spatial nonuniformity $h_i^A, h_i^B, \dots$, conjugate to $A_i(\sigma), B_i(\sigma), \dots$. A generalized Hamiltonian is obtained by adding terms like

$$-\sum_i \left(h_i^A A_i + h_i^B B_i + \cdots\right) \tag{5.100}$$

to the Hamiltonian (5.61); so we can construct the generating functionals for the correlation functions of the type

$$G^{(2)}_{A_iB_j} = \langle A_i(\sigma)B_j(\sigma)\rangle$$

$$G^{(2)}_{A_iA_j} = \langle A_i(\sigma)A_j(\sigma)\rangle \tag{5.101}$$

or the irreducible correlation functions

$$\begin{aligned}\chi^{(2)}_{A_iB_j} &= \langle (A_i(\sigma) - \langle A_i(\sigma)\rangle)(B_j(\sigma) - \langle B_j(\sigma)\rangle)\rangle \\ &= \langle\langle A_iB_j\rangle\rangle.\end{aligned} \tag{5.102}$$

Sometimes the functions of σ like $\varphi(\sigma), A(\sigma), \ldots$ are called *composite functions* or, when σ is a quantum mechanical operator — *composite operators.* Such composite functions (or operators) are the energy densities $\mathcal{H}_i(\sigma)$ and $\mathcal{H}_h^{(i)}(\sigma)$. The Hamiltonians $\mathcal{H}(\sigma)$ and $\mathcal{H}_h(\sigma)$ can always be written as sums of the densities $\mathcal{H}_i(\sigma)$ and $\mathcal{H}_h^{(i)}(\sigma)$ over the sites i. For example,

$$\mathcal{H}(\sigma) = \sum_i \mathcal{H}_i(\sigma). \tag{5.103}$$

Then we can write the expressions

$$C_\varphi = \frac{1}{T^2}\sum_{ij}\chi_{ij}^{(U)} \tag{5.104}$$

and

$$C_h = \frac{1}{T^2}\sum_{ij}\chi_{ij}^{(W)} \tag{5.105}$$

for the specific heats C_φ, see Eq. (5.12); $C_h = -T(\partial^2\Phi/\partial T^2)$. These specific heats are expressed by the irreducible energy–energy correlation function

$$\chi_{ij}^{(U)} = \langle (\mathcal{H}_i - \langle \mathcal{H}_i\rangle)(\mathcal{H}_j - \langle \mathcal{H}_j\rangle)\rangle \tag{5.106}$$

and the irreducible "enthalpy–enthalpy" correlation function

$$\chi_{ij}^{(W)} = \langle \left(\mathcal{H}_h^{(i)} - \langle \mathcal{H}_h^{(i)}\rangle)(\mathcal{H}_h^{(j)} - \langle \mathcal{H}_h^{(j)}\rangle\right)\rangle, \tag{5.107}$$

where $\langle \mathcal{H}_h^{(i)}\rangle)$ is equal to the local enthalpy W_i (see the end of the previous section). The field conjugate to the Hamiltonian ($\mathcal{H}$ or $\mathcal{H}_h$) is the reciprocal temperature $(1/T)$.

5.6 Ising Model

The main task of the statistical mechanics is to calculate the partition and correlation functions for a given Hamiltonian $\mathcal{H} = \mathcal{H}(\sigma, h)$. Whatever the mathematical expression for $\mathcal{H}$ is, it is never an exact reflection of the real system. So writing down the Hamiltonian we always introduce a modelling of the real system or the physical problem of interest. The relatively complex Hamiltonians, those which include several terms, showing the real nature of the inter-particle interactions, are too difficult to be treated within the mathematical scheme of the statistical mechanics. Such complex models are investigated approximately, i.e., by using appropriate approximations at some suitable stages of the statistical calculation. In many cases the approximate study of a complex model corresponds (or is equivalent) to the exact treatment of simplified models. So the modelling of the system finally comes from two sources: the choice of the Hamiltonian and the approximations in its statistical treatment. Usually, we call "simple models" those Hamiltonians that can, at least in particular cases, be exactly calculated. In this and the next sections we shall enumerate several relatively simple models together with their main properties.

5.6.1 *Definition of the Ising model*

Consider a solid with localized spin$-1/2$ angular momenta S_i at sites i of the crystalline lattice (note that the symbol S_i and similar symbols of angular moments and magnetic moments, used below denote vectors). If the crystal field anisotropy is strong enough, it will influence the exchange spin interaction and in certain cases the spins S_i and, respectively, the spin magnetic moments m_i will tend to align parallel to one of the lattice axes, for example, the $\hat{z}$-axis. In this case we can neglect the vector components m_i^x and m_i^y (S_i^x and S_i^y, as well) along the $\hat{x}$- and $\hat{y}$-axes and focus on the behaviour of the component m_i^z (extreme uniaxial anisotropy; Section 3.8.3). Before introducing the Ising model, which describes such systems, we will elucidate this point in more details.

The localized magnetic moments m_i at the sites i of the lattice are produced by the non-compensated (effective) spin angular momentum S_i and the orbital angular momentum L_i of particles, localized at these sites. One atom (molecule) per site is usually considered. When the orbital angular momenta L_i are not taken into consideration, the magnetic momenta m_i at the sites i result from the non-compensated electron spin angular momenta

S_i of the atoms (sites) i. We shall assume that at any site i there is one non-compensated spin angular momentum S_i. Recall the known relation $m_i = \gamma_e S_i = -(g_s\mu_B/\hbar)S_i$, where $\gamma_e = -g_s\mu_B/\hbar$ is the electron gyromagnetic ratio, $\mu_B = |e|\hbar/2mc$ is the Bohr magneton, and $g_s \approx 2$ is the electron g-factor.[1] Owing to the negative sign of the electron electric charge e, the vectors m_i and S_i have opposite directions.

In uniaxial ferromagnets, where a preferred axis of spin orientation exists, for example, the $\hat{z}$-axis, we consider the possible values of the $\hat{z}$-projections S_i^z of the spins. In case of extreme anisotropy, it is expected that other components of the spin vector are equal to zero. In such systems the magnetic moment at site i is given by $m_i = (0, 0, m_i^z)$, where $m_i^z = -(g_s\mu_B/\hbar)S_i^z$. As the electron spin quantum number is $S = 1/2$, the vector spin component may take two possible values: $S_i^z = \pm\hbar/2$. So for spin–1/2 systems we have $m_i^z = \mu_B\sigma_i$, where the dimensionless variable σ_i takes the values $\sigma_i = \pm 1$, and these values exactly correspond to the values $S_i^z = \mp\hbar/2$ of the spin component S_i^z (the *up-down* order of the signs should be kept in mind). The respective σ_i-vectors, $(0, 0, \sigma_i)$, are parallel to the magnetic moment vectors m_i and anti-parallel to the spin angular momentum vectors S_i (here and below we use the notation σ_i for the component σ_i^z, too). More generally, one may use the relation $m_i^z = g_s\mu_B S\sigma_i$ which is approximately equivalent to $m_i = \mu_B\sigma_i$ provided the small fine-constant correction to the value of g_s is ignored.

Now we should introduce the exchange interaction between the magnetic moments and the interaction of the same moments with the external magnetic field H, oriented along the $\hat{z}$-axis; see, e.g., Kittel (1971) and White (1970); Getzlaff (2008). These interactions can be described in terms of the dimensionless variable σ_i. For example, the Zeeman energy $(-m_i H)$ at site i can be written in the form $-h\sigma_i$, where $h = H/g_s\mu_B S \approx H/\mu_B$. The simplest (original) form of the Ising model, defined on a regular spin lattice (Ising, 1925) is given by the spin Hamiltonian

$$\mathcal{H} = -\frac{1}{2}J_{\text{ex}}\sum_{\langle ij\rangle}\sigma_i\sigma_j - h\sum_i \sigma_i, \tag{5.108}$$

where the first sum $\langle ij\rangle$ extends over *the nearest neighbour* (*nn*) *pairs* of lattice spin variables σ_i and simulates the exchange interaction in one-constant (J_{ex}) approximation (J_{ex} is proportional to the integral (constant)

[1] Note, that within the Dirac mechanics $g_s = 2$ but taking into account the fine-structure constant $\alpha = 1/137$, we have $g_s = 2(1 + \alpha/2\pi + \cdots) \approx 2.002319$. Here we use $g_s = 2$.

of the direct exchange interaction between two atoms). The factor 1/2 in the front of the first term on the r.h.s. of Eq. (5.108) is introduced for convenience. Certainly this is the simplest way to define an extremely short range (i.e., *nn*) exchange coupling in a homogeneous system, described by the variables σ_i which take discrete values on the lattice vertices i.

Despite of the fact that the lattice field $\sigma = \{\sigma_i\}$ has a quantum origin, within the Ising model it behaves as a classical field, and in all further treatments of the Ising model we may forget about quantum mechanics rules of calculation. To clarify this point, let us recall that $\sigma_i = -(g_s/\hbar)S_i^z$ and $S_i^x = S_i^y = 0$. As the spin components S_i^x and S_i^y are equal to zero for the strong uniaxial anisotropy, as well as an external magnetic field H along the $\hat{z}$-axis does not generate any nonzero value of these components, the quantum mechanical commutation relations are trivially satisfied and the spin component S_i^z, apart from the quantization, behaves as a classical quantity (a total lack of commutation conditions).

The Ising model may be applied to the study of various problems beyond magnetism and quantum mechanics. This model is always useful when one has to describe any collective phenomenon in a system consisting of many constituents which are localized on the sites i of a regular (or irregular) lattice and take two or more discrete values (see also the discussion in the beginning of Section 5.6.3 and the lattice gas model formulation in Section 5.7.2); see also numerous applications in various fields of physics discussed in the review article by Pelissetto and Vicari (2002).

In a perfect (regular) lattice the *nn* sites are at the same distance and obviously, the exchange forces acting on a given spin site i from its *nn*, are equal. This is why, a single exchange constant J_{ex} is sufficient to describe such simple spin system. Even in this simple form, the *nn* interaction is sufficient to determine the major collective properties of the spin ensemble.

The second (Zeeman) term in Eq. (5.108) describes the interaction of the spin-1/2 particles i with the external magnetic field $H = h/\mu_{\text{B}}$. When the σ_i-vector is parallel to the field vector H, that is when $m_i \parallel H$, the Zeeman energy is minimal, and the magnetic moments tend to align along the external field direction. This means that the spin angular momentum vectors S_i tend to be anti-parallel to the field H.

If we define the vectors $(0, 0, \sigma_i)$ along the direction of the spin vectors S_i (by the change $\sigma_i \to -\sigma_i$), the first term on the r.h.s. of Eq. (5.108) will remain invariant but the Zeeman term will change sign and the sign of this term will be positive. In this representation of the Ising model, the statistical variable σ_i simulates the behaviour of the spins S_i rather than the behaviour of the magnetic moments m_i. This alternative representation of the Ising model is often used, too.

To complete the formulation of the Ising model we have to specify the lattice structure. Except for some special cases, we shall assume a regular lattice: one-dimensional chain (1D lattice) square (2D) lattice, cubic or another 3D lattice. In certain cases we may study Ising models in dD lattices, i.e., hyper-cubic or other lattices of general spatial dimensionality $d \geq 0$. In many cases, one may consider dD lattices of non-integer dimensionality $d \geq 0$; see Craco et al (1999). In studies of disordered systems, the lattice is usually irregular, which brings about a lack of (quasi)translational invariance of the Ising model (see Sections 5.6.2 and 9.5).

If the number of sites i in the lattice is N, the system defined by the Ising model Eq. (5.108) will have 2^N possible microstates (configurations) $\sigma = \{\sigma_i\}$. For each configuration σ we can define the overall (nonequilibrium) magnetization

$$\mathcal{M}(\sigma) = \mu_{\rm B} \sum_i \sigma_i = \sum_i m_i. \tag{5.109}$$

In a more general form, valid for extensions of the Ising model mentioned in Section 5.6.2, one may change the factor $\mu_{\rm B}$ to $g_s \mu_{\rm B} S$, or, $g_J \mu_{\rm B} J$, where J is the quantum number of the total angular momentum of atoms. Equilibrium states are characterized by $M(T, h) = \langle \mathcal{M}(\sigma) \rangle$, where the mean magnetization M is an extremum of the generalized potential $\Phi(T, h, \mu)$. Any equilibrium magnetization M can be realized through a large number of microscopic configurations $\sigma = \{\sigma_i\}$.

The structure of the ordered states ($M \neq 0$) depends on the sign of the exchange constant J_{ex} It is readily seen from the model (5.108) that the positive ($J_{ex} > 0$) *nn* exchange interactions will tend to align the spins parallel to each other as these (ferromagnetic type) configurations correspond to a lower exchange energy $\mathcal{H}(h = 0)$. The Hamiltonian (5.108) with $J_{ex} > 0$ describes a simplified version of anisotropic ferromagnets, that is *the Ising model ferromagnets.* The case $J_{ex} < 0$ corresponds to anisotropic (uniaxial) antiferromagnets (for further notes about Ising antiferromagnets see Section 5.8.2).

5.6.2 *General exchange, general values of spin and angular momentum*

One may digress from the one-constant approximation and to define the *nn* exchange coupling as depending on the sites i and j : $J_{\rm ex} \to J_{ij}$. This may correspond to some irregularity (inhomogeneity) of the lattice and/or the bonds J_{ij}, i.e., to magnets with some types of disorder (see Section 9.5). In this case we need an extension of the Ising model which is provided by the notation J_{ij}. But very often the notation J_{ij} is used for simple systems, described by Eq. (5.108) provided the *nn* interaction is stipulated. We can also introduce the next-nearest-neighbour (*nnn*) and etc., longer range exchange in a regular (perfectly ordered) lattice, and in this case J_{ij} will depend on the distance $|i-j|$ ($\equiv |\mathrm{r}_i - \mathrm{r}_j|$) between the sites i and j. Sometimes, the uniform field h can be replaced by a spatially dependent external field h_i, in particular, in considerations of disordered systems described by the so-called random external field (see Section 9.5.3). For other types of disordered systems, the elements J_{ij} of the exchange matrix $\hat{J} = \{J_{ij}\}$ cannot be represented as dependent on the difference $|i-j|$ only and, hence, the one constant (J_{ex}-) representation in Eq. (5.108) cannot be applied even for *nn* interactions in a regular lattice. This is the case of randomly distributed ("disordered") bonds J_{ij} (see also Section 9.5).

So there are several important reasons to write the Ising model in a more general form:

$$\mathcal{H} = -\frac{1}{2}\sum_{i\neq j} J_{ij}\sigma_i\sigma_j - \sum_i h_i\sigma_i, \tag{5.110}$$

where $J_{ii} = 0$ for any i. Bearing in mind that $J_{ii} = 0$ for any i (a lack of "self-interaction"), one may omit the explicit indication that the summation over i and j in the first term in Eq. (5.110) is performed under the condition $i \neq j$.

For perfect (regular) lattices one may use that $J_{ij} = J(|i-j|) \equiv J(|\boldsymbol{r}_i - \boldsymbol{r}_j|)$, so the system possesses a (quasi)translational invariance: $J(|i'-j'|) = J(|i-j|)$, where the sites $i' = i + R_n$ and $j' = j + R_n$ are given by any lattice vector $R_n = (n_1a_1, \cdots, n_da_d)$; $a_1, \ldots, a_d$ are the lattice constants along the axes of the dD lattice, and $n_1, \ldots, n_d$ are integer numbers. For simple hyper-cubic lattice ($a_1 = a_2 = \cdots = a_d \equiv a_0$ – the lattice constant), the *nn* model of the interaction yields $J_{ij} = J_{\text{ex}}$ provided $|i-j| = a_0$, and $J_{ij} = 0$ otherwise. Then any site i is located in an exchange potential $J_0 = zJ_{\text{ex}}$, where z is the *nn* number; for a hyper cube of dimensionality d (dD cube), we have $z = 2d$, provided the lattice is simple (Fourier transformations of the *nn* exchange interaction function

$J(|i-j|)$ for dD simple hypercubic (sc) and dD body-centered tetragonal (bct) lattices are presented in Section F). The translational invariance is broken in disordered systems, where the exchange parameters J_{ij} exhibit an additional local dependence, for example, $J_{ij} = J(i)f(|i-j|)$; here the function $f(x)$ is equal to unity for $x = a_0$ and tends to zero for $x \to \infty$.

Sites i with a total spin quantum number $S > 1/2$ have $(2S+1)$ projections S_i^z of spin angular momentum S_i on the $\hat{z}$-axis, given by $\hbar(-S, -S+1, \ldots, S-1, S)$. In this case we can easily write a more general spin-S Ising model in which the variable σ_i will take $(2S+1)$ values at any site i of lattice. If we set $m_i^z = g_s\mu_{\rm B}S\sigma_i$, σ_i will run the values $(1, 1-1/S, \ldots, -1+1/S, -1)$; once again the (maximal) value of σ_i is normalized to unity and the values of σ_i differ in sign from the values of the spin component S_i^z (see also Section 5.6.1).

Alternatively, one may work with the quantity $\sigma_i' = S\sigma_i$ which is related with the magnetic moments m_i by $m_i^z = g_s\mu_{\rm B}\sigma_i'$. When the variable σ_i is used the field parameter h should be written as $h = g_s\mu_{\rm B}S$ whereas for the alternative field σ_i' the choice $h = g_s\mu_{\rm B}$ should be made. Besides, in these two cases the exchange matrices $\hat{J} = \{J_{ij}\}$ differ by a factor S^2.

When the orbital angular momentum vectors L_i of particles (with orbital momentum quantum number L) cannot be neglected we must consider the total magnetic moments (denoted again by m_i) as sums $m_i = -(\mu_{\rm B}/\hbar)(2S_i + L_i)$, or, alternatively, we may use the total angular momentum vectors $J_i = (S_i + L_i)$ of particles (with quantum number J of the total angular momentum), in order to write $m_i = -(g_J\mu_{\rm B}/\hbar)J_i$, where g_J is the Landé g-factor $g_J = (L+2S)/(L+S)$ which is usually referred to as:

$$g_J = 1 + \frac{J(J+1) + S(S+1) - L(L+1)}{2J(J+1)}. \tag{5.111}$$

For $L = 0$, $J = S$, and $g_J = 2$.

The extreme anisotropy or the presence of external field H along the $\hat{z}$-axis gives the opportunity to fix the $\hat{z}$-direction and once again to consider only the z-component of the magnetic moment: $m_i^z \equiv m_i = -(g_J\mu_{\rm B}/\hbar)J_i^z$. Now the Ising model (5.108) undergoes another generalization, namely, the variables $\sigma_i = m_i^z/g_J\mu_{\rm B}J = -J_i^z/\hbar J$ at each site i of the lattice can take $(2J+1)$ values $(1, 1-1/J, \ldots, -1+1/J, -1)$. The external field parameter h will be given by $h = g_s\mu_{\rm B}S$. Alternatively, one may use the relation $m_i^z = g_J\mu_{\rm B}\sigma_i'$ and write the respective Ising model in the terms of the variable $\sigma_i' = J\sigma_i$. Here the parameter h is given by $h = g_s\mu_B$ and exchange parameters J_{ij} will differ by a factor $1/J^2$ from exchange parameters in the model, written in terms of the lattice field σ_i; for a comprehensive discussion of magnetic Hamiltonians and, in particular, of spin Hamiltonians see, e.g., Mattis (1965), White (1970), Kittel (1971), and Getzlaff (2008). We shall often refer to the lattice models with lattice variable $\sigma_i \sim -J_i$ as "spin models", and to σ_i as "spin variable", although the total angular momentum is a more complex physical object than the spin S_i and here the field σ_i has a sign opposite to that of the total angular momentum J_i.

The consistent treatment of magnetic models is made in the framework of quantum mechanics, where S_i, L_i, J_i, and m_i are operators. In the *semi-classical approximation* these quantities are considered as usual vectors; see, e.g., Landau and Lifshitz (1958), and Rubinowicz (1968). For such (semi)classical treatment the magnetic moments m_i are vectors that lie along the direction of the magnetic polarization at site i and may take continuous values. The projection of the vector m_i on the h_i–direction is given by $|m_i| \cos\theta_i$, where $0 \leq \theta_i \leq \pi$. The field term in Eq. (5.108) can be written in the form

$$g_J\mu_B \sum_i H_i \cdot J_i = -m \sum_i h_i \cos\theta_i, \tag{5.112a}$$

where $m = g_J\mu_B J$ is the magnitude of magnetic moment, corresponding to the total angular momentum J of particles. Such description is valid in the "infinite spin" *semi-classical limit* [$J \to \infty$ so that $g_J J$ and, hence, m remains constant; see, e.g., Pathria (1972)]. The generalization of Ising model from ($S = 1/2$)–system to systems with $S > 1/2$ does not introduce qualitatively new features of the cooperative many-particle behaviour and, therefore, no new phase transition properties. That is why we shall often use the spin–1/2 Ising model.

5.6.3 *Brief notes*

The Ising model describes an effective potential energy of the spin particles. The electron spin J_{ij}–exchange is obviously a potential energy. The electron exchange interaction is relatively strong and allows for the description of magnetic order up to temperatures of several 100 K, whereas the other magnetic interactions (dipole–dipole, etc.) are weak and cannot affect the cooperative behaviour above temperatures of order 1 K.

The Zeeman term including only the spin angular momentum, $-h_iS_i$, has also a pure potential origin (external potential). But, as known from the theory of magnetism, the part of Zeeman term containing the orbital angular momentum, $-h_iL_i$, has a kinetic origin (White, 1970). This term is a part of the gauge-invariant kinetic energy

$$K = \frac{1}{2m}\sum_{\alpha=1}^{Z}\left(\boldsymbol{p}_\alpha + \frac{e}{c}\boldsymbol{A}\right)^2 \tag{5.112b}$$

of electrons of an atom of atomic number Z, which includes the vector potential $\boldsymbol{A}$ of the external magnetic field $\boldsymbol{H} = \nabla \times \boldsymbol{A}$ (the Coulomb gauge $\nabla.\boldsymbol{A} = 0$ should is satisfied).

For the sake of simplicity we have written Eq. (5.112b) in the continuum limit ($i \to \boldsymbol{r}$). The kinetic energy term (5.112b) is usually treated for an uniform magnetic field $\boldsymbol{H}$. In this particular case, the choice $\boldsymbol{A} = \frac{1}{2}\boldsymbol{H}\times\boldsymbol{r}$ is consistent with the Coulomb gauge $\nabla.\boldsymbol{A} = 0$. Inserting this expression of the vector potential in Eq. (5.112b) one can represent the energy K as a sum of three terms: the simple kinetic energy $p^2/2m = \sum_\alpha p_\alpha^2/2m$, the orbital momentum term, $\mu \mathrm{B}\boldsymbol{L}.\boldsymbol{H}$, where $\boldsymbol{L} = \sum_\alpha \boldsymbol{L}_\alpha$, and the known diamagnetic term $(e^2/8m)\sum_\alpha(\boldsymbol{H}\times\boldsymbol{r}_\alpha)^2$ (the latter is neglected in the Ising model).

Now we can put this atom at the site i of the lattice and consider the lattice as a whole. In such lattice the two kinetic terms are labeled by i: $p_i^2/2m$ and $\mu_{\mathrm{B}}\boldsymbol{L}_i.\boldsymbol{H}$. The dynamic variables $\boldsymbol{p}_i$ can be easily integrated out from the partition function of the lattice. They give a contribution to the free energy which is unessential for the cooperative phenomena and can be ignored. So, a portion of kinetic energy enters in the Ising Hamiltonian only when we take into account the orbital angular momentum.

The Ising model and similar lattice models, introduced in Section 5.7, describe only the potential energy of the spin particles. The kinetic energy is neglected when the particles are rigidly localized at the vertices i of the lattice. One may accept this picture or assume that the kinetic energy term, always quadratic in the particle momenta $\boldsymbol{p}_i$, can be integrated in the partition function. Such procedure always gives a standard temperature dependent contribution to the thermodynamic potential, corresponding to classical noninteracting particles.

If the external field H is not uniform, we cannot deduce the orbital momentum term in this simple form $(-h_iL_i)$, and for this reason, Ising models with spatially dependent external fields (h_i) and include terms of type h_iL_i are not justified. In such cases the site dependent field h_i should be considered as a theoretical tool rather than as describing a real system and should be taken as uniform $h_i = h$ at the end of the calculation and in interpretations of real systems. For systems with $L_i = 0$ such a restriction does not exist. An external field h_i, which varies on scales of the order of the lattice constant a_0 is difficult to realize in practice. As a rule, the external fields have larger characteristic scales of spatial variations, and the theoretical study of real systems is usually performed for uniform field parameter h.

The Ising models (5.108) and (5.110) initially introduced in the theory of magnetic systems are now widely used to describe various many-body systems (see also Section 5.6.1, where we point out the possibility to apply the model beyond quantum mechanical restrictions). In fact, one may forget the spin origin of the variables (σ_i) and assume that they describe any physical quantity that can take two or more possible values at the vertices of a spatial lattice; see, e.g., Section 5.7.2, where the lattice gas model is formulated on the basis of the Ising model.

Interesting information about the history of the Ising model is given by Mattis (1965), Brush (1967, 1983), Stanley (1971). The model (5.108) was at first proposed by *Wilhelm Lenz* in 1920 and then solved for 1D systems by his diploma thesis student *Ernst Ising* (Ising, 1925). Thus this model is sometimes referred to as *the Lenz–Ising model* . The Ising solution for 1D systems is exact and shows that the 1D "Ising ferromagnet" does not undergo a phase transition up to $T_c = 0$. The exact solution of the same model for 2D systems was performed by Onsager (1944); see also Temperley (1972), and Baxter (1982).

5.7 Other Lattice Models

5.7.1 *Heisenberg and XY model*

While the Ising model describes magnets with an extreme uniaxial anisotropy, the Heisenberg model is intended to describe entirely isotropic spin systems. Assuming that the external field H_i is along one of the spatial directions, the Heisenberg model can be written again with the help of Eq. (5.108) or Eq. (5.110) in the case of the generalized exchange J_{ij}. Within this model however the components m_i^x and m_i^y of the magnetic moments $m_i = (m_i^x, m_i^y, m_i^z)$ are taken into account on equal grounds with m_i^z. So, in the classical approximation, the variables σ_i in Eqs. (5.108) and (5.110) are three dimensional vectors $\sigma_i = (\sigma_i^x, \sigma_i^y, \sigma_i^z)$ and the terms $\sigma_i \sigma_j$

and $h_i\sigma_i$ are scalar products; $h_i = (h_i^x, h_i^y, h_i^z)$. Within the quantum mechanical treatment the classical components σ_i^x, σ_i^y, and σ_i^z are replaced with the corresponding quantum mechanical operators.

The XY model describes magnets with easy (XY)-plane of magnetization: $\sigma_i = (\sigma_i^x, \sigma_i^y, 0) \equiv (\sigma_i^x, \sigma_i^y)$; $\sigma_i^z = 0$ for $h_i^z = 0$. It is therefore a special case of the Heisenberg model. There are two possibilities for the XY model depending on the spatial orientation of the external field: i) a field parallel to the $\hat{z}$-axis (*XY model in a transverse field* or, *transverse XY model*) and, ii) a field parallel to the XY plane (*XY model in a longitudinal field*).

The Heisenberg model and its particular cases, the Ising and XY models, can be represented by the interpolating Hamiltonian:

$$\mathcal{H} = -\frac{1}{2}\sum_{\langle ij\rangle} J_{ij}\left[\sigma_i^x\sigma_j^x + \sigma_i^y\sigma_j^y + \Delta\sigma_i^z\sigma_j^z\right] - \sum_i h_i\sigma_i, \tag{5.113}$$

where $\Delta \geq 0$. When $\Delta = 1$, this Hamiltonian, which is often referred to as "XXZ model", coincides with the isotropic Heisenberg model; $\Delta = 0$ and $\Delta \gg 1$ describe to the XY and Ising models, respectively. The intermediate values of Δ make possible to investigate the anisotropic Heisenberg model.

The quantum mechanical treatment of the Heisenberg and XY models is rather difficult because there are terms in the Hamiltonian expressed by operators, which do not commute with each other. Because of the absence of transverse components (σ_i^x and σ_i^y) of the operator σ_i no commutation problems exist in the quantum Ising model. Such a problem arises in the so-called transverse quantum Ising model (when the external field is perpendicular to the operator σ_i^z).

Within the classical treatment of the spin–1/2 Heisenberg and XY models the vector σ_i can rotate under the normalization condition

$$\sigma_i^2 = 1. \tag{5.114}$$

In contrast to the Ising model where σ_i can take two discrete values (± 1) in the Heisenberg model the values of σ_i belong to the 3D unit sphere (5.114). The values of the σ_i variables in the classical XY model lie on the circumference $\sigma_{ix}^2 + \sigma_{iy}^2 = 1$ in the two dimensional spin space $(\sigma_{ix}, \sigma_{iy})$.

Like in the Ising model, when the quantum number J of the total angular momenta $J_i \sim -\sigma_i$ is different from 1/2, i.e., for $J > 1/2$, the value of σ_i is not fixed and depends on the particular quantum state. Sometimes, the classical lattice models like (5.113) and are called "Ising-like" or "Ising-type" models because they have many common features with the original Ising model (5.108).

5.7.2 *Lattice gas model*

The "lattice gas" provides a somewhat oversimplified model of fluid system. Mathematically, this model is isomorphous to the Ising model (5.108) and is used to describe the vapour-liquid phase transition. Imagine that the Lennard–Jones–like potential in Fig. 3.8(a) is replaced with the interparticle potential shown in Fig. 5.2.

We divide the volume V of the gas in cells of a size r_0 and label the center of each cell by the index i. As the effective size of each particle is r_0, a particle can occupy a cell i but no other particles can be placed in the same cell. The potential $u(r)$ from Fig. 5.2 can be written in the form

$$u_{ij} = \begin{cases} \infty, & \text{for } i = j, \\ -u_0, & \text{if } i \text{ and } j \text{ are } nn, \\ 0, & \text{in all other cases.} \end{cases} \tag{5.115}$$

where $u_0 > 0$ is an interaction parameter.

Obviously this lattice gas model will have the energy

$$\mathcal{H}(\tau) = -u_0 \sum_{i \neq j} \tau_i \tau_j, \tag{5.116}$$

where the variable τ_i will be zero, if the site i is empty and will be equal to unity when the site i is occupied. Each spatial distribution of the molecules in the gas can be given by a set of numbers $\tau = \{\tau_i, i = 1, \ldots, \widehat{N}\}$; $\widehat{N}$ is the number of the cells i. The number of particles $N < \widehat{N}$ is

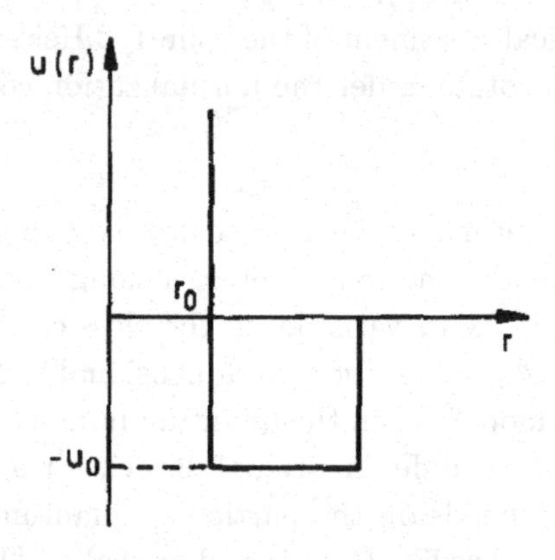

Fig. 5.2 Model potential of a lattice gas ($u_0 > 0$).

$$N = \sum_i \tau_i. \tag{5.117}$$

Now we can compare the Ising Hamiltonian (5.108) with the energy $\mathcal{H}(\mu)$ corresponding to the grand canonical ensemble $(T, V = \widehat{N}r_0^3, \mu)$ of the lattice gas:

$$\mathcal{H}(\mu) = -\mu N - u_0 \sum_{i \neq j} \tau_i \tau_j. \tag{5.118}$$

Using the relation

$$\tau_i = \frac{\sigma_i + 1}{2} \tag{5.119}$$

between the numbers $\tau_i = 0, 1$ and $\sigma_i = \pm 1$, we replace in Eqs. (5.117) and (5.118) τ_i with σ_i. The subsequent simple calculations reveal how important the number z of the nearest neighbours of the site i is:

$$z = \frac{1}{2} \sum_{j \neq i} 1.$$

The Hamiltonian (5.118) then becomes:

$$\mathcal{H}(\mu) = -\frac{u_0}{4} \sum_{(ij)} \sigma_i \sigma_j - h_{\text{eff}} \sum_i \sigma_i - E_0, \tag{5.120}$$

where $E_0 = \widehat{N}(4\mu + u_0 z)/4$, and $h_{\text{eff}} = (2\mu + u_0 z)/4$. Note that the substitution $u_0 \to u_0'/2$ makes possible to relate u_0' to the single coupling $\tau_i \tau_j$. If the model is given in terms of u_0', instead of z we must write $z' = 2z$. The expressions for E_0 and h_{eff} are invariant under the change $(u_0, z) \to (u_0', z')$.

The similarity between the models (5.108) and (5.118) has been particularly fruitful for comparing the properties of Ising magnets and gas–liquid systems (Yang and Lee, 1952). The lattice gas model was introduced in the late thirties by Peierls (1936), Fowler R. H. (1938), Rushbrooke (1938), and Guggenheim (1939). More information about the lattice gas is given by Fisher (1967), Runnels (1972), and Baxter (1982).

5.7.3 *n-component vector models*

A theoretical generalization of the Heisenberg model is obtained by extending the number of the "spin" components $(\sigma_i^x, \sigma_i^y, \sigma_i^z)$ to an arbitrary value $n \geq 3$. Then the variables σ_i will be n-component vectors. Classical models of this type are often called *n-component* (or *vector*) *models* $(n > 1)$. The reader could notice that the Heisenberg model is a 3-component vector model and XY is a two-component vector model. These models are used to describe on a phenomenological level magnets with complex structure or structural phase transitions; see Bruce (1980), Cowley (1980), and Izyumov and Syromiatnikov (1984).

5.7.4 *The s-state Potts model*

Describes a lattice, each site i of which can have any of S different states (Potts, 1952). The interaction energy is of the form

$$\mathcal{H} = -J \sum_{\langle ij \rangle} \delta(\sigma_i, \sigma_j), \tag{5.121}$$

where δ is the Kronecker symbol and the variables σ_i take s different values. This model can be used when the configurational energy of each two states (σ_i and σ_j) does not depend on the states themselves (i.e., on the values of σ_i and σ_j) but depends on whether they are equivalent ($\sigma_i = \sigma_j$; $\delta = 1$) or not ($\delta = 0$). Clearly, the Potts model is a generalization of the Ising model and coincides with it for $s = 2$.

A possible interpretation of the Potts model is given by considering a set of rods, one at each site i in an s-dimensional space (Hertz, 1985). According to the Potts model these rods will interact, only if they are parallel. Alternative formulations of this model are discussed by Zia and Wallace (1975), Amit (1976), and Lubensky (1979). The statistical studies by integral transformations as given in Section 6.6, demonstrate that for $s < 2$ and $s > 2$ this model describes first order transitions (the Landau expansion gains a third order invariant; see Lubensky (1979), Mitag and Stephen (1974), and Hertz (1985).

5.7.5 *The spherical model*

In 1952 Berlin and Kac introduced a simplified version of the Ising model that can be solved exactly for an arbitrary spatial dimensionality d

(Berlin and Kac, 1952). The constraint $\sigma_i = \pm 1$ or $\sigma_i^2 = 1$ in the Ising model is replaced here with the weaker condition

$$\sum_{i=1}^{N} \sigma_i^2 = N, \tag{5.122}$$

where N is the number of vertices of the lattice. The *spherical* condition (5.122) means that the square σ_i^2 of each variable σ_i can take any value provided the *spherical* constraint (5.122) is obeyed. It has been shown by Stanley (1968, 1969) that *the spherical* or, *the Berlin–Kac model* is a particular case of the n-vector model with nearest-neighbour exchange interaction and corresponds to the limit $n \to \infty$; see also Kac and Thompson (1971), Pearce and Thompson (1977); reviews are given by Stanley (1971), Joyce (1972), and Baxter (1982).

5.7.6 *Biquadratic exchange and single spin energy*

The spin-1 model

$$\mathcal{H} = -J \sum_{\langle ij \rangle} \sigma_i \sigma_j \left[1 + a\sigma_i \sigma_j\right] + \Delta \sum_{\langle ij \rangle} \sigma_i^2, \tag{5.123}$$

was introduced by Blume, Emery and Griffiths (BEG) in 1971 to describe the phase separation in He^3–He^4 mixtures (Blume, Emery and Griffiths , 1971). In Eq. (5.123), J is the usual Ising coupling, the parameter $a > 0$ gives the constant aJ of the biquadratic exchange and the parameter Δ stands for the interaction of the $(S = 1)$–"spins", $\sigma_i = 0, \pm 1$, with the crystal field at each point i. If in Eq. (5.123) we set $a = 0$, the BEG Hamiltonian will become the spin–1 Ising model with crystal field interaction, often referred to as the *Blume–Capel model* (Blume, 1966; Capel, 1966). The spin term proportional to Δ gives simple-spin energy levels with a $\sigma_i = 0$ singlet lowest level and a $\sigma_i = \pm 1$ twice degenerate energy level at an energy Δ above the singlet (provided $\Delta > 0$).

Systems like (5.123) are relatively complex and cannot be treated exactly. Blume, Emery and Griffiths (1971) have applied the molecular field approximation (Section 5.8) and have obtained the corresponding Landau expansion; analogous approach to the Blume-Capel model is presented by Blume (1966), Capel (1966), and Blume and Watson (1967).

5.7.7 *About the exact solutions of lattice models*

We shall not discuss how the partition function of the Ising model and the models enumerated in this section can be calculated exactly. This is a large field of investigation beyond the scope of the present book; see Stanley (1971), and Baxter (1982) for detailed information and references to original sources). We shall focus our attention on two general methods that make possible to derive the (quasi)macroscopic properties of a system from its microscopic Hamiltonian. One of these methods is based on the MF approximation (see also Chapter 1 and Section 5.8). Except for very special cases this approximation can be applied to any system or, precisely speaking, to microscopic model Hamiltonians of real systems. In this way the Landau expansion can be obtained from the corresponding microscopic Hamiltonian (Section 5.8). The second method includes an integral transformation of the lattice models to new models, expressed in variables which take continuous values. Making after that a transition to the continuum limit we go to the field theoretic counterpart of the microscopic lattice model (Section 6.6). So we obtain a theory formulated in terms of fluctuating fields, which gives opportunity to apply the fluctuation theory of phase transitions.

5.8 Mean-Field Approximation

Here we shall illustrate how the mean field (MF) approximation is applied for Ising models like (5.110). The MF method is quite general and our present consideration can be straightforwardly colligated for more complex Hamiltonians. The discussion in this Section can easily be extended to Heisenberg, XY, and other magnetic models by considering σ_i and h_i as vectors and $\sigma_i\sigma_j$ and $h_i\sigma_i$ as scalar products. We have discussed the ferromagnetic order but the same MF treatment is easily extended to antiferromagnets, where the exchange bonds J_{ij} are negative and one is forced to introduce the staggered magnetization as an order parameter (see Sections 3.8.4 and 5.8.2 for brief discussions of this topic).

It would be more instructive to carry out our discussion for an arbitrary "spin" variable ($J > 1/2$; see Section 5.6.2), because we can demonstrate the transition from quantum to semi-classical limit ($J \to \infty$) (often referred to as "classical limit"). We assume that the variables σ_i in Eq. (5.110) take values $(1, 1-1/J, \dots, -1+1/J, -1)$ corresponding to the total angular momentum of the particles i as explained in Section 5.6.2. More generally, we may digress from the magnetic analogy and think of σ_i as any physical quantity that can take $(2J+1)$ equally spaced values. In this way the Ising model, as well as other magnetic models, can be widely used in description of systems, consisting of many constituents in other fields of physics, chemistry, biology, economics and even in sociology. In view of certain applications it seems reasonable to demonstrate the MF treatment of inhomogeneous systems (Section 5.3).

5.8.1 *Mean field*

The difficulty in treating the Ising models lies in the presence of the $\sigma\sigma$ interaction. If we look at Eq. (5.110) carefully, we will notice that the $\sigma\sigma$ term can be written in the form

$$-\sum_i \left(\frac{1}{2}\sum_j J_{ij}\sigma_j\right)\sigma_i, \tag{5.124}$$

which resembles the $h\sigma$ interaction. Recall the note below Eq. (5.110) that the matrix $\hat{J}$ satisfies the conditions $J_{ii}=0$ for any i (self-interaction in the lattice sites is often forbidden as "unphysical" within these type of models[2]). This is why, the special indication that the summation over i and j is performed for $i \neq j$ is redundant and might be omitted, as in Eq. (5.124) and below.

Clearly, the term in the brackets of Eq. (5.124) acts on the σ_i-variable as an i-dependent *internal (molecular) field* created by the interactions of $j(\neq i)$–spins with the ith ones. If the number of nn is z and no other (nnn etc) interactions are taken into account, we have a picture where each variable σ_i is in the (*exact*) molecular field (potential) generated by its z nearest neighbours σ_j. The (exact) internal field however depends on the local microscopic states at sites i, i.e., on the configuration $\{\sigma_j\}_{nn}^i$ of the nn at the sites i. Our construction of an internal field acting like the external (h_i) one will be useless, unless we are able to free it from the σ_i-dependence. An appropriate choice for the molecular field will be the most probable value of the bracketed sum in Eq. (5.124), i.e.,

$$-\frac{1}{2}\sum_j J_{ij}\langle\sigma_j\rangle. \tag{5.125}$$

[2]Models, like the Hamiltonian (5.123), which contain "self-interaction", can also be considered in the MF approximation outlined in this Section.

This is the known *mean field* (MF) *approximation* of the molecular field.

Using the MF (5.125) we certainly neglect some real effects. To understand what is lost in this approximation we will apply the MF procedure more consistently, writing the Ising variables σ_i in the form

$$\sigma_i = \langle\sigma_i\rangle + \delta\sigma_i, \tag{5.126}$$

where $\delta\sigma_i$ is the deviation (fluctuation) of σ_i from the mean value $\langle\sigma_i\rangle$ (the statistical average of σ_i). The $\sigma_i\sigma_j$ term in Eq. (5.110) becomes

$$\sigma_i\sigma_j = \sigma_i\langle\sigma_j\rangle + \langle\sigma_i\rangle\sigma_j - \langle\sigma_i\rangle\langle\sigma_j\rangle + \delta\sigma_i\delta\sigma_j. \tag{5.127}$$

At this stage we will neglect the fluctuation term $\delta\sigma_i\delta\sigma_j$ in Eq. (5.127). Further we shall replace the $\sigma\sigma$ interaction in Eq. (5.110) with the first three terms on the r.h.s. of Eq. (5.127). This is the only approximation to be made the neglecting of the fluctuation part of the scalar product $\sigma_i\sigma_j$ in Eq. (5.127).

The Ising model (5.110) can now be written in the reduced form

$$\mathcal{H} = \frac{1}{2}\sum_{ij} J_{ij}\langle\sigma_i\rangle\langle\sigma_j\rangle - \sum_{ij}\left(J_{ij}\langle\sigma_j\rangle + h_i\right)\sigma_i, \tag{5.128a}$$

where we have taken in account that the matrix $\hat{J} = \{J_{ij}\}$ is symmetric and for this reason the contributions of the first and the second terms on the r.h.s. of Eq. (5.127) are equal. The Eq. (5.128a) can be written in the following way

$$\mathcal{H} = A(\langle\sigma\rangle) - \sum_i h_i^{\text{eff}}\sigma_i, \tag{5.128b}$$

where

$$h_i^{\text{eff}} = h_i + h_i^{\text{MF}} \tag{5.129}$$

with the MF ("mean molecular field")

$$h_i^{\text{MF}} = \frac{1}{2}\sum_j (J_{ij} + J_{ji})\langle\sigma_j\rangle = \sum_{ij} J_{ij}\langle\sigma_j\rangle \tag{5.130}$$

and

$$A(\langle\sigma\rangle) = \sum_i A_i(\langle\sigma\rangle), \qquad A_i(\langle\sigma\rangle) = \frac{1}{2}\sum_j J_{ij}\langle\sigma_j\rangle\langle\sigma_i\rangle, \tag{5.131}$$

and $\langle\sigma\rangle = \{\langle\sigma_i\rangle\}$ denotes the statistical average of any particular lattice field configuration.

The original Hamiltonian (5.110) with the bilinear exchange interaction $\sigma_i\sigma_j$ has been reduced to the Hamiltonian (5.128b), which describes free (noninteracting) variables σ_i in an external effective field h_i^{eff} given by Eq. (5.129). The MF, h_i^{MF}, acts on the variables σ_i in the same way as the external field h_i does.

5.8.2 *Thermodynamic potential in mean-field approximation and symmetry breaking*

The reader is well aware that problems in theoretical physics concerning the behaviour of free particles in external fields are often easy to solve. Using the *trace* symbol in the form

$$\prod_i \sum_{\sigma_i=-1}^{1} \tag{5.132}$$

the partition function (5.74) with the MF Hamiltonian (5.128b) is

$$\mathcal{Z} = e^{-A(\langle\sigma\rangle)/T} \prod_i \left[\sum_{\sigma_i=-1}^{1} \exp\left(\frac{1}{T} h_i^{\text{eff}} \sigma_i\right) \right]. \tag{5.133a}$$

or, equivalently,

$$\mathcal{Z} = e^{-A(\langle\sigma\rangle)/T} \prod_i \left[\sum_{\sigma_i'=-J}^{J} \exp\left(\frac{1}{JT} h_i^{\text{eff}} \sigma_i'\right) \right]. \tag{5.133b}$$

Note that, according to the definition of the lattice field σ_i in Section 5.6.2, the sums in Eq. (5.132) and Eq. (5.133a) are over the sequence of $(2J+1)$ values of σ_i: $-1, -1+1/J, \ldots, 1-1/J, 1$. $\sigma_i' = J\sigma_i$ is however a more suitable summation variable and, in practice, one may use Eq. (5.133b).

The sum over σ_i is a geometrical progression. Performing this summation and using Eq. (5.29) we obtain the Gibbs potential

$$\Phi(T, J_{ij}, \langle\sigma\rangle) = \sum_l \Phi_l(T, J_{ij}, \langle\sigma\rangle), \tag{5.134}$$

where l runs the N lattice sites, and the local potentials

$$\Phi_l = A_l\left(\langle\sigma\rangle\right) - T\ln\left[\frac{\sinh\left(1+\dfrac{1}{2J}\right)a_l}{\sinh\left(\dfrac{a_l}{2J}\right)}\right], \tag{5.135}$$

depend on

$$a_l = \frac{1}{T}h_l^{\rm eff}. \tag{5.136}$$

The MF approximation has led us to a representation of the total Gibbs potential Φ as a sum over local (single-particle) potentials Φ_i. The reason is clear enough: we are dealing with a system consisting of free (noninteracting) degrees of freedom.

From Eqs. (5.134), (5.135), and the relation

$$\langle\sigma_i\rangle = -\frac{\partial\Phi}{\partial h_i} = -\frac{\partial\Phi_i}{\partial h_i}, \tag{5.137}$$

cf. Eq. (5.84), we obtain

$$\langle\sigma_i\rangle = B_J(a_i), \tag{5.138}$$

and

$$B_J(y_i) = \left(\frac{2J+1}{2J}\right)\coth\left(\frac{2J+1}{2J}a_i\right) - \frac{1}{2J}\coth\frac{a_i}{2J}, \tag{5.139}$$

is the *Brillouin function* of order J, known from the theory of magnetism; see White (1970), Stanley (1971), and Getzlaff (2008). The system of transcendental equations (5.138) describes strongly couples spins; see the dependence of a_i on $h^{\rm MF}$ and the expression (5.130) for $h_i^{\rm MF}$. For translational invariant systems $[h_i = h,\ J_{ij} = J(|i-j|)]$ we have a system of N equivalent equations and, hence, N identical solutions $(\langle\sigma_i\rangle = \langle\sigma_j\rangle)$. But the equations remain strongly coupled.

For $a_i \to \infty$, $B_J(a_i) \to 1$. This limit occurs either for very large external fields h_i or for very low temperatures ($T \approxeq 0$). Then we have $\langle\sigma\rangle = 1$ and, hence, *a saturated magnetization* $M_i^{(s)}/N = \langle m_i \rangle_s = g_J \mu_B J \langle \sigma \rangle = g_J \mu_B J$. Sometimes, the same quantity (M_s) is denoted by M_0 (and its density – by m_0) to indicate that this is the magnetization at $T = 0$ (we shall use both notations; see, e.g., Section 5.8.5). The function $B_J(a_i)$ becomes zero only in the limit $a_i \to 0$ and, hence, the zero solution ($\langle \sigma_i \rangle = 0$) of Eq. (5.138) exists only for $h_i = 0$.

Up to this stage the sign of the exchange constants J_{ij} has not been of significance, and our consideration is therefore valid for both positive and negative signs of J_{ij} and, hence, of h_i^{MF}. The Brillouin function $B_J(y)$ is odd $[B_J(y) = -B_J(-y) \geq 0]$ and the change of the sign of a_i leads to a change of the sign of the order parameter configuration $\langle \sigma \rangle = \{\langle \sigma_i \rangle\}$. For free (non-interacting) spins ($J_{ij} = 0$, $h_i^{\text{MF}} = 0$) this merely indicates that the magnetic moments follow the external field orientation (*pure paramagnetic effect*). For interacting spins the role of external field is played by the effective field h_i^{eff} which includes the molecular field h_i^{MF}. In this extremely interesting case, the change of sign of $a_i \sim h_i^{\text{eff}}$ depends on the signs of h_i and h_i^{MF} and the ratio of their magnitudes. The field ($\langle \sigma_i \rangle$) will take positive values only if $h_i > -h^{\text{MF}}$; otherwise the spins will align along the external field direction (*paramagnetic effect*). The field $\langle \sigma_i \rangle$ will be negative for $h_i < -h^{\text{MF}}$.

In zero external field ($h_i = 0$), Eq. (5.138) is homogeneous and always has the trivial (paramagnetic) solution ($\langle \sigma_i \rangle = 0$) (paramagnetic solution corresponding to disordered spin orientations). Besides, again for $h_i = 0$, and if $J_{ij} > 0$ (*ferromagnetic case*), under certain conditions Eq. (5.138) has nontrivial solutions of type $\pm\langle \sigma_i \rangle$, which describe the ferromagnetic order. This important result is due to the positive sign of J_{ij}. To elucidate this point we stress that the sign of the argument h_i^{MF}/T of odd function $B_J(h_i^{\text{MF}}/T)$ is the same as the sign of the order parameter $\langle \sigma_i \rangle$ and, hence, one can write Eq. (5.138) in the form $B_J(\pm|h^{\text{MF}}|) = \pm\langle \sigma_i \rangle$, which is equivalent to $B_J(|h^{\text{MF}}|) = \langle \sigma_i \rangle$. The even function $[B_J(|h^{\text{MF}}|)/\langle \sigma_i \rangle - 1]$ may have real nontrivial ($\langle \sigma_i \rangle$) solutions in couples of the type $\pm\langle \sigma_i \rangle$.

If $J_{ij} < 0$ (*antiferromagnetic case*) and $h_i = 0$, $\mp|h^{\text{MF}}|$ will correspond to $\pm\langle \sigma_i \rangle$, and the Eq. (5.138) will read $\mp B_J(|h^{\text{MF}}|) = \pm\langle \sigma_i \rangle$ (here the odd property of the Brillouin function and the negative sign of J_{ij} are essential). In this case, again for $T \geq 0$, there is only a trivial solution ($\langle \sigma_i \rangle = 0$), which means that the system cannot exhibit any ferromagnetic order for non-negative temperatures. But we still have an option of learning from Eq. (5.138). We may suppose that the mean spins $\langle \sigma_j \rangle$ in Eq. (5.130), which are the interaction neighbours of σ_i and create the mean field $h_i^{\rm MF}$ at site i, have a direction (sign) opposite to that of $\langle\sigma_i\rangle$ on the l.h.s. of Eq. (5.138). In fact, this is the idea of a simple antiferromagnetic order, provided the spins σ_j are *nn* of σ_i. To describe the antiferromagnetic order of simplest type one should divide the crystal lattice in two equivalent interpenetrating sublattices (Section 3.8.4) and consider the magnetizations (M_1 and M_2) of these sublattices as different order parameters – a problem which will not be discussed in this book; see, e.g., Mattis (1985), Getzlaff (2008), and references therein.

Let we continue the discussion for the zero external magnetic field ($h_i = 0$). We have already shown that in the ferromagnet ($J_{ij} > 0$) – the system which will be considered henceforth, there may be a couple $\langle\sigma_i\rangle_\pm = \pm|\langle\sigma_i\rangle| \neq 0$ of nontrivial solutions which are expected to describe the ferromagnetic order. In contrast with the case of ferromagnet in external field, where odd number of nontrivial solutions may occur (see, e.g., Section 4.10), for $h_i = 0$, these even number solutions are symmetric, they appear in couples ($\langle\sigma_i\rangle_\pm$) of type $\langle\sigma_i\rangle_+ = -|\langle\sigma_i\rangle_-|$, and the solutions in any couple are absolutely equivalent in their physical meaning and the description of the system thermodynamics. Here we shall discuss this point in more details. In Section 5.8.6 we will truncate the series expansion of $B_J(y)$ for small y to the third order in y. Then we shall remain with the possibility to have only one couple of nontrivial solutions of Eq. (5.138) but namely this couple will describe a proper ferromagnetic phase.

The original field configuration $\sigma(J) = \{\sigma_i(J)\}$ is an odd function of the quantum number J and, therefore, the quadratic form in the Hamiltonian (5.110) is an even (lattice) functional of this lattice field configuration. So, for $h = 0$ the Ising Hamiltonian (5.110) is invariant under the change of sign of the field configuration, namely, with respect to a "parity transformation" in the N-dimensional σ-space. The same symmetry is duplicated in the even function (5.134)–(5.135), representing the free energy $\Phi(\langle\sigma\rangle)$, which directly results from the form of the initial Hamiltonian. Further, the same symmetry property is imitated by the odd function $B_J(y)$ which enters in the equation of state (5.138). Now we consider functions of the averaged field configuration $\langle\sigma\rangle = \{\langle\sigma_i\rangle\}$ which has not the same J-dependence as that in the initial field configuration σ. When $h_i = 0$, the statistical averages $\pm\langle\sigma_i\rangle \neq 0$ follow the symmetry of the initial spin variables σ_i: "$\pm$" symmetry, or, which is the same, *up-down* orientational symmetry.

Recall that two thermodynamic states are equivalent if they have equal free energies. As the thermodynamic potential (5.134)–(5.135) is an even function of a_i, the energies of any couple of antisymmetric mean configurations , $\{\langle\sigma\rangle\}_+ = \{\langle\sigma_i\rangle\}$ and $\{\langle\sigma\rangle\}_- = \{-\langle\sigma_i\rangle\}$, are equal at any given $T \geq 0$ and $h_i = 0$. So, we can describe the system thermodynamics by taking one of the solutions of Eq. (5.138) for $h_i = 0$, for example, $\langle\sigma_i\rangle > 0$ and forget about the second one, provided we have no further interest in the symmetry analysis. This equivalence is called *discrete (up-down) symmetry* and is sometimes referred to as $Z_{(2J+1)}$ *symmetry* (for $J = 1/2$, Z_2 *symmetry*).

The fact that the original spin states σ_i are quantized, is not the reason to call this symmetry "discrete". As we see from Eq. (5.138) the mean spin $\langle\sigma_i\rangle$ is not discrete but continuously vary in the interval $[-1, 1]$. The symmetry remains discrete even in the semi-classical limit ($J \to \infty$) when it transforms to Z_∞ *symmetry.* The reason is that σ_i is the $\hat{z}$-component of the vector $(0, 0, \sigma_i)$ and behaves as a pseudoscalar. As the order parameter is a component of a vector and the latter is oriented along the positive $\hat{z}$-direction, it cannot take the opposite direction by a continuous rotation, i.e., without abrupt "up"–"down" jump, because the other vector components are constrained to zero by an extreme magnetic anisotropy. In systems where the order parameter is an actual scalar the discrete symmetry is not present.

This purely geometric symmetry is broken by the external field ($h_i \neq 0$), conjugate to both the original field σ_i and the order parameter field $\langle\sigma_i\rangle$. For the original spins σ_i this is obvious from the Hamiltonian (5.110). For $h_i \neq 0$, Eq. (5.138) clearly demonstrates that the different spins "feel" the direction of the field h_i as more suitable for their alignment. Thus the two possible (*up* and *down*) directions become nonequivalent for the spins. In such cases we speak for the phenomenon of *breaking of the discrete symmetry.*

When $h_i \equiv 0$ but $\langle\sigma_i\rangle \neq 0$ (for instance, we choose $\langle\sigma_i\rangle > 0$) and the molecular field $h^{\mathrm{MF}} > 0$, the Zeeman energy in the MF Hamiltonian (5.128b) is different for the different signs of σ_i (note, that now σ_i and $\langle\sigma_i\rangle$ are different variables and the sign of σ_i does not depend on the sign of $\langle\sigma_i\rangle \neq 0$ and, hence, on the sign of h_i^{MF}). Therefore, in the absence of h_i, the spins σ_i "feel" the directions of h_i^{MF} as their preferred directions of orientation. The spins may take these directions, at least, partially (by their mean statistical portions $\langle\sigma_i\rangle$ only), provided the ratios h_i^{MF}/T in the respective Zeeman terms $\langle\sigma_i\rangle h_i^{\mathrm{MF}}/T$ are enough large in order the molecular field ordering effect to prevail over the destructive kinetic energy $k_{\rm B}T$ ($k_{\rm B} = 1$). When this order occurs as a result of the molecular field action in the system, without the external help, exercised by h_i, we speak about *a spontaneous breaking of the symmetry*, in our case, a spontaneous breaking of the discrete ("up-down"; $\pm$) symmetry.

Usually this discussion is performed for an uniform external field $h_i = h$ and uniform molecular field $h_i^{\rm MF} = h^{\rm MF}$, and then all spins σ_i are in equivalent conditions. But the present generalized picture within which we consider a possible i-dependencies of h and $h^{\rm MF}$, is also self-consistent within the MF concept and, in particular, in the interpretation of Eq. (5.138). The possible solutions of Eq. (5.138) clearly show that below some temperature, the molecular field configuration, which depends on the equilibrium (mean) spins will be either predominantly positive or predominantly negative.

The concept of the spontaneous symmetry breaking has a direct generalization for vector models like the XY and the Heisenberg models. When the vector order parameter field, $(\langle\sigma_i^{(x)}\rangle, \langle\sigma_i^{(y)}\rangle)$ for the XY model and $(\langle\sigma_i^{(x)}\rangle, \langle\sigma_i^{(y)}\rangle, \langle\sigma_i^{(z)}\rangle)$ for the Heisenberg model, may take any (2D or 3D) spatial direction within the same thermodynamic state, i.e., without change of the free energy, we say that the system possesses "a continuous symmetry". This symmetry can be broken by some external field, conjugate to the vector order parameter field of the system (*breaking of the continuous symmetry*), or, by the internal molecular field (*spontaneous breaking of the continuous symmetry*). The remarkable phenomenon of spontaneous symmetry breaking usually produces continuous phase transitions because the latter are described by thermodynamic potentials which are even functions of the order parameter and produce equations of state which contain only odd functions of the order parameter components.

Another well-known function $L(y)$, first introduced by Langevin on the basis of classical statistics of free (noninteracting) magnetic moments, is just the limiting case $B_\infty(y_i)$ and describes the semi-classical limit; see our discussion in Section 5.6.2. Taking the limit $J \to \infty$ in Eq. (5.139) we obtain the *Langevin function*

$$L(a_i) = \coth a_i - \frac{1}{a_i}. \tag{5.140a}$$

The usual ($J = 1/2$)–Ising model corresponds to the particular Brillouin function

$$B_{\frac{1}{2}}(a_i) = \tanh a_i, \qquad J = \frac{1}{2}. \tag{5.140b}$$

This result is easily derived repeating the calculation of the partition function sum (5.133b) for $J = 1/2$. Both $B_J(y)$ and $L(y)$ are odd functions of y and are expanded in odd powers of y (for $y \ll 1$). Φ_i from Eq. (5.135) is an even function of a_i.

5.8.3 *Self-consistency of the mean-field approximation*

So far we have demonstrated an essential part of the MF approximation. Within the MF theory we have assumed that one of the spin-like variable σ_i in the coupling term $\sigma_i \sigma_j$ can be replaced by $\langle \sigma_i \rangle$ — an unspecified quantity to be determined at a later stage of investigation. So what we make by the replacement $\sigma_i \sigma_j \to \langle \sigma_i \rangle \langle \sigma_j \rangle$ is to exclude $\langle \sigma_j \rangle$ from the degrees of freedom and to represent the spin-like variables σ_i as independent on each other. That is, we disregard the direct correlations (interactions) between the spins but we allow for correlations between a spin σ_i at site i and a "mean spin" $\langle \sigma_j \rangle$ at site j. This substitution of the initial spin interactions allows for a straightforward calculation of the partition function.

Up to some advanced stage of consideration, the unspecified quantity $\langle \sigma_j \rangle$ in the MF Hamiltonian (5.128b) can be interpreted as a nonequilibrium order parameter. This loose treatment of the quantity $\langle \sigma_i \rangle$ can be extended up to the stage of the derivation of Eq. (5.134) and Eq. (5.135) for the thermodynamic potential $\Phi(\langle \sigma \rangle)$. In fact, we often draw the shape of functions of type $\Phi(\langle \sigma \rangle)$, where the field $\langle \sigma \rangle$ takes a wide range of values. Of course, this is practically possible when the field $\langle \sigma \rangle$ is uniform; see Section 5.8.7 and examples in Chapter 4. Such an arbitrary order parameter $\langle \sigma \rangle$ defines the thermodynamic potential $\Phi(\langle \sigma \rangle)$ in Eqs. (5.134) and (5.135) as *the nonequilibrium thermodynamic potential* of the system, because almost all values of this function do not describe stable or metastable equilibrium states (Section 2.5). Exceptions are only the minimal values of $\Phi(\langle \sigma \rangle)$ which correspond to equilibrium states of physical interest (see Section 2.5).

At the next stage of the MF procedure, when the equilibrium equation of state (5.137) is applied to determine $\langle \sigma_i \rangle$, we have no choice but to interpret $\langle \sigma_i \rangle$ on the l.h.s. of Eq. (5.138) as equal to the equilibrium order parameter ($\langle \sigma_i \rangle = \langle \sigma_i \rangle_{\text{eq}}$). This is so, because the general thermodynamic relation Eq. (5.137) results from the second thermodynamic law and, hence, describes only equilibrium states.

The result Eq. (5.138) can be derived directly from the definition for statistical average in the statistical ensemble defined by the MF Hamiltonian

Eq. (5.128b). The calculation is quite simple for $J = 1/2$. Such calculation merely confirms that the quantity $\langle \sigma_i \rangle$ on the l.h.s. of the equation of state Eq. (5.138) is the supposed statistical average and that this average belongs to the equilibrium field configuration $\langle \sigma \rangle_{\text{eq}} = \{\langle \sigma_i \rangle_{\text{eq}}\}$. But we are still not convinced that the quantities $\langle \sigma_j \rangle$, which enter in the argument a_i of the Brillouin function $B_J(a_i)$ also belong to the same equilibrium configuration; even, we are not certain that these are equilibrium quantities at all.

The problem is how to convince ourselves that the quantity a_i on the r.h.s. of the equation of state (5.138), which depends on variables $\langle \sigma_j \rangle$, also takes equilibrium values. This problem can be solved by deriving the explicit form of the variational equation of state $\delta\Phi(\langle \sigma \rangle)/\delta\langle \sigma_i \rangle = 0$, where $\delta\Phi(\langle \sigma \rangle)$ is given by Eqs. (5.134) and (5.135). The equilibrium states $\langle \sigma \rangle_{\text{eq}}$ are obtained as solutions of this equation. The calculation is straightforward and, in result, we obtain the same mathematical form of the equation of state as that given by Eq. (5.138):

$$\langle \sigma_i \rangle_{\text{eq}} = B_J \left(\frac{h_i}{T} + \frac{1}{2T} \sum_j J_{ij} \langle \sigma_j \rangle_{\text{eq}} \right) . \tag{5.141}$$

The fact that the form of this equation is the same as the form of Eq. (5.138) contains the prove of the self-consistency of our MF approach. As the l.h.s. of Eq. (5.138) consists of $\langle \sigma_i \rangle = \langle \sigma_i \rangle_{\text{eq}}$ and both sides of Eq. (5.141) contain averages from the same field configuration, we may be certain that the argument a_i of $B_J(a_i)$ on the r.h.s. of Eq. (5.138) also involves equilibrium values $\langle \sigma_j \rangle_{\text{eq}}$ of the averaged field variables rather than arbitrary ones $(\langle \sigma_j \rangle)$.

Now we are sure that $\langle \sigma_i \rangle$ on both sides of the equation of state (5.138) belong to the same averaged field configuration and that this configuration is one of the possible equilibria of the system: $\sigma_{\text{eq}} = \{\langle \sigma_i \rangle_{\text{eq}}\}$. In particular, the same equilibrium configuration may be one of the possible global minima of the equation of state (5.141) which describes stable phases. We have to solve the equation of state in order to obtain the possible equilibria: unstable, (metastable, if available), and stable phases (see Section 2.5). In this way we have proven the self-consistency of the MF analysis in a quite general case when the system is inhomogeneous due the presence of spatially varying external field.

The MF approximation neglects the fluctuations. The latter are defined as deviations from the exact statistical averages $\langle \sigma_i \rangle_{\text{ex}}$ – the averages over the probability distribution function $\exp(-H/T)$, where H is the exact (initial) Hamiltonian (in our case, the Ising Hamiltonian (5.110). Obviously, our statistical averages $\langle\sigma_i\rangle$, calculated by the reduced MF Hamiltonian (5.128b) are not exact statistical averages but rather approximate $\langle\sigma_i\rangle_{\text{MF}}$ and, hence, the (neglected) deviations $\delta\sigma_i = (\sigma_i - \langle\sigma_i\rangle_{\text{MF}})$ from the averages in MF approximation — $\langle\sigma_i\rangle_{\text{MF}}$, are not the proper fluctuations $\delta\sigma_i^{\text{ex}} = (\sigma_i - \langle\sigma_i\rangle_{\text{ex}})$.

In fact we have demonstrated a lowest order approximation — the MF approximation within the general self-consistency approach in statistical physics (Uzunov, 1996, 2008). If the fluctuation term indicated in Eq. (5.127) is properly taken into account (Uzunov, 1996, 2008), the MF approximation will be well justified for finite value of the parameter $J_0 = zJ_{\text{ex}}$, in which the number of interacting neighbours (see Section 5.6.2) tends to infinity ($z \to \infty$) and the nn exchange constant J_{ex} tends to zero ($J_{\text{ex}} \to 0$). This is the simplest way to simulate extremely long range interaction. In the present case we use "infinitesimally small and infinitely ranged interaction" in a way similar to classic studies of the Van der Waals condensation (note, that the Van der Waals equation is an equation of MF type); see Section 5.9.1 and references therein.

5.8.4 *Free spins and Curie law*

If the spins do not interact ($J_{ij} = 0$), $h_i^{\text{MF}} = 0$, $a_i = h_i/T$, and the nonzero average $\langle\sigma_i\rangle$ is produced only by the Zeeman term in the Hamiltonian (5.128a)–(5.128b), where we must set $A(\langle\sigma\rangle) = 0$, and $h_i^{\text{eff}} = h)i$. The equation of state (5.138) is still valid for $a_i = h_i/T$, and we can use it in the calculation of the magnetic susceptibility $\chi_0(T) = (\partial M/\partial H)_T$ in zero external magnetic field ($H = 0$); here $M = \sum_i M_i = g_J\mu_{\text{B}}J\sum_i\langle\sigma_i\rangle$ (see Section 5.6.2).

We deal with an inhomogeneous external field H_i and we will continue the consideration within this variant of the theory. In the present case $h_i^{\text{MF}} = 0$ and because of that, the equations of state (5.138) are completely decoupled. Then $\partial\langle\sigma_i\rangle/\partial h_j = \delta_{ij}(\partial\sigma_i/\partial h_i$ and, hence, $\chi = \sum_i \chi_i$, where $\chi_i = \partial\langle M_i\rangle/\partial H_i$. Bearing in mind that $H_i = h_i/g_J\mu_{\text{B}}J$ (see Section 5.6.2) and using Eq. (5.138), we can easily obtain $\chi_i = (g_J\mu_{\text{B}}J)^2 B'_J(a_i)/T$, where $B'_J(a_i)$ is the first derivative of $B_J(a_i)$ with respect to a_i. We need the values $\chi_0(T) \equiv \chi(T, H_i = 0)$ and $\chi_i^{(0)} = \chi_i(T, H_i = 0)$ of χ and χ_i at zero external field ($H_i = 0$), so we need to calculate $B'_J(a_i)$ in the limit $a_i \to 0$. Taking this limit with the help of Eq. (5.139) we obtain $B'_J(0) = (J+1)/3J$. Now we readily see that $\chi_i^{(0)}$ does not depend on the site i and is given by

$$\chi_0 = \frac{C}{T}, \qquad C = \frac{1}{3} N (g_J \mu_B)^2 J(J+1) \tag{5.142}$$

(*Curie law* and *Curie constant* C); in cgs system one should change the factor 1/3 to $1/3k_B$ (recall, that here we work with $k_B = 1$). Sometimes, this law is written in the following form $\chi_0 = N\mu_{\text{eff}}^2/3T$, where $\mu_{\text{eff}} = g_J \mu_B \sqrt{J(J+1)}$ is the so-called effective spin moment of the atom (site i).

The same "inverse temperature dependence" is derived for homogeneous system of free spins, where the field H_i is uniform ($H_i = H$, $h_i = h$), $a_i = a$ and $\langle \sigma_i \rangle = \langle \sigma \rangle$, and Eq. (5.138) describes the single variable $\langle \sigma_i \rangle = \langle \sigma \rangle$. Having in mind that $M = N g_J \mu_B J \langle \sigma \rangle$ and $H = h / g_J \mu_B J$, the Eq. (5.142) is obtained directly from $\chi_0 = (\partial M / \partial H)_T$, where one should use again Eq. (5.138) for $\langle \sigma \rangle$. The Curie law (5.142) is valid only for noninteracting spins. In real ferromagnets the magnetic susceptibility has the form $\chi_0 \sim |T - T_c|^{-1}$, where T_c is the critical temperature (see Sections 5.8.5 and 5.8.6).

5.8.5 *Uniform systems and equation of state*

Our consideration is somewhat complicated because the external field h_i is nonuniform and no assumptions have been made about the translational invariance of the system in a zero field ($h_i = 0$), namely, about the properties of the exchange matrix $\hat{J} = \{J_{ij}\}$ (Section 5.6.2). Usually, the MF approximation is demonstrated for uniform external fields ($h_i = h$) and translational invariant exchange interaction ($J_{ij} \equiv J(|i - j|)$) in a perfect (regular) "infinite" lattice, where the surface effects do not exist, or, for the spin behaviour in the bulk of a large lattice, where the surface effects can be ignored. Here we shall restrict ourselves to this case but, as is easy to see, the subsequent considerations can be directly generalized for inhomogeneous systems.

Setting $h_i = h$ in Eq. (5.110) together with the assumption for the translational invariance of the exchange interaction do not directly lead to the conclusion that the equilibrium field configurations should be uniform ($\langle \sigma_i \rangle = \langle \sigma_i \rangle \equiv \bar{\sigma}$). One may recall the phenomenological concept that the equilibrium order in a homogeneous system should be macroscopically uniform but this generally true statement is not entirely convincing in our microscopic MF theory, where the averages $\langle \sigma_i \rangle$ are locally defined.

The reliable justification of the uniformity of stable equilibria is given at a next stage of the theory when we consider the effective fluctuation Hamiltonian of the system in the continuum limit ($i \to x$) (see, e.g., Eq. (7.1) in

Section 7.1). This effective Hamiltonian is the counterpart of the thermodynamic potential (5.134)–(5.135) in the field theory of phase transitions. Equation (7.1) contains a positive gradient term which raises the energy of the non-uniform field configuration. Therefore, to any possible nonuniform field configuration, there may be found a uniform configuration with a lower energy. As the stable equilibrium states are minima of the thermodynamic potential (5.134)–(5.135) and the effective Hamiltonian (7.1), we may conclude that these states (phases) are given by uniform field configurations (both $\bar{\sigma} \neq 0$ and $\bar{\sigma} = 0$).

Note, that reasons for a non-uniformity of the equilibrium MF solutions for the order parameter field $\langle \sigma_i \rangle$ in the bulk of the system could be:

(1) Inhomogeneous external field h_i, conjugate to the order parameter.
(2) Surface effects in small systems.
(3) Inhomogeneities due to irregular composition and lattice defects.
(4) Spatially varying gauge fields interacting with the order parameter (for example, the vector potential of the magnetic field in type II superconductors).
(5) Fluctuations.

Some of these cases are reviewed in Chapter 9. In studies beyond the MF approximation (Sections 7–9), when the fluctuations $\delta\sigma_i$ are taken into account, the total order parameter field $\sigma = \langle \sigma \rangle + \delta\sigma_i$ always depends on the site i (the fluctuation field part $\delta\sigma_i$ always contains uniform ($\delta\sigma$) and spatially varying ($\delta\sigma_i'$) components, i.e., $\delta\sigma_i = \delta\sigma + \delta\sigma_i'$, where $\delta\sigma \neq 0$ at any site i).

The sources of non-uniformity of $\langle \sigma_i \rangle$ within the framework of the relatively simple Ising model (5.110), defined on a regular lattice are only three: the inhomogeneous field h_i, a presumable lack of translational invariance of the exchange interaction, i.e., irregular J_{ij} bonds, and surface effects, i.e., the cases (1)–(3), enumerated above. Excluding these options, we set $h_i = h$ — uniform external field, assume that the exchange is translational invariant ($J_{ij} = J(|i - j|)$), and consider only the behaviour in the bulk of a large regular lattice.

The translational invariance of J_{ij} is not sufficient to ensure uniformity of the molecular field h_i^{MF} given by Eq. (5.130). But following an argument already given in this Section, we may set $\langle \sigma_j \rangle = \langle \sigma \rangle$ in Eq. (5.130) and then the sum over j will give

$$J_0 = \sum_j J_{ij} = zJ_{\text{ex}}, \tag{5.143}$$

where z is the number of nn sites, and J_{ex}, like in Eq. (5.108), is the exchange constant of any nn couple of spins. For a simple hyper-cubic dD lattice, $z = 2d$ (Section 5.6.2). But in what follows we may not be involved in any specification of the lattice structure and dimensionality d, and work with the effective exchange parameter J_0, which does not depend on the sites i. Now we have $a = h^{\text{eff}}/T$ and $h^{\text{eff}} = h + J_0\bar{\sigma}$; cf. Eq. (5.136), Eq. (5.129), and Eq. (5.130).

The assumptions made up to now make possible to consider the mean value of any thermodynamic quantity as independent of the site i and reduce the system of N coupled equations of state (5.138) to a single equation for the uniform field (order parameter) $\bar{\sigma}$. For example, Eq. (5.131) for the energy $A(\langle\sigma\rangle)$ becomes

$$A = \frac{1}{2}N\hat{J}_0\bar{\sigma}^2. \tag{5.144}$$

The single-particle potentials Φ_i, given by Eq. (5.135), are equal and the Gibbs potential (5.134) will be

$$\Phi = \frac{1}{2}NJ_0\bar{\sigma}^2 - NT\ln\left[\frac{\sinh\left(1+\frac{1}{2J}\right)a}{\sinh\frac{a}{2J}}\right], \tag{5.145a}$$

for $(J > 1/2)$–systems and

$$\Phi = \frac{1}{2}NJ_0\bar{\sigma}^2 - NT\ln\left[2\cosh\left(\frac{h+J_0\bar{\sigma}}{T}\right)\right], \tag{5.145b}$$

for a two-state $(J = 1/2)$–systems.

The mean overall magnetization is given by

$$M = g_J\mu_{\text{B}}J\sum_i\langle\sigma_i\rangle = N(g_J\mu_{\text{BJ}}\bar{\sigma}). \tag{5.146a}$$

By using Eqs. (5.137) and (5.145a), or, directly from (Eq. (5.138)), we obtain

$$M(T,H) = M_0\bar{\sigma}, \tag{5.146b}$$

with

$$\bar{\sigma} = B_J(a) \equiv B_J\left(\frac{h + J_0\bar{\sigma}}{T}\right) \tag{5.147}$$

and

$$M_0 \equiv M(0,H) = m_0 N, \qquad m_0 = g_J\mu_B J, \qquad h = m_0 H. \tag{5.148}$$

The magnetization M_0 at $T = 0$ describes the total spin polarization. In Section 5.8.2 we pointed out the same quantity as the saturated magnetization ($M_s = M_0$, $m_s = m_0$), corresponding to $B_J(\infty) = 1$ and $\bar{\sigma}(T \to 0) \to 1$. When this limit is achieved by very large external magnetic field H we usually speak for a saturated magnetization. When this limiting case is achieved by lowering the temperature to the absolute zero this entirely spin-polarized state is often referred to as "the ferromagnetic ground state". The usual form of the equation of state (5.146b)–(5.147) is

$$M = M_0 B_J\left[\frac{m_0}{T}(H + \lambda M)\right], \tag{5.149}$$

where the parameter

$$\lambda = \frac{J_0}{m_0^2 N} = \frac{J^2}{m_0^2 N} J_0', \tag{5.150}$$

with $J_0' = J_0/J^2$ is the actual exchange constant. To elucidate this point, remember that in the Ising model, Eq. (5.108) and Eq. (5.110), we use the variable $\sigma_i = \sigma_i'/J$; see Section 5.6.2. If we take the variable σ_i', which is often used in representations of spin models, the exchange parameters J_{ij} in Eq. (5.110) and J_{ex} in Eq. (5.108) should be multiplied by $1/J^2$. Then we shall have the actual exchange parameters: $J_{ij}' = J_{ij}/J^2$ and $J_{\text{ex}}' = J_{\text{ex}}/J^2$. In Ising models, represented in terms of σ_i', the external field h_i is given by $(g_J\mu_B)H_i$; instead of $m_0 H_i = (g_J\mu_B J)H_i$. Having in mind that $m_0 \sim J$ we see that the parameter λ does not depend on J.

The magnetic susceptibility in zero external magnetic field $\chi_0 = (\partial M/\partial H)_T$ is obtained from the equation of state (5.149). Differentiating this equation with respect to H (the argument λM on the r.h.s. should be differentiated, too), setting $H = 0$ in the resulting equality, and substituting λ from Eq. (5.150), we obtain

$$\chi_0 = \frac{N(g_J \mu_B J)^2 B'_J(J_0\bar{\sigma}/T)}{T - J_0 B'_J(J_0\bar{\sigma}/T)}, \tag{5.151}$$

where the derivative $B'_J(y) = \partial B_J(y)/\partial y$ is taken at $y = J_0\bar{\sigma}/T$.

Equation (5.151) will become equivalent to the Curie law (5.142) only if the exchange interaction is turned off ($J_0 = 0$). In the paramagnetic phase, where $\bar{\sigma} = 0$, we have $B'_J(0) = (J+1)/3J$ (see also Section 5.8.4 and Eq. (5.153c) in Section 5.8.6) and, hence, Eq. (5.151) takes the simple form

$$\chi_0 = \frac{C}{T - T_c}, \qquad T_c = \frac{J+1}{3J} J_0, \tag{5.152}$$

where C is the Curie constant given by Eq. (5.142) and T_c is the critical temperature of the ferromagnetic phase transition in MF approximation (often referred to as *the Curie temperature*).

Note, that the MF approximation yields phase transition temperatures which are much higher than the real ones. The MF phase transition temperatures for various systems are at the upper bound of the possible theoretical predictions for these quantities because the order parameter fluctuations are completely ignored within the MF approximation. This explanation is quite plausible because the fluctuations have a substantial destructive effect on the order. So, the fluctuation contributions to T_c have a negative sign; note, that a more detailed investigation within the framework of the general self-consistency approach yields satisfactory results for the critical temperature T_c; see Uzunov (1996, 2008).

In order to describe the shape of the magnetic susceptibility in the ferromagnetic phase (for $T < T_c$) we need to know the equilibrium value of the order parameter $\bar{\sigma}$; see Section 5.8.6, where we obtain this value in Landau expansion of equation of state. Note, the total angular momentum quantum number and the interaction constant J_0 are effective parameters of the atoms at the sites of lattice. The values of these material parameters depend on the electronic properties. The latter depend on thermodynamic parameters as T, P, and chemical composition c, and this justifies the observed dependencies $C(T, P, c)$ and $T_c(P, c)$ in some materials; see also our discussion of the magnetic susceptibility in Section 3.8.2, where general phenomenological arguments have been presented.

5.8.6 *Expansion of the equation of state and the critical point*

Here we shall show that the MF theory confirms the phenomenological Landau expansion and the thermodynamics of second order phase transitions presented in Sections 4.6 and 4.10.1. We expand the Brillouin function in Eq. (5.147) for small a to order a^3 and neglect the terms of order $O(a^5)$. Besides, we shall assume that $h \ll J_0\bar{\sigma}$ and keep only the term linear in h, whereas terms of order $O(h^2/T^2)$ and $O(hJ_0^2\bar{\sigma}^2/T^2)$ will be ignored as small. The expansion is straightforward and yields

$$B_J \approx \frac{J+1}{3JT}(h + J_0\bar{\sigma}) - u_0\left(\frac{J_0\bar{\sigma}}{T}\right)^3 + O[(J_0\bar{\sigma}/T)^5] \tag{5.153a}$$

where

$$u_0 = \left(\frac{J+1}{90J^3}\right)(2J^2 + 2J + 1)\,. \tag{5.153b}$$

From Eq. (5.153a) we obtain the derivative $B_J'(J_0\bar{\sigma}/T)$ in Eq. (5.151):

$$B_J'\left(\frac{J_0\bar{\sigma}}{T}\right) = \frac{J+1}{3J} - 3u_0\left(\frac{J_0}{T}\bar{\sigma}\right)^2 + O[(J_0\bar{\sigma}/T)^4] \tag{5.153c}$$

From Eq. (5.147) and Eq. (5.153a) we obtain the following form of the equation of state:

$$-\frac{J+1}{3JT}h + \left(\frac{T_c}{T}\right)t\bar{\sigma} + u\left(\frac{T_c}{T}\right)^3\bar{\sigma}^3 + O(\bar{\sigma}^5) = 0, \tag{5.153d}$$

or,

$$-\bar{h} + t\bar{\sigma} + u\bar{\sigma}^3 = 0, \tag{5.154}$$

where the critical temperature T_c is given by Eq. (5.152), the parameter t is introduced by Eq. (4.67b), $\hat{h} = (J+1)/3JT_c$, and $u(J) = [(3J/(J+1)]^3u_0$ is a material parameter which monotonically varies from $u(1/2) = 1/3$ to $u(\infty) = 3/5$. The Eq. (5.154) was obtained by setting all factors of type T_c/T in Eq. (5.153d) equal to unity; for example, the factor $1/T$ in the first term of Eq. (5.153d) can be written as $(T_c/TT_c) \approx 1/T_c$ and, in result, the field $\bar{h}$ has a factor $1/T_c$.

Note, that the substitution of all factors T_c/T in Eq. (5.153d) with unity was an obligatory step in the derivation of the correct equation of state Eq. (5.154). To elucidate this point we must remind that our expansion is valid for small $a \sim J_0\bar{\sigma}/T \sim |t| < 1$; (note, we have assumed $h \ll J_0\bar{\sigma}$). For $h = 0$, the Eq. (5.153d) has two types of solutions: $\bar{\sigma}_{\rm P} = 0$ (the para phase), and $\bar{\sigma}_{\rm F}^2 = (T/T_c)^2(-t)/u$ (the ferromagnetic phase order parameter for $T < T_c$). If we do not substitute $T = T_c(1+t)$ with T_c in the factor $(T/T_c)^2$, $\bar{\sigma}_{\rm F}^2$ will contain small corrections, $2(-t)^2/u$ and $(-t)^3/u$, to the leading term $(-t)/u$. But these corrections are beyond the accuracy of our theory and for this reason we should not take them into account. Rather, we should accept the expression $\bar{\sigma}_{\rm F}^2 = (-t)/u$ as the correct result for the equilibrium value of $\bar{\sigma}^2$.

To justify this point of view, consider the neglected terms in the series for the equation of state (5.153d). The first neglected term in this infinite series is equal to zero in the para phase and is of order $\bar{\sigma}_{\rm F}^5 \sim (-t)^{5/2}$ in the ferromagnetic phase. Therefore, in our approximation for the equation of state we ignore terms of order $O(|t|^{5/2})$. The factors (T_c/T) in Eq. (5.153d) generate such terms and should be excluded.

So, we have formulated the general rule of dealing with expansions like (5.153a), (5.153c), and (5.153d). The calculation of the quantity $\bar{\sigma}_{\rm F}^2 = (-t)/u$ is an example of a consistent calculation of any thermodynamic quantity within this series expansion approach. Besides, it became evident that Eq. (5.153d) is just an intermediate step in the derivation of the equation of state Eq. (5.154).

The free energy Φ can be obtained either by a direct expansion of Eq. (5.145a), or, by performing the integration of the l.h.s. of Eq. (5.154) with respect to $\bar{\sigma}$. The first variant needs simple but lengthy calculations. The second variant is shorter and more instructive. The integration of the l.h.s. of Eq. (5.154) yields the dimensionless free energy density

$$\bar{\Phi} = \bar{h}\bar{\sigma} + \frac{t}{2}\bar{\sigma} + \frac{u}{4}\bar{\sigma}^4, \tag{5.155a}$$

(the integration constant is zero because we take into account only energies which depend on $\bar{\sigma}$, i.e. we evaluate only the free energy of the ordered phase). Now the relation between $\bar{\Phi}$ and the free energy Φ is given by $\Phi = \Theta\bar{\Phi}$, where Θ is some energy factor. This is so because Φ is the overall free energy of the N-particle system. Therefore we can write Θ as $\Theta = N\bar{\theta}$ and now the choice of the energy density $\bar{\theta}$ depends on the choice of our energy units. If we measure the energy in temperature units

($k_{\rm B} = 1$) we shall have $\bar{\theta} = T$, which is compatible with Eq. (5.145a) for Φ. Now we remember our "($T_c/T \approx 1$) rule" and set $\Theta = T \approx T_c$. Thus the thermodynamic potential is given by

$$\Phi = NT_c \left(\bar{h}\bar{\sigma} + \frac{t}{2}\bar{\sigma} + \frac{u}{4}\bar{\sigma}^4 \right). \tag{5.155b}$$

This is a $\bar{\sigma}^4$-theory of the type of the phenomenological ϕ^4–theory; cf. Eq. (4.58) and Eq. (4.134).

All results obtained in Sections 4.6 and 4.8.1 for the phenomenological φ^4-theory can be ascribed to the present MF theory. For this aim we have to compare the parameters and the field $\bar{\sigma}$ in Eq. (5.155b) with the respective quantities in Eq. (4.134). In order to make $\bar{\sigma}$ an extensive order parameter like φ we will perform the transformation $N\bar{\sigma} = \eta$. Now the factors in the front of the linear, quadratic and fourth-order terms in our MF theory become: $-\bar{h}T_c$, $tT_c/2N$, and $uT_c/4N^3$. The thermodynamic potential takes the form

$$\Phi = -\frac{(J+1)}{3J}h\eta + \frac{tT_c}{2N}\eta^2 + \frac{uT_c}{4N^3}\eta^4. \tag{5.155c}$$

Using the label "L" for the Landau parameters in Eq. (4.134) and having in mind Eq. (4.67b) and Eq. (4.68), we may easily show that $h_{\rm L} = (J+1)h/2J$, $u_{\rm L} = uT_c/4N^3$, and $r_{\rm L} = \alpha_0 t = tT_c/2N$; $\alpha_0 = T_c/2N$. The order parameter η can be identified with the order parameter φ in Eq. (4.134).

To see how the theory works we will consider the magnetic susceptibility (5.151) in the domain of the ferromagnetic phase ($T < T_c$). The equilibrium value of B'_J is easy to obtain by setting $\bar{\sigma}_{\rm F}^2 = (-t)/u$ in Eq. (5.153c):

$$B'_J(J_0\bar{\sigma}_{\rm F}/T) = \left(\frac{J+1}{3J}\right)(1+3t) \approx \frac{(J+1)}{2J}. \tag{5.156}$$

Note, that the neglected terms in Eq. (5.153c) are of order of order $O(|t|^2)$ and, therefore, the correction $3|t| \ll 1$ in Eq. (5.157) is consistent in this series expansion approach.

With the help of Eq. (5.156) for B'_J we can write the Eq. (5.151) in the following simple form

$$\chi_0(T) = \frac{C(1+3t)}{2(T_c - T)} \approx \frac{C}{2(T_c - T)}, \qquad T < T_c, \qquad |t| \ll 1. \tag{5.157}$$

In a close vicinity of the critical point T_c this magnetic susceptibility is twice smaller than that in the paramagnetic phase, given by Eq. (5.152)(cf. Eq. (4.142) and Eq. (4.143) for the respective results of the phenomenological theory). The $3t$-correction in the nominator of Eq. (5.156) might be significant for $0.1 < 3|t| < 1$, but the check of the reliability of this correction required a calculation of the denominator of Eq. (5.157) to second order in $|t|$, which is beyond the scope of this theory. It may turn out from some more precise calculation that the number factor 3 in this ($3|t|$-) correction should be changed, or nullified.

Alternatively, the Eq. (5.157) can be obtained from Eq. (4.140) and Eq. (4.143). Using the relations between the parameters of the phenomenological theory Eq. (4.134) and the present MF theory, we obtain that $\chi_0' = (\bar{\sigma}/\partial h)_T = (J+1)/6JT(-t)$. Applying the relation $\chi_0 = (g_J \mu_B J)^2 \chi_0'$ as well as our rule that we must set $T = T_c$ everywhere except in t, directly reproduces the result Eq. (5.157) (without the $3|t|$–correction).

When the exchange parameter $\hat{J}_0$ is site-dependent $[J_{0i} = z_i J_{\rm ex}(i)]$ the system will not possess the property of translational invariance. The exchange parameters J_{ij} will produce a site dependence of the exchange interaction if we choose them in the form $J_{ij} = J_{\rm ex}(i)f(|i-j|)$, where $J_{\rm ex}$ is i-dependent, and function $f(x)$ is defined by $f(f(|i-j|) = f(a_0) = 1$ (a_0 is the lattice constant) and $f(|i-j| \neq 0) = 0$. In this way we may introduce a form of irregular exchange (see also Section 5.6.2); a definition of the same problem beyond the nn exchange is straightforward. Then T_c will depend on the site $i : T_c \to T_{ci} \sim J_{0i}$. However the coefficients in the expansion of Φ_i for small $\bar{\sigma}_i$ will give the same linear dependence of the coefficient t, which can now be written as $t_i = (1 - T_{ci}/T)$ and $u_i > 0$. One may try to obtain these results with the help of the equations presented in Section 5.8.2. These results are essentially used in studies of systems with disorder of type "random critical temperature" (see Sections 9.5.1 and 9.5.2).

5.8.7 *Helmholtz free energy and generalized potential*

The equilibrium magnetization M discussed so far is a function of T and H. The supplementary variables Y are: N, λ, z, and J. Having the equilibrium Gibbs potential $\Phi(T, H)$ from Eqs. (5.134) and (5.135), or from Eqs. (5.145a)–(5.145b) and (5.146a)–(5.146b) , we can calculate the equilibrium Helmholtz potential $F(T, M)$ and the generalized Gibbs function $\Phi(T, H/\mathcal{M})$, where $\mathcal{M}$ is the nonequilibrium counterpart of M. We shall demonstrate how this can be performed. The equation of state corresponding to the potential (5.145b) is

$$\bar{\sigma} = \tanh(K\bar{\sigma} + \bar{h}), \tag{5.158}$$

where $K = z\lambda/T$ and $\bar{h} = h/T$. Solving this equation with respect to $\bar{h}$,

$$\bar{h} = \tanh^{-1}\bar{\sigma} - K\bar{\sigma} \tag{5.159}$$

and substituting $\bar{h}(\bar{\sigma})$ in the thermodynamic relation $F(T, M) = \Phi(T, M) + HM$, where $H = \bar{h}/\mu_B T$ and $M = \mu_B N\sigma$ for $J = 1/2$, we have

$$F(T, \bar{\sigma}) = NT\left[-\frac{1}{2}K\bar{\sigma}^2 + \frac{1}{2}\ln(1 - \bar{\sigma}^2) + \bar{\sigma}\tanh^{-1}\bar{\sigma} - \ln 2\right]. \tag{5.160}$$

The generalized Gibbs free energy $\Phi(T, \bar{h}/\sigma)$ with $\sigma = \mathcal{M}/\mu_B N$ is then given by the relation $\Phi(T, \bar{h}/\sigma) = F(T, \sigma) - \sigma h$. The algebraic calculations yield:

$$\Phi(T, \bar{h}/\sigma) = NT\left[-\bar{h}\sigma - \frac{1}{2}K\sigma^2 + \frac{1}{2}\ln(1 - \sigma^2) + \sigma\tanh^{-1}\sigma - \ln 2\right]. \tag{5.161}$$

For the reader will be instructive to minimize this potential, and to confirm the results for the critical behaviour presented in Table 4.1. Often it is very convenient to work in variables $(\sigma, \bar{h})$. The expansion of $\Phi(\sigma)$ from Eq. (5.161) gives the Landau theory in powers of σ^2.

5.9 Real Space Condensation and Statistical Mechanics

5.9.1 *On the derivation of the Van der Waals equation*

The oldest example of the MF approximation is the Van der Waals (1873) theory of condensation. The Van der Waals equation of state $P = P(V, T)$ of a gas–liquid system is

$$P = \frac{NT}{V - Nb} - \frac{aN^2}{V^2}, \tag{5.162}$$

where the positive parameters a and b describe the inter-particle interactions. The parameter b accounts for the effect of the short-range repulsive part of the intermolecular potential and a is related with the attractive long-range tail of the interaction; see Fig. 3.8(a).

The Eq. (5.162) together with Eqs. (2.62) and (3.4) yield the coordinates of the critical point:

$$T_c = \frac{8a}{27b}, \quad P_c = \frac{a}{27b^2}, \quad V_c = 3Nb. \tag{5.163}$$

The parameters a and b are never zero in real systems. The theoretical demonstration that the gas–liquid transition terminates at an isolated critical point as it comes about in real systems is the great success of the Van der Waals theory. Moreover, Eq. (5.162) gives a quite good description of the first order gas–liquid transition. The difference $(\rho_L - \rho_G)$ between the liquid and gas densities below T_c is the main result from the specific action of the Van der Waals intermolecular forces represented by the interaction parameters a and b. The same density difference vanishes when the temperature is increased up to a value, for which the thermal energy prevails over the internal potential energy.

We shall not enter into details of the statistical derivation of the Eq. (5.162). Instead we shall remind how the macroscopic parameters a and b are connected with the microscopic interaction.

Usually, to illustrate how the Van der Waals equation is obtained an expansion of the grand canonical potential $\Omega = -PV$ is made in power series of the density $\rho = N/V$ up to the second virial term; see Uhlenbeck and Ford (1963), Landau and Lifshitz (1980), Ma (1985). Then an appropriate modelling of the interaction potential $u(r)$ shown in Fig. 3.8(a) is applied. This model interaction consists of a hard-core part acting at short distances and a relatively weak long-range attractive part $u_a(r)$. The parameter b turns out proportional to r_0^3, where the hard-core radius r_0 is approximately equal to the molecular size (Fig. 3.8a), and the parameter a is related with the weak attractive potential $u_a(r)$:

$$a \sim \int_{r_0}^{\infty} r^2 |u_a(r)| dr. \tag{5.164}$$

The actual result from the expansion of Ω to the order ρ^2 and from the above mentioned modelling of the interaction is the equation:

$$P = \frac{NT}{V} + \frac{N^2T}{V^2}\left(b - \frac{a}{T}\right). \tag{5.165}$$

The Eq. (5.162) can be obtained from Eq. (5.165) after making the approximation

$$\frac{NT}{V}\left(1+\frac{N}{V}b\right) \approx \frac{NT}{V-Nb}. \tag{5.166}$$

It is important to note that Eq. (5.165) does not give a critical point for any $T > 0$, whereas the Van der Waals equation does. The parameter b appears in the denominator $(V - Nb)$ of Eq. (5.162) and such a result cannot be obtained by a finite series perturbation expansion. The approximation (5.166) rather corresponds to the summation of an infinite geometrical progression. So the Van der Waals idea about the way, in which the intermolecular forces act and the resulting equation (5.162) cannot be deduced by simple perturbation calculations.

A derivation of the Van der Waals formula (5.162) based on the MF approximation is given by Ornstein in 1908; see Uhlenbeck and Ford (1963), Fisher (1965), de Boer (1974). Within the Ornstein approach, Eq. (5.162) is obtained for 1D gases where the effect of the strong repulsive potential can be calculated exactly. For $(d > 1)$-systems however one is forced to apply the low-density limit $V \gg Nb$, i.e., the approximation (5.166). Thus we cannot be certain about the reliability of the formula (5.162) for the domains $(T, P) \leq (T_c, P_c)$ of the phase diagram (T, P, V), where the substance is either a dense gas or a liquid; see also Section 3.6.1.

Although the mentioned statistical derivations of the Van der Waals theory are not mathematically rigorous, they confirm the early Van der Waals idea about the way, in which the intermolecular forces act. Exact theoretical studies can be carried out for 1D and 2D model systems with a two-particle interaction consisting of a hard-core repulsion and an infinitely small attraction of almost-infinite range. It is proven that the condensation of such 1D gases is correctly described by the Van der Waals equation (Kac, Uhlenbeck and Hemmer , 1963, 1964); see also Stanley (1971). This result is further generalized for 3D systems; see Lebowitz and Penrose (1966), Penrose and Lebowtz (1971), and Lebowitz (1974).

The physical picture, in which the realistic interactions are approximated by an infinitesimally small two-particle interaction $u_a(r)$ of an infinite range (i.e., $u_a(r)$ is almost constant) is very similar to the replacement of the two-particle Ising interaction in magnets with the interaction of a particle in a MF. The result of numerous theoretical studies by using models with weak long-range interactions is that now such models in the limit of infinitesimally small-infinite range interaction are considered physically equivalent to the MF Hamiltonians.

5.9.2 *Corresponding states and universality*

Now we shall draw the reader's attention to some important features of the phase transition properties described by Eq. (5.162).

Firstly, the parameters a and b vary from one substance to another and, therefore the critical point coordinates (5.163) as well as the interaction strength itself are not universal features of the critical phenomena. It is a common rule in the phase transition theory that the location of the critical point or, more generally, of the phase transition line(s) depends on the interaction parameters and on certain features of the microscopic degrees of freedom like, for example, the value J of the total angular momentum in magnets (see, e.g., Eq. (5.152) for the Curie temperature T_c).

Secondly, the thermodynamic behaviour near the critical points is universal, that is it remains one and the same for different substances. Now remember the law of the corresponding states; see, e.g., Pippard (1957), Landau and Lifshitz (1980). This law has a very general meaning but for the sake of clarity we shall discuss it for the Van der Waals fluids described by Eq. (5.162). It is more convenient to write this equation with the help of the dimensionless variables $T' = T/T_c, P' = P/P_c$, and $V' = V/V_c$: $P' = P'(T', V')$. Then the function $P'(T', V')$ will not depend on a and b. The reduced parameters of state T', P', and V' in this way give the thermodynamics for all systems satisfying the equation $P' = P'(T', V')$ same as that of the Van der Waals fluid. So *the Van der Waals fluid* is a term introduced to denote the *universality class* for all such systems. It is more suitable to use from now on other dimensionless variables:

$$t = \frac{T - T_c}{T_c}, \quad p = \frac{P - P_c}{P_c}, \quad v = \frac{V - V_c}{V_c}. \tag{5.167}$$

The universal form of Eq. (5.162) is

$$\left[(p+1)(v+1)^2 + 3\right](3v+2) = 8(1+t)(v+1)^2. \tag{5.168}$$

The law of the corresponding states means that if we know two of the parameters p, v, and t, the third one can be always found from Eq. (5.168). If two different systems belong to the Van der Waals class of universality and say, their p and t are equal, the v - s will be equal too. Moreover they will behave in one and the same way at the state (p, v, t) — *the corresponding state.* The law of corresponding states was first suggested by Van der Waals in 1880; see de Boer (1974). But the same law can be applied to other universality classes as well. For example, one can define the universality class of the systems satisfying the φ^4-theory within the Landau expansion or the universality class defined by the MF lattice models discussed in Section 5.8.

It can be demonstrated by simple considerations that as far as the states near the critical point(s) are concerned, all these universality classes coincide. Then we say that all these universality classes fall into the *MF universality class* of the second order phase transitions (ordinary critical points). Although the MF theories give a quite good qualitative description of the critical properties, the real systems never follow exactly the MF predictions. So the universality class coming from the MF theories is a theoretical approximation of the real critical behaviour.

5.9.3 *Critical properties of Van der Waals fluids*

We have already mentioned that the main features of the critical behaviour of the Van der Waals fluids are the same as those of the Landau theory of second order phase transitions and MF Ising magnets. To illustrate this we shall analyze briefly the vicinity of the gas–liquid critical point, where $v \ll 1$.

The expansion of the Eq. (5.168) in power series of v is

$$p = 4t - 6tv + 9v^2 t - \frac{3}{2}(1 + 9t)v^3 + \frac{21}{4}v^4 + \dots \tag{5.169}$$

As in Section 2.6.4 and Section 4.4.5 we shall calculate the dependence of the thermodynamic potentials on the order parameter v. Replacing $P(V)$ from Eq. (5.162) in the relation $P = -(\partial F/\partial V)_{T,N}$ and integrating over V we get the Helmholtz free energy

$$F(V,T) = -NT\ln(V - Nb) - \frac{aN^2}{V} + f(T,N), \tag{5.170}$$

where $f(T,N)$ is an undeterminate function of T and N. It is unsubstantial to our consideration, because it does not depend on V. The generalized Gibbs potential $\Phi(V) = PV + F$ is

$$\Phi(P,T/V) = PV - NT\ln(V - bV) - \frac{aN^2}{V}, \tag{5.171}$$

or, expanding in terms of v,

$$\Delta\Phi(t, p/v) = \frac{3}{8}NT_c \Big[p + pv - 4tv + 3tv^2 - 3tv^3 + \frac{3}{8}(1+9t)v^4 + O(v^5)\Big], \tag{5.172}$$

where $\Delta\Phi = \Phi(t,p,v) - \Phi(t,p,0)$. The effective field conjugate to v is not p but rather $h = (4t - p)$.

Remember that the gas (G)–liquid (L) coexistence conditions are $t_G = t_L$, $p_G = p_L$ and $\Phi_G = \Phi_L$. Imposing these conditions for $T \leq T_c$ and performing lengthy but simple calculations we obtain that

$$v_G = -v_L = 2|t|^{1/2}, \tag{5.173}$$

where

$$v_G = \frac{V_G - V_c}{V_c}, \qquad v_L = \frac{V_L - V_c}{V_c}, \tag{5.174}$$

are the equilibrium values $\bar{v}$ of v for the gas and the liquid states on the coexistence surface. The difference $\Delta v = (v_G - v_L)$ from Eq. (5.173) is given by $\Delta v = 2v_G$ so that $\Delta v = 4|t|^{1/2}$. For $t > 0$ there is a trivial result $\bar{v} = v_G = v_L = 0$.

The relation between p and v on the critical isotherm ($t = 0$) is obtained from the equation of state (5.169). It reads

$$p = \frac{3}{2}|v|^3 + \frac{21}{4}|v|^4 + O(v^5). \tag{5.175}$$

The v^4-term is a correction to the main dependence $p \sim v^3$; cf. Eq. (4.173).

The critical behaviour of the isothermal compressibility K_T is found from Eqs. (5.169) and (2.56):

$$K_T = \frac{1}{P_c\left[6t + \frac{9}{2}\bar{v}^2 + O(|t|^{3/2})\right]}. \tag{5.176}$$

For $t > 0$ the equilibrium value $\bar{v}$ is zero so that $K_T^{>} \sim t^{-1}$. For $t < 0$ the stable states are given by $\bar{v}^2 = v_L^2 = 4|t|$ and, consequently, $K_T^{<} = K_T^{>}(-t)/2 \sim |t|^{-1}$. Clearly, the fluid susceptibility K_T exhibits the same critical behaviour as the magnetic susceptibility χ given by Eqs. (4.172) and (4.173).

The entropy difference $\Delta S = -\partial \Delta\Phi/\partial T$ is calculated from Eq. (5.172):

$$\Delta S = \frac{9}{2} N t. \tag{5.177}$$

The specific heat difference $\Delta C_V = T(\partial \Delta S/\partial T)$ near T_c is

$$\Delta C_V(T) = \frac{9}{2} N \frac{T}{T_c}. \tag{5.178}$$

From Eq. (5.178) the jump of the specific heat at T_c is $\Delta C_V(T_c) = 9N/2$.

The MF analysis makes possible to establish the microscopic mechanism of the phase transition and the relationship between the microscopic and macroscopic parameters of the systems.

The above results can be also derived by using the density difference $\Delta\rho = (\rho - \rho_c)$ as an order parameter. The term tv^3 in Eq. (5.172) is of the order $|t|^{5/2}$ and actually does not contribute to the main t-dependence of the thermodynamic quantities.

As far as the dependence of the thermodynamic functions on the temperature and external field h are concerned, the MF theories confirm the general phenomenological approach (Chapter 4).

Chapter 6

Fluctuations and Fields

6.1 Description of Macroscopic Fluctuations

The notion of the fluctuations of the physical quantities has been widely used in our previous considerations. In Chapter 2 we have studied the stability conditions with respect to macroscopic fluctuations like δV^{α}, δN^{α}, δT^{α} of the volume V^{α}, the number of particles N^{α} and the temperature T^{α} for macroscopic subsystems (cells) α of a given thermodynamic system; see Section 2.7.

The quasi-macroscopic treatment of the fluctuations is formulated within the generalized thermodynamics where both equilibrium and near-to-equilibrium states are described by an extension of the entropy maximum principle. Once again we shall consider a total isolated system $\sigma^t(\varphi^t)$ consisting of the system under consideration $\sigma(\varphi)$ and its surroundings $\sigma^0(\varphi^0)$; see Sections 2.2.4 and 4.2. While the stable equilibrium states $\bar{\varphi}$ are those for which the entropy is maximal $S^t_{\text{max}} = S^t(\bar{\varphi})$, the fluctuation states $\varphi = \bar{\varphi} + \delta\varphi$ correspond to entropy $S^t(\varphi) < S^t_{\text{max}}$. It can be accepted as a postulate in the generalized thermodynamics that the thermodynamic probability of a nonequilibrium state φ is given by the macroscopic distribution function

$$w(\varphi) \sim e^{S^t(\varphi) - S^t_{\text{max}}} \tag{6.1}$$

(Callen, 1960; Landau and Lifshitz, 1980). If the subsystems $\sigma(\varphi)$ and $\sigma^0(\varphi^0)$ of $\sigma^t(\varphi^t)$ have a macroscopic size, we can apply the thermodynamic additivity property and investigate the difference $\Delta S^t = [S^t(\varphi) - S^t_{\text{max}}] \sim O(\delta\varphi^2)$ in the same way as we have done in Section 2.4.2 for the set of extensive variables $a = \{a_i\}$.

The same scheme for the study of the extrema of ΔS^t can be applied to a lattice of macroscopic cells α as we have done in Section 2.7.3. The new point here is that ΔS^t is in the exponent of the thermodynamic probability (6.1) and this implies the use of the methods of statistical physics. We shall not restrict our discussion to macroscopic cells only but we shall also consider quasi-macroscopic subsystems (see, e.g., Section 5.4). Then the additivity property is not necessarily fulfilled and the entropy $S^\alpha(a)$ of a quasi-macroscopic cell α may depend on the variables a_β of cells β, $\beta \neq \alpha$; in particular, $a_i^\alpha = \varphi^\alpha$ according to our notations in Chapter 2.

For reasons explained in Section 4.2 our main interest will be concentrated on the behaviour of the system $\sigma(\varphi)$ at fixed (T, Y, h). Using Eqs. (2.94) we can write the relation $\Delta S^t = -\Delta\Phi(\varphi)/T$, which is valid for fixed T, h and supplementary variable(s) Y. The Gibbs potential difference $\Delta\Phi(\varphi) = \Delta\Phi(\bar{\varphi}, \delta\varphi)$ is given by Eq. (5.22). One may think of $\Delta\Phi(\varphi)$ as the difference between the generalized potential $\Phi(\varphi)$ and its equilibrium value $\bar{\Phi}(\bar{\varphi})$ at the stationary point $\bar{\varphi}(h)$; see, e.g., Eq. (4.30) with $h = \bar{h}$. It seems out of sense to investigate fluctuations around unstable equilibria (Section 2.5). In any vicinity of such equilibria states φ always exist, for which $\Delta S > 0$ (or $\Delta\Phi < 0$) and for these states $w(\varphi)$ is not well-defined ($w \geq 1$). From now on we shall denote by $\bar{\varphi}$ only stable or metastable phases $\bar{\varphi}$ (minima of Φ).

For a system at fixed (T, Y, h) the thermodynamic probability $w(\varphi)$ of the state φ is written in the form

$$w(\varphi) = Ce^{-\Delta\Phi(\varphi)/T} \tag{6.2}$$

in accordance with the relation (6.1). The factor C is independent on φ and is obtained from the normalization condition

$$\int_{-\infty}^{+\infty} w(\varphi) D\varphi = 1, \tag{6.3}$$

where $D\varphi = d\varphi_1 d\varphi_2 \cdots d\varphi_n$ for a n-component order parameter $\varphi = \{\varphi_\alpha, \alpha = 1, \ldots, n\}$. Now we shall employ our considerations from Sections 5.2 and 5.4, which make clear that the postulate (6.2) has a statistical basis.

We have to recall Eq. (5.29), in which the partition function

$$\mathcal{Z}(h) = \int_{-\infty}^{+\infty} D\varphi e^{-\Phi(\varphi)/T} \tag{6.4}$$

is expressed by the nonequilibrium potential (effective Hamiltonian)

$$\Phi(\varphi) = \bar{\Phi}[\bar{\varphi}(h)] + \Delta\Phi(\bar{\varphi}, \delta\varphi), \tag{6.5}$$

where $\Delta\Phi(\bar{\varphi}, 0) = 0$. From Eq. (6.4) and Eq. (6.5) we get

$$\mathcal{Z}(h) = e^{-\Phi(\bar{\varphi})/T}\mathcal{Z}_f, \tag{6.6}$$

where

$$\mathcal{Z}_f = \int_{-\infty}^{+\infty} D\delta\varphi e^{-\Delta\Phi(\bar{\varphi},\delta\varphi)/T} \tag{6.7}$$

is *the fluctuation partition function;* the integration variables φ_α has been changed: $\varphi_\alpha \to \delta\varphi_\alpha$, $d\varphi_\alpha = d\delta\varphi_\alpha$.

The integration over φ (or $\delta\varphi$) in Eqs. (6.4) and (6.7) is between $-\infty$ and ∞ and, hence, states φ far from the stability hypersurface $\varphi = \bar{\varphi}$ shall also contribute to the integrals. The nonequilibrium entropy $S^t(\varphi)$ and the nonequilibrium potential $\Phi(\varphi)$ are certainly well defined for near-to-equilibrium states ($\varphi \sim \bar{\varphi}$). The phenomenological treatment of a state φ far from $\bar{\varphi}$ may be however unreliable. Therefore, expanding the integration over all possible values $(-\infty, \infty)$ of the real quantity φ we suppose that there is a sharp minimum of $\Phi(\varphi)$ at $\bar{\varphi}$, which means the near-to-equilibrium states ($\delta\varphi \ll 1$) to give the main contribution to the statistical integrals.

The total equilibrium Gibbs potential $\Phi(h)$ can be written as a sum

$$\Phi(h) = \bar{\Phi}[\bar{\varphi}(h)] + \Phi_f(\bar{\varphi}), \tag{6.8}$$

where the fluctuation part Φ_f is

$$\Phi_f = -T \ln \mathcal{Z}_f, \tag{6.9}$$

in accordance with Eq. (5.29). In Chapter 5 we have denoted the potential (6.8) by $\widetilde{\Phi}(h)$; see, e.g., Eqs. (5.29) and (5.59). As $\bar{\Phi}(\bar{\varphi})$ is a minimum of $\Phi(\varphi)$, $\Phi_f > 0$ — the fluctuation contribution to the total Gibbs potential $\Phi(h)$ is always positive.

To complete the brief summary of our previous studies of fluctuations we shall note that the postulate (6.1) of the fluctuation probability theory within the generalized thermodynamics can be accepted as a "theorem" in statistical mechanics. Namely, the distribution (6.1) of an isolated system σ^t or the distribution (6.2) of an open system $\sigma(\varphi)$ at fixed T, Y, h can in principle be obtained by a coarse graining of the microscopic statistical distribution $w(\sigma)$. Such problems have already been discussed in Section 5.4. Apart from particular examples in Chapter 5 we have neither calculated the fluctuation contribution to the thermodynamic quantities nor we have given methods for such calculations. This and the next two Chapters are intended to present the systematical way for calculations in the framework of the fluctuation theory of phase transitions.

6.2 Spatially Independent Fluctuations and Anisotropy

Here we shall discuss the simplest case of spatially independent fluctuations. Within this picture:

(i) the quasi-macroscopic cells ν of the lattice considered in Section 5.4 are so large that the correlations between the microscopic degrees of freedom $\sigma_\nu = \{\sigma_i\}_\nu$ and $\sigma_\mu = \{\sigma_i\}_\mu$ of each couple (ν, μ) of cells can safely be ignored, and

(ii) the cells ν are sufficiently small so that the value of the nonequilibrium order parameter φ_ν for each cell ν can be considered uniform, that is spatially independent.

Now we can treat the cell ν as an independent thermodynamic system. It can be shown within the field fluctuation theory (Section 7.6) that a macroscopic body, for which φ is spatially independent represents a "zero-dimensional system" ("mote" or little sphere).

The problem is to calculate the multiple integral (6.4) over the spatially independent components φ_α of the order parameter φ and then to obtain $\Phi(h) = -T \ln \mathcal{Z}(h)$; see Eq. (6.8). We can try to perform this with the help of the expansion (4.30) for $\Phi(\varphi)$, where $\bar{\Phi}(\bar{\varphi}(h)) = \bar{\Phi}(h); h = \bar{h}$. For anisotropic systems with a multi-component order parameter $\varphi = \{\varphi_\alpha, \alpha = 1, \ldots, n\}$ this expansion can be written in the following general form

$$\Delta\Phi(\bar{\varphi}, \delta\varphi) = \sum_{l=2}^{\infty} \sum_{\alpha_1,\ldots,\alpha_l} u^{(l)}_{\alpha_1\ldots\alpha_l}(\bar{\varphi})\delta\varphi_{\alpha_1} \cdots \delta\varphi_{\alpha_l}. \tag{6.10}$$

Such an approach implies an integration in the partition function (6.4) by the saddle point method, in which the stable phase $\bar{\varphi} = \{\bar{\varphi}_\alpha\}$ is the saddle point of $\Phi(\varphi) \equiv \Phi(\bar{\varphi}, \delta\varphi)$, obtained from the equations $(\partial\Phi/\partial\varphi_\alpha)_{\bar{\varphi}_\alpha} = 0$. Alternatively we can use the Landau expansion of $\Phi(\varphi)$ for $\varphi = 0$ to calculate $\mathcal{Z}(h)$ and $\Phi(h) = -T \ln \mathcal{Z}(h)$ from Eq. (6.4) and finally to obtain the stationary phase $\bar{\varphi}(h) = \{\bar{\varphi}_\alpha(h)\}$ with the help of the equations

$[\partial\Phi(h)/\partial h_\alpha] = -\bar{\varphi}_\alpha(h)$. Here we shall accept the first variant. So we focus our attention on the problem how to find out $\mathcal{Z}_f$ from Eq. (6.7).

The explicit calculations are carried out by an appropriate truncation of the expansion (6.10). As the phase $\bar{\varphi}(h)$ is stable, $\Delta\Phi > 0$ for all fluctuation vectors $\delta\varphi = \{\delta\varphi_\alpha\}$. Let the series (6.10) be abbreviated in a way to include the terms of order less or equal to $\delta\varphi^m$; all terms containing higher powers of fluctuations than $O(\delta\varphi^m)$ are neglected. For large vectors $\delta\varphi$, $\Delta\Phi$ must be positive and this puts restrictions on the variations of the $u^{(m)}$-coefficients. The phase $\bar{\varphi}$ will be stable, if the lowest order non-vanishing term in Eq. (6.10) is even, $l = 2k$, and positively definite:

$$\sum_{\alpha_1,\dots,\alpha_{2k}} u^{(2k)}_{\alpha_1\dots\alpha_{2k}}(\bar{\varphi})\delta\varphi_{\alpha_1}\cdots\delta\varphi_{\alpha_{2k}} > 0. \tag{6.11}$$

It is convenient at this stage to consider the following auxiliary example:

$$\begin{aligned}\Delta\Phi = &\sum_{\alpha_1,\dots,\alpha_{2k}} u^{(2k)}_{\alpha_1\dots\alpha_{2k}}(\bar{\varphi})\delta\varphi_{\alpha_1}\cdots\delta\varphi_{\alpha_{2k}} \\ &+ \sum_{\alpha_1,\dots,\alpha_m} u^{(m)}_{\alpha_1\dots\alpha_m}(\bar{\varphi})\delta\varphi_{\alpha_1}\cdots\delta\varphi_{\alpha_m},\end{aligned} \tag{6.12}$$

where $m > 2k, k = 1, 2, \dots$. The coefficients $u^{(2k)}(\bar{\varphi})$ are related to the first non-vanishing inverse susceptibility of the phase $\bar{\varphi}$:

$$\begin{aligned}u^{(2k)}_{\alpha_1\dots\alpha_{2k}} &= \frac{1}{(2k)!}\left[\chi^{(2k)}_{\alpha_1\dots\alpha_{2k}}\right]^{-1} \\ &= \frac{1}{(2k)!}\left[\frac{\partial^{2k}\Phi(\varphi)}{\partial\delta\varphi_{\alpha_1}\cdots\partial\delta\varphi_{\alpha_{2k}}}\right]_{\bar{\varphi}}.\end{aligned} \tag{6.13a}$$

When there is an overall isotropy in the n-dimensional space of the order parameter components φ_α, namely,

$$\left[\frac{\partial^{2k}\Phi(\varphi)}{\partial\delta\varphi_{\alpha_1}\cdots\partial\delta\varphi_{\alpha_l}}\right]_{\bar{\varphi}} = \delta_{\alpha\alpha_1}\delta_{\alpha_1\alpha_2}\cdots\delta_{\alpha_{l-1}\alpha_l}\left[\frac{\partial^l\Phi(\varphi)}{\partial\varphi^l_\alpha}\right]_{\bar{\varphi}}, \tag{6.13b}$$

where $l = 2k$, $k \geq 1$, the l-th order derivative on the *r.h.s.* of this equation does not depend on α. Then Eq. (6.12) becomes

$$\Delta\Phi = u_{2k}(\bar{\varphi})\delta\varphi^{2k} + u_m(\bar{\varphi})\delta\varphi^m, \tag{6.14a}$$

where $u_l(\bar{\varphi}) = (\partial^l\Phi/\partial\varphi^l_\alpha)_{\bar{\varphi}} > 0$, $(l = 2k, m)$,

$$\delta\varphi^{2k} = (\delta\varphi_1^2 + \cdots + \delta\varphi_n^2)^k = (\delta\varphi^2)^k, \tag{6.14b}$$

and $\delta\varphi^m = (\delta\varphi^2)^{m/2}$.

The states $\bar{\varphi}$ of the system, which are near to an ordinary critical point T_c :, namely, corresponding to temperatures $T \sim T_c$, are described by $k = 1$. For $k > 1$ the models (6.10), (6.12), and (6.15) describe the fluctuations at the ordinary critical point $\{u_2[\bar{\varphi}(T_c) = 0] = u_3 = 0, u_4(0) > 0\}$ or at a higher-order critical point $(u_2 = u_3 = \cdots = u_{2k-1} = 0, u_{2k} > 0; k > 2)$.

6.2.1 *Gaussian fluctuations*

In the simple case of $u^m_{\alpha_1 \dots \alpha_m} = 0$ and $k = 1$, the model (6.12) contains a single quadratic form and the partition function (6.7) is a Gaussian integral (see, e.g., Eq. (B.1) for $l = 1$ and $m = 2$). For this reason these fluctuations are often called *Gaussian fluctuations.* In the present simplified consideration these are *uniform Gaussian fluctuations*; see also the discussion below, as well as Eq. (5.21) in Section 5.2.1.

The quadratic form (6.11) for $k = 1$ can always be diagonalized by a rotation ($\varphi_\alpha \to \varphi'_\alpha$) in the order parameter space. It can be therefore expressed as a sum over the squares φ'^2_α of the new (rotated) components φ'_α. As $\hat{u}^{(2)} = u^{(2)}_{\alpha_1\alpha_2}$ is a symmetric positively definite matrix, the quadratic fluctuation term of an anisotropic system can always be written in the form

$$\sum_\alpha u_2^{(\alpha)} \delta\varphi_\alpha^2, \tag{6.14c}$$

where $u_2^{(\alpha)} > 0$ are the eigenvalues of $\hat{u}^{(2)}$; see our calculation of Φ_f in Section 5.4, Eqs. (5.58a) and (5.58b), where $u_2^{(\alpha)} = \chi^{(\alpha)}/2$.

In the isotropic case, all $u_2^{(\alpha)}$ are equal and the sum (6.14c) will be $u_2\delta\varphi^2$.

6.2.2 *Gaussian-like and non-Gaussian fluctuations*

We note that, except for very special unrealistic cases, $(2k)$-order fluctuation terms like those given by Eq. (6.11) with $k > 1$ cannot be diagonalized in the above described way. Moreover one cannot simultaneously diagonalize both terms in Eq. (6.12) even in the simplest case when $k = 1, m = 3$.

A glance on Eq. (6.14a) and Eq. (6.14b) shows that the mixed terms like $\delta\varphi_\alpha^2\delta\varphi_\beta^2, \alpha \neq \beta$, will be always present in $\Delta\Phi$ for $k \geq 2$ and $n \geq 2$, even if $m = 0$. So $\Delta\Phi$ from Eq. (6.12) can be represented in the following way:

$$\Delta\Phi = \sum_{\alpha=1}^{n} u_{2k}^{(\alpha)}(\bar{\varphi})\delta\varphi_\alpha^{2k} + O(\delta\varphi_\alpha^m; m > 2k), \tag{6.15}$$

if only

$$n = 1, \qquad k \geq 1, \tag{6.16a}$$

or

$$n \geq 1, \qquad k = 1. \tag{6.16b}$$

The form (6.15) of the first non-vanishing fluctuation term is of principal interest as it gives $\Delta\Phi$ as a sum over independent (decoupled) fluctuation terms $\delta\varphi_\alpha^{2k}$. So the fluctuations $\delta\varphi_\alpha$ can be considered as independent degrees of freedom in the partition integral $\mathcal{Z}_f$. Note, that the fluctuation term $\delta\varphi_\alpha^m$ cannot be represented in a decoupled form and the term $O(\delta\varphi^m)$ always describes fluctuation couplings (fluctuation interactions). We have already explained that the important case is $k = 1$ and systems with $k > 1$ may arise rather pure academic than real interest.

Regardless of the value of k, $\Delta\Phi$ given by Eq. (6.15) corresponds to a multiple (n-fold) *Gaussian-like integral* for the partition function (6.7); see Section B.1. In fact, the form (6.15) for $\Delta\Phi$ leads to a partition integral of the type, given by Eq. (B.1). The latter can be transformed to a derivative of a standard Gaussian integral, as shown in Section B. So, the fluctuations described by the form (6.15) of $\Delta\Phi$ could be referred to as *Gaussian-like fluctuations*, or, *pseudo*-Gaussian fluctuations, provided the $\delta\varphi_\alpha^m$-terms are neglected.

Setting the interaction coefficients u_m in Eq. (6.12) equal to zero and taking anyone of the conditions (6.16a) and (6.16b), it is easy to solve the partition integral (6.7) with $\Delta\Phi$ given by Eq. (6.15). In this case we have to solve a simple Gaussian-like integral for $\mathcal{Z}_f$; see, e.g., Eq. (B.1). Mathematical details of the calculation of the partition function $\mathcal{Z}_f$ in Gaussian approximation are summarized in Section B.

The study of fluctuation interactions ($u_m > 0$) requires the calculation of *non-Gaussian integrals.* In this case we are faced with the problem of treatment of *non-Gaussian fluctuations*, or, alternatively, *interacting fluctuations.* When the interaction between the fluctuations is taken into account ($u_m > 0$), we can achieve an exact solution of the fluctuation partition integral $\mathcal{Z}_f$ only for the relatively simple case of uniform fluctuations. In the next three subsections we shall discuss the calculation of $\mathcal{Z}_f$ for uniform (spatially independent) Gaussian-like and non-Gaussian fluctuations.

6.3 Partition Function of Spatially Independent Fluctuations

We have already used the Gaussian approximation for $\Delta\Phi$ in Sections 5.2 and 5.4, where we have calculated $\Phi_f = -T\ln \mathcal{Z}_f$ for $k = 1$. In this section we shall discuss both Gaussian and non-Gaussian (interacting) fluctuations.

Taking the Gaussian (noninteracting or free) part of $\Delta\Phi$,

$$\Delta\Phi_0 = \sum_{\alpha=1}^{n} u_{2k}^{(\alpha)}(\bar{\varphi})\delta\varphi_{\alpha}^{2k}, \tag{6.17}$$

see Eq. (6.15), and the interacting part

$$\Delta\Phi_{\rm int}(\delta\varphi) = \Delta\Phi(\delta\varphi) - \Delta\Phi_0(\varphi), \tag{6.18}$$

as given by the series (6.10) with terms of order $l > 2k$, we can write Eq. (6.7) in the form

$$\mathcal{Z}_f = \int_{-\infty}^{+\infty} D\delta\varphi e^{-(\Delta\Phi_0+\Delta\Phi_{\rm int})/T}. \tag{6.19}$$

The average value $\langle A\rangle$ of any function $A(\delta\varphi)$ is

$$\langle A\rangle = \mathcal{Z}_f^{-1}\int_{-\infty}^{+\infty} D\delta\varphi A(\delta\varphi) e^{-(\Delta\Phi_0+\Delta\Phi_{\rm int})/T}. \tag{6.20}$$

Neglecting the interaction term $\Delta\Phi_{\rm int}$ we have the Gaussian average:

$$\langle A\rangle_0 = \mathcal{Z}_{0f}^{-1}\int_{-\infty}^{+\infty} D\delta\varphi A(\delta\varphi) e^{-\Delta\Phi_0/T}, \tag{6.21}$$

where the Gaussian partition function $\mathcal{Z}_{0f}$ is given by Eq. (6.19) for $\Delta\Phi_{\rm int} = 0$. $\mathcal{Z}_f$ and $\mathcal{Z}_{0f}$ as well as the total average $\langle A\rangle$ and the Gaussian average $\langle A\rangle_0$ are related by very useful formulae. With the help of the expression (6.21) for the Gaussian averages, Eq. (6.19) is represented in the form

$$\begin{aligned}\mathcal{Z}_f =& \mathcal{Z}_{0f}\left[\frac{\int_{-\infty}^{\infty} D\delta\varphi\left(e^{-\Delta\Phi_0/T}\right)\left(e^{-\Delta\Phi_{\rm int}/T}\right)}{\mathcal{Z}_{0f}}\right]\\ =& \mathcal{Z}_{0f}\langle e^{-\Delta\Phi_{\rm int}/T}\rangle_0.\end{aligned} \tag{6.22}$$

So, $(\mathcal{Z}_f/\mathcal{Z}_{0f}) = 1$, if only $\Delta\Phi_{\rm int} = 0$. In the same way, Eq. (6.20) becomes

$$\langle A\rangle = \left(\frac{\mathcal{Z}_{0f}}{\mathcal{Z}_f}\right)\frac{\int_{-\infty}^{\infty} D\delta\varphi\left[A(\delta\varphi)\,e^{-\Delta\Phi_{\rm int}/T}\right]\left(e^{-\Delta\Phi_0/T}\right)}{\int_{-\infty}^{\infty} D\delta\varphi e^{-\Delta\Phi_0/T}}$$
$$= \frac{\langle A(\delta\varphi)\,e^{-\Delta\Phi_{\rm int}/T}\rangle_0}{\langle e^{-\Delta\Phi_{\rm int}/T}\rangle_0}. \tag{6.23}$$

Clearly, our first task is to learn how the Gaussian averages $\langle\ldots\rangle_0$ can be calculated. Having these averages we can in principle find the total average $\langle\ldots\rangle$ and the total partition function $\mathcal{Z}_f$.

6.3.1 *Treatment of Gaussian-like fluctuations*

The partition function $\mathcal{Z}_{0f} = \mathcal{Z}_f(\Delta\Phi_{\rm int} = 0)$ can be represented as a multiple Gaussian integral

$$\mathcal{Z}_{0f} = \int_{-\infty}^{+\infty}\left(\prod_{\alpha=1}^{n} d\delta\varphi_\alpha\right)\exp\left[-\frac{1}{T}\sum_{\alpha=1}^{n} u_{2k}^{(\alpha)}\delta\varphi_\alpha^{(2k)}\right]$$
$$= \prod_{\alpha=1}^{n}\int_{-\infty}^{+\infty} d\delta\varphi_\alpha \exp\left[-\frac{1}{T}u_{2k}^{(\alpha)}\delta\varphi_\alpha^{(2k)}\right] \tag{6.24}$$

(hereafter, in our notations, $\int\left(\prod_{\alpha=1}^{n} dX_n\right) \equiv \prod_{\alpha=1}^{n}\int dX_n$). Employing Eq. (B.1) we obtain

$$\mathcal{Z}_{0f} = \left(\prod_{\alpha=1}^{n} a_k^\alpha\right)[J_0(k)]^n\,, \tag{6.25}$$

where

$$a_k^\alpha = \left[\frac{T}{u_{2k}^{(\alpha)}}\right]^{1/2k}, \tag{6.26}$$

and

$$J_0(k) = \int_{-\infty}^{+\infty} e^{-z^{2k}}dz = \frac{1}{k}\Gamma\left(\frac{1}{2k}\right); \tag{6.27}$$

$\Gamma(y)$ is the gamma function (Section A.2). For $k = 1$, $\Phi_{0f} = -T\ln\mathcal{Z}_{0f}$ as given by Eq. (5.58b); $a_k^{(\alpha)} \equiv a_1^{(\alpha)} = 2\chi_\alpha$.

The averages $\langle A\rangle_0$ can be easily calculated, if $A(\delta\varphi)$ has the form

$$A = \delta\varphi_{\alpha_1} \cdots \delta\varphi_{\alpha_j}. \tag{6.28}$$

From the properties of the Gaussian integrals, see Eq. (B.2), we find that

$$\langle \delta\varphi_{\alpha_1} \cdots \delta\varphi_{\alpha_j} \rangle_0 = \begin{cases} \langle \delta\varphi^2_{\alpha_1} \cdots \delta\varphi^2_{\alpha_l} \rangle_0, & j = 2l; \quad l = 1, 2, \dots \\ 0, & \text{for odd } j \end{cases} \tag{6.29}$$

This expression is a result of the direct calculation of the mean $\langle \delta\varphi_{\alpha_1} \cdots \delta\varphi_{\alpha_j} \rangle_0$. Using Eq. (6.21) we have

$$\langle \delta\varphi_{\alpha_1} \cdots \delta\varphi_{\alpha_j} \rangle_0$$
$$= \frac{\int_{-\infty}^{+\infty} d\delta\varphi_{\alpha_1} \cdots d\delta\varphi_{\alpha_j} \left(\delta\varphi_{\alpha_1} \cdots \delta\varphi_{\alpha_j} \right) e^{-\left[u_{2k}^{(\alpha_1)} \delta\varphi^{2k}_{\alpha_1} + \cdots + u_{2k}^{(\alpha_j)} \delta\varphi^{2k}_{\alpha_j} \right]/T}}{\left(\int_{-\infty}^{+\infty} d\delta\varphi_{\alpha_1} e^{-u_{2k}^{(\alpha_1)} \delta\varphi^{2k}_{\alpha_1}/T} \right) \cdots \left(\int_{-\infty}^{+\infty} d\delta\varphi_{\alpha_j} e^{-u_{2k}^{(\alpha_j)} \delta\varphi^{2k}_{\alpha_j}/T} \right)}. \tag{6.30}$$

The nominator in this equation will be not equal to zero, if only $\delta\varphi_{\alpha_1} \cdots \delta\varphi_{\alpha_j}$ can be represented as a product of squares $\delta\varphi^2_{\alpha_1} \cdots \delta\varphi^2_{\alpha_l}$; see Eq. (6.28). Of course some of the indices $\alpha_1, \dots, \alpha_l$ in this product of squares may coincide. If $A(\delta\varphi)$ is not a product of the fluctuation components $\delta\varphi_\alpha$, the Gaussian integral (6.21) cannot be straightforwardly calculated and we shall be forced to develop a perturbation expansion; see Sections 6.3.2 and 6.3.3.

The averages (6.29) are called Gaussian cumulants of order j or, shortly, *cumulants*. For example, the simplest cumulant of order $2l$ is

$$\langle (\delta\varphi_\alpha)^{2l} \rangle_0 = (a_k^\alpha)^{2l} \frac{\Gamma\left(\dfrac{2l+1}{2k} \right)}{\Gamma\left(\dfrac{1}{2k} \right)}. \tag{6.31}$$

For $l = k$, Eq. (6.29) yields

$$\langle (\delta\varphi_\alpha)^{2k} \rangle_0 = \frac{T}{2k u_{2k}} \sim \chi_{2k}^{(\alpha)}, \tag{6.32}$$

where we have used Eq. (6.13a), Eq. (6.13b), and the property Eq. (A.3) of the gamma function. In our model all susceptibilities χ_l of order $l < 2k$ are infinite ($\chi_l \sim u_l^{-1}, u_l = 0$ for $l < 2k$). For this reason the cumulants of order $l < 2k$ can be considered as effective susceptibilities related to the actual susceptibility in the Gaussian approximation $\chi_{2k}^{(\alpha)} \sim a_k^\alpha$. For example, $\left\langle (\delta\varphi_\alpha)^2 \right\rangle_0 \sim [\chi_{2k}^{(\alpha)}]^{1/k}$.

6.3.2 *Beyond the Gaussian approximation. Weak coupling*

In the remaining part of this section we shall study the case $k = 1$. We shall assume that $n = 1; \varphi = \varphi_1$. The reader could try to rederive the subsequent formulae for $k, n > 1$.

The partition function of the non-Gaussian model (6.15) with $n = k = 1$ is

$$\mathcal{Z}_f = aJ(g), \qquad a = \left(\frac{T}{u_2}\right)^{1/2}, \tag{6.33a}$$

$$J(g) = \int_{-\infty}^{\infty} dz e^{-z^2 - gz^m}. \tag{6.33b}$$

The dimensionless coupling constant g is given by

$$g = T^{m/2-1}\left(\frac{u_m}{u_2^{m/2}}\right); \tag{6.34}$$

$z = \delta\varphi/a$ is a dimensionless integration variable.

Now it is convenient to write Eq. (6.22) in the form

$$\mathcal{Z}_f(g) = \mathcal{Z}_f(0)\langle e^{-gz^m}\rangle_0, \tag{6.35}$$

where $\mathcal{Z}_f(0) = \mathcal{Z}_{0f}(n = k = 1) = \sqrt{\pi}a$; see Eqs. (6.24)–(6.27), where $\delta\varphi = \delta\varphi_1$, $a_1^1 = a$, $J_0(1) = J(0) = \sqrt{\pi}$. Expanding the exponent from Eq. (6.35) in power series,

$$\mathcal{Z}_f(g) = (\sqrt{\pi}\, a) \sum_{l=0}^{\infty} \frac{(-1)^l}{l!} g^l \langle z^{ml}\rangle_0, \tag{6.36}$$

and solving the integral (6.21) for the cumulants $\langle z^{ml}\rangle_0$ we obtain

$$\mathcal{Z}_f(g) = \begin{cases} a \sum_{l=0}^{\infty} \frac{(-g)^l}{l!} \Gamma\left(\frac{ml+1}{2}\right), \text{ for even } m, & \text{(a)} \\ a \sum_{l=0}^{\infty} \frac{(g)^{2l}}{(2l)!} \Gamma\left(\frac{2ml+1}{2}\right), \text{ for odd } m. & \text{(b)} \end{cases} \tag{6.37}$$

It seems at first sight that these perturbation series for $\mathcal{Z}_f(g)$ are valid when

$$g \sim \left(\frac{u_m}{u_2^{m/2}} \right) \ll 1, \qquad \text{even} \quad m, \tag{6.38a}$$

or

$$g^2 \sim \left(\frac{u_m^2}{u_2^m} \right) \ll 1, \qquad \text{odd} \quad m. \tag{6.38b}$$

In fact, certain qualitative predictions about the effect of the interaction can be received from the first terms ($l = 1, 2, \ldots$) of this expansion. For example, Eq. (6.38a) is obtained by comparing $\mathcal{Z}_{0f}$ and $\mathcal{Z}_f(g) \approx \mathcal{Z}_f^1(g)$ to the order g^1; $\mathcal{Z}_f^1$ is given by the terms $l = 0, 1$ in (6.37a), neglecting all terms $O(g^2)$. So $g \ll 1$ corresponds to the ratio $(\mathcal{Z}_{0f}/\mathcal{Z}_f^1) \sim 1$. When Eqs. (6.38a) and (6.38b) are satisfied the qualitative results connected with the fluctuation effect can be calculated in the Gaussian approximation ($g = 0$).

Although this conclusion is correct we could not truncate the infinite series (6.37) for some l, say $l = l_1$, and neglect all terms $O(\delta\varphi^{l_1+1})$. To see this let us evaluate the ratio $R = A_{l+1}/A_l$ of two neighbouring terms in (6.37a) for $l \gg 1$. Using the asymptotic formula (Abramowitz and Stegun, 1965),

$$\frac{\Gamma(y+a)}{\Gamma(y+b)} \sim y^{a-b} \left[1 + O\left(\frac{1}{y}\right) \right], \tag{6.39}$$

valid for $y \gg 1$, we obtain

$$R \sim g l^{-1+m/2}, \qquad l \gg 1. \tag{6.40}$$

If $m > 2$, the expansion (6.37a) is divergent for all $g > 0$. The convergence appears at $g = 0$. Therefore, the expansion (6.37a) represents the function $\mathcal{Z}_f(g)$ asymptotically, i.e., in the limit $g \to 0$ (*asymptotic expansion*). The ratio $R = A_{2l+2}/A_{2l}$ for odd m and $l \gg 1$, see Eq. (6.37b), is $R \sim g^2 l^{m-2}$ and, hence, the expansion (6.37b) is valid only asymptotically. The divergence of the non-Gaussian partition function $\mathcal{Z}_f(g)$ and its asymptotic properties are discussed in details by Kazakov and Shirkov (1980) for $k = n = 1$, and by Tuszyński, Clouter and Kiefte (1986) in the more general cases.

Let us consider the averages $\langle A(z) \rangle \equiv \langle A(az) \rangle$ in the framework of the same fluctuation model. The Eq. (6.23) now yields

$$\langle A\rangle = \frac{\langle A(z)e^{-gz^m}\rangle_0}{\langle e^{-gz^m}\rangle_0}, \tag{6.41}$$

or, expanding in power series of g,

$$\langle A\rangle = \frac{\langle A\rangle_0 + L_A}{1 + L_1}, \tag{6.42}$$

where

$$L_A = \sum_{l=0}^{\infty} \frac{(-g)^l}{l!} \langle A(z)\,(z^m)^l\rangle_0, \tag{6.43}$$

and $L_1 \equiv L_{(A=1)}$. If $A(z)$ can be represented as a power of $z = \delta\varphi/a$, say z^n, or by finite or infinite polynomials of z, the sum L_A can be calculated as a sum of cumulants. Therefore, there is no principal difficulty to calculate any total mean value $\langle A\rangle$ by expanding the non-Gaussian integral (6.22) in power series of the interaction term(s) $\Delta\Phi_{\rm int}/T$; the term $\Delta\Phi_{\rm int}$ may consist of several interaction terms with different coupling constants: $g_m z^m$, $g_\nu z^\nu$, etc. From one side this is an expansion in powers of the coupling constant(s) and from the other side the expansion coefficients are proportional to cumulants of $z \sim \delta\varphi$. Such series are often referred to as *the cumulant expansion*. The Eq. (6.36) is an example of a cumulant expansion for $\mathcal{Z}_f$. The calculations of the averages $\langle A\rangle$ can be made in a more compact form. Instead of Eq. (6.42) the equivalent representation can be used

$$\langle A\rangle = \langle A\rangle_0 + \sum_{l=1}^{\infty} \frac{(-g)^l}{l!} \langle A(z)\,(z^m)^l\rangle_{0c}, \tag{6.44}$$

where the averages $\langle \ldots \rangle_{0c}$ are given by the formula

$$\langle A(z)B(z)\rangle_{0c} = \langle A(z)B(z)\rangle_0 - \langle A(z)\rangle_0 \langle B(z)\rangle_0, \tag{6.45}$$

and are called *connected Gaussian averages* (or *irreducible cumulants*); $B(z) = (z^m)^l$. All possible products of fractional averages, like $\langle A\rangle$ and $\langle B\rangle$ in Eq. (6.45) are subtracted from the connected averages $\langle \ldots \rangle_{0c}$. At first we shall discuss the expansion (6.44) and after that we shall prove it.

As $B(z)$ is a power of z^m, the first term ($l = 1$) in the expansion (6.44) is obtained by substituting $B(z) = z^m$ and calculating the ordinary averages on the r.h.s. of Eq. (6.45) with the help of Eq. (6.31); the last equation yields $\langle z^{2l}\rangle_0 = \Gamma(l + 1/2)/\sqrt{\pi}$ for $k = n = 1, a_1^\alpha = a$. Let $A(z) = z^n$. If n is odd,

$\langle A\rangle_0 = 0$ and the cumulant in Eq. (6.45) for $l = 1$ is $\langle z^{n+m}\rangle_{0c} = \langle z^{n+m}\rangle_0$. When $(n+m)$ is odd this cumulant is zero but if $(n+m)$ is an even number, the same cumulant will be equal to $\Gamma(n+m+1/2)/\sqrt{\pi}$. When n is even but m is odd the cumulant $\langle z^{n+m}\rangle_{0c}$ in Eq. (6.44) is zero. For even n and m, the first order ($l = 1$) term in Eq. (6.44) is

$$-\frac{g}{\sqrt{\pi}}\left[\Gamma\left(n+m+\frac{1}{2}\right)-\Gamma\left(n+\frac{1}{2}\right)\Gamma\left(m+\frac{1}{2}\right)\right]. \tag{6.46}$$

Under the requirement this lowest order interaction contribution to $\langle A\rangle$ to be small compared with $\langle A\rangle_0 = \langle z^n\rangle_0 = \Gamma(n+1/2)/\sqrt{\pi}$ we can re-derive the criterion (6.38a). If $A(z) = A(\delta\varphi) = \delta\varphi^n$ as it is usually, $\langle\delta\varphi^n\rangle = a^n\langle z^n\rangle$ and all results should be multiplied by a^n. It is more convenient to continue with the notation $A(z) = z^n$. The second term ($l = 2$) in the expansion (6.44) is

$$\begin{aligned}\langle A(z)\,(z^m)^2\rangle_{0c} = \langle A(z)\,(z^m)^2\rangle_0 - \langle A(z)\rangle_0\langle z^{2m}\rangle_0\\ - 2\langle A(z)z^m\rangle_0\langle z^m\rangle_0 + 2\langle A(z)\rangle_0\langle z^m\rangle_0^2;\end{aligned} \tag{6.47}$$

cf. Eqs. (5.91) and (5.77), where $\varphi_i = A$, $\varphi_j = \varphi_k = z^m$. The calculation of this term is somewhat more complicated but it can be straightforwardly performed for monomials $A(z) = z^n$, or polynomials of z.

The main problem for the higher-order perturbation terms ($l \geq 3$) is to find the connection between the irreducible $\langle\dots\rangle_{0c}$ and the ordinary (reducible) $\langle\dots\rangle_0$ Gaussian cumulants of order $l \geq 3$. These relations can be in principle derived simultaneously with the derivation of the expansion (6.44) from Eq. (6.42). It is easy to find out the second or third perturbation terms ($l = 2, 3$); see Section 7.6 where this is made for field models. After direct calculations we obtain relations like

$$\begin{aligned}\langle A(z)\,(z^m)^2\rangle = \langle A(z)\,(z^m)^2\rangle_{0c} + \langle A(z)\rangle_{0c}\langle z^m z^m\rangle_{0c}\\ + 2\langle A(z)z^m\rangle_{0c}\langle z^m\rangle_{0c} + 2\langle A(z)\rangle_{0c}\langle z^m\rangle_{0c}^2,\end{aligned} \tag{6.48}$$

where $\langle A(z)\rangle_0 \equiv \langle A(z)\rangle_{0c}$, and $\langle z^m\rangle_0 \equiv \langle z^m\rangle_{0c}$; Eq. (6.48) is the inverted Eq. (6.47).

It is obvious from equations such as (6.45), (6.47), and (6.48) that the relation between the cumulants $\langle(z^m)^l\rangle_0$ and $\langle z^{ml}\rangle_{0c}$ is quite similar to that between the ordinary and irreducible correlation functions $G^{(l)}$ and $\chi^{(l)}$ introduced in Section 5.5. In our present example the Gaussian averages

$\langle (z^m)^l \rangle_0$ and $\langle (z^m)^l \rangle_{0c}$ are spatially independent and this is the difference between them and the correlation functions ($G^{(l)}$ and $\chi^{(l)}$). Another difference is that in our present discussion Gaussian cumulants $\langle \dots \rangle_0$ are involved. However, it is very easy to define total cumulants with the help of the rule (6.20) for the total averages $\langle \dots \rangle$. For example,

$$\begin{aligned}\langle (z^m)^2 \rangle_c &= \mathcal{Z}_f^{-1} \int_{-\infty}^{\infty} d\delta\varphi \, (z^{2m}) \, e^{-\Delta\Phi/T} \\ &\quad - \mathcal{Z}_f^{-2} \left(\int_{-\infty}^{\infty} d\delta\varphi z^m e^{-\Delta\Phi/T} \right)^2 \\ &= \frac{\partial}{\partial(-g)} \langle z^m \rangle = \frac{\partial \ln \mathcal{Z}_f(g)}{\partial(-g)^2}. \end{aligned} \tag{6.49}$$

The relations between the total (or Gaussian) cumulants can be obtained from the relations between $G^{(l)}$ and $\chi^{(l)}$ after neglecting the indices $(i, j, \dots)$ of the spatial dependence in the correlation functions and identifying $G^{(l)}$ with $\langle (z^m)^l \rangle_0$ and $\chi^{(l)}$ with $\langle (z^m)^l \rangle_{0c}$; for example, compare Eq. (6.47) for $A = z^m$ with Eq. (5.91), where $G^{(l)} = \langle (z^m)^l \rangle_0$. The fact that the cumulants in the expansions (6.42) and (6.4) are a simplified version of the correlation function raises the idea to apply the technique of the generating functionals (Section 5.5). The functional

$$L(g, \lambda) = \ln \mathcal{Z}_f(g, \lambda) = \ln \int_{-\infty}^{\infty} dz e^{-z^2 - gz^m - \lambda A(z)}, \tag{6.50}$$

generates the irreducible averages

$$\langle A^\mu(z) \, (z^m)^\nu \rangle_c = \left[\frac{\partial^{\mu+\nu} L(g, \lambda)}{\partial(-\lambda)^\mu \partial(-g)^\nu} \right]_{\lambda=0}. \tag{6.51}$$

The ordinary averages $\langle A^\mu \, (z^m)^\nu \rangle$ are proportional to the derivatives of $\mathcal{Z}_f(g, \lambda)$.

For example, $\langle A \rangle_c = \langle A \rangle$ is given by

$$\langle A \rangle = \left[\frac{\partial L}{\partial(-\lambda)} \right]_{\lambda=0} = \frac{1}{\mathcal{Z}_f(g)} \left[\frac{\partial \mathcal{Z}_f(g, \lambda)}{\partial(-\lambda)} \right]_{\lambda=0}. \tag{6.52}$$

The expansion of $\langle A \rangle$ in powers of g is

$$\langle A \rangle = \sum_{l=0}^{\infty} \frac{(-g)^l}{l!} \left[\frac{\partial}{\partial(-\lambda)} \frac{\partial^l}{\partial(-g)^l} L(g, \lambda) \right]_{\lambda=g=0}. \tag{6.53}$$

This expansion should be identical to Eq. (6.44) as $\langle A \rangle$ has an unique expansion in powers of g. The derivatives of $L(g, \lambda)$ in Eq. (6.53) are the connected Gaussian averages from Eq. (6.44). So we prove the validity of the expansion (6.44) for all l. Doing the mathematical operations shown in Eq. (6.53) we can obtain any average $\langle A \rangle$ to any order in g.

The expansion of the partition function $\mathcal{Z}_f(g)$ with the help of Eqs. (6.32), (6.33a) and (6.33b) is given in the form

$$\mathcal{Z}_f(g) = \int_{-\infty}^{\infty} dz e^{-z^2} \left(\sum_{l=0}^{\infty} \frac{(-g)^l}{l!} z^{ml} \right)$$
$$= \left[e^{-g(\partial/\partial g)} \mathcal{Z}_f(g) \right]_{\mathcal{Z}_f = \mathcal{Z}_{0f}} . \tag{6.54}$$

The fluctuation potential $\Phi_f(g) = -T \ln \mathcal{Z}_f(g)$ then will be

$$\Phi_f(g) = -T \left[e^{-g(\partial/\partial g)} \ln \mathcal{Z}_f(g) \right]_{\mathcal{Z}_f = \mathcal{Z}_{0f}} , \tag{6.55}$$

or

$$\Phi_f g) = \Phi_f(0) + \Phi_f^{\text{int}} , \tag{6.56}$$

where

$$\Phi_f^{\text{int}} = \sum_{l=1}^{\infty} \frac{(-g)^l}{l!} \langle (z^m)^l \rangle_{0c} . \tag{6.57}$$

These formulae are often used in theoretical studies when perturbation expansions are applied.

The perturbation theory developed here is a prototype of the perturbation theory in the field theory of fluctuations, where the fluctuations $\delta\varphi$ depend on the spatial point x (Chapter 7). From the above simplified version of the perturbation theory one can understand the basic rules, on which the perturbation theories are built up. Moreover the approximation of spatially independent fluctuations $\delta\varphi$ gives the opportunity to demonstrate quite easily the perturbation theory in its strong coupling limit.

6.3.3 *Strong coupling and closed-form expression of the partition integral*

The series (6.37) is divergent for $g > 0$. The reason is that m in Eq. (6.40) is bigger than 2. In order to avoid this feature of the weak coupling expansion

($g \ll 1$) we shall perform an expansion in inverse non-integer powers of g. To escape from calculational complications we shall use the even number $m = 2p, p = 2, 3, \ldots$. If we introduce the notation

$$\tilde{g} = g^{-1/2p} = \left[\frac{T^{(1-p)/p} u_2}{u_{2p}^{1/p}} \right]^{1/2}, \tag{6.58}$$

$\mathcal{Z}_f$ will take the form

$$\begin{aligned} \mathcal{Z}_f =& (a\tilde{g}) \int_{-\infty}^{\infty} dy e^{-y^{2p} - (\tilde{g}y)^2} \\ =& \frac{\Gamma(1/2p)}{p} \left(\frac{T}{u_{2p}} \right)^{1/2p} \langle e^{-(\tilde{g}y)^2} \rangle_0, \end{aligned} \tag{6.59}$$

where we have used the new integration variable $y = z/\tilde{g}$ and the equality $a\tilde{g} = (J/u_{2p})^{1/2p}$. The factor $\Gamma(1/2p)/p$ is equal to the integral $J_0(p)$; see Eq. (6.27).

The average in Eq. (6.59) corresponds to the Gaussian distribution $e^{(-y^{2p})}$. Now the picture is inverted. We do not expand in powers of the interaction $(\Delta\Phi^{\text{int}}/T) \sim y^m$, rather the interaction is used as a basis (zero approximation) for a new expansion in powers of $(\tilde{g}y)^2 \sim (\Delta\Phi_0/T)$:

$$\mathcal{Z}_f = \frac{1}{p} \left(\frac{T}{u_{2p}} \right)^{1/2p} \sum_{l=0}^{\infty} \frac{(-\tilde{g}^2)^l}{l!} \Gamma\left(\frac{2l+1}{2p} \right). \tag{6.60}$$

The ratio $R = A_{2(l+1)}/A_{2l}$ of two given nn terms in this sum for $l \gg 1$ will be $R \sim \tilde{g}^2/l$ and will tend to zero for $l \to \infty$ provided $\tilde{g}^2$ is finite. Consequently, the expansion (6.60) will be convergent, if u_m is different from zero and u_2 does not tend to infinity. The domain of convergence of the expansion (6.60) is obviously evaluated from the condition $\tilde{g}^2 \leq 1$, which yields

$$T^{(1-p)/p} u_2 \leq u_{2p}^{1/p}. \tag{6.61}$$

The expansion of $\langle A(\delta\varphi) \rangle$ in terms of $\tilde{g}$ is also convergent in the domain (6.61) of the parameters u_2 and u_m. It will be a good exercise for the reader to obtain this expansion. The inequality (6.61) is the inverse inequality (6.38a). The Eq. (6.61) represents the line $T_1(Y)$ near the critical line $T_c(Y)$ where the system passes from the critical region of the strong interaction

$(u_2 < u_{2p}^{1/p})$ to the region of the weak fluctuation interaction $(u_2 > u_{2p}^{1/p})$; cf. the Ginzburg–Levanyuk criterion (see Sections 7.5.2, 7.11.1, and 7.11.2).

For one who is interested in the compact representation of the total partition function $\mathcal{Z}_f(g)$ we shall recall the integral Eq. (B.3); see also the original papers by Witschel (1980, 1981). With the help of the formula Eq. (B.3) we can write the solution of the partition integral $\mathcal{Z}_f(g)$ for the quite general model (6.15) with $m = 4k$. Choosing in Eq. (B.2) $\nu = (1/2k)$ we obtain

$$\begin{aligned} \mathcal{Z}_f =& a_k \int_{-\infty}^{\infty} dz e^{-z^{2k} - g_k z^{4k}} \\ =& \frac{\Gamma(1/2k)}{k} \frac{a_k e^{1/(8g_k)}}{(2g_k)^{1/(4k)}} D_{-1/2k}(1/\sqrt{2g_k}), \end{aligned} \tag{6.62}$$

where $a_k = (T/u_{2k})^{1/(2k)}$, $g_k = (Tu_{4k}/u_{2k}^2)$, and $D_{-\nu-1/2}(z)$ is the parabolic cylinder function. The cumulants are given by $\langle z^{2n+1}\rangle = 0$, because the distribution $e^{-z^{2k} - g_k z^{4k}}$ is symmetric $(z \leftrightarrow -z)$, and

$$\langle z^{2n}\rangle = \frac{\Gamma\left(\dfrac{2n+1}{2k}\right)}{\Gamma\left(\dfrac{1}{2k}\right)(2u_{4k})^{n/2k}} \frac{D_{-(2n+1)/2k}\left(1/\sqrt{2g_k}\right)}{D_{-1/2k}\left(1/\sqrt{2g_k}\right)}. \tag{6.63}$$

When $k = 1$,

$$\begin{aligned} \mathcal{Z}_f =& \left(\frac{\pi^2 T}{2u_4}\right)^{1/4} e^{(u_2^2/8u_4T)} D_{-1/2}\left(u_2/\sqrt{2u_4T}\right) \\ =& \frac{ae^{1/8g}}{(4g)^{1/2}} K_{1/4}(1/8g); \end{aligned} \tag{6.64}$$

for a see Eq. (6.33a), g is given by Eq. (6.34) for $m = 4$, and $K_\nu(z)$ is the modified Bessel function.

Now we may use the expansions of $D_{-\nu}(z)$ for $z \ll 1$ and $z \gg 1$ to derive the strong and weak coupling expansion of $\mathcal{Z}_f$; for the properties of $D_{-\nu-1/2}(z)$ and $K_\nu(z)$; see Abramowitz and Stegun (1965). The case of spatially independent fluctuations of an one-component order parameter presents the rare possibility to sum up the infinite perturbation series by functions like $D_{-\nu}(z)$. Within the field theory of fluctuations this possibility does not exist.

6.4 Fluctuation Field

The notion of spatially dependent fluctuation $\delta\varphi$ has been used in our discussion of discrete systems (lattices; see Section 5.4). Now we pass to the continuum limit for the lattice systems and, consequently, the order parameter φ_i defined for cells i becomes an order parameter depending on a continuous spatial vector $\boldsymbol{x}$ as discussed in Sections 6.5 and 6.6. Physical quantities, which depend on the continuous spatial vector $x(\equiv \boldsymbol{x})$ are called *fields* (remember the electric $E(x)$ or the magnetic $H(x)$ fields). Sometimes the order parameter φ_i defined on a lattice is referred to as a *discrete field* (or, a lattice field). There are two approaches to the description of fields. In this section we shall outline the general phenomenological approach. It can be verified with the help of integral transformations applied to the microscopic models (see Sections 6.6 and 6.7).

In order to develop the phenomenological field theory of phase transitions we must replace the ordinary function $\Phi(\varphi)$ of n variables, $\varphi = \{\varphi_\alpha; \alpha = 1, \ldots, n\}$ with the functional $\Phi[\varphi(x)]$, where the vector order parameter $\varphi(x) = \{\varphi_\alpha(x)\}$ is itself a real function of the spatial vector x. Now we can consider the generalized Gibbs potential $\Phi[\varphi(x)]$ as a function of an infinite number of variables, namely, $\varphi_\alpha(x)$ for $x \in V$, where $V \sim L^d$ is the volume of a d-dimensional system. The field theory is then a generalization of the contents in Chapter 4, Section 6.2, and Section 6.3. We shall work with a n-component real order parameter vector $\varphi(x)$. The results can be straightforwardly applied for $(n/2)$-component complex vectors $\varphi(x)$ provided n is an even number.

We start with Eq. (4.14). When there is a spatial dependence, it becomes:

$$\Phi(\varphi) = F(\varphi) - \sum_\alpha \int d^d x\, h_\alpha(x)\varphi_\alpha(x), \tag{6.65}$$

where $F[\varphi(x)]$ is the nonequilibrium Helmholtz functional, and $h(x) = \{h_\alpha(x)\}$ is the field conjugate to $\varphi(x)$; $d^d x \equiv d\boldsymbol{x}$. We shall discuss the series expansion of $\Phi(\varphi)$ at the stationary solution $\bar{\varphi}(x) = \{\bar{\varphi}_\alpha(x)\}$, where $\bar{\varphi}_\alpha(x) \equiv \bar{\varphi}_\alpha[x/h_\alpha(x)]$ are themselves functionals of $h_\alpha(x)$; for the definition of functional series see Section C.2. The stationary point (function) $\bar{\varphi}(x)$ of $\Phi(\varphi)$ for a given $h(x)$ is determined by the equations

$$\left[\frac{\delta\Phi(\varphi)}{\delta\varphi_\alpha(x)}\right]_{\bar{\varphi}} = 0; \tag{6.66}$$

see the definition Eq. (C.11) of the functional differentiation. The same equations of state can be expressed by F,

$$\left[\frac{\delta F}{\delta\varphi_\alpha(x)}\right]_{\bar{\varphi}} = \bar{h}_\alpha(x); \tag{6.67}$$

cf. Eq. (4.7). If we denote with $\delta\varphi_\alpha(x)$ the difference

$$\delta\varphi_\alpha(x) = \varphi_\alpha - \bar{\varphi}_\alpha(x), \tag{6.68}$$

the functional Taylor series in powers of *the fluctuation field* $\delta\varphi(x) = \{\delta\varphi_\alpha(x)\}$ will be the analogue of the expansion (4.30); for elementary information about the properties of functionals and functional series see Sections C.2–C.5.

Expanding $F(\varphi)$ in Eq. (6.65) and applying Eq. (6.67) we have

$$F(\varphi) = F(\bar{\varphi}) + \sum_\alpha \int d^dx\, \bar{h}_\alpha(x)\delta\varphi_\alpha(x) + \delta^{(2)}F + \delta^{(3)}F + \cdots, \tag{6.69}$$

where

$$\delta^{(l)}F = \sum_{\alpha_1,\ldots,\alpha_l} \int d^dx\, u^{(l)}_{\alpha_1,\ldots,\alpha_l}(x_1,\ldots,x_l/\bar{\varphi})\delta\varphi_{\alpha_1}(x_1)\cdots\delta\varphi_{\alpha_l}(x_l), \tag{6.70}$$

are the variations of F of order $l \geq 2$; here we define

$$d^dx \equiv \prod_{j=1}^{l} d^dx_j.$$

The coefficient functions $u^{(l)}$ are given by

$$u^{(l)}(\bar{\varphi}) = \frac{1}{l!}\left[\frac{\delta^{(l)}F}{\delta\varphi_{\alpha_1}(x_1)\cdots\delta\varphi_{\alpha_l}(x_l)}\right]_{\bar{\varphi}}, \quad l \geq 2, \tag{6.71}$$

(we shall very often omit the suffices α_i, x_i or $\bar{\varphi}$ of $u^{(l)}$). The coefficients $u^{(l)}_{\alpha_1,\ldots,\alpha_l}(x_1,\ldots,x_l/\bar{\varphi})$ are invariant under the change $(\alpha_i, x_i) \leftrightarrow (\alpha_j, x_j)$; $i, j = 1, = \ldots, l$. The relation (6.65) shows that the functional derivatives of F and Φ of order $l \geq 2$ are equal so that F and Φ in Eq. (6.71) can be used on equal grounds. Putting $F(\varphi)$ from Eq. (6.69) in Eq. (6.65) we obtain

$$\Phi(\varphi) = \bar{\Phi} - \sum_{\alpha} \int d^d x \left[h_\alpha(x) - \bar{h}_\alpha(x) \right] \varphi_\alpha(x) + \sum_{l \geq 2}^{\infty} \delta^{(l)} F(\bar{\varphi}, \delta\varphi), \quad (6.72)$$

with $\bar{\Phi}$ the saddle point potential:

$$\bar{\Phi} = F(\bar{\varphi}) - \sum_{\alpha} \int d^d x \bar{h}_\alpha(x) \bar{\varphi}_\alpha(x), \quad (6.73)$$

cf. Eqs. (4.30) and (4.12). The Landau expansion for $\bar{\varphi} = 0$ will be

$$\Phi(\varphi) = F(0) - \sum_{\alpha} \int d^d x \left[h_\alpha(x) - h_\alpha^{(0)}(x) \right] \varphi_\alpha(x) + \sum_{l \geq 2}^{\infty} \Phi^{(l)}(\varphi). \quad (6.74)$$

Here

$$\Phi^{(l)}(\varphi) = \sum_{\alpha_1, \ldots, \alpha_l} \int d^d x \, u^{(l)}_{\alpha_1, \ldots, \alpha_l}(x_1, \ldots, x_l / 0) \varphi_{\alpha_1}(x_1) \cdots \varphi_{\alpha_l}(x_l), \quad (6.75)$$

and $h_\alpha^{(0)}$ and $u^l(0)$ are given by the derivatives (6.67) and (6.71) for $\bar{\varphi} = 0$.

Following the arguments presented in Section 4.4.2 we shall consider $h(x)$ equal to the equilibrium field $\bar{h}(x)$ and the difference $\Delta h = [h - h^{(0)}]$ will be further denoted by $h^{\text{ext}}(x)$. Now the second term on the r.h.s. of Eq. (6.72) becomes zero. For $\Delta h = h^{\text{ext}} = 0$ the second term on the r.h.s. of Eq. (6.74) is also zero and this means that $\bar{\varphi}(x) = 0$ is a stationary solution of the equations of state (6.66), or, equivalently, (6.67). If h^{ext} is different from zero, $\bar{\varphi}(x) = 0$ cannot be a saddle point of $\Phi(\varphi, h^{\text{ext}})$; in this case the saddle point will be given by another function of x : $\bar{\varphi}[x/h^{\text{ext}}]$. As in our next considerations, we shall deal only with the field difference $\Delta h = h^{\text{ext}}$, it is convenient to introduce the short notation

$$h(x) \equiv h^{\text{ext}}(x). \quad (6.76)$$

Further on we shall employ the functional $\Phi(\varphi)$ in the form:

$$\Phi(\varphi) = \bar{\Phi}(\bar{\varphi}) + \sum_{l=2}^{\infty} \sum_{\alpha_1, \ldots, \alpha_l} \int d^d x \, u^{(l)}_{\alpha_1, \ldots, \alpha_l}(x_1, \ldots, x_l / \bar{\varphi}) \delta\varphi_{\alpha_1}(x_1) \cdots \delta\varphi_{\alpha_l}(x_l), \quad (6.77)$$

or

$$[\Phi(\varphi) - \Phi(0)] \longrightarrow \Phi(\varphi) = -\sum_{\alpha} \int d^d x\, h_\alpha(x) \varphi_\alpha(x)$$
$$+ \sum_{l=2}^{\infty} \sum_{\alpha_1,\dots,\alpha_l} \int d^d x\, u^{(l)}_{\alpha_1,\dots,\alpha_l}(x_1,\dots,x_l) \varphi_{\alpha_1}(x_1) \cdots \varphi_{\alpha_l}(x_l), \tag{6.78}$$

where we have subtracted $\Phi(0) = F(0)$ from $\Phi(\varphi)$, see Eq. (6.74). It is easy to derive the connection between these two forms of Φ. Substituting $\varphi_\alpha(x)$ from Eq. (6.68) in Eq. (6.78), $\bar{\Phi}(\bar{\varphi})$ in Eq. (6.77) becomes

$$\bar{\Phi}(\bar{\varphi}) = \Phi(0) - \sum_{\alpha} \int d^d x\, h_\alpha(x) \bar{\varphi}_\alpha(x) + \sum_{l \geq 2}^{\infty} \Phi^{(l)}(\bar{\varphi}), \tag{6.79}$$

The terms $\delta\Phi^{(l)} = \delta F^{(l)}$ in the series (6.77) for $l \geq 2$ are given by Eq. (6.70) and the coefficients $u^{(l)}(\bar{\varphi})$ are expressed by the recursive relations:

$$u^{(2)}_{\alpha\beta}(x, y/\bar{\varphi}) = u^{(2)}_{\alpha\beta}(x, y/0)$$
$$+ 3 \sum_{\gamma} \int d^d z\, u^{(3)}_{\alpha\beta\gamma}(x, y, z/0) \bar{\varphi}_\gamma(z) + \dots$$
$$+ \left[\frac{(l+2)!}{(l-3)!}\right] \sum_{\gamma_1 \dots \gamma_l} \int d^d x\, u^{(l+2)}_{\alpha\beta\gamma_1 \dots \gamma_l} \bar{\varphi}_{\gamma_1}(z_1) \cdots \bar{\varphi}_{\gamma_l}(z_l) + \dots, \tag{6.80a}$$

$$u^{(3)}_{\alpha\beta\gamma}(x, y, z/\bar{\varphi}) = u^{(3)}_{\alpha\beta\gamma}(x, y, z/0)$$
$$+ 4 \sum_{\gamma_1} \int d^d z_1\, u^{(4)}_{\alpha\beta\gamma\gamma_1}(x, y, z, z_1/0) \bar{\varphi}_{\gamma_1}(z_1) + \dots, \tag{6.80b}$$

etc. We have made use of the symmetry of $u^{(l)}$ with respect to the change $(\alpha_i, x_i) \leftrightarrow (\alpha_j, x_j)$ and the equation of state (6.66), which can be explicitly written as infinite series. With the help of Eq. (6.66) and Eq. (6.78) we obtain n equations for the fields $\bar{\varphi}_\alpha(x)$:

$$0 = -h_\alpha(x) + \sum_{l=1}^{\infty} l \left[\sum_{\alpha_1,\dots,\alpha_l} \int d^d x\, u^{(l+1)}_{\alpha\alpha_1,\dots,\alpha_l}(x, x_1, \dots, x_l/0) \right.$$
$$\left. \times \bar{\varphi}_{\alpha_1}(x_1) \cdots \bar{\varphi}_{\alpha_l}(x_l) \right]. \tag{6.81}$$

A lot of studies are performed in the reciprocal space built up on the wave vectors $\boldsymbol{k}$; below called *the momentum space.* For a large but finite system the Fourier transformation of the fields $\varphi_\alpha(x)$ will be

$$\varphi_\alpha(x) = \sum_k \varphi_\alpha(k) e^{ikx}, \tag{6.82a}$$

where

$$kx \equiv \boldsymbol{k} \cdot \boldsymbol{x}, \qquad k \equiv \boldsymbol{k}; \tag{6.82b}$$

note, once again, that we shall often avoid bold-face mathematical symbols when this does not introduce any confusion.

For a system of volume $V = L^d$, we shall assume the periodic boundary conditions $\varphi_\alpha(x) = \varphi_\alpha(x + L)$. With the help of this condition and the transformation (6.82a), we find out that the momentum vector $\mathrm{k} = (k_1, \dots, k_d) \equiv \{k_i\}$ takes the values $k_i = 2\pi l_i/L$, where $l_i = 0, \pm 1, \dots$. The inverse transformation from the Fourier amplitudes $\varphi_\alpha(k)$ to $\varphi_\alpha(x)$ is

$$\varphi_\alpha(k) = \frac{1}{V} \int d^d x \, \varphi_\alpha(x) e^{-ikx}. \tag{6.83}$$

Useful formulae for making the transformation $\varphi_\alpha(k) \leftrightarrow \varphi_\alpha(x)$ are:

$$\delta(k) = \frac{1}{V} \int d^d x \, e^{-ikx}, \qquad \delta(x) = \frac{1}{V} \sum_k e^{ikx}. \tag{6.84}$$

Note, that in our notations, $\delta(x) = \delta(x_1) \dots \delta(x_d)$ and $\delta(k) = \delta(k_1) \dots \delta_(k_d)$ represent the Kronecker symbol, $\delta(\mu, 0) \equiv \delta(\mu)$ with ($\mu = k$ or x), as an integral over $x \in V$ and as a sum over the discrete vectors k in the corresponding reciprocal (k-)space. For a large but finite volume V, we must use the Fourier summation over k rather than the integration (see Section 6.5). If $\varphi_\alpha(x)$ is x-independent, $\varphi_\alpha(x) = \varphi_\alpha^0$ is the density of the uniform component α. Then Eq. (6.83) yields $\varphi_\alpha(k) = \varphi_\alpha^0 \delta(k)$ — the nonzero amplitude $\varphi_\alpha(k)$ is $\varphi_\alpha(0) = \varphi_\alpha^0$.

The Fourier expansion of the coefficients $u^{(l)}$ is

$$u^{(l)}_{\alpha_1, \dots, \alpha_l}(x_1, \dots, x_l) = \frac{1}{V^{l-1}} \sum_{k_1 \dots k_l}' u^{(l)}_{\alpha_1, \dots, \alpha_l}(k_1, \dots, k_l) e^{i(k_1 x_1 + \dots + k_l x_l)}. \tag{6.85}$$

where $l \geq 1$, $u_\alpha^{(1)} \equiv h_\alpha$ and the factor $(1/V^{l-1})$ is introduced to fix a suitable dimension of the amplitudes $u_{\alpha_\nu}^{(l)}(k_\nu)$. Making use of Eqs. (6.82a) and (6.82b) with $\varphi_\alpha \to \delta\varphi_\alpha$, Eq. (6.84) and Eq. (6.85), the Eq. (6.77) takes the form

$$\frac{\Phi(\varphi)}{V} = \bar{\phi}(\bar{\varphi}) + \sum_{l=2}^{\infty} \sum_{\alpha_1,\dots,\alpha_l} \sum_{k_1,\dots,k_l} u_{\alpha_1,\dots,\alpha_l}^{(l)}(-k_1,\dots,-k_l)\delta\varphi_{\alpha_1}(k_1)\cdots\delta\varphi_{\alpha_l}(k_l); \tag{6.86}$$

$\bar{\phi}(\bar{\varphi}) = \Phi(\bar{\varphi})/V$ is the density of the equilibrium potential. In the same way we obtain the transformed Eq. (6.78):

$$\frac{\Phi(\varphi)}{V} = -\sum_{\alpha,k} h_\alpha(-k)\varphi_\alpha(k) + \sum_{l=2}^{\infty} \sum_{\alpha_i,k_i} u_{\alpha_1,\dots,\alpha_l}^{(l)}(-k_1,\dots,-k_l)\varphi_{\alpha_1}(k_1)\cdots\varphi_{\alpha_l}(k_l), \tag{6.87}$$

where we have introduced the Fourier transform $h_\alpha(k)$ of the intensive variable $h_\alpha(x)$; see Eq. (6.82a), where $\varphi_\alpha \to h_\alpha$.

Note that $\varphi_\alpha(x)$ and $h_\alpha(x)$ are real field (components): $\varphi_\alpha^*(x) = \varphi_\alpha(x)$, where φ_α^* is the complex conjugate to φ_α. With the help of this property and Eq. (6.82a), it is easy to prove that the Fourier amplitudes $a(k)$ of any real field $a(x)$ obey the relation: $a^*(k) = a(-k)$. So, for a real field $\varphi_\alpha(x)$, we have

$$\varphi_\alpha^*(k) = \varphi_\alpha(-k), \tag{6.88a}$$

and, hence,

$$\varphi_\alpha(k)\varphi_\alpha(-k) = \varphi_\alpha(k)\varphi_\alpha^*(k) \equiv |\varphi_\alpha(k)|^2 \tag{6.88b}$$

Besides, Eq. (6.88a) indicates that $\varphi_\alpha^*(0) = \varphi_\alpha(0)$, which means that the amplitude $\varphi_\alpha(0)$ is a real quantity. The same relations are valid for Fourier amplitudes of the real field components $h_\alpha(x)$. We shall widely use these relations.

As $h_\alpha(x)$ and $\varphi_\alpha(x)$ are real fields, the first term in Eq. (6.87) can be written in the form

$$-\sum_{k>0}\left[h_\alpha^*(k)\varphi_\alpha(k)+h_\alpha(k)\varphi_\alpha^*(k)\right]; \tag{6.89}$$

the summation here is over the positive momenta ($k > 0$).

The representations (6.77) and (6.86) of Φ are used for the calculation of the fluctuation effects. The Landau expansion (6.78) and (6.87) are more convenient when the investigation is carried out for the total field $\varphi = \bar{\varphi} + \delta\varphi$.

6.5 From Lattice to Field Models and Vice Versa

The field theory comes as a result of applying the continuum limit to lattice models.

6.5.1 *Continuum limit*

Let us return to the lattice of cells i and ν as described in Section 5.4. The volume $v_i \sim a_i^d$ of the microscopic cells i is fixed by the characteristic microscopic scale $a_0 \sim a_i$ (a_0 is the mean inter-particle distance; for crystals, this is the lattice constant). The value φ_i^α of the order parameter field component φ^α corresponding to the cell i is an extensive quantity proportional to the volume v_i. The corresponding density is denoted by $\varphi^\alpha(i) = (\varphi_i^\alpha/v_i)$. The order parameter component φ_ν^α of a given cell ν can be written as a sum over the cells $i_\nu \in V_\nu$ — the volume of the cell ν :

$$\varphi_\nu^\alpha = \sum_{i_\nu=1}^{N_\nu}\varphi_{i_\nu}^\alpha = \sum_{i_\nu=1}^{N_\nu}\varphi^\alpha(i_\nu)v_{i_\nu}. \tag{6.90a}$$

Because the volume $V_\nu \sim L_\nu^d$ of the quasi-macroscopic cell ν is much larger than any v_{i_ν}, $L_\nu \gg a_{i_\nu}$, the sum (6.90a) can be approximated by the integral

$$\varphi_\nu^\alpha = \lim_{v_{i_\nu}\to 0}\left[\sum_{i_\nu=1}^{N_\nu} v_{i_\nu}\varphi^\alpha(i_\nu)\right] = \int_{V_\nu} d^d x_\nu\, \varphi_\alpha(x_\nu), \tag{6.90b}$$

where we have denoted $v_{i_\nu} \to 0$ by $d^d x$, and the density $\varphi^\alpha(i) = (\varphi_i^\alpha/v_i)$ in the limit $v_i \to 0$ — by $\varphi_\alpha(x)$. So $\varphi_\alpha(x)$ is the density of φ_α at the point x ($\equiv \boldsymbol{x}$). Now we can pass to the continuum limit for the cells ν provided their volume L_ν^d is small compared to the volume $V \sim L^d$ of the total system ($L_\nu \ll L$). The order parameter component φ_α of the total system is then written by the integral

$$\varphi_\alpha = \int_V d^d x \, \varphi_\alpha(x), \tag{6.91a}$$

or

$$\frac{\varphi_\alpha}{V} \longrightarrow \varphi_\alpha = \frac{1}{V} \int d^d x \, \varphi_\alpha(x). \tag{6.91b}$$

Of course, the continuum limit can be performed directly from φ_i^α for the microscopic cells i to the field $\varphi_\alpha(x)$ in Eq. (6.91a) and Eq. (6.91b). In deriving the above formulae we have supposed that the variable φ_i^α is proportional to the volume v_i of the initial cells i. In some studies, this is quite convenient and leads to a field variable $\varphi_\alpha(x)$, the dimension $[\varphi_\alpha(x)]$ of which differs from that of the original variable φ_i^α by a factor $[V]$: $[\varphi_i^\alpha] = [\varphi_\alpha(x)][V]$; see, e.g., Eq. (6.90b).

Alternatively a slightly different way can be used to perform the continuum limit. Let us consider that the variable φ_i^α is defined per unit particle as it is, for example, for the Ising model. Instead of using the quantity $\varphi^\alpha(i) = \varphi_i^\alpha / v_i$, we shall take the continuum limit directly for φ_i^α. Having in mind that $v_{i_\nu} \sim N^\nu / V_\nu$, we can write φ_ν^α in the form

$$\varphi_\nu^\alpha = \frac{N_\nu}{V_\nu} \sum_{i_\nu}^{N_\nu} v_{i_\nu} \varphi_{i_\nu}^\alpha \longrightarrow \frac{N_\nu}{V_\nu} \int_{V_\nu} d^d x_\nu \, \varphi'_\alpha(x_\nu), \tag{6.92a}$$

where $\varphi'_\alpha(x_\nu)$ corresponds to φ_i^α :

$$\varphi_i^\alpha \longrightarrow \varphi'_\alpha(x_\nu). \tag{6.92b}$$

Obviously $\varphi_\alpha(x_\nu) = (N_\nu / V_\nu) \varphi'_\alpha(x_\nu)$. If the limit is taken directly from the cells i to the macroscopic volume $V = L^d$ of the total system, the suffices ν should be omitted in the above formulae. So we can use either the rule of correspondence

$$\sum_i \longrightarrow \int d^d x, \tag{6.93a}$$

or the alternative,

$$\sum_i \longrightarrow \frac{N}{V}\int d^d x. \tag{6.93b}$$

The correspondence (6.93b) gives the field $\varphi_\alpha(x)$ in the same dimension as the initial lattice variable φ_i^α.

Two physical ideas are intuitively supposed when passing to the continuum limit of a system with many (but not infinite) degrees of freedom. The first one is to fix the microscopic lengths $a_0 \sim a_i \neq 0$ and to consider them small compared to all other characteristic lengths (L_ν or L) of the system. Then fixing $v_i \neq 0$, the continuum limit will correspond to the thermodynamic limit ($N, V \to \infty$), where $N \sim (L/a_0)^d \sim V/v$ is proportional to the number of degrees of freedom; for cells ν, $N_\nu \sim (L_\nu/a_0)^d \sim (V_\nu/v)$.

In statistical mechanics the transition to this limit is always justified as the total system of a volume V is of a macroscopic size: $(V/v) \geq 10^{22}$. The condition for the macroscopic treatment $(V/v) \geq 10^{22}$ is usually fulfilled so its mathematical approximation by the limit $(V/v) \to \infty$ is quite plausible. The total system is finite ($V \neq \infty$), the microscopic cell i is also finite ($v_i \neq 0$) but we can take $v_i \to 0$ in virtue of the expression $v_i = (V/N) \ll V$; in the case of the cell ν, $v_i = (V_\nu/N_\nu) \ll V_\nu$.

The second possibility is to fix the macroscopic length scale L as finite ($L \neq \infty$) and simultaneously to assume the limit $a_0 \to 0$. This precisely corresponds to the real fact that a macroscopic system has a finite size but involves problems connected with the microscopic size, which is now mathematically zero for $(V/v) \to \infty$ (another interpretation of the limit $a_0 \to 0$ is briefly discussed at the end of Section 6.5.4).

From a physical point of view we can assume that a does not approach the zero, rather it is small of the order $10^{-22}V$. The continuum limit gives an exact representation of an infinite system ($V, N = \infty$) or of a finite system where the physical quantities can be defined at any point x of the continuous medium; both cases are a form of theoretical idealization. Certainly, in statistical mechanics we can use two cutoff lengths — the microscopic characteristic length a_0, which is of order of the mean inter-particle distance and the macroscopic length L, $(L/a_0) \geq 10^{22/d}$. The physical phenomena will be then: (i) microscopic (of scale a_0), (ii) quasi-macroscopic ($a_0 \ll L_\nu \ll L$), and (iii) macroscopic (of scale L). The phenomena (ii) and (iii) can be approximately investigated in the continuum (field) limit.

6.5.2 *The origin of fields in the generalized thermodynamics*

It is very useful to compare the cell expansion (5.51) with the functional series (6.77). In order to present this example on a general basis let us consider the cell-ν expansion of $\Phi(\varphi)$,

$$\Phi(\varphi) = -\sum_{\alpha;\nu} h_\nu^\alpha \varphi_\nu^\alpha + \sum_{\alpha_1\alpha_2;\nu_1\nu_2} u^{(2)\alpha_1\alpha_2}_{\nu_1\nu_2} \varphi^{\alpha_1}_{\nu_1}\varphi^{\alpha_2}_{\nu_2} + O\left[(\varphi_\nu^\alpha)^3\right]. \qquad (6.94)$$

This expansion looks like the entropy expansion in Section 2.7, where we have studied macroscopic cells ν (denoted there by α). If the cells ν are macroscopic systems themselves, they must obey the condition of thermodynamic additivity (Section 2.2.3). As a result, the potential Φ will be separable $\Phi = \sum \Phi_\nu$; Φ_ν will depend *only* on φ_ν (see also Section 5.4.3):

$$\left(\frac{\partial \Phi_\nu}{\partial \varphi_\mu}\right)_{\nu \neq \mu} = 0.$$

In this case the coefficients in the expansion (6.94) will obey the relations

$$u^{(l)\alpha_1\ldots\alpha_l}_{\nu_1\ldots\nu_l} = \delta_{\nu_1\nu_2}\delta_{\nu_2\nu_3}\cdots\delta_{\nu_{l-1}\nu_l} u^{(l)\alpha_1\ldots\alpha_l}_{\nu_1}. \qquad (6.95)$$

Then $\Phi(\varphi)$ will be a sum of powers like $(\varphi_\nu^\alpha)^l$ and no correlation between the cells will exist. Now we have a simple sum of independent cells ν, which act as thermodynamic systems, placed in different fields h_ν. Certainly in this case the series (6.94) will describe spatially independent fluctuations φ_ν^α within each thermodynamic system ν. The expansion (6.94) will contain mixed (inter-cell) terms $\varphi_{\nu_1}\varphi_{\nu_2}$, $\nu_1 \neq \nu_2$, $\varphi_{\nu_1}\varphi_{\nu_2}\varphi_{\nu_3}$, etc., if only the size of the macroscopic cells ν is lowered to the quasi-macroscopic level ($L_\nu \ll L$), which means the breaking of the thermodynamic additivity. Clearly, there will be correlations when L_ν is less than some characteristic length of the system — the correlation length, which at this stage of consideration remains unspecified (see Sections 7.1 and 7.2).

A glance on the first two terms in Eq. (6.94) shows a similarity to the structure of the Ising model (5.110). The inter-cell correlation coefficients $u^{(2)}_{\nu_1\nu_2}$ play the role of the exchange constants J_{ij} in Eq. (5.110). If we suppose the existence of short-range correlations between the cells ν, we may accept the approximation of the nearest-neighbour inter-cell correlation ($u_{\nu,\nu+1} \neq 0$, $u_{\nu,\nu+l} = 0, l > 1$) which is usual for microscopic lattice models. We shall not go into details of the similarities between microscopic and quasi-macroscopic lattice models because the differences are also quite transparent (for example, φ_ν in Eq. (6.94) is a continuous variable). Moreover expansions over quasi-macroscopic cells like (6.94) require a treatment by phenomenological arguments and, hence, it is convenient from both mathematical and physical point of view to investigate them in the continuum limit.

The functional $\Phi(\varphi)$ is obtained by substituting $\varphi_\nu^\alpha = V_\nu \varphi^\alpha(\nu)$ in Eq. (6.94) and taking the continuum limit $V_\nu \to 0$; see also the definition Eq. (C.5) of a functional derivative. Thus we directly arrive at the functional series (6.78). Alternatively we can use the correspondence rule (6.92b). Then,

$$\Phi(\varphi) = -\frac{\tilde{N}}{V}\sum_\alpha \int d^d x\, h_\alpha(x)\varphi'_\alpha(x) + \frac{\tilde{N}^2}{V^2}\sum_{\alpha_1\alpha_2}\int d^d x_1 d^d x_2\, u^{(2)}_{\alpha_1\alpha_2}(x_1,x_2)\varphi'_{\alpha_1}(x_1)\varphi'_{\alpha_2}(x_2) + \cdots, \quad (6.96)$$

where $\tilde{N} = (L/L_\nu)^d$ is the number of cells ν in the total volume $V = L^d$. In the limit $(\tilde{N}, V) \to \infty$ the density $\rho = \tilde{N}/V \neq (0,\infty)$, can be attached to $\varphi' : \rho\varphi'_\alpha \to \varphi_\alpha$.

6.5.3 *Discretization of field models*

Some theoretical studies are performed by the inverse transformation — by substituting the field model with its lattice counterpart. So we must recall how an integral can approximately be expressed as a sum over discrete variables. As an example we shall use the simple φ^4-theory for a one-component isotropic field $\varphi(x)$:

$$\Phi = \int d^d x_1 d^d x_2\, u_2(x_1,x_2)\varphi(x_1)\varphi(x_2) + \int d^d x_1 \cdots d^d x_4\, u_4(x_1,\ldots,x_4)\varphi(x_1)\cdots\varphi(x_4). \quad (6.97)$$

Let there be a lattice of cells i; we shall consider $\varphi_i \equiv \varphi(x_i)$, $x_i \in v_i$. Here the index i does not necessarily denote a microscopic (or quasi-macroscopic) cell. The important point is that $v_i \ll V$. Now we approximate all values $\varphi(x)$, $x \in v_i$ with one and the same value $\varphi(x_i)$ of $\varphi(x)$ at the site i. So the integrals in Eq. (6.97) can be written with the help of the sums:

$$\Phi = \sum_{ij} K^{(2)}_{ij} \varphi_i \varphi_j + \sum_{ijkl} K^{(4)}_{ijkl} \varphi_i \varphi_j \varphi_k \varphi_l, \tag{6.98}$$

where $\varphi_i = \varphi(x_i) v_i$, $K^{(2)}_{ij} = u_2(i,j)$, and $K^{(4)}_{ijkl} = u_4(i,j,k,l)$. The lack of correlations between the cells will be equivalent to the choice $K^{(2)}_{ij} = \delta_{ij} K^{(2)}_i$ and $K^{(4)}_{ijkl} = \delta_{ij}\delta_{ik}\delta_{il} K^{(4)}_i$.

Alternatively, we can attach the volumes v_i to the coefficients $K^{(2)}$ and $K^{(4)}$. For example, $K^{(2)}_{ij} = u_2(i,j) v_i v_j$, and then, $\varphi_i = \varphi(x_i)$ are the actual lattice variables at the centers (or the vertices) of the lattice. For any function $f(x)$, we can define a sum over the corresponding discrete values $f(x_i) \equiv f_i$ by the set of transformations

$$\int d^d x\, f(x) \longrightarrow \sum_i v_i f(x_i) \longrightarrow v \sum_i f(x_i) \longrightarrow \frac{V}{N} \sum_i f(x_i),$$

where we have used the equality $v_i = v$, which holds true for a regular lattice and can be considered as a limiting case of the relation $v_i \sim (L/a_i)^d$ for $L \gg a_i$.

6.5.4 *Continuum limit in the momentum space*

The continuum limit in the x-space is consistent with the continuum limit in the reciprocal (k-) space. Let the system have again volume $V = L^d$. The corresponding reciprocal volume in the reciprocal space of the vectors $k(= \boldsymbol{k})$ is $v_L = (2\pi/L)^d$. This is the minimal volume in the k-space; the maximal volume is given by $v_a = (2\pi/a_0)^d$ with a_0 the lattice spacing (for simplicity we restrict ourselves to a regular cubic lattice; $a_i = a_0$). A sum over k will be transformed to an integral:

$$\sum_k f(k) \longrightarrow \frac{1}{v_L} \sum_k v_L f(k) \longrightarrow \frac{V}{(2\pi)^d} \int d^d k f(k),$$

where we have substituted the factor v_L into the sum by $d^d k$, which is valid for $(L/a_0) \to \infty$. If $f(k) = 1$, the integral over k will be

$$\int d^d k \cdot 1 = \int_0^{2\pi/a_0} dk_1 \cdots \int_0^{2\pi/a_0} dk_d = \left(\frac{2\pi}{a_0}\right)^d, \tag{6.99}$$

so the sum $\sum_k .1$ is exactly equal to $(V/a_0^d) = N$ — the number of cells. The rule of correspondence now will be

$$\frac{1}{V}\sum_{k} \longrightarrow \int_{k} \equiv \int \frac{d^d k}{(2\pi)^d}. \tag{6.100}$$

The integral over k can be written in spherical coordinates

$$\int d^d k = \int_0^{\Lambda} dk k^{d-1} \int d\Omega, \tag{6.101}$$

where Λ is the upper cutoff for the magnitude $|k|$ of the momentum: $0 < k < \Lambda$; $|k| = k$ (see also Section D.1). The value of Λ in spherical coordinates can be found by imposing the condition that the volume (6.99) of the d-dimensional cube with an edge $(2\pi/a)$ in the k-space is equal to the volume of the d-dimensional sphere with radius Λ : $(2\pi/a_0)^d = \Omega_d \Lambda^d/d$, where Ω_d is the area of the unit ($\Lambda = 1$) sphere; see Eq. (D.6). Thus we obtain $\Lambda = [d\Gamma(d/2)/2]^{1/d}(2\sqrt{\pi}/a_0) \sim 1/a_0$.

The quantities written previously by Fourier sums can now be expressed by integrals with the help of the rule (6.100). For example, Eq. (6.82a) will take the form

$$\varphi_\alpha(x) = V \int \frac{d^d k}{(2\pi)^d} \varphi_\alpha(k) e^{ikx}, \tag{6.102a}$$

where we have used notations introduced in Eq. (6.82b). The Eq. (6.85) becomes

$$u^{(l)}_{\alpha_1,\ldots,\alpha_l}(x_1,\ldots,x_l) = V \int \frac{d^d k_1}{(2\pi)^d} \cdots \frac{d^d k_l}{(2\pi)^d} u^{(l)}_{\alpha_1,\ldots,\alpha_l}(k_1,\ldots,k_l) e^{i(k_1 x_1 + \cdots + k_l x_l)}. \tag{6.102b}$$

The continuum limit must be also performed in the expressions (6.84) for the Kronecker symbols $\delta(k)$ and $\delta(x)$. In the "true" (mathematically exact) continuum limit ($L/a_0 \to \infty$), these symbols turn into the Dirac δ-function. From Eq. (6.84) and the rule (6.100) we obtain

$$\delta(k) = \int \frac{d^d x}{(2\pi)^d} e^{-ikx}, \qquad \delta(x) = \int \frac{d^d k}{(2\pi)^d} e^{ikx}. \tag{6.103}$$

If the rule (6.100) is applied to Eq. (6.87), the expansion terms should be multiplied by the factors $V^{l+1}, l \geq 1$. This is related to the initial definition of the Fourier series (6.83) and (6.85) for $\varphi_\alpha(x)$ and $u^{(l)}$. In the continuum limit these definitions are no longer convenient and ought to be changed by introducing the Fourier amplitudes $\varphi_\alpha(k) \to (1/V)\varphi_\alpha(k)$ and $u^{(l)}(k_i) \to (1/V)u^{(l)}(k_i)$ having another dimension. Now the sums (6.83) and (6.85) are multiplied by a volume dependent coefficient $(1/V)$. The corresponding terms in the expansion of Φ will have volume dependent coefficients $(1/V)^l$, which change to $(2\pi)^{-dl}$ when we pass to the continuum limit (6.100). Therefore,

$$\begin{aligned}\Phi = &-\sum_\alpha \int \frac{d^dk}{(2\pi)^d} h_\alpha(-k)\varphi_\alpha(k)\\ &+\sum_{l=2}^{\infty}\sum_{\alpha_i,k_i} \int \frac{d^dk_1}{(2\pi)^d}\cdots\frac{d^dk_l}{(2\pi)^d} u^{(l)}_{\alpha_1,\dots,\alpha_l}(-k_1,\dots,-k_l)\\ &\times \varphi_{\alpha_1}(k_1)\cdots\varphi_{\alpha_l}(k_l).\end{aligned} \tag{6.104}$$

The same result will be obtained, if the transformations

$$\varphi(x) = \int \frac{d^dk}{(2\pi)^d}\varphi(k)e^{ikx}, \tag{6.105}$$

$$\begin{aligned}&u^{(l)}_{\alpha_1,\dots,\alpha_l}(x_1,\dots,x_l)\\ &\quad= \int \frac{d^dk_1}{(2\pi)^d}\cdots\frac{d^dk_l}{(2\pi)^d} u^{(l)}_{\alpha_1,\dots,\alpha_l}(k_1,\dots,k_l)e^{i(k_1x_1+\cdots+k_lx_l)},\end{aligned} \tag{6.106}$$

and (6.103) are directly applied to the x-representation (6.78) of Φ in the thermodynamic limit $(V \to \infty)$.

For finite systems $(V \neq \infty)$ the integrals (6.103), (6.105), and (6.106) are approximate. It has been already mentioned that we can keep the lattice spacing a_0 finite and to consider the size L of the system much larger than a_0. The availability of a finite upper cutoff $(\Lambda \neq \infty)$ points that $a_0 > 0$. The approximation is connected with the lack of a lower momentum cutoff $\Lambda' \sim 1/L$. In finite systems $(\Lambda' \neq 0)$ the small momenta $(0 < k < \Lambda')$ are not present and setting $\Lambda' = 0$, as we have made above, the real finite system is substituted by the corresponding infinite system. In fact, neither Λ nor Λ' can be neglected in real systems. While at the field model is looked as an approximation, which describes very well macroscopic systems we can write Fourier sums instead of integrals and we can work with a finite value of the volume V. Moreover we can safely take the continuum limit from sums to integrals when this is convenient from a calculational point of view. In most of the cases we shall use this (quasi-continuum) version of the theory.

If the mean inter-particle distance (lattice constant in crystals) a_0 tends to zero, the density $\rho = (N/V) \sim a_0^{-d}$ will tend to infinity (extremely dense matter). In this quite rare case the continuum limit is entirely exact even for the case of finite size systems.

6.5.5 *Functional integral*

The partition function of the field functional $\Phi(\varphi)$ will be again expressed by the relation (6.4) but now the integration is over an infinite number of variables represented by the fields $\varphi(x)$ at any point x :

$$D\varphi \equiv \prod_x \prod_{\alpha=1}^{n} d\varphi_\alpha(x) = \lim_{v_i \to 0} \prod_i \prod_{\alpha=1}^{n} d\varphi_i^\alpha. \tag{6.107}$$

Note that φ_i^α from Eq. (6.99) is substituted with $\varphi'_\alpha(x)$; see Eq. (6.92b). We have already explained that the fields $\varphi_\alpha(x)$ and $\varphi'_\alpha(x)$ differ from one another by the factor $(N/V) \sim 1/v_i$, which in the continuum limit yields the density factor $\rho(x) = 1/v(x)$. Then $\prod_x \rho(x) = N/V$ will give a trivial contribution, $-T\ln(N/V)$ to the equilibrium thermodynamic potential $\Phi(\bar{\varphi}, h)$. It can be always subtracted from the para-phase potential $\Phi(0, h)$. Such a term is completely irrelevant to our consideration.

The continuum limit has led to the necessity of treating the partition function as a functional integral. The main problem in the phase transition theory is how to solve the functional integrals $\mathcal{Z}$ and $\mathcal{Z}_f$. The techniques for dealing with such integrals rapidly developed when Feynman introduced the functional integrals in quantum mechanics and quantum field theory (Feynman and Hibbs, 1965; Bogoliubov and Shirkov, 1959; Parisi, 1988) and soon after in statistical physics; see, e.g., Popov (1976, 1987). We shall not enter into details of the functional analysis because the relevant solutions of the functional integral rather come from certain physical arguments than from applying special mathematical methods; see Sections C.2–C.5. The reader interested in mathematical details is referred to the books by Volterra (1959), Taylor and Coy (1980), Vassilev (1976) and the review by Tarski (1968).

Other functional integrals are defined by applying the continuum limit to the correlation functions $G^{(l)}$ and $G_c^{(l)} = \chi^{(l)}$ (Section 5.5). For example the continuum limit of Eq. (5.78) defines $G^{(l)}(x_1, \ldots, x_l)$ — the l-point correlation function of the one-component field $\varphi(x)$ as proportional to the l-th order functional derivative of the partition integral; $h(i_j) \to h(x_j)$. We shall not re-derive the field version of the results presented in Section 5.4 because all these results are valid for fields $\varphi(x)$ and $h(x)$. Moreover they are straightforwardly generalized for n-component fields: $h = \{h_\alpha\}$ and $\varphi = \{\varphi_\alpha\}$.

6.5.6 *Truncated series*

Often, it is more easy to calculate the functional integrals for the correlation functions than the integral for the partition function. However the series expansion (6.78) is rather general and when we make an attempt to calculate the functional integrals for $\mathcal{Z}$ and $G^{(l)}$ we shall meet difficulties. It is possible to develop the general phenomenological approach by appropriate assumptions about the properties of the coefficient functions $u^{(l)}$.

For example, the equations of state show that stationary solutions $\bar{\varphi}_\alpha(x)$ will be independent on x provided h is uniform and the coefficients $u^{(l+1)}(x, x_1, \ldots, x_l)$ depend on l vector differences $(x_1 - x), \ldots, (x_l - x)$ rather than on $(l+1)$ independent vectors (translational invariance). Further one must keep in mind that $u^{(l)}$ are functions of the thermodynamic parameters T and Y as well. In this respect they have the properties of the coefficients in the usual Landau expansion (4.32); see, e.g., Landau and Lifshitz (1980).

Using general stability and symmetry requirements about the behaviour of the functional Taylor series for $\Phi(\varphi)$, it is possible to develop a quite general theory of phase transitions and critical phenomena. The opposite approach is the investigation of particular functionals $\Phi(\varphi)$, which describe essential aspects of real systems (models). For example, abbreviating the series expansion one may investigate the φ^2-theory (Gaussian approximation), φ^3-, φ^4-, or φ^6-theories; the arguments for such a truncation have been presented in Sections 4.4.4 and 4.8.3.

The most important problems of the theory are revealed from the particular field models such as the φ^4-theory. This is the reason, for which it seems quite reasonable to reduce our general treatment and to focus our attention on the φ^4-field theory that describes the behaviour near the ordinary critical points. The properties of the φ^4-functional; see, e.g., Eq. (6.97), related with the description of translationally noninvariant systems or other secondary effects are not among the main features of critical phenomena. In

Chapter 7 we shall discuss a simple variant of φ^4-field theory having in mind that the obtained results can be generalized for more complex systems.

6.6 Integral Transformation of Microscopic Models

A wide class of microscopic models can be investigated with the help of the identity

$$e^{\frac{1}{2}\sum_{ij} K_{ij}\sigma_i\sigma_j} = A\int_{-\infty}^{\infty} D\varphi e^{-\frac{1}{2}\sum_{ij}^{N} K_{ij}^{-1}\varphi_i\varphi_j + \sum_i \sigma_i\varphi_i}, \tag{6.108}$$

where we use notations explained in Eq. (5.110), $K_{ij}^{-1} \equiv [\hat{K}^{-1}]_{ij}$ are the elements of the inverse matrix $\hat{K}^{-1}$,

$$A = \frac{(2\pi)^{-N/2}}{\mathrm{Det}(\hat{K}^{1/2})}, \quad D\varphi = \prod_{i=1}^{N} d\varphi_i\,, \tag{6.109}$$

N is the number of cells (or vertices) i of the lattice, on which the microscopic variables σ_i and the interaction $\sigma_i\sigma_j$ through the symmetric matrix $\hat{K}$ are defined. For a vector lattice field $\sigma_i \equiv \{\sigma_\alpha\}$, as in Heisenberg and XY models, the conjugate field φ_i should also be a vector from the same vector space, and in this case the products $\sigma_i\sigma_j$, $\varphi_i\varphi_j$, $\sigma_i\varphi_i$ are understood as scalar products, for example, $\sum_\alpha \sigma_\alpha\sigma_\alpha$. From a pure mathematical point of view the identity (6.108) is well defined when $\mathrm{Det}\hat{K}^{1/2} = (\mathrm{Det}\hat{K})^{1/2}$ is a finite real number (different from zero) but the same identity has a quite wider, and quite well justified application in theoretical physics.

The expression (6.108), which relates the bilinear exponent $\sigma\hat{K}\sigma$ of variables σ_i which take discrete values on the sites i of a lattice with a multiple integral over another lattice field φ_i which takes continuous values ($-\infty < \varphi < \infty$) on the same lattice sites i is called *the Hubbard–Stratonovich transformation* (Hubbard, 1958; Stratonovich, 1957); see also Berlin and Kac (1952), Baker (1962), Zinn–Justin (1996), and Herbut (2007).

From a mathematical point of view this is a transformation of a multiple Gaussian integral and is often called *the Gaussian transformation.* The way of the derivation of the identity (6.108) is outlined in Section B.2; see Eq. (B.5) and Eq. (B.5), where replacement of h by σ, $\hat{K}$ by $\hat{K}^{-1}$ and n by N should be made; see also Micnas (1979). If $\hat{K}$ is a diagonal matrix:

$\hat{K} = \delta_{ij}K_j$, $A^{-1} = \prod_i (2\pi K_i)^{1/2}$. The identity (6.108) becomes very simple for a single variable ($\sigma_i = x, \varphi = y$):

$$e^{ax^2/2} = \frac{1}{\sqrt{2\pi a}} \int_{-\infty}^{\infty} dy e^{-y^2/(2a)+xy}. \tag{6.110}$$

The relation (6.110) is valid only for $a > 0$. The general identity (6.108) obviously is fulfilled for positively definite matrices $\hat{K}$: $(\mathrm{Det}\hat{K}) > 0$. In this case the denominator of A in (6.109) is well defined: $\mathrm{Det}(\hat{K}^{1/2}) = (\mathrm{Det}\hat{K})^{1/2}$.

If the identity (6.108) is taken in the continuum limit (Section 6.5), in which $(i, j) \to (x, y)$ the requirement for the positive definiteness of $\hat{K}$ seems to be not necessary. This is easy to see by taking the logarithm of Eq. (6.108). When the continuum limit is performed the quantity $\ln(\mathrm{Det}\hat{K})$ becomes an integral over the continuous variables x and y. This integral will be finite for a wide class of infinitely-rowed matrices $K(x, y)$. Such a matrix $K(x, y)$, the integral of which has a finite value, may correspond to a divergent quantity $\ln(\mathrm{Det}\hat{K})$ of the original lattice model. For example, if we suppose that the matrix K can be diagonalized, $K_{ij} = \delta_{ij}K_j$, the corresponding integral in the continuum limit will be $\int d^dx K(x)$. Although some eigenvalue(s) $K_{i_1} = 0$ may give rise to a singularity in $\ln(\mathrm{Det}\hat{K})$ of the lattice model, the corresponding integral over x may have a finite value.

Now we should mention that the identity used in statistical mechanics is just the logarithm of Eq. (6.108). The lack of an exact correspondence between the analytical properties of the lattice model and its field counterpart should be taken into account in the treatment of models with non-positively definite exchange interaction matrices $\hat{K}$. When the transformation (6.108) is performed for the investigation of the initial (σ_i) model in terms of new lattice variables (φ_i), it is very important the transformed matrix $\hat{K}$ to be positively definite (or to be chosen of this type). In the cases, for which the identity (6.108) is applied in order to transform the initial model to a field model $\varphi_i \to \varphi(x)$ and to investigate the behaviour of the system in terms of the field $\varphi(x)$, one can use all matrices $\hat{K}_{ij}$ which being taken in the continuum limit $K(x, y)$ give well defined functional integrals. The loss of an accuracy in this "lattice-field" correspondence is too small in comparison with the uncontrolled approximation for the study of real phenomena with the help of simple lattice models.

6.6.1 *Integral transformation of the Ising model*

The partition function of the spin ($S \geq \frac{1}{2}$) Ising model (5.110) is

$$\mathcal{Z} = \prod_{i=1}^{N} \sum_{\sigma_i' = -S}^{S} e^{\frac{1}{2}\sum_{i \neq j} \sigma_i' K_{ij} \sigma_j' + \sum_i \tilde{h}_i \sigma_i'}, \tag{6.111}$$

where $K_{ij} = J_{ij}/TS^2$ and $\tilde{h}_i = h_i/TS = g_s \mu_B/T$, and the extra factor $1/S^2$ in K_{ij} comes from the fact that in Eq. (6.111) we use the variables $\sigma_i' = S\sigma_i$ (see Section 5.6.2 for a comprehensive discussion of the spin angular momentum S_i and the variables σ_i and σ_i').

Note, that the definition of h_i accepted in Eq. (5.110) for a general quantum number S is $h_i = g_s \mu_B S$ so that the change of σ_i to σ_i' introduces an extra factor $1/S$ in the external field parameter h_i, too. Now we may assume that the factor $1/S^2$ in K_{ij} is absorbed in the exchange parameters J_{ij} ($J_{ij}/S^2 = J_{ij}'$, and hereafter the superscript (*prime*) will be omitted). In Eq. (6.111), S can be substituted by J, the total angular momentum, if the orbital angular momentum (L) is to be taken into account (see Section 5.6.2). While $K_{ii} = J_{ii} = 0$ (see Section 5.6.2) we may forget about the (ij)-summation rule $i \neq j$ in Eq. (6.111).

Note, that for two-state variables $\sigma_i' = \pm 1/2$, as in the usual spin-1/2 Ising model, $(\sigma_i')^2 = 1/4$ at each site i and a constant term $(\mathcal{A}/2)\sum_i \sigma_i^2 = \mathcal{A}N/2$ can always be added to the exponent in Eq. (6.111) (Berlin and Kac, 1952). This trick is used to ensure the positive definiteness of the matrix $\hat{K}$ replacing it by $(\hat{K} + \mathcal{A}\hat{I})$, where $\mathcal{A} > 0$ and $\hat{I} = \{\delta_{ij}\}$. The same procedure cannot be applied for spin-($S > 1/2$) models where $(\sigma_i')^2$ changes with σ_i'; for example, for $S = 1$, $\sigma_i' = 0, \pm 1$ and $\sigma_i^2 = 0, 1$.

Substituting the exponent $\frac{1}{2}\sigma' \hat{K} \sigma'$ in Eq. (6.111) by the r.h.s. of Eq. (6.108) and having in mind that the variables σ_i' do not depend on φ_i, we obtain

$$\mathcal{Z} = A \int_{-\infty}^{\infty} D\varphi e^{-\frac{1}{2}\sum_{ij}^{N} \varphi_i K_{ij}^{-1} \varphi_j} \left\{ \prod_{i=1}^{N} \left[\sum_{\sigma_i = -S}^{S} e^{(\tilde{h}_i + \varphi_i)\sigma_i'} \right] \right\}. \tag{6.112}$$

The sum over σ_i' in Eq. (6.112) has already been calculated in Section 5.8.2; see Eqs. (5.113) and (5.135) for $J = S$. Representing $\prod_i(\cdots)$ in the integrand in Eq. (6.112) by $\exp[\sum_i \ln(\cdots)]$ and performing the shift transformation

$$(\varphi_i + \tilde{h}_i) \longrightarrow \varphi_i, \tag{6.113}$$

we obtain

$$\mathcal{Z} = A \int_{-\infty}^{\infty} D\varphi e^{-\mathcal{H}_{\text{eff}}/T}. \tag{6.114}$$

The effective Hamiltonian $\mathcal{H}_{\text{eff}}$ is given by

$$\mathcal{H}_{\text{eff}}(\varphi) = \frac{T}{2} \sum_{ij} K_{ij}^{-1} [\tilde{h}_i \tilde{h}_j - 2\tilde{h}_i \varphi_j + \varphi_i \varphi_j] - T \sum_i \ln \left[\frac{\sinh(S + \frac{1}{2})\varphi_i}{\sinh(\varphi_i/2)} \right]. \tag{6.115}$$

The transformation (6.108) yields the exact Hamiltonian (6.115) of the Ising systems in terms of the new continuous ($-\infty < \varphi_i < \infty$) variables φ_i. These variables, like σ_i', are defined on a lattice so $\mathcal{H}_{\text{eff}}$ in Eq. (6.115) represents a spatially discrete system.

Up to this stage of investigation, no approximations have been made so the name "effective Hamiltonian" points out the coming approximations. We shall see in the remaining part of our derivation of the field counterpart of the Ising model that some of the calculations are also exact and the approximation appears only when the properties of the matrix $\hat{K}$ in the continuum limit are Determined and the respective functional series is abbreviated.

The next step is to expand in power series the ln(sinh) in Eq. (6.115) for $\varphi_i \ll S^{-1}$ and to obtain a lattice version of the Landau series. Alternatively one may use the expansion Eq. (5.153a) of the Brillouin function $B_S(S\varphi_i)$ and the fact that this function is the first derivative of the logarithm in Eq. (6.115). This logarithm is equal to $\ln(2S+1)$ for $\varphi_i = 0$. The integration of the series (5.153a) when the integration constant $\ln(2S+1)$ is taken into account yields the desired expansion. It is

$$\ln \left[\frac{\sinh(S + \frac{1}{2})\varphi_i}{\sinh(\varphi_i/2)} \right] \approx \ln 2S + 1) + \frac{S(S+1)}{6} \varphi_i^2 - \frac{S(S+1)(2S^2 + 2S + 1)}{360} \varphi_i^4 + O(\varphi_i^6). \tag{6.116}$$

This expression is substituted in Eq. (6.115) and thus an expansion of $\mathcal{H}_{\text{eff}}$ in terms of φ_i is obtained. Such a series may be of use in some particular investigations, but it will not lead us to the desired result — the generalization of the mean field analysis from Section 5.8. The reason is that φ_i in Eq. (6.116) is not the right field.

Our task is to represent the first term $\varphi K^{-1}\varphi$ in Eq. (6.115) in the Ising-like form $\varphi'\hat{K}\varphi'$, where φ' is some field obtained by a transformation of φ. The appropriate transformation is obviously given by

$$\varphi_i \longrightarrow \sum_j K_{ij}\varphi_j. \tag{6.117}$$

Inserting (6.116) and (6.117) in Eq. (6.115) we get

$$\begin{aligned}\mathcal{H}_{\text{eff}} &= \frac{1}{2}\sum_{ij}\left[J_{ij}^{-1}h_ih_j + J_{ij}\varphi_i\varphi_j - 2\delta_{ij}h_i\varphi_j\right] \\ &\quad - NT\ln(2S+1) - \frac{1}{2}a_2\sum_i\left(\sum_{j_1j_2}J_{ij_1}J_{ij_2}\varphi_{j_1}\varphi_{j_2}\right) \\ &\quad + a_4\sum_i\left(\sum_{j_1\dots j_4}J_{ij_1}\cdots J_{ij_4}\varphi_{j_1}\cdots\varphi_{j_4}\right) - \cdots,\end{aligned} \tag{6.118}$$

where $a_2 = S(S+1)/3T$ and $(-a_4T^3) < 0$ is equal to the coefficient before the φ^4-term in Eq. (6.116). The terms $\frac{1}{2}h_iJ_{ij}^{-1}h_j$ and $-NT\ln(2S+1)$ in Eq. (6.118) automatically go out from the functional integral (6.114) and give direct contributions to the potential $\Phi = -T\ln\mathcal{Z}$.

The reason is that these terms do not depend on φ_i. The factor A in Eq. (6.114) yields the same type of contribution to the thermodynamic potential. Its contribution to Φ is however through the term $(-T\ln\tilde{A})$, where $\tilde{A} = A\text{Det}(\hat{K}) = [\text{Det}(\hat{K})]^{1/2}/2\pi)^{N/2}$. The reason is in the transformation Eq. (6.117) which has a Jacobian equal to $\text{Det}(\hat{K})$ and namely this Jacobian changes the quantity A to $\tilde{A}$ under the logarithm of the partition function (see also Section B.2 where the Jacobian $\text{Det}(\hat{K}^{1/2})$ of the transformation (B.10) by the matrix $\hat{K}^{1/2}$ is discussed).

Note, that the lattice "field" theories (6.115) and (6.118) can be used for the development of perturbation (cumulant) expansions beyond the standard MF approximation (see, e.g., Uzunov (1996, 2008) and references therein). That is why one should choose the zero approximation of the Hamiltonian $\mathcal{H}_{\text{eff}}^{(0)}$ appropriate to the particular problem of study and treat the remaining interaction part $\mathcal{H}_{\text{eff}}^{\text{int}} = (\mathcal{H}_{\text{eff}} - \mathcal{H}_{\text{eff}}^{(0)})$ as a perturbation; see, e.g., Section 6.4. Usually the "free" field part $\mathcal{H}_{\text{eff}}^{(0)}$ is chosen among the terms of the type $\varphi\varphi$ or $h\varphi$; see, e.g., (Hertz, 1985), where this approach is demonstrated for the example of disordered (spin-glass) systems.

6.6.2 *Continuum limit for the Ising model and relation to the mean-field theory*

Now we are ready to perform the continuum limit for the effective Hamiltonian (6.118) . As in Section 6.5 we introduce the density $\varphi(x_i) \to \varphi_i/v_i$ and the notations $v_i \to d^d x$, $\varphi(x_i) \to \varphi(x)$, which correspond to infinitesimal cell volumes v_i. Besides an additional factor $v_i \to d^d x$ arises in the sums over i in the a_2- and a_4-terms in Eq. (6.118), because of the set of substitutions $J_i(y_1)J_i(y_2)/v_i \to J(x_i, y_1)J(x_i, y_2) \to J(x, y_1)J(x, y_2)$ made simultaneously with passing to the continuum limit $(j_1, j_2) \to (y_1, y_2)$ for the variables describing φ_i in the a_2-term; the procedure is the same for the a_4-term. Omitting the φ-independent terms in Eq. (6.118) we obtain

$$\begin{aligned}\mathcal{H}_{\text{eff}} = &- \int d^d x h(x)\varphi(x) + \frac{1}{2}\int d^d x d^d y J(x,y)\varphi(x)\varphi(y) \\ &- \frac{1}{2}a_2 \int d^d x d^d y_1 d^d y_2 J(x,y_1)J(x,y_2)\varphi(y_1)\varphi(y_2) \\ &+ a_4 \int d^d x d^d y_1 \cdots d^d y_4 J(x,y_1)\cdots J(x,y_4)\varphi(y_1)\cdots\varphi(y_4) \\ &- \ldots \end{aligned} \tag{6.119}$$

The fundamental properties of the Ising model are exhibited in the translationally invariant version $J(i-j)$ of the coupling J_{ij}. Concentrating our attention on this example, we shall use $J(x,y)$ in the form $J(|x-y|)$. Our further analysis can continue in the x-representation but the results are more clear and more easy to derive in the momentum (k-) representation, so we shall perform the Fourier transformation in $\mathcal{H}_{\text{eff}}$. The Fourier amplitudes of $\varphi(x)$ and $h(x)$ are obtained from Eq. (6.82a) and Eq. (6.85) for $l = 1$. The transform $J(k)$ of $J(x-y)$ is

$$J(x-y) = \frac{1}{V}\sum_k J(k)e^{i(x-y)k}. \tag{6.120}$$

The effective Hamiltonian (6.119) written by Fourier amplitudes is

$$\frac{\mathcal{H}_{\text{eff}}}{V} = -\sum_k h(-k)\varphi(k) + \frac{1}{2}\sum_k \left[J(k) - a_2|J(k)|^2\right]|\varphi(k)|^2$$
$$+ a_4 \sum_{k_1k_2k_3k_4} \delta(k_1+k_2+k_3+k_4)J(-k_1)\cdots J(-k_4)$$
$$\times \varphi(k_1)\cdots\varphi(k_4) - \dots. \tag{6.121}$$

Here we have applied Eqs. (6.88a) and (6.88b), and the Kronecker δ-symbol stands for the momentum conservation ($\text{k}_1 + \cdots + \text{k}_4 = 0$); in accord with our notations after Eq. (6.82a).

The interesting problem is how the Fourier amplitudes $J(k)$ behave. These amplitudes as well as $J(x-y)$ are related to the definition of the interaction in the system. The physical arguments, which lead to the appropriate form of $J(k)$ will be presented below; see also Section 6.7. They consist of some general requirements for the behaviour of the system. If we do not take into account the k-dependence of φ, J and h in Eq. (6.121) or the x-dependence of the same quantities in Eq. (6.118), we shall merely confirm the Landau expansion (4.32) in terms of $\varphi(0) \equiv \varphi(k=0)$. Within this approximation after neglecting the terms $O(\varphi^6)$ we can compare the present φ^4-theory with Eq. (4.134): $r = J_0(1-a_2J_0)/2$ and $u = a_4J_0^4$; $J_0 \equiv J(0)$. Consequently, $u \sim J_0^4/T^3$ and r takes the form Eq. (4.68): $r = \alpha_0'\Delta T_c$ with $\alpha_0' = (J_0/2T)$ and $\Delta T_c = T - T_c'$, where $T_c' = S(S+1)J_0/3$.

To compare this result for T_c' with that for a ferromagnet in MF approximation, given by Eq. (5.152) we ought to remember that we have redefined the exchange constants from J_{ij} to J_{ij}/S^2; see the discussion below Eq. (6.111). Now we restore the original exchange parameters J_{ij} which yields an extra factor $1/S^2$ in J_0. So in our final results we can substitute J_0 with J_0/S^2. Thus we see that $T_c' = S(S+1)J_0/3$ becomes equal to $T_c = (S+1)J_0/3S$ as given from Eq. (5.152) for $J = S$. Therefore we have re-derived the result for the critical temperature $T_c = (S+1)J_0/3S$ for the spin-S Ising model (MF approximation). One can establish a full conformity between these results for the parameters r and u and those obtained in Section 5.8.6. For this aim the difference in the definitions of the order parameter fields for these two cases should be taken in mind.

Consider the auxiliary example of extremely short range forces when $J(|x-y|) = J_0\delta(x-y)$ and, hence, $J(k) = J_0$. Then the expansion (6.119) will be simplified:

$$\mathcal{H}_{\text{eff}} = \int d^d x \left[-h(x)\varphi(x) + r\varphi^2(x) + u\varphi^4(x) + \cdots\right]$$
$$= \int d^d x \mathcal{H}_{\text{eff}}(x), \tag{6.122a}$$

where $\mathcal{H}_{\text{eff}}(x)$ is the "density" of $\mathcal{H}_{\text{eff}}$. The corresponding equation of state

$$\left[\frac{\delta \mathcal{H}_{\text{eff}}}{\delta \varphi(x)}\right]_{\bar{\varphi}} = 0 \tag{6.122b}$$

yields the equilibrium phase $\bar{\varphi}(x)$. The order parameter $\bar{\varphi}(x)$ depends on x only because of the x-dependence of h. However no correlations between different spatial points x and y exists.

The idealized system, described by Eq. (6.122a), is a sum of an infinitely large number of "point subsystems" parameterized by the spatial vector x. For each $h(x_1)$ fixed at x_1, the system $\varphi(x_1)$ is described by the theory presented in Section 4.10. For an uniform h, the "point systems" $\varphi(x)$ are in the same conditions but the Eq. (6.122a) does not straightforwardly yield the intuitively supposed result $\varphi = const$, even, it does not yield uniform φ at all. Rather we remain with terms of type $\int d^d x \varphi^l(x)$, where $l = 1, 2, 4$. However, the solutions of Eq. (6.122b) will be the same for any x and from here we conclude that for uniform h, the equilibrium order parameter configuration(s) $\bar{\varphi}(x)$ will be uniform $[\varphi(x) = \varphi = \text{const}]$. While the equilibrium states are uniform, the fluctuations $\delta\varphi(x)$ around these equilibrium states are definitely arbitrary (both uniform and x-dependent) and do not interact one another and, hence, they could not experience statistical correlations $\langle\varphi(x_l)\cdots\varphi(x_k)\rangle$ because in this scheme these averages are always decoupled: $[\langle\varphi(x_l)\rangle\cdots\langle\varphi(x_k)\rangle]$.

In order to clarify this result we shall try to establish the interrelationship between the MF analysis (Section 5.8 and the present consideration. Firstly, let us remind the correspondence $J_0 = zJ_{\text{ex}} = \sum_j \hat{J}$; see Eqs. (5.143) and (5.152) and the discussion below Eq. (5.110). If the radius of the interaction $J(|i-j|)$ in the initial Ising lattice is small, say, equal to n inter-site distances a_0, the sum (5.143) will extend over the finite distance $a_0 n$. The continuum limit for this sum ($\sum_j \hat{J} \to \sum_j v_j J(i, x_j)$, $v_j \to a_0^d \to 0, i \to x$) will lead to the function $J(x-y) = J_0\delta(x-y) + f(x-y)$, where $J_0 = zJ_{\text{ex}}$; $f(x-y)$ is a slowly varying function (tail), which is neglected in Eq. (6.122a).

In this continuum model J_0 plays a role similar to that of J_{ex} in the lattice model. Since no other physical mechanisms of correlation between the variables σ_i except the interaction J_{ij} are available, the system separates in an infinite number of independent subsystems $\varphi(x)$ (independent fields). The Hamiltonian (6.122a) corresponds to a factorized partition function and, hence, to an additive potential Φ (a sum of the contributions of the independent order parameters $\varphi(x)$). The density of the equilibrium order parameter will be given by the integral

$$\bar{\varphi} = \frac{1}{V}\int d^d x \bar{\varphi}(x), \qquad \bar{\varphi}(x) = \langle\varphi(x)\rangle. \tag{6.123}$$

The equilibrium density $\bar{\varphi}(x)$ will be x-independent (spatially uniform), when $h(x)$ is spatially uniform. Then we can apply the fluctuation theory presented in Section 6.3 and obtain absolutely uncorrelated fluctuation contributions to the thermodynamic potential of the system.

To obtain the theory in this form we have taken the continuum limit in the variant in which the lattice constant a tends to zero ($a_0 \to 0$). In Section 6.5.4 we have mentioned that this could be justified only for some cases of dense matter, where the density ρ is unusually large, or, infinite. Below we shall see that this is an approximation to a more general consideration in which the lattice constant a is different from zero.

The sum (5.143) can also be considered in the limit of "infinitely small-ranged ($J_{\text{ex}} \to 0$) – infinitely long-ranged ($z \to \infty$) interaction" between the "spins" σ_i [$zJ_{\text{ex}} \to J_0 \neq (0 \text{ or } \infty)$]. Then the continuum limit for the Eq. (5.143) with $J_{ij} = J(i-j)$ will be the integral $J_0 = \int d^d R J(R) = J(0); R = (x-y)$ and $J(0) = J(k=0)$, where $J(k)$ are the Fourier amplitudes of $J(|x-y|)$. Assuming that the values $J(R) \sim J_{\text{ex}}$ are infinitesimally small, it will not be a big error to suppose that $J(R) \approx const$ and, hence, $VJ_{\text{ex}} \approx J(0) = J_0$. The function $J(x-y)$ for this type of interactions differs from that for short-range (neighbour) interactions by the factor $\delta(x-y)$. Obviously the expansion (6.119) in the present case, $J(x,y) = J_0$, is made in terms of $(J_0\varphi/T)^2$, that is, in even powers of φ — the nonequilibrium order parameter of the Landau potential (φ is given by the integral (6.123), after substituting $\bar{\varphi}$ by $\varphi \neq \langle\varphi(x)\rangle$).

The mean field theory developed in Section 5.8 is an exact description of systems with extremely small — almost infinitely — ranged interactions; for more details, see Stanley (1971), Kac (1968). Spatially dependent fluctuations are not available in this variant of the theory because Eq. (6.123) defines an overall density of the order parameter. The latter may exhibit only the uniform fluctuation $\delta\bar{\varphi} = (\bar{\varphi} - \bar{\varphi}_{\text{eq}})$ as a deviation from the equilibrium value $\bar{\varphi}_{\text{eq}}$, given as a solution of the equation of state $[\delta\mathcal{H}_{\text{eff}}/\delta\bar{\varphi}]_{\bar{\varphi}_{\text{eq}}} = 0$.

After these two limiting cases of extremely short range interactions and extremely long-range ones, we shall consider a more general variant of the theory, which is appropriate for real systems. We shall consider slow variations of the exchange function $J(|x-y|)$. As $J(|x-y|)$ is a real function, the Eq. (6.120) yields $J(k)^* = J(-k)$. Besides, for the even property, $J(R) = J(-R)$, the same Eq. (6.120), which can be written for $J(|x-y|)$, yields $J(k) = J(-k)$, namely, $J(k)$ does not depend on the direction of the wave vector k, and can be written as $J(k^2)$. From these two properties of $J(k)$ we immediately obtain that $J^*(k) = J(k)$, i.e., that $J(k)$ is a real function. The supposed slow variation of $J(|x-y|) \approx J_0\delta(x-y) + f(x-y)$ at large distances corresponds to the small momentum limit,[1] $(ka_0)^2 \ll 1$, where $k = |\boldsymbol{k}|$, and a_0 is the lattice constant of the initial lattice (or, $(kL)^2 \sim 1$, where L is the size of the system, $V \sim L^d$). Then we can expand $J(k^2)$ for small k :

$$J(k^2) = J_0 + J_2 k^2 + \cdots\,, \qquad J_2 = \left(\frac{\partial J}{\partial k^2}\right)_{k=0}, \tag{6.124}$$

and to substitute this result in Eq. (6.121); $k^2 < |J_0/J_2|$. Now we take in mind that the change $J_0 \to J_0/S^2$ in the final results should be performed. The first two terms in Eq. (6.124) yield the following form for the second term in Eq. (6.121):

$$\frac{1}{2}\sum_k (r' + ck^2)|\varphi(k)|^2, \tag{6.125}$$

where

$$r' = 2r = \frac{J_0^2(S+1)}{3ST}t, \qquad t = \frac{T - T_c}{T_c}, \tag{6.126a}$$

and

$$c = -J_2\left(\frac{2T_c}{T} - 1\right), \qquad T_c = \frac{(S+1)J_0}{3S}. \tag{6.126b}$$

We focus now our attention on the term (6.125) and neglect the effect of all other terms in Eq. (6.121) on the formation of the phases and their stability properties. Such a simple Gaussian theory will describe only one

[1] Often called "the long wavelength limit".

phase — the disordered phase $\varphi(x) = 0$, so $\varphi(k)$ in Eq. (6.125) are the fluctuations of φ around the value $\varphi = 0$ (see also Section 7.2). Alternatively we may imagine that the system is in the disordered state and the term (6.125) gives the essential (but not the total) fluctuation effect. Anyway the paraphase ($\varphi = 0$) will be stable, if the function

$$f(k^2) = J(k^2) - a_2 J^2(k^2), \tag{6.127}$$

in Eq. (6.121) and its reduced version (6.125) are positive for any k. Note that the Eq. (6.127) imposes the requirement that $J(k^2) > 0$, because otherwise $f(k^2)$ is always negative; the condition $(1 + a_2|J(k^2)|) < 0$ for $J(k^2) < 0$ cannot be fulfilled for any $T > 0$.

The case $J(k^2) = 0$ corresponds to an ideal (noninteracting) system, it is not of interest here. The quadratic form Eq. (6.125) is positively definite for $T > T_c$ provided $c > 0$. If J_2 in Eq. (6.126b) is negative, $c > 0$ for all $T < 2T_c$, i.e., $t < 1$. The fact that c changes sign at $T = 2T_c$ is not supposed to give any physical effect; it seems that this is an artifact of the expansions (6.121) and (6.124).

The change of the sign of c is just at $t = 1$ — the value of t, below which the Landau expansion for small φ can be applied. So we are interested in temperatures $T < 2T_c$, i.e., $t < 1$ and we exclude the possibility $J_2 > 0$, which is related to the instability of the disordered phase towards spatially dependent fluctuations $\varphi(k), k \neq 0$ (the stability with respect to spatially independent fluctuations depends on the sign of r').

It is obvious that the stability of the disordered phase at $T > T_c$ is ensured by the fact that $J(k)$ has a maximum at $k = 0$ ($J_2 < 0$). This maximum of $J(k)$ corresponds to a minimum of $f(k)$ for $t < 1$ (a distinct difference between the functions $f(k)$, $J(k)$ and $f(k^2)$, $J(k^2)$ should be made). The condition $c > 0$, which ensures the stability of the system with respect to spatially dependent fluctuations of the order parameter is called *Lifshitz criterion* (see also Section 4.11.5). This criterion is widely discussed in the book by Landau and Lifshitz (1980); for further developments see Dzyaloshinskii (1964), Goshen, Mukamel and Shtrikman (1974).

The stability properties of the amplitudes $\varphi(k)$ have to be studied more carefully and we shall briefly present the corresponding calculation for states near the critical point ($t < 1; h = 0$). The ordered phase(s) $\varphi(k) \neq 0$ now appear for temperatures $T < T_c(k_0)$, where the critical temperature $T_c(k_0)$ is defined by the equation $f(k_0^2) = 0$.

From this equation we get

$$T_c(k_0) = T_c\frac{J_1}{J_0}, \qquad T_c \equiv T_c(0), \tag{6.128}$$

$J_1 = J(k_0)$ and J_0 is given by Eq. (6.124). So there is a possibility of appearance of the usual phase $\varphi(0) \neq 0$ at T_c as well as other phases with characteristic amplitudes $\varphi(k_0), k_0 \neq 0$; an example of the last case is the antiferromagnetic state where $|k_0|$ is the momentum at the boundary of the Brillouin zone. All these cases can be simultaneously analyzed by setting $k_0^2 \geq 0$. Having in mind that now $t = [T - T_c(k_0)]/T_c(k_0)$, we obtain that the disordered phase for $0 < t < 1$ will be stable, if $J_2(k_0) = [\partial J/\partial k^2]_{k_0} < 0$. This condition is analogous to the requirement $f'(k_0^2) = (\partial f/\partial k^2)_{k_0} > 0$ and, hence,

$$f(k^2) = f(k_0^2) + f'(k_0^2)(k - k_0)^2 > 0, \tag{6.129}$$

for all k near k_0.

The relation (6.129) means that no states with energy $\mathcal{H}_{\text{eff}}$ lower than $\mathcal{H}_{\text{eff}} = 0$ exist (in this qualitative analysis $h\varphi$ and $O(\varphi^4)$ terms in Eq. (6.121) are neglected). Moreover the most important fluctuations in the disordered phase ($t > 0$) will be given by the amplitudes $\varphi(k_0)$ as they correspond to a minimal contribution to the energy $\mathcal{H}_{\text{eff}}$; see Eq. (6.129). Because of this the qualitative analysis of the ordered states $\varphi(k_0)$ below $T_c(k_0)$ can be carried out by neglecting all amplitudes except $\varphi(k_0)$ in the φ^4-theory that result from the φ^2- and φ^4-terms in the expansion (6.121):

$$\mathcal{H}_{\text{eff}} = V\left[\frac{1}{2}f(k_0)\varphi^2(k_0) + a_4 J^4(k_0)\varphi^4(k_0)\right]. \tag{6.130}$$

Below $T_c(k_0)$, the equilibrium (minimal) potential $\Phi = \mathcal{H}_{\text{eff}}(\bar{\varphi})$ that corresponds to the equilibrium order parameter $\bar{\varphi}^2 = |f|/4a_4J^4$ is $\mathcal{H}_{\text{eff}} = -|f|\bar{\varphi}^2/4$; $f \equiv f(k_0)$ and $J \equiv J(k_0)$. Simple calculations yield $\mathcal{H}_{\text{eff}}(\bar{\varphi})$ in the form

$$\mathcal{H}_{\text{eff}} = -\frac{(a_2 - \frac{1}{J})^2}{16a_4}, \qquad a_2 > \frac{1}{J}. \tag{6.131}$$

Here both a_2 and a_4 do not depend on $J = J(k_0)$. Now one can check out from the derivative $[\partial\mathcal{H}_{\text{eff}}/\partial k_0^2]$ that $\mathcal{H}_{\text{eff}}(k)$ will have a minimum at k_0, if $(\partial J/\partial k^2)_{k_0} = 0$. This corresponds to $f'(k_0^2) = 0$ and, hence, it is consistent with the approximation $f(k^2) \approx f(k_0^2)$ and $k \approx k_0$ in the Hamiltonian. It depends on the values of $\mathcal{H}_{\text{eff}}(k_0)$ for different k_0 which one of the minima $\bar{\varphi}(k_0)$ will give the stable phase for $T < T_c(k_0)$. It is easy to find from Eq. (6.131) that the stable state $\varphi(k_0^{st})$ will correspond to those k_0^{st} that give maximal $J(k_0^2)$.

6.6.3 *Transformation of other models*

Consider the model

$$\mathcal{H} = -\frac{1}{2}\sum_{ij;\alpha} J_{ij}\sigma_i^\alpha \sigma_j^\alpha, \tag{6.132}$$

where σ_i^α are the components of the classical vectors $\sigma_i = \{\sigma_i^\alpha, \alpha = 1, \ldots, n\}$. These vectors can change in direction but their magnitudes $|\sigma_i|$ obey the condition

$$\sigma_i^2 = \sum_\alpha (\sigma_i^\alpha)^2 = n. \tag{6.133}$$

Equations (6.132) and (6.133) define a type of vector models, which describe some phase transitions (see the brief discussion in Section 5.7.3). In fact each $(\sigma_i^\alpha)^2$ can vary from 0 to n but taking into account the condition (6.133) in the partition function we can think about $(\sigma_i^\alpha)^2$ as being in the interval $(0, \infty)$. This is very suitable from a calculational point of view.

The partition function will be

$$\mathcal{Z} = \int D\sigma \prod_i \delta(\sigma_i^2 - n) e^{-\mathcal{H}(\sigma)/T}, \tag{6.134}$$

where

$$\int D\eta = \int_{-\infty}^{\infty} \prod_{i=1}^{N} \prod_{\alpha=1}^{n} D\eta_i^\alpha, \qquad \eta = \sigma, \varphi. \tag{6.135}$$

With the help of the transformation (6.108), $\mathcal{Z}$ becomes

$$\mathcal{Z} = A \int D\varphi e^{-\frac{1}{2}\sum_{ij;\alpha} K_{ij}^{-1}\varphi_i^\alpha \varphi_j^\alpha + f(\varphi)}. \tag{6.136}$$

The exponent $f(\varphi)$ is given by the expression

$$f(\varphi) = \ln I(n, \varphi), \tag{6.137}$$

with

$$\begin{aligned} I(n, \varphi) &= \int_{-\infty}^{\infty} D\sigma \prod_{i=1}^{N} \delta(\sigma_i^2 - n) \exp\left[\sum_i \varphi_i \cdot \sigma_i\right] \\ &= \prod_{i=1}^{N} I_i(n, \varphi_i), \end{aligned} \tag{6.138}$$

where

$$I_i(n, \varphi_i) = \int_{-\infty}^{\infty} \prod_{\alpha=1}^{n} d\sigma_i^\alpha \delta(\sigma_i^2 - n) e^{\varphi_i \cdot \sigma_i}. \tag{6.139}$$

The reader must realize that in the exponent Eq. (6.139) $\varphi_i \cdot \sigma_i$ denotes the scalar product $\sum_\alpha \varphi_i^\alpha \sigma_i^\alpha$; so the exponent in Eq. (6.138) is a sum of such scalar products.

Now our task is to calculate the integral (6.139). This is not a difficult problem because the integral is over the Cartesian coordinates σ_i^α of the n-dimensional vectors σ_i. $I_i(n, 0)$ is nothing else but the area of the n-dimensional sphere of radius $R = \sqrt{n}$. It is very convenient to use the spherical coordinates of the vector σ_i; see Eq. (D.14). As the magnitude $|\sigma_i|$ of σ_i is always fixed at the value $\sqrt{n}$, at each point i, σ_i, can rotate in the n-dimensional space of its components (for this reason these types of models are referred to as *models of two-*, *three-* and so on *dimensional rotators*). In this picture φ_i is fixed in both magnitude and direction.

Let φ_i be along the axis $\hat{\varphi}_i^{(1)}$ of the n-component space: $\varphi_i = (\varphi_i^{(1)}, 0, \cdots)$; see Eq. (D.14). The product $\varphi_i \cdot \sigma_i$ then will be $\sqrt{n}|\varphi_i| \cos\theta$, where θ is the angle between φ_i and σ_i. Making use of Eq. (D.1), Eq. (D.2), and Eq. (D.8) the integral I_i becomes

$$\begin{aligned} I_i &= \Omega_{n-1} \int_0^{\infty} d\sigma_i \sigma_i^{n-1} \delta(\sigma_i^2 - n) \int_0^{\pi} d\theta \sin^{n-2}\theta e^{\sqrt{n}|\varphi_i|\cos\theta} \\ &= \Omega_{n-1} n^{(n-1)/2} I_0. \end{aligned} \tag{6.140}$$

Here I_0 denotes the integral over θ. The integral I_0 is calculated by expanding the integrand exponent in power series. The odd powers in $\cos\theta$ will give zero after integration so that the expression for I_0 will contain only the powers of $(n|\varphi_i|^2)$:

$$I_0 = \sum_{p=0}^{\infty} \frac{(n\varphi_i^2)^p}{(2p)!} 2 \int_0^{\infty} \sin^{n-2}\theta \cos^{2p}\theta d\theta. \tag{6.141}$$

Calculating the integrals in Eq. (6.141) with the help of Eq. (D.18) and substituting the result in Eq. (6.140) we obtain

$$I_i = S_n(\sqrt{n}) \sum_{p=0}^{\infty} \frac{1}{p!} \left(\frac{n\varphi_i^2}{4}\right)^p \frac{\Gamma(n/2)}{\Gamma(p+n/2)}. \tag{6.142}$$

$S_n(\sqrt{n}) = \Omega_n n^{(n-1)/2}$ is the area of the n-dimensional sphere of radius $\sqrt{n}$; it is easy to be convinced after replacing $d = n$ and $R = \sqrt{n}$ in $S_d(R) = [dV_d(R)/dR]$ with V_d given by Eq. (D.4).

Now we can expand $f(\varphi)$ in powers of φ^2

$$f(\varphi) = \sum_i \ln I_i(n, \varphi_i), \tag{6.143}$$

and to substitute the result in Eq. (6.136). Thus we find the effective potential in the following form

$$\mathcal{H}_{\text{eff}} = \frac{T}{2} \sum_{ij;\alpha} K_{ij}^{-1} \varphi_i^\alpha \varphi_j^\alpha - T \sum_i \ln I_i(n, \varphi_i). \tag{6.144}$$

It differs from the integrand exponent (6.136) by the factor $(-1/T)$; see Eq. (6.104). This effective potential taken to the order φ_i^4 will be

$$\mathcal{H}_{\text{eff}} = \frac{T}{2} \left\{ \sum_{ij;\alpha} K_{ij}^{-1} \varphi_i^\alpha \varphi_j^\alpha - \sum_i \left[\varphi_i^2 - \frac{\varphi_i^4}{2(n+2)} \right] \right\}. \tag{6.145}$$

It is not difficult to calculate the next terms $(\varphi_i^{2m}, m \geq 3)$ in $\mathcal{H}_{\text{eff}}$. For example, the coefficient u_6 of the $u_6\varphi_i^6$-term is $u_6 = T(3n^2+9n+4)/48(n+2)(n+4)$; as $n \to \infty$, $u_6 \to T/16$, whereas $u_4 \to 0$. The transformation of φ_i^α with the help of Eq. (6.117) yields:

$$\begin{aligned} \mathcal{H}_{\text{eff}} = &\frac{1}{2} \sum_{ij;\alpha} J_{ij} \varphi_i^\alpha \varphi_j^\alpha - \frac{1}{2T} \sum_{ij_1j_2;\alpha} J_{ij_1} J_{ij_2} \varphi_{j_1}^\alpha \varphi_{j_2}^\alpha \\ &+ \frac{1}{4(n+2)T^3} \sum_{i;\alpha\beta} \sum_{j_1 \dots j_4} J_{ij_1} \cdots J_{ij_4} \varphi_{j_1}^\alpha \varphi_{j_2}^\alpha \varphi_{j_3}^\beta \varphi_{j_4}^\beta, \end{aligned} \tag{6.146}$$

(the term $[-TN \ln S_n(\sqrt{n})]$ has been omitted); cf. Grinstein and Luther (1976). Now we may perform the continuum limit ($i \to x$) receiving as a result the φ^4-field theory of a n-component classical field $\varphi(x) = \{\varphi_\alpha(x)\}$. The behaviour of the exchange matrix $J_{ij} \to J(x,y)$ can be considered on the basis of the arguments presented in Sections 6.6.1 and 6.7.

The two spin-1 Ising systems (s_i and σ_i) with a biquadratic coupling ($s_i^2\sigma_j^2$) are described by the Hamiltonian

$$\frac{\mathcal{H}}{T} = -\frac{1}{2}\sum_{i\neq j}\left\{J_{ij}^{(s)} s_i s_j + J_{ij}^{(\sigma)}\sigma_i\sigma_j + L_{ij}s_i^2\sigma_j^2\right\}, \tag{6.147}$$

where s_i and σ_i at any site i of the lattice take the values 0 and ± 1; the factor $(1/T)$ is absorbed in the exchange constants $J^{(s)}, J^{(\sigma)}$ and L. After applying the identity (6.98) to both terms (s_is_j) and $(\sigma_i\sigma_j)$ of the Hamiltonian two types (ϕ and φ) of new continuous fields are introduced. Note that in the case of "spin"-$\frac{1}{2}$ variables, i.e., $\sigma_i = \pm 1$, $s_i = \pm 1$, the coupling of the type $s_i^2\sigma_j^2$ does not exist because $s_i^2 = \sigma_i^2 = 1$ and, hence, this term is a constant. At the intermediate stage of the derivation of the continuum model we obtain the expression:

$$\begin{aligned}\frac{\mathcal{H}}{T} &= \frac{1}{2}\phi J_s^{-1}\phi + \frac{1}{2}\varphi J_\sigma^{-1}\varphi \\ &\quad - \sum_i \ln\left[1 + 2\cosh\phi_i + 2(1 + e^{\mu_i}\cosh\phi_i)\cosh\varphi_i\right].\end{aligned} \tag{6.148}$$

Here

$$\mu_i = \frac{1}{2}\sum_j L_{ij}, \tag{6.149}$$

and the additive constants have been omitted. In Eq. (6.148), φ_i and ϕ_i are one-component ($n = 1$) continuous variables corresponding to s_i and σ_i, respectively.

Further calculations are analogous to that presented in Sections 6.6.1 and 6.6.2. Imposing the requirement for the translational invariance of the couplings J_s, J_σ and L, we obtain the following field model

$$\begin{aligned}\mathcal{H}_{\text{eff}} &= \frac{1}{2}\int d^dx d^dy \left[a_s(x-y)\phi(x)\phi(y) + a_\sigma(x-y)\varphi(x)\varphi(y)\right] \\ &\quad + \int d^dx_1 \cdots d^dx_4 \left[b_s(x_i)\phi(x_1)\cdots\phi(x_4) + b_\sigma(x_i)\varphi(x_1)\cdots\varphi(x_4)\right. \\ &\quad \left. + b_{s\sigma}(x_i)\phi(x_1)\phi(x_2)\varphi(x_1)\varphi(x_2)\right],\end{aligned} \tag{6.150}$$

where

$$a_{s,\sigma}(x-y) = J_{s,\sigma}(x-y) - a'_{s,\sigma}\int d^dzJ(z-x)J(y-z); \qquad (6.151)$$

b_s, b_σ, and $b_{s\sigma}$ depend on $(x_1 - x_4)$, $(x_2 - x_4)$, and $(x_3 - x_4)$ and are given by

$$b_{s,\sigma} = b'_{s,\sigma}\int d^dzJ_{s,\sigma}(x_1 - z)\cdots J_{s,\sigma}(x_4 - z), \qquad (6.152)$$

and

$$b_{s\sigma} = b'_{s\sigma}\int d^dzJ_s(x_1-z)J_s(x_2-z)J_\sigma(x_3-z)J_\sigma(x_4-z). \qquad (6.153)$$

The parameters $a'_{s,\sigma}$, $b'_{s,\sigma}$, and $b'_{s\sigma}$ are simple functions of the temperature similar to those of a_2 and a_4 in Eq. (6.118) and Eq. (6.119). The parameter $b'_{s\sigma}$ depends on μ_0, which is independent on i : $2\mu_0 \equiv 2\mu_i = \sum_j L(i - j)$. When $\mu_0 = 0$, $b'_{s\sigma} = 0$ and the system decouples in two independent subsystems (ϕ and φ).

The symmetry of the initial Hamiltonian (6.147), $s_i \leftrightarrow -s_i$, $\sigma_i \leftrightarrow -\sigma_i$ is conserved in the resulting field theory as it should be. The above obtained type of the φ^4-theory describes the interaction between two systems (ϕ and φ) or, which is the same, the interaction between the order parameters of two phase transitions. We shall see in Chapter 9 that the model (6.149) gives the bicritical and tetracritical points.

Another lattice model of two interacting "spin"-1/2 variables ($s_i = \pm1$ and $\sigma_i = \pm1$) can be obtained by substituting the biquadratic term $s_i^2\sigma_j^2$ in Eq. (6.147) by the bilinear term $L_{ij}s_i\sigma_j$. Now the effective Hamiltonian has the structure

$$\mathcal{H}_{\text{eff}} = \int d^dxd^dy\{A(x-y)\varphi(x)\varphi(y) + B(x-y)\phi(x)\phi(y) + C(x-y)\varphi(x)\phi(y)\} + \mathcal{H}^{(4)}_{\text{eff}} + \mathcal{H}^{(6)}_{\text{eff}} + \cdots, \qquad (6.154)$$

where $\mathcal{H}^{(4)}_{\text{eff}}, \ldots$ are of fourth, sixth, etc order in φ and ϕ. For example, $\mathcal{H}^{(4)}_{\text{eff}}$ will contain terms of the type φ^4, ϕ^4, $\varphi\phi^3$, $\phi\varphi^3$, and $\varphi^2\phi^2$. All these terms are allowed by the symmetry of the system; they are a result of direct calculations. The coefficient functions A, B, and C in Eq. (6.154) as well as the coefficient functions before the higher-order terms are functionals of $J^{(\sigma)}$, $J^{(s)}$, and L.

The n-vector model (6.132) can be straightforwardly generalized for two coupled (n- and m-) vector variables. In this way field models of coupled n- and m-component fields φ and ϕ can be derived. A special case of the Blume–Emery–Griffiths model with bilinear and biquadratic couplings has been studied by Carneiro, Henriques and Salinas (1987). The path integral representation of the Hubbard model (White, 1970; Smart, 1966) by the means of the Hubbard–Stratonovich transformation is given by Hubbard (1979), Moriya (1979), Hertz (1976), and Izyumov and Skryabin (1987).

6.7 Short-Range and Long-Range Translationally Invariant Interactions

To derive the asymptotic properties of the exchange $J(k)$ for small momenta k we have used several general requirements about the behaviour of the system near the phase transition point. These phenomenological arguments can be avoided by performing the Fourier transformation of the exchange coupling $J(R)$, $R = |x - y|$. The inverse transformation (6.120) in the continuum limit (6.100) yields

$$J(k) = \int d^d R J(R) e^{-i\boldsymbol{k}\cdot\boldsymbol{R}}. \tag{6.155}$$

This approach of obtaining $J(k)$ is based on the assumption that $J(R)$ could be taken from experiments or given by model arguments. The same is true for the inverse problem — if there is a possibility to propose a reliable form for $J(k)$, $J(R)$ will be obtained with the help of the formula

$$J(R) = \int \frac{d^d k}{(2\pi)^d} J(k) e^{i\boldsymbol{k}\cdot\boldsymbol{R}}. \tag{6.156}$$

In fact one may suggest correctly the type of the inter-cell interaction $J(R)$, or $J(k)$, but one can be never sure about the precise form of this function. Certainly $J(R)$ will decrease with R according to a power or an exponential (screened power) law, so this function can be written as follows:

$$J_1(R) = \frac{\tilde{J}_1}{R^{d+\theta}}, \tag{6.157}$$

or,

$$J_2(R) = \tilde{J}_2 \frac{e^{-\lambda R}}{R^{d+\theta}}, \tag{6.158}$$

where $\tilde{J}_1$, $\tilde{J}_2$, and $\lambda > 0$ are constants. The Fourier transforms $J(k)$ of the functions (6.157) and (6.158) are presented in Section D.2. Strictly speaking Eq. (6.155) will be a Fourier transformation, if the inverse transformation (6.156) also exists and converges to the initially chosen function $J(R)$ at each point of its domain of definition. For $J(R) = J_0\delta(R)$, $J(k) = J_0 =$ *const*. This kind of extremely short-range interactions has already been discussed in Section 6.6.2.

The investigation of the functions (6.157) and (6.158) is more complicated. The function (6.157) is not absolutely integrable because of the divergence at small distances, when $\theta > 0$ and because of the slow decay at long distances, when $\theta < 0$. That is why when we are making an attempt to regularize the transformation we are forced to use mathematical tricks like the introduction of a cutoff of the integral over R at short distances or the substitution of the original function $J_1(R)$ by $\tilde{J}(R) = J(R)e^{-\delta y}$; $\delta > 0$. In the last case, the limit $\delta \to +0$ is taken after the transformation is performed see also Eq. (D.24). Here we shall not dwell upon details, rather we shall enumerate the main results.

The behaviour of the function $J_1(k)$ corresponding to $J_1(R)$ from Eq. (6.157) is qualitatively different for $\theta > 0$, $\theta < 0$, and $\theta = 0$. For $\theta < 0$, the leading term in the asymptotic behaviour of $J(k)$ is of the type k^θ:

$$J_1(k) = ck^\theta + O(k^2). \tag{6.159}$$

Therefore, no phase transition point $T_c \sim J_0$ can exist in systems with $\theta < 0$. This instability of the system is connected with the impossibility to perform the thermodynamic (continuum) limit or, in other words the thermodynamic limit for such systems does not exist; see, e.g., Pfeuty and Toulouse (1975). The coupling $J(R)$ given by Eq. (6.157) with $\theta < 0$ can be referred to as the extremely long-range interaction limit. For $\theta \geq 0$ we have to distinguish between two cases: (i) *long-range interactions* described by $0 < \theta < 2$, and (ii) *short-range interactions* described by $\theta > 2$. For these two cases there is a distinct difference between the small-momentum behaviour of $J(k)$ and this can be seen with the help of the expressions given in Section D.2. For long-range interactions and $(k/\Lambda) \ll 1$, where $\Lambda \sim 1/a$ is the upper cutoff,

$$J_1(k) \approx J_0 - \rho^2 k^\theta + O(k^2), \qquad 0 < \theta < 2; \tag{6.160}$$

the parameters $J_0, \rho^2 > 0$ can be obtained explicitly from Eq. (D.27). Unlike the k^2-dependence considered in Section 6.6.2, here we are faced with a leading term of the type $k^\theta > k^2$. Neglecting all terms of the order $O(k^2)$ we obtain that the k^2-dependence in $f(k)$, see Eq. (D.27), should be changed to k^θ. For short range forces, $\theta > 2$, the first non-vanishing k-term of $J(k)$ is of $\rho^2 k^2$-type and, hence, $f(k)$ remains quadratically dependent on k; see Eq. (6.124) with $J_2 < 0$. The crossover from short-range to long-range interactions at $\theta = 2$ is signalled by a logarithmic term of the type $k^2 \ln(\Lambda/k)$; see Eq. (D.29). Another logarithmic term, $\sim \ln(\Lambda/k)$, that appears for $\theta = 0$ signals the crossover from long-range ($0 \leq \theta < 2$) to extremely long-range ($-d < \theta < 0$) interactions; see Eq. (D.28). We are not aware that there are systems with θ strictly 2 or 0 and these cases give a special fluctuation behaviour; see also Joyce (1966, 1972), Lacour–Gayet and Toulouse (1974). The Fourier transform $J_2(k)$ of the interaction (6.157) is always of the form $J_2(k) = (J_0 - \rho^2 k^2)$ and hence describes short-range interactions; see Eq. (D.21).

It becomes clear from our present discussion and the contents of Section D.2 that the total form of the exchange amplitude $J(k)$ can hardly be used in a field theory because the rather complicated coefficients $u^{(2)}$ and $u^{(4)}$ would be a source of great calculational difficulties. Furthermore all consistent physical arguments convince us that the phase transition properties near T_c are completely described by the small momentum ("infrared") asymptotic expression for $J_1(k)$. The main results for the short-range interactions can always be adapted for systems with long-range interactions so we shall focus our attention mainly on the quadratic k-dependence of $J(k)$ and $f(k)$. Clearly, the above consideration does not add anything new to our knowledge of the form of $J(k)$ for the realistic short-range interactions previously suggested by phenomenological arguments. These arguments have been used for translational invariant coupling $J(R)$ but, on the same grounds, they can be straightforwardly applied to the coefficient $u^{(2)}(-k_1, -k_2)$ in the expansion (6.87). When there is a translational invariance, $k_2 = -k_1 \equiv k$ and $u^{(2)}$ is a function of $|k|$. The appropriate form of $u^{(2)}(|k|)$ for short-range microscopic interactions will be

$$u^{(2)}(k) = \frac{1}{2}(r + ck^2), \tag{6.161}$$

where $r = \alpha_0 t$ and $c > 0$. The φ^2-term in Eq. (6.87) will be then

$$\frac{V}{2}\sum_{\alpha;k} u^{(2)}(k)|\varphi_\alpha(k)|^2. \tag{6.162}$$

The transformation of this term to the x-representation can be made with the help of Eq. (6.82a) and the obvious relation

$$\nabla\varphi_\alpha(x) = \sum_k i\boldsymbol{k}\varphi_\alpha(k)e^{i\boldsymbol{k}\cdot\boldsymbol{x}}. \tag{6.163}$$

The inverted transformation (6.163) is

$$i\boldsymbol{k}\varphi_\alpha(k) = \frac{1}{V}\int d^dx \nabla\varphi_\alpha(x)e^{-i\boldsymbol{k}\cdot\boldsymbol{x}}. \tag{6.164}$$

The term $k^2|\varphi_\alpha(k)|^2$ in Eq. (6.162) can be represented as

$$[i\boldsymbol{k}\varphi_\alpha(k)].[-i\boldsymbol{k}\varphi_\alpha(-k)].$$

Making use of Eq. (6.83) and Eq. (6.164), we obtain that Eq. (6.162) in a x-representation will be

$$\frac{1}{2}\sum_\alpha \int d^dx \left[r\varphi_\alpha^2(x) + c[\nabla\varphi_\alpha(x)]^2\right], \tag{6.165}$$

where the gradient term is

$$(\nabla\varphi)^2 = (\nabla\varphi_\alpha)(\nabla\varphi_\alpha) = \left(\frac{\partial}{\partial x_1}\varphi_\alpha\right)^2 + \cdots + \left(\frac{\partial}{\partial x_d}\varphi_\alpha\right)^2. \tag{6.166}$$

The form of the terms $u^{(l)}$ with $l > 2$ in Eq. (6.78) and Eq. (6.87) is deduced from general arguments or with the help of our experience with series like (6.118), (6.145), (6.150), and (6.154) for particular models. From a general point of view, the translational invariant coupling coefficients $u^{(l)}$ in Eq. (6.78) should obey the relation

$$u^{(l)}(x_1 + x, \ldots, x_l + x) = u^{(l)}(x_1, \ldots, x_l), \tag{6.167}$$

where x is an arbitrary vector (in the continuum limit the invariant translations with a vector $\boldsymbol{x} = n\boldsymbol{a}$ where $\boldsymbol{a}$ is the vector of the primitive lattice, extend over all vectors $\boldsymbol{x}$). With the help of Eq. (6.167) it is readily seen that the vectors k_i in the transformation (6.85) obey the condition

$k_1 + \cdots + k_l = 0$. The expansion (6.87) for translationally invariant interactions becomes then

$$\frac{\Phi}{V} = -\sum_{\alpha} h_\alpha(-k)\varphi_\alpha(k) + \frac{1}{2}\sum_{\alpha;k} u^{(2)}(k)|\varphi_\alpha(k)|^2$$
$$+\sum_{l=3}^{\infty}\sum_{\alpha_i;k_i} \delta(k_i)u^{(l)}_{\alpha_i}(k_i)\varphi_{\alpha_1}(k_1)\cdots\varphi_{\alpha_l}(k_l), \qquad (6.168)$$

or, in a x-representation

$$\Phi = -\sum_{\alpha}\int d^d x h_\alpha(x)\varphi_\alpha(x) + \frac{1}{2}\sum_{\alpha}\int d^d x\left[r\varphi_\alpha^2 + c(\nabla\varphi_\alpha)^2\right]$$
$$+\sum_{l=3}^{\infty}\sum_{\alpha_i}\int d^d x\left[\frac{1}{V}\int d^d y\, u^{(l)}_{\alpha_i}(x_i - y)\right]\varphi_{\alpha_1}(x_1)\cdots\varphi_{\alpha_l}(x_l). \quad (6.169)$$

The integral in the square bracket of the last term in Eq. (6.169) is quite similar to the corresponding integral over x in the expansion (6.119) provided the $J(x, y)$ factors are taken in the form $J(x - y)$. The integration over y in Eq. (6.169) will certainly give a coefficient function of the type $u^{(l)}(x_1 - x_2, \ldots, x_1 - x_l)$. To see this the change $y = x_l + y'$ should be done. Then the integration over y' yields a function of $(l - 1)$ vector differences $(x_i - x_l)$, $i = 1, \ldots, (l - 1)$. The same result is obtained from Eq. (6.167) by setting $x = -x_l$.

A less general consideration of the short-range interactions can be performed on the basis of the *nn* interaction picture. As shown in Section F for dD simple hypercubic (sc) and dD body-centered tetragonal lattices, the Fourier amplitudes of *nn* exchange interactions can be represented in the form (6.124) with $J_2 \sim -J_{\rm ex}a_0^2$, where a_0 is the lattice constant. This means that the coefficient c in Eq. (6.125) will be given by $c \sim J_{\rm ex}a_0^2$; see also Eq. (6.126b).

Chapter 7

Perturbation Theory of Fluctuation Fields

7.1 Basic Fluctuation Hamiltonian

In this Chapter, we shall represent the perturbation theory for interacting fluctuation fields. The principles of the perturbation expansion have been already illustrated in the Section 6.3 on the example of spatially independent (macroscopic) fluctuations described by the model potential (6.15). At this point we shall consider the basic effective Hamiltonian $\mathcal{H}$ of the fluctuating order parameter near the usual critical points:

$$\mathcal{H} = \int d^d x\, \mathcal{H}(x)$$
$$= \int d^d x \left[-h(x)\varphi(x) + \frac{r_0}{2}\varphi^2(x) + \frac{c}{2}\left[\nabla\varphi(x)\right]^2 + u_0\varphi^4(x) \right]. \tag{7.1}$$

Here $r_0 = \alpha_0 t$, c, u, and α_0 are positive parameters, and $\mathcal{H}(x)$ is the Hamiltonian density at a point x (x is the d-dimensional spatial vector: $x = (x_1, \dots, x_d)$; or, in another notation: $x = \{x_\alpha; \alpha = 1, \dots, d\}$). In Eq. (7.1), $\varphi = \{\varphi_\alpha\}$ is the n-component order parameter and $h = \{h_\alpha\}$ is the conjugate external field. The terms in Eq. (7.1) are scalar products: $h\varphi = h_\alpha \varphi_\alpha$, $\varphi^2 = \varphi_\alpha\varphi_\alpha$, $(\nabla\varphi)^2 = (\nabla\varphi_\alpha)(\nabla\varphi_\alpha)$, and $\varphi^4 = (\varphi^2)^2$; a summation over the repeated indices is implied.

Since the statistical weight $\exp(-\mathcal{H}/T)$ is often used in our considerations, we shall assume that the factor $(1/T)$ is absorbed into $\mathcal{H}$:

$$\frac{\mathcal{H}}{T} \longrightarrow \mathcal{H}, \tag{7.2}$$

which leads to a change of the other parameters: $h \to h/T$, $r \to r/T$, $c \to c/T$ and $u \to u/T$. This is equivalent to the introduction of a dimensionless

Hamiltonian ($\mathcal{H} \equiv \mathcal{H}/T$) and, hence, of a dimensionless thermodynamic potential $\Phi \to (\Phi/T)$,

$$\Phi(h) = -\ln \mathcal{Z}(h), \tag{7.3}$$

where

$$\mathcal{Z}(h) = \int D\varphi \, e^{-\mathcal{H}(\varphi,h)}, \tag{7.4}$$

is the partition function. It is very suitable for the calculations to make Φ and $\mathcal{H}$ dimensionless but this may become a source of confusion in certain thermodynamic considerations. In such cases the temperature factors of h and Φ have to be restored. The factor $(1/T)$ of r, c, u, and h can be replaced by $(1/T_c)$ because the correction $O(t)$ is small near the phase transition point ($t = 0$) (this point is discussed in details in Section 5.8.6). One should restore the original factor $(1/T)$, if the task is to evaluate a correction of type $O(t^{a+1})$ to the leading t-dependence ($\sim t^a$) of some particular thermodynamic quantity. The best variant is to work with the original factor $(1/T)$ and substitute it with $(1/T_c)$ at the end of the calculation everywhere in our final result except for the leading t-dependence (for an explicit example see Section 5.8.6).

One important result of our previous considerations is that the φ^4-theory with the n-vector order parameter φ describes the most essential features of the usual second order phase transitions. Let us recall the main arguments. Firstly, neglecting the spatial dependence of $\varphi(x)$ leads to the usual Landau theory of second order phase transitions (Sections 4.6 and 4.10) or, which is the same, to the MF approximation of the basic microscopic models (see Sections 5.6 and 5.8). Secondly, the model (7.1) makes possible to investigate both x-independent and x-dependent (field) fluctuations giving in this way the corrections to the MF theory.

It has been demonstrated in Chapter 6 that the Hamiltonian (7.1) certainly stands for the simplest nontrivial theory of interacting fluctuations in isotropic systems. The form of the φ^2-terms in Eq. (7.1) has been justified by both phenomenological and model considerations (Sections 4.6, 6.6, and 6.7). The spatially independent (self)interaction constant u_0 describes the limiting case for the fluctuation interactions. It corresponds to a special substitution

$$u(x_1, \ldots, x_4) = \delta(x_2 - x_1)\delta(x_3 - x_1)\delta(x_4 - x_1)u_0, \quad x_1 \equiv x, \tag{7.5}$$

in the φ^4-term of the generalized Landau expansion of the potential $\Phi(\varphi)$ in even powers of φ for symmetry breaking transitions; see, e.g., Eq. (6.78) with $l = 2m$; $m = 1, 2$. So we neglect the spatial "dispersion" of $u(x_i)$ reducing it to a spatial independent constant u_0 that describes the *"self-interaction"* of the field $\varphi(x)$, that is the interaction of four fields $\varphi^4(x)$ at the same point x. This approximation is justified *a posteriori*, from the obtained theoretical predictions and their comparison with the experimental data.

The form (7.5) of $u(x_i)$ gives the relevant φ^4-term in Eq. (7.1) while the more complex expressions for $u(x_i)$ introduce features of the φ^4-fluctuation interaction, which are irrelevant in the close vicinity of the critical points. The physical meaning of the interaction constant u_0 has become clear from our discussion of the lattice models. For example, the substitution of $J(x-y)$ by $J_0\delta(x-y)$ in Eq. (6.119) or $J(k_i)$ by $J(0)$ in Eq. (6.121) corresponds to the simple form of the φ^4-term in Eq. (7.1). From this point of view the φ^4-interaction in Eq. (7.1) describes extremely short-range "inter-cell" interactions or, according to the alternative interpretation, it corresponds to the fluctuation interactions only in the cells of volume ξ^d, where ξ is *the correlation length* of the system (see Section 7.2). For very large correlation lengths ($\xi \gg a_0$, a_0 is the inter-particle distance) the essential part of the spatial fluctuation "dispersion" is included into the gradient term of the effective Hamiltonian.

When a spatial anisotropy or an anisotropy in the order parameter space are present, the Hamiltonian (7.1) should be generalized in an appropriate way (*complex Hamiltonians*). The results obtained from Eq. (7.1), which are presented in this and the next Chapters, can be generalized for complex Hamiltonians without principal difficulties. So we can discuss the complex field models that describe anisotropic systems or multicritical phenomena, as an application of the theory developed for the isotropic φ^4-Hamiltonian (7.1).

The effective Hamiltonian (7.1) will be called shortly the "Hamiltonian". For reasons that will become clear below it is also referred to as *the Landau–Ginzburg* (LG) or even, *Landau–Ginzburg–Wilson* (LGW) Hamiltonian.

Now our task is to calculate within the field model (7.1) the functional integral (7.4) for $\mathcal{Z}$. We shall obtain the potential (7.3) as a function of the parameters of $\mathcal{H}$: c, r_0, u_0, and the volume V; see Eq. (7.1). Another problem is to find the mean value of the physical quantities $A(\varphi)$, which are functionals of $\varphi(x)$:

$$\langle A \rangle = \mathcal{Z}^{-1}(h) \int D\varphi \, A \, e^{-\mathcal{H}}. \tag{7.6}$$

If $A(\varphi) = [\varphi_{\alpha_1}(x_1) \cdots \varphi_{\alpha_l}(x_l)]$, the average $\langle A \rangle$ will be the lth order correlation function defined by

$$G^{(l)}_{\alpha_1 \dots \alpha_l}(x_1, \dots, x_l/h) = \langle \varphi_{\alpha_1}(x_1) \cdots \varphi_{\alpha_l}(x_l) \rangle_h = \frac{1}{\mathcal{Z}(h)} \frac{\delta^{(l)} \mathcal{Z}(h)}{\delta h_{\alpha_1}(x_1) \cdots \delta h_{\alpha_l}(x_l)}; \tag{7.7}$$

cf. Eq. (5.78), where $\varphi(i)$ is a discrete one-component ($n = 1$) field. The correlation functions $G^{(l)}_{\alpha_i}(x_i/h)$ in a nonzero field are general physical quantities but in most of the cases we need the corresponding functions $G^{(l)}_{\alpha_i}(x_i) \equiv G^{(l)}_{\alpha_i}(x_i/0)$ in a zero field. The reason is that the systems have critical points (T_c, h_c) at a zero field ($h = h_c = 0$). From Eq. (7.7) follows that

$$G^{(l)}_{\alpha_1 \dots \alpha_l}(x_1, \dots, x_l) = \frac{1}{\mathcal{Z}(0)} \left[\frac{\delta^{(l)} \mathcal{Z}(h)}{\delta h_{\alpha_1}(x_1) \cdots \delta h_{\alpha_l}(x_l)} \right]_{h=0}. \tag{7.8}$$

The irreducible correlation functions $\chi^{(l)} = G^{(l)}_{(c)\alpha_i}(x_i/h)$ are obtained from the functional derivatives of $(-\Phi) = \ln \mathcal{Z}(h)$:

$$G^{(l)}_{(c)\alpha_1 \dots \alpha_l}(x_1, \dots, x_l/h) = \frac{\delta^{(l)} \ln \mathcal{Z}(h)}{\delta h_{\alpha_1}(x_1) \cdots \delta h_{\alpha_l}(x_l)}. \tag{7.9}$$

The corresponding zero-field irreducible correlation functions $G^{(l)}_{(c)}$ can be found by setting $h = 0$ in Eq. (7.9). All formulae, written for the correlation functions $G^{(l)}$ and $\chi^{(l)} = G^{(l)}_{(c)}$ in Section 5.5 where the lattice field $\varphi(i)$ is discrete, are straightforwardly transformed for field models in the continuum limit ($i \to x$); see Section 6.5.

It is difficult to calculate Φ and $G^{(l)}$ because the corresponding functional integrals for $\mathcal{Z}(h)$ and its derivatives cannot be solved exactly. In this situation we are forced to develop a perturbation theory for the model (7.1). The first step is to divide the Hamiltonian $\mathcal{H}$ in two parts: $\mathcal{H} = \mathcal{H}_0 + \mathcal{H}_{\text{int}}$. The part $\mathcal{H}_0$ should satisfy two important requirements: (i) to include in itself as much as possible parts of the total Hamiltonian $\mathcal{H}$ and (ii) to give an exactly solvable functional integral for $\mathcal{Z}(h)$. For $\mathcal{H} = \mathcal{H}_0$,

$$\mathcal{Z}_0(h) = \int D\varphi \, e^{-\mathcal{H}_0}; \tag{7.10}$$

the corresponding potential Φ and correlation functions will be denoted by Φ_0 and $G_0^{(l)}$. The Hamiltonian $\mathcal{H}_0$ describes an idealized (auxiliary) system, but the knowledge of its properties gives the opportunity to make an expansion of the total partition integral (7.4) in terms of $\mathcal{H}_{\rm int}$, i.e., in powers of the interaction constant u_0.

7.2 Free Field

Our next task is to investigate $\mathcal{H}_0 \equiv \mathcal{H}(u_0 = 0)$ — the Hamiltonian of the noninteracting field $\varphi(x)$. It is often called a *free field model* or *Gaussian model* because it leads to simple φ^2-Gaussian integrals for the partition and correlation functions of the noninteracting (free) fields $\varphi_\alpha(x)$. From Eq. (7.1) we have

$$\mathcal{H}_0 = \frac{1}{2}\int d^d x \left\{r_0\varphi^2(x) + c[\nabla\varphi(x)]^2 - 2h(x)\varphi(x)\right\}, \tag{7.11}$$

or, in Fourier amplitudes,

$$\varphi_\alpha(x) = \frac{1}{\sqrt{V}}\sum_k \varphi_\alpha(k)e^{ikx}, \tag{7.12a}$$

(kx is a scalar product) and

$$h_\alpha(x) = \frac{1}{\sqrt{V}}\sum_k h_\alpha(k)e^{ikx}, \tag{7.12b}$$

$$\mathcal{H}_0 = \frac{1}{2}\sum_{\alpha,k}\left[(ck^2 + r_0)|\varphi_\alpha(k)|^2 - 2h_\alpha(-k)\varphi_\alpha(k)\right]. \tag{7.13}$$

The volume factors before the sums over k in Eq. (7.12a) and Eq. (7.12b) are different from those used in Chapter 6 and, hence, the amplitudes $\varphi_\alpha(k)$ and $h_\alpha(k)$ have another dimension. The inverse transformation is given by

$$\eta_\alpha(k) = \frac{1}{\sqrt{V}}\int d^d x\, \eta_\alpha(x)e^{-ikx}, \qquad \eta \equiv (\varphi, h). \tag{7.14}$$

The formulae (6.84) for $\delta(x)$ and $\delta(k)$ can be used here too.

In investigations of the bulk properties of crystal bodies, we usually assume periodic boundary conditions. In this quite common case, the wave vector k lies in the first Brillouin zone which can be defined by

$\pi/a_0 \le k_l < \pi/a_0$, or, alternatively, by $\pi/a_0 < k_l \le \pi/a_0$, where a_0 is the lattice constant of a hyper-cubic lattice, $k_l = 2\pi n_l/L$ are the components of the wave vector, $k = \{k_l; l = 1, \ldots, d; n_l = 0, \pm 1, \pm 2, \ldots, [L/a_0]\}$; $[L/a_0]$ denotes the maximal integer which is either less or equal to the ratio L/a_0 (Kittel, 1963). Now we see that the number of quantum states k in the first Brillouin zone is equal to the number of particles N. The quantum states k will be symmetrically located in this zone if the particle number $N > 1$ is odd. If N is however even, the l.h.s. of the zone (negative k_l), or, alternatively, the r.h.s. of the zone (positive k_l) will contain one more quantum state. For macroscopic systems, considered in this book, we may set absolutely symmetric borders of the Brillouin zone ($\pi/a_0 \le k_l \le \pi/a_0$) and forget about the minor asymmetry for even particle numbers.

7.2.1 *Equation of state*

The equation of state will give the equilibrium states $\bar{\varphi}(x) = \{\bar{\varphi}_\alpha(x)\}$ of the system (below the *bar* of $\bar{\varphi}_\alpha$ will be omitted). It is derived from the stability condition

$$\delta\mathcal{H}_0 = 0. \tag{7.15}$$

Since $\mathcal{H}_0$ depends on the field $\varphi(x)$ we must use the Euler–Lagrange equations for the calculations of variations. From Eq. (7.11) and Eq. (7.15) we obtain the set of n equations:

$$\int d^d x \; [(r_0\varphi_\alpha - h_\alpha)\delta\varphi_\alpha + c(\nabla\varphi_\alpha)(\nabla\delta\varphi_\alpha)] = 0. \tag{7.16}$$

Now we must exclude the variations $\delta\varphi_\alpha$ and this problem is connected with the appropriate choice of the boundary condition for the behaviour of $\varphi(x)$ at the surface S of the system. Let the system be very large so we can ignore the surface effects. Therefore we focus our attention on the bulk properties of $\varphi(x)$. Mathematically our description will correspond to an infinite system, which extends over the whole d-dimensional space.

The gradient term in Eq. (7.16) can be integrated by parts. The corresponding surface integral

$$\int_S \delta\varphi_\alpha(\nabla\varphi_\alpha)\cdot \boldsymbol{e}_n dS, \qquad |\boldsymbol{e}_n| = 1, \tag{7.17}$$

where $e_n(\equiv \boldsymbol{e}_n)$ is the normal to the surface S, is neglected as small in comparison with the bulk integrals in Eq. (7.16). As a result, Eq. (7.16) becomes:

$$\int d^d x \left[r_0\varphi_\alpha - c\nabla^2\varphi_\alpha - h_\alpha\right]\delta\varphi_\alpha = 0. \tag{7.18}$$

Here the "gradient" term $\nabla^2\varphi_\alpha$ appears when the integration by parts is done.

Since Eq. (7.18) is true for any variation $\delta\varphi_\alpha$, the Euler–Lagrange equations are obtained in the simple form:

$$\left(r_0 - c\nabla^2\right)\varphi_\alpha = h_\alpha. \tag{7.19}$$

The solutions of the linear differential equations (7.19) depend on the form of the external field $h = \{h_\alpha(x)\}$. For $h_{\alpha_i} \neq h_{\alpha_j}$, the solutions $\varphi_\alpha(x)$ will be different; when $h_{\alpha_i} = h_{\alpha_j}$, Eq. (7.19) consists of n identical equations. The most simple solution has been given in Section 4.10, where h_α and φ_α have been considered spatially independent.

Let us introduce the notations

$$\varphi_\alpha = \varphi_\alpha^0 + \Delta\varphi_\alpha(x), \tag{7.20a}$$

and

$$h_\alpha = h_\alpha^0 + \Delta h_\alpha(x); \tag{7.20b}$$

φ_α^0 and h_α^0 are the spatially independent parts of φ_α and h_α. Neglecting $\Delta\varphi_\alpha$ and Δh_α as small we obtain

$$\varphi_\alpha^0 = \frac{h_\alpha^0}{r_0} = \chi_0 h_\alpha^0. \tag{7.21}$$

The susceptibility χ_0 is given by Eq. (4.142) with $r \to r_0/2$.

The present Gaussian model has nontrivial solutions $\varphi_\alpha \neq 0$ of the type $\varphi_\alpha \sim h_\alpha$, i.e., field-induced solutions but it does not exhibit states of spontaneously broken symmetry ($\varphi_\alpha \neq 0, h_\alpha = 0$). So the free field Hamiltonian describes neither a phase transition, nor the behaviour of real systems below the critical point ($r \sim t < 0$), where the spontaneous symmetry breaking $(\varphi_\alpha, 0)$ occurs. However, there is a real situation which can quite approximately be described by the free field model $\mathcal{H}_0$. The solution ($\bar{\varphi} = 0$) of the total Hamiltonian (7.1) for $r_0 > 0$ and $h = 0$ gives a disordered phase, for which the order parameter $\varphi = \bar{\varphi} + \delta\varphi$ is represented only by its fluctuations ($\varphi_\alpha = \delta\varphi_\alpha$). Neglecting the fluctuation interaction ($u_0 = 0$), we get the behaviour of the free fluctuations $\delta\varphi_\alpha$ from the model $\mathcal{H}_0$.

The equations (7.19) have no x-dependent solutions $\Delta\varphi_\alpha(x)$ for uniform external fields ($h_\alpha = h_\alpha^0$). The reason is that the coefficient c is positive and the nonuniform φ-states correspond to a lower statistical weight $\exp(-\mathcal{H}_0)$ than the uniform ones ($\nabla\varphi_\alpha^0 = 0$). The same result follows directly from the theory of the partial differential equations. Besides, it is valid for the total model ($u_0 > 0$) too.

7.2.2 *Green function*

Substituting φ_α and h_α from Eqs. (7.20a) and (7.20b) in the Eq. (7.19) and having in mind Eq. (7.21), we obtain the equations for $\Delta\varphi_\alpha$:

$$(r_0 - c\nabla^2)\Delta\varphi_\alpha = \Delta h_\alpha. \tag{7.22}$$

The solution of these equation(s) can be discussed for particular functions $\Delta h_\alpha(x)$. However the *Green function* G_0 of the equations can be always obtained. It corresponds to the point-field source

$$\Delta h_\alpha(x) = \gamma_\alpha \delta(x), \tag{7.23}$$

of intensity $\gamma_\alpha = 1$ located at the point $x = 0$ in the volume of the system. Although the solution of Eq. (7.22) with $\Delta h_\alpha(x)$ in the form (7.18) is well known, it will be useful to discuss this solution in more details. Applying the transformations (7.12a)–(7.12b) to $\Delta\varphi_\alpha$ and Δh_α we obtain Eq. (7.22) as a simple algebraic equation for the amplitudes $\Delta\varphi_\alpha(k)$:

$$(r_0 + ck^2)\Delta\varphi_\alpha(k) = \Delta h_\alpha(k). \tag{7.24}$$

The formula (7.23) for $\Delta h_\alpha(x)$ corresponds to $\Delta h_\alpha(k) = \gamma_\alpha$ and, hence, the equation

$$(r_0 - c\nabla^2)G_0(x) = \delta(x), \tag{7.25}$$

for the Green function $G_0(x)$ of the operator $(r_0 - c\nabla^2)$ yields the Fourier amplitude $G_0(k)$:

$$G_0(k) = \frac{1}{r_0 + ck^2}. \tag{7.26}$$

If the explicit form of the amplitudes $\Delta h_\alpha(k)$ is available, $\Delta\varphi_\alpha(k)$ can also be found from the relation:

$$\Delta\varphi_\alpha(k) = G_0(k)\Delta h_\alpha(k). \tag{7.27}$$

The inverse Fourier transformation of this equation gives

$$\Delta\varphi_\alpha = \frac{1}{V}\sum_k G_0(k)\Delta h(k)e^{-ikx} \tag{7.28}$$

$$= \int \frac{d^dk}{(2\pi)^d}\frac{\Delta h(k)}{r_0 + ck^2}e^{-ikx}. \tag{7.29}$$

Here we have applied the continuum limit; see Eq. (6.100). For the point source in Eq. (7.23),

$$\Delta\varphi_\alpha(x) = \gamma_\alpha G_0(x), \tag{7.30}$$

with

$$G_0(x) = \int \frac{d^dk}{(2\pi)^d}\frac{e^{-ikx}}{r_0 + ck^2}. \tag{7.31}$$

The above consideration demonstrates the principal role of the Green function $G_0(x)$, when we investigate the response $\Delta\varphi$ of the system to the fields Δh conjugate to the order parameter. Besides, this function describes also the fluctuation properties of the field $\varphi(x)$. With the help of Eq. (7.26), the Eq. (7.13) takes the form

$$\mathcal{H}_0 = \frac{1}{2}\sum_{\alpha,k}\left[G_0^{-1}(k)|\varphi_\alpha(k)|^2 - 2h_\alpha(-k)\varphi_\alpha(k)\right]. \tag{7.32}$$

Let us remember that $\mathcal{H}_0$ coincides with the generalized Gibbs potential $\Phi_0(\varphi, h)$, in which the interaction terms $O(\varphi^3)$ are neglected. The function $G_0^{-1}(k)$ is then the second stability coefficient (u_2) in the expansion of $\Phi(\varphi, h)$ in powers of the Fourier amplitudes $\varphi_\alpha(k)$; see, e.g., Eq. (6.87). For $h_\alpha \equiv 0$, $G_0^{-1}(k)$ will describe the stability property of the disordered state ($\bar{\varphi} = 0$) with respect to fluctuations; $\varphi_\alpha(k) = \delta\varphi_\alpha(k)$; see also Sections 7.4 and 7.5.

The Green function $G_0(x)$ in Eq. (7.31) is calculated with the help of the identity Eq. (D.21). The result is:

$$G_0(x) = \frac{(2\pi)^{-d/2}}{c} \left(\xi_0 |x|\right)^{1-d/2} K_{\frac{d}{2}-1}\left(\frac{|x|}{\xi_0}\right), \tag{7.33}$$

where

$$\xi_0(T) = \left(\frac{c}{r_0}\right)^{1/2}, \tag{7.34}$$

is the correlation length of the free fluctuations and $K_\nu(z)$ is the modified Bessel function of second kind (Abramowitz and Stegun, 1965).

For $T = 0$, i.e., $|t| = 1$, $\xi_0(0) \equiv \xi_{00} = (c/\alpha_0)^{1/2}$ is the so-called zero-temperature correlation length. Taking in mind Eqs. (6.125)–(6.126b), Eq. (F.4) and Eq. (F.6), as well as that $T_c \sim J_0 \sim J_{\text{ex}}$, we conclude that for models of Ising type with nn interactions, $\xi_{00} \sim a_0$ (a_0 is the lattice constant).

The expression (7.33) is rather complicated, that is why we shall briefly discuss the asymptotic cases when $z = (x/\xi_0) \ll 1$, ($x \equiv |x|$) and $z \gg 1$. For $z \ll 1$,

$$K_\nu(z) \sim z^{-\nu} \qquad (\nu \neq 0, \ \Re\, \nu > 0), \tag{7.35a}$$

and

$$K_0(z) \sim \ln \frac{1}{z}, \tag{7.35b}$$

so that

$$G_0(x) \sim \frac{1}{cx^{d-2}}, \quad \text{for} \quad d \neq 2 \tag{7.36a}$$

and

$$G_0(x) \sim \frac{1}{c} \ln\left(\frac{\xi_0}{x}\right), \quad \text{for} \quad d = 2. \tag{7.36b}$$

In the above expressions the reader must pay more attention to the special role of the dimensionality $d = 2$ for systems with quadratic energy spectrum $\varepsilon_0(k) = ck^2$ of the fluctuations; $G_0(k) = r_0 + \varepsilon_0(k)$. The spatial dependence of the order parameter φ brings about different correlation properties for different dimensionalities d. For $d > 2$, $G_0(x) \to \infty$ when $(x/\xi_0) \to 0$. When $d < 2$, $G_0(x) \to 0$ as $(x/\xi_0) \to 0$. The *dimensional crossover* between the behaviour for low ($d < 2$) and "usual" ($d > 2$) dimensions d is at the *borderline* dimension $d = 2$, where $G_0(x)$ is logarithmically divergent at small distances $(\xi_0/x) \gg 1$.

The most important quantity is the correlation length ξ_0. As T approaches T_c from the high-temperature side ($T \to T_c^{(+)}$) all distances $x \leq L$ in the volume L^d of the system become small in the sense that $x \leq L < \xi_0$. In the close vicinity of T_c, Eqs. (7.36a) and (7.36b) are true for any real system ($L < \infty$). The divergence of $G_0(x)$ described by Eqs. (7.36a) and (7.36b) is closely connected with the corresponding divergence of $G_0(k, r_0)$ for $r_0 \to 0$ and $k \to 0$; see Eq. (7.26).

It is obvious from Eq. (7.32) that *the critical fluctuations* ($r_0 = h_\alpha = 0$) with small energies $\varepsilon_0(k) \sim k^2$ ($k \ll \Lambda$) will have an anomalously high statistical weight $\exp(-\mathcal{H}_0)$, which tends to unity for the normalization condition $|\varphi(0)|^2 = 1$. These critical fluctuations are often referred to as *anomalous fluctuations.* For reasons explained above here we do not discuss the behaviour of the system below T_c. The correlation length ξ_0 for $T < T_c$ has similar properties and the critical fluctuations appear as T approaches T_c from the low-temperature side (see Section 7.3).

When $T > T_c$, $\xi_0 < L$ and there are distances x, for which $\xi_0 < x < L$. The asymptotic properties of $G_0(x)$ for $(x/\xi_0) \gg 1$, (long distances) are obtained from the expansion of $K_\nu(z)$ for $z \gg 1$:

$$K_\nu(z) = \sqrt{\frac{\pi}{2z}}e^{-z}\left[1 + O\left(\frac{d-4}{z}\right)\right]. \tag{7.37a}$$

Taking only the leading term from Eqs. (7.36a) and (7.36b) we obtain that

$$G_0(x) = \left(\frac{\xi_0}{2c}\right)(2\pi\xi_0 x)^{(1-d)/2}\, e^{-x/\xi_0}, \tag{7.37b}$$

or, for $d = 3$,

$$G_0(x) = \frac{e^{-x/\xi_0}}{4\pi c x}, \qquad (x/\xi_0) \gg 1. \tag{7.37c}$$

The exponential decrease of $G_0(x)$ with the distance x is a characteristic feature of the *noncritical* (usual) fluctuations, i.e., the fluctuations far from the critical point. Since ξ_0 varies with the temperature from finite values to infinity (for $T = T_c$), the behaviour of $G_0(x)$ changes from the rapid exponential decrease to the anomalous (critical) power law, see Eq. (7.36a) for $d > 2$. This is a signal for the appearance of critical fluctuations; for a detailed discussion; see also Fisher (1962), Stanley (1971).

The function $G_0(k)$ in the form (7.26), or, in the form Eq. (7.37c) has been first introduced in the theory of critical opalescence (Landau and Lifshitz, 1960) by Ornstein and Zernicke (1914); a comprehensive representation of the Ornstein–Zernicke theory and its interrelationship with the older Rayleigh's, Smoluchowskii's and Einstein's theories of fluctuations is given in the book by Kociński and Wojtczak (1978). In order to understand the meaning of the Ornstein–Zernicke form of $G_0(k)$ we should remember that the gradient φ^2-term enters in the theory as a result of the inter-cell exchange J_{ij} in the original lattice model (see Section 6.6).

Certainly the Ornstein–Zernicke approach makes possible to investigate the fluctuation correlations between the states in small neighbouring volumes (cells) of the system, while the older theories do not take into account such correlations which corresponds to the approximation $G_0(k, r_0) \approx G_0(0, r_0) = 1/r_0$. The Ornstein–Zernicke theory ignores the fluctuation interactions ($\delta\varphi_\alpha^l, l \geq 3$). This theory describes correctly the fluctuation behaviour far from the critical point but it breaks down in the vicinity of T_c where the fluctuation interaction, say the $\delta\varphi^4$-interaction in Eq. (7.1) is relatively strong and cannot be neglected.

In a close vicinity of the critical point one has to consider the total Green function $G(x)$ rather than $G_0(x)$. Using the scheme of calculation presented in this section the reader can show that the Green function $G(x)$ of the interacting fluctuations obeys a non-linear differential equation in which the fluctuation interaction is taken into account.

7.3 Thermodynamic Properties of Free Fields

The partition function (7.10) of the free field model is given by the Gaussian integral
indexGaussian integral

$$\mathcal{Z}_0(h) = \int D\varphi \exp\left\{-\frac{1}{2}\int d^dx \left[r_0\varphi^2 + c(\nabla\varphi)^2 - 2h\varphi\right]\right\}, \qquad (7.38a)$$

or, after the Fourier transformation of $\mathcal{H}_0$, see Eq. (7.32),

$$\mathcal{Z}_0(h) = \int \tilde{D}\varphi \exp\left\{ -\frac{1}{2} \sum_{\alpha,k} \left[G_0^{-1}(k)|\varphi_\alpha(k)|^2 - 2h(-k)\varphi_\alpha(k) \right] \right\}, \quad (7.38b)$$

where the integration symbol $\int \tilde{D}\varphi$ is related to the Fourier amplitudes $\varphi_\alpha(k)$ and may considerably differ from the definition (6.107). This point will be discussed at a next stage of the calculation of $\mathcal{Z}_0(h)$.

Now we shall calculate $\mathcal{Z}_0(h)$ and the thermodynamic potential of the free field model:

$$\Phi_0(h) = -\ln \mathcal{Z}_0(h), \quad (7.39)$$

cf. Eq. (7.3). In order to avoid the non-diagonal gradient term appearing in the exponent in Eq. (7.38a) in the x-representation, we shall use the k-representation (7.38b), where $\mathcal{H}_0$ is diagonal. It is useful to apply the shift transformation to the fields $\varphi_\alpha(k)$ to new fields $\varphi'_\alpha(k)$,

$$\varphi_\alpha(k) = \varphi'_\alpha(k) + A_\alpha(k), \quad (7.40a)$$

which has the same form in the x-space,

$$\varphi_\alpha(x) = \varphi'_\alpha(x) + A_\alpha(x). \quad (7.40b)$$

(similar transformations have been used in our previous considerations; see, e.g., Eq. (6.113) and Section B.2). The shift transformation is intended to eliminate the linear term in the integrand exponent in Eq. (7.38b). The "shift field" $A(x) = \{A_\alpha(x)\}$ is real because $\varphi(x)$ and $h(x)$ are real.

After substituting $\varphi_\alpha(k)$ from Eq. (7.40a) in Eq. (7.38b) we should choose the shift variables $A_\alpha(k)$ in such way that the linear terms $h_\alpha\varphi'_\alpha$ vanish. We also use the fact that the fields $\varphi_\alpha(x)$, $h_\alpha(x)$ and $A(x)$ are real and, therefore, their Fourier amplitudes satisfy conditions like Eq. (6.88a) for $\varphi_\alpha(k)$. Further, we take in mind the important circumstance that the limits of the summation over k are symmetric with respect to $k = 0$ (as explained below Eq. (7.14), in case of a macroscopic crystal, the vector k lies in the symmetric Brillouin zone; for a non-crystalline substance, k varies from $-\infty$ to ∞). Thus we can write the term $2h(-k)\varphi_\alpha(k)$ under the sum in the exponent of Eq. (7.38b) as a sum of two terms,

$$h(-k)\varphi(k) + h(k)\varphi(-k),$$

where we substitute $\varphi_\alpha(k)$ according to the prescription (7.40a). Making the same substitution in the first term of the exponent of Eq. (7.38b) and having in mind that $G_0(k) = G_0(-k)$, after simple algebra we obtain

$$A_\alpha(k) = G_0(k)h_\alpha(k),$$
$$A_\alpha(-k) = G_0(k)h_\alpha(-k), \tag{7.41}$$

and the corresponding expression for the partition function

$$\mathcal{Z}_0(h) = \mathcal{Z}_0(0) \exp\left[\frac{1}{2}\sum_{\alpha,k} G_0(k)h_\alpha(k)h_\alpha(-k)\right]. \tag{7.42}$$

Here

$$\mathcal{Z}_0(0) = \int \tilde{D}\varphi' \exp\left[-\frac{1}{2}\sum_{\alpha,k} G_0^{-1}(k)|\varphi'_\alpha(k)|^2\right] \tag{7.43}$$

is the partition integral of the free field φ'_α (henceforth, the superscript (*prime*) of φ'_α will be omitted).

The factor $|\varphi_\alpha(k)|^2$ in the integrand exponent in Eq. (7.43) can be written in the form

$$|\varphi'_\alpha(k)|^2 \longrightarrow |\varphi_\alpha(k)|^2 = a^2_{\alpha k} + b^2_{\alpha k}, \tag{7.44}$$

where

$$a_{\alpha k} = \Re[\varphi_\alpha(k)] \quad \text{and} \quad b_{\alpha k} = \Im[\varphi_\alpha(k)] \tag{7.45a}$$

are the real and imaginary parts of the complex amplitudes $\varphi_\alpha(k)$ and we can write

$$\varphi_\alpha(k) = a_{\alpha k} + ib_{\alpha k},$$
$$\varphi^*_\alpha(k) = a_{\alpha k} - ib_{\alpha k}. \tag{7.45b}$$

In order to perform the integration in Eq. (7.43) we must know the complete set of independent fields $\varphi_\alpha(k)$. Because of the relations (6.88a) $b_{\alpha 0} = \Im[\varphi_\alpha(0)]$ is zero and, moreover, the following relations $a_{\alpha,-k} = a_{\alpha,k}$ and $b_{\alpha,-k} = -b_{\alpha,k}$ are valid. Therefore, the complete set of independent fields $\varphi_\alpha(k)$ can be given by $a_{\alpha 0}$ and all $a_{\alpha k}$ and $b_{\alpha k}$ with positive wave vectors ($k > 0$); or, alternatively, again by $a_{\alpha 0}$, and all $a_{\alpha k}$ and $b_{\alpha k}$ with negative wave vectors ($k < 0$).

For a macroscopic number of particles we can approximate the first Brillouin zone by the full-sphere $|k| \leq \Lambda \sim \pi/a_0$ (a_0 is the mean inter-particle distance) containing the wave vectors k. Thus our choice of the complete set of independent fields could be either the fields $a_{\alpha k}$ and $b_{\alpha k}$ with wave vectors in the upper half ($k \geq 0$) of the sphere $|k| \leq \Lambda \sim \pi/a_0$, or, the same fields with wave vectors in lower half ($k \leq 0$) of this momentum sphere.

In virtue of the relations (7.45b) we may define the set of independent fields in terms of $\varphi_\alpha(k)$ and $\varphi^*_\alpha(k)$. Then the two equivalent variants of the complete set of independent fields are given by the series [$\varphi_\alpha(k)$, $\varphi^*_\alpha(k)$, for $k \geq 0$] and [$\varphi_\alpha(k)$, $\varphi^*_\alpha(k)$, for $k \leq 0$]. The work with the real fields $a_{\alpha k}$ and $b_{\alpha k}$ has calculational advantages and we shall use the notations (7.45a).

The next step is to write the exponent and the integration symbol $\int \tilde{D}\varphi$ in Eq. (7.43) by the independent fields. The exponent in Eq. (7.43) takes the form

$$\begin{aligned} &-\frac{1}{2}G_0^{-1}(0)a_{\alpha 0}^2 - \frac{1}{2}\sum_{\alpha;k\neq 0} G_0^{-1}(k)\left(a_{\alpha k}^2 + b_{\alpha k}^2\right) \\ &= -\frac{1}{2}G_0^{-1}(0)a_{\alpha 0}^2 - \sum_{\alpha;k>0} G_0^{-1}(k)\left(a_{\alpha k}^2 + b_{\alpha k}^2\right). \end{aligned} \tag{7.46}$$

The choice of the symbol $\int \tilde{D}\varphi$ in Eq. (7.43) is not unique but rather we have several choices. In order to cover all possibilities we shall use an integration of the type

$$\int \tilde{D}\varphi = \aleph \int_{-\infty}^{\infty} \prod_{\alpha=1}^{n} da_{\alpha 0} \prod_{k>0} da_{\alpha k} db_{\alpha k}, \tag{7.47}$$

where $\aleph$ is a constant with respect to the relevant temperature parameter r_0. The factor $\aleph$ tunes the zero level of the energy scale by the respective contribution ($-T \ln \aleph$) to the free energy (6.9).

In order to achieve a physically justified picture in which the free energy (7.39) exhibits divergencies only at the critical temperature ($r_0 = 0$) we should choose the factor $\aleph$ in such way that the irrelevant (often divergent) terms produced by the integration vanish. Here we have two options: to use the constant $\aleph$ in order to compensate the above-mentioned terms produced by the integration and in this way to achieve a self-consistent description, or, alternatively, disregarding this constant, we ought to ignore the respective integration products as unessential to our consideration (such terms could be relevant only in studies of generic divergencies of the functional integral; Section 6.5.5). Below we shall show an example of divergent term which is compensated by the factor $\aleph$.

Using Eqs. (7.46), (7.47) and (B.1) we calculate the integral in Eq. (7.43):

$$\mathcal{Z}_0(0) = \aleph \tilde{I}_0^n \prod_{k>0} I_k^{2n}, \tag{7.48}$$

where

$$\tilde{I}_0 = \int_{-\infty}^{\infty} dy \exp\left[-\frac{G_0(0)}{2} y^2\right] = [2\pi G_0(0)]^{1/2}, \tag{7.49a}$$

and

$$I_k = \int_{-\infty}^{\infty} dy \exp\left[-G_0(k) y^2\right] = [\pi G_0(k)]^{1/2}. \tag{7.49b}$$

Now one can easily find a more compact form of $\mathcal{Z}_0(0)$:

$$\begin{aligned} \mathcal{Z}_0(0) &= \aleph\,[2\pi G_0(0)]^{n/2} \prod_{k>0} [\pi G_0(k)]^n \\ &= \aleph 2^{n/2} \prod_k [\pi G_0(k)]^{n/2} \\ &= \aleph 2^{n/2} \prod_k \left(\frac{\pi}{r_0 + ck^2}\right)^{n/2}, \end{aligned} \tag{7.50}$$

where Eq. (7.26) for $G_0(k)$ has been used. By the last equality (7.50) the possibility of variations of the vector k in the whole volume $V_d(\Lambda)$ of the dD full-sphere $|k| \leq \Lambda$ in the k-space is restored (see, e.g., Eq. (D.4) in Section D.1). This is convenient in calculations of integrals over the wave vector k.

Now we can find the potential (7.39):

$$\Phi_0(h) = \Phi_{0h} + \Phi_{0f}, \tag{7.51}$$

where

$$\Phi_{0h} = -\frac{1}{2}\sum_{\alpha,k} G_0(k)|h_\alpha(k)|^2 \tag{7.52}$$

is the h-dependent part of Φ_0 and

$$\Phi_{0f} = -\frac{n}{2}\sum_{k} \ln[G_0(k)] + [\Upsilon\ (=0)] \tag{7.53a}$$

is the fluctuation part. In Eq. (7.53a), a term of type

$$\Upsilon(\aleph, N, n) = -\ln \aleph \left(2^{1/N}\pi\right)^{nN/2} \tag{7.53b}$$

also appears. This term can be reduced to zero by the particular choice $\aleph = \left(2^{1/N}\pi\right)^{-nN/2}$ of the arbitrary factor $\aleph$ in Eq. (7.47). Therefore, with the help of the factor $\aleph$, we have been able to remove an irrelevant term, $-\ln(2^{1/N}\pi)^{nN/2}$, in the free energy (7.53a); this term is divergent in the thermodynamic limit ($N \to \infty$).

We have performed the calculation of $\mathcal{Z}_0(0)$ in one of the possible variants, let us say, the most direct one. Now we shall show other options.

Let us "symmetrize" the expression (7.46) by the simple transformation: $a_{0k} = a'_{0k}/\sqrt{2}$, $b_{0k} = b'_{0k}/\sqrt{2}$ of all fields with $k \neq 0$. This is equivalent to a quite common transformation of type (7.45b), where the r.h.s. of the two equations are multiplied by $1/\sqrt{2}$. If we work in this way, by the fields a'_{0k} and b'_{0k}, the second term on the r.h.s. of Eq. (7.46) will acquire a factor 1/2 as it is for the first term. The same factor (1/2) will appear in the exponent of the integrand in Eq. (7.49b) and, hence, the integral (7.49b) will become exactly equal to the integral $\bar{I}_0$, given by Eq. (7.49a). Now we denote the arbitrary factor in Eq. (7.47) by another symbol, $\tilde{\aleph}$, because our independent integration fields $(a_{\alpha 0}, a'_{\alpha k}, b'_{\alpha k})$ are somewhat different. Following the next steps in this alternative variant of calculation we immediately arrive at the result (7.53a) for Φ_{0f}, where Υ should be changed to $\tilde{\Upsilon} = -\ln[\tilde{\aleph}(2\pi)^{nN/2}]$. Now we must set $\tilde{\aleph} = (2\pi)^{-nN/2}$ in order to nullify the redundant term $\tilde{\Upsilon}$.

This result is quite plausible and can be deduced in another way. The ratio $\tilde{\aleph}/\aleph = 2^{-n(N-1)/2}$ is exactly equal to the number which we should obtain as an extra factor in the front of the integral (7.47) if we perform the transformation of the integration variables $(a_{\alpha k}, b_{\alpha k;k\neq 0})$ to the new ones $(a'_{\alpha k}, b'_{\alpha k}; k \neq 0)$. In fact we have $n(N-1)$ independent real modes (fields) with $k \neq 0$ which play the role of integration variables, and this yields the additional factor $(1/\sqrt{2})^{n(N-1)} = 2^{-n(N-1)/2}$ on the r.h.s. of Eq. (7.47), provided the transformation of the fields is performed.

A third variant of integration variable choice is also possible, namely, when we introduce a new integration variable $a'_{\alpha 0} = \sqrt{2} a_{\alpha 0}$ for the $(k = 0)$–field, too. But we already understand the interrelationship between the normalization factor $\aleph$ and the field normalization and the demonstration of a third example seems redundant.

Certainly, we vary the possible normalization of our field variables and this leads to a variation of the norm of the statistical distribution (6.1), i.e., the partition function $\mathcal{Z}$. The multi-variant solution of this problem is a direct consequence of the fact that the energy scale and the energy zero are loosely definite and can be changed according to our convenience. But if we fix them, the proper normalization of the fields and the norm of the integration (7.47) can be identified. We have used these relatively simple examples in order to demonstrate certain peculiarities of the functional integrals (Section 6.5.5) in k-representation.

A glance on Eqs. (7.27) and (7.41) shows that $A_\alpha(k)$ is the same as the solution $\bar{\varphi}_\alpha(k)$ of the equation of state (7.24) for $\Delta h_\alpha(k) = h_\alpha(k)$, i.e., for the variations h_α around the value $h^0_\alpha = 0$. The same equation of state is given by the first derivative of $\Phi_0(h)$ with respect to $h_\alpha(-k)$. From Eqs. (7.51)–(7.53a) we have

$$\frac{\delta \Phi_0}{\delta h_\alpha(-k)} = -\langle \varphi_\alpha(k) \rangle_{0h} = -\bar{\varphi}_\alpha(k). \tag{7.54}$$

In fact $A_\alpha(k)$, $\bar{\varphi}_\alpha(k)$, and $\langle \varphi_\alpha(k) \rangle_{0h}$ are different symbols for one and the same quantity — the equilibrium value of the field $\varphi_\alpha(k)$ at the external field h_α. For spatially independent fields, $h^0_\alpha(x) = const$, Eq. (7.14) yields $h_\alpha(k) = \sqrt{V} h_\alpha \delta(k)$ and, accordingly,

$$\Phi_{0h} = -\frac{V}{2r_0} h_0^2, \tag{7.55}$$

where $r_0^{-1} = \chi_0$ is the susceptibility per unit volume of the state $\bar{\varphi}_0 = 0$ with respect to variations of the uniform field $h_0 = \{h^0_\alpha\}$ around the value $h_0 = 0$; see also Eq. (7.21). The susceptibility $(V\chi_0)$ corresponds to the volume V.

The term Φ_{0f} of $\Phi_0(h)$ has a pure fluctuation origin. Since $A_\alpha(k)$ is equivalent to the solution $\bar{\varphi}_\alpha(k)$ of the equation of state, φ'_α in Eq. (7.40a) will be the fluctuation part of $\varphi_\alpha(k)$. Therefore, the integral $\mathcal{Z}_0(0)$ and, hence, Φ_{0f} are related to the fluctuations $\delta\varphi_\alpha(k) = \varphi'_\alpha(k)$ of the free field. For $h_\alpha = 0$, $\bar{\varphi}_\alpha = 0$ and $\varphi_\alpha(k) = \delta\varphi_\alpha(k)$. Then $\Phi_0(h)$ coincides with its fluctuation part Φ_{0f} or, in other words, it describes a free (fluctuation) field in a zero external field.

Having Eqs. (7.51)–(7.53a) we can obtain any thermodynamic quantity of interest. The entropy S and the specific heat C_h at fixed h are found by an appropriate differentiation of $\Phi_0(h)$ with respect to the temperature. The consistent approach requires to restore the original notations for the quantities in Eqs. (7.51)–(7.53a). For example, $\Phi_0(h)$ must be substituted by $\Phi_0(h)/T$, $G_0(k) \to TG_0(k)$,, etc.; see the discussion of Eq. (7.2). After making this procedure it can be easily demonstrated by direct calculations that near the critical point ($0 < t \ll 1$) the essential dependence on T, i.e., on $t = (T - T_c)/T_c$ is given by the parameter r_0. As we are interested in the behaviour of the physical quantities in a close vicinity of T_c we can substitute the factor $(1/T)$ in all thermodynamic parameters by $(1/T_c)$, then we can investigate the dependence of Φ_0 only on r_0.

As an example we shall calculate the specific heat C_h near T_c when the external field is zero. In this case $\Phi_{0h} = 0$ and the fluctuation specific heat C_{0f} is given by,

$$C_{0f} = -T_c \left[\frac{\partial^2 (T_c \Phi_{0f})}{\partial r_0^2}\right] \left(\frac{\partial r_0}{\partial T}\right)^2 . \tag{7.56}$$

Here we have used (Eq. (7.53a) for Φ_{0f} with the substitution $\Phi_{0f} \to \Phi_{0f}/T_c$. The final expression for C_{0f} is obtained by taking the continuum limit for the sum in Eq. (7.53a) according to the prescription (6.100) and differentiating the corresponding integral with respect to r_0. Making use of Eq. (D.11) and Eq. (D.12) we find that

$$C_{0f} = \left(\frac{n}{2}\right) \alpha_0^2 V K_d \int_0^\Lambda \frac{k^{d-1} dk}{(r_0 + ck^2)^2} . \tag{7.57}$$

The integral in Eq. (7.57) is calculated in Section E.1. The result for the asymptotic behaviour of C_{0f} near the critical point ($\xi_0 \gg \Lambda \sim 1/a$) is

$$C_{0f} = \frac{C_0}{t^{2-d/2}} , \tag{7.58}$$

where

$$C_0 = \frac{n}{2}\Gamma(2-d/2)\left(\frac{\alpha_0}{4\pi c}\right)^{d/2} V. \tag{7.59}$$

From Eq. (7.58) and Eq. (7.59), we obtain C_{0f} as a function of ξ_0 : $C_{0f} \sim \xi_0^{4-d}$. This calculation is valid for $d < 4$ and gives the highest power divergent term of C_{0f} at $t = 0$ (see Section E.1). Certainly a temperature $T_1 > T_c$ exists, below which $T_c < T < T_1$ the fluctuation contribution C_{0f} becomes bigger than the mean field jump ΔC of C_h at the critical point; see, e.g., Eq. (4.74). The dependence $C_{0f}(t)$ for $d \geq 4$ is discussed in Section E.1; see also Eq. (E.3) and Eq. (E.4), where the behaviour of $C_{0f}(t)$ is given for $d = 4$ and $d = 5$. For $d = 4$ the integral in Eq. (7.57) and, hence, $C_{0f}(t)$ exhibits a logarithmic divergence when $t \to 0$. For $d = 5$ and, in general, for $d > 4$, C_{0f} tends to a finite constant as $t \to 0$ ("MF-like" behaviour). The contribution to the specific heat of the term (7.52) can also be evaluated in the presence of an external field. For a constant field $h(x) = h$ this contribution is very strong near $t = 0$. From Eq. (7.55) we have that $C_{0f} \sim t^{-3}$. For a point source, when $h(x) \sim \delta(x)$ and $h_\alpha(k)$ in Eq. (7.52) does not depend on k, the sum $\sum_k G_0(k)$ is proportional to the derivative of Φ_{0f} with respect to r_0. The corresponding contribution to the specific heat at small t will therefore be of the type $(\partial C_{0f}/\partial t) \sim t^{d/2-3}$. This singularity merely indicates the strong field limit near T_c; see also Section 4.10.1, Eq. (4.148).

Eqs. (7.51)–(7.53a) can be represented by the Green function of the system $G_0(k)$. Since $G_0(k)$ has the simple form Eq. (7.26), we are able to calculate any physical quantity in the theory of free fluctuations. This is true also for the φ^4-model (7.1). But in this case we have to find the so-called full (or total) "Green function", which corresponds to the nonlinear differential equation of state of the system. To do this we must at first investigate the correlation functions of the free fluctuations.

7.4 Correlation Function of the Free Fields

The correlation functions $G_0^{(l)h}$ and $G_{0c}^{(l)h}$ of the free fluctuations in an external field h are obtained with the help of Eqs. (7.7) and (7.9), in which $\mathcal{Z}(h)$ is replaced with $\mathcal{Z}_0(h)$ given by Eqs. (7.38a) and (7.38b). We can also use the results from Section 7.3, in particular, Eqs. (7.42) and (7.43) for $\mathcal{Z}_0(h)$ and Eqs. (7.51)–(7.53a) for $\Phi_0(h)$.

It has been already shown by Eq. (7.54) that the lowest order correlation function $G_0^{(1)h} = G_{0c}^{(1)h}$ is equal to the stable solution $\bar{\varphi}_\alpha \sim h_\alpha$ of the equation of state, therefore $G_0^{(1)h} = G_{0c}^{(1)h} = 0$. The second variation of $\ln \mathcal{Z}_0(h) = [-\Phi_0(h)]$ is calculated from Eqs. (7.51)–(7.53a). The result is

$$G_{0c;\alpha\alpha'}^{(2)h}(k,k') = \delta_{\alpha\alpha'}\delta(k+k')G_0(k), \tag{7.60}$$

where $G_0(k)$ is the Green function. In order to compare $\Phi_0(h)$ from Eqs. (7.51)–(7.53a) with the general expression (5.85) we must substitute there $\widetilde{\Phi}(h)/T$ by $\Phi_0(h)$ and $\chi^{(l)}$ by $G_0^{(l)h}$ (the k-representation is assumed). The explicit form of the expansion (5.85) in the k-representation for the free fluctuations is very simple. It can be found by putting the relation $h_\alpha(k) = \bar{h}_\alpha(k) + \delta h_\alpha(k)$ in Eq. (7.52); the field $\bar{h}_\alpha$ corresponds to the stable state: $\bar{\varphi}_\alpha(k) \equiv \bar{\varphi}_\alpha(k/\bar{h})$. In this way the above results for $G_0^{(1)h}$ and $G_{0c}^{(2)h}$ are re-derived. The correlation functions for $l \geq 3$ obey the condition:

$$G_{0c}^{(l)h} = 0, \qquad l \geq 3. \tag{7.61}$$

This result for the free fluctuations is obvious because the potential (7.51) has a simple quadratic dependence on the external field h_α.

The function (7.60) can be obtained also by a direct differentiation of $\ln \mathcal{Z}_0(h)$; see Eq. (7.9) for $l = 2$ with $\mathcal{Z}_0(h)$ taken from Eq. (7.38b). Therefore, for the 2nd order correlation function we have:

$$\begin{aligned} G_{oc;\alpha\alpha'}^{(2)h}(k,k') &= \langle \varphi_\alpha(k)\varphi_{\alpha'}(k')\rangle - \langle \varphi_\alpha(k)\rangle_0 \langle \varphi_{\alpha'}(k')\rangle_0 \\ &= \langle [\varphi_\alpha(k) - \langle \varphi_\alpha(k)\rangle_0]\,[\varphi_{\alpha'}(k') - \langle \varphi_{\alpha'}(k')\rangle_0]\rangle \\ &= \langle \varphi'_\alpha(k)\varphi'_{\alpha'}(k')\rangle_0 \\ &= \langle\langle \varphi_\alpha(k)\varphi_{\alpha'}(k')\rangle\rangle_0, \end{aligned} \tag{7.62}$$

where the field $\varphi'_\alpha(k)$ is defined by Eqs. (7.40a) and (7.41), and $\langle\langle \cdots \rangle\rangle$ denotes the irreducible averages; cf. Eq. (5.83). The irreducible function $G_{0c}^{(2)h}$ of the field $\varphi_\alpha(k)$ is equal to the reducible correlation function of the field $\varphi'_\alpha(k) = \varphi_\alpha(k) - \bar{\varphi}_\alpha(k)$. The solution $\bar{\varphi}_\alpha = A_\alpha$ of the equation of state (7.54) is subtracted from the field $\varphi_\alpha(k)$ so $G_{0c}^{(2)h}$ describes the pure fluctuation contribution $\varphi'_\alpha(k) = \delta\varphi_\alpha(k)$. Imagine the phase diagram (φ, h) of the system, which describes the behaviour of the field φ in the presence of an external field h. This picture is equivalent to the phase diagram $(\varphi', 0)$, where the h-axis is shifted by the value of h. Therefore the field φ' represents a system in a zero external field.

The function (7.60) is equal to zero except when $\alpha = \alpha'$ and $k = -k'$. In this case it coincides with the Green function

$$G_0(k) = \langle|\varphi_\alpha(k)|^2\rangle_0. \tag{7.63}$$

We can write $G_0(k)$ in a different way using the transformation (7.14) for $\varphi_\alpha(k)$:

$$G_0(k) = \frac{1}{V}\int d^d x d^d y \, \langle\varphi_\alpha(x)\varphi_\alpha(y)\rangle_0 e^{-ik(x-y)}. \tag{7.64}$$

From this relation we obtain that

$$\sum_k G_0(k)e^{ikR} = \int d^d x \, \langle\varphi_\alpha(x)\varphi_\alpha(x+R)\rangle_0, \tag{7.65}$$

where $R = (x - y)$, i.e., according to our notations, here $R \equiv \boldsymbol{R} = (\boldsymbol{x} - \boldsymbol{y})$. The Fourier transformation of the average $\langle\varphi_\alpha(x)\varphi_\alpha(x+R)\rangle_0$ can be found with the help of Eqs. (7.12a) and (7.60). It is

$$\langle\varphi_\alpha(x)\varphi_\alpha(x+R)\rangle_0 = \frac{1}{V}\sum_k G_0(k)e^{ikR}.$$

Therefore, this average does not depend on x; it is equal to $G_0(R)$:

$$G_0(R) = \langle\varphi_\alpha(x)\varphi_\alpha(x+R)\rangle_0. \tag{7.66}$$

Substituting the integrand in Eq. (7.65) with $G_0(R)$ from Eq. (7.66) we have

$$G_0(R) = \frac{1}{V}\sum_k G_0(k)e^{ikR}, \tag{7.67}$$

which in the continuum limit (6.100) coincides with Eq. (7.31); $R = x$.

Up to now we have considered the irreducible correlation functions $G_{0c}^{(l)h}$ for a free field in an external field h. The reducible functions $G_o^{(l)h}$ are treated in a similar way. They can also be calculated exactly with the help of Eqs. (7.7) and (7.8) and the generating functional $\mathcal{Z}_0(h)$ in the form (7.38b) or (7.42). It is easy to see from Eqs. (7.7) and (7.38a)–((7.38b) that the Fourier amplitudes of the functions $G_{0;\alpha_i}^{(l)h}(x_i) \equiv G_0(x_1, \ldots, x_l/h)$ are expressed by the derivatives of $\mathcal{Z}_0(h)$ with respect to $h_{\alpha_i}(-k_i)$, so we can write

$$G^{(l)h}_{(0);\alpha_1\ldots\alpha_l}(k_1,\ldots,k_l/h) = \langle\varphi_{\alpha_1}(k_1)\cdots\varphi_{\alpha_l}(k_l)\rangle_{0h}$$
$$= \frac{1}{\mathcal{Z}_0(h)}\frac{\delta^{(l)}\mathcal{Z}_0(h)}{\delta h_{\alpha_1}(-k_1)\cdots\delta h_{\alpha_l}(-k_l)}. \tag{7.68}$$

The reducible functions $G^{(l)}_0$ in a zero field ($h = 0$) have a more restricted meaning, but they are the most important correlation functions for the description of the second order phase transitions, where the transition point is at $h = 0$. These functions are given by

$$G^{(l)}_{(0);\alpha_1\ldots\alpha_l}(k_1,\ldots,k_l) = \langle\varphi_{\alpha_1}(k_1)\cdots\varphi_{\alpha_l}(k_l)\rangle_0$$
$$= \frac{1}{\mathcal{Z}_0(0)}\left[\frac{\delta^{(l)}\mathcal{Z}_0(h)}{\delta h_{\alpha_1}(-k_1)\cdots\delta h_{\alpha_l}(-k_l)}\right]_{h=0}. \tag{7.69}$$

The properties of the functions $G^{(1)}_0$ and $G^{(2)}_0$ have been discussed above. For $h = 0$, $G^{(2)}_0$ and $G^{(2)}_{0c}$ coincide; see Eqs. (7.60) and (7.63). So we shall focus our attention on $G^{(l)}_0$ for $l \geq 3$. They are represented by sums of the products consisting of the Green functions $G^{(2)} = G_0(k)$ (Wick's theorem); see, e.g., Amit (1978) for the classical fluctuation fields, and Fetter and Walecka (1971) for the averages of quantum operators. In the remaining part of this section we shall prove the Wick's rule, which is very important for the perturbation expansions. The reader can easily retrace the prototype of the Wick's rule within the perturbation theory of macroscopic fluctuations (Section 6.3). This rule can be deduced, at least to some order $l = 6$ or 8, by a direct calculation of the averages (7.69) with the help of the formulae (7.6) and (7.38a)–(7.38b); the reader may perform this calculation as an useful exercise. $G^{(l)}_0$ for $l \geq 3$ are found by making an expansion of the exponent in Eq. (7.42).

$$\frac{\mathcal{Z}(h)}{\mathcal{Z}(0)}$$
$$= 1 + \sum_{m=1}^{\infty}\frac{1}{m!\,2^m}\sum_{\substack{k_1\ldots k_m\\ \alpha_1\ldots\alpha_m}} G_0(k_1)\cdots G_0(k_m)|h_{\alpha_1}(k_1)|^2\cdots|h_{\alpha_m}(k_m)|^2, \tag{7.70}$$

and comparing it with the series

$$\frac{\mathcal{Z}(h)}{\mathcal{Z}(0)} = 1 + \sum_{l=1}^{\infty} \frac{1}{l!} \sum_{\substack{k_1 \dots k_l \\ \alpha_1 \dots \alpha_l}} G^{(l)}_{(0);\alpha_1 \dots \alpha_l}(k_1, \dots, k_l) h_{\alpha_1}(-k_1) \cdots h_{\alpha_l}(-k_l), \tag{7.71}$$

which follows from the definition (7.69) of $G_0^{(l)}$ and the rule for the expansion of the functionals. The trivial result

$$G^{(2l+1)} = 0, \tag{7.72}$$

is obvious from the properties of the Gaussian integrals for the averages (7.69); see Eq. (B.2). The next task is to obtain the expression for $G_0^{(2l)}$. The comparison between the term ($l = 2$) in Eq. (7.71) and ($m = 1$) in Eq. (7.70) yields:

$$G^{(2)}_{0\alpha_1\alpha_2}(k_1, k_2) = \delta_{\alpha_1\alpha_2}\delta(k_1 + k_2) G_0(k_1), \tag{7.73}$$

cf. Eq. (7.60).
In order to continue this procedure for the higher order terms in Eq. (7.70) and Eq. (7.71), we must perform a suitable rearrangement of the series (7.70). For example the term for $m = 2$ can be written in the form

$$\begin{aligned} &\frac{1}{4!} \sum_{\substack{k_1 \dots k_4 \\ \alpha_1 \dots \alpha_4}} [h_{\alpha_1}(-k_1) h_{\alpha_2}(-k_2) h_{\alpha_3}(-k_3) h_{\alpha_4}(-k_4)] \\ &\times \left\{ \frac{4!}{2! 2 \cdot 3} [\delta_{\alpha_1\alpha_2}\delta_{\alpha_3\alpha_4}\delta(k_1 + k_2)\delta(k_3 + k_4) G_0(k_1) G_0(k_3) \right. \\ &+ \delta_{\alpha_1\alpha_3}\delta_{\alpha_2\alpha_4}\delta(k_1 + k_3)\delta(k_2 + k_4) G_0(k_1) G_0(k_2) \\ &\left. + \delta_{\alpha_1\alpha_4}\delta_{\alpha_2\alpha_3}\delta(k_1 + k_4)\delta(k_2 + k_3) G_0(k_1) G_0(k_2)] \right\}. \end{aligned} \tag{7.74}$$

The numerical factor before the curly brackets is equal to unity because the sum of the three products of the G_0-functions is compensated by the factor (1/3). Comparing this term with the ($l = 4$)-term in Eq. (7.71) we obtain that $G_0^{(4)}$ is exactly equal to the sum in the curly brackets of Eq. (7.74). Taking into account Eq. (7.73) we can write

$$\begin{aligned} G^{(4)}_{\alpha_1 \dots \alpha_4}(k_1, \dots, k_4) &= G^{(2)}_{(0)\alpha_1\alpha_2}(k_1, k_2) G^{(2)}_{(0)\alpha_3\alpha_4}(k_3, k_4) \\ &+ G^{(2)}_{(0)\alpha_1\alpha_3}(k_1, k_3) G^{(2)}_{(0)\alpha_2\alpha_4}(k_2, k_4) \\ &+ G^{(2)}_{(0)\alpha_1\alpha_4}(k_1, k_4) G^{(2)}_{(0)\alpha_2\alpha_3}(k_2, k_3). \end{aligned} \tag{7.75}$$

Now the same consideration can be applied to the terms of the order h^6, h^8 etc. in Eq. (7.70) and Eq. (7.71) and this is a pure combinatorial problem.

Alternatively, we can use the expression (7.68) for the function $G_0^{(l)h}$ and the relation

$$\frac{\delta \mathcal{Z}_0(h)}{\delta h_\alpha(-k)} = \frac{h_\alpha(k)}{r_0 + ck^2}\mathcal{Z}_0(h), \tag{7.76}$$

which ensues from Eqs. (7.26) and (7.42). For example, the four-point function $G_0^{(4)h}$ is transformed to a sum of products of $G_0^{(2)}$ functions, represented by the following set of equalities

$$\begin{aligned}
G_0^{(4)h} &= \langle \varphi_{\alpha_1}(k_1)\cdots\varphi_{\alpha_4}(k_4)\rangle_{0h} \\
&= \mathcal{Z}_0^{-1}(h)\frac{\delta^3}{\delta h_{\alpha_1}(-k_1)\delta h_{\alpha_2}(-k_2)\delta h_{\alpha_3}(-k_3)}\left[\frac{h_{\alpha_4}(k_4)}{r_0 + ck_4^2}\mathcal{Z}_0(h)\right] \\
&= \mathcal{Z}_0^{-1}(h)\frac{\delta^2}{\delta h_{\alpha_1}(-k_1)\delta h_{\alpha_2}(-k_2)} \\
&\times \left\{\mathcal{Z}_0(h)\left[\frac{\delta_{\alpha_3\alpha_4}\delta(k_3+k_4)}{(r_0+ck_3^2)} + \frac{h_{\alpha_3}(k_3)h_{\alpha_4}(k_4)}{(r_0+ck_3^2)(r_0+ck_4^2)}\right]\right\} \\
&= \mathcal{Z}_0^{-1}(h)\frac{\delta}{\delta h_{\alpha_1}(-k_1)}\{\cdots\} \\
&= \frac{\delta_{\alpha_1\alpha_2}\delta_{\alpha_3\alpha_4}\delta(k_1+k_2)\delta(k_3+k_4)}{(r_0+ck_1^2)(r_0+ck_3^2)} \\
&+ \frac{\delta_{\alpha_1\alpha_3}\delta_{\alpha_2\alpha_4}\delta(k_1+k_3)\delta(k_2+k_4)}{(r_0+ck_1^2)(r_0+ck_2^2)} \\
&+ \frac{\delta_{\alpha_1\alpha_4}\delta_{\alpha_2\alpha_3}\delta(k_1+k_4)\delta(k_2+k_3)}{(r_0+ck_1^2)(r_0+ck_2^2)} + O(h_\alpha^2).
\end{aligned} \tag{7.77}$$

Setting $h = 0$ in the above expressions, we can re-derive Eq. (7.75). The main result is that the average of the product of four fields is a sum of all possible products of the averages of the field couples $(\varphi_{\alpha_i}\varphi_{\alpha_j})$:

$$\begin{aligned}
\langle \varphi_{\alpha_1}(k_1)\cdots\varphi_{\alpha_4}(k_4)\rangle_0 &= \langle \varphi_{\alpha_1}(k_1)\varphi_{\alpha_2}(k_2)\rangle_0\langle \varphi_{\alpha_3}(k_3)\varphi_{\alpha_4}(k_4)\rangle_0 \\
&+ \langle \varphi_{\alpha_1}(k_1)\varphi_{\alpha_3}(k_3)\rangle_0\langle \varphi_{\alpha_2}(k_2)\varphi_{\alpha_4}(k_4)\rangle_0 \\
&+ \langle \varphi_{\alpha_1}(k_1)\varphi_{\alpha_4}(k_4)\rangle_0\langle \varphi_{\alpha_2}(k_2)\varphi_{\alpha_3}(k_3)\rangle_0.
\end{aligned} \tag{7.78}$$

The number of the terms in this sum is given by the following rule. We fix one of the fields on the l.h.s. of Eq. (7.78), say, $\varphi_{\alpha_1}(k_1)$ and count the possible ways, in which this field can be linked in a couple with one of the other three fields: $\varphi_{\alpha_2}, \varphi_{\alpha_3}$ and φ_{α_4}. There are three possible ways to do this:

$$\begin{aligned}
&1.\quad [\varphi_{\alpha_1}(k_1)\varphi_{\alpha_2}(k_2)]\varphi_{\alpha_3}(k_3)\varphi_{\alpha_4}(k_4),\\
&2.\quad [\varphi_{\alpha_1}(k_1)\varphi_{\alpha_3}(k_3)]\varphi_{\alpha_2}(k_2)\varphi_{\alpha_4}(k_4),\\
&3.\quad [\varphi_{\alpha_1}(k_1)\varphi_{\alpha_4}(k_4)]\varphi_{\alpha_2}(k_2)\varphi_{\alpha_3}(k_3).
\end{aligned} \tag{7.79}$$

For each of these possibilities, there is only one way to form the second couple. For example, the variant 1 in Eq. (7.79) yields one product of couples $(\varphi_{\alpha_1}\varphi_{\alpha_2}) \times (\varphi_{\alpha_3}\varphi_{\alpha_4})$, where the brackets $(\cdots)$ show the way of averaging, $\langle\varphi_{\alpha_1}\varphi_{\alpha_2}\rangle_0 \times \langle\varphi_{\alpha_3}\varphi_{\alpha_4}\rangle_0$; so we obtain the first terms on the r.h.s. of Eqs. (7.78), (7.77) and (7.75). Using the variant 2 we obtain the second term in Eq. (7.78). All different ways of making couples must be taken into account. The number 3 appears when some of the indices α_j or the momenta k_j coincide. For example,

$$\langle|\varphi_\alpha(k)|^4\rangle = 3\langle|\varphi_\alpha(k)|^2\rangle_0, \tag{7.80a}$$

and

$$\begin{aligned}
\langle|\varphi_\alpha(k)|^2\varphi_{\alpha_1}(k_1)\varphi_{\alpha_2}(k_2)\rangle &= 2\langle\varphi_\alpha(k)\varphi_{\alpha_1}(k_1)\rangle_0\langle\varphi_\alpha(k)\varphi_{\alpha_2}(k_2)\rangle\\
&\quad + \langle|\varphi_\alpha(k)|^2\rangle\langle\varphi_{\alpha_1}(k_1)\varphi_{\alpha_2}(k_2)\rangle.
\end{aligned} \tag{7.80b}$$

Obviously, this rule which is very useful in the perturbation theory can be applied to any product of fields

$$\langle\varphi_{\alpha_1}(k_1)\cdots\varphi_{\alpha_{2l}}(k_{2l})\rangle_0. \tag{7.81}$$

For $l = 3$, the sum for $G_0^{(6)}$ consists of products like

$$\langle\varphi_{\alpha_1}(k_1)\varphi_{\alpha_2}(k_2)\rangle_0\langle\varphi_{\alpha_3}(k_3)\varphi_{\alpha_4}(k_4)\rangle_0\langle\varphi_{\alpha_5}(k_5)\varphi_{\alpha_6}(k_6)\rangle_0. \tag{7.82}$$

The number of products in the sum will be $5 \cdot 3 \cdot 1 \equiv 5!!$. In general, the average (7.81) will be a sum of products of m averaged couples in number $(2m-1)(2m-3)\cdots 3 \cdot 1 \equiv (2m-1)!!$. The average $\langle A\rangle$ of any functional $A(\varphi)$ of free fields, which is represented by a monomial (or polynomial) of φ can be calculated with the help of the above rules.

We have expressed all correlation functions $G_0^{(l)}$ by the Green function $G_0(k)$ or, equivalently, by the two-point correlation function (7.73) in a k-representation. In order to obtain the correlation functions in a x-representation we have to do the corresponding Fourier transformations. It will be quite instructive to receive the x-representation of the exponent in Eq. (7.42). From Eqs. (7.14) and (7.64), we have

$$\mathcal{Z}_0(h) = \mathcal{Z}_0(0) \exp\left[\frac{1}{2}\sum_{\alpha\beta}\int d^dx d^dy\, G^{(2)}_{(0);\alpha\beta}(x-y) h_\alpha(x) h_\beta(y)\right], \quad (7.83)$$

where

$$G^{(2)}_{(0)}(x,y) = \frac{1}{V}\sum_{k_1,k_2} G^{(2)}_{(0)\alpha\beta}(k_1,k_2)\, e^{-ik_1x-ik_2y}, \quad (7.84)$$

calculated with the help of Eq. (7.73) to becomes

$$G^{(2)}_{(0);\alpha\beta}(x-y) = \delta_{\alpha\beta} G_0(x-y). \quad (7.85)$$

Now it is very easy to show that the averages like

$$\langle \varphi_{\alpha_1}(x_1)\cdots\varphi_{\alpha_l}(x_l)\rangle_0 \quad (7.86)$$

are represented as sums of products of $G^{(2)}_{(0);\alpha\beta}$ functions in the x-representation, i.e., as products of pair averages of the type

$$G^{(2)}_{(0);\alpha\beta}(x_1-x_2) = \langle \varphi_{\alpha_1}(x_1)\varphi_{\alpha_2}(x_2)\rangle_0. \quad (7.87)$$

In fact the differentiation in Eq. (7.8) for $\mathcal{Z}_0(h)$ in the form (7.37a)–(7.37b) yields the functions $G^{(l)}_0$ in the x-representation, i.e., the averages (7.86). The same differentiation with the help of $\mathcal{Z}_0(h)$ in the form (7.83) makes possible to confirm the above statement (*Wick's theorem* in a x-representation). For example, the average

$$\begin{aligned}\langle \varphi_{\alpha_1}(x_1)\cdots\varphi_{\alpha_4}(x_4)\rangle_0 = {} & G^{(2)}_{(0)\alpha_1\alpha_2}(x_1-x_2) G^{(2)}_{(0)\alpha_3\alpha_4}(x_3-x_4) \\ & + G^{(2)}_{(0)\alpha_1\alpha_3}(x_1-x_3) G^{(2)}_{(0)\alpha_2\alpha_4}(x_2-x_4) \\ & + G^{(2)}_{(0)\alpha_1\alpha_4}(x_1-x_4) G^{(2)}_{(0)\alpha_2\alpha_3}(x_2-x_3), \end{aligned} \quad (7.88)$$

can be further represented by $G_0(x-y)$ from Eq. (7.85). The same result can be obtained by a direct Fourier transformation of Eq. (7.75).

We have demonstrated that the Wick theorem is valid for the φ^4-theory. This very important theorem is valid also for the most relevant classical and quantum theories (Hamiltonians, actions) in Physics, and this makes possible the investigation of high-order perturbation expansions. A rigorous proof of the Wick theorem is presented, for example, by Fetter and Walecka (1971).

7.5 Gaussian Approximation

Here we shall consider the φ^4-model (7.1) in the Gaussian approximation for the fluctuations. This is the most simple treatment of fluctuations within nontrivial models that describe phase transitions. The most important property of the Gaussian approximation is that it can be easily applied to any complex Hamiltonian of phase transitions. So it is possible to receive:

(i) a fairly good information about the fluctuation properties far from the critical point, and

(ii) a notion about the size of the fluctuation effects in a close vicinity of the critical points.

Making use of the equation $\delta\mathcal{H} = 0$ and treating it as explained in Section 7.2.1, we obtain that the equilibrium order parameter $\bar{\varphi}_\alpha(x)$ satisfies the following equation of state

$$(r_0 + 4u\bar{\varphi}^2 - c\nabla^2)\bar{\varphi}_\alpha(x) = h_\alpha(x), \tag{7.89}$$

cf. Eq. (7.19) for the free field. Now we substitute the total order parameter $\varphi_\alpha = \bar{\varphi}_\alpha + \delta\varphi_\alpha$ in Eq. (7.1) and we take into account the equation of state (7.89). Neglecting the surface integral, similar to that given by Eq. (7.17), we obtain

$$\mathcal{H}(\varphi) = \mathcal{H}(\bar{\varphi}) + \mathcal{H}_G(\bar{\varphi}, \delta\varphi) + \mathcal{H}_{\text{int}}(\bar{\varphi}, \delta\varphi), \tag{7.90}$$

where $\mathcal{H}(\bar{\varphi})$ is the stationary (or mean field) Hamiltonian,

$$\mathcal{H}(\bar{\varphi}) = \frac{1}{2}\int d^dx \left[r_0\bar{\varphi}^2 + c(\nabla\bar{\varphi})^2 + 2u\bar{\varphi}^4 - 2h\bar{\varphi}\right]. \tag{7.91}$$

$\mathcal{H}_G$ is the Hamiltonian part in the Gaussian (quadratic) approximation for the fluctuations,

$$\mathcal{H}_G = \frac{1}{2}\int d^dx \left[r_0\delta\varphi^2 + c(\nabla\varphi)^2 + 8u(\bar{\varphi}\delta\varphi)^2 + 4u\bar{\varphi}^2\delta\varphi^2\right], \tag{7.92}$$

and $\mathcal{H}_{\text{int}}$ represents the fluctuation interactions,

$$\mathcal{H}_{\text{int}} = \int d^d x \left[u(\delta\varphi^2)^2 + 4u(\delta\varphi)^2 \bar{\varphi} \cdot \delta\varphi \right]. \tag{7.93}$$

The term $\mathcal{H}(\bar{\varphi})$ gives the stationary value of the generalized thermodynamic potential ($\mathcal{H} \equiv \Phi$). Without a loss of generality we can choose the field $h = \{h_\alpha\}$ to be along the first axis in the space of the order parameter components $\varphi_\alpha : h = (h_1, 0, \dots, 0)$. Then all $\bar{\varphi}_\alpha$ except $\bar{\varphi}_1$ will be equal to zero (see Section 4.10.2. Since h and φ point in one and the same direction at each point x, the fluctuation field $\delta\varphi(x)$ can be represented by the sum

$$\delta\varphi = \delta\varphi_\parallel + \delta\varphi_\perp, \tag{7.94}$$

where $\delta\varphi_\parallel = (\delta\varphi_1, 0, \dots)$ is the component parallel to h and $\bar{\varphi}$ and $\delta\varphi_\perp = (0, \delta\varphi_2, \dots)$ is the transverse component.

We shall assume that $\bar{\varphi}_\alpha(x)$ are known from the solution of Eq. (7.89) and we shall investigate the fluctuation contribution in the Gaussian approximation:

$$\mathcal{H} \approx \mathcal{H}(\bar{\varphi}) + \mathcal{H}_G(\bar{\varphi}, \delta\varphi). \tag{7.95}$$

Written by $\delta\varphi_\parallel$ and $\delta\varphi_\perp$, Eq. (7.92) takes the form

$$\mathcal{H}_G = \frac{1}{2} \int d^d x \left[r_\parallel \delta\varphi_\parallel^2 + r_\perp \delta\varphi_\perp^2 + c(\nabla\varphi_\parallel)^2 + c(\nabla\varphi_\perp)^2 \right]. \tag{7.96}$$

Here

$$r_\parallel = r_0 + 12u\bar{\varphi}^2, \qquad r_\perp = r_0 + 4u\bar{\varphi}^2. \tag{7.97}$$

The quantities $r_\parallel$ and $r_\perp$ are connected with the longitudinal ($\chi_\parallel$) and the transverse ($\chi_\perp$) susceptibilities of the system (Section 4.10.2).

The calculation of the fluctuation contribution of $\mathcal{H}_G$ to the thermodynamic quantities is not easy to be made because $r_\parallel$ and $r_\perp$ depend on x : $\bar{\varphi}^2 = \bar{\varphi}_1^2$ is x-dependent. We shall consider the simple case, when $h(x) = 0$. The stable phase $\bar{\varphi}$ is then x-independent. It is obtained from Eq. (4.61), where $u_2 = r_0/2$; see also Eq. (7.89) with $(\nabla^2\varphi_\alpha) = 0$ and $h_\alpha \equiv 0$. Since $\bar{\varphi}^2 = (-r_0/4u)$, the susceptibilities of the stationary phase $\bar{\varphi} \neq 0$ will be in the form:

$$\chi_{\parallel}^{-1} = r_{\parallel} = 2|r_0|, \tag{7.98a}$$

and

$$\chi_{\perp}^{-1} = r_{\perp} = 0, \qquad r_0 < 0, \tag{7.98b}$$

(for $r_0 > 0$, $\bar{\varphi} = 0$ and $\chi_{\parallel}^{-1} = \chi_{\perp}^{-1} = r_0$). This result is remarkable because it shows that the transverse fluctuation modes $\delta\varphi_{\perp}$, which always exist for $n > 1$, correspond to an infinite susceptibility $\chi_{\perp} = \infty$ for any $T \leq T_c$, i.e., in the ordered phase $\bar{\varphi} \neq 0$. The Fourier transformation of Eq. (7.96) yields

$$\mathcal{H}_G = \frac{1}{2} \sum_{\alpha,k} \left[r_1 \delta_{\alpha,1} + ck^2\right] |\delta\varphi_\alpha(k)|^2, \tag{7.99}$$

where we have used the fact that $\delta\varphi_{\parallel} = (\delta\varphi_1, 0, \dots)$, $\delta\varphi_{\perp} = (0, \delta\varphi_2, \dots)$; $r_1 \equiv r_{\parallel}$.

7.5.1 *Goldstone modes*

The correlation function of the fluctuation modes $\varphi_\alpha(k)$, $1 < \alpha < n$ is

$$G_{\alpha\alpha'} = \delta_{\alpha\alpha'} \delta(k + k') \frac{1}{ck^2}. \tag{7.100}$$

It tends to infinity for $k \to 0$, i.e., the self energy $G_0^{-1}(k) = \varepsilon_0(k) = ck^2$ of these modes is equal to zero for $k = 0$. Therefore the modes $\varphi_\alpha(0)$ always appear in the spontaneously broken states. In the phase $\bar{\varphi} = \bar{\varphi}_1$, in which the symmetry is broken along the axis 1 in the space of the n-component order parameter, the fluctuation correlations transverse to the axis 1 have an infinite range $\xi_{\perp} = (c/r_{\perp})^{1/2} = \infty$. In the quantum field theory the quantity which corresponds to the inverse susceptibility $\chi^{-1} \sim r$ is the square of the particle mass (m^2). In the language of quantum field theory the above result means that the spontaneous breaking of the symmetry is always followed by the appearance of $(n - 1)$ massless fields or, *Goldstone bosons* (in our case, classic Goldstone modes); see Goldstone (1961), Lange (1965, 1966), Wagner (1966), and Brout (1976). We have demonstrated the existence of the Goldstone modes within the simple Gaussian approximation below T_c but they come out in the theory of interacting fluctuations too; see, e.g., Amit (1978). These massless fluctuations occur in systems where the effective Hamiltonian possesses a continuous global symmetry in the space of the order parameter components φ_α; for the relationship between the breaking of the continuous symmetry and the appearance of Goldstone modes; see also Brout (1976), Parisi (1988). In contrast to the Goldstone fluctuations ($\delta\varphi_\perp$) the massless critical fluctuations appear when $r_0 \to 0$ (i.e., $T \to T_c$).

7.5.2 *Ginzburg–Levanyuk criterion and critical region*

The Gaussian approximation (7.95) gives the results of the free field theory for $T > T_c$, where $\bar{\varphi} = 0$ and it makes possible to evaluate the fluctuation corrections to the ordered state $\bar{\varphi} \neq 0$ below T_c. We shall again discuss the simple phase $\bar{\varphi} = -r_0/4u$ corresponding to a homogeneous system in a zero external field.

For both cases $T \geq T_c$ and $T \leq T_c$, the thermodynamic potential $\Phi_0(h = 0)$ will be

$$\Phi_G = -V\frac{r_0^2}{16u} - \ln \int D\delta\varphi e^{-\mathcal{H}_G}; \tag{7.101}$$

the second term represents the fluctuation part Φ_{Gf}. Obviously, Φ_{Gf} coincides with Φ_{0f}, see Eq. (7.53) for $\bar{\varphi}(T \geq T_c) = 0$ (see Eq. (7.92) for $\bar{\varphi} = 0$). The fluctuation contribution Φ_{Gf} for $T \geq T_c$ will be negligible, if

$$V\frac{r_0^2}{16u} \gg \Phi_{Gf}. \tag{7.102}$$

In the opposite case

$$\Phi_{Gf} \gtrsim V\frac{r_0^2}{16u}, \tag{7.103}$$

the thermodynamics of the system will be essentially affected by the fluctuations. Obviously, this will happen for sufficiently small $r_0 \sim t$, i.e., near the critical point T_c where the usual Landau expansion is no longer valid. This criterion for the validity of the usual Landau theory near the second order phase transition points was first introduced by Levanyuk (1959) and Ginzburg (1960) — (*Levanyuk-Ginzburg criterion*, or, as it is most often referred to — *the Ginzburg criterion*). This criterion is very general and it can be used to evaluate the fluctuation effects and the size of *the critical region* of rather complex models describing the phase transitions.

The size of the critical region, that is, the domain around T_c where the mean field solution $\bar{\varphi}$ does not determine correctly the phase transition can be obtained from the inequality (7.103) or from the corresponding inequality for any of the thermodynamic quantities (derivatives of Φ). In Section 7.11 the Ginzburg criterion will be obtained as a criterion for the validity of the perturbation expansion. Actually, Levanyuk and Ginzburg have compared the fluctuation contribution C_{Gf} to the specific heat in the Gaussian approximation with the specific heat jump ΔC at T_c predicted by the mean field theories; see Eq. (4.74) or, in our notations $\Delta C = (V\alpha_0^2/16uT_c)$. From

$$\frac{C_f(t>0)}{\Delta C} \gtrsim 1 \tag{7.104}$$

which corresponds to the inequality (7.103) and the result given by Eqs. (7.58)–(7.59), we get for the specific heat $C_f(t>0)$ above the transition point T_c the following expression:

$$t^{2-d/2} \lesssim 4n\Gamma\left(2-\frac{d}{2}\right)\left(\frac{\alpha_0}{4\pi c}\right)^{d/2}\left(\frac{uT_c}{\alpha_0^2}\right). \tag{7.105}$$

Since the fluctuation integral (7.57) yields Eq. (7.58) for $d < 4$, the above result is applicable for these dimensions. The equality will be fulfilled at some temperature $T_G > T_c$, which defines the size $(T_G - T_c)$ of the critical region above T_c. The numerical factor in Eq. (7.105) is not essential because the above criterion has a qualitative sense. The dimensionless quantity (*Ginzburg number*)

$$G_i = \left(\frac{u^2T_c^2}{c^d\alpha_0^{4-d}}\right), \tag{7.106}$$

is often introduced in the discussion of the size of the critical region. Then the critical interval above T_c will be

$$\Delta T_G = (T_G - T_c) \sim G_i^{1/(4-d)}T_c. \tag{7.107}$$

For $T < T_c$, the integral in Eq. (7.101) should be calculated separately for the fluctuations $\delta\varphi_1$ and the Goldstone modes $\delta\varphi_\alpha, \alpha > 1$; see the calculation of the integral (7.43). Then the $(n-1)$ Goldstone modes will give a contribution to Φ_G of the type

$$-\frac{(n-1)}{2}V\int\frac{d^dk}{(2\pi)^d}\ln\left(\pi ck^2\right). \tag{7.108}$$

This is a temperature independent term that does not affect the thermodynamics of the system (a characteristic feature of these modes connected with their zero energy at $k = 0$). Therefore we have to take into account only the contribution from the longitudinal fluctuations $\delta\varphi_1$. The essential part of the fluctuation integral in Eq. (7.101) will be

$$-\frac{V}{2}\int \frac{d^dk}{(2\pi)^d} \ln\left(\frac{\pi}{2|r_0| + ck^2}\right). \tag{7.109}$$

In this case the criterion (7.105) below T_c will be slightly different because the constant C_0 changes to $C_0' = 2^{d/2}C_0/n$, i.e., the constant C_0 for $n = 1$ and $\alpha_0 \to 2\alpha_0$. The ratio

$$\frac{C_0}{C_0'} = \frac{n}{2^{d/2}}, \tag{7.110}$$

between the t-dependent amplitudes of the specific heat above and below T_c is often used in discussions of experiments. If we know this ratio with a high precision, the number of the order parameter components (symmetry index of the ordering) can be determined.

Having in mind the condition (4.183), the mean field expansions in power series of the order parameter as well as the Landau phenomenological expansion will be valid in the temperature interval

$$t_G = \frac{\Delta T_G}{T_c} \lesssim t \lesssim 1. \tag{7.111}$$

In most of the real systems, $t_G < 1$, and therefore such an interval exists.

The size ΔT_G or $t_G = G_i^{1/(4-d)}$ of the critical region is very sensible to the dimension d. For $d < 4$ the critical region always exists and its size increases with the decrease of d (for $d = 3$, $t_G = G_i$). The border dimensionality is $d = 4$. From the logarithmic dependence of the integral (7.57) for $d = 4$, see Eq. (E.3), we obtain the size of the critical region:

$$t_G \sim \frac{c\Lambda^2}{\alpha_0} \exp\left[-\frac{Const}{uT_c}\right]. \tag{7.112}$$

The positive *const* in the above expression is easily found during the process of calculation. This critical region is extremely small because of the exponential dependence on (uT_c). For $d > 4$, see Eq. (E.4), the value of C_f tends to a finite constant for $t \to 0$ and this constant is much smaller than the MF jump ΔC. Thus, the critical region does not exist.

This is what the Gaussian approximation can tell us about the critical phenomena. To examine the effect of the fluctuation interactions we should go beyond the Gaussian approximation by applying the perturbation theory.

7.6 Perturbation Expansion

Let us consider a fluctuation Hamiltonian $\mathcal{H}$, the statistical treatment of which cannot be performed exactly. This can be done at most for its non-interacting part $\mathcal{H}_0$. If all the averages

$$\langle A \rangle_0 = \mathcal{Z}_0^{-1} \int D\varphi \, A \, e^{-\mathcal{H}_0}, \tag{7.113}$$

can be calculated exactly, in particular, the correlation functions $\langle A \rangle \rightarrow G_0^{(l)}$, we can consider the remainder $\mathcal{H}_{\text{int}} = (\mathcal{H} - \mathcal{H}_0)$ as a perturbation. The perturbation theory in this case is based on the same principle as that for the macroscopic fluctuations in Section 6.2.1. In almost all the cases, the part of $\mathcal{H}$ which brings about the calculation difficulties and describes the interaction(s); it is often called the interaction part, or interaction Hamiltonian — $\mathcal{H}_{\text{int}}$. In this section we shall discuss a general Hamiltonian where $\mathcal{H}_0$ and $\mathcal{H}_{\text{int}}$ are not specified. The "free" part, $\mathcal{H}_0$, satisfies the requirement to give the correlation functions $G_0^{(l)}$ exactly in terms of the products of two-point functions $G_0^{(2)}$ as shown in Section 7.4. No doubt that $\mathcal{H}_0$ consists only of φ^2-terms. The interaction part $\mathcal{H}_{\text{int}}$ may be formed from one or more interaction terms like the φ^4-interaction in the model (7.1). At some stages of our discussion we shall illustrate the results by using the example of the φ^4-Hamiltonian (7.1). For simplicity we shall consider it for $h = 0$ and $r_0 > 0$.

7.6.1 *Perturbation expansion for the partition function and the thermodynamic potential*

Let $A(\varphi)$ in Eq. (7.113) be equal to $\exp(-\mathcal{H}_{\text{int}})$. Then the integral in Eq. (7.113) is equal to the partition function $\mathcal{Z}$ of the system described by $\mathcal{H}$, i.e.,

$$\mathcal{Z} = \mathcal{Z}_0 \langle e^{-\mathcal{H}_{\text{int}}} \rangle, \tag{7.114}$$

where, as below, the suffix (0) of the averages $\langle \cdots \rangle_0$ is omitted. The expansion of the exponent in Eq. (7.114) yields

$$\mathcal{Z} = \mathcal{Z}_0(1 + L_1). \tag{7.115}$$

Here

$$\begin{aligned} L_1 &= \sum_{l=1}^{\infty} \frac{(-1)^l}{l!} \langle \mathcal{H}_{\rm int}^l \rangle \\ &= -\langle \mathcal{H}_{\rm int} \rangle + \frac{1}{2!} \langle \mathcal{H}_{\rm int}^2 \rangle - \frac{1}{3!} \langle \mathcal{H}_{\rm int}^3 \rangle + \dots . \end{aligned} \tag{7.116}$$

In order to take into account the interaction corrections to the partition function $\mathcal{Z}_0$ of the corresponding noninteracting system ($\mathcal{H}_0$) we have to find out the terms of the infinite series L_1. Usually it is possible to calculate the first few terms; the first order perturbation correction — $\langle \mathcal{H}_{\rm int} \rangle$, the second order correction $\langle \mathcal{H}_{\rm int}^2 \rangle$, etc. $\mathcal{H}_{\rm int}$ is built up by fields $\varphi_\alpha(x)$ subject to mathematical operations like integration and summation as, for example, the φ^4-term in Eq. (7.1) is constructed. So the perturbation terms in Eq. (7.116) could be represented by averages of the type (7.81) or (7.86), which will be later decomposed in sums of products of pair averages $\langle \varphi_{\alpha_i} \varphi_{\alpha_j} \rangle$. The averages within the free theory ($\mathcal{H}_0$) can be always obtained which means that we should not meet principal difficulties in the calculation of the perturbation terms in Eq. (7.116). For example, the term $\langle \mathcal{H}_{\rm int} \rangle$ for the φ^4-theory (7.1) will be

$$\begin{aligned} \langle \mathcal{H}_{\rm int} \rangle &= u_0 \sum_{\alpha\beta} \int d^d x \, \langle \varphi_\alpha^2(x) \varphi_\beta^2(x) \rangle \\ &= u_0 \sum_{\alpha\beta} \int d^d x \, \{ \langle \varphi_\alpha^2(x) \rangle \langle \varphi_\beta^2(x) \rangle + 2 \langle \varphi_\alpha(x) \varphi_\beta(x) \rangle^2 \} \\ &= u_0 \sum_{\alpha\beta} \int d^d x \, \{ G_{\alpha\alpha}(0) G_{\beta\beta}(0) + 2 G_{\alpha\beta}^2(0) \}; \end{aligned} \tag{7.117}$$

here the suffices (2) and (0) of $G^{(2)}_{(0);\alpha\beta}$ have been omitted, $G^{(2)}_{(0);\alpha\beta} \equiv G_{\alpha\beta}$. In Eq. (7.117), $G_{\alpha\beta}(0) \equiv G_{\alpha\beta}(x, x) = G_{\alpha\beta}(x - x)$; see Eq. (7.85).

The terms higher than the first in the series (7.116) can be classified, and this is of principal importance for further discussion. Let us consider the second order term. The average $\langle \mathcal{H}_{\rm int}^2 \rangle$ is represented by two series of terms:

$$\langle \mathcal{H}_{\text{int}}(\varphi)\mathcal{H}_{\text{int}}(\varphi')\rangle = \langle \mathcal{H}_{\text{int}}(\varphi)\rangle\langle \mathcal{H}_{\text{int}}(\varphi')\rangle + \langle \mathcal{H}_{\text{int}}(\overbrace{\varphi'_\alpha \dots \varphi'_\gamma)\mathcal{H}_{\text{int}}(\varphi_\alpha}\dots \varphi_\mu)\rangle_c, \tag{7.118}$$

where we have denoted φ_α in one of the Hamiltonian parts $\mathcal{H}_{\text{int}}$ by φ'_α in order to distinguish between the fields describing different interaction Hamiltonians. The first term on the r.h.s. of Eq. (7.118) is the square $\langle \mathcal{H}_{\text{int}}\rangle^2$ of the first order perturbation term in Eq. (7.116). In this part of $\langle \mathcal{H}^2_{\text{int}}\rangle$ there are no mixed pair averages like $\langle \varphi_\alpha \varphi'_\beta\rangle$ — all pair averages are formed from the fields φ_α belonging to the same Hamiltonian: $\langle \varphi_\alpha\varphi_\beta\rangle$ or $\langle \varphi'_\alpha\varphi'_\beta\rangle$. The averages like $\langle \varphi_\alpha\varphi'_\beta\rangle$ are included in the second term in Eq. (7.118), which is symbolically marked by a bar and an index "c". For example, the second order term $\langle \mathcal{H}^2_{\text{int}}\rangle$ for the φ^4-Hamiltonian (7.1) will be

$$\begin{aligned}\langle \mathcal{H}^2_{\text{int}}\rangle &= u_0^2 \sum_{\alpha\beta\gamma\delta}\int d^dx d^dy \langle \varphi^2_\alpha(x)\varphi^2_\beta(x)\varphi^2_\gamma(y)\varphi^2_\delta(y)\rangle \\ &= u_0^2\left[\sum_{\alpha\beta}\int d^dx\,\langle \varphi^2_\alpha(x)\varphi^2_\beta(x)\rangle\right]^2 \\ &\quad + u_0^2\sum_{\alpha\beta\gamma\delta}\int d^dx d^dy \langle [\varphi^2_\alpha(x)\varphi^2_\beta(x)]\,[\varphi^2_\gamma(y)\varphi^2_\delta(y)]\rangle_c.\end{aligned} \tag{7.119}$$

The suffix "c" of the average in the second term means that at least one pair average must be of the type $\langle \varphi_\alpha(x)\varphi_\beta(y)\rangle$; the dependence of φ_α on x or y clearly indicates whether the fields belong to the same Hamiltonian part or they are from different Hamiltonian parts ($x \neq y$). The pair averages $\langle \varphi_\alpha(x)\varphi_\beta(y)\rangle$ are called connected averages because they give the correlation between the fields from different Hamiltonian parts. All disconnected averages are included in the first term $\langle \mathcal{H}_{\text{int}}\rangle^2$ of $\langle \mathcal{H}^2_{\text{int}}\rangle$; see Eq. (7.119).

Therefore, the second order perturbation term in Eq. (7.116) is divided in two parts: (i) a sum of products of "disconnected" pair averages, which exactly reproduce $\langle \mathcal{H}_{\text{int}}\rangle^2$ — the square of the first perturbation term $\langle \mathcal{H}_{\text{int}}\rangle$ and (ii) a sum of products of "connected" averages. This is illustrated in Fig. 7.1, where the Hamiltonian part $\langle \mathcal{H}_{\text{int}}\rangle$ is represented by a *diagram* (*graph*).

Further we shall often make a diagrammatic representation of the perturbation terms and this will be very helpful for both the classification and calculation of the perturbation contribution to the physical quantities.

The diagrammatic representation of perturbation series has been first introduced by Mayer for the linked-cluster expansion in statistical mechanics (Isihara, 1971; Pathria, 1972) and further this technique has been developed in the quantum field theory (quantum electrodynamics) by Feynman (Bjorken and Drell, 1965; Bogoliubov and Shirkov, 1959). The graphs we are going to use in our studies are the Feynman diagrams corresponding to classic (non-quantum) fluctuation fields $\varphi_\alpha(x)$.

Any perturbation term can be represented by a diagram and the success of the diagrammatic technique is guaranteed by the appropriate choice of the correspondence between mathematical symbols and graph elements. The reader need not have the knowledge of the diagrammatic rules, the only necessary thing is to learn how to perform independently the calculations up to the first and the second order of the perturbation theory on the example of one or two simple models like Eq. (7.1). In this way the understanding of the simple mechanism according to which the perturbation terms are described by figures (diagrams) can be acquired. Having the experience from the lowest order perturbation calculations, we can also treat successfully the higher-order terms in the perturbation expansion.

Let us go back to Fig. 7.1(a). In this figure, the straight lines (*legs*) represent the fields $\varphi_\alpha(x)$ in $\mathcal{H}_{\text{int}}(\varphi)$. The shaded circle (*vertex*) denotes the mathematical operations included in $\mathcal{H}_{\text{int}}$ (integration over x, summation over α, etc.). The averaging is given by all possible connections between the legs (φ_α) which lead to pair averages like $G_{\alpha_i\alpha_j} = \langle\varphi_{\alpha_i}\varphi_{\alpha_j}\rangle$; see Fig. 7.1(b). The lines with ends terminating at one and the same vertex, as in Figs. 7.1(b, c), are a picture of the correlation functions G formed by the legs (fields φ_α and φ_β, or φ'_α and φ'_β) of one and the same Hamiltonian part. The lines that connect two vertices describe the correlation functions of the type $G = \langle\varphi_\alpha\varphi'_\beta\rangle$; see Figs. 7.1(d–f).

The diagrams like that in Fig. 7.1(d), which can be separated in two parts by cutting a single G-line are often called one particle reducible diagrams (reducible in two parts). This terminology comes from the field theory of the elementary particles where G is the particle propagator (Bjorken and Drell, 1965). The graph given in Fig. 7.1(e) is one-particle irreducible or, equivalently, two-particle reducible diagram. The graph in Fig. 7.1(f) with $m > 2$ links is the two-particle irreducible diagram. According to the above classification the disconnected graphs, such as in Fig. 7.1(c), are always reducible.

Apart from a numerical factor the third order term in Eq. (7.116) is given by the average $\langle\mathcal{H}^3_{\text{int}}\rangle$, which can be written as a sum

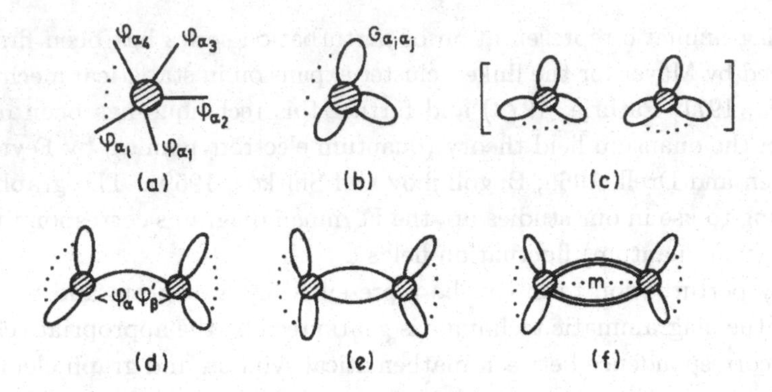

Fig. 7.1 Graphs for: (a) $\mathcal{H}_{\rm int}$, (b) $\langle\mathcal{H}_{\rm int}\rangle$, (c) $\langle\mathcal{H}_{\rm int}\rangle^2$, (d) one-particle reducible diagrams in $\langle\mathcal{H}^2_{\rm int}\rangle$, (e) two-particle reducible diagrams, (f) two-particle irreducible diagrams.

$$\langle\mathcal{H}^3_{\rm int}\rangle = \langle\mathcal{H}_{\rm int}\rangle^3 + 3\langle\mathcal{H}^2_{\rm int}\rangle_c\langle\mathcal{H}_{\rm int}\rangle + \langle\mathcal{H}^3_{\rm int}\rangle_c. \tag{7.120}$$

The numerical factor 3 before the second term in this equation is equal to the ways, in which two Hamiltonian parts can be chosen from three ones to form a connected average of the type $\langle\mathcal{H}^2_{\rm int}\rangle_c$. Alternatively, there are three ways to separate one of the three factors $\mathcal{H}_{\rm int}$ on the l.h.s. of Eq. (7.120) to form the factor $\langle\mathcal{H}_{\rm int}\rangle$ in the second term on the r.h.s. of the equation. The suffix "c" of two of the averages in Eq. (7.120) denotes those diagrams, for which all three factors $\mathcal{H}_{\rm int}$ are linked to each other; all disconnected diagrams, see Figs. 7.2(a, b), have been already taken into account in the first two terms on the r.h.s. of Eq. (7.120). So Figs. 7.2(a–c) illustrate the type of diagrams arising from the first, second and third terms on the r.h.s. of Eq. (7.120), respectively.

We have shown the principle of constructing diagrams for any Hamiltonian part $\mathcal{H}_{\rm int}$ with, say, l legs: $\mathcal{H}_{\rm int} \sim \varphi^l$. It is evident that the higher-order perturbation terms will generate a big number of diagrams. Bearing in mind that the perturbation series (7.116) is infinite and moreover, the number of diagrams will increase with the order l of the perturbation expansion we meet the first difficulty — the impossibility to take into account all perturbation (diagrammatic) contributions to $\mathcal{Z}$.

The number of perturbation terms can be reduced for the thermodynamic potential $\Phi = -\ln\mathcal{Z}$. Despite the number of perturbation terms remains infinite, when we take the logarithm of $\mathcal{Z}$ we automatically

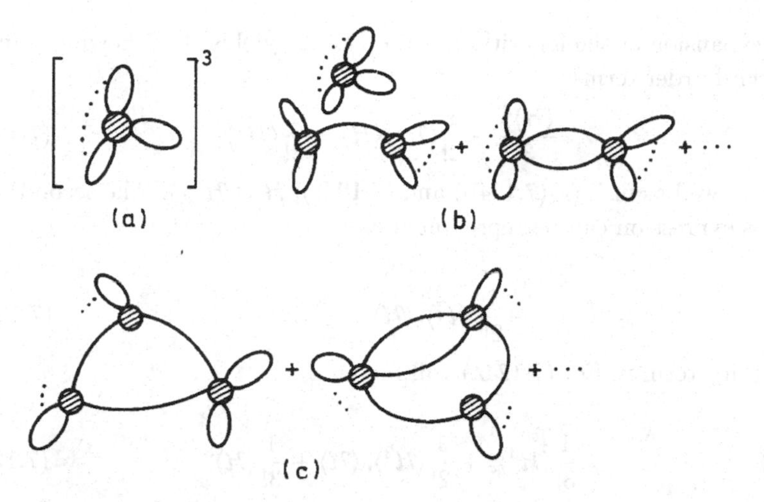

Fig. 7.2 Diagrammatic representation of the third-order perturbation terms for $\mathcal{Z}$: (a) $\langle\mathcal{H}_{\rm int}\rangle^3$, (b) disconnected diagrams from the second term in Eq. (7.120), (c) connected diagrams for the third term in Eq. (7.120).

remove the necessity to consider the disconnected diagrams as those shown in Fig. 7.1(c), Fig. 7.2(a), and Fig. 7.2(b). We shall show this to the third order in $\mathcal{H}_{\rm int}$. Taking the logarithm of Eq. (7.115) we obtain

$$\Phi = \Phi_0 + \Phi_{\rm int}, \tag{7.121}$$

where $\Phi_0 = -\ln \mathcal{Z}_0$ and

$$\Phi_{\rm int} = -\ln(1 + L_1). \tag{7.122}$$

The expansion to the order $\mathcal{H}^3_{\rm int}$ gives

$$\Phi_{\rm int} = \langle\mathcal{H}_{\rm int}\rangle - \frac{1}{2!}\langle\mathcal{H}^2_{\rm int}\rangle_c + \frac{1}{3!}\langle\mathcal{H}^3_{\rm int}\rangle_c + O(\mathcal{H}^4_{\rm int}); \tag{7.123}$$

the averages $\langle\ldots\rangle_c$ have the same meaning as explained above — disconnected diagrams cannot appear from these averages. As far as our aim is to calculate Φ rather than $\mathcal{Z}$, the absence of disconnected diagrams for $\Phi_{\rm int}$ is an advantage of this perturbation calculation. We shall illustrate the mechanism of excluding all disconnected diagrams from the perturbation series for Φ by calculating explicitly the third order term in Eq. (7.123).

The expansion of the logarithm in Eq. (7.122) yields the following sum for the third order terms

$$\frac{\langle\mathcal{H}\rangle^3}{3} - \frac{1}{2!}\langle\mathcal{H}^2\rangle\langle\mathcal{H}\rangle + \frac{1}{3!}\langle\mathcal{H}^3\rangle, \tag{7.124a}$$

(here as well as in Eqs. (7.124b) and (7.124c), $\mathcal{H} \equiv \mathcal{H}_{\text{int}}$). The second term in this expression can be represented as

$$-\frac{1}{2!}\langle\mathcal{H}^2\rangle_c\langle\mathcal{H}\rangle - \frac{1}{2!}\langle\mathcal{H}\rangle^3. \tag{7.124b}$$

The third term in Eq. (7.124a) will be

$$\frac{1}{3!}\langle\mathcal{H}^3\rangle_c + \frac{1}{2!}\langle\mathcal{H}^2\rangle_c\langle\mathcal{H}\rangle + \frac{1}{3!}\langle\mathcal{H}\rangle^3. \tag{7.124c}$$

Putting these expressions in Eq. (7.124a) we obtain the connected average from Eq. (7.123). This direct calculation of the perturbation series (7.123) is quite inconvenient when the terms $O(\mathcal{H}_{\text{int}}^4)$ are taken into account. There is a straightforward way to demonstrate that the total series for Φ_{int} can be written in terms of connected averages, i.e.,

$$\Phi_{\text{int}} = \sum_{l=1}^{\infty} \frac{(-1)^{l-1}}{l!}\langle\mathcal{H}_{\text{int}}^l\rangle_c. \tag{7.125}$$

The way, in which this expression can be proven is shown in Section 6.3.2. Let $\mathcal{H}_{\text{int}}$ contain a single term with an interaction constant like u_0 in Eq. (7.1). The coefficients in the expansion of Φ at the point $u_0 = 0$ in the parameter space (r_0, c, u_0) of the Hamiltonian $\mathcal{H}$ will give the connected averages in Eq. (7.125). In fact,

$$\Phi = \Phi_0 + \sum_{l=1}^{\infty} \frac{1}{l!}\left(\frac{\partial^l \Phi_{\text{int}}}{\partial u_0^l}\right)_{u_0=0} u_0^l, \tag{7.126}$$

and from

$$\Phi = -\ln \int D\varphi \exp\left[-\mathcal{H}_0 - u_0(\mathcal{H}_{\text{int}})\right], \tag{7.127}$$

with $(\mathcal{H}_{\text{int}}) \equiv \mathcal{H}_{\text{int}}/u_0$, we obtain Eq. (7.125), where $\langle\mathcal{H}_{\text{int}}^l\rangle_c \equiv \langle\langle\mathcal{H}_{\text{int}}^l\rangle\rangle$ are the lth order irreducible correlation functions (irreducible cumulants) of $\mathcal{H}_{\text{int}}$. If $\mathcal{H}_{\text{int}}$ consists of several interaction terms, for example: $u_4\varphi^4$,

$u_6\varphi^6$, etc. we can formally multiply all interaction constants with a number $\lambda \neq 0$; then we can expand Φ in powers of λ at $\lambda = 1$ and we can demonstrate that the derivatives $(\partial^l \Phi_{\text{int}}/\partial\lambda^l)_{\lambda=1}$ correspond to $\langle\langle \mathcal{H}^l_{\text{int}} \rangle\rangle$. These considerations prove the validity of the series (7.125) for complex Hamiltonians with several fluctuation interactions. The practical calculation of the terms in Eq. (7.125) requires to choose an explicit form of the Hamiltonian (see Section 7.7).

7.6.2 *Perturbation expansion for the averages*

The perturbation series for any average $\langle A(\varphi)\rangle$, see Eq. (7.6) is formed in the way explained in Section 6.3: see (Eqs. (6.20)–(6.23). We can write Eq. (7.6) in the following way:

$$\langle A \rangle = \frac{\langle A(\varphi) e^{-\mathcal{H}_{\text{int}}} \rangle_0}{\langle e^{-\mathcal{H}_{\text{int}}} \rangle_0}, \tag{7.128}$$

cf. Eq. (6.23). We shall distinguish between the Gaussian averages $\langle \dots \rangle_0$ corresponding to $\mathcal{H} = \mathcal{H}_0$ and the full averages $\langle \dots \rangle$ corresponding to the total Hamiltonian $\mathcal{H} = \mathcal{H}_0 + \mathcal{H}_{\text{int}}$. Supposing that all $\langle \dots \rangle_0$ are known or that they can be in principle calculated, we expand the formula (7.128) in power series of $\mathcal{H}_{\text{int}}$ with the hope to achieve some progress in the calculation of $\langle A \rangle$. In result we obtain

$$\langle A \rangle = \frac{\langle A \rangle_0 + L_A}{1 + L_1}, \tag{7.129}$$

where

$$L_A = \sum_{l=1}^{\infty} \frac{(-1)^l}{l!} \langle A(\varphi) \mathcal{H}^l_{\text{int}}(\varphi) \rangle_0, \tag{7.130}$$

and $L_1 \equiv L_{(A=1)}$; cf. Eqs. (6.42) and (6.43). Now we can use our experience from Section 6.3 and Section 7.6.1 to show that

$$\langle A \rangle = \langle A \rangle_0 + \sum_{l=1}^{\infty} \frac{(-1)^l}{l!} \langle A \mathcal{H}^l_{\text{int}} \rangle_{0c}, \tag{7.131}$$

i.e., the perturbation series for any $\langle A \rangle$ consists of an infinite sum of irreducible averages. The connected averages $\langle \dots \rangle_{0c}$ in Eq. (7.131) ought to be correctly understood, in particular, that the suffix "c" in Eq. (7.131) does not necessarily mean connected diagrams. The topology of these diagrams will depend on the particular type of the functionals $A(\varphi)$ and $\mathcal{H}_{\text{int}}(\varphi)$. For example, if $A = A_1(\varphi)A_2(\varphi)$ and $\langle A_1(\varphi)\rangle_0 \neq 0$ averages of the type $\langle A_2\mathcal{H}_{\text{int}}\rangle\ \langle A_1\rangle$ or $\langle A_2\mathcal{H}_{\text{int}}\rangle\ \langle A_1\mathcal{H}_{\text{int}}\rangle$ should also be included into the sum (7.131), because such averages which connect A with the factors $\mathcal{H}_{\text{int}}$ are not eliminated by the expansion of the denominator in Eq. (7.129) or from the definition of the irreducible functions of the type $\langle A\mathcal{H}^l_{\text{int}}\rangle_c$; see also the treatment of the four-point function $A = [\varphi_{\alpha_1} \cdots \varphi_{\alpha_4}]$ in Section 7.7.2 and the diagrams in Fig. 7.10(a). For $h = 0$ and $T > T_c$, since this is the case studied below, the functionals A like $A = \varphi_{\alpha_1}\varphi_{\alpha_2}$ do not give disconnected diagrams. We may continue our further studies of the perturbation series for $\langle A\rangle$ without specifying the Hamiltonian $\mathcal{H}$. But it seems more instructive to choose the φ^4-Hamiltonian (7.1) with $h = 0$ and $r_0 > 0$ and do some explicit calculations together with the further presentation of the general scheme of the perturbation theory.

7.7 φ^4-theory. Coordinate Representation

Here we shall demonstrate how the scheme developed in Section 7.6 works for the φ^4-Hamiltonian (7.1) in the x-representation. Setting $h = 0$ and $r_0 > 0$ in Eq. (7.1) means that the average $\langle\varphi_\alpha(x)\rangle$ is equal to zero and we have to consider the fluctuation field ($\varphi_\alpha = \delta\varphi_\alpha$) above T_c. We can choose between two variants of the diagrammatic representation of the interaction part $\mathcal{H}_{\text{int}} \sim \varphi^4$; see Fig. 7.3. The picture in Fig. 7.3(a) precisely corresponds to the original φ^4-interaction included in Eq. (7.1). The point, u_0, where the four straight lines describing the fields $\varphi_\alpha(x)$ terminate is expressed by the "mathematical" operation:

$$\bullet \quad \triangleq \quad u_0 \sum_{\alpha\beta}^{n} \int d^d x\,; \tag{7.132}$$

the symbol $\triangleq$ denotes the equivalence of the graph elements and mathematical symbols. If we imagine that Fig. 7.3(a) is placed in the volume of the system, the vertex will be located at the point x — the coordinate of the fields φ_α and φ_β.

Alternatively we may suggest a slightly different diagrammatic representation Fig. 7.3(b) arising from the equivalent mathematical expression for $\mathcal{H}_{\text{int}}$:

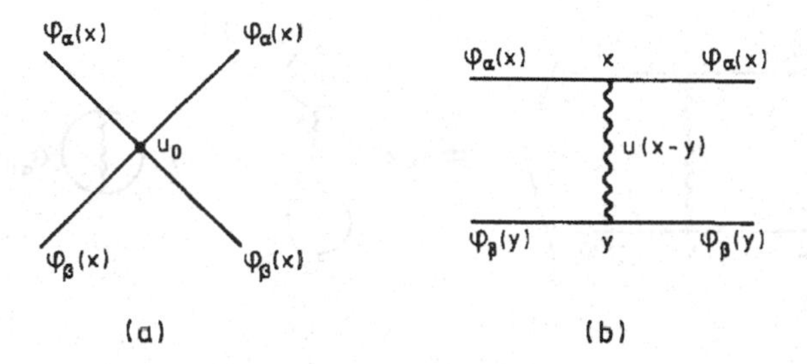

Fig. 7.3 Diagrams for the φ^4-interaction; see the text.

$$\mathcal{H}_{int} = \sum_{\alpha\beta}^{n} \int d^d x d^d y u(x-y) \varphi_\alpha^2(x) \varphi_\beta^2(y), \tag{7.133}$$

where

$$u(x-y) = u_0 \delta(x-y). \tag{7.134}$$

The mathematical meaning of the wavy line in Fig. 7.3(b) is given by the symbolic equality:

$$x \bullet\!\!\sim\!\sim\!\!\bullet y \triangleq \sum_{\alpha\beta}^{n} \int d^d x d^d y \, u(x-y), \qquad (y \leftrightarrow x); \tag{7.135}$$

the line $x \bullet\!\!\sim\!\!\bullet y$ in virtue of Eq. (7.134) is identical to the simple vertex in Fig. 7.3(a); in some cases this representation may be very useful. We must however keep in mind that in our case the point x coincides with y $(y \leftrightarrow x)$. The remaining part of the diagrammatic rules will be deduced in the course of the calculations up to the first or second order of the perturbation expansion.

7.7.1 *First order perturbation contribution to the thermodynamic potential*

Consider the first term in Eq. (7.123). It has already been calculated, see Eq. (7.117), so we have to discuss only its graph representation by the correlation functions $G^{(2)}_{(0)\alpha\beta} \equiv G_{\alpha\beta}$. The three terms in Eq. (7.117) correspond to the all possible different links between the legs $\varphi_\alpha(x)$ and $\varphi_\beta(y \to x)$ in Fig. 7.3(b) (or Fig. 7.3a).

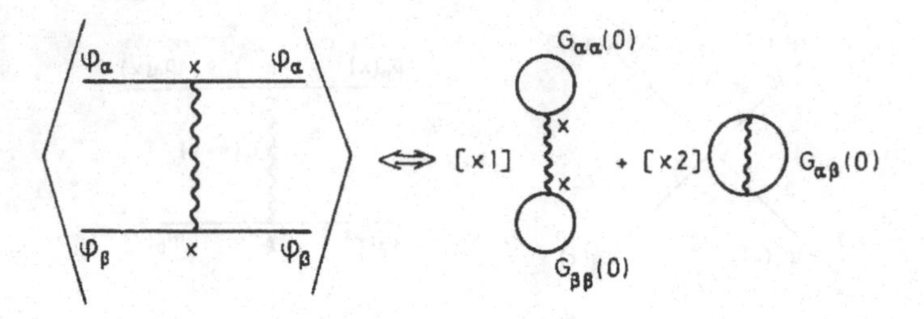

Fig. 7.4 First-order diagrams for the potential Φ.

Let us choose one of the four legs in Fig. 7.3(a), or, Fig. 7.3(b), say one of the two legs $\varphi_\alpha(x)$. It can form one pair with the other φ_α, $\langle\varphi_\alpha^2(x)\rangle_0$, then the second pair will be $\langle\varphi_\beta^2(x)\rangle_0$; or $\langle\varphi_\beta^2(y)\rangle_0$ in Fig. 7.3(b). So we obtain the product $G_{\alpha\alpha}(0)G_{\beta\beta}(0)$ in Eq. (7.117) with a multiplying numerical factor 1 because the way in which this product can be obtained is only one. However $\varphi_\alpha(x)$ can participate in another two pair averages provided it is coupled to the fields φ_β. From this coupling we have the pair $\langle\varphi_\alpha\varphi_\beta\rangle$ and another factor $\langle\varphi_\alpha\varphi_\beta\rangle$ comes from the remaining two fields (legs in the diagrams 7.3). The above explanation is depicted in Fig. 7.4, where the multiplying factors [×1] and [×2] of the diagrams relevant to the terms in Eq. (7.117) are also shown. Certainly, we have established a full correspondence between Eq. (7.117) and the diagrams in Fig. 7.4.

We leave it to the reader to make again all possible connections between the legs on the l.h.s. of Fig. 7.4(a) taking into account the respective multiplying factors. Further, having in mind that the loops in the first diagram on the r.h.s. of Fig. 7.4 as well as the half-loops in the second diagram describe the G-functions and that the wavy line means the mathematical operation (7.132), it is easy to re-derive Eq. (7.117) in the new diagrammatic way. Usually those, who have a lot of experience with the diagrammatic rules discuss the properties of the perturbation theory directly in terms of diagrams. The experience from familiar cases can be easily adapted to any model different from the above one; the slight differences become obvious from a simple analysis of the first and second order perturbation terms.

Now we shall calculate the contribution from the diagrams in Fig. 7.4. Denoting the first-order perturbation correction to Φ by $\Phi_{\text{int}}^{(1)} \equiv \langle\mathcal{H}_{\text{int}}\rangle$ and taking the sum and integral in Eq. (7.117) we have

$$\Phi_{\text{int}}^{(1)} = u_0 V n(n+2)G^2(0); \tag{7.136}$$

$G(0)$ is the Green function $G_0(x = 0)$ of the free ($u_0 = 0$) theory; see Eq. (7.31). Therefore we can formulate the next diagrammatic rule: a factor n corresponds to any closed loop of G-lines. The first diagram in Fig. 7.4 has two closed loops and so it gives a factor n^2 in Eq. (7.136). The second diagram has one closed loop of G-lines and gives a factor n. From Eqs. (7.136) and (7.67) we finally obtain

$$\frac{\Phi_{\rm int}^{(1)}}{V} = n(n+2)u_0 I_1^2(r_0), \tag{7.137}$$

where

$$I_1(r_0) = \int_k \frac{1}{r_0 + ck^2}, \qquad \int_k \equiv \int_{0<|k|<\Lambda} \frac{d^d k}{(2\pi)^d}, \tag{7.138}$$

is the first and the most simple perturbation integral, with which we meet in our discussion. The contribution $\Phi_{\rm int}^{(1)}/V$ to the potential density (Φ/V) is a function of the Hamiltonian parameters (r_0, c, u_0), the symmetry index n and the cutoff Λ. When we calculate $G(x)$ for $(|x|/\xi) \ll 1$ from Eqs. (7.33)–(7.36), we use an infinite cutoff ($\Lambda = \infty$), so the results cannot be applied for $x = 0$. However $G(0)$ or, which is the same, the integral $I_1(r_0, c, \Lambda)$ can always be found for a finite cutoff.

The higher-order perturbation terms in the expansion (7.123) can also be calculated with the help of diagrams. For example, the second-order term in Eq. (7.123) corresponds to connected diagrams which can be obtained from the connected average in the second term of Eq. (7.119). The reader may try to make this calculation and to persuade himself that a number of diagrams with different *combinatorial* (multiplying) factors appear in them. There is no reason to carry out such calculations here and there is a more simple way to do this with the help of the perturbation series for the correlation functions.

7.7.2 *Perturbation contribution to the correlation functions*

Let $A(\varphi)$ in Eq. (7.131) be equal to $\varphi_{\alpha_1}(x_1)\varphi_{\alpha_2}(x_2)$, then the expansion (7.131) will be

$$G_{\alpha_1\alpha_2}(x_1, x_2) = G_{\alpha_1\alpha_2}^{(0)}(x_1, x_2) + \sum_{l=1}^{\infty} \frac{(-1)^l}{l!} \langle \varphi_{\alpha_1}(x_1)\varphi_{\alpha_2}(x_2)\mathcal{H}_{\rm int}^l \rangle_{0c}, \tag{7.139}$$

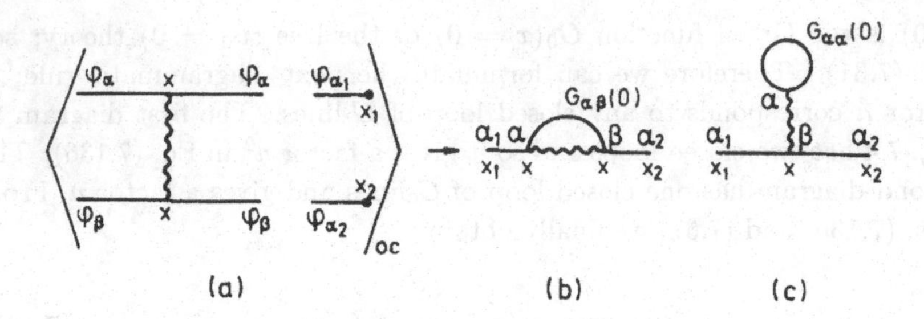

Fig. 7.5 First-order diagrams for G. The combinatorial coefficients $[\times\cdots]$ include the factor $1/l!$, see Eq. (7.139), and the loop factor n.

where $G_{\alpha\alpha} \equiv G^{(2)}_{\alpha\alpha}$ is the two-point correlation function of the total (interacting) system and $G^{(0)}_{\alpha_1\alpha_2} \equiv G^{(2)}_0$ is the two-point correlation function of the corresponding ideal system ($\mathcal{H}_{\text{int}} = 0$). The first perturbation contribution to G will be

$$-u_0 \sum_{\alpha\beta} \int d^d x \, \langle \varphi_{\alpha_1}(x_1)\varphi_{\alpha_2}(x_2)\varphi^2_\alpha(x)\varphi^2_\beta(x)\rangle_{0c}. \tag{7.140}$$

Apart from the factor (-1), this term is represented diagrammatically in Fig. 7.5.

The free legs in Fig. 7.5(a) stand for the fields φ_{α_1} and φ_{α_2} in Eq. (7.140). All three graph elements in the brackets of Fig. 7.5(a) ought to be linked to each other in connected diagrams and this conditions yield the two types of diagrams in Figs. 7.5(b, c) together with their combinatorial factors; see Section 7.7.3. The diagrams (b) and (c) in Fig. 7.5 have external legs which represent the correlation functions $G^{(0)}(x_1-x)$ and $G^{(0)}(x-x_2)$; remember that $G^{(0)}(x,x') = G^{(0)}(x-x')$. The internal $G^{(0)}$-functions correspond to coinciding coordinates $(x = x')$, $G^{(0)}(x-x) = G^{(0)}(0)$. Now we have to do the summation over α and β, the indices of the vertices (α, x) and (β, x), and the integration over x. The wavy line is the graph symbol of the factor $(-u_0)$. So the expression (7.140) becomes equal to

$$-u_0 \sum_{\alpha\beta} \int d^d x \left[8G^{(0)}_{\alpha\beta}(0) + 4G^{(0)}_{\alpha\alpha}(0)\right] G^{(0)}_{\alpha_1\alpha}(x_1 - x)G^{(0)}_{\beta\alpha_2}(x - x_2), \tag{7.141}$$

or bearing in mind that $G^{(0)}_{\alpha\beta} = \delta_{\alpha\beta}G^{(0)}$, we obtain

$$-4(n+2)u_0 G^{(0)}(0)\delta_{\alpha_1\alpha_2} \int d^d x G^{(0)}(x_1 - x)G^{(0)}(x - x_2). \tag{7.142}$$

In this way we confirm the rule that a factor n always appears from the closed loops of $G^{(0)}$-lines like that of the diagram (c) in Fig. 7.5. Because of this rule we can omit the indices α, β etc. of the $G^{(0)}$-functions and we can accept that there is a factor $\delta_{\alpha_1\alpha_2}$ for any diagram with two external legs φ_{α_1} and φ_{α_2}. It is more easy to deduce all these rules from direct calculations with the expression (7.140). Since $G^{(0)}_{\alpha_1\alpha_2}(x_1, x_2)$ in Eq. (7.139) depends on $(x_1 - x_2)$ and $\delta_{\alpha_1\alpha_2}$ and the same is true for the expression Eq. (7.142) we can write

$$G_{\alpha_1\alpha_2}(x_1, x_2) = \delta_{\alpha_1\alpha_2} G(x_1 - x_2). \tag{7.143}$$

It ensues from Eqs. (7.139)–(7.143) that the nonzero full correlation function $G(x_1 - x_2)$ is

$$\begin{aligned} G(x_1 - x_2) &= G^{(0)}(x_1 - x_2) \\ &\quad - 4(n+2)u_0 G^{(0)}(0) \int d^d x G^{(0)}(x_1 - x) G^{(0)}(x - x_2) + O(u_0^2). \end{aligned} \tag{7.144}$$

This equation is approximate, it gives G to the first order in u_0 (lowest-order interaction contribution). This approach can be straightforwardly applied to the second order perturbation contribution of u_0. While the second term in Eq. (7.144) can be easily calculated without the help of diagrams, the treatment of the next order terms of the perturbation theory is greatly facilitated by the diagrammatic technique. To see this and to get an idea of the practical way for obtaining the perturbation series for G, we shall briefly consider the second term in the expansion (7.139):

$$\frac{1}{2!} u_0^2 \sum_{\alpha\beta\gamma\delta} \int d^d x d^d y \langle \varphi_{\alpha_1}(x_1) \varphi_{\alpha_2}(x_2) \varphi_\alpha^2(x) \varphi_\beta^2(x) \varphi_\gamma^2(y) \varphi_\delta^2(y) \rangle_{0c}, \tag{7.145}$$

which is represented by the graph in Fig. 7.6.

We can calculate in the same way as above the set of the second-order diagrams. The topologically different diagrams are of two types: one-particle reducible and one-particle irreducible diagrams; see, e.g., Fig. 7.7 and Fig. 7.8.

The combinatorial coefficients $[\times \cdots]$ are shown for all topologically different diagrams. In order to find the coefficients $[\times \cdots]$ the number of the topologically equivalent diagrams should be multiplied by the coefficient $(1/2!)$ for the second-order term in Eq. (7.139) and the loop factor n^2. For

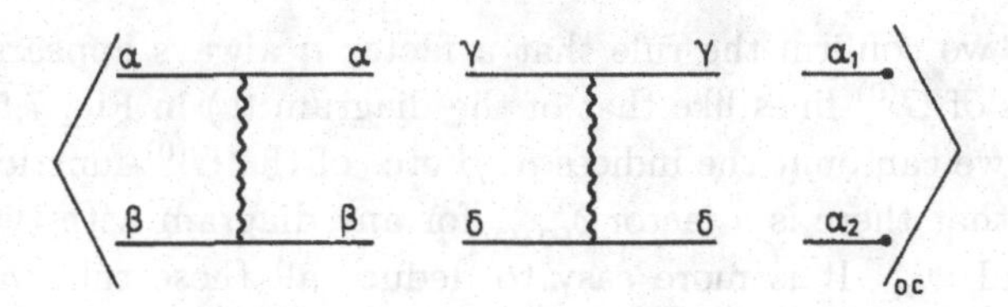

Fig. 7.6 The graph for the expression (7.145).

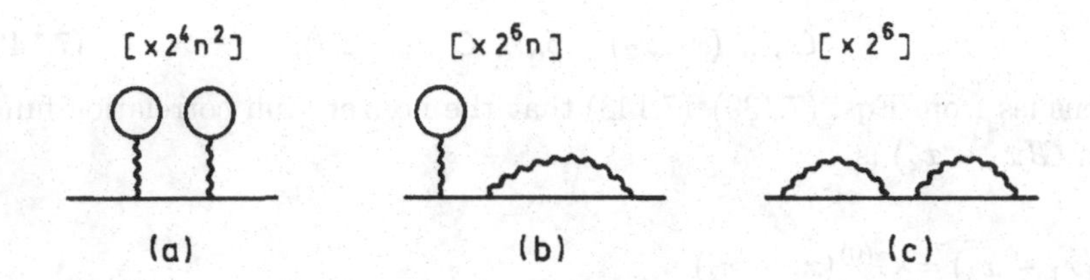

Fig. 7.7 One-particle reducible diagrams for G of order u_0^2.

example, there are 2^5 diagrams of the type 7.8a and, hence, $[\times 2^4 n^2] = n^2(2^5/2!)$. We shall show in the next section that the calculation of the one-particle reducible diagrams can be avoided when the equation for G is transformed to an equation for G^{-1}. The irreducible diagrams (7.8) will be often used in our discussion in Chapter 8. We shall not present the mathematical expressions corresponding to each diagram in Fig. 7.7 and Fig. 7.8. This may be a good exercise for the reader; see also Section 7.8 where these expressions are given in the k-representation. Rather we shall estimate the contribution to $G_{\alpha_1\alpha_2}$ from, say, the diagrams (b) and (e) in Fig. 7.8. They are shown separately in Fig. 7.9.

Following the above explained diagrammatic rules and the meaning of the symbols (x, y, α, etc) in Fig. 7.9, we obtain

$$16u_0^2 \sum_{\alpha\beta\gamma\delta} \int d^d x d^d y G^{(0)}_{\alpha_1\alpha}(x_1 - x) G^{(0)}_{\alpha\gamma}(x - y) \times G^{(0)}_{\gamma\alpha_2}(y - x_2)[G^{(0)}_{\beta\delta}(x - y)]^2 \tag{7.146}$$

for diagram 7.9a and

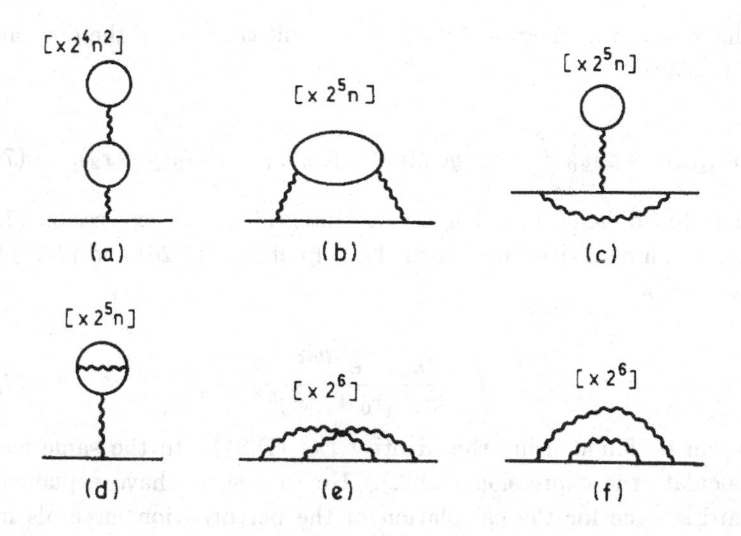

Fig. 7.8 Second-order (one-particle) irreducible diagrams for G. The factor $(1/2!)$ from the second term in Eq. (7.139) is included in the combinatorial coefficients $[\times 2^m]$.

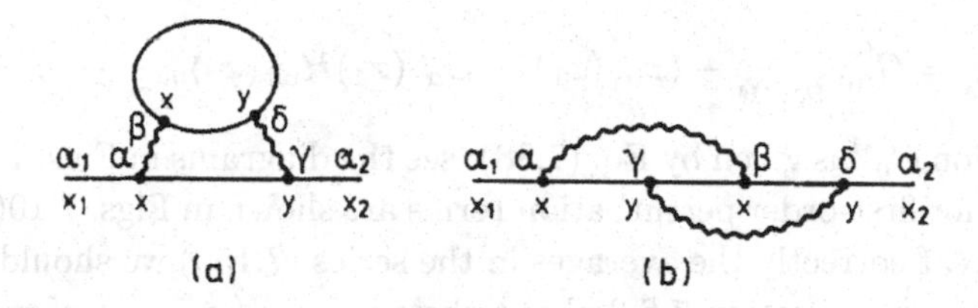

Fig. 7.9 Second order diagrams, which are expressed by equivalent mathematical relations with the only exception of the combinatorial coefficient.

$$32u_0^2 \sum_{\alpha\beta\gamma\delta} \int d^dx d^dy G^{(0)}_{\alpha_1\alpha}(x_1 - x) G^{(0)}_{\alpha\gamma}(x - y) G^{(0)}_{\gamma\beta}(y - x)$$

$$\times G^{(0)}_{\beta\delta}(x - y) G^{(0)}_{\delta\alpha_2}(y - x_2) \qquad (7.147)$$

for the diagram 7.9b. The $\delta_{\alpha\beta}$-factors of the $G^{(0)}_{\alpha\beta}$-functions in the expression (7.147) eliminate completely the four-fold summation over the symmetry indices while the same factors in the expression (7.146) are such that the summation over a single index, say β, remains. But, only the factor $\delta_{\beta\delta}[G^{(0)}_{\beta\delta}]^2$ in (7.146) depends on β and this summation yields the factor n

(see the loop in the diagram 7.9b). As a result the sum of the diagrams in Fig. 7.9 will be

$$16\delta_{\alpha_1\alpha_2}(n+2)u_0^2\int d^dxd^dyG_0^3(x-y)G_0(x_1-x)G_0(y-x_2). \qquad (7.148)$$

It is not too difficult to calculate the integral in the expression (7.148) because it can be transformed with the help of Eqs. (7.26) and (7.67) to an integral over k,

$$\int\frac{d^dk}{(2\pi)^d}\frac{e^{-ikR}}{(r_0+ck^2)^5}, \qquad (7.149)$$

which can be found using the identity Eq. (D.21). In the same way we can calculate the expression (7.142). Up to now we have explained the standard scheme for the calculation of the perturbation integrals in the x-representation.

If A from Eq. (7.131) has the form $[\varphi_{\alpha_1}(x_1)\cdots\varphi_{\alpha_4}(x_4)]$, Eq. (7.131) will be

$$G^{(4)}_{\alpha_1\dots\alpha_4}=G^{(4)}_{(0)\alpha_1\dots\alpha_4}-\langle\varphi_{\alpha_1}(x_1)\cdots\varphi_{\alpha_4}(x_4)\mathcal{H}_{\text{int}}(\varphi)\rangle_{0c}+\cdots. \qquad (7.150)$$

The function $G_0^{(4)}$ is given by Eq. (7.88); see the diagrams in Figs. 7.10(a–c). Diagrams for first-order perturbation terms are shown in Figs. 7.10(d–f). In order to treat correctly the averages in the series (7.150) we should remember the remark in Section 7.6.2 about the averages $\langle\varphi_{\alpha_1}\cdots\rangle$ of more than two fields. The diagrams (d) and (e) in Fig. 7.10 describe reducible averages $\langle\varphi_{\alpha_1}\cdots\varphi_{\alpha_4}\rangle$ of type G_0G whereas the connected diagram (f) describes the first-order contribution to the irreducible four-point correlation function $G_c^{(4)}$. Subtracting all disconnected diagrams from the diagrammatic series for $G^{(4)}$ we obtain $G_c^{(4)}$. The combinatorial factor of the diagram 7.10(f) is 24.

To the first order in u_0 $G_c^{(4)}$ is given by

$$G^{(4)}_{(c)\alpha_1\dots\alpha_4}(x_1,\dots,x_4)$$
$$=-u_0\sum_{\beta\gamma}\int d^dx\,\langle\varphi_{\alpha_1}(x_1)\cdots\varphi_{\alpha_4}(x_4)\varphi_\beta^2(x)\varphi_\gamma^2(x)\rangle_{occ}, \qquad (7.151)$$

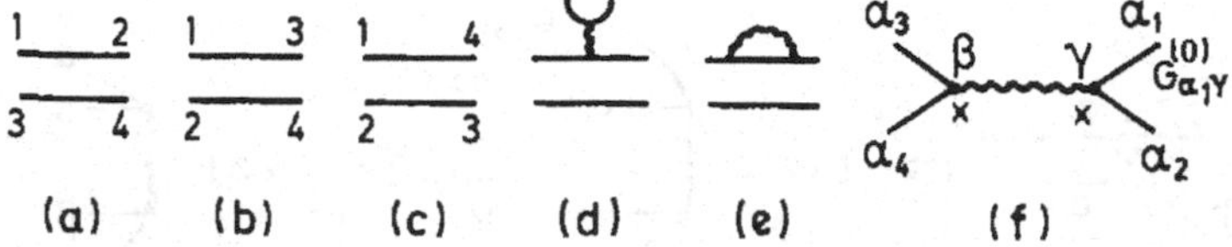

Fig. 7.10 Diagrams for $G^{(4)}$.

where the double suffix "cc" stands for the above remark. From the direct calculation of the four-point correlation function (7.151), and taking into account the properties of the functions $G_0^{(2)}$ we can easily obtain that the diagram shown in Fig. 7.10(f) corresponds to the following mathematical expression

$$G_c^{(4)} = -24\Delta_{\alpha_1\alpha_2\alpha_3\alpha_4} u_0 \times \int d^d x G_0(x - x_1) G_0(x - x_2) \times G_0(x - x_3) G_0(x - x_4); \tag{7.152a}$$

here

$$\Delta_{\alpha_1 \ldots \alpha_4} = \frac{1}{3}(\delta_{\alpha_1\alpha_2}\delta_{\alpha_3\alpha_4} + \delta_{\alpha_1\alpha_3}\delta_{\alpha_2\alpha_4} + \delta_{\alpha_1\alpha_4}\delta_{\alpha_2\alpha_3}). \tag{7.152b}$$

Clearly, the irreducible correlation functions are represented by connected diagrams.

7.7.3 *Remarks and calculation of the combinatorial factors*

The perturbation series are represented by a number of diagrams which are classified according to their topology. The suitable choice of the graph elements makes equivalent mathematical expressions for the topologically equivalent diagrams possible. So our task is to describe all possible topologically equivalent diagrams and to determine their multiplying (combinatorial) factors. Here we shall show how these numerical factors can be calculated. Let us do this for the simple diagram (b) in Fig. 7.5. There are four possible ways in which one of the legs, say φ_{α_1}, in Fig. 7.5(a) can be connected with one of the four legs (φ_α and φ_β) of the diagram for $\mathcal{H}_{\text{int}}$; see Fig. 7.11(a).

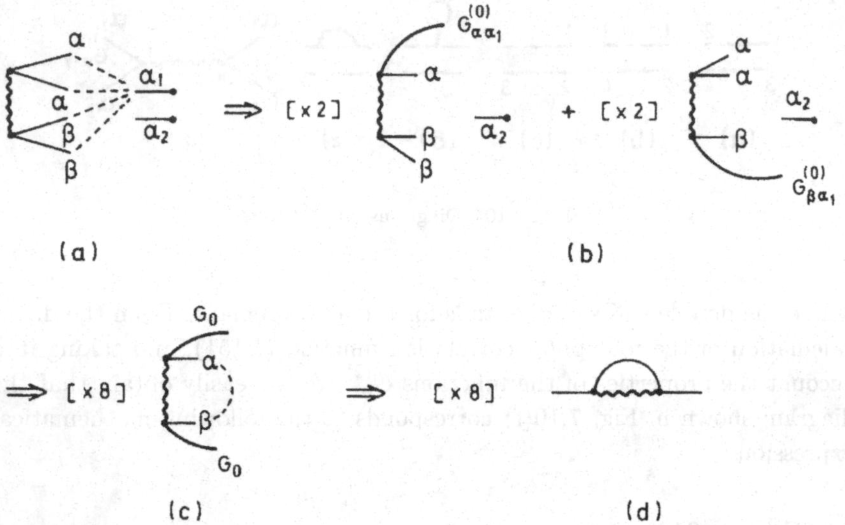

Fig. 7.11 Construction of the first-order diagrams.

For each of these four ways, there are three ways of linking the leg φ_{α_2} with one of the remaining three free legs of $\mathcal{H}_{\text{int}}$ to form the first external $G^{(0)}$-function, say, the line $x_1 - x$ in Fig. 7.5(b). We must be careful, because one of these ways of linking will give the diagram 7.5(c) but not 7.5(b). Therefore, if the leg α_1 is linked to one of the α-legs we should connect the leg α_2 with one of the two β-legs; see also Fig. 7.11. The remaining two legs in $\mathcal{H}_{\text{int}}$ must be joined together and there is only one way to do this. Therefore the combinatorial factor of the diagram (b) in Fig. 7.5 is $4 \cdot 2 = 8$.

As a second example we shall consider the diagram (a) in Fig. 7.8. This diagram is obtained from the graph elements in Fig. 7.6. Both legs α_1 and α_2 should be joined to the legs with equal indices ($\alpha\alpha, \beta\beta$, $\gamma\gamma$ or $\delta\delta$). From this choice ($\alpha_1\alpha_2\alpha\alpha$) a factor 4 appears because there are another two equivalent variants ($\alpha_1\alpha_2\beta\beta$) ... The possible links between the legs ($\alpha_1\alpha_2\alpha\alpha$) are 2, so the combinatorial factor increases to $4 \cdot 2 = 8$. In this way we obtain the intermediate diagram shown in Fig. 7.12(a). Performing the next linking we arrive either at the diagram 7.8(a) or at the diagram 7.8(d).

We shall focus our attention on the ways of linking, which lead to the diagram 7.8a. Their number is 4, so we obtain that the number of these diagrams is $8 \cdot 4 = 2^5$ and, hence, the coefficient $[\times 2^4 n^2]$ in Fig. 7.8 is equal to $(2^5/2!)n^2$; $(1/2!)$ is the corresponding factor coming from Eq. (7.139). Very often the factors $(1/l!)$ and n^m are not included in the combinatorial coefficients; they are taken into account separately. Sometimes the numbers $[\times \cdots]$ are called *weight factors* of the diagrams.

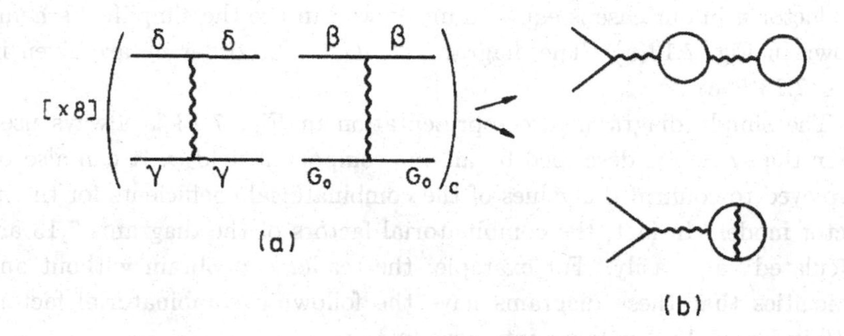

Fig. 7.12 Construction of second-order diagrams.

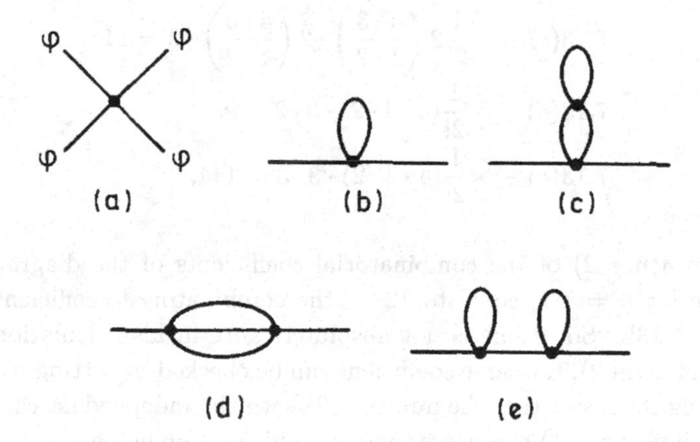

Fig. 7.13 Diagrams for scalar fields.

It is not difficult to make a mistake in the calculation of the combinatorial factors. So we shall point out how the combinatorial factors for some types of diagrams can be checked. We consider that $n = 1$, i.e., Eq. (7.1) defines a scalar-field Hamiltonian. In this case we shall not use the expression for $\mathcal{H}_{\rm int}$ with two vertices connected by a wavy line; see Fig. 7.3(a) and Fig. 7.13(a). The point vertex in Fig. 7.3(a) has been substituted by a wavy line in Fig. 7.3(b) in order to distinguish the graph representation of the averages $\langle\varphi_\alpha^2\rangle$ and $\langle\varphi_\alpha\varphi_\beta\rangle$, $\alpha \neq \beta$, and hence, to introduce correctly the factors n corresponding to the closed loops of the G_0-functions. As the factor n in our case is equal to unity, we can use the simplified scheme shown in Fig. 7.13(a). The diagrams for G to the order u_0^2 are given in Figs. 7.13(b–e).

The simple diagrammatic representation in Fig. 7.13 is always used when the system is described by an one-component field φ. It can also be employed to confirm the values of the combinatorial coefficients for the n-vector model. In fact, the combinatorial factors of the diagrams 7.13 are calculated very easily. For example, the reader can obtain without any difficulties that these diagrams have the following combinatorial factors ($1/l!$ in Eq. (7.139) is taken into account):

$$\begin{aligned}
&7.13(b) \longrightarrow \frac{1}{1!}\left(\frac{4\cdot 3}{1\cdot 2}\right)\cdot 2 = 12\\
&7.13(c) \longrightarrow \frac{1}{2!}2\left(\frac{4\cdot 3}{1\cdot 2}\right)\cdot 2\left(\frac{4\cdot 3}{1\cdot 2}\right)\cdot 2 = 144\\
&7.13(d) \longrightarrow \frac{1}{2!}(4\cdot 4\cdot 2)\cdot 3\cdot 2 = 96\\
&7.13(e) \longrightarrow \frac{1}{2!}(4\cdot 4\cdot 2)\cdot 3\cdot 3 = 144.
\end{aligned} \tag{7.153}$$

The sum $4(n+2)$ of the combinatorial coefficients of the diagrams 7.5b and 7.5c for $n = 1$ is equal to 12 — the combinatorial coefficient of the diagram 7.13b. So, if one is not absolutely sure in his calculation of the coefficient $4(n+2)$, the same coefficient can be checked by setting $n = 1$ and comparing the result (i.e. the number 12) with the independent calculation for the simple ($n = 1$)-component model. This is often helpful in the higher-order perturbation calculations. Certainly, the comparison with the scalar field diagrams does not remove all possible errors rather it can make easier to understand the right way of calculation.

Having in mind that for $n > 1$ the diagrams 7.5b and 7.5c on one side and the diagram 7.13b on the other side are represented by exactly equal mathematical expressions, see Eq. (7.142), we can write

(7.154)

The meaning of this graphical equation is that its l.h.s. denotes the sum of all diagrams of the same type (below we shall omit the brackets). In a big number of the presentations of the theory we can use the simple graph 7.13(b) rather than the two graphs 7.5(b) and 7.5(c). Note that the diagrams of one and the same type are obtained by reducing the wavy line to a point in all diagrams; see Figs. 7.5, 7.7, and 7.8. This procedure yields one type of the first order diagrams, following from Eq. (7.154). There are three different types of second-order diagrams:

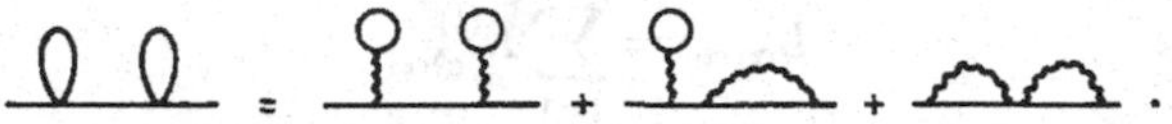

(7.155a)

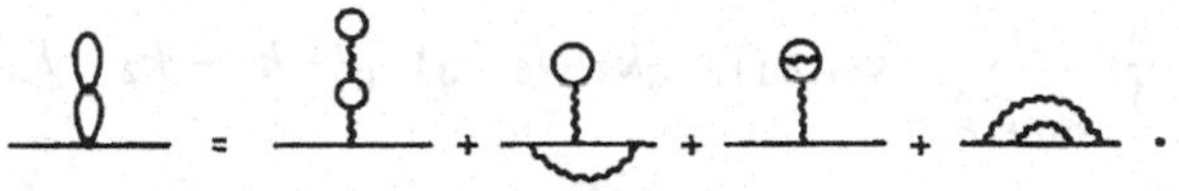

(7.155b)

and

(7.155c)

We can check that the sum of the combinatorial coefficients ($n = 1$) of the diagrams on the r.h.s. of each of these equations is equal to the corresponding combinatorial coefficient of the diagram on the l.h.s.; see Eqs. (7.153), Fig. 7.7 and Fig. 7.8.

It is obvious from Fig. 7.13 and Eqs. (7.154)–(7.155c) that the first-order perturbation terms of the φ^4-theory for a ($n = 1$)-component field are represented by diagrams with one closed loop of the internal G_0-function, the second-order terms give two-loop diagrams, etc. The order in u_0 is equal to the number of loops. That is why this expansion is often called *the loop expansion*; of course, we can use the term "loop expansion" for the n-vector models, too (see Section 8.3).

7.8 φ^4-theory. Momentum Representation

The Fourier transformation of Eq. (7.139) yields

$$G_{\alpha_1\alpha_2}(k_1,k_2) = G^{(0)}_{\alpha_1\alpha_2}(k_1,k_2) + \sum_{l=1}^{\infty} L^{(l)}_{(c)\alpha_1\alpha_2}(k_1,k_2), \tag{7.156}$$

where

$$L^{(l)}_{(c)\alpha_1\alpha_2} = \frac{(-1)^l}{l!}\langle \varphi_{\alpha_1}(k_1)\varphi_{\alpha_2}(k_2)\mathcal{H}^l_{\text{int}}[\varphi_\alpha(k)]\rangle_{0c}; \tag{7.157}$$

the notation

$$L_{\alpha_1\alpha_2} = \sum_{l=1}^{\infty} L^{(l)}_{(c)\alpha_1\alpha_2} \tag{7.158}$$

is introduced for the sum in Eq. (7.156). In Eq. (7.157),

$$\mathcal{H}_{\text{int}} = \frac{u_0}{V}\sum_{\alpha\beta;k_1k_2k_3} \varphi_\alpha(k_1)\varphi_\alpha(k_2)\varphi_\beta(k_3)\varphi_\beta(-k_1-k_2-k_3). \tag{7.159}$$

7.8.1 *First-order perturbation terms*

The diagrammatic technique for $G_{\alpha_1\alpha_2}(k_1,k_2)$ is quite similar to that in the coordinate space. We can find $L^{(1)}_c$ by the Fourier transformation of the expression Eq. (7.142). The same can be made for the other results previously obtained in the x-representation (Section 7.7). But it will be more instructive to re-derive the results to the first order in u_0 directly in the k-representation demonstrating in this way how the diagrammatic rules naturally arise from the lowest-order perturbation calculations.

The first-order perturbation contribution to $G_{\alpha_1\alpha_2}$ is

$$L^{(1)}_{(c)\alpha_1\alpha_2} = -\frac{u_0}{V}\sum_{\alpha\beta;p_1\dots p_4}\delta(p_1+p_2+p_3+p_4)$$
$$\times\langle\varphi_{\alpha_1}(k_1)\varphi_{\alpha_2}(k_2)\varphi_\alpha(p_1)\varphi_\alpha(p_2)\varphi_\beta(p_3)\varphi_\beta(p_4)\rangle_{0c}, \tag{7.160}$$

where we have used the δ-symbol for the momenta p of the Hamiltonian fields $\varphi_\alpha(p)$. The fields $\varphi_\alpha(k)$ and $\mathcal{H}_{\text{int}}$ are depicted in Fig. 7.14(a). They represent the connected average in Eq. (7.160) with the help of all possible

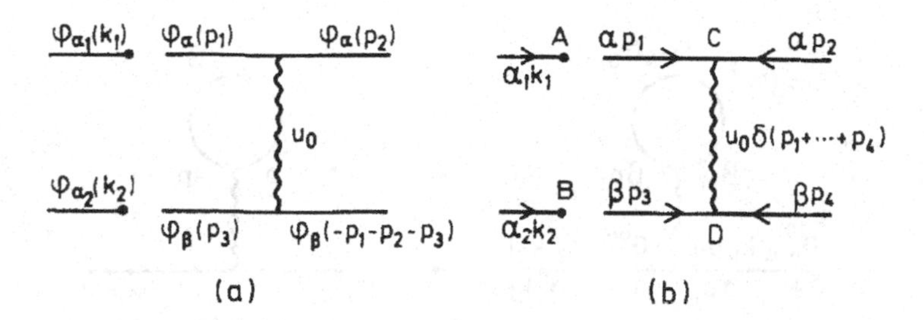

Fig. 7.14 Illustration of the diagrammatic representation of $\varphi_\alpha(k)$ and $\mathcal{H}_{\rm int}[\varphi_\alpha(k)]$; (a) and (b) are explained in the text.

products of the pair averages. This means to join the legs $(\alpha_1 k_1)$ and $(\alpha_2 k_2)$ in Fig. 7.14(a) to the legs $\varphi_\alpha(p)$ of the Hamiltonian in all possible ways, as shown (by A and B) in Fig. 7.14(b). The wavy line stands for the parameter $(-u_0/V)$ and a summation over the indices $\alpha\beta$ and the momenta p_i.

The diagrams are the same as those shown in Fig. 7.5 but now the diagrammatic rules are slightly different; see Fig. 7.15. $G^{(0)}_{\alpha\beta}(k_1, k_2)$ is zero for $\alpha \neq \beta$ or $k_1 \neq -k_2$ and the δ-factor in Eq. (7.160) is an additional restriction on the momenta of the nonzero diagrams. This allows the notation for the internal G_0–lines and the external G_0-legs to be as shown in Fig. 7.15(b). A $G^{(0)}_{\alpha\alpha}(k)$-line carries an index α and a momentum $k(\equiv \boldsymbol{k})$. The sum of the momenta of the external legs of each diagram is equal to zero as it is for $\mathcal{H}_{\rm int}$. A summation over the internal indices and an integration over the internal momenta are assumed; see, e.g., the index β and the momentum p in Fig. 7.15(b). Alternatively one may use the property $\varphi^*_\alpha(k) = \varphi_\alpha(-k)$ of the real fields $\varphi_\alpha(x)$ in order to introduce oriented lines for $\varphi_\alpha(k)$ and $G_0(k)$. This version of the diagrammatic technique is shown in Figs. 7.14(b), and 7.15(c, d). An arrow is attached to the fields $\varphi_{\alpha_1}(k_1)$, $\varphi_{\alpha_2}(k_2)$ and the fields of the Hamiltonian $\mathcal{H}_{\rm int}$ so as the direction of the arrows to reflect the directions of the momenta (k_i and p_i). In Fig. 7.14(b) all oriented lines (momenta) enter in the points A, B, C and D and terminate there. If we change the direction of a line, say that for $\varphi_{\alpha_1}(k_1)$, it will be a "going out" line and it will describe $\varphi_{\alpha_1}(-k_1)$ or, which is the same $\varphi^*_\alpha(k)$. No doubt that we can change the direction of the arrows having in mind that when doing this we change the sign of k ($k \to -k$). This variant of diagrammatic representation by oriented lines is very convenient in the practical calculations. Now we can represent the first-order diagrams by oriented lines and this is shown in Fig. 7.15(c) and Fig. 7.15(d) for one of the first-order diagrams.

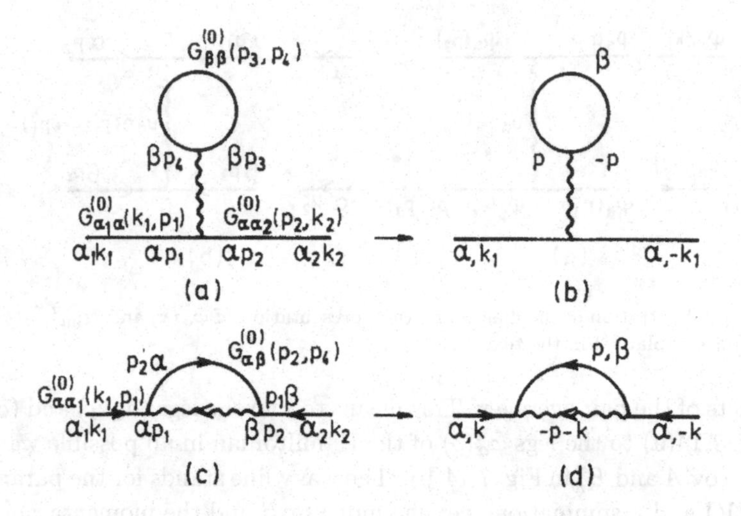

Fig. 7.15 **First-order diagrams in the k-representation.**

To each G_0 and wavy lines we attach momenta so as the sum of the "entering" and "going out" momenta for each vertex point to be equal to zero. Since in our model the interaction parameter u_0 does not depend on the momenta of the fields in $\mathcal{H}_{\text{int}}$ we can always attach a momentum vector to the wavy line of the diagrams. In some field models, the four-point interaction constant is not so simple and depends on the momentum. In such cases we must take into account the precise momenta carried by the wavy lines. In our example the momenta of the wavy lines determine the conservation law ($\sum_i k_i = 0$) for the momenta k_i which enter or go out of each vertex.

Calculating directly or using the above mentioned diagrammatic rules we obtain the expression (7.160) for $L_c^{(1)}$ in the form:

$$L^{(1)}_{(c)\alpha_1\alpha_2}(k_1, k_2) = \delta_{\alpha_1\alpha_2}\delta(k_1 + k_2)L_c^{(1)}(k_1), \tag{7.161a}$$

with

$$L_c^{(1)}(k) = -4(n+2)u_0 G_0^2(k) I_1(r_0), \tag{7.161b}$$

where the integral

$$I_1(r_0) = \frac{1}{V}\sum_p G(p) \longrightarrow \int_p G_o(p) \tag{7.162}$$

has already appeared in Eq. (7.137); cf. Eq. (7.142), where $I_1(r_0) \equiv G_0(R = 0)$, see also Eq. (7.67). The above discussion shows the basic diagrammatic rules in the k-space. Any diagrammatic rule is usually obtained by calculating relatively low-order terms of the perturbation expansion and after that the result is applied for the calculation of the higher-order terms. Accepting this principle the reader may deduce the rules of the diagrammatic technique for any field model of interest. We shall use in our further discussion several field models of phase transitions which are different from the basic φ^4-Hamiltonian (7.1). Once the diagrammatic rules are known for the model (7.1) they can be easily adapted to more complex models and we shall not dwell upon the details of this technical problem.

7.8.2 *Second-order perturbation terms*

With the help of diagrams, the reader can easily derive the expressions for the second-order perturbation terms included in $L_c^{(2)}$; see Eq. (7.157). This could be a good exercise for the diagrammatic calculation within the perturbation theory. Here we shall present only the results. The one-particle (OP) reducible diagrams in Fig. 7.7 are now shown in Fig. 7.16 (the indices α of the G_0-lines are omitted). They are explicitly separated from the other terms in Eq. (7.155a) and their contribution to $L_c^{(2)}$ will be denoted by $L_c^{(2a)}$:

$$L^{(2a)}_{(c)\alpha_1\alpha_2}(k) = \delta_{\alpha_1\alpha_2} L_c^{(2a)}(k), \tag{7.163a}$$

where

$$L_c^{(2a)}(k) = 16(n+2)^2 G_0^3(k) I_1^2(r_0). \tag{7.163b}$$

The remaining part of $L_c^{(2)}$ consists of the OP-irreducible terms in Fig. 7.8(a), which are classified in two groups — see Eqs. (7.155b) and (7.155c), and below their contributions are denoted by $L_c^{(2b)}$ and $L_c^{(2c)}$ respectively. So

$$L^{(2b)}_{(c)\alpha_1\alpha_2}(k) = \delta_{\alpha_1\alpha_2} L_c^{(2b)}(k), \tag{7.164a}$$

and

$$L^{(2c)}_{(c)\alpha_1\alpha_2}(k) = \delta_{\alpha_1\alpha_2} L_c^{(2c)}(k), \tag{7.164b}$$

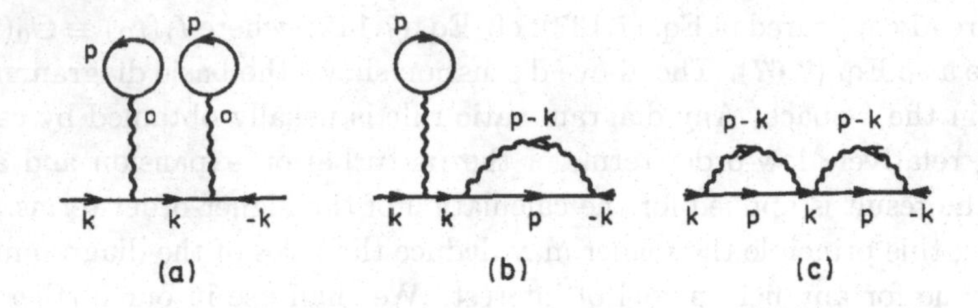

Fig. 7.16 OP-reducible diagrams in the k-representation.

where

$$L_c^{(2b)}(k) = 16(n+2)^2 u_0^2 G_0^2(k) I_1(r_0) I_2(r_0), \tag{7.165a}$$

and

$$L_c^{(2c)}(k) = 32(n+2) u_0^2 G_0^2(k) J(r_0, k). \tag{7.165b}$$

In the above expressions, we meet two other typical perturbation integrals:

$$I_l(r_0) = \int_k \frac{1}{(r_0 + ck^2)^l}, \qquad l = 1, 2, \ldots \tag{7.166}$$

and

$$J(r_0, k) = \int \frac{d^d p}{(2\pi)^d} \frac{d^d q}{(2\pi)^d} \frac{1}{(p^2 + r_0)(q^2 + r_0)[(p+q+k)^2 + r_0]}; \tag{7.167}$$

see also Eq. (7.138) and Eq. (D.10). In J the integration is over two internal momenta (p and q). The integrand in Eq. (7.167) depends on two angles because $(p+q+k)^2$ is the scalar product of the sum of the vectors p, q and k; see also Fig. 7.17. The diagrams in Fig. 7.17 exhibit a new property — their internal G_0-lines and hence the integral J depend on the external momentum. Even if the wavy lines are reduced to points, the external legs terminate at different vertices and this is the reason that the dependence of one of the internal G_0-lines on k cannot be avoided. Because of their different topology these diagrams have another analytical properties.

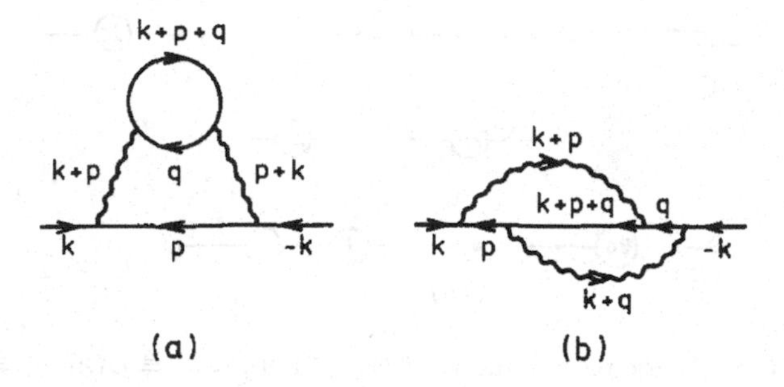

Fig. 7.17 The diagrams in Fig. 7.9 in the k-representation.

We can summarize our results by writing the full two-point correlation function $G_{\alpha\beta}(k, k')$ in the form

$$G_{\alpha\beta}(k, k') = \delta_{\alpha\beta}\delta(k + k')G(k), \tag{7.168}$$

where $G(k)$ is the "Green function" of the interacting system. From Eqs. (7.156)–(7.158), (7.161a, b), (7.163a)–(7.163b), and (7.164a)–(7.167), we have

$$\begin{aligned} G(k) = G_0(k) &- 4(n+2)u_0 G_0^2(k) I_1(r_0) \\ &+ 16(n+2)^2 G_0^3(k) I_1^2(r_0) + 16(n+2)^2 u_0^2 G_0^2(k) I_1(r_0) I_2(r_0) \\ &+ 32(n+2)u_0^2 G_0^2(k) J(r_0, k) + O(u_0^3). \end{aligned} \tag{7.169}$$

The diagrammatic representation of this equation is shown in Fig. 7.18. The OP-reducible terms $L_c^{(2a)}$ in $L_c^{(2)}$ are given by the square of $L_c^{(1)}$.

We must keep in mind that the first-order interaction contribution, i.e., the second term in Eq. (7.168) should be small compared to $G_0(k)$; the second-order terms ($\sim u_0^2$) should be small with respect to the first order in u_0, etc. Therefore the Eq. (7.168) can be treated by iterations. For example, let us neglect the terms $O(u_0^2)$, then,

$$\begin{aligned} G(k) &\approx \frac{G_0(k)}{1 + 4(n+2)u_0 G_0(k) I_1(r_0)} \\ &= \frac{1}{r_1(u_0) + ck^2}, \end{aligned} \tag{7.170a}$$

where

$$r_1(u_0) = r_0 + 4(n+2)u_0 I_1(r_0), \tag{7.170b}$$

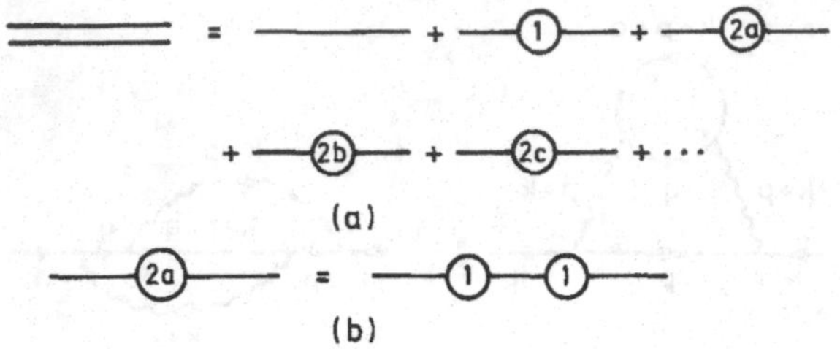

Fig. 7.18 Diagrammatic representation of Eq. (7.169); ═══ $\triangleq G(k)$, Ⓢ $\triangleq L_c^{(S)}$; $S = (1, 2a, 2b, 2c)$.

which is the first order correction to the "bare" (undressed) value of r_0. This value is "dressed" by the field interaction (a folklore commonly used in the perturbation theory). Now the function $G(k)$ has the same form as $G_0(k)$ but one of the parameters, r_0, is dressed, i.e., it is corrected by the interaction. If we accept the form (7.170a) of $G(k)$ we have to introduce a correcting term of order u_0^2 equal to the third term in Eq. (7.169) — the contribution from the OP-reducible diagrams. In fact the second term in the expansion of the geometrical progression in Eq. (7.170a) will be exactly compensated by the third term on the r.h.s. of Eq. (7.169). In the next section we shall present a more convenient way of finding the corrections to the bare parameters of the theory.

7.9 Dyson Equation and Self-energy Function

The fact that the thermodynamic potential Φ is equal to the logarithm of the partition function $\Phi = -\ln \mathcal{Z}$ has helped us to eliminate the infinite set of disconnected diagrams for the partition function $\mathcal{Z}$ (Section 7.6.1). This is a methodic advantage because we are actually rather interested in the calculation of Φ than of $\mathcal{Z}$. The series (7.139) for $G_{\alpha\beta} = \delta_{\alpha\beta} G$ consists only of connected diagrams. The mechanism of removing of the disconnected diagrams is ensured by the denominator in Eq. (7.128). In spite of this success, the series (7.139) of connected diagrams remains infinite and moreover the subsets of different types of diagrams are also infinite. Our task is to seek the way in which some type of weakly-linked diagrams can be excluded from the expansion for $G_{\alpha\beta}$ so as to simplify the further calculations. We have already mentioned in Section 7.2.2 that the OP-reducible diagrams can be removed, if we consider $G_{\alpha\beta}^{-1}$ instead of $G_{\alpha\beta}$ itself. The matrix $G_{\alpha\beta}$ is diagonal, see Eq. (7.168) and we can study directly the full

Green function $G(k)$. Moreover we have shown at the end of Section 7.8 that the series for $G(k)$ does not give in a straightforward way the effect of interactions on the initial parameters of the theory. In order to find the interaction corrections to these parameters we must do additional calculations to obtain results that are consistent with the perturbation expansion. We can avoid this and the calculation of OP-irreducible terms by performing the perturbation expansion of $G^{-1}_{\alpha\beta}$. This perturbation expansion is often called *the Dyson equation*; see also Abrikosov, Gor'kov, and Dzyaloshinskii (1962), Fetter and Walecka (1971), and Amit (1978). Here we shall restrict our discussion to the k-representation which is more convenient for calculations.

7.9.1 *Dyson equation*

The matrix equation (7.156) can be inverted multiplying it on the right by $G^{-1}_{\alpha_2\beta}$ and on the left by $G^{(o)-1}_{\alpha_1\alpha}$. After summing up over α_1 and α_2 we obtain

$$G^{-1}_{\alpha\beta}(k_1,k_2) = G^{(o)-1}_{\alpha_1\alpha}(k_1,k_2) - \Sigma_{\alpha\beta}(k_1,k_2), \tag{7.171}$$

where

$$\Sigma_{\alpha\beta} = \sum_{\alpha_1\alpha_2} G^{(o)-1}_{\alpha_1\alpha} L_{\alpha_1\alpha_2} G^{-1}_{\alpha_2\beta} \tag{7.172}$$

is called *the self-energy function* (or, the mass operator). We can add Eqs. (7.157) and (7.158) to Eqs. (7.171), and (7.172) which yields the way of the calculation of $\Sigma_{\alpha\beta}$ to any order (l) of the perturbation theory. For the model (7.1) we can represent the Dyson equation (7.171) in a more simple form. It is clear from the calculations in Section 7.8 to the order u_0^2 and the general arguments related with the symmetry of the φ^4-Hamiltonian that $\Sigma_{\alpha\beta}$ will have the form (7.168) of $G_{\alpha\beta}$. Defining $\Sigma_{\alpha\beta}(k_1,k_2)$ by

$$\Sigma_{\alpha\beta}(k,k') = \delta_{\alpha\beta}\delta(k+k')\Sigma(k), \tag{7.173}$$

and bearing in mind the analogous representation of $G_{\alpha\beta}$ and $G^{(0)}_{\alpha\beta}$, we obtain the Dyson equation (7.171) in the form

$$G^{-1}(k) = G_0^{-1}(k) - \Sigma(k), \tag{7.174}$$

where

$$\Sigma(k) = G_0^{-1}(k)L_c(k)G^{-1}(k). \tag{7.175}$$

In Eq. (7.175), $L_c(k)$ is determined from $L_{\alpha_1\alpha_2} = \delta_{\alpha_1\alpha_2}L_c$; see Eq. (7.158).

7.9.2 *Expansion of the self-energy function to order u_0^2*

Obviously Eqs. (7.171)–(7.172) and Eqs. (7.174)–(7.175) do not give the perturbation expansion for G^{-1} in powers of u_0. Rather $L(k)$ and $G(k)$ in $\Sigma(k)$ are themselves infinite series in u_0. Let us expand Eq. (7.174) in powers of u_0. To do this we should write the perturbation series for $G^{-1}(k)$ in the following way:

$$\begin{aligned} G^{-1}(k) = G_0^{-1}(k) - G_0^{-1}(k)L_c(k)\left\{G_0^{-1}(k)\right. \\ \left. - G_0^{-1}(k)L_c(k)\left\{G_0^{-1}(k) - G_0^{-1}(k)L_c(k)\left\{G_0^{-1}(k) - \cdots\right.\right.\right. , \end{aligned} \tag{7.176}$$

or, which is the same,

$$\begin{aligned} G^{-1}(k) &= G_0^{-1}(k) - G_0^{-1}(k)L_c(k)G_0^{-1}(k) \\ &+ G_0^{-1}(k)L_c(k)G_0^{-1}(k)L_c(k)G_0^{-1}(k) - \cdots \\ &(\mp)^m G_0^{-m-1}(k)L_c^m(k) + \cdots . \end{aligned} \tag{7.177}$$

The same procedure can be performed in the general case — the only difference with Eq. (7.171) is that instead of the usual products in Eq. (7.177), there shall be products of the matrices $G_{\alpha\beta}^{(0)}$ and $L_{\alpha\beta}$. Certainly, $\Sigma(k)$ is represented by the infinite series in powers of $L(k)$ from Eq. (7.177). In order to obtain the contributions $\Sigma^{(l)}(k)$ of order u_0^l to $\Sigma(k)$,

$$\Sigma(k) = \Sigma^{(1)}(k) + \Sigma^{(2)}(k) + \cdots , \tag{7.178}$$

we have to use the partial terms $L_c^{(l)}$ of L_c; see Eq. (7.158). Therefore,

$$\Sigma^{(1)}(k) = G_0^{-1}(k)L_c^{(1)}(k)G_0^{-1}(k), \tag{7.179}$$

and

$$\Sigma^{(2)}(k) = \Sigma_1^{(2)}(k) + \Sigma_2^{(2)}(k), \tag{7.180}$$

where

$$\Sigma_1^{(2)}(k) = G_0^{-1}(k)L_c^{(1)}(k)G_0^{-1}(k)L_c^{(1)}(k)G_0^{-1}(k), \qquad (7.181)$$

and

$$\Sigma_2^{(2)}(k) = -G_0^{-1}(k)L_c^{(2)}(k)G_0^{-1}(k). \qquad (7.182)$$

As $\Sigma^{(2)}$, the self-energy parts $\Sigma^{(l)}$ for $l > 2$ can be written as sums of different types of terms. Here we shall restrict our consideration to the second-order perturbation contributions to $G_0^{-1}(k)$. We shall use the results from Section 7.8 for $L_c(k)$ in order to clarify the properties of $\Sigma(k)$. Eq. (7.174) in u_0^2 order is

$$G^{-1}(k) = G_0^{-1}(k) - \left[\Sigma^{(1)}(k) + \Sigma^{(2)}(k)\right]. \qquad (7.183)$$

The self-energy contributions $\Sigma^{(1)}$ and $\Sigma^{(2)}$ are obtained with the help of Eqs. (7.179), (7.180), and Eqs. (7.161b), (7.163b), (7.164a)–(7.164b), (7.165a)–(7.165b) for $L_c^{(1)}$ and $L_c^{(2)} = L_c^{(2a)} + L_c^{(2b)} + L_c^{(2c)}$, respectively. The term $\Sigma_2^{(2)}(k)$ is equal to

$$\begin{aligned}\Sigma_2^{(2)}(k) &= -G_0^{-2}(k)\left[L_c^{(2a)} + L_c^{(2b)} + L_c^{(2c)}\right] \\ &= \Sigma_2^{(2a)}(k) + \Sigma_2^{(2b)}(k) + \Sigma_2^{(2c)}(k). \qquad (7.184)\end{aligned}$$

From the above equations and Eq. (7.163b) we have

$$\Sigma_2^{(2a)} = -16(n+2)^2 G_0(k) I_1^2(r_0). \qquad (7.185)$$

The calculation of $\Sigma_1^{(2)}(k)$ using Eqs. (7.181) and (7.161b) yields

$$\Sigma_1^{(2)} = 16(n+2)^2 G_0(k) I_1^2(r_0). \qquad (7.186)$$

So,

$$\Sigma_1^{(2)} + \Sigma_2^{(2a)} = 0, \qquad (7.187)$$

namely, the OP-reducible contribution $\Sigma_2^{(2a)}$ to $\Sigma^{(2)}$ is completely cancelled by the term $\Sigma_1^{(2)}$. Eq. (7.187) is shown graphically in Fig. 7.19, where the graphs for $L^{(1)}(k), L^{(2a)}(k)$ and $G_0(k)$ are given in accordance with Eqs. (7.181), (7.182) and (7.185). The mechanism of the compensation of

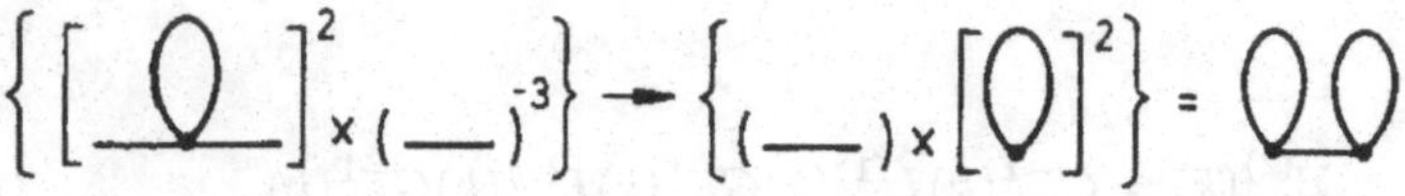

Fig. 7.19 Diagrammatic illustration of the relation (7.187).

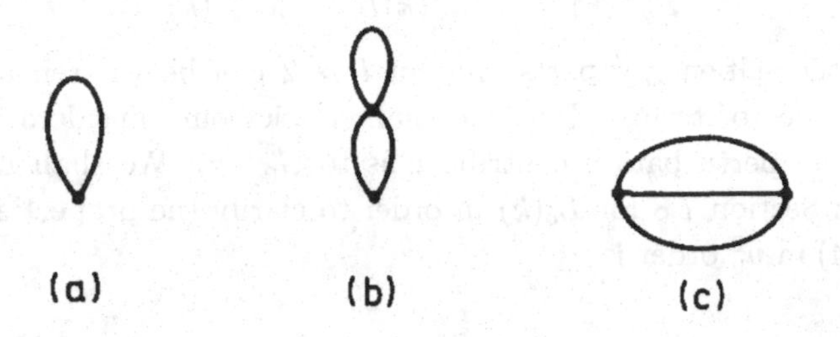

Fig. 7.20 Compact diagrams for: (a) $\Sigma^{(1)}$, (b) $\Sigma^{(2)}_{21}$, and (c) $\Sigma^{(2)}_{22}$.

the OP-reducible terms acts for any order of the perturbation expansion of $\Sigma(k)$; see Section 7.9.3.

It is seen from the diagrammatic relations in Fig. 7.19 and Eqs. (7.177)–(7.182) that the diagrams for $\Sigma(k)$ are obtained by cutting the external legs of the diagrams for $L(k)$ and removing the OP-reducible diagrams like those in Fig. 7.19. We shall introduce the notation

$$\Sigma^{(2)}_2 = \Sigma^{(2)}_{21} + \Sigma^{(2)}_{22}; \tag{7.188}$$

$\Sigma^{(2)}_{21}$ is the momentum-independent part of $\Sigma^{(2)}_2$, and $\Sigma^{(2)}_{22}$ is the momentum-dependent part; see Fig. 7.20, where $\Sigma^{(2)}_2$ is represented by compact diagrams.

Equation (7.183) can now be written in the form

$$G^{-1}(k) = (r_0 + ck^2) - \left[\Sigma^{(1)}(r_0) + \Sigma^{(2)}_{21}(r_0) + \Sigma^{(2)}_{22}(r_0, k)\right]. \tag{7.189}$$

Here

$$\Sigma^{(1)}(r_0) = -4(n+2)u_0 I_1(r_0), \tag{7.190a}$$

$$\Sigma^{(2)}_{21}(r_0) = 16(n+2)^2 u_0^2 I_1(r_0) I_2(r_0), \tag{7.190b}$$

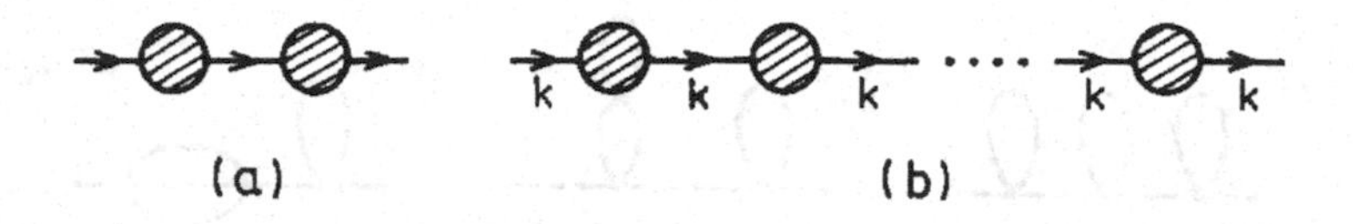

Fig. 7.21 Reducible diagrams with irreducible bubbles.

and

$$\Sigma_{22}^{(2)}(r_0, k) = 32(n+2)u_0^2 J(r_0, k). \tag{7.190c}$$

Equations (7.189)–(7.190c) will be widely used in our further considerations.

7.9.3 *Remarks on the reducibility of the diagrams*

Eq. (7.187) reveals one general property of the perturbation series for $\Sigma(k)$. The self-energy part $\Sigma^{(2)}(k)$ does not include in itself OP-reducible diagrams. In virtue of the same mechanism of compensation demonstrated for $\Sigma^{(2)}(k)$, it can be shown that the total $\Sigma(k)$ does not contain *reducible diagrams.* One example of reducible diagrams which we have met in our discussion are the OP-reducible diagrams (generally presented in Fig. 7.21(a). The two-particle (TP)-reducible perturbation terms correspond to diagrams which can be separated in three parts by cutting two internal G_0-lines. One l-particle reducible diagram for $L(k)$ is drawn in Fig. 7.21(b); there the internal G_0-lines connect the irreducible bubbles.

The bubbles can be of different type; see Fig. 7.22. Figures 7.22(a)–7.22(c) display all possible reducible diagrams of order u_0^3 and Figs. 7.22(d,e) give examples of the fourth-order reducible diagrams. The third order term

$$G_0^{-1}(k) L_c^{(3)}(k) G_0^{-1}(k) \tag{7.191}$$

is a part of the second term on the r.h.s. of Eq. (7.177), it can be represented by the diagrams 7.22a–7.22c provided the external legs have been cut; these legs are cancelled out by the factor $G_0^{-2}(k)$ in Eq. (7.191). It is a combinatorial problem to prove that these diagrams are compensated by the diagrams corresponding to the third order terms, which appear in the series (7.177):

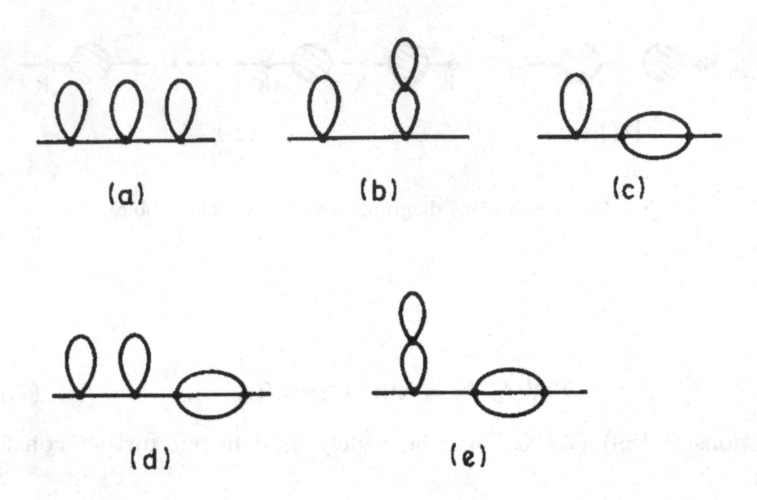

Fig. 7.22 Examples of reducible diagrams for $L_c^{(3)}$ and $L_c^{(4)}$.

$$G_0^{-4}(k)\left[L_c^{(1)}(k)\right]^3 \tag{7.192}$$

and

$$G_0^{-3}(k)L_c^{(2)}(k)L_c^{(1)}(k). \tag{7.193}$$

Thus the reducible part $L_R^{(3)} = L_c^{(3)} - L_I^{(3)}$ will be completely compensated and we have to investigate the irreducible part $L_I^{(3)}$. The diagrams for

$$\Sigma_I^{(3)}(k) = G_0^{-2}(k)L_I^{(3)}(k), \tag{7.194}$$

are given in Figs. 7.23(a, b). The irreducible diagrams of order u_0^4 for $\Sigma(k)$ are shown in Figs. 7.23(c–f).

The irreducible diagrams that describe $\Sigma(k)$ are generated by the second term in the expansion (7.177), which has the following form:

$$-G_0^{-2}(k)(L_I + L_R), \tag{7.195}$$

where $L_I = L_c - L_R$ is the irreducible part of L_c and L_R is the reducible part. The role of the remaining infinite series of terms $O(L_c^2)$ in Eq. (7.177) is to cancel the reducible part $-G_0^{-2}(k)L_R$. Therefore the total $\Sigma(k)$ will be:

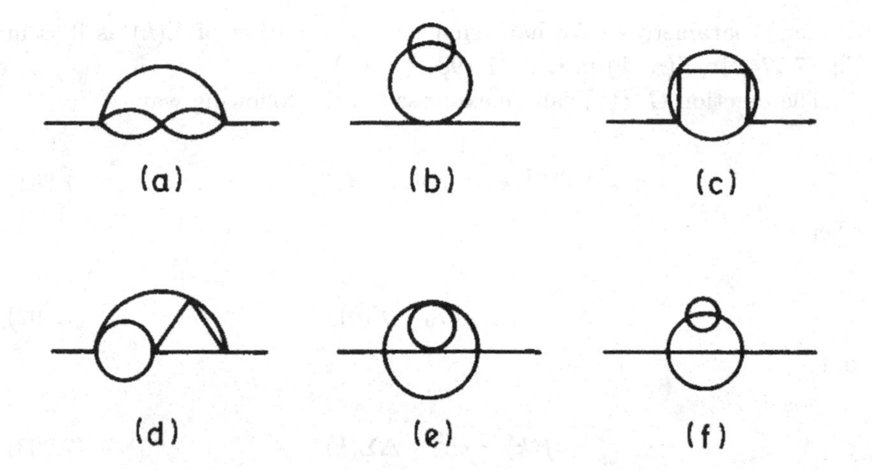

Fig. 7.23 Third-order (a)–(b) and fourth-order (c)–(f) irreducible diagrams for $\Sigma(k)$.

$$\Sigma(k) = G_0^{-2}(k) L_I(k), \tag{7.196}$$

or, for any contribution of order $l - \Sigma^{(l)}$,

$$\Sigma^{(l)} = G_0^{-2}(k) L_I^{(l)}(k). \tag{7.197}$$

7.10 Notion of Renormalization

Let us return to Eq. (7.174) and write it in the form

$$G^{-1}(k) = (r_0 + ck^2) - \Sigma(0) - \Delta\Sigma(k), \tag{7.198}$$

where

$$\Sigma(0) = \Sigma(r_0, 0), \tag{7.199}$$

and

$$\Delta\Sigma(k) = \Sigma(r_0, k) - \Sigma(r_0, 0). \tag{7.200}$$

The total function $G(k)$ depends on the parameters (r_0, c, and u_0) of the Hamiltonian and the momentum k. The self-energy $\Sigma(k)$ depends on the same parameters. We have changed the notation of $\Sigma(k)$ as it is in Eq. (7.174) to $\Sigma(r_0, k)$ in Eqs. (7.198)–(7.200).

The function $G^{-1}(k)$ can be expressed in the following way:

$$G^{-1}(k) = r(u_0) + f(k), \tag{7.201}$$

where

$$r(u_0) = r_0 - \Sigma(0), \tag{7.202}$$

and

$$f(k) = ck^2 - \Delta\Sigma(k). \tag{7.203}$$

With the help of the above rearrangement of the Dyson equation we divide $\Sigma(k)$ in two parts: (i) the k-dependent part, $\Delta\Sigma(k)$, which is zero for $k = 0$, it gives contributions to the k-dependent part of $G^{-1}(k)$, and (ii) the k-independent part, $\Sigma(0)$, which describes the perturbation contributions to the k-independent part of $G^{-1}(k)$. No doubt that $\Sigma(0)$ and $\Sigma(k)$ will alter the simple form of the bare (initial) values r_0 and c of the parameters of the theory, even they may alter the simple form of the quadratic energy spectrum, $\varepsilon_0(k) = ck^2$ of the fluctuations. Such changes which are a result from the fluctuation interaction or, from the interaction of other modes with the field $\varphi(x)$ are called *renormalization*. It has been shown at the end of Section 7.8.2 that the bare parameter r_0 gains a first-order perturbation correction and thereby its value in the interacting system changes from r_0 to $r(u_0)$. This is the simplest example of the *mass renormalization*, which we shall discuss in more details. In addition the interaction leads to a renormalization of the bare interaction constant u_0 and the field $\varphi(x)$. This problem will be examined in Chapter 8 within the framework of the general renormalization group approach to the critical phenomena. The preliminary consideration of the mass renormalization gives us the opportunity to analyze the perturbation series in a more consistent way. There are two equivalent ways to present the mass-renormalization and we shall present both of them.

We shall concentrate our attention on Eq. (7.202); Eq. (7.203) will be discussed in Section 7.11. Equations (7.189)–(7.190c) yield

$$r(u_0) = r_0 - \left[\Sigma^{(1)}(0) + \Sigma^{(2)}(0)\right], \tag{7.204}$$

where

$$\Sigma^{(1)}(0) = -4(n+2)u_0 I_1(r_0) \tag{7.205}$$

and

$$\Sigma^{(2)}(0) = 16(n+2)^2 u_0^2 I_1(r_0) I_2(r_0) + 32(n+2)u_0^2 J(r_0, 0). \tag{7.206}$$

In this section it will be convenient to change the standard notation $r_0 = \alpha_0(T - T_c^0)/T_c^0$ to $r_0 = \alpha_0(T - T_c^0)$; $(\alpha_0/T_c^0) \to \alpha_0$. Usually, we denote the critical temperature by T_c. Here and sometimes further on we shall use the notation T_c^0 for the bare (initial) critical temperature, i.e., the critical temperature of the corresponding ideal system where the fluctuation interaction is neglected. T_c^0 is just the critical temperature used in our discussion of the Landau expansion, the MF and Gaussian approximations. It becomes clear from Eq. (7.201) that $G^{-1}(0) \to 0$ provided the renormalized parameter $r = r(u_0) \to 0$. This happens as we shall see, at $T_c \neq T_c^0$, i.e., at the renormalized (true) critical temperature defined by $r(T_c) = 0$. The shift $(T_c^0 - T_c) > 0$ or, which is the same, the reduction of T_c due to the fluctuation interaction has been first discussed by Vaks, Larkin and Pikin (1966); see also Section 7.11.1.

7.10.1 *Mass renormalization from the perturbation series*

Eqs. (7.204)–(7.206) are written in a form which is not very suitable for the analysis of the perturbation effects and the critical behaviour of the interacting systems. One of the reasons is that the perturbation integrals depend on the initial parameter r_0 and, consequently, on the initial (bare) temperature. That is why we shall rearrange Eqs. (7.204)–(7.206) so that the parameter r_0 in the perturbation integrals to be changed to r. Neglecting the term $\Sigma^{(2)}(0)$ in Eq. (7.204) we have

$$r = r_0 + 4(n+2)u_0 \int_k \frac{1}{r_0 + ck^2}. \tag{7.207}$$

This expression can be presented also in the form:

$$r = r_0 + 4(n+2)u_0 \int_k \frac{1}{r + ck^2} + \Delta r_1, \tag{7.208}$$

where

$$\begin{aligned}\Delta r_1 &= 4(n+2)u_0 \int_k \left[\frac{1}{r_0 + ck^2} - \frac{1}{r + ck^2}\right] \\ &= 4(n+2)u_0 \left[(r - r_0) \int_k \frac{1}{(r + ck^2)(r_0 + ck^2)}\right] .\end{aligned} \tag{7.209}$$

In our first-order perturbation calculation, $(r - r_0) \sim u_0$, see Eq. (7.207). If the higher-order perturbation terms are included into consideration, $(r - r_0) \sim O(u_0)$. Therefore, to the first order in u_0 we can neglect $\Delta r_1 \sim O(u_0^2)$ and we can use the equation

$$r = r_0 + 4(n+2)u_0 I_1(r), \tag{7.210}$$

instead of Eq. (7.207). The above calculation can also be made with the help of $\Sigma^{(1)}(0) = \Sigma^{(1)}(r_0, 0)$. Then Eq. (7.207) will be

$$r = r_0 - \Sigma^{(1)}(r_0, 0); \tag{7.211}$$

Eq. (7.210) becomes

$$r = r_0 - \Sigma^{(1)}(r, 0), \tag{7.212}$$

and the term

$$\Delta\Sigma^{(1)} = \Sigma^{(1)}(r, 0) - \Sigma^{(1)}(r_0, 0) \sim O(u_0^2) \tag{7.213}$$

is omitted. The term $\Delta\Sigma^{(1)}$ should be taken into account when the higher orders in u_0 are considered. With an accuracy to the order u_0^2, the correction Δr_1 is obtained from Eqs. (7.209) and (7.207):

$$\begin{aligned}\Delta r_1 &\approx 16(n+2)^2 u_0^2 I_1(r) I_2(r) \\ &\approx 16(n+2)^2 u_0^2 I_1(r_0) I_2(r_0).\end{aligned} \tag{7.214}$$

It is exactly equal to the first term of $\Sigma^{(2)}(0)$ in Eq. (7.206), which enters in Eq. (7.204) with a minus; therefore Δr_1 and the first term of $\Sigma^{(2)}(0)$ are exactly cancelled with an accuracy to terms of order u_0^2. The term $\Sigma^{(2)}(0)$ is called the mass-renormalization part of $\Sigma(k)$ to the second order in u_0. The respective diagrams (a), (c), (d), and (f) in Fig. 7.8, or in compact notations the diagram (b) in Fig. 7.20 are named the mass-renormalization diagrams of second order. This way of renormalization can continue to the second and so on orders of the perturbation expansion for $G^{-1}(k)$.

7.10.2 *Mass-renormalization counter-term*

The mass renormalization stands for the effect of the interaction, which appears in the lowest (first-) order of the perturbation theory. Mathematically an essential part of the phenomena described by $\mathcal{H}_{\rm int}$ are represented by transforming r_0 into r. So we can include from the very beginning the part of $\mathcal{H}_{\rm int}$, from which the mass-renormalization arises in the initial non-perturbed Hamiltonian $\mathcal{H}_0 = \mathcal{H} - \mathcal{H}_{\rm int}$. For this purpose we shall use the technique of the counter-terms. Instead of considering the free field Hamiltonian

$$\mathcal{H}_0(r_0) = \frac{1}{2}\sum_{\alpha,k}\{r_0 + ck^2\}|\varphi_\alpha(k)|^2, \tag{7.215}$$

we shall "dress" the "free" fields φ_α with the difference $(r_0 - r)$ between the bare r_0 and the renormalized r values of the inverse susceptibility. This is equivalent to the change of the initial Hamiltonian $\mathcal{H}_0(r_0)$ to $\mathcal{H}_0(r)$,

$$\mathcal{H}_0(r) = \frac{1}{2}\sum_{\alpha,k}\{r + ck^2\}|\varphi_\alpha(k)|^2. \tag{7.216}$$

Then the interaction Hamiltonian will gain the counter-term $\mathcal{H}_c = \mathcal{H}'_{\rm int} - \mathcal{H}_{\rm int}$; here

$$\mathcal{H}'_{\rm int} = \mathcal{H}_{\rm int} + \frac{1}{2}(r_0 - r)\sum_{\alpha,k}|\varphi_\alpha(k)|^2. \tag{7.217}$$

The counter-term $\mathcal{H}_c(r_0 - r)$ is shown in Fig. 7.24.

The next step is to do the perturbation expansion for $G^{-1}(r,k)$ in terms of the new interaction Hamiltonian $\mathcal{H}'_{\rm int}$ and the new initial Green function

$$G_0(r,k) = \frac{1}{r + ck^2}. \tag{7.218}$$

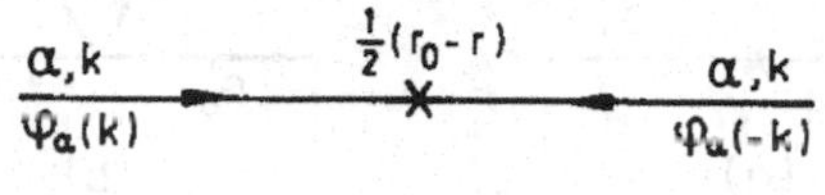

Fig. 7.24 Diagram for $\mathcal{H}_c$; see Eq. (7.217).

As a result, the Dyson equation becomes

$$G^{-1}(r,k) = r + ck^2 - \widetilde{\Sigma}(r,k), \tag{7.219}$$

where the modified self-energy $\widetilde{\Sigma}$ is given by the Eq. (7.175), in which $L_c(k)$ is substituted by

$$\widetilde{L}_c(k) = \sum_{l=1}^{\infty} \frac{(-1)^l}{l!} \langle |\varphi_\alpha(k)|^2 \mathcal{H}'^l_{\text{int}} \rangle_{0c}, \tag{7.220}$$

and $G_0(r_0, k)$ is substituted by G_0 from Eq. (7.218). In this way a part of the interaction effects are taken into account in the zero approximation of the theory. Of course the perturbation expansion for $\mathcal{Z}$, Φ, $G^{(2)}$, and $G^{(4)}$ discussed previously in terms of $\mathcal{H}_{\text{int}}$ can be straightforwardly re-derived for $\mathcal{H}'_{\text{int}}$. We shall examine Eq. (7.219) further. Owing to the properties of the infinite series (7.177) explained in Section 7.9, we can write $\widetilde{\Sigma}$ in the following way

$$\widetilde{\Sigma}(k) = G_0^{-2}(k)\widetilde{L}_I(k); \tag{7.221}$$

cf. Eq. (7.196); from now on the "prime" of $\mathcal{H}'_{\text{int}}$ and the "tilde" of $\widetilde{\Sigma}$ and $\widetilde{L}$ will be omitted. The calculation of $\Sigma^{(1)}(r)$ and $\Sigma^{(1)}(r,k)$ is somewhat lengthy but the reader can make it without any difficulties.

A part of the diagrams for $L_I^{(1)} \equiv L^{(1)}$ and $L_I^{(2)}$ is shown in Fig. 7.20. The new diagrams for the counter-terms of the perturbation expansion are depicted in Fig. 7.25.

The diagram 7.25(a) gives the contribution $(r - r_0)$ to $\Sigma(r,k)$ and, hence, it restores the initial parameter r_0 in Eq. (7.219). The diagram 7.25(b) describes the term

$$(n+2)(r_0 - r)u_0 I_2(r) \sim O(u_0^2), \tag{7.222}$$

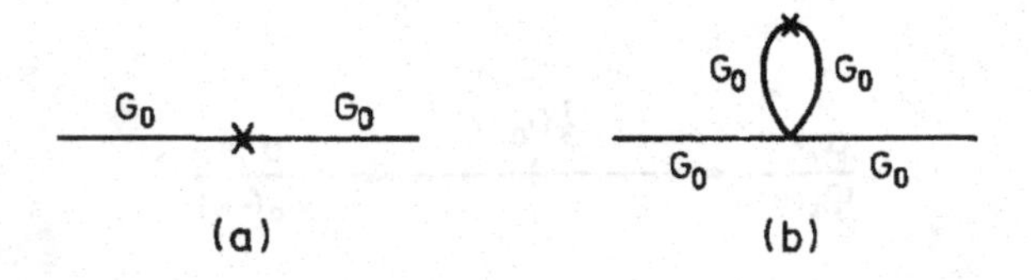

Fig. 7.25 Diagrams for counter-terms: (a) of order $O(u_0)$, (b) of order $O(u_0^2)$.

of $\Sigma(r,k)$. This results in G^{-1} becomes

$$G^{-1} = r_0 + ck^2 + 4(n+2)u_0 I_1(r) - 16(n+2)^2 u_0^2 I_1(r) I_2(r) \\ - 4(n+2)(r_0 - r)u_0 I_2(r) - 32(n+2)u_0^2 J(r,k). \qquad (7.223)$$

All terms in G^{-1} but the first depend on the renormalized parameter r. The k-dependent part of G^{-1} is given by Eq. (7.203), in which

$$\Delta\Sigma(k) = 32(n+2)u_0^2 \left[J(r,k) - J(r,0)\right]. \qquad (7.224)$$

The renormalized parameter r is defined by the $(k=0)$-part of G^{-1}, $r = G^{-1}(0)$. Then

$$r = r_0 + 4(n+2)u_0 I_1(r) - 16(n+2)^2 u_0^2 I_1(r) I_2(r) \\ - 4(n+2)(r_0 - r)u_0 I_2(r) - 32(n+2)u_0^2 J(r,0). \qquad (7.225)$$

If we neglect the u_0^2-terms we shall obtain Eq. (7.210). Putting this first-order result in the fourth term on the r.h.s. of Eq. (7.225), we see that this term is exactly cancelled by the third term in the same equation. We have already mentioned that such a procedure is correct because we work with an accuracy to the order of u_0^2; terms of order equal or higher than $O(u_0^3)$ are omitted. Therefore,

$$r = r_0 + 4(n+2)u_0 I_1(r) - 32(n+2)u_0^2 J(r,0) + O(u_0^3). \qquad (7.226)$$

The method of successive renormalization of the perturbation integrals presented in Section 7.10.1 in the lowest order of the perturbation theory and the method of Hamiltonian counter-terms discussed in this subsection lead to one and the same result — the mass-renormalized theory does not contain mass-renormalization diagrams. In fact, the expression (7.226) represents the inverse susceptibility of the interacting system; it is written as a perturbation expansion, in which the mass-renormalization diagrams are excluded.

To draw a general conclusion from the above discussion we shall write the full Green function $G(r,k)$ and the self-energy $\Sigma(k)$ to the second order in u_0 :

$$G^{-1}(r,k) = r + k^2 - \Delta\Sigma(k), \qquad (7.227a)$$

$$r = r_0 - \Sigma(r, 0), \tag{7.227b}$$

and

$$\Delta\Sigma(k) = \Sigma(r, k) - \Sigma(r, 0). \tag{7.227c}$$

This is the general form of the Dyson equation (7.227a) valid to any order of the perturbation theory.

7.11 Properties of the Standard Perturbation Series

The main effect of the fluctuation interaction on the critical behaviour can be understood from the first non-vanishing perturbation correction to r; see Eq. (7.210).

7.11.1 *How to determine the critical region from the perturbation expansion*

We shall write the integral $I_1(r)$ in the form

$$I_1(r) = I_1(0) + \Delta I_1(r), \tag{7.228a}$$

where

$$\Delta I_1(r) = -r \int_k \frac{1}{ck^2(r + ck^2)}. \tag{7.228b}$$

The integrals $I_1(0)$ and $\Delta I_1(r)$ are calculated with the help of Eq. (D.12):

$$I_1(0) = \frac{K_d \Lambda^{d-2}}{(d-2)c} \tag{7.229}$$

and

$$\Delta I_1(r) = -K_d \left(\frac{r}{c}\right) \int_0^\Lambda \frac{k^{d-3} dk}{r + ck^2}. \tag{7.230}$$

Changing the integration variable in the last integral to $y = (c/r)k^2$ and approximating the new cutoff $\Lambda' = (\Lambda^2 c/r)$ by infinity ($r \ll \alpha_0 T_c$), we obtain by making use of Eq. (E.2) that

$$\Delta I_1(r) = -C_1(d)r^{d/2-1}, \qquad 2 < d < 4; \tag{7.231}$$

here

$$C_1(d) = c^{-d/2} K_d \Gamma\left(\frac{d}{2} - 1\right) \Gamma\left(2 - \frac{d}{2}\right) > 0. \tag{7.232}$$

So Eq. (7.210) for r becomes

$$r = r_{0c} - 4(n+2)C_1 u_0 r^{d/2-1}, \tag{7.233}$$

with

$$r_{0c} = r_0 + 4(n+2)u_0 I_1(0) = \alpha_0(T - T_c), \tag{7.234a}$$

and a renormalized critical temperature equal to

$$T_c = T_c^0 - 4(n+2)\left(\frac{u_0}{\alpha_0}\right) I_1(0). \tag{7.234b}$$

It is obvious from the expression (7.229), that $T_c(u_0)$ depends on the cutoff Λ too. Of course, our study is valid for $T \geq T_c$ and $2 < d < 4$. The quantity r_{0c} is equal to zero at $T = T_c$ — the true (renormalized) transition temperature. From Eq. (7.233) we can obtain r at least for some particular values of $2 < d < 4$. For example, if $d = 3$, we have to solve a quadratic equation for $\sqrt{r}$ $(r > 0)$. We prefer not to do this simple calculation, but to focus our attention on the properties of the perturbation term $\sim 1/r^{1-d/2}$. The main question is whether $G^{-1}(r, 0)$ has the simple Ornstein–Zernicke form $G^{-1}(r, 0) \sim (T - T_c) \sim r_{0c}$, or in other words whether the susceptibility $\chi \sim r^{-1}$ diverges linearly at T_c. This will be possible, if

$$r \approx r_{0c}, \tag{7.235a}$$

i.e.,

$$r \gg 4(n+2)C_1(d)u_0 r^{d/2-1}. \tag{7.235b}$$

The condition (7.235b) is identical to the Ginzburg (–Levanyuk) criterion (Section 7.5.2). Let us remember that the original Ginzburg criterion defines the temperature interval where the MF approximation is valid and the fluctuation contribution to the specific heat (and, certainly, to the other thermodynamic quantities) can be neglected as small compared to the MF values. The region around T_c where this condition is violated is the critical region. The criterion (7.235b) has another meaning. It defines the temperature interval above T_c where the fluctuation interaction can be neglected, i.e., where the Gaussian approximation works well. The opposite condition,

$$r^{(4-d)/2} \lesssim 4(n+2)C_1(d)u_0, \tag{7.236}$$

will be therefore satisfied in the critical region, in which the fluctuation interaction is relatively strong and the Gaussian approximation works no more. The mathematical expressions (7.236) and (7.105) for the above mentioned criteria coincide (apart from an unimportant numerical factor) which means that the critical region is this vicinity of T_c, where both the MF and the Gaussian approximations break. Therefore, these approximations are valid in one and the same temperature interval ($|t| > |t_G|$). There is no doubt that the analogous calculations for $t < 0$ will give the criterion (7.236) with $r \to 2|r|$. Often, the critical region is estimated by perturbation calculations similar to that presented here] see, e.g., Amit (1974); for bicritical and tetracritical point, see, e.g., Tonchev and Uzunov (1981a). The advantage of this approach is that the critical region is calculated for the true critical temperature T_c.

The criterion for the validity of the Ornstein–Zernicke form of $G^{-1}(k)$ can also be applied to the k-dependent part $f(k)$; see Eq. (7.203). In this case the requirement will be

$$ck^2 \gg |\Delta\Sigma(k)|; \tag{7.237}$$

$\Delta\Sigma(k)$ can be calculated from Eqs. (7.224) and (7.167). Let us set $r = 0$ and consider $\Delta\Sigma(0,k)$. The integral (7.167) for $r_0 = 0$ can be estimated qualitatively. Here we are interested in the leading dependence of $J(0,k)$ on k. To find it we should count up the powers of k in the integrand of $J(0,k)$. This analysis yields

$$J(0,k) \sim \frac{[k]^{2d}}{[k]^6} \tag{7.238}$$

or

$$J(0,k) \sim \frac{k^{2d-6}}{c^3}. \tag{7.239}$$

Note that this estimate is valid for $r = 0$, i.e., at the critical point ($T = T_c$). At this point $J(r, k)$ is divergent for $d < 3$; conclusions when $d = 3$ cannot be given straightforwardly because the above analysis does not take into account the weak dependencies like $\ln k$. However this divergence does not appear in $\Delta\Sigma(k)$, where it is totally compensated by the term $J(0, 0)$. From Eq. (7.237) and Eq. (7.239) we find the inequality

$$c^4 \gg u_0^2 k^{2d-8} \qquad (k \neq 0). \tag{7.240}$$

For $d > 4$ the criterion is always fulfilled provided $c^2 > (u_0 \Lambda^{d-4})$. The difficulties exist for the spatial dimension $d < 4$, where the integral $J(0, k)$ is divergent. Clearly, the simple quadratic fluctuation spectrum described by the Ornstein–Zernicke correlation function may change to another type of k-dependence for small k. In particular, this can be expected for small r, i.e., in the critical region where $\Delta\Sigma(k)$ has large values for small k.

At this stage the reader must pay a special attention to the following property of the interacting fluctuations $\varphi_\alpha(k)$; namely, their influence on the critical behaviour is very strong for small momenta k, *infrared region (or, limit)* or, *infrared limit* ($k \to 0$) in analogy with the red color region of the light spectrum at relatively small photon frequencies. Obviously, the integral $J(r, k)$ will not exhibit divergences, even it may become negligible for $r \neq 0$. The criterion (7.240) breaks down for small values of both r and k. Therefore, the fluctuation interactions become important in the critical region and their main effect is given by their "infrared" asymptote.

7.11.2 *Effective expansion parameter*

Up to now we have discussed a perturbation expansion in powers of the coupling constant u_0. The power series in u_0 however have another effective expansion parameter which is expressed by the ratio of two neighbouring terms of the expansion of any physical quantity. The criterions (7.235b) and (7.237) for the validity of the Ornstein–Zernicke form of $G^{-1}(k)$ have been obtained under the condition that the first nonvanishing perturbation corrections to $G^{-1}(k)$ are small compared to the zero order approximation ($u_0 = 0$). The effective expansion parameter for $r \neq 0$ is found from this requirement. From Eq. (7.235b) we have

$$u_{\text{eff}} = u_0 r^{d/2-2}, \tag{7.241}$$

where the numerical factor $4(n+2)C_1$ is neglected. Therefore the effective expansion parameter itself defines the critical region. In this region $u_{\text{eff}} \geq 1$ and, hence, the standard perturbation expansion is not valid. In a close vicinity of T_c the perturbation calculation of the fluctuation-interaction effects is unreliable. Clearly, we can use the above perturbation theory to calculate the fluctuation effects outside the critical region or, to receive some information about the qualitative behaviour of the interacting system near the critical region ($u_{\text{eff}} \sim 1$). We shall show how the effective parameter u_{eff} is obtained from the ratio of any two neighbouring terms of order u_0^l and u_0^{l+1}. For example the ratio

$$\left|\frac{\Sigma^{(2)}(r,0) - \Sigma^{(2)}(0,0)}{\Sigma^{(1)}(r,0) - \Sigma^{(1)}(0,0)}\right| \tag{7.242}$$

yields

$$\frac{u_0|J(r,0) - J(0,0)|}{|\Delta I_1(r)|} \sim u_{\text{eff}}. \tag{7.243}$$

The integral difference $\Delta J(r) = J(r,0) - J(0,0)$ in Eq. (7.231) is evaluated in a way similar to that for $J(0,k)$; see Eqs. (7.238) and (7.239). In the present case we have to replace k^2 in Eq. (7.239) with (r/c), and so $J(r,0) \sim r^{d-3}$ for any $r > 0$. Then $|\Delta J(r)| \sim r^{d-3}$ $(r > 0)$. $\Delta I_1(r)$ is taken from Eq. (7.231).

In general, any new order of the perturbation theory gives an extra factor u_0. Moreover two additional Green functions G_0 and one extra integration over the momenta should be taken into account. So the general expression for the effective parameter will be

$$u_{\text{eff}}(r,q) \sim u_0 \int_k \frac{1}{(r+ck^2)[r+c(k+q)^2]}. \tag{7.244}$$

For $q = 0$, the integral in Eq. (7.244) is equal to $I_2(r)$; its estimation gives $I_2(r) \sim r^{2-d/2}$ and as a result u_{eff} coincides with the quantity from Eq. (7.241). For $r = 0$,

$$u_{\text{eff}}(q) \sim u_0 q^{d-4}; \tag{7.245}$$

this dependence is responsible for the peculiar behaviour of all momentum-dependent integrals at the critical point discussed in the previous subsection. The perturbation integral (7.244) appears in the expansion for the four-point correlation function $G^{(4)}$.

The properties of the perturbation series discussed in this section can be determined from the perturbation expansion of other physical quantities such as, for example, the thermodynamic potential Φ or the ($l > 2$)-point correlation functions $G^{(l)}$ (see Section 7.4 and Section 7.7.2). The qualitative treatment of the perturbation integrals makes it possible to obtain the right power dependence on r or the external momentum k. The systematical way for the calculation of the perturbation integrals is presented in Section E.2.

7.11.3 *Dependence on the symmetry index n and the spatial dimension d. Universality classes*

The dependence of the thermodynamic and correlation functions on the number n of the order parameter components φ has been first established for the Gaussian approximation discussed in Section 7.5; see, e.g., Eq. (7.110). For interacting fluctuations, this dependence is more essential because the perturbation terms gain numerical factors which are functions (polynomials) of n. In contrast to the MF approximation, the fluctuation theory of phase transitions gives the possibility to distinguish between systems with different symmetry index n. For example, the n-vector φ^4-Hamiltonian (7.1) describes a variety of isotropic systems, whose fluctuation critical behaviour can be classified according to the value of n. Systems with different n belong to different symmetries of the ordering below T_c.

The infinite number of perturbation terms and the difficulties we have run upon in their calculation within the standard perturbation theory developed in this Chapter do not allow a systematic investigation of the critical behaviour for the different *universality classes* (this point is explained in more details in Chapter 8 and Chapter 9). The perturbation series also depend on the spatial dimension d. Systems with equal n but different spatial dimension will belong to different universality classes. Therefore we should define an universality class with the help of two values — n and d.

The classification of the critical behaviour in universality classes determined by the couple (n, d) will be useful if we know the main properties of the critical behaviour within each of the possible classes of universality. Unfortunately, little can be said about this problem by means of the standard perturbation theory. Another remark concerns the Hamiltonian of the system. For different Hamiltonians the perturbation series will be different and, hence, the universality classes (n, d) should be defined for each Hamiltonian separately. The way, in which the fluctuation behaviour in the vicinity of critical and multicritical points is described and classified in universality classes will be discussed in Chapter 8 and Chapter 9.

7.11.4 *Borderline dimensions*

Our calculation of the lowest-order perturbation terms for $G^{-1}(r, k)$ has been done for the spatial dimensions $2 < d < 4$. We have revealed that in this case a nontrivial fluctuation behaviour occurs in the critical region of the second-order phase transitions.

Let now consider the cases $d \leq 2$ and $d \geq 4$. Our brief discussion is based on the Eq. (7.210) for r. The integral

$$I_1(r) = K_d \int_0^{\Lambda} \frac{k^{d-1} dk}{(r + ck^2)} \tag{7.246}$$

can be calculated for a finite cutoff Λ. Obviously for $d \leq 2$, $I_1(r) \to \infty$ as $r \to 0$. This divergence is of the type

$$I_1(r) \sim \ln(1/r) \quad \text{for} \quad d = 2, \tag{7.247a}$$

and

$$I_1(r) \sim r^{d/2-1} \quad \text{for} \quad d < 2. \tag{7.247b}$$

Then the integral $I_1(0)$ for $d \leq 2$ is infinite and the reason lies in the strong fluctuation effect at low spatial dimensions. For these dimensions ($d \leq 2$), the true critical temperature T_c follows the divergence of $I_1(0)$; see Eq. (7.234b). For $d = 2$, $T_c \to -\infty$ logarithmically while T_c tends to $-\infty$ by the power law (7.247b) for $d < 2$. Of course this result makes no sense and it merely indicates that the critical temperature and, hence, the macroscopic ordering does not exist for $d \leq 2$. At low dimensions the ordering is destroyed by the fluctuation effects. This effect has been obtained in somewhat different way by Mermin and Wagner (1966) and Hohenberg (1967); see also Mermin (1968). The fluctuation instability of the ordering in low-dimensional ($d = 1, 2$) crystal lattices is known from the early works by Peierls (1934, 1935) and Landau (1937b); see also Peierls (1979) and Landau and Lifshitz (1980). Therefore we can determine the lower *borderline* (or *critical*) dimension

$$d_L = 2, \tag{7.248}$$

below which the ordering does not exist. In the interval $2 < d < 4$ the behaviour of the integral $I_1(r)$ is explained in Section 7.11.1. For $d = 4$, $I_1(0)$ is given again by Eq. (7.229) but

$$\Delta I_1(r) = -\frac{K_d}{4c} r \ln\left(\frac{c\Lambda^2 + r}{r}\right)$$
$$\sim \frac{r}{c} \ln\left(\frac{c\Lambda^2}{r}\right), \quad \text{for} \quad c\Lambda^2 \gg r. \tag{7.249}$$

From Eq. (7.249) and the condition (7.235a) we can obtain the critical temperature interval (7.112) in this case. For $d > 4$, $\Delta I_1(r)$ is again proportional to $r^{-1+d/2}$, see Eq. (7.231) but for these higher dimensions, $\Delta I_1(r) \to 0$ for $r \to 0$ and the critical region does not exist. Therefore we can define the *upper borderline dimension*

$$d_U = 2, \tag{7.250}$$

for the φ^4-model (7.1). With the increase of d from $d < 4$ to $d > 4$ the nontrivial critical behaviour undergoes a dimensional crossover to the simple Gaussian critical behaviour in which the fluctuation interactions can be neglected. Why $d_L = 2$ and $d_U = 4$? If there are long-range interactions (Section 6.7) the quadratic fluctuation spectrum $\varepsilon(k) = ck^2$ will be replaced with $\varepsilon(k) = ck^\theta$, $0 < \theta < 2$. Then

$$d_L = \theta \quad \text{and} \quad d_U = 2\theta. \tag{7.251}$$

If the φ^4-interaction is changed to another, for example, φ^m-interaction the critical dimensions d_L and d_U will take another values; see, e.g., Pfeuty and Toulouse (1975), where the value

$$d_U = \frac{m\theta}{m-2} \tag{7.252}$$

is presented. It will be useful to derive d_L, d_U and the Ginzburg criterion for the φ^3- and φ^6- theories. For this purpose, it is necessary to consider the Eq. (7.227a) and to find the first perturbation terms of $\Sigma(r, 0)$. The perturbation series for $G^{-1}(k)$ within the φ^3- and φ^6- theories are quite different from that for the model (7.1). For example the first nonvanishing perturbation term to $\Sigma(r, k)$ for the φ^3-model is of order u_0^2 and depends both on r and the external momentum k.

7.11.5 *Breakdown of the standard perturbation expansion*

By means of the field theory of fluctuations and the perturbation technique presented in this Chapter we have obtained important information about the fluctuation effects on the critical behaviour. Our discussion of the perturbation expansion was limited to the two-loop approximation but the reader may take advantage from the books by Zinn–Justin (1996), and Kleinert and Schulte-Frohlinde (2001), and learn more about the higher orders of the loop expansion especially regarding the ϕ^4-model.

We have determined the condition for the validity of the MF theories in the vicinity of the critical points. There exists a fluctuation (critical) region in a close vicinity of any critical point in which both the MF theory and the Gaussian approximation for the fluctuations break. The treatment of this region requires sophisticated theoretical methods describing strongly interacting fields. The MF and the Gaussian approximations can be considered as reliable in the whole temperature interval outside the critical region (for the Landau expansion one should have in mind the additional condition $t < 1$). Clearly, the perturbation calculations can be used for obtaining the corrections to the results from these theories.

The description of the critical phenomena inside the critical region of systems having spatial dimensions $d_L < d < d_U$ is a serious theoretical problem (below, the usual case of $2 < d < 4$ will be discussed). The calculations in the second order of the standard perturbation theory demonstrate that the perturbation terms become infinite in a close vicinity of the critical temperature ($r \to 0$). Apart from the asymptotic case ($r \to 0$), the fluctuation interaction inside the critical region gives corrections to the thermodynamic and correlation quantities which are much larger than the corresponding MF values or the values in the Gaussian approximation. It is then clear why the critical behaviour is often interpreted as the behaviour of the interacting fluctuations. The large effect of the fluctuation interactions certainly follows from the instability of the zero-order solution (in the Gaussian approximation) for the possible states of the system near the critical point. The stability of the thermodynamic states is determined in Gaussian approximation by the inverse Green function $G_0^{-1}(k) = (r_0 + ck^2)$. As $r \to 0$, the correlation length $\xi_0 = (c/r_0)^{1/2}$ of the free noninteracting fluctuations tends to infinity which means that the fluctuation correlations extend over large (macroscopic) scales. The large scale (infrared $k \to 0$) fluctuations are the reason for the "infrared divergence" of the perturbation integrals. When r_0 is near to zero, the response G_0 of the system to small momentum free fluctuations $\varphi_\alpha(k)$, $k \sim 0$, is too large which signals about the weakened stability of the thermodynamic state of the system. Surely the relatively higher-momentum fluctuations $\varphi_\alpha(k)$ those having k in the interval $\Lambda < k < \Lambda'$, where Λ' is some momentum different from zero, will not be essential for the behaviour of the system in small intervals around T_c. On the contrary, the instability towards the small momentum fluctuations may lead to the reconstruction of the fluctuation spectrum $G_0^{-1}(k) = (r_0 + ck^2)$. We have already obtained that the bare temperature T_c^0 and the bare parameter $r_0 \sim (T - T_c^0)$ are renormalized ("dressed" by interaction) and this is the first sign of the possibility for more drastic reconstruction of the free field fluctuation distribution. With the help of the mass-renormalization ($r_0 \to r$) we have eliminated a part of the divergent perturbation integrals but we have not been able to take into account the divergences systematically.

Further studies of the critical fluctuations can follow different ways. The first idea which may arise is to investigate the higher orders of the perturbation theory and try to perform the summation of the most divergent (leading) perturbation terms. There exist different methods for the partial summation of the diagram series. For the Dyson equation one can use the so-called skeleton technique which can be deduced from the general perturbation formulae; see, e.g., Fetter and Walecka (1971) and Popov (1976, 1987). Up to now such an approach to the critical states has not led to a substantial progress because the attempts to select partial infinite subsets of leading perturbation terms reveal that the number of these subsets is itself infinite.

Chapter 8

Renormalization Group

8.1 Modified Perturbation Scheme

We have seen in Chapter 7 that the divergences of the perturbation terms arise from the small momenta k, i.e., from the "infrared" asymptote of the perturbation integrals. According to the experimental and theoretical results (see Chapter 1 and Chapter 7) the field amplitudes $\varphi_\alpha(k)$ with relatively large wave vectors k ($\equiv \boldsymbol{k}$) describe effects of having short characteristic lengths ($L \ll \xi$). They are less important than the (quasi)macroscopic phenomena ($k \sim \xi^{-1} \sim 0$) in a close vicinity of the critical temperature T_c.

So we have enough arguments in favour of the reconstruction of the perturbation theory. We shall try to divide the relatively short-range effects from the long-range ones. Let the Hamiltonian be $\mathcal{H}(\varphi)$ with a momentum cutoff $\Lambda \sim 1/a_0$, where a_0 is the mean inter-particle distance (of the order of the lattice spacing in a crystalline solid). We shall separate the Fourier sum for the field components $\varphi_\alpha(x)$ of the vector order parameter $\varphi(x) = \{\varphi_\alpha(x)\}$,

$$\varphi_\alpha(x) = \frac{1}{\sqrt{V}} \sum_{k=0}^{\Lambda} \varphi_\alpha(k) e^{ikx}, \qquad kx \equiv \boldsymbol{k} \cdot \boldsymbol{x}, \tag{8.1}$$

in two parts

$$\varphi_\alpha(x) = \frac{1}{\sqrt{V}} \sum_{k=0}^{\Lambda'} \varphi_\alpha(k) e^{ikx} + \frac{1}{\sqrt{V}} \sum_{k>\Lambda'}^{\Lambda} \varphi_\alpha(k) e^{ikx}, \tag{8.2}$$

where $\Lambda' < \Lambda$ is another (auxiliary) cutoff; $\Lambda' \sim 1/a_0'$, corresponds to another effective minimal distance a_0'.

The first term in Eq. (8.2) describes an effective fluctuation field $\varphi_0(x)$ in a system where the minimal distance taken into account is of the order $a_0' > a_0$. The remaining part $\varphi_1(x) = \varphi(x) - \varphi_0(x)$ of $\varphi(x)$ is related to the behaviour of a finite system ($k > \Lambda'$) where no phenomena of interest to us can occur at distances larger than $(\Lambda')^{-1}$.

It is convenient to introduce "small" and "large" momentum amplitudes

$$\varphi_{0\alpha}(k) = \begin{cases} \varphi_\alpha(k), & \text{if } 0 < k < b^{-1}\Lambda, \\ 0, & \text{otherwise,} \end{cases} \tag{8.3a}$$

and

$$\varphi_{0\alpha}(k) = \begin{cases} \varphi_\alpha(k), & \text{if } b^{-1}\Lambda < k < \Lambda, \\ 0, & \text{otherwise,} \end{cases} \tag{8.3b}$$

The number $b = (\Lambda/\Lambda') > 1$ is *the rescaling factor.* Then the total field $\varphi_\alpha(k)$ can be written as a sum of the small momentum (or slow-varying) fields $\varphi_{0\alpha}(k)$ and the large momentum (or fast-varying) fields $\varphi_{1\alpha}(k)$:

$$\varphi_\alpha(k) = \varphi_{0\alpha}(k) + \varphi_{1\alpha}(k). \tag{8.4}$$

Using this relation the sum (8.1) is naturally divided in the parts shown in Eq. (8.2). One may think of the borderline mode $\varphi_\alpha(\pm\Lambda')$ as included in $\varphi_{0\alpha}(k)$, see Eq. (8.2) or, equivalently, as included in $\varphi_{1\alpha}$ (this is of no importance to the discussion in this Chapter). If $\Lambda' \to 0$, $\varphi_{0\alpha}(k) \to \varphi_\alpha(0)$ — the zero momentum (x-independent) mode. The mode $\varphi_\alpha(0)$ gives the order parameter of homogeneous systems.

The exclusion of the zero-momentum (condensate) mode $\varphi(0)$ from all amplitudes $\varphi(k)$ has been well-known for a long time. One of the famous examples is the description of the lower-momentum (superfluid) spectrum of the almost-degenerate nonideal Bose gas (Bogoliubov, 1947); see also Abrikosov, Gor'kov, and Dzyaloshinskii (1962), Fetter and Walecka (1971), and Lifshitz and Pitaevskii (1980). The present division in slow and fast fields follows the idea of the Kadanoff block transformation in the momentum (k-) space (Chapter 1).

If we are able to build a perturbation theory, in which the momentum integration is only over the fast variables in the shell ($b^{-1}\Lambda < k < \Lambda$; $k \equiv |\boldsymbol{k}|$) we may hope to avoid the infrared divergences of the perturbation integrals. Moreover, instead of a total functional integration of the partition integral, see Eq. (7.4), we may try to receive an useful information from the functional integration only over the large-momentum amplitudes $\varphi_{1\alpha}$. From this integration we do not obtain the partition integral, rather it will give an effective Hamiltonian $\mathcal{H}'(\varphi_0)$ which depends only on the slow fields $\varphi_{0\alpha}(k)$. The relation between $\mathcal{H}'(\varphi_0)$ and the total initial Hamiltonian $\mathcal{H}(\varphi) \equiv \mathcal{H}(\varphi_0 + \varphi_1)$ is

$$e^{-\mathcal{H}'(\varphi_0)} = \int D\varphi_1 \, e^{-\mathcal{H}(\varphi)}, \tag{8.5}$$

where the integration is over the independent degrees of freedom $\varphi_{1\alpha}(k)$ in the momentum interval $b^{-1}\Lambda < k < \Lambda$:

$$\int D\varphi_1 = \prod_{\alpha} \prod_{b^{-1}\Lambda<k<\Lambda} d\,[\Re\, \varphi_{1\alpha}(k)]\, d\,[\Im\, \varphi_{1\alpha}(k)]\,; \tag{8.6}$$

cf. Eq. (7.47). As in Chapter 7, the factor $(1/T)$ in the statistical exponent $(-\mathcal{H}/T)$ is absorbed in $\mathcal{H}$.

Now our task is to put Eq. (8.4) in $\mathcal{H}(\varphi)$ and to perform the functional integration in Eq. (8.5). According to our experience with the perturbation series gained in Chapter 7, we may expect that this integration will give an effective Hamiltonian $\mathcal{H}(\varphi_0)$, in which the bare parameters of the original Hamiltonian $\mathcal{H}(\varphi)$ will be renormalized. In addition, terms which are not present in the initial Hamiltonian $\mathcal{H}$ may appear in $\mathcal{H}(\varphi_0)$. Having these remarks in mind we shall try to use the integral transformation (8.5) in order to learn new information about the properties of the fluctuation fields $\varphi_\alpha(x)$.

The above scheme can be obviously applied to any Hamiltonian but, in order to obtain explicit results we have to work with some particular fluctuation model. As in Chapter 7 we choose the φ^4-effective Hamiltonian of the system in a zero external field above the critical point $T \geq T_c$ where $\langle\varphi\rangle = 0$:

$$\mathcal{H}(\varphi) = \mathcal{H}_0(\varphi) + \mathcal{H}_{\text{int}}(\varphi). \tag{8.7a}$$

$\mathcal{H}_0$ and $\mathcal{H}_{\text{int}}$ are given by the following expressions

$$\mathcal{H}_0(\varphi) = \frac{1}{2}\sum_{\alpha,k}(r_0 + ck^2)|\varphi_\alpha(k)|^2, \tag{8.7b}$$

and

$$\mathcal{H}_{\text{int}}(\varphi) = \frac{u_0}{V} \sum_{\alpha\beta;k_1k_2k_3} \varphi_\alpha(k_1)\varphi_\alpha(k_2)\varphi_\beta(k_3)\varphi_\beta(-k_1-k_2-k_3). \quad (8.7c)$$

Eq. (8.7b) can be rewritten in an equivalent way by substituting $\varphi_\alpha(k)$ according to Eq. (8.4)

$$\mathcal{H}_0(\varphi) = \mathcal{H}_0(\varphi_0) + \mathcal{H}_0(\varphi_1). \quad (8.8)$$

Here the quadratic Hamiltonian parts $\mathcal{H}_0(\varphi_0)$ and $\mathcal{H}_0(\varphi_1)$ have the same form as in Eq. (8.7b) because the bilinear term $\sim \varphi_{1\alpha}(k)\varphi_{0\alpha}(-k)$ is equal to zero in accordance with Eq. (8.3a) and Eq. (8.3b) ; the momentum k cannot belong simultaneously to the outer $(0 < k < b^{-1}\Lambda)$ and the inner $(b^{-1}\Lambda < k < \Lambda)$ intervals.

For $\mathcal{H}_{\text{int}}(\varphi_0 + \varphi_1)$ after the same replacement we have

$$\mathcal{H}_{\text{int}}(\varphi_0, \varphi_1) = \mathcal{H}_{\text{int}}(\varphi_0) + \mathcal{H}_{\text{int}}(\varphi_1) + \mathcal{H}_{\text{int}}^{(0,1)}. \quad (8.9)$$

$\mathcal{H}_{\text{int}}^{(0,1)}$ represents the $(\varphi_0 - \varphi_1)$-interactions:

$$\begin{aligned}\mathcal{H}_{\text{int}}^{(0,1)} = \frac{2u_0}{V} \sum_{\alpha\beta;k_1k_2k_3} & [\varphi_{0\alpha}(k_1)\varphi_{0\alpha}(k_2)\varphi_{1\beta}(k_3)\varphi_{1\beta}(-k_1-k_2-k_3) \\ & + 2\varphi_{0\alpha}(k_1)\varphi_{0\beta}(k_2)\varphi_{1\alpha}(k_3)\varphi_{1\beta}(-k_1-k_2-k_3) \\ & + 2\varphi_{0\alpha}(k_1)\varphi_{1\alpha}(k_2)\varphi_{1\beta}(k_3)\varphi_{1\beta}(-k_1-k_2-k_3) \\ & + 2\varphi_{0\alpha}(k_1)\varphi_{0\alpha}(k_2)\varphi_{0\beta}(k_3)\varphi_{1\beta}(-k_1-k_2-k_3)]. \end{aligned} \quad (8.10)$$

The two terms of the type $(\varphi_0^2\varphi_1^2)$ coincide for $(n = 1)$-component systems (the numerical factor of such term is 6). The factor 2 of the first term in Eq. (8.10) and the factor 4 of the other terms result from the invariance of $\mathcal{H}_{\text{int}}$ under the changes $\alpha \leftrightarrow \beta$ and $k_i \leftrightarrow k_j$; $i, j = 1, 2, 3$.

The division of $\mathcal{H}_{\text{int}}$ in four types of φ^4-interactions included in $\mathcal{H}_{\text{int}}^{(0,1)}$ and other two φ_0^4 and φ_1^4 included in $\mathcal{H}(\varphi_0)$ and $\mathcal{H}(\varphi_1)$ is shown diagrammatically in Fig. 8.1(a).

The field $\varphi_{0\alpha}(k)$ is represented by a line with a little bubble. These graphs stand for the external legs $\varphi_{0\alpha}$ of the perturbation diagrams (further on the bubbles will be not drawn). The terms in Eq. (8.10) correspond to the diagrams (b)–(e). The combinatorial factors [×2] and [×4] give the number of the equivalent terms. For example, there are two equivalent ways to obtain the diagram (b): $\varphi_{0\alpha}^2\varphi_{1\beta}^2$ and $\varphi_{1\alpha}^2\varphi_{0\beta}^2$. Substituting $\varphi_\alpha =$

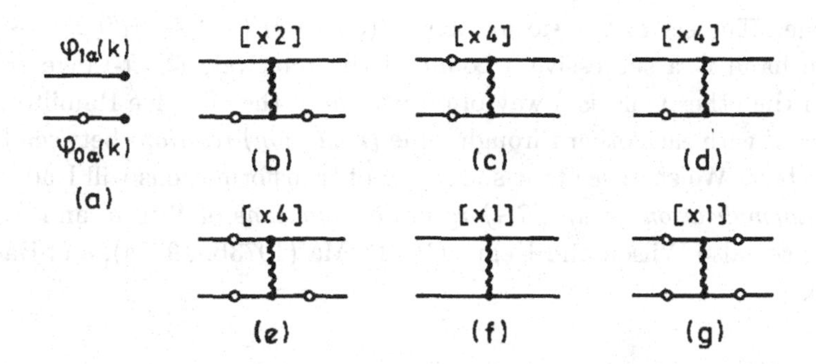

Fig. 8.1 Diagrams for: (a) φ_0, φ_1 and (b)–(e) the interaction terms in the φ^4-model.

$\varphi_{0\alpha}+\varphi_{1\alpha}$ in Eq. (8.7c) we receive the four equivalent diagrams (c) describing terms of the type $\varphi_{0\alpha}\varphi_{1\alpha}\varphi_{0\beta}\varphi_{1\beta}$. Instead of making this substitution, we can write all possible diagrams with legs φ_1 and φ_0 as shown in Fig. 8.1. We shall explain how the numerical factors can be found on the example of the diagram (d). There are four possibilities of the legs to be of the type φ_0 (then the other three will be of the type φ_1). Therefore, the numerical factor of the diagram (d) is 4.

Any functional $A(\varphi) = A(\varphi_0 + \varphi_1)$ can be averaged over the fields $\varphi_{1\alpha}$ as follows

$$\langle A\rangle_1 \equiv \bar{A}(\varphi_0) = \frac{\int D\varphi_1\, e^{-\mathcal{H}(\varphi)} A(\varphi_0+\varphi_1)}{\int D\varphi_1\, e^{-\mathcal{H}(\varphi)}}, \tag{8.11a}$$

or, using Eq. (8.5),

$$\bar{A}(\varphi_0) = e^{\mathcal{H}'(\varphi_0)} \int D\varphi_1\, e^{-\mathcal{H}(\varphi)} A(\varphi). \tag{8.11b}$$

For reasons which will become clear subsequently, we shall investigate the case $A(\varphi) = 1$, i.e., the transformation (8.5). Let us assume that the integration (8.5) can be carried out so that we have the explicit form of $\mathcal{H}'(\varphi_0)$,

$$\mathcal{H}'(\varphi_0) = -\ln \int D\varphi_1\, e^{-\mathcal{H}(\varphi_0+\varphi_1)} \tag{8.12}$$

Further we can integrate over one next set of amplitudes $\varphi'_{0\alpha}(k)$, say, those in the momentum shell $\Lambda b_1^{-1} < k < \Lambda b^{-1}$ $(b_1 > b)$ and this procedure may continue. The set of transformations $\mathcal{H}(\varphi) \to \mathcal{H}'(\varphi_0) \to \mathcal{H}''(\varphi_0') \to \cdots$, on one hand is a successive account of the relatively short-range effects and on the other hand is a way of derivation of the effective Hamiltonians connected with each other through some *(recursion) relations* between their parameters. We shall see that such type of transformations will lead us to *the renormalization group (RG) recursion relations* of Wilson and Fisher (1972); see also Wilson and Kogut (1974), Ma (1973b, 1976a), and Barber (1977).

8.2 Perturbation Expansion for the Effective Hamiltonian

The Eq. (8.12) can be written in another way

$$\mathcal{H}'(\varphi_0) = -\ln\left[\frac{\int D\varphi_1\, e^{-\mathcal{H}(\varphi)}}{\int D\varphi_1\, e^{-\mathcal{H}_0(\varphi_1)}}\right] + \Phi_{01}, \tag{8.13}$$

where

$$\Phi_{01} = -\ln \mathcal{Z}_{01} = -\ln \int D\varphi_1\, e^{-\mathcal{H}_0(\varphi_1)}, \tag{8.14}$$

is the thermodynamic potential corresponding to the free fluctuations $\varphi_1 = \{\varphi_{1\alpha}\}$ described by the quadratic Hamiltonian $\mathcal{H}_0(\varphi_1)$. The difference between this (Gaussian) part and the free fluctuation contribution given by Eq. (7.53a) is that the integration over k now is restricted:

$$\Phi_{01} = -\frac{nV}{2}\int_k^1 \ln\left(\frac{\pi}{r_0 + ck^2}\right), \tag{8.15a}$$

$$\int_k' \equiv \int_{b^{-1}\Lambda<|k|<\Lambda} \frac{d^dk}{(2\pi)^d}; \tag{8.15b}$$

see also Eq. (D.10).

Below we shall leave out terms, like Φ_{01}, which do not depend on the field φ_0. Our task is to obtain the functional dependence $\mathcal{H}(\varphi_0)$. The numerous φ_0-independent finite terms which can arise from the integration (8.12) will merely shift the level, from which the energy in the system is measured. One can imagine that these terms are subtracted from $\mathcal{H}(\varphi_0)$ and the energy scale is changed respectively.

After these remarks we can write Eq. (8.13) in the form

$$\mathcal{H}'(\varphi_0) = -\ln\langle e^{-[\mathcal{H}(\varphi)-\mathcal{H}_0(\varphi_1)]}\rangle_{(01)c}, \tag{8.16}$$

where the averaging is made with the help of the statistical distribution

$$\mathcal{P}(\varphi_1) = e^{-\mathcal{H}_0(\varphi_1)} \tag{8.17}$$

of the free φ_1-fluctuations (the index "(01)c" of the connected averages $\langle \dots \rangle_{(01)c}$ will be omitted hereafter). $\mathcal{H}(\varphi_0) = \mathcal{H}_0(\varphi_0) + \mathcal{H}_{\text{int}}(\varphi_0)$ in the exponent $[\mathcal{H}(\varphi) - \mathcal{H}_0(\varphi_1)]$ from Eq. (8.16) does not depend on the $\varphi_{1\alpha}$-fields. Its remaining part will be denoted by

$$\mathcal{H}_4(\varphi_1) = \mathcal{H} - \mathcal{H}_0(\varphi_1) - \mathcal{H}(\varphi_0). \tag{8.18a}$$

This interaction part is equal to $[\mathcal{H}_{\text{int}}^{(0,1)} + \mathcal{H}_{\text{int}}(\varphi_1)]$; see Eq. (8.7c) with $\varphi_\alpha \to \varphi_{1\alpha}$, and Eqs. (8.9)–(8.10). The diagrammatic representation of $\mathcal{H}_4$ is given in Figs. 8.1(b–f). Now $\mathcal{H}'(\varphi_0)$ from Eq. (8.16) becomes

$$\mathcal{H}'(\varphi_0) = \mathcal{H}(\varphi_0) - \ln\langle e^{-\mathcal{H}_4}\rangle, \tag{8.18b}$$

or,

$$\mathcal{H}'(\varphi_0) = \mathcal{H}(\varphi_0) - \sum_{l=1}^{\infty} \frac{(-1)^l}{l!} \langle \mathcal{H}_4^l(\varphi_1, \varphi_0)\rangle_c. \tag{8.18c}$$

8.2.1 *First order calculation*

Let us find out the first-order ($l = 1$) perturbation contribution to $\mathcal{H}_0'(\varphi_0)$. It will be useful for the unexperienced reader to perform the direct averaging of $\langle \mathcal{H}_4 \rangle$ over the fields $\varphi_{1\alpha}$. The averages $\langle \mathcal{H}_4^l \rangle$ are calculated in the same way as explained in Section 7.6 for the standard perturbation series of the thermodynamic potential. Here we shall briefly explain the diagrammatic way of calculation.

The φ_0-legs of the diagram in Fig. 8.1(b) may be of two types: $\varphi_{0\alpha}\varphi_{0\alpha}$ and $\varphi_{0\beta}\varphi_{0\beta}$. So this diagram has a combinatorial factor 2 as shown in the figure. The $\varphi_{1\alpha}$-legs of the same diagram can be connected in only one way to form the average $\langle \varphi_{1\alpha}(k_1)\varphi_{1\alpha}(k_2)\rangle$. Thus we obtain Fig. 7.5(c), or, in a momentum representation Fig. 7.15(b) with the only difference that the external legs correspond to φ_0-fields rather than to G_0-functions. The closed loop of the internal G_0-line yields an extra factor n so that the combinatorial factor of this diagram becomes $[\times 2n]$.

The internal (G_0)-lines of the diagrams represent the fast-field correlation functions

$$G_{0\alpha\alpha'}(k,k') = \langle \varphi_{1\alpha}(k)\varphi_{1\alpha'}(k')\rangle = \frac{\delta_{\alpha\alpha'}\delta(k+k')}{r_0+ck^2}, \tag{8.19}$$

where $b^{-1}\Lambda < |k| < \Lambda$. Obviously, all external legs of the diagrams will correspond to $\varphi_{0\alpha}$-fields and all $\varphi_{1\alpha}$-fields in $\langle \mathcal{H}_4^l\rangle$ will participate in the nonzero products of the pair averages $G_0(k) = \langle |\varphi_\alpha(k)|^2\rangle$ with opposite momenta $\varphi^*(k) = \varphi(-k)$.

As a result Fig. 8.1(c) will give the diagram from Fig. 7.5(b), or, in a momentum representation, that from Fig. 7.15(d) with the modification of the correspondence rules as explained above. The averaging of (d) and (e) diagrams in Fig. 8.1 yields zero because these Hamiltonian parts consist of an odd number of $\varphi_{1\alpha}$-fields. The diagram (f) in Fig. 8.1 will give φ_0-independent diagrammatic contributions to $\mathcal{H}_0'(\varphi_0)$ as shown in Fig. 7.4 and according to our convention such φ_0-independent contributions will be neglected. Clearly, the modification of the diagrammatic rules with respect to those in Chapter 7 is in the different interpretation of the external legs and in the performing of the integration over the internal momenta in the restricted domain ($b^{-1}\Lambda < |k| < \Lambda$).

The Hamiltonian $\mathcal{H}'(\varphi_0)$ to the first order in u_0 will be

$$\begin{aligned}\mathcal{H}'(\varphi_0) &= \frac{1}{2}\sum_{\alpha,k}{}' \left[r^{(1)} + c^{(1)}k^2\right]|\varphi_\alpha(k)|^2 \\ &\quad + \frac{1}{V}\sum_{\alpha\beta;k_i}{}' u_4^{(1)}(k_i)\varphi_\alpha(k_1)\varphi_\alpha(k_2)\varphi_\beta(k_3)\varphi_\beta(-k_1-k_2-k_3),\end{aligned} \tag{8.20}$$

where we have substituted the fields $\varphi_{0\alpha}$ with φ_α in accordance with Eqs. (8.3a) and (8.3b). The "prime" of the sums denotes the restricted summation ($b^{-1}\Lambda < |k|, |k_i| < \Lambda$) over $|k|$ and $|k_i|$, $i = (1,2,3,4)$, $k_4 = -k_1-k_2-k_3$; for example, $b^{-1}\Lambda < |k_1+k_2+k_3| < \Lambda$. In Eq. (8.20), $r^{(1)}$ and $c^{(1)}$ are the "renormalized" parameters to the order u_0^1 :

$$r^{(1)} = r_0 + 4(n+2)u_0 A_1(r_0), \tag{8.21a}$$

$$c^{(1)} = c, \tag{8.21b}$$

and

$$u_4^{(1)} = u_0, \tag{8.21c}$$

where

$$A_1(r_0) \equiv I_1(r_0; b) = \int_p' \frac{1}{r_0 + cp^2}, \tag{8.22}$$

is the lowest-order perturbation integral already obtained in our previous considerations; see, e.g., Eq. (7.138) and Eq. (7.162). In Eq. (8.21a), we have taken into account the factor 1/2 of the φ_0^2-term of $\mathcal{H}'(\varphi_0)$. The "prime" of the integral $A_1(r_0)$ indicates the difference from the ordinary perturbation integral $I_1(r_0)$ — the integration in A_1 is bounded in the inner momentum shell ($b^{-1}\Lambda < |p| < \Lambda$).

The relations (8.21b) and (8.21c) are trivial — the "renormalized" values $c^{(1)}$ and $u_4^{(1)}$ are equal to the bare (initial) ones. The parameter r gains the usual mass-renormalization to the first order in u_0 ensured by the one-loop self-energy diagrams represented by the compact graph (a) in Fig. 7.20. The simplicity of the renormalization group (recursion) relations (8.21a)–(8.21c) is due to the lowest order of our calculation. We shall see that these relations become rather complex to the second, third, etc. orders in u_0.

8.2.2 *Vertex functions*

Now we can use Eqs. (8.21a)–(8.21c) as an example of the vertex functions to the first order in u_0. In general, the two-point $\Gamma^{(2)}$ and the four-point $\Gamma^{(4)}$ vertex functions are defined by

$$\Gamma^{(2)}(k) = \tilde{r} + \tilde{c}k^2 \tag{8.23}$$

and

$$\Gamma^{(4)}(k_1, k_2, k_3) = \delta(k_1 + k_2 + k_3 + k_4)\tilde{u}_4(k_1, k_2, k_3), \tag{8.24}$$

where $\tilde{r}$, $\tilde{c}$ and $\tilde{u}_4$ denote the effective (renormalized) values of the Hamiltonian parameters to any order in u_0, for instance, to the order $(u_0)^0$, $\tilde{r} = r_0$, $\tilde{c} = c$ and $\tilde{u}_4 = u_0$, to the order u_0^1, $\tilde{r} = r^{(1)}$, $\tilde{c} = c^{(1)} = c$ and $\tilde{u}_4 = u_4^{(1)} = u_0$, and so on. We shall see in the next section that the higher-order terms in the series (8.18c) generate the respective vertex functions $\Gamma^{(6)}$, $\Gamma^{(8)}$, etc.

which are proportional to the nonzero coefficients of the φ_0^6, – φ_0^8 – etc. terms in $\mathcal{H}'(\varphi_0)$.

Usually $\mathcal{H}'(\varphi_0)$ is written as a functional power series in φ_0 :

$$\mathcal{H}'(\varphi_0) = \frac{1}{2}\sum_{\alpha,k}{}' \Gamma^{(2)}(k)|\varphi_\alpha(k)|^2 + \frac{1}{V^{m-1}} \sum_{\substack{\alpha_1\dots\alpha_m \\ k_1\dots k_{2m}}}{}' \Gamma^{(2m)}(k_1,\dots,k_{2m})\varphi_{\alpha_1}(k_1)\cdots\varphi_{\alpha_m}(k_{2m}), \quad (8.25)$$

$m = 2, 3, \dots$ The function $\Gamma^{(2)}(k)$ coincides with the inverse irreducible correlation function $G^{-1}_{(2)}(k)$; for the bare values, $\Gamma_0^{(2)}(k) = [G_0^{(2)}(k)]^{-1}$.

Alternatively, the vertex functions can be defined so as the Kronecker factors for the indices $(\alpha, \beta, \dots)$ to be included into them, for example, $\Gamma^{(2)}_{\alpha\beta}(k, k') = \delta_{\alpha\beta}\delta(k + k')\Gamma^{(2)}(k)$; see, e.g., Amit (1978). The important point is that the effective potential $\mathcal{H}'(\varphi_0)$ is represented by the series (8.25) where the explicit expressions for $\Gamma^{(2m)}$ are determined from the order of calculation with respect to u_0.

Within the φ^4-theory for $T > T_c$, i.e., $\langle\varphi\rangle = 0$, the series (8.25) will consist of even powers in $\varphi_{0\alpha}(k) \equiv \varphi_\alpha(k; k < b^{-1}\Lambda)$. This result indicates that the symmetry of $\mathcal{H}(\varphi)$ and $\mathcal{H}'(\varphi_0)$ with respect to rotations in the order parameter space is one and same. In other models odd-order vertex functions $\Gamma^{(2m-1)}$, $m \geq 1$ may appear. Some of the main properties of the vertex functions $\Gamma^{(m)}$ will become clear from further discussion in this Chapter.

We have to underline that the definition of $\Gamma^{(2m)}$ presented above is somewhat unusual, it actually does not correspond to the standard definition (Amit, 1978). The commonly accepted definition of $\Gamma^{(2)}$ and $\Gamma^{(4)}$ is given by Eqs. (8.13) and (8.24), provided $b \to \infty$, i.e., for $\varphi_{0\alpha} \equiv \varphi_\alpha(0)$. The same is valid for every $\Gamma^{(2m)}$ in the expansion (8.25) so that $\mathcal{H}'(\varphi_0)$ will be equivalent to $\mathcal{H}'[\varphi(0)]$ — the effective thermodynamic potential of a homogeneous system.

8.2.3 *Second order calculation*

It is not difficult to investigate the second-order perturbation contributions to the functions $\Gamma^{(2)}$ in the series (8.25) because the analogous calculation for $G^{-1}(k, r_0)$ has been already done in Section 7.9.

Here we should take into account the modifications in the correspondence rules between the graphs and the analytical expressions owing to the division in slow (φ_0) and fast (φ_1) fields.

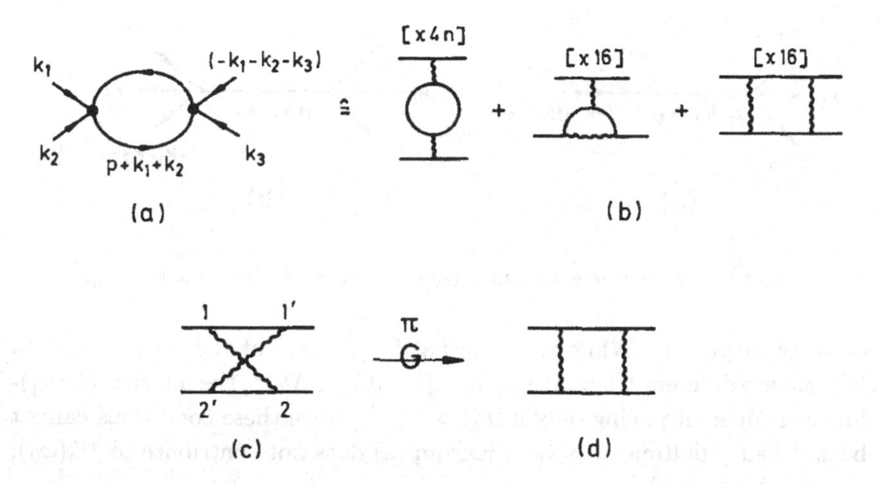

Fig. 8.2 OP-irreducible diagrams of type φ_0^4: (a) a compact notation, (b) equivalent representation by a sum of three diagrams with wavy lines standing for the φ^4-interaction. For diagrams (c) and (d) see the text.

Diagrammatically the contributions of order u_0^2 to $\mathcal{H}'(\varphi_0)$ of the type φ_0^2, i.e., quadratic in the fields $\varphi_{0\alpha}$ are given by the OP-irreducible diagrams for $G^{-1}(k, r_0)$, see Fig. 7.8. The OP-reducible diagrams in Fig. 7.7 should be excluded from the φ_0^2-term of $\mathcal{H}'(\varphi_0)$ because within the present perturbation scheme there exists strong restriction over the values of the internal momentum of the internal (OP) reducible line. In fact, the diagrams in Fig. 7.7 are of the same type as those in Figs. 7.21(a, b). If we look carefully at Fig. 7.21(b), we shall see that the momentum k have to satisfy simultaneously two conditions $b^{-1}\Lambda < |k| < \Lambda$ and $0 < |k| < b^{-1}\Lambda$. Since this is impossible, the OP-reducible diagrams do not enter in the self-energy (φ_0^2-) terms of $\mathcal{H}'(\varphi_0)$.

Therefore, the diagrams for $\Gamma^{(2)}(k)$ have the same topology as those for the self-energy function $\Sigma(k)$ discussed in Section 7.9; see also Fig. 7.20 for the OP-irreducible diagrams in compact graph representation. Using the compact graph representation (Section 7.7.3 and Fig. 7.13) we can write the OP-irreducible diagrams corresponding to the terms φ_0^4 in $\mathcal{H}'(\varphi_0)$, see Figs. 8.2(a, b). The OP-reducible diagrams having contribution to the φ_0^4- and φ_0^6-invariants of $\mathcal{H}'(\varphi_0)$ are to be discussed too; see Fig. 8.3 for the compact graph representation of these contributions. Let us consider the momenta $|k_i|$, $(i = 1, \cdots, 5)$, $k_4' = -k_1 - k_2 - k_3$, and $k_6' = (-k_1 - \cdots - k_5)$

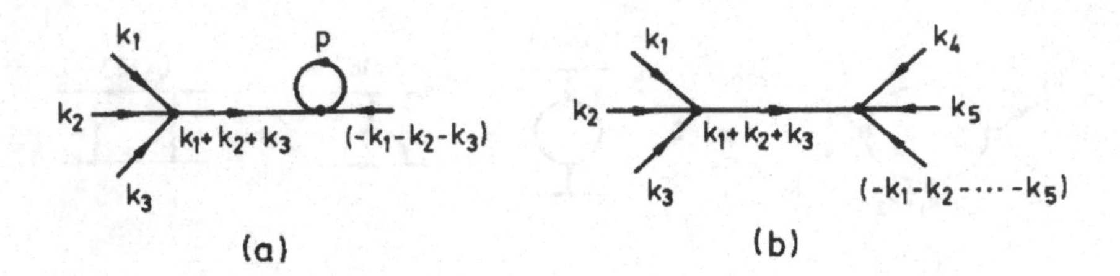

Fig. 8.3 OP-reducible diagrams: (a) of the type φ_0^4, (b) of the type φ_0^6.

of these diagrams. While the external legs $\varphi_0(k_i)$ of the diagram (a) in Fig. 8.3 are different from zero for $0 \le |k_i|, |k_4'| < \Lambda b^{-1}$, the internal $G_0(k_4')$-line can come into being only if $|k_4'| > \Lambda b^{-1}$. Since these conditions cannot be satisfied simultaneously the diagram (a) does not contribute to $\mathcal{H}_0'(\varphi_0)$; merely, this term does not exist.

The same arguments can be applied to Fig. 8.3(b). It is readily seen that a very restricted region $|k_i| \le b^{-1}\Lambda$, $|k_4'| > \Lambda b^{-1}$ exists, in which the integration over the internal momentum k_4' should be done, namely:

$$\int_{k_4'}' G_0(k_4'), \qquad (b^{-1}\Lambda < |k_4'| < \Lambda, 0 < |k_i| \le b^{-1}\Lambda)\,. \tag{8.26a}$$

In fact, this is an angular integration over two independent angles formed between the momenta k_1, k_2, and k_3; we shall not discuss k_4, k_5, and k_6' and the conditions on them which may bring about new restrictions.

Our aim is to show, that the integral (8.26a) has a relatively small value, for example, that it is much smaller than the integrals $\mathcal{A}_1(r_0)$ and $\mathcal{A}_1(0)$. Thus we shall demonstrate that the φ_0^6-term described by Fig. 8.3(b) can be neglected. To see this we shall replace the integral (8.26a) with a larger integral

$$K_d \int_{b^{-1}\Lambda}^{3b^{-1}\Lambda} k^{d-1} dk, \tag{8.26b}$$

which corresponds to the Eq. (8.26a) in the limiting case $|k_i| = b^{-1}\Lambda$ and where the vectors k_1, k_2, and k_3 are supposed mutually parallel so that $k_4' = 3|k_i| = 3b^{-1}\Lambda$. This integral is proportional to $4(\Lambda/b)^d$; it vanishes as $b \to \infty$. For large $b \gg 1$ it is small and can be neglected. In comparison, $\mathcal{A}_1(r_0)$ is finite for large b; see also Eq. (8.22).

Such considerations can be applied to the other OP-reducible diagrams like φ_0^6, φ_0^8, etc. appearing in the higher orders of the perturbation theory.

For $b \to \infty$, i.e., for $\varphi_{0\alpha} \equiv \varphi_\alpha(0)$, the case for which the vertex functions $\Gamma^{(2m)}$ are usually defined, the OP-reducible diagrams are equal to zero (they do not appear in $\Gamma^{(2m)}$). In our case ($b \neq \infty$) we should give additional arguments for the irrelevance of the OP-reducible terms in $\mathcal{H}'(\varphi_0)$. Further we shall use only OP-irreducible diagrams. At this point of our description, we shall not present the explicit expressions of $\tilde{r}$, $\tilde{c}$, and $\tilde{u}_4$ to the second order in u_0. The reader may easily carry out the calculation with the help of the diagrams for $\Sigma^{(2)}$ (see Chapter 7) and the diagrams in Fig. 8.2 for the $u_0^2\varphi_0^4$-terms; see also the discussion in the next sections.

8.3 Loop Expansion

Here we shall discuss the expansion in the number of loops for the coefficient ($\Gamma-$) functions of $\mathcal{H}'(\varphi_0)$; see Eq. (8.25). The loop expansion is nothing new but a slightly different rearrangement of the terms in the u_0-expansion. For example, let us consider *the one-loop approximation.* In order to clarify this approximation we shall use the compact graph representation when the wavy line of the φ^4-interaction is reduced to a point vertex (as in the ($n = 1$)-component field theory).

The one-loop approximation for $\mathcal{H}'(\varphi_0)$ is given by all perturbation diagrams consisting only of *one* closed loop of internal G_0-functions. The one-loop approximation for the self-energy Σ or, which is the same, for $\Gamma^{(2)}$ or the self-energy parameters $\tilde{c}$ and $\tilde{r}$ coincides with the contribution from the diagrams to the first order in u_0. But there are other cases, for example, the contributions to the $\tilde{u}_4\varphi_0^4$-term or, which is the same to $\Gamma^{(4)}$ or $\tilde{u}_4$, where the one-loop approximation means to take into account the terms up to the second order in u_0; see Fig. 8.2. The two-loop approximation for the $\tilde{u}_4\varphi_0^4$-term in $\mathcal{H}'(\varphi_0)$ includes terms up to the order u_0^3, etc.

So, in general, the approximation in the number of loops does not coincide with the expansion in powers of the bare interaction constant u_0. Within the renormalization-group (RG) transformation considered in the remainder of this Chapter the loop expansion will be widely used. Here we shall determine the effective parameters ($\tilde{r}, \tilde{c}$, $\tilde{u}_4$, $\tilde{u}_6$, etc.) in the two-loop approximation; the OP-reducible diagrams for $\tilde{u}_6$, $\tilde{u}_8$, etc. are neglected.

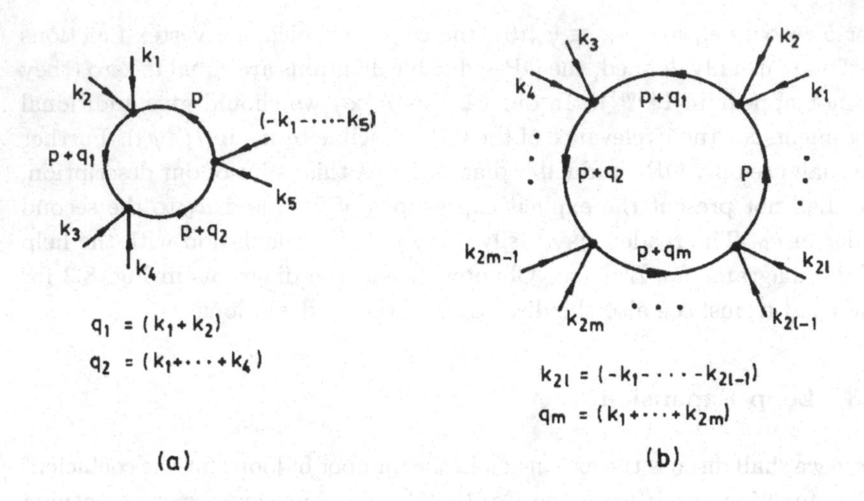

Fig. 8.4 One-loop diagrams: (a) of the type φ_0^6, (b) of the type φ_0^{2l}; $l \geq 1$.

8.3.1 *One-loop approximation*

The diagrams representing the one-loop approximation are shown in Fig. 8.4; see also Fig. 8.2(a) for the φ_0^4-term and Fig. 7.13(a) for the φ_0^2-term. The corresponding analytic expressions for $\tilde{r}$ and $\tilde{c}$ are given by Eqs. (8.21a) and (8.21b) written to the first order in u_0.

So, we have to obtain the one-loop approximation for $\tilde{u}_{2l}$; $l > 1$. With the help of Fig. 8.4 we find that

$$\tilde{u}_4 = u_0 - 4(n+8)u_0^2 \mathcal{A}^{(2)}(r_0, q_1), \tag{8.27a}$$

$$\tilde{u}_6 \sim u_0^3 \mathcal{A}^{(3)}(r_0, q_1, q_2), \tag{8.27b}$$

$$\vdots$$

etc., with

$$\mathcal{A}^{(l)}(r_0, k_1, \ldots, k_{2l-1}) = \int_p' \frac{1}{(r_0 + cp^2)[r_0 + c(p+q_1)^2] \cdots [r_0 + c(p+q_{l-1})^2]}, \tag{8.28}$$

where $b^{-1} < |q_i + p| < \Lambda$. The connection between the momenta $q_i(i = 1, \ldots, l-1)$ and the external momenta $k_i(i = 1, \ldots, 2l-1)$ is seen from Fig. 8.4. The integrals $\mathcal{A}^{(l)}$ for $k_i \equiv 0$ coincide with the integrals

$$\mathcal{A}_l(r_0) = \int_p' \frac{1}{(r_0 + ck^2)^l}, \qquad l \geq 1, \tag{8.29}$$

which satisfy the relation

$$\mathcal{A}_{l+1}(r_0) = -\frac{1}{l}\left[\frac{\partial}{\partial r_0}\mathcal{A}_l(r_0)\right]. \tag{8.30}$$

The numerical factor $4(n+8)$ in (Eq. (8.27a) is obtained from the combinatorial numbers of the diagrams in Fig. 8.2(b); the reader should bear in mind that the factor $(-1/2!)$ appearing before the second-order perturbation term in the series (8.18c) is also taken into account. Besides, diagrams like that in Fig. 8.2(c) are to be taken into account too. After a rotation around the wavy line 1-2 (or 1′-2′), see Fig. 8.2(c,d); the diagram (c) becomes the same as the last diagram in the graph sum (b). The numerical factor of the expression (8.27b) for $\tilde{u}_6$ can be found in the same way. To do this the reader has to write down all topologically different diagrams corresponding to φ_0^6 with an interaction presented by a wavy line and to calculate their combinatorial factors. Now the factor, standing before the third-order terms in (Eq. (8.18c) will be $(1/3!)$.

8.3.2 *Two-loop approximation*

The two-loop diagrams for the self-energy parameters $\tilde{r}$ and $\tilde{c}$ are obtained from both the k-independent and the k-dependent second-order perturbation terms for $\Sigma(k, r_0)$ taking into account the modified conditions for the integration over the internal momenta; see the OP-irreducible diagrams (b) and (c) in Fig. 7.20. Alternatively one may use the OP-irreducible second-order diagrams for the Green function, see Fig. 7.8, where the external G_0-lines are replaced with φ_0-lines.

The two-loop diagrammatic contributions to the φ_0^4-term in $\mathcal{H}'(\varphi_0)$, presented by compact graphs, are shown in Fig. 8.5. Each of the three compact diagrams (a), (b), and (c) in this figure are equivalent, not including a numerical factor, to the respective finite sets of diagrams given in Fig. 8.6. When writing the topologically different diagrams with a wavy line standing for the φ^4-interaction we have not made difference between

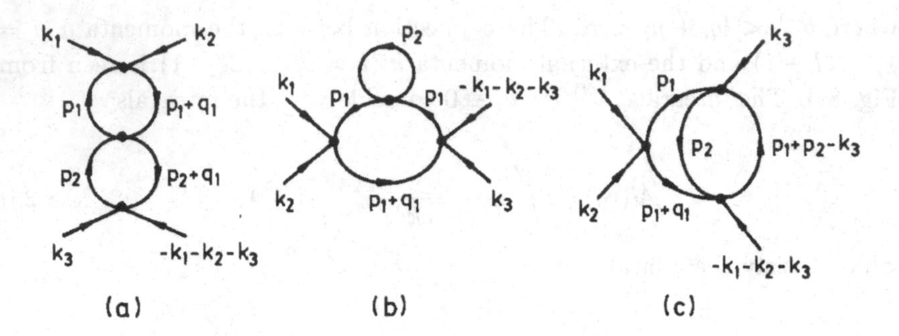

Fig. 8.5 Two-loop (u_0^3-) diagrams of type φ_0^4. The momentum q_1 is given in Fig. 8.4.

diagrams which are obtained from each other by a rotation like that shown in Figs. 8.2(c, d). The sum of the six diagrams (a) in Fig. 8.6 together with the respective combinatorial factors yields an analytical expression of order u_0^3 corresponding to the diagram (a) in Fig. 8.5. The same is valid for the diagram (b) and (c) in Fig. 8.5 and Fig. 8.6.

It is readily seen from the momentum notations in Fig. 8.5 that the three sets of diagrams (a), (b), and (c) result in different analytic expressions for the perturbation integrals, so these diagrams are of a different type. Here we have only three sets of different diagrams but there are models, for which diagrams of the same order and graph topology could be divided in subsets bringing about different analytic expressions.

In such cases, one is forced to distinguish between the different subsets of diagrams by the directions of the arrows of the G_0-lines (see, e.g., the treatment of the nonideal Bose gas in the limit $T \to 0$; Section 9.3). The combinatorial numbers of the self-energy diagrams are known from Chapter 7. In a similar way one can find the combinatorial factors of the diagrams in Fig. 8.5 and Fig. 8.6.

After these explanations we can write for $\tilde{r}$, $\tilde{u}$ and $\tilde{c}$ the following expressions:

$$\tilde{r} = r_0 + 4(n+2)u_0\mathcal{A}_1(r_0) - 16(n+2)^2u_0^2\mathcal{A}_1(r_0)\mathcal{A}_2(r_0) \\ - 32(n+2)u_0^2B(r_0, 0), \tag{8.31a}$$

$$\tilde{u} = u_0 - 4(n+8)u_0^2\mathcal{A}^{(2)}(r_0, q_1) + 16(n^2+6n+20)u_0^3[\mathcal{A}^{(2)}(r_0, q_1)]^2$$
$$+ 64(5n+22)u_0^3 C(r_0, k_i)$$
$$+ 32(n+2)(n+8)u_0^3 \mathcal{A}_1(r_0)\mathcal{A}^{(3)}(r_0, q_1, 0), \tag{8.31b}$$

and

$$\tilde{c} = c - 32(n+2)u_0^2\left\{\frac{1}{k^2}\left[B(r_0,k) - B(r_0,0)\right]\right\} + O(k^\tau), \quad \tau > 0, \tag{8.31c}$$

where the integral

$$B(r_0, k; b) = \int_{p_1}' \int_{p_2}' \frac{1}{(r_0 + cp_1^2)(r_0 + cp_2^2)[r_0 + c(p_1 + p_2 + k)^2]} \tag{8.32}$$

coincides with the integral (7.167) for $b \to \infty$. The integral $\mathcal{A}^{(3)}$ is given by Eq. (8.28) for $l = 3$ and $q_2 = 0$, $q_1 = (k_1 + k_2)$, and the integral C is

$$C(r_0, k_i; b) = \int_{p_1}' \int_{p_2}' \frac{1}{(r_0 + cp_1^2)[r_0 + c(p_1 + q_1)^2]}$$
$$\times \frac{1}{(r_0 + cp_2^2)[r_0 + c(p_1 + p_2 - k_3)^2]}. \tag{8.33}$$

In Eq. (8.31c), the last ("τ"-) term merely indicates that k-dependent insertions from the B-integral difference of order higher than k^2, are ignored as small and irrelevant (note, that the external wave number k obeys the condition $|k| \ll \Lambda$). In addition to the conditions $b^{-1}\Lambda < |p_i| \leq \Lambda$ and $0 < |k_i| \leq b^{-1}\Lambda$ for the external momenta we have to take into account in the above integrals the restrictions on the momenta of the internal G_0-functions; the "prime" in the above integrals stands for all these requirements. So the square of each momentum participating in the integrands should be bigger than $(b^{-1}\Lambda)^2$ and less than Λ^2, for example, the momentum $(p_1 + p_2 - k_3)$ in the integral (8.33) is subject to the condition $(b^{-1}\Lambda)^2 < (p_1 + p_2 - k_3)^2 < \Lambda^2$. We shall not discuss the φ_0^6-, φ_0^8-, etc. invariants of $\mathcal{H}'(\varphi_0)$ because such terms will be not further used in our considerations.

8.3.3 *Large-b limit and one-component field*

The large $b \gg 1$ and the large-b limit ($b \to \infty$) can be used for the study of the infrared asymptote of $\mathcal{H}'(\varphi_0)$. When $b \gg 1$, $\varphi_{0\alpha} \approx \varphi_\alpha(0)$ and very often

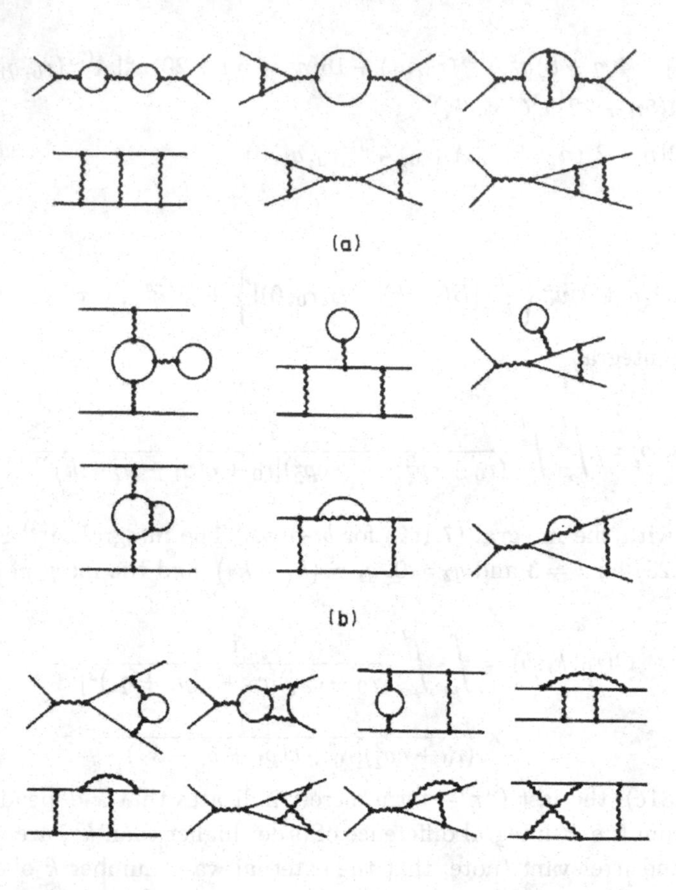

Fig. 8.6 Two-loop diagrams corresponding to the compact representation in Fig. 8.5.

the external momenta k_i in the perturbation integrals can be neglected as small compared to the "mass" parameter r_0 and the internal momenta (p_i). In the limiting case $b \to \infty$, all $|k_i| \to 0$ and $\varphi_{0\alpha} \to \varphi_\alpha(0)$. In our further discussion both $b \gg 1$ and $b \to \infty$ are referred to as the *large-b limit.* In this limit the integrals $\mathcal{A}^{(l)}$ defined by Eqs. (8.28) are approximately (or exactly) equal to the integrals $\mathcal{A}_l$ from Eqs. (8.29). Under the same condition $(k_i \approx 0)$,

$$C(r_0, 0; b) = -\frac{1}{3}\left[\frac{\partial B(r_0, k)}{\partial r_0}\right]_{k=0}. \tag{8.34}$$

For $b \to \infty$, the analytical expressions corresponding to the diagrams in Fig. 8.2(a), Fig. 8.4(a) and Fig. 8.4(b) are simplified because all external momenta (k_i) are zero. $\mathcal{H}'[\varphi_\alpha(0)]$ in the one-loop approximation is easily calculated to any order in $\varphi^{2l}(0)$ for $(n = 1)$-component systems. So we have to find the analytical expressions of the diagrams in Fig. 8.4(b) for any $l \geq 1$. The result for $\mathcal{H}'[\varphi(0)]$ will be

$$\begin{aligned}\mathcal{H}'[\varphi(0)] &= \frac{r_0}{2}\varphi^2(0) + \frac{u_0}{V}\varphi^4(0) + 6u_0 I_1(r_0)\varphi^2(0) \\ &- \frac{36u_0^2}{V} I_2(r_0)\varphi^4(0) + \frac{288u_0^3}{V^2} I_3(r_0)\varphi^6(0) - \cdots ,\end{aligned} \tag{8.35}$$

where $I_l(\varphi_0) \equiv \mathcal{A}(\varphi_0; b = \infty)$ are the integrals (7.166). Introducing the notation $\varphi(0) = \sqrt{V}\bar{\varphi}$, we obtain that

$$\mathcal{H}'(\bar{\varphi}) = V\left[\frac{r_0}{2}\bar{\varphi}^2 + u_0\bar{\varphi}^{\,4}\right] + \frac{1}{2}\int_k \ln\left[1 + \frac{12u_0\bar{\varphi}^2}{(r_0 + ck^2)}\right] \tag{8.36}$$

to any order in $\bar{\varphi}^2$. The coefficient $288 = (4.2/3!) \cdot (4.3/1.2)^3$ in Eq. (8.35) is the combinatorial factor of the diagram (a) in Fig. 8.4; the factor (1/3!) before the third-order term in Eq. (8.18c) is included too. It is not difficult to verify by direct calculations the logarithmic expression (8.36) in the one-loop approximation for $\mathcal{H}'$.

8.3.4 *Remarks on the renormalization*

In Section 7.10 we have introduced the notion of the renormalization of field models. Moreover the mass-renormalization ($r_0 \to r$) has been demonstrated up to the second order in u_0. Within the present modified perturbation technique with upper (Λ) and lower ($b^{-1}\Lambda$) cutoffs, no divergent integrals appears but the divergences arise immediately when we make an attempt to consider the infrared ($b \to \infty$) and the ultraviolet ($\Lambda \to \infty$; $b^{-1}\Lambda < \infty$) asymptotics of the theory. The infrared divergences occur below the borderline dimensionality $d_U = 4$, whereas the ultraviolet divergences ($k \to \infty$) take place above d_U. At $d = 4$ the perturbation integrals give terms proportional to $b^m(\ln b)^{m'}$ which are logarithmically divergent ($m = 0$, $m' > 0$) for $b \to \infty$; see also Section 7.11.

The mass-renormalization procedure from Section 7.10 can be rederived in the present perturbation scheme in both its versions by including a counter term $\mathcal{H}_c$ in the Hamiltonian or, by introducing a successive set of perturbation counterterms for the renormalized ($r_0 \to r$) perturbation series. In this way one can also do the coupling constant renormalization ($u_0 \to u$) and the field renormalization ($\varphi_{0\alpha} \to \varphi'_{0\alpha}$). In the limit

$b \to \infty$, this yields the conventional renormalization procedure explained in detail in many books; see, e.g., Amit (1978). Of course, the result for $\tilde{r}$, $\tilde{c}$, and $\tilde{u_4}$ following from the two-loop approximation can be employed to find the renormalized values of the respective parameters. This might be done for $b \neq \infty$ within the Wilson-Fisher renormalized scheme presented in Sections 8.5–8.7. In some particular studies the necessity to take the limit $b \to \infty$ may arise; then we obtain the commonly used renormalization scheme within the standard perturbation theory.

8.4 Large-n Limit and Hartree Approximation

When the symmetry index n is large enough the n-dependent polynomial coefficients in the expansions (8.31a) and (8.31b) can be approximated by the leading power in n. Therefore, the perturbation series is reduced to the diagrams with a maximal number of closed loops. Here we shall discuss the infinite series for large n.

8.4.1 *Hartree diagrams*

The perturbation series for $\Gamma^{(2)}(k)$ and $\Gamma^4(k_i)$ in the large-n limit are shown in Fig. 8.7, where the wavy-line representation of the interaction u_0 is used. For reasons that will become clear in the next subsection these loop diagrams are often called *Hartree diagrams.* It is not difficult to see that they describe the main part of the perturbation corrections to $\tilde{r}$, $\tilde{c}$, and $\tilde{u}_4$ (below denoted by r, c, u, respectively). For example, in Eq. (8.31a) we can approximately replace $(n+2) \approx n$ for $n \gg 2$, in addition we can neglect the last term $\sim nu_0^2$ as small compared to the terms $\sim nu_0$ and $\sim n^2u_0^2$. Thus, in the large-n limit $(n \to \infty)$ the parameter c will not gain perturbation corrections.

The diagrammatic series in Fig. 8.7(a) is analytically expressed as an infinite geometrical progression:

$$\begin{aligned} r = r_0 &+ 4n\mathcal{A}_1(r_0) - 16n^2u_0^2\mathcal{A}_1(r_0)\mathcal{A}_2(r_0) \\ &+ 48n^3u_0^3\mathcal{A}_1(r_0)\mathcal{A}_2^2(r_0) + \cdots . \end{aligned} \tag{8.37a}$$

This can be rewritten in the following way

$$r = r_0 - \Sigma(r_0), \tag{8.37b}$$

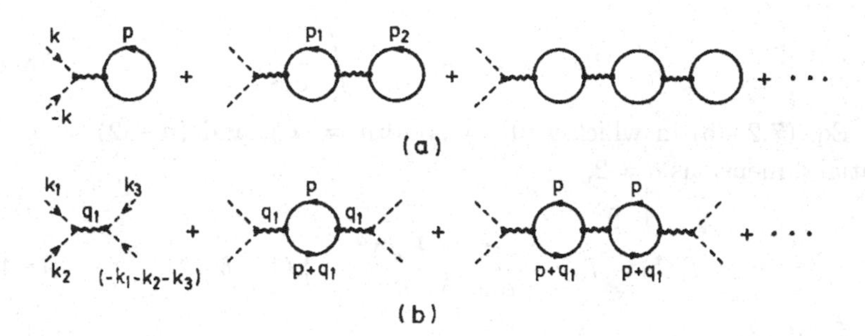

Fig. 8.7 Hartree diagrams: (a) for r, (b) for u; $q_1 = (k_1 + k_2)$; the same diagrams can be applied for the φ_0^2- and φ_0^4-terms in $\mathcal{H}'(\varphi)$, provided the dotted legs are taken into consideration.

where

$$\Sigma(r_0) = -\frac{4nu_0\mathcal{A}_1(r_0)}{1 + 4nu_0\mathcal{A}_2(r_0)} \tag{8.38}$$

is the k-independent self-energy function. Since no k-dependent contribution to $\Gamma^{(2)}(k, r_0) = G_{(2)}^{-1}(k, r_0)$ appear to any order in u_0, the self energy function is given by the large-n mass renormalization diagrams (a) in Fig. 8.7.

There are another two equivalent expressions for r:

$$r = r_0 + 4nu_0 \int_p' \frac{1}{cp^2 + r_0 + 4nu_0\mathcal{A}_1(r_0)}, \tag{8.39a}$$

and

$$r = r_0 + 4nu_0 \int_p' \frac{1}{(cp^2 + r)}. \tag{8.39b}$$

The last result can be obtained either by performing a successive mass-renormalization ($r \to r_0$), or by introducing a counter term

$$\mathcal{H}_c = \frac{1}{2}\left[\Sigma(r)\right] \sum_{\alpha,k} |\varphi_\alpha(k)|^2. \tag{8.40}$$

This is a simplified version of the renormalization procedure presented in the Section 7.10.

The critical temperature $T_c(b)$ is found from the equation $r(T_c) = 0$ and Eq. (8.38b). Remembering that $r_0 = \alpha_0(T - T_{c0})$, where T_{c0} is the bare critical temperature, we have

$$T_c(b) = T_{c0} - \frac{4nu_0}{\alpha_0} \int_p' \frac{1}{cp^2}; \tag{8.41a}$$

cf. Eq. (7.234b) in which $I_i(0) = \mathcal{A}_1(0; b = \infty)$, and $(n+2) \approx n$. For spatial dimensions $d \neq 2$,

$$T_c(b) = T_{c0} - \frac{4nu_0}{\alpha_0 c} \left(\frac{K_d \Lambda^{d-2}}{d-2} \right) (1 - b^{2-d}), \tag{8.41b}$$

and for d=2,

$$T_c(b) = T_{c0} - \frac{4nu_0}{\alpha_0 c} K_d \ln b. \tag{8.41c}$$

When $b \to \infty$, $T_c(b)$ of $(d \leq 2)$-dimensional systems tends to $-\infty$ as discussed previously in Section 7.11. The number $b^{-1} \ll 1$ in the limit of large b can be thought of as proportional to a dimensionless quantity like $t = (T - T_c)/T_c$. Then the divergence $(-\infty)$ of T_c for $d \leq 2$ is expressed by t.

8.4.2 *Large-n critical behaviour*

The susceptibility exponent γ, see Table 1.1, is obtained from Eq. (8.39b) and the relation $r = r'(T - T_c)^\gamma$, where r' does not depend on $(T - T_c)$. Let us denote the quantity $\alpha_0(T - T_c)$ by $\tilde{r}_0$. Then Eq. (8.39b) yields

$$\tilde{r}_0 = r + 4nu_0 r K_d \int_{b^{-1}\Lambda}^{\Lambda} \frac{p^{d-1} dp}{cp^2(cp^2 + r)}. \tag{8.42}$$

For $b \neq 0$, the integral in Eq. (8.42) will be always finite. It tends to a nonzero constant for $r \to 0$. Therefore, for small r, $\tilde{r}_0 \sim r$ and, hence, $\gamma = 1$. When $b \to \infty$, the behaviour of the parameter r depends on the spatial dimension d. From now on the notation for the critical temperature T_c will be $T_c(\infty)$. For $2 < d < 4$, the integral in Eq. (8.42) is proportional to $r^{-2+d/2}$. This result is obtained when a change of the variables $(c/r)^{1/2} k = k'$ and an approximation $\Lambda(c/r)^{1/2} \to \infty$ for the upper cutoff of $|k'|$ are introduced; see a similar calculation presented in Section E.1. Therefore, Eq. (8.42) become

$$\alpha_0 \left(\frac{r}{r'} \right)^{1/\gamma} = r + 4nu_0 C_1(d) r^{d/2-1}, \tag{8.43}$$

where $C_1(d)$ is given by Eq. (7.232).

For $d < 4$, $\dfrac{d}{2} - 1 < 1$ so that the second term in Eq. (8.43) tends to zero as $T \to T_c$ more slowly than the term linear in r. Eq. (8.43) will be valid for $T \to T_c$ provided

$$\gamma = \frac{2}{d-2} \tag{8.44}$$

and $\alpha_0 r' = [4nu_0 C_1(d)]^\gamma$; when r is small, Eq. (8.43) cannot be satisfied for values $\gamma \leq 1$. When $d > 4$, the second term in Eq. (8.43) tends to zero at T_c faster than the linear one and we get $\gamma = 1$.

Since the quadratic fluctuation spectrum $\varepsilon_0(k) = ck^2$ is not changed in the large-n limit, the Fisher exponent η will be equal to zero; see Chapter 1, Eq. (1.15). The correlation length exponent ν will be determined from the relation $\xi \sim (c/r)^{1/2}$ and, hence, $\nu = \gamma/2$. If $\gamma = 1$, as it is for $d > 4$, $\nu = 1/2$, the classical value of ν. The exponents α, β, and δ will certainly differ from their classical values for $2 < d < 4$; see Table 1.2.

In order to calculate the other exponents we have to find the effective potential $\mathcal{H}'(\varphi_0)$. Now we shall use the diagrams in Fig. 8.7(b). Thus we obtain another geometrical progression:

$$u = u_0 - 4nu_0^2 \mathcal{A}^{(2)}(r_0, q_1) + 16n^2 u_0^3 \left[\mathcal{A}^{(2)}(r_0, q_1)\right]^2 - \cdots \tag{8.45}$$

In this series the bare value r_0 can be replaced by r to any order in u_0. This is made without approximation because to each term in Eq. (8.45) corresponds a term in the general expansion (8.31b) for $u(\equiv \tilde{u})$ which compensates the difference. For example, the difference

$$-4nu_0^2 \left[\mathcal{A}^{(2)}(r_0, q_1) - \mathcal{A}^{(2)}(r, q_1)\right]$$

is exactly cancelled by the $n^2u_0^3$-term in Eq. (8.31b):

$$32n^2 u_0^3 \mathcal{A}_1(r_0) \mathcal{A}^{(3)}(r_0, q_1, q_1).$$

Therefore we can substitute r_0 by r in the integral $\mathcal{A}^{(2)}$ in Eq. (8.45). For the $n^2u_0^3$-term in the series (8.45) a compensating term of order u_0^4 exists and so on. The necessity of such a successive compensation procedure does not arise when the large–n limit is investigated with the help of the counter term $\mathcal{H}_c$.

Table 8.1 Critical exponents for the spherical model.

α	β	γ	δ	ν	η
$\frac{4-d}{d-2}$	$\frac{1}{2}$	$\frac{2}{d-2}$	$\frac{d+2}{d-2}$	$\frac{1}{d-2}$	0

No matter what method we use, the result is

$$u(r, q_1) = \frac{u_0}{1 + 4nu_0 \mathcal{A}^{(2)}(r, q_1)}. \tag{8.46}$$

For $b \to \infty$, $\mathcal{H}'(\bar{\varphi})$ becomes

$$\mathcal{H}'(\bar{\varphi}) = V\left[\frac{r}{2}\bar{\varphi}^2 + \bar{u}\bar{\varphi}^4\right], \tag{8.47}$$

where

$$\bar{\varphi}^2 = V \sum_{\alpha=1}^{n} \varphi_\alpha^2(0) \tag{8.48}$$

and $\bar{u} = u(r, 0)$; see Eq. (8.46). For $r \sim 0$, $\mathcal{A}^{(2)}(r, 0) \sim r^{-2+d/2}$, and

$$\bar{u} \sim \frac{1}{n} r^{2-d/2} \sim 0 \tag{8.49}$$

It is now easy to obtain the exponents α, β, and δ in the way explained in Chapter 4 with $r \sim (T - T_c)^\gamma$ and

$$\bar{u} \sim r^{2-d/2} \sim (T - T_c)^{(4-d)/(d-2)}. \tag{8.50}$$

The minimization of $\mathcal{H}'(\bar{\varphi})$ with respect to $\bar{\varphi}$ yields $\bar{\varphi} \sim (T - T_c)$, that is $\beta = 1/2$. The second derivative of $\mathcal{H}'[\bar{\varphi}(T - T_c)]$ with respect to T gives the specific heat exponent α. Finally, by introducing an external field h conjugate to $\bar{\varphi}$ we obtain $h \sim \bar{u}\,\bar{\varphi}^3$ or, using $(T - T_c) \sim \bar{\varphi}^2$ and Eq. (8.50), $h \sim \bar{\varphi}^{(d+2)/(d-2)}$, therefore $\delta = (d+2)/(d-2)$. The results for the critical exponents are shown in Table 8.1.

The large-n critical behaviour coincides with the critical behaviour described by the Berlin-Kac spherical model (Section 5.7.1). It happens often that the results following from an exactly solvable model as that of Berlin and Kac are the same as those by summation of infinite subsets of perturbation terms arising from a more general model. The fluctuation interaction effects described by the infinite geometric progressions in Fig. 8.7 are only a part of the fluctuation interaction effects given by the φ^4-interaction; see also Busiello and Uzunov (1987a).

The "spherical" critical behaviour is quite interesting because it establishes a rather intriguing dependence of the critical behaviour on the spatial dimension d but the "spherical" values of the critical exponents do not explain the experiments in real systems. Clearly, the accounting for the fluctuation effects in the large-n limit does not present adequately the fluctuation phenomena near T_c.

8.4.3 *Hartree approximation*

We have seen that in the large-n limit the φ^4-Hamiltonian exhibits properties analogous to those of the Berlin-Kac spherical model (Berlin and Kac, 1952); see also Section 5.7.5. Here we shall show that the same limit corresponds to the famous Hartree approximation.

The Hartree approximation for the φ^4-Hamiltonian is done by replacing the φ^4-interaction part $\mathcal{H}_{\rm int}$ with

$$\mathcal{H}_{\rm int} \approx \frac{u_0}{V} \sum_{\alpha\beta;k_i} \left(\langle \varphi_\alpha \varphi_\alpha \rangle \varphi_\beta \varphi_\beta + \langle \varphi_\beta \varphi_\beta \rangle \varphi_\alpha \varphi_\alpha + 4 \langle \varphi_\alpha \varphi_\beta \rangle \varphi_\alpha \varphi_\beta \right); \tag{8.51}$$

the momenta k_i of the fields φ_α, φ_β in the curly brackets are not explicitly written. The coefficient 4 stands for the four possible ways of forming the pair average $\langle \varphi_\alpha \varphi_\beta \rangle$ from the product $\varphi_\alpha^2 \varphi_\beta^2$. Performing the respective summation and integration in Eq. (8.51) we obtain

$$\mathcal{H}_{\rm int} \approx 2(n+2) u_0 I_1(r_0) \sum_{\alpha,k} |\varphi_\alpha(k)|^2. \tag{8.52}$$

The total "Hartree" Hamiltonian is

$$\mathcal{H}_H = \frac{1}{2} \sum_{\alpha,k} \{ck^2 + r_H\} |\varphi_\alpha(k)|^2, \tag{8.53}$$

where

$$r_H = r_0 + 4(n+2) u_0 I_1(r_0). \tag{8.54}$$

This expression for the inverse ("Hartree") susceptibility $r_H \sim \chi_H^{-1}$ can be improved by introducing the counter term $\mathcal{H}_c$ provided the averages in

Eq. (8.51) are taken over the probability distribution of the free Hamiltonian $\mathcal{H}_2 = (\mathcal{H}_0 - \mathcal{H}_c)$. Then r_0 in the integral $I_1(r_0)$ will become r and, hence,

$$r = r_0 + 4(n+2)u_0 I_1(r); \tag{8.55}$$

cf. Eq. (8.39b) for $b \to \infty$.

From the results in Section 8.4.1 we can make the conclusion that the Hartree approximation is equivalent to the zero order ($n \approx \infty$) approximation within the large-n limit; so we can write $(n+2) \approx n$ in Eq. (8.55). The Hartree approximation can be carried out within the scheme of slow and fast fields as well. The result is expressed by Eqs. (8.54) and (8.55) where instead of I_1 we must write $\mathcal{A}_1(r_0)$ or $\mathcal{A}_1(r_1)$. The Hartree approximation is a good starting point for the $(1/n)$–expansion; see Section 8.10.3, and Busiello and Uzunov (1987a), where a field counterpart of the lattice Berlin–Kac model is discussed.

8.5 Renormalization-group Recursion Relations

Except for simple cases, such as the large-n limit considered in Section 8.4, the effective Hamiltonian $\mathcal{H}'(\varphi_0)$ cannot be calculated to an arbitrary order of the loop expansion for the vertex functions $\Gamma^{(2m)}$. Certainly this calculational problem seems to be as hard as the direct calculation of the partition function by the standard perturbation technique. The modified perturbation theory presented in Sections 8.1–8.4 has however an advantage connected with the absence of infrared divergences ($k \to \infty$) for $r \to 0$.

In certain cases the calculation of the perturbation integrals is simplified. For example, when $(b^{-1}\Lambda)^2 \gg q_i^2$ one can neglect the external momenta q_i or to take them into account by expanding the corresponding integrands in powers of q_i^2. Within the standard perturbation theory, an analogous expansion can be performed in powers of (q_i^2/r). When $r \sim 0$ as it is near T_c, this expansion cannot be done. Within the modified perturbation scheme however the perturbation integrals can be expanded both in powers of $(r/b^{-1}\Lambda)^2$ provided $T \sim T_c$ and $q_i^2 \ll (b^{-1}\Lambda)^2$. Bearing this in mind, we can calculate approximately the vertex functions $\Gamma^{(2m)}$ or the vertex parameters $\tilde{c}, \tilde{r}, \tilde{u}_4, \ldots$ for small r and q_i^2. This opportunity will be used in our further discussion.

The problem for the investigation of the critical state can now be reformulated as a problem for obtaining the relation between the "old" ($\mathcal{H}$)

and the "new" ($\mathcal{H}'$) effective Hamiltonians. This relationship has been discussed at a qualitative level by Kadanoff (1966); the successful treatment of the problem has been demonstrated by Wilson (1971); see also Wilson and Fisher (1972), and the discussion in Chapter 1.

Let us write the new effective Hamiltonian $\mathcal{H}'(\varphi_0)$ in the form

$$\begin{aligned}\mathcal{H}'(\varphi_0) = \frac{1}{2}\sum_{\alpha,k}{}' \left(\tilde{r} + \tilde{c}k^2\right)|\varphi_\alpha(k)|^2 \\ + \frac{1}{V}\sum_{\alpha\beta;k_i}{}' \tilde{u}\varphi_\alpha(k_1)\cdots\varphi_\beta(-k_1-k_2-k_3) \\ + \frac{1}{V^2}\sum_{\alpha\beta\gamma;k_1\ldots k_5}{}' \tilde{u}_6\varphi_\alpha(k_1)\cdots\varphi_\gamma(-k_1-k_2-\cdots-k_5) + O(\varphi^8).\end{aligned} \tag{8.56}$$

Here the fields $\varphi_{0\alpha}$ are replaced with φ_α according to Eqs. (8.3a)–(8.3b) and the "prime" of the sums means a restricted summation over the cells $0 < |k_i| < b^{-1}\Lambda; i = 1, 2, \ldots$. The effective parameters $\tilde{r}$ and $\tilde{c}$ describe the effective fields $\varphi_{0\alpha} = \varphi_\alpha(k < b^{-1}\Lambda)$ when the interaction is not taken into account ($\tilde{u}_4 = \tilde{u}_6 = \cdots = 0$). The interaction parameters (functions)-$\tilde{u}_4(k_i, r)$, $\tilde{u}_6(k_i, r)$, etc. determine the slow-field interactions of type $\varphi_0^4, \varphi_0^6, \ldots$.

We have established the relations between the effective and initial (r_0, c, u_0) parameters of the systems up to the two-loop approximation; see Eqs. (8.31a)–(8.31c). However these relations are not sufficient for setting up the way of the mapping of $\mathcal{H}(\varphi)$ to $\mathcal{H}'(\varphi_0)$ because the upper cutoffs of the momenta in these Hamiltonians are different. In order to restore the initial cutoff Λ in $\mathcal{H}'(\varphi)$ we shall introduce the following transformations of the momenta (k) in $\mathcal{H}'(\varphi_0)$ to new momenta k':

$$k = b^{-1}k', \tag{8.57}$$

where $b > 1$ is a real number: a dimensionless quantity that is varying over the following range $1 < b < \infty$. The new momenta k' satisfy the condition $0 < |k'| < \Lambda$, as it is readily seen from Eq. (8.57), and the inequality $0 < |k| < b^{-1}\Lambda$. According to the length-scale change (8.57), any length L will become L' through the relation $L = bL'$. The number $b > 1$ alters the units of measurement of lengths, that is why it is sometimes called the rescaling factor; see also Section 8.1. For example, the volume $V \sim L^d$ of a d-dimensional systems will be transformed to V':

$$V = b^d V', \tag{8.58}$$

i.e., the volume shrinks by a factor of b^{-d}.

The scale transformation (8.57) of k to k' will give rise to the corresponding transformations of the fields

$$\varphi_\alpha(k) = \varphi_\alpha(b^{-1}k') = \zeta(b)\varphi'_\alpha(k'), \tag{8.59}$$

where $\zeta(b)$ is a function of b. Whether the fields φ_α really obey the scaling transformation (8.59) will be discussed below. For certain models of anisotropic systems the rescaling factor $\zeta(b)$ depends on $\alpha; \zeta(b) \to \zeta_\alpha(b)$.

Introducing the relations (8.57)–(8.59) in Eq. (8.56) for $\mathcal{H}'(\varphi)$, we obtain the following expression for the transformed Hamiltonian

$$\begin{aligned}\mathcal{H}'(\varphi') &= \frac{1}{2}\sum_{\alpha,k'}\{r' + c'k'^2\}|\varphi'_\alpha(k')|^2 \\ &+ \frac{1}{V'}\sum_{\alpha\beta;\, k'_i} u'\varphi'_\alpha(k'_1)\cdots\varphi'_\beta(-k'_1 - k'_2 - k'_3) \\ &+ \frac{1}{V'^2}\sum_{\alpha\beta\gamma;\, k'_i} u'_6\varphi'_\alpha(k'_1)\cdots\varphi_\gamma(-k'_1 - \cdots - k'_5) + \ldots,\end{aligned} \tag{8.60}$$

where the summations are over k' and k'_i in the interval $0 < |k'|, |k'_i| < \Lambda$. The expressions of $c', r', u', u'_6, \ldots$ depend on those for $\tilde{c}, \tilde{r}, \tilde{u}_4, \ldots$, i.e., on the order in the loop-expansion taken into consideration. Let us first introduce some general notations and then discuss the form of the functions $c', r', u', \ldots$

It is convenient to represent the parameters $c, r, \ldots$ by a vector $\mu = (c, r, \ldots)$ in *the parameter space* of the Hamiltonian. For the initial φ^4-Hamiltonian this space is three-dimensional. Let us denote the parameters of the initial φ^4- Hamiltonian $\mathcal{H}(\varphi)$ by the vector $\mu = (c, r_0, u_0) \equiv (\mu_1, \mu_2, \mu_3)$. The new Hamiltonian $\mathcal{H}'(\varphi')$ will be given by the vector $\mu' = \{\mu_i\} = (c', r', u', u'_6, \ldots)$ in an extended (many-dimensional) parameter space. The new Hamiltonian parameters $\mu' = \{\mu'_i\}$ can be expressed by the "old" ones $\{\mu_i\}$ through the transformation

$$\mu'_i = \mu'_i(\{\mu_i\}, b). \tag{8.61}$$

These functional relations for the parameters μ_i and Eq. (8.59) for the field components φ_α are called *renormalization-group (RG) recursion relations.* The RG theory is concerned with the problem to find an appropriate receipt for obtaining such RG recursion relations and after that to investigate them.

In the following discussion, we shall use the results for the effective parameters $\tilde{c}, \tilde{r}, \tilde{u}_4, \ldots$ from Section 8.3 in one- and two-loop approximations. Certainly the mathematical form of the RG equations (8.61) depends on the form of the initial Hamiltonian and the accuracy to which the integration over the fast field $\varphi_{1\alpha}$ has been performed. For simplicity, we shall at first consider the RG recursion relations in the one-loop approximation. From Eqs. (8.21a), (8.21b) and (8.27a)–(8.27b) we have

$$c' = \zeta^2(b)b^{-2}c, \tag{8.62a}$$

$$r' = \zeta^2(b)\left[r_0 + 4(n+2)u_0\mathcal{A}_1(r_0)\right], \tag{8.62b}$$

$$u_0' = \zeta^4(b)b^{-d}\left[u_0 - 4(n+8)u_0^2\mathcal{A}_2(r_0, q_1b^{-1})\right], \tag{8.62c}$$

$$u_6' = \zeta^6(b)b^{-2d}\tilde{u}_6(q_ib^{-1}), \tag{8.62d}$$

$$\vdots$$

$$u_{2l}' = \zeta^{2l}(b)b^{(l-1)d}\tilde{u}_{2l}(q_ib^{-1}), \tag{8.62e}$$

where $\tilde{u}_{2l}(q_ib^{-1})$ is the $(2l)$-th effective interaction parameter and $q_i(\equiv q_i')$ are the new external momenta; in Eqs. (8.62c) – (8.62e) the "prime" of q_1' and q_i' has been omitted.

Eqs. (8.62a)–(8.62e) show a quite complicated RG (transformation); below the abbreviation "RG" will be used to denote both the "renormalization group" (RG) and the "renormalization-group transformation". The RG recursion relations (8.62a)–(8.62e), or, more generally the RG, Eqs. (8.61) present a scheme for relating phenomena having different length scales ($L > \Lambda^{-1} \sim a$, and $L' \sim (b^{-1}\Lambda)^{-1} \sim a' > a$) which are described by different effective Hamiltonians, $\mathcal{H}(\varphi)$ and $\mathcal{H}'(\varphi)$, referred to one and the same systems.

The RG is defined by two successive mathematical operations:

(i) the integration (I_b) over the degrees of freedom $\varphi_{1\alpha}$ corresponding to relatively short-range phenomena, and

(ii) the rescaling (S_b) of the effective Hamiltonian $\mathcal{H}'(\varphi_0)$ which restores the initial cutoff Λ in the system.

The product $I_b S_b$ of these operations gives the RG connecting $\mathcal{H}(\varphi)$ and $\mathcal{H}'(\varphi')$:

$$\mathcal{H}' = R_b \mathcal{H}, \tag{8.63}$$

where

$$R_b = S_b I_b \tag{8.64a}$$

or, more explicitly,

$$R_b = -\ln \int D\varphi_1 \exp\left[-(\cdots)\right]\Big|_{\substack{k=k'/b \\ \varphi_\alpha(k)=\zeta(b)\varphi'_\alpha(kb)}} . \tag{8.64b}$$

The definition (8.63)–(8.64b) of RG is not restricted to particular fluctuation Hamiltonians. It can be applied to a variety of field models describing critical phenomena.

The successive application of the RG, i.e., of the R_b-operation (8.64a)–(8.64b),

$$\mathcal{H}'' = R_{b_2}\mathcal{H}' = R_{b_2}R_{b_1}\mathcal{H} \tag{8.65}$$

will give a new Hamiltonian $\mathcal{H}''(\mu_i'')$ in which the phenomena corresponding to the "fast" fields $\varphi_\alpha(k)$ with $(b_1 b_2)^{-1}\Lambda < |k| < \Lambda$ are taken into account through the functional relations

$$\mu_i'' = \mu_i''\left[\mu'(\mu)\right]. \tag{8.66}$$

However the same Hamiltonian $\mathcal{H}''(\mu'')$ should be obtained from the single transformation

$$\mathcal{H}'' = R_{b_1 b_2}\mathcal{H}. \tag{8.67}$$

Both the integration I_b and the scaling transformation S_b are independent on whether the fields $\varphi_{1\alpha}(k)$, $(b_1 b_2)^{-1}\Lambda < |k| < \Lambda$, are integrated out and then rescaled by a single $R_b = R_{b_1 b_2}$-operation ($b = b_1 b_2$) or the same is performed by two successive RG transformations, R_{b_1} and R_{b_2}. Therefore, we can write down the important (group) property of the generator R_b of RG:

$$R_b = R_{b_1} R_{b_2}, \tag{8.68}$$

for any $b_1 b_2 = b$. The above rule is trivially generalized for more than two transformations: $R_{b_1} R_{b_2} R_{b_3} \ldots = R_{b_1 b_2 b_3 \ldots}$.

If the group property (8.68) does not hold the successive RG transformations will depend on the way in which they are performed. The operations I_b and S_b themselves cannot be a cause for such a possibility to occur. But there exists a possibility for breaking down of the property (8.68), for example, when the RG is constructed in a quite oversimplified manner. It is clear from the physical arguments presented above that in order to apply the RG scheme in a sensible way it is absolutely necessary that the group property (8.68) to be fulfilled; for details, see Uzunov (1987).

Because of the integration I_b the operation R_b cannot be inverted and, hence, the inverse element R_b^{-1} cannot be defined. The absence of the R_b^{-1}-operation determines RG rather as a semi-group than as a group of transformations. Another group properties following from general physical requirements for the behaviour of the systems under RG are given by the relations $I_b S_b = S_b I_b$ and $R_{b_1} R_{b_2} = R_{b_2} R_{b_1}$.

The application of RG to the fluctuation Hamiltonian of the system gives the opportunity to define an infinite set of Hamiltonians: $\mathcal{H}(\mu, \varphi)$, $\mathcal{H}'(\mu', \varphi')$, $\mathcal{H}''(\mu'', \varphi'')$, ... corresponding to the points $\mu, \mu', \ldots$ in the parameter space (the space of the vectors $\mu = \{\mu_i\}$). Since $\mu, \mu', \mu'', \ldots$ are points in the Hamiltonian parameter ($\mu-$)space, the RG recursion relations (8.61) will give the (RG) trajectories in this space; sometimes called RG flow lines or Hamiltonian flow lines. The geometrical representation of the RG equation by RG flow lines in the μ-space is very applicable to the RG studies.

The parameters $\mu = \{\mu_i\}$ are functions of the fixed thermodynamic quantities (T, h, Y); see, e.g., Section 4.2. Let two different physical systems be described by a fluctuation Hamiltonian $\mathcal{H}(\varphi)$. For example, these can be two ferromagnets with the same symmetry of ordering but with different chemical composition so that the initial parameters $\mu_1 = \{\mu_i^{(1)}\}$ and $\mu_2 = \{\mu_i^{(2)}\}$ are different and despite of the same structure of the functional $\mathcal{H}(\varphi)$ for these systems, $\mathcal{H}(\varphi, \mu_1) \neq \mathcal{H}(\varphi, \mu_2)$. Since $\mu_1 \neq \mu_2$ these systems will have different critical temperatures, T_{c_1} and T_{c_2}. Such a family of different Hamiltonians corresponding to different physical systems can also be represented either by the whole μ-space or, at least by a domain in this space. In addition the variations of T and h will create variations of the parameters $\mu_1, \mu_2, \ldots$. This is the standard mapping of the points in the phase diagram (T, h, Y) to the points in the μ-space of the effective

Hamiltonian. As far as this effective Hamiltonian has a fixed structure, the only thing that can vary are the values of its initial parameters, which correspond to different physical systems belonging to one and the same class.

In contrast to this standard representation of a family of different physical systems as points in the μ-space, the RG family $\mathcal{H}, \mathcal{H}', \mathcal{H}'', \ldots$ of Hamiltonians or, that is the same, of Hamiltonian flow lines $\mu_i'(\mu)$ represent one and the same system at different levels of description. The number of degrees of freedom of the "new system" $(')$ corresponding to $\mathcal{H}'$ is less than that of the original one $(\mathcal{H})$ at the expense of introducing the renormalization of the parameters $\mu \to \mu'$ and the appearance of new effective interactions in it (like $u_6\varphi^6, u_8\varphi^8$, etc. in the φ^4-model). Suppose that the initial Hamiltonian determined by the parameters $\mu_1 = \{\mu_i\}$ describes a real ferromagnetic or superconducting sample and suppose that another real system (μ_2) of the same type exists for which $\mu_1' = \mu_2$. Obviously, if the points μ_1 and μ_2 represent different ferromagnets having equal thermodynamic parameters T and h or two identical ferromagnets at different phase diagram points (T, h), (T_1, h_1) and (T_2, h_2), the transformed system μ_1' is formally equivalent to the initial one (μ_1) but taken at another point in the phase diagram (T, h, Y).

The operation S_b for the φ^4-model has been carried out with the help of the linear relation (8.59), where $\zeta(b)$ is the rescaling factor for the field φ_α. Sometimes the RG with a linear rescaling, $\varphi \sim \zeta\varphi'$, is called *linear RG* (for a discussion of the nonlinear RG; see, e.g., Wegner (1976).

In the remaining part of this Chapter as well as in Chapter 9 our considerations will be based on the linear RG which has a large practical application. Even in this case the RG Eqs. (8.61) display a rather complicated nonlinear connection between the parameters (vectors) μ and μ'; see, e.g., Eqs. (8.62a)–(8.62c) which give the relatively simple example of RG equations in the one-loop approximation. Obviously relations like (8.62a)–(8.62e) can hardly be used in the studies of the critical behaviour unless additional arguments ensuing from some basic features of the critical state are involved into consideration. Following Wilson and Fisher (1972) we shall demonstrate in the next section how the Eqs. (8.62a)–(8.62e) can be analyzed in a close vicinity of the critical point.

8.6 Renormalization Group in the One-loop Approximation

8.6.1 *General features of the renormalization group equations*

Two features of the Eqs. (8.62a)–(8.62e) create difficulties for the RG analysis. The first is their nonlinearity and the second is that these relations are not in a diagonal form — the parameter r_0' depends on r_0, u_0, and c rather than on r_0 only and the same is valid for u_0', u_6', etc. If the RG relations were linear it would be not difficult to diagonalize them by some rotation in the μ-space. Here we shall analyze Eqs. (8.62a)–(8.62e) with the only purpose to arrive at their linearized version (*linearized* RG).

We begin with the group property (8.68). Certainly, Eqs. (8.62a)–(8.62e) and, in particular the simple Eq. (8.62a) for c satisfy it. Note that the last equation holds also when $u_0 = 0$, i.e., when there is no interaction between the fields φ_α. Applied to Eq. (8.62a)–(8.62e) the group property (8.68) reads

$$\zeta(b_1)\zeta(b_2) = \zeta(b_1 b_2). \tag{8.69}$$

If the function $\zeta(b)$ is not identically zero, as it is supposed here, it will obey Eq. (8.69) only under the assumption that

$$\zeta(b) = b^y. \tag{8.70}$$

Here the exponent y does not depend on b; for the proof see, e.g., Aczél (1966, 1969) and Stanley (1971). It will be more convenient to denote the exponent y by $1 - (\eta/2)$, because as we shall see later on $\eta = 2(1 - y)$ exactly coincides with the Fisher exponent discussed in Chapter 1.

It becomes clear that the group property (8.68) is not automatically fulfilled by the RG Eqs. (8.62a)–(8.62e) for the φ^4-model in one-loop approximation. Rather one has to impose the property (8.68) and to deduce the corresponding consequences. This is valid for the application of RG to all models of phase transitions. In the practical applications which are based on the ϵ-expansion and the differential RG equations (Section 8.7) the group property (8.68) is always fulfilled; see also Uzunov (1987).

Using the new notation for y, we can rewrite Eqs. (8.62a)–(8.62c) in an equivalent form:

$$c' = b^{-\eta}c, \tag{8.71a}$$

$$r_0' = b^{2-\eta}\left[r_0 + 4(n+2)u_0\mathcal{A}_1(r_0)\right], \tag{8.71b}$$

and

$$u_0' = b^{4-d-2\eta}\left[u_0 - 4(n+8)u_0^2\mathcal{A}_2(r_0, q_1 b^{-1})\right], \tag{8.71c}$$

where $q_1 = (k_1' + k_2')$ and k_i' are the transformed external momenta; Eqs. (8.62d)–(8.62e) are discussed at the end of this Section and in Section 8.7.

The bare parameters $r_0 = \alpha_0(T - T_{c0})$ becomes zero above the (renormalized) critical temperature $T_c < T_{c0}$; see, e.g., Eqs. (8.41) and (7.234b); that is why we shall introduce a new parameter

$$\begin{aligned} r &= \alpha_0(T - T_c) \\ &= r_0 + 4(n+2)u_0\mathcal{A}_1(0), \end{aligned} \tag{8.72}$$

which is quite similar to the parameter r_{0c} introduced in Eq. (7.233) and the parameter $\tilde{r}_0$ introduced in Section 8.4.2. Obviously $r(T_c) = 0$, so the critical temperature will be given by

$$T_c(b) = T_{c0} - \frac{4(n+2)u_0}{\alpha_0}\mathcal{A}_1(0), \tag{8.73}$$

where

$$\mathcal{A}_1(0) = \frac{K_d\Lambda^{d-2}}{c(d-2)}(1 - b^{2-d}), \qquad d \neq 2; \tag{8.74}$$

cf. Eq. (7.234b) and Eqs. (8.41a)–(8.41c). In fact, $T_c(b)$ is the critical temperature of the system described by the transformed Hamiltonian $\mathcal{H}$ while the real temperature is $T_c = T_c(\infty)$. The difference $(r_0 - r)$ is of order u_0.

The recursion relation (8.71b) is valid to the first order in u_0 and the relation (8.71c) is valid to the second order in u_0. Therefore we can replace r_0 with r in the integrals $\mathcal{A}_1$ and $\mathcal{A}_2$. In consequence the recursion relations (8.71b) and (8.71c) become

$$r' = b^{2-\eta}f_r r \tag{8.75a}$$

and

$$u_0' = b^{4-d-2\eta} f_u u_0 \tag{8.75b}$$

with

$$f_r(r, u_0; b) = 1 + 4(n+2)u_0 \left[\frac{\Delta\mathcal{A}(r)}{r}\right], \tag{8.76}$$

where

$$\begin{aligned}\Delta\mathcal{A}(r) &= \mathcal{A}(r) - \mathcal{A}(0)\\ &= K_d \int_{b^{-1}\Lambda}^{\Lambda} dk \cdot k^{d-1}\left[\frac{1}{r+ck^2} - \frac{1}{ck^2}\right]\end{aligned} \tag{8.77}$$

and

$$f_u(r, u_0, q_1; b) = 1 - 4(n+8)u_0\mathcal{A}_2\left(r, q_1 b^{-1}\right). \tag{8.78}$$

8.6.2 *Renormalization group at the critical point*

Let us consider the critical state $r(T_c) = 0$. At the critical point T_c the relation (8.75a) is trivial $(0 = 0)$. When the initial system $\mathcal{H}(\varphi)$ is at the critical point it cannot be taken away from this state by the RG transformations. The correlation length $\xi = (c/r)^{1/2}$ at $r = 0$ is infinite and it cannot become finite under any scale transformation

$$\xi' = b^{-1}\xi, \tag{8.79a}$$

where

$$\xi' = (c'/r')^{1/2}. \tag{8.79b}$$

At this point we should remember that the stable critical state is described by positive parameters c and u_0 (other values of c and u correspond to multicritical phenomena or first-order transitions which require special consideration (see, e.g., Sections 4.8, 6.6, and 9.6). When the initial parameter c takes a finite positive value, the transformed parameter c' will remain in the region $0 < c < \infty$, corresponding to some real physical situation (the so-called "physical region") if only the exponent η in Eq. (8.71a) is equal to zero. Thus the physical requirements lead to the equalities

$$c' = c, \qquad \eta = 0, \tag{8.80}$$

i.e., to the invariance of the parameter c under the RG; note, that Eqs. (8.80) is valid within the one-loop approximation.

While at the critical state $r(= 0)$ and $c > 0$ remain unchanged under RG this is not so for the interaction parameter u_0, which obeys the relation

$$u_0' = b^{4-d}\left[u_0 - 4(n+8)u_0^2\mathcal{A}_2(0, q_1 b^{-1})\right] \tag{8.81}$$

with

$$\mathcal{A}_2(0, q_1 b^{-1}) = \frac{K_{d-1}}{2\pi c^2}\int_{b^{-1}\Lambda}^{\Lambda} dk \cdot k^{d-3}\int_0^{\pi}\frac{\sin^{d-2}\theta d\theta}{k^2 + 2kq_1 b^{-1}\cos\theta + (q_1 b^{-1})^2}, \tag{8.82}$$

where $q_1 (\equiv |\boldsymbol{q}_1|)$ varies in the interval $0 < q_1 < \Lambda$; see also Eqs. (D.8)–(D.11), Eq. (E.32), and Eq. (E.33). Because of the factor b^{-1} the integrand in Eq. (8.82) can be expanded in powers of $(q_1/kb)^2 < (q_1/\Lambda)^2 < 1$ to give

$$\mathcal{A}_2(0, q_1 b^{-1}) = \mathcal{A}_2(0,0) - \frac{(d-4)K_d}{dc^2}\left(\frac{q_1}{b}\right)^2\int_{b^{-1}\Lambda}^{\Lambda} k^{d-7}dk + O(q_1^4 b^{-4}). \tag{8.83}$$

Then Eq. (8.81) takes the form

$$u_0' = b^{4-d}u_0\left[1 - 4(n+8)u_0\mathcal{A}_2(0,0) + O(q_1^2 b^{-2})\right], \tag{8.84}$$

or, equivalently, using the function (8.78), Eq. (8.75b), Eq. (8.80) for $\eta = 0$ and the effective exponent

$$\lambda_u^{(1)}(b) = \frac{\ln f_u}{\ln b}, \tag{8.85}$$

we obtain

$$u_0' = u_0 b^{4-d+\lambda_u^{(1)}(b)}. \tag{8.86}$$

At the critical point the RG transformation of the interaction parameter u does not depend on $r(= 0)$ and $c(= c')$; r and c are fixed. Then the RG is performed only through the parameter u: $u' = u'(u_0)$. The group property will be fulfilled if the exponent $\lambda_u^{(1)}(b)$ does not depend on b. Eqs. (8.78) and (8.85) show that this will be so provided $\ln f_u(r=0)$ is proportional to $\ln b$. The corresponding solutions of the integral (8.82) exist for four-dimensional systems ($d=4$) if the corrections of order $q_1 b^{-1}$ in the integral (8.83) are neglected. These q_1-corrections become smaller with the repeating of the R_b-operation. Therefore, we can accept the approximation $\mathcal{A}_2 \approx \mathcal{A}_2(0,0)$. From Eq. (8.82) we have

$$\mathcal{A}_2(0, q_1 b^{-1}) = \frac{K_{d-1}}{2\pi c^2}\int_{b^{-1}\Lambda}^{\Lambda} dk \cdot k^{d-5}\int_0^{\pi} d\theta \sin^{d-2} \equiv \int_k' \frac{1}{(ck^2)^2}, \quad (8.87\text{a})$$

in an entire conformity with Eq. (8.15b), Eq. (8.28) for $l=2$ and $r_0 = q_1 = 0$, and Eq. (D.13) for $l=1$. For $d=4$ we obtain from Eq. (8.87a) that

$$\mathcal{A}_2(0,0) = \frac{K_d}{c^2}\int_{b^{-1}\Lambda}^{\Lambda} dk \cdot k^{d-5} = \frac{K_d}{c^2}\ln b, \qquad (\text{for } d=4). \qquad (8.87\text{b})$$

A solution of the type *const.*$\ln b$ of the same integral can be obtained near four dimensions ($d \sim 4$) too. For $d \neq 4$ but $d \sim 4$ the integral $\mathcal{A}_2 \equiv \mathcal{A}_2(0,0)$ will be

$$\mathcal{A}_2 = \frac{K_d}{c^2}\Lambda^{d-4}\left(\frac{1-b^{4-d}}{d-4}\right); \qquad (8.88)$$

cf. Eq. (E.22). Assuming that

$$|d-4|\ln b \ll 1, \qquad (8.89)$$

and expanding $b^{4-d} \equiv \exp[(4-d)\ln b]$ in Eq. (8.88) we find

$$\mathcal{A}_2 = \frac{K_d}{c^2}\Lambda^{d-4}(\ln b)\,\{1 + O[(d-4)\ln b]\}; \qquad (8.90)$$

see also Section E.3 and Eq. (E.27). Neglecting the terms $O[(d-4)\ln b]$ in Eq. (8.90) and assuming that

$$4(n+8)u_0\frac{K_d}{c^2}\Lambda^{d-4}\ln b \ll 1 \qquad (8.91)$$

we can represent $f_u(0, u_0, 0, b)$ in Eq. (8.78) in the form $b^{\lambda_u^{(1)}}$, where

$$\lambda_u^{(1)} = -4(n+8)u_0 \frac{K_d}{c^2} \Lambda^{d-4}. \tag{8.92}$$

Unlike c and r, the interaction parameter can vary at the critical point ($r = 0$). The case $u_0 \equiv 0$ is trivial. For $u_0 \neq 0$, u_0' will tend to a fixed value and this is implied by the nonlinearity of the transformation (8.86); $\lambda_u^{(1)}$ depends on u_0. To show this it is convenient to write the recursion relation for u' in the following way

$$u_0' = b^{4-d} \left[u_0 - 4(n+8)u_0^2 \frac{K_d \Lambda^{d-4}}{c^2} \ln b \right] \tag{8.93}$$

and to define that the repeated RG leads to *a fixed point* (FP) u^* at which u does not change anymore. So after applying the next R_b-transformation $b^{4-d+\lambda_u^{(1)}(u^*)}$ to u^* we can see that

$$u' = b^{4-d+\lambda_u^{(1)}(u^*)} u^* = u^*. \tag{8.94a}$$

Therefore, the equation for u^* will be

$$u^* = b^{4-d} u^* \left[1 - 4(n+8)u^* \frac{K_d \Lambda^{d-4}}{c^2} \ln b \right]. \tag{8.94b}$$

One solution of this equation is

$$u_0^* = 0. \tag{8.95a}$$

The second solution is obtained with the help of the relation $b^{4-d} \approx 1 + (4-d) \ln b$:

$$u_1^* = \frac{(4-d)c^2 \Lambda^{4-d}}{4(n+8)K_d}. \tag{8.95b}$$

Now we can investigate the relation (8.93) in the neighbourhood of the fixed point (FP) values u^*, see Eqs. (8.95a) and (8.95b) for u_0^* and u_1^*, by writing $u = u^* + \delta u$ and taking into account the variations of u': $\delta u' = u' - u^*$. The linearization of the relation (8.93) at u^* yields

$$\delta u' = b^{4-d} \left[1 - 8(n+8) \frac{u^* K_d \Lambda^{d-4}}{c^2} \ln b \right] \delta u. \tag{8.96}$$

For $u^* = 0$,

$$\delta u' = b^{4-d}\delta u, \tag{8.97}$$

while for $u_1^* \neq 0$,

$$\delta u' = b^{d-4}\delta u. \tag{8.98}$$

For any $u > 0$, the $(d > 4)$-dimensional systems will tend to the FP $u^* = 0$ while the $(d < 4)$-dimensional systems will tend to the FP $u_1^* \neq 0$.

The existence of two different FPs provokes two types of critical behaviour. The RG transformation eliminates the unessential (fast) modes $\varphi_{1\alpha}$ and preserves the small momentum fluctuations of the order parameter φ. The fluctuations at the critical point $(r = 0)$ with $c \neq 0$ and $u^* = 0$ are of the Gaussian type and, therefore, the FP (8.59a) describes the Gaussian (free-field) critical behaviour. If the fluctuation interaction in the system is neglected the RG analysis begins with $u_0 \equiv 0$ (in the restricted $\mu = (c, r)$-space). Then the system will always sustain this free field behaviour. Moreover the same behaviour is stable towards the appearance of a fluctuation interaction $(u_0 \neq 0)$ for $d > 4$; the exponent $(4 - d)$ in Eq. (8.97) is negative and for any $\delta u = u_0 - u^* = u_0$, $\delta u'$ will tend to zero, i.e., $u_0 \to u_0^* = 0$. For $d < 4$, the free field critical behaviour is unstable with respect to the δu-perturbations of the system. When $u_0 \neq 0$ and $d < 4$ the stable critical behaviour will be described by Eq. (8.98) which corresponds to $u_1^* \neq 0$.

8.6.3 *Critical Hamiltonian and fixed points*

The φ^4-Hamiltonian with $r = 0$ is called *the critical Hamiltonian* ($\mathcal{H}_c$). We have seen that the application of the RG to the critical Hamiltonian

$$\mathcal{H}_c = \frac{1}{2}\int d^dx \left[c(\nabla\varphi)^2 + 2u\varphi^4\right]. \tag{8.99}$$

always leads to the FP Hamiltonian,

$$\mathcal{H}_{\rm FP} = \frac{1}{2}\int d^dx \left[c(\nabla\varphi)^2 + 2u^*\varphi^4\right]. \tag{8.100}$$

This will be either the free field FP Hamiltonian $\mathcal{H}_{(0)FP} = \mathcal{H}_{\rm FP}(u^* = 0)$ which describes the Gaussian critical behaviour or the interaction field FP Hamiltonian $\mathcal{H}_{\rm FP}(u_1^*)$ which describes the interacting fluctuations at T_c.

The FP $u_0^* = 0$ is called *the trivial FP* or *the Gaussian* FP (in short, GFP). Its coordinates are denoted by the index (0) as above or by the index (G), for example, $u_G(\equiv u_0^*)$. This GFP always exists for models which adequately describe critical phenomena produced by non-interacting fluctuations and, therefore, include the Gaussian type of critical behaviour as an adequate description of the fluctuation phenomena outside the critical region T_c. Since the Gaussian critical behaviour always occurs as an approximation to the behaviour of the system near (and at) the critical point, the GFP (or, shortly, G) will always describe this Gaussian approximation.

The FP $u_1^* \neq 0$ is called *the nontrivial FP* or *the Heisenberg FP* (HFP; often denoted by $u_H(\equiv u_1^*)$. It gives a picture of the critical behaviour of $(n \geq 1)$-component systems with interacting fluctuations which resemble the Heisenberg $(n = 3)$ model of a ferromagnet. When $n = 1$ the same FP is sometimes called *Ising* FP (IFP). We shall use also the shorter notations: G for GFP, H for HFP and I for IFP.

In $\mathcal{H}_c$, Eq. (8.99), u should vary in the physical domain $u > 0$ (u is a general notation for the interaction constant whereas u_0 stands for the initial value of the same constant). Thus the positive axis u in the parameter space (r, c, u) of the Hamiltonian represents all critical states (all critical Hamiltonians) of the systems corresponding to the model under consideration. In fact, the critical states lie on the critical plane $r = 0$ or, equivalently, on the family of lines defined by $(r = 0, c = const, u)$. This subspace of the μ-space is called the "critical subspace" or the "critical manifold". In the case of the relatively simple φ^4-Hamiltonian we are dealing with the critical ray $u > 0$; see Fig. 8.8. The arrows in Fig. 8.8(a) show that the critical Hamiltonians for $(d < 4)$-dimensional systems will tend always to H; i.e., H is stable and G is unstable. For $d > 4$, G is stable (all critical Hamiltonians tend to the G-Hamiltonian where $u_G = 0$); see Fig. 8.8(b). In this case H lies in the unphysical (instability) interval $u < 0$. For $d = 4, u_H = u_G = 0$ so G and H coincide. AT $d = 4$, both G and H exhibit a marginal stability and the study of the stability properties of these FPs requires a more careful analysis within the two-loop approximation (Section 8.9).

In order to complete the RG analysis of the critical state let us mention that G and H have the following coordinates in the parameter space (c, r, u):

$$\mathrm{G} : (c, 0, 0), \tag{8.101a}$$

$$\mathrm{H} : (c, 0, u_H). \tag{8.101b}$$

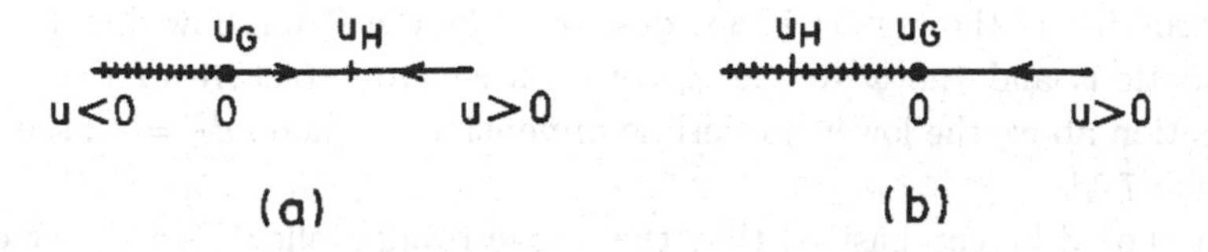

Fig. 8.8 The critical ray of the φ^4-model: (a) $(d<4)$-dimensional systems, (b)$(d>4)$-dimensional systems. The ray $u<0$ has no physical meaning (instability domain of the φ^4-model).

Since we usually know the initial value of r_0, we can easily calculate from Eqs. (8.72), (8.74), and (8.95a)–(8.95b) that

$$r_{0G} = 0, \tag{8.102}$$

and

$$\begin{aligned} r_{0H} &= -\frac{4(n+2)u_H K_d \Lambda^{d-2}(1-b^{2-d})}{c(d-2)} \\ &= -\frac{(n+2)}{(n+8)}\frac{(4-d)}{(d-2)}c\Lambda^2(1-b^{2-d}). \end{aligned} \tag{8.103}$$

The value r_{0H} depends on b as it is in the approximate RG formula derived by Wilson (1971). The reason is that we have chosen as a parameter in our study $r = \alpha_0[T - T_c(b)]$. In fact, one should take $r = \alpha_0[T - T_c(\infty)]$, where $T_c(\infty)$ is the actual critical temperature of the system. Of course the results from our RG analysis within the one-loop approximation will be not changed if we instead of r from Eq. (8.72) use $r = \alpha_0[T - T_c(\infty)]$. The difference $\Delta T_{cb} = [T_c(b) - T_c(\infty)]$ is given by

$$\Delta T_{cb} = \frac{4(n+2)u_0}{\alpha_0}\frac{K_d\Lambda^{d-2}}{c(d-2)}b^{2-d}$$

and, therefore, the shift of r by the quantity $\alpha_0\Delta T_{cb}$ will give the precise FP value

$$r_{0H} = -\frac{(n+2)(4-d)}{(n+8)(d-2)}c\Lambda^2. \tag{8.104}$$

Taking the true critical temperature $T_c = T_c(\infty)$ we have to bear in mind that our analysis is valid for $(d>2)$-dimensional systems, where the fluctuation shift $(T_{c0} - T_c)$ of the critical temperature is finite. In this case the ordering in the system is not destroyed by the strong low-dimensional fluctuations and the φ^4-model gives a satisfactory picture of the phase transition above the lower borderline dimension d_L (here, $d_L = 2$); see also Section 7.11.

It should be emphasized that the above results follow as a direct consequence of the fact that the RG analysis of the critical state has been performed in the "nearly" four-dimensional ($d \sim 4$) space. We have used the small parameter $\epsilon = (4-d)$ and our ϵ-analysis is correct for small $|\epsilon \ln b|$; see Eq. (8.89). All terms of order higher than $|\epsilon| \ln b \ll 1$ have been neglected; remember, for example, the approximation $\mathcal{A}_2 \sim \ln b$ for $\mathcal{A}_2$ from Eq. (8.88).

The effective interaction terms like u_6, u_8, etc. which are not present in the initial φ^4-Hamiltonian are of order $u_0^2, u_0^3, \dots$, respectively and, therefore, they corresponds to zero FP values within the $u_0 \sim \epsilon^1$-order analysis ($u_6^*, u_8^* \sim \mathrm{O}(\epsilon^2) \sim 0$). These parameters turn out completely inessential for the description of the main features of the critical behaviour (for more details, see Section 8.7).

The results from this ϵ-expansion (Wilson and Fisher, 1972) are consistent with those from the loop expansion. In fact we have obtained that u_H and r_{0H} are proportional to $\epsilon = (4-d)$. The one-loop expansion in powers of the interaction constant u_0 is now effectively substituted by the ϵ-expansion in nearly four-dimensional space. Having notion about the way, in which the critical state and the RG analysis within the ϵ-expansion are related to each other we shall present in the next section the systematic RG calculations in the neighbourhood of the critical point.

8.7 Linearized Renormalization Group in the One-loop Approximation

Here we shall use the concept of FP introduced in Section 8.6 in order to make a complete investigation of the linearized version of the RG recursion relations (8.62a)–(8.62e). Unlike the analysis in Section 8.6, where our aim has been to clarify the relation between the FPs of RG and the critical state here we shall examine the system away from the critical point ($r \neq 0$). Moreover we shall present the RG analysis in an explicit form suitable for the study of the (asymptotic) critical behaviour in the Ginzburg region described by a number of fluctuation models.

To simplify the notations and to avoid clumsy mathematical expressions we shall introduce units, in which $c = \Lambda = 1$. The parameter c is equal to unity when the shift $\sqrt{c}\varphi_\alpha \to \varphi_\alpha$ of the field components in the φ^4-model is made; the accompanying change in the other parameters is $(r_0/c) \to r_0$ and $(u_0/c^2) \to u_0$. To make $\Lambda = 1$, the respective change of the momenta (k, k_i) in the $\mathcal{H}(\varphi)$ should be: $k \to k\Lambda$; as a result $(r_0/\Lambda^2) \to r_0$, $\Lambda\varphi_\alpha \to \varphi_\alpha$ and $(u_0/\Lambda^4) \to u_0$. So, to have $c = 1$ and $\Lambda = 1$ simultaneously, we set $k \to k\Lambda$, $\sqrt{c}\Lambda\varphi_\alpha \to \varphi_\alpha$, $(r/c\Lambda^2) \to r_0$, and $(u_0/c^2\Lambda^4) \to u_0$.

8.7.1 *Recursion relations and fixed points*

Having in mind these remarks, we can write Eq. (8.62a) in the form

$$1 = b^{-\eta} \cdot 1. \tag{8.105a}$$

It shows that the Fisher exponent η in this case is equal to zero. Equations (8.62b) and (8.62c) for r' and u' then become

$$r' = b^2 \left[r_0 + 4(n+2)u_0\mathcal{A}_1(r_0)\right] \tag{8.105b}$$

and

$$u_0' = b^\epsilon \left[u_0 - 4(n+8)u_0^2\mathcal{A}_2(r)\right], \tag{8.105c}$$

where

$$\mathcal{A}_m(r) = \int_p' \frac{1}{(p^2+r_0)^m}, m = 1, 2, \ldots, \tag{8.106}$$

and $\int'$ denotes the integration over p in the shell $b^{-1} < |p| < 1$. Our task is to investigate the behaviour of Eqs. (8.105a)–(8.105c) near the FPs generally denoted by (r^*, u^*). The FP equations will be

$$r^* = b^2 \left[r^* + 4(n+2)u^*\mathcal{A}_1(r^*)\right], \tag{8.107a}$$

and

$$u^* = b^\epsilon \left[u^* - 4(n+8)u^{*2}\mathcal{A}_2(r^*)\right]. \tag{8.107b}$$

With the help of these equations we can verify the existence of the Gaussian and the Heisenberg fixed points (GFP and HFP, or G and H). In fact, from Eqs. (8.107b) we find $u_G = 0$ and introducing this value of u^* in Eq. (8.107a) we obtain $r^* = r_G = 0$. The HFP is a solution of the equation

$$1 = b^{\epsilon}\left[1 - 4(n+8)u_H \mathcal{A}_2(r_H)\}\right] \tag{8.108}$$

which follows from (8.107b) for $u^* \neq 0$. Using that $b^{\epsilon} \approx 1 + \epsilon \ln b$ we have

$$u_H = \frac{\epsilon \ln b}{4(n+8)\mathcal{A}_2(r_H)}. \tag{8.109}$$

This result is put back in Eq. (8.107a) to give

$$r_H(b^{-2} - 1) = \frac{(n+2)}{(n+8)}\frac{\mathcal{A}_1(r_H)}{\mathcal{A}_2(r_H)}\epsilon \ln b. \tag{8.110}$$

Obviously r_H will be proportional to ϵ. The calculation of integrals of the type $\mathcal{A}_m(r_H)$ is made by expanding them in powers of r_H and r_H and ϵ (see Section E.3).

As far as we do the ϵ-analysis here with an accuracy to the first order in ϵ, we can set $\epsilon = r_H = 0$ in $\mathcal{A}_1$ and $\mathcal{A}_2$; see Eq. (8.110). Thus, we obtain

$$\mathcal{A}_1 = K_d \int_{b^{-1}}^{1} kdk = \frac{K_d}{2}(1 - b^{-2}) + \mathrm{O}(\epsilon), \tag{8.111}$$

and

$$\mathcal{A}_2 = K_d \ln b + \mathrm{O}(\epsilon). \tag{8.112}$$

In fact, $K_d = 2^{1-d}\pi^{-d/2}\Gamma^{-1}(d/2)$ can also be expanded in powers of $\epsilon = (4-d)$ but this is not always convenient. Bearing in mind that in Eq. (8.111) and Eq. (8.112) one must set $K_d = K_4 = 1/8\pi^2$ we shall keep the notation K_d for K_4.

From Eqs. (8.109)–(8.112) we obtain

$$r_H = -\frac{(n+2)}{(n+8)}\epsilon \tag{8.113}$$

and

$$u_H = \frac{\epsilon}{4(n+8)K_d} = \frac{2\pi^2\epsilon}{(n+8)} \tag{8.114}$$

for the coordinates of H in the plane (r, u); see Fig. 8.9, where G and H are depicted. So we have just derived Eq. (8.95b) and Eq. (8.104) in a different way.

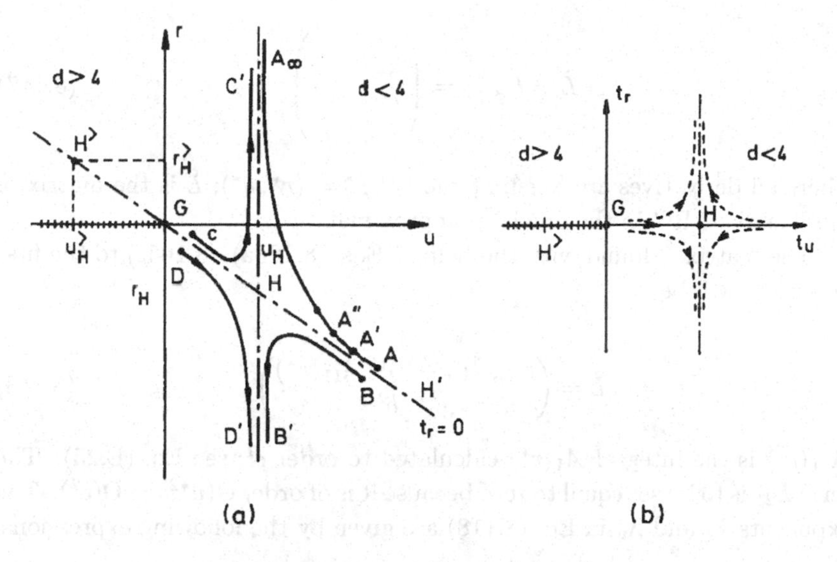

Fig. 8.9 G and H FPs and flow lines (RG trajectories): (a) in the (r_0, u) diagram, (b) in the (t_r, t_u)-diagram. $H^>$ is the unphysical HFP ($d > 4$).

8.7.2 *Linearization and exponents*

Let us now set

$$\mu' = \mu^* + \delta\mu', \qquad \mu = \mu^* + \delta\mu, \tag{8.115}$$

where $\mu = (r_0, u)$.

With the help of these expressions we shall linearize Eqs. (8.105a)–(8.105c) around the FP vector μ^*; $\mu^* = \mu_G \equiv (0, 0)$ or $\mu^* = \mu_H \equiv (r_H, u_H)$. To do this, we have to take into account the FP Eqs. (8.107a) and (8.107b) and to write down the equations obtained for $\delta r'$ and $\delta u'$ to the first-order in δr_0 and $\delta u (\equiv \delta u_0)$. It is however more convenient to perform the linearization by a direct variation of r_0' and u' with respect to r_0 and u (below we shall often omit the index (0) of r_0', u_0' and u_0). The result will be

$$(\delta r', \delta u') - \hat{L} \begin{pmatrix} \delta r_0 \\ \delta u \end{pmatrix} \tag{8.116}$$

with

$$\hat{L} = \{L_{ij}\} = \begin{pmatrix} \frac{\delta r'}{\delta r_0} & \frac{\delta r'}{\delta u} \\ \frac{\delta u'}{\delta r_0} & \frac{\delta u'}{\delta u} \end{pmatrix}, \tag{8.117}$$

where all derivatives are taken at the FP $\mu^* = (r^*, u^*)$; $\hat{L}$ is the matrix of the linearized RG in the one-loop approximation.

The matrix $\hat{L}$ found with the help of Eqs. (8.105a)–(8.105c) to the first order in ϵ will be

$$\hat{L} = \begin{pmatrix} b^{\lambda_r}, & 4(n+2)b^2\mathcal{A}_{11}(r^*) \\ 0, & b^{\lambda_u} \end{pmatrix}; \tag{8.118}$$

$\mathcal{A}_{11}(r^*)$ is the integral $\mathcal{A}_1(r^*)$ calculated to order ϵ^1; see Eq. (E.25). The term L_{21} is to be set equal to zero because it is of order $\mathrm{O}(u^{*2}) \sim \mathrm{O}(\epsilon^2)$. The exponents λ_r and λ_u in Eq. (8.118) are given by the following expressions:

$$\lambda_r = 2 - 4(n+2)K_d u^* \tag{8.119a}$$

and

$$\lambda_u = \epsilon - 8(n+8)K_d u^*. \tag{8.119b}$$

For G, $u_G = 0$ and therefore,

$$\lambda_r^G = 2, \lambda_u^G = \epsilon. \tag{8.120}$$

For H, we obtain from Eqs. (8.114) and (8.119a)–(8.119b) that

$$\lambda_r^H = 2 - \frac{(n+2)}{(n+8)}\epsilon \tag{8.121a}$$

and

$$\lambda_u^H = -\epsilon. \tag{8.121b}$$

Now we have to find the eigenvalues τ and eigenvectors $\boldsymbol{e}$ of the matrix $\hat{L}$. It is convenient to represent τ in the form b^λ so that the eigenvalues $(\tau_i = b^{\lambda_i}; i = 1, 2)$ of $\hat{L}$ will be determined by the eigenvalue exponent λ; λ_1 and λ_2. Within the approximation assumed above,

$$b^{\lambda_r} \approx b^2[1 + (\lambda_r - 2)\ln b], \tag{8.122a}$$

$$b^{\lambda_u} \approx 1 + \lambda_u \ln b \tag{8.122b}$$

and

$$b^{\lambda} \approx 1 + \lambda \ln b. \tag{8.122c}$$

These relations give the opportunity the eigenvalue equation

$$|\hat{L} - b^{\lambda}\hat{I}| = 0 \tag{8.123}$$

with $\hat{I}$ being the unit matrix, to be reduces to a single algebraic equation for λ:

$$(\lambda - \lambda_r)(\lambda - \lambda_u) = 0. \tag{8.124}$$

Therefore, the eigenvalues λ_1 and λ_2 of $\hat{L}$ are

$$\lambda_1 = \lambda_r, \quad \lambda_2 = \lambda_u. \tag{8.125}$$

The associated eigenvectors $\boldsymbol{e} = (e_1, e_2)$, in our notations, $\boldsymbol{e}_r = (e_{r1}, e_{r2})$ and $\boldsymbol{e}_u = (e_{u1}, e_{u2})$, are obtained from the eigenvector equation

$$(\hat{L} - b^{\lambda}\hat{I})\begin{pmatrix} e_1 \\ e_2 \end{pmatrix} = 0. \tag{8.126}$$

From here we can easily calculate that

$$\boldsymbol{e}_r = c_r(1, 0) \tag{8.127a}$$

and

$$\boldsymbol{e}_u = c_u[-4(n+2)\mathcal{A}_1, 1], \tag{8.127b}$$

where $\boldsymbol{e}_r$ corresponds to the parameter r and the eigenvectors λ_r and $\boldsymbol{e}_u$ correspond to u and λ_u; $\mathcal{A}_1$ in Eq. (8.127b) is the value of the integral (8.111) for $\epsilon = 0$ (zero order in ϵ). The arbitrary constants c_r and c_u are set for convenience equal to unity.

The vector $\delta\mu = (\delta r_0, \delta u)$ can be represented as a linear combination of $\boldsymbol{e}_r$ and $\boldsymbol{e}_u$:

$$(\delta r_0, \delta u) = t_r \boldsymbol{e}_r + t_u \boldsymbol{e}_u. \tag{8.128}$$

Making use of Eqs. (8.127a) and (8.127b) , we find the connection of the coefficients t_r and t_u with δr_0 and δu:

$$t_r = \delta r_0 + 4(n+2)\mathcal{A}_1 \delta u, \tag{8.129a}$$

$$t_u = \delta u. \tag{8.129b}$$

Since the variations $\delta\mu \approx \mu^* - \mu_0$, where $\mu_0 = (r_0, u)$, we can easily calculate, taking into account the FP Eqs. (8.107a) and (8.107b), that

$$t_r = r_0 + 4(n+2)\mathcal{A}_1 u = r \tag{8.130a}$$

and

$$t_u = (u_0 - u^*). \tag{8.130b}$$

The quantity $t_r \equiv r = \alpha_0(T - T_c)$ in Eq. (8.130a) is the shifted value of r_0 (see, e.g., Eqs. (8.72) and (8.73)) corresponding to the critical temperature shift $(T_{c0} - T_c)$. Thus the linearized RG is represented by new parameters t_r and t_u which describe the variations around the FPs; see also Fig. 8.9(b). The parameters t_r and t_u are sometimes called *critical "fields"* or, in a more general context, *linear scaling fields*; see, e.g., Ma (1976a), Barber (1977), and Fisher (1983).

The profit from the representation of RG as a transformation of the scaling fields t_i, $i = (r, u)$ is obvious. The transformed vector $\mu' = R_b\mu$ is now given by

$$(\delta r', \delta u') = \left(b^{\lambda_r}\delta r, b^{\lambda_u}\delta u\right), \tag{8.131}$$

where $\delta r = t_r$ and $\delta u = t_u$, in accordance with Eq. (8.130a), Eq. (8.130b) and $r^* = 0$, i.e., $\delta r = r$. The RG is now completely diagonalized so the group property is easily demonstrated:

$$\begin{aligned}(t_r^{(2)}, t_u^{(2)}) &= [b_2^{\lambda_r} t_r^{(1)}, b_2^{\lambda_u} t_u^{(1)}] \\ &= [(b_2 b_1)^{\lambda_r} t_r, (b_2 b_1)^{\lambda_u} t_u] \\ &= (b^{\lambda_r} t_r, b^{\lambda_u} t_u), \qquad (b = b_2 b_1).\end{aligned} \tag{8.132}$$

The above results are valid for both G and H. The explicit expressions can be obtained by substituting the corresponding values of $\mu^* = (r^*, u^*)$ or $t^* = (t_r^* = 0,\ t_u^* = u^*)$ and $\lambda_i(\mu^*)$ in the above formulae.

8.7.3 *Renormalization group flows and classification of the scaling fields*

The results from the analysis of the RG equations are shown graphically in Fig. 8.9. For ($d < 4$)-dimensional systems, i.e., where a nontrivial critical behaviour generated by the φ^4-fluctuation interaction occurs near T_c, the HFP is stable for $r = t_r = 0$, i.e., at the true critical temperature ($T = T_c$); see the line GH in Fig. 8.9(a) and the t_u-axis in Fig. 8.9(b). This is in accordance with our conclusions in Section 8.6. For $T \neq T_c$, $r \sim (T - T_c)$ is a *relevant variable* (relevant scaling field) in the sense that its scaling exponent λ_r is positive and, therefore, for any initial value $r \neq 0$, $r' = R_b \cdot r$ tends to infinity. The points A', A'', ... lying on the oriented line AA_∞ in Fig. 8.9(a) represent the successive RG steps — $A' = R_b A$, $A'' = R_b A'$, etc. The RG flow on the line GHE of a ($d < 4$)-dimensional system at T_c is represented by the two possible RG trajectories terminating at H. On the contrary, the lines like AA_∞ with initial points A, B, C, D, ..., as shown in Fig. 8.9(a), first approach the point $H = (r_H, u_H)$ because $u' \to u^* = u_H$ (λ_n^H is negative) but after a sufficient number of RG steps, $A - A'$, $A' - A''$, ... go away from the FP and tend to an infinite value of the relevant variable (r_0, or which is the same, the shifted value r). The corresponding picture in the (t_r, t_u)-diagram of the system Fig. 8.9(b) is more symmetrical. Both *FPs*, G, and H are unstable with respect to the relevant variable r (or r_0). As r_0 depends on T, it is often said that T is the relevant thermodynamic parameter. This fact directly relates the RG behaviour of the relevant scaling field ($r = t_r$) with the simple thermodynamic result that an ordinary critical point is approached when $T \to T_c$; the field h conjugate to the order parameter φ is equal to zero. The HFP is unstable with respect to δr-variations around the value $r\,(T_c) = 0$ but it exhibits a stability with respect to the variations δu of the fluctuation interaction u. The GFP has a double instability — towards both δr and δu (*complete instability*). So, the GFP is twice unstable whereas the HFP has a single instability. The critical point of the φ^4-Hamiltonian occurs for $r = 0$. The Gaussian critical behaviour at $r = 0$ represents an idealized system with zero φ^4-Hamiltonian or, which is the same, a real system, where the φ^4-fluctuation interaction can be neglected (outside the Ginzburg region). The Gaussian critical behaviour is well understood without any application of RG arguments and, therefore, the GFP is of little interest to our attempts at describing the critical behaviour in a close (asymptotic) vicinity of T_c. The nontrivial behaviour certainly corresponds to the HFP.

The investigation of the critical behaviour is not connected only with the variation of the temperature. Another relevant thermodynamic parameter which can take a system away from its ordinary critical point is the external field $h = \{h_\alpha\}$ conjugate to the order parameter $\varphi = \{\varphi_\alpha\}$. The RG recursion relation

$$\{h'_\alpha\} = R_b\{h_\alpha\} \tag{8.133}$$

is easily derived, when RG is applied to the fluctuation Hamiltonian, in which the field h is included. Let, as usual, $h = \{h_\alpha\}$ be an uniform field so that the additional term in $\mathcal{H}(\varphi, h)$ will be

$$-\sum_\alpha h_\alpha \int d^d x \varphi_\alpha(x), \tag{8.134a}$$

or, in Fourier components $\varphi_\alpha(k)$,

$$-\sqrt{V}\sum_\alpha h_\alpha \varphi_\alpha(0). \tag{8.134b}$$

The recursion relation (8.133) for h_α, within the approximation we use, becomes

$$h'_\alpha = b^{1+d/2} h_\alpha, \tag{8.135a}$$

or,

$$h'_\alpha = b^{\lambda_h} h_\alpha, \tag{8.135b}$$

where

$$\lambda_h = 3 - \epsilon/2 \tag{8.136}$$

for $d = (4-\epsilon) \sim 4$. The external field exponent $\lambda_h > 0$; for $\epsilon = 0$ it coincides with the MF exponent ($\delta = 3$). Therefore, the external parameters for an ordinary critical point is at the same time the second relevant scaling field ($t_h = h$) of the RG transformation.

An external field $h(x)$ will not have the behaviour described above if its zero momentum ($k \sim 0$) asymptote $h(k = 0) = 0$ is lacking. The field h, for which all $h(k)$ with k in the interval $0 < |k| < \Lambda_h \ll 1$ are zero, will be integrated out up to a *const* in the free energy by the successive RG "steps";

so it has no effect on the asymptotic (small $k \sim 0$) behaviour of the system. Such a field is the simplest example of a finite-scale effect that is irrelevant in the study of the scaling critical behaviour. Here we shall not enter into a detailed discussion of the influence of a slowly-varying external fields, for which $h(k) \neq 0$ when k is small ($0 < |k| < \Lambda_h$). The perturbation expansion for a nonzero h or for $T < T_c$ should be somehow modified because in both cases $\langle \varphi \rangle \neq 0$ and one has to separate the fluctuations $\delta\varphi = \varphi - \langle \varphi \rangle$ from the MF value $\bar{\varphi} \equiv \langle \varphi \rangle$ of the order parameter; the reader who is interested in this topic is referred to Ma (1976a), Wallace (1976), Rudnick (1978). The behaviour of the system in the plane $(0, t_r, t_u)$ in the scaling-field space $(t_h, t_r, t_u,)$ for $T < T_c$ is given in Fig. 8.9(b); see also the lines BB' and DD' in Fig. 8.9(a) which correspond to $t_r < 0$, i.e., $T < T_c$.

8.7.4 *Remarks on the effect of the high-order interaction terms*

Here we shall briefly discuss the effect of the interactions like φ^6, $\varphi^8, \dots$ near four dimensions ($d = 4 - \epsilon; \epsilon \ll 1$). In the end of the previous section we have already mentioned that the interaction parameter u_6 will be of order $O(u^2) \sim O(\epsilon^2)$, $u_8 \sim O(\epsilon^3)$, etc. Then $u_6^* = u_8^* = \cdots = 0$ to the ϵ^1-order of the RG analysis. Moreover such interactions near $d \sim 4$ have negative scaling exponents; see Eqs. (8.62a)–(8.62e) for $d = (4 - \epsilon)$. The above arguments set up such interactions to be considered *irrelevant* to the main scaling behaviour. They do not lead to the appearance of another FPs and, hence, the critical behaviour is described within the scaling fields t_i, the scaling exponents λ_i and the RG flows in the parameter space (h, r, u) of the φ^4-Hamiltonian.

We should mention that the consistent RG study of interactions of type φ^6, φ^8, etc. is made by taking them into accounts in the initial Hamiltonian $\mathcal{H}(\varphi)$. Let us briefly discuss the effect of the term

$$\lambda_h = 3 - \epsilon/2, \tag{8.137a}$$

or, in Fourier amplitudes $\varphi_\alpha(k)$,

$$\frac{u_6}{V^2} \sum_{\alpha\beta\gamma;k_i} \varphi_\alpha(k_1)\varphi_\alpha(k_2)\varphi_\beta(k_3)\varphi_\beta(k_4)\varphi_\gamma(k_5)\varphi_\gamma(-k_1 - \cdots - k_5). \tag{8.137b}$$

The above term will give corrections to r', u' and u_6' to the first order in u_6; see Fig. 8.10.

Including the term (8.137a), or, equivalently, (8.137b), leads to the following recursion relations

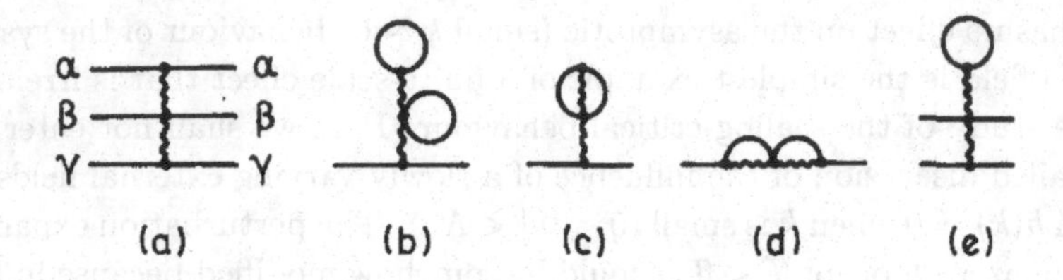

Fig. 8.10 (a) $u_6\varphi^6$-term; the contribution to u_6' of the type $u_6\varphi_0^6$ is given by the same diagram. (b)–(d) diagrams for r' of order u_6; e) a diagram for u' of order u_6. Note that not all possible topologically different diagrams for r' and u' are shown.

$$r' = b^2 f_r(r_0, u, u_6), \tag{8.138a}$$

$$u' = b^{4-d} f_u(u, u_6) \tag{8.138b}$$

and

$$u_6' = b^{6-2d} f_6(u, u_6), \tag{8.138c}$$

where we have neglected the u_6-dependence of the functions f_u and f_6 as being irrelevant to the RG analysis to the first order in ϵ. It is not difficult to obtain the explicit form of the Eqs. (8.138a)–(8.138c). The function f_r contains all terms considered previously together with a term proportional to u_6. The function f_u has an additional term of the type uu_6 (the term $\sim u_6^2$ appears in the two-loop approximation). The function f_6 contains a term of u_6^1-order, and terms proportional to uu_6 and u_6^2. The main problem now is to determine the upper critical dimension d_U. When $d_U = 4$, as it is for the φ^4-model, the parameter u_6 will be irrelevant. Clearly, in the zero order in RG, i.e., $R_b \equiv S_b$,

$$u_6' = b^{6-2d} \cdot u_6 \tag{8.139}$$

and choosing $d_U = 4$ we should do an $\epsilon = (4-d)$–expansion. In this case, for $d_U = 4$, $u_6' = b^{-2}u_6$ so that the exponent λ_6 of the scaling field u_6 is negative to the order ϵ^0. The ϵ-correction to the value $\lambda_6 = -2$ in the one- or higher-loop approximations cannot change the negative sign of λ_6 because they are small enough. Therefore,

$$\lambda_6 = -2 + O(\epsilon) < 0$$

to any order of RG and, consequently, the parameter u_6 is irrelevant to the G and H FP of the φ^4-model; thus to the respective critical phenomena.

The example discussed above is intended to demonstrate the way, in which the relevant and irrelevant scaling fields (parameters) are selected. To the zero order in ϵ, u is a *marginal parameter* for $(d = 4)$-dimensional systems; $f_u = u$, $\epsilon = 4 - d = 0$; see Eqs. (8.138a)–(8.138c). The next order in RG, i.e., the RG in the one-loop approximation reveals that this marginality is quite artificial and u is an irrelevant rather than marginal parameter.

An alternative way for treating Eqs. (8.138a)–(8.138c) is to choose $d_U^{(6)} = 3$ which means that the interaction u_6 is marginal to the order ϵ^0. After that we have to perform an $\epsilon = (3 - d)$–expansion and to investigate Eqs. (8.138a)–(8.138c) near the borderline dimension $d_U^{(6)} = 3$. Then both r and u will be relevant to $\lambda_r = 2+O(\epsilon)$ and $\lambda_u = 1+O(\epsilon)$. We should remember that the parameter u is very essential in the study of tricritical phenomena (Section 4.8). If we consider the manifold $u(T_{\rm tr}, Y_{\rm tr}) = r(T_{\rm tr}, Y_{\rm tr}) = 0$ in the parameter space (r_0, u, u_6) the system will be at the tricritical point $(T_{\rm tr}, Y_{\rm tr})$ and then the role of u_6 will be similar to that of u as discussed in Section 8.6. The departure of the system from the tricritical point is described by the variations δr and δu of the relevant parameters r and u (for the RG studies of tricritical phenomena see the references cited in Section 9.7).

8.7.5 *Dimensional analysis*

The above example of a complex Hamiltonian demonstrates that it is very convenient to begin the RG study after the analysis of the dimension of all fields and parameters included in the Hamiltonian of interest is accomplished (*dimensional analysis*). When a theoretical study of the (multi)critical state is done within the framework of RG the dimensional analysis is equivalent to the RG to the zero order in $\epsilon = d_U - d$, i.e., when only the rescaling (S_b-) operation of RG is considered.

Let us study the auxiliary Hamiltonian

$$\mathcal{H}(\varphi) = \frac{1}{2}\int G_0^{-1}(R, r_0)\varphi(x)\varphi(y)dxdy + u_m\int \varphi^m(x)dx, \tag{8.140}$$

where $R = |x - y|$ and

$$G_0^{-1}(R, r_0) = \frac{1}{V}\sum_k e^{ikR}G_0^{-1}(k, r_0) \tag{8.141a}$$

with

$$G_0^{-1}(k, r_0) = (k^\theta + r_0). \tag{8.141b}$$

is the inverse correlation function. The exponent θ, $0 < \theta \leq 2$, describes short ($\theta = 2$) and long-range interactions, respectively (Section 7.11). The interaction φ^m in Eq. (8.140) stands for either the φ^4-interaction ($m = 4$) or for anyone else with $m > 2$. The Fourier transform of Eq. (8.140) will be

$$\mathcal{H}(\varphi) = \frac{1}{2}\sum_k (k^\theta + r_0)|\varphi(k)|^2 + \frac{u_m}{V^{m/2-1}} \sum_{k_1 \ldots k_{m-1}} \varphi(-k_1 \cdots - k_{m-1}). \tag{8.142}$$

After applying the rescaling S_b with $\zeta(b) = b^y$, we obtain that the k^θ-term will be invariant ($k^\theta = k'^\theta$), provided $y = \theta/2$. Therefore, the upper borderline dimension $d_U^{(m)}$ will be the dimension d for which the exponent of the scale transformation of the parameter u_m,

$$u'_m = b^{(1-\frac{m}{2})d+\frac{\theta}{2}m} u_m, \tag{8.143}$$

is equal to zero, i.e.,

$$d_U^{(m)} = \frac{m\theta}{m-2}; \tag{8.144}$$

this is exactly the Eq. (7.252) mentioned in Section 7.11. If more than one interaction is included in $\mathcal{H}(\varphi)$, say, $u_m\varphi^m + u_l\varphi^l + \cdots$, $m \neq l$, we have to choose this upper borderline dimension $d_U^{(i)}$, $i = (m, l, \ldots)$ which corresponds to the problem under investigation. At the upper borderline, dimension d_U the perturbation terms are always proportional to powers of $u \ln b$, $u^{(l)}(\ln b)^{l'}$, $(l, l') = 1, 2, \ldots$, so that the $\epsilon = (d_U - d)$-expansion can be carried out in powers of small $\epsilon^l(\ln b)^{l'}$; note that $l > l'$ is valid for $l > 1$ only (see, e.g., Section 8.9).

8.7.6 *Differential renormalization group relations*

It is worth nothing that the RG Eqs. (8.105a)–(8.105c) can be cast in a differential form, sometimes more convenient for calculations (in particular, for complex models $\mathcal{H}(\varphi)$) with competing interactions). The differential RG equations within the Wilson–Fisher RG scheme have been introduced by Wegner and Houghton (1973). Here we shall illustrate them by the φ^4-model to the first order in $\epsilon = (4-d)$.

Choosing the rescaling factor b in Eqs. (8.105a)–(8.105c) in the form

$$b = e^{\delta l} \approx 1 + \delta l, \tag{8.145}$$

where $0 < \delta l \ll 1$, and calculating the respective integrals according to this quasi-infinitesimal RG step δl,

$$A_m(r_0) = K_d \int_{1-\delta l}^{1} dk \frac{k^{d-1}}{(r_0 + k^2)^m} = \frac{K_d}{(1+r_0)^m}\delta l + O(\delta l^2), \tag{8.146}$$

we obtain

$$r(\delta l) = r_0 + \left[2r_0 + \frac{4(n+2)u_0 K_d}{1+r_0}\right]\delta l \tag{8.147}$$

and

$$u(\delta l) = u_0 + \left[\epsilon u_0 - \frac{4(n+8)u_0^2 K_d}{(1+r_0)^2}\right]\delta l. \tag{8.148}$$

Here we have introduced the notations $r = r'(\delta l)$ and $u' = u(\delta l)$; the terms $O(\delta l^2)$ have been dropped. It is not difficult to see that the double application of the RG transformation, $b = b_1 b_2$, $b_i = (1 + \delta l_i)$, gives again Eqs. (8.147) and (8.148) with $\delta l = \delta l_1 + \delta l_2$; $i = (1, 2)$ and the semi-group property (8.68) is satisfied within the framework of this differential transformation. So we can write the differential RG equations in the limit $\delta l \equiv dl \to 0$ in the form

$$\frac{dr(l)}{dl} = 2r + \frac{4(n+2)uK_d}{1+r} \tag{8.149}$$

and

$$\frac{du(l)}{dl} = u\left[\epsilon - \frac{4(n+8)uK_d}{(1+r)^2}\right], \tag{8.150}$$

where $l = \ln b$ and $dl = (l - l_0)$; l_0 is a fixed number $l > 0$. The initial conditions for these differential equations are $r(dl = 0) = r(l_0)$, $u(dl = 0) =$

$u(l_0)$ or, $r(l_0) = r_0$, $u(l_0) = u_0$ when $l_0 \to 0$. The denominator $(1+r)^2$ in Eq. (8.150) can be set equal to unity, $r \sim 0$, and this approximation does not change anything in the first-order ϵ-analysis. The denominator $(1+r)$ in Eq. (8.150) can be replaced by unity after carrying out the variation of the same equation with respect to $r \equiv r(l)$.

The G and HFP's are obtained from the FP conditions

$$\left(\frac{dr}{dl}\right)_{\mu^*} = \left(\frac{du}{dl}\right)_{\mu^*} = 0. \tag{8.151}$$

Thus one rederives the FP coordinates: (i) $r_G = uG = 0$ for G, and (ii) Eqs. (8.113) and (8.114) for H. The variation of Eqs. (8.149)–(8.150) with respect to r and u yields

$$\frac{d\delta r}{dl} = \lambda_r(l)\delta r + \frac{4(n+2)K_d}{1+r}\delta u \tag{8.152a}$$

and

$$\frac{d\delta u}{dl} = \lambda_u(l)\delta u; \tag{8.152b}$$

$\lambda_r(l)$ and $\lambda_u(l)$ are effective exponents given by Eqs. (8.119a) and (8.119b), in which u^* should be substituted by $u(l)$. The parametric form of the RG trajectories (see Fig. 8.9) is represented with the rescaling parameter $l = \ln b$. At the point $l = l_0$, $\lambda_r(l_0)$ and $\lambda_u(l_0)$ are connected with $u(l_0)$ and δr and δu describe the infinitesimal RG step: $r(l_0) + \delta r = r(l_0 + dl)$, $u(l_0) + \delta u = u(l_0 + dl)$; see the method develop by Rudnick (1975), Rudnick and Nelson (1976) and applied by Rudnick (1978) to systems with a cubic anisotropy.

8.8 Renormalization Group and Thermodynamics

The thermodynamic potential $\Phi(T,h,Y)$ is expressed by the (T,h,Y)-dependent Hamiltonian parameters $\mu = \{\mu_i(T,h,Y)\}$; $Y = \{Y_i\}$ is the set of additional thermodynamic variables which are necessary for the description of the phase diagrams in complex system (Section 4.2). The RG operation R_b transforms the Hamiltonian $\mathcal{H}(\mu)$ to $\mathcal{H}'(\mu')$ but it does not change the potential, i.e., we have $\Phi(\mu) = \Phi'(\mu')$. This is the basic relation in the "scaling" thermodynamics which shows the invariance of Φ with respect to R_b:

$$\Phi(\mu) = \Phi(\mu'), \tag{8.153}$$

where

$$\mu' = R_b\mu, \qquad \mu = \{\mu_i\}, \tag{8.154}$$

is an expression for the RG transformation equivalent to the RG Eqs. (8.61) and (8.63). Alternatively, Eq. (8.153) can be written with the help of the Hamiltonian $\mathcal{H}(\varphi, h)$ describing a system in an external magnetic field h:

$$\Phi(\mathcal{H}) = \Phi(R_b\mathcal{H}). \tag{8.155}$$

It is convenient to express the transformed potential $\Phi' = \Phi(\mu')$ by the transformed length scale $L' = b^{-1}L$. Since Φ is an extensive quantity it is proportional to some fixed extensive parameter, for example, the volume V (or the total number of particles N when the thermodynamic ensemble is chosen so that the volume V varies and N is fixed). The potential density $\phi(\mu) = \Phi[\mu(T, h, Y)]/V$ is transformed under RG according to Eqs. (8.153) and (8.58):

$$\phi(\mu) = b^{-d}\phi(\mu'), \tag{8.156}$$

where

$$\phi(\mu') = \frac{\Phi(\mu')}{V'}, \tag{8.157}$$

or, equivalently, $\phi(\mu') = \phi'(\mu') = \Phi'(\mu')/V'$ which, in virtue of the equality $\Phi'(\mu') = \Phi(\mu')$, yields Eq. (8.157).

Our discussion can be carried out on the basis of a general Hamiltonian having the form

$$\mathcal{H}(\varphi) = \sum_i \mu_i O_i(\varphi), \tag{8.158}$$

where $\mu = \{\mu_i\}$ a vector in the Hamiltonian parameter (μ)-space; μ_i is a general notation, representing, for example, the external field $h = \{h_\alpha\}$ conjugate to the order parameter field $\varphi = \{\varphi_\alpha\}$; $r_0 \sim (T - T_c^0)$ or $r \sim (T - T_c)$; the interaction parameters like u, u_6, u_8, etc. In the Eq. (8.155) for Φ the vector quantities like $h = \{h_\alpha\}$ will enter only through their magnitudes, $|h|$, which we shall often denote by $h(\equiv |h|)$.

We shall assume that all new terms generated by RG are taken into account in the initial Hamiltonian (8.158). Alternatively one may imagine that the initial Hamiltonian is chosen to be one of the transformed Hamiltonians, for example, the effective Hamiltonian $\mathcal{H}'(\varphi')$ after the first RG step. That is why, it is convenient to include in the initial Hamiltonian all the field invariants $O_i(\varphi)$; sometimes called *scaling operators.*

For example, terms which do not dependent on the fields φ_α are usually excluded from $\mathcal{H}(\varphi)$ by a change of the zero of the energy scale or, equivalently such terms are included in the potential of the disordered phase. The appearance of this type of terms in $\mathcal{H}'(\varphi')$ is a result from the integration (I_b) of the "fast" fields $\varphi_\alpha(k), |k| > b^{-1}\lambda$. They are presented by regular integrals which do not exhibit infrared divergences, of course, if only $b \neq \infty$; see Ma (1976a) and the discussion in Section 8.8.2. So we shall assume that the term $\mu_0 O_0$, where O_0 does not depend on φ_α is included in Eq. (8.158). Since μ_0 is an intensive variable and $\mathcal{H}$ is measured in energy units, we have that $O_0 \sim V$ (or N). The field independent term in Eq. (8.158) can be therefore written as $C(\mu) = \mu_0(\mu)V$ with μ_0 being the energy density parameter, which depends on $\mu = \{\mu_i\}$. We shall see below that this term requires a special treatment.

8.8.1 *Scaling form of the thermodynamic potential and classification of the scaling fields*

Suppose that we have performed the RG transformation of $\mathcal{H}(\varphi)$ with the subsequent diagonalization so that the transformed Hamiltonian $\mathcal{H}'(\varphi')$ is represented by the scaling fields $\{t_i\}$; denoted by the scaling vector $\boldsymbol{t} = \{t_i\}$. The components $t_i(\mu)$ of the scaling vector $\boldsymbol{t}$ are connected with the original parameters $\mu = (r_0, u, \ldots)$ or their transformed counterparts $\mu' = (r', u', \ldots)$ through a rotation in the μ-space (see Section 8.7). The RG transformation of μ,

$$\mu' = L_b \cdot \mu \tag{8.159}$$

with $L_b \approx R_b$ — the linearized R_b transformation that can be expressed by the scaling fields $\{t_i\}$. By analogy with Eqs. (8.129) the vector μ can be presented as a linear combination of the set of eigenvalues $\boldsymbol{e}_i$ of the matrix L_b corresponding to the eigenvalues $\{\lambda_i\} = (\lambda_r, \lambda_u, \ldots)$, i.e.,

$$\boldsymbol{\mu} = \boldsymbol{\mu}^* + \sum_i t_i \boldsymbol{e}_i \qquad (8.160)$$

(below the bold face of the vector will be avoided). In Eq. (8.160) the coefficients t_i are the (*canonical*) scaling fields. The eigenvalues λ_i of L_b are sometimes denoted by y_i.

This particular scheme of the RG method is a straightforward generalization of the analysis of the φ^4–model in Section 8.7. The L_b–transformation of μ, see Eq. (8.159) yields

$$\mu' = \mu^* + \sum_i b^{\lambda_i} t_i e_i, \qquad (\mu, e_i) \equiv (\boldsymbol{\mu}, \boldsymbol{e}_i), \qquad (8.161)$$

and, with the help of Eq. (8.157) we obtain the following form for $\phi(\mu)$:

$$\phi(\mu) = b^{-d}\phi\left(\mu^* + b^{\lambda_1} t_1 e_1 + b^{\lambda_2} t_2 e_2 + \cdots\right). \qquad (8.162)$$

Clearly, Eqs. (8.161) and (8.162) obey the group property, after m transformations with equal rescaling factor b,

$$\phi(\mu) = b^{-md}\phi(\mu^* + b^{m\lambda_1} t_1 e_1 + b^{m\lambda_2} t_2 e_2 + \cdots). \qquad (8.163)$$

Let the first scaling field $t_1 \sim t_r$ be identified with $t = (T - T_c)/T_c$, the usual notation for the "temperature distance" from the critical point T_c; then $\lambda_1 = \lambda_r \equiv \lambda_t$. The scaling field t_2 is identified with the external field $h = |h|$; $t_2 \equiv t_h \approx h$, and $\lambda_2 \equiv \lambda_h$. Then we shall denote by $\{t_j\}$ and $\{\lambda_j\}$ the subsets of all other scaling fields and associated exponents λ_j (like μ_0 and λ_0, $t_u \sim u$ and λ_u, $t_6 \sim u_6$ and λ_6, etc.). As the dependence on

$$\mu' = \mu^* + \sum_i b^{\lambda_i} t_i e_i \qquad (8.164)$$

in Eq. (8.162) is not explicit, it rather illustrates general theoretical arguments, we shall change nothing if we write it in the following way

$$\phi(\mu) = b^{-d}\phi\left(b^{\lambda_t} t, b^{\lambda_h} h, b^{\lambda_j} t_j\right); \qquad (8.165)$$

the term $t'_j = b^{\lambda_j} t_j$ in (8.164) is intended to represent the dependence on the whole set $\{t_j\}$ of fields $t_j \neq t, h$. The form (8.165) of $\phi(\mu)$ is valid on both sides of T_c; $T < T_c$ and $T \geq T_c$.

In Eq. (8.165) $\mu = \mu^*$ corresponds to $t_1 = t = 0$ or $T \equiv T_c$ and, therefore, $t < 0$ describes the case $T < T_c$ and $t \geq 0$ stands for $T \geq T_c$ (provided all other relevant fields, like h, are neglected). The vectors e_i are obtained to an indeterminate factor; so we can choose e_i such that the scaling field $t_r \sim t$ is always positive (for example, te_1 for $t > 0$ and $te_1' = |t|e_1$ for $t < 0$; $e_1' = -e_1$).

Further we should notice that the RG calculations and the resulting scaling lead to one and the same critical exponents above and below the critical point $(T_c, h = 0)$; see also Chapter 1. For this reason we shall assume that the exponents λ_i of the scaling fields t_i are identical for $t \lessgtr 0$ (the generalization to different exponents above and below T_c is straightforward). Therefore we rewrite Eq. (8.165) in the form

$$\phi(\mu) = b^{-d}\phi_{\pm}\left(b^{\lambda_t}t, b^{\lambda_h}h, b^{\lambda_j}t_j\right) \tag{8.166}$$

where

$$\phi_{\pm} \equiv \phi\left(\pm|t|', h', t_j'\right). \tag{8.167}$$

The arbitrary rescaling factor b in (8.166) can be fixed in a way to explain the purposes of the particular study. Let $t \neq 0$, so we set

$$b^{\lambda_t}|t| = 1, \tag{8.168}$$

i.e., $b = |t|^{-1/\lambda_t}$; $\lambda_t > 0$. For this choice of b,

$$\phi(\mu) = |t|^{d/\lambda_t}\phi\left(\pm 1, |t|^{-\Delta_h}h, |t|^{-\Delta_j}t_j\right), \tag{8.169}$$

where the ratios

$$\Delta_h = \frac{\lambda_h}{\lambda_t}, \qquad \Delta_j = \frac{\lambda_j}{\lambda_t}, \tag{8.170}$$

are called crossover exponents (sometimes these exponents are denoted by ϕ_i; ϕ_h and ϕ_j).

We have already demonstrated, see Eq. (8.136), that $\lambda_h > 0$ and, hence, Δ_h is positive, too. This fact together with the condition $\lambda_t > 0$ defines t and $t_h \sim h$ as relevant scaling fields. The classification of the scaling fields as relevant, irrelevant and marginal has been already introduced in Section 8.7 for the φ^4-Hamiltonian. Here it naturally follows from the scaling properties of the scaling fields within a quite general scheme of the RG analysis.

The relevant parameters μ_i (or scaling fields t_i) are those with positive exponents λ_i, the irrelevant parameters are those with negative scaling exponents λ_i and the marginal ones have exponents $\lambda_i = 0$. The corresponding scaling "operators" O_i in Eq. (8.158) are classified in the same way.

All these fields and operators are directly connected with the properties of the potential density $\phi(\mu)$ or other observable thermodynamic and correlation properties. On the contrary, the *redundant operators*, $O_i^{(r)}$ which can also appear in special cases and the corresponding (redundant) parameters $\mu_i^{(r)}$ have no effect on the potential $\phi(\mu)$. We shall not discuss these parameters because they are usually of no importance in practical theoretical calculations and to the interpretation of the experimental situations; for more information, see Wegner (1974, 1976), and Fisher (1983).

In Section 8.7 we have mentioned that there are two relevant scaling fields of an ordinary critical point, i.e., t and h. They are directly related to the two relevant thermodynamic parameters — the temperature T and the external field $h = |h|$ which drive the system into and take it away from the critical point $(T_c, h = 0)$.

If we suggest that the scaling fields relevant to the description of an ordinary critical point are not two but three, one of them should be identically equal to zero, at least, in that part of the (T, h, Y) diagram of the system, where the phase transition is of second order. For example, although we may assume that a term $u_3\varphi^3$ exists in $\mathcal{H}(\varphi)$, the critical point $(T_c, h = 0)$ or the line $[T_c(Y), h = 0]$ of critical points will appear in the manifold (T, h, Y), where $u_3(T, h, Y) \equiv 0$.

Another example is the tricritical point. When we consider the tricritical scaling behaviour we should admit that the $u\varphi^4$-term in $\mathcal{H}(\varphi)$ changes sign (see Section 4.6) under the variation of T, h, and Y. Then $u(T, h, Y)$ becomes a new relevant scaling field with a positive exponent $\lambda_u = 1 + O(\epsilon)$ near three dimensions ($\epsilon = 3 - d$); see Section 8.7.4. In order to obtain the coordinates $(T_{\rm tr}, O)$ of tricritical point in the (T, h)-diagram from the set of equations $t = u = 0$, it is necessary that this system to be degenerate. If this is not so, we should try to expand the analysis including a supplementary variable Y in the $(T, Y, h = 0)$-diagram. Then the equations $T_{\rm tr} = T_c(Y)$ and $u(T_{\rm tr}, Y_{\rm tr}) = 0$ may have a solution giving the coordinates $(T_{\rm tr}, Y_{\rm tr})$ of the tricritical point. Of course this is out of sense if the system is such that Y does not influence the phase transition under condition. So there are three relevant thermodynamic parameters (T, Y, h) which drive the system in tricritical point and three associated relevant Hamiltonian parameters (r, u, h).

Surely, several (or even an infinite number) of irrelevant scaling fields, t_j, with exponents $|\lambda_j| = -\lambda_j$ will always be present and the reason lies in the complexity of the effective Hamiltonian $\mathcal{H}'(\varphi'_0)$ obtained after the integration (I_b) of the "fast" fields $\varphi_{1\alpha}$. The complicated form of $\mathcal{H}'(\varphi'_0)$ and therefore of $\mathcal{H}'(\varphi')$ is of no importance to the main (asymptotic) scaling properties of the critical (or multicritical) behaviour because the irrelevant fields t_j tend to zero when $T \to T_c$; see Eq. (8.169) with $\Delta_j < 0$. These variables may affect the thermodynamics at temperatures T which are not close enough to T_c.

We shall see below that the irrelevant variables give corrections to the asymptotic $(T \sim T_c)$ scaling laws (Wegner, 1972; see also Chapter 1). Another source of corrections to the scaling laws may arise from the non-linearity of the RG. These effects are studied with the help of the so-called nonlinear scaling fields; for details, see Wegner (1976), Barber (1977), and Fisher (1983).

In order to describe the scaling behaviour of the potential density $\phi(t_i)$ together with the corrections to the main scaling we shall separate the subset $\{t_k\}$ of irrelevant variables from the set $\{t_j\}$ of supplementary variables t_j; see Eq. (8.169). To make the discussion thorough let us assume that subsets of marginal scaling fields, $\{t_m\}$, and supplementary relevant fields $\{t_p\}$ also exist. Supposing that the marginal fields, (t_m), are small we shall expand $\phi(\{t_i\})$, where $\{t_i\} = (t, h, \{t_p\}, \{t_m\}, \{t_k\})$ to the first order in $t'_k = |t|^{-\Delta_k} t_k$ and $t'_m = t_m(\lambda_m = 0)$. Dropping the terms of order $\mathcal{O}(t'^2_k)$ and $\mathcal{O}(t^2_m)$ we obtain

$$\begin{aligned}\phi(t_i) = |t|^{d/\lambda_t} \Big[\phi\left(\pm 1, |t|^{-\Delta_h} \cdot h, |t|^{-\Delta_p} \cdot t_p\right) \\ + \sum_k \left(\frac{\partial \phi_\pm}{\partial t'_k}\right)_{t'_k = 0} |t|^{|\Delta_k|} \cdot t_k \\ + \sum_m \left(\frac{\partial \phi_\pm}{\partial t_m}\right)_{t_m = 0} \cdot t_m \Big] . \end{aligned} \tag{8.171}$$

This is the total scaling form of the potential $\phi(t_i)$, from which the scaling laws together with the corrections to them can be found. The contributions to $\phi(t_i)$ of the marginal (t_m) and the irrelevant (t_k) parameters are proportional to $|t|^{d/\lambda_y}$ and $|t|^{|\Delta_k| + d/\lambda_t}$, respectively. Therefore, the second and third terms in Eq. (8.171) are small compared to the first term which gives the effect of the relevant parameters t, h and t_m.

We need, of course, knowledge of the precise form of $\phi(t_i)$. In Chapter 4 we have discussed the thermodynamic potential following the usual Landau expansion which yields a simple form of $\phi(t_i)$. Taking into account either the free fluctuation in the Gaussian approximation or the integral fluctuations within the RG, it is possible to obtain such a form of $\phi(t_i)$ which describes more adequately the critical thermodynamics in the Ginzburg region. In order to find the explicit scaling form of $\phi(t_i)$ above T_c of the systems described by the φ^4-Hamiltonian, one should use the RG recursion relations, presented in Section 8.7 and the standard functional integral relating $\Phi(\mu)$ and $\mathcal{H}(\varphi; \mu)$; see, e.g., Ma (1976a).

In order to obtain the scaling form of $\phi(t_i)$ either below T_c or in the presence of an external field h conjugate to the order parameter $\langle\varphi\rangle$, above and below T_c, the RG transformation of $\mathcal{H}(\varphi, h)$ has to be carried out within the perturbation expansion corresponding to a nonzero equilibrium order parameter $\langle\varphi\rangle$; see, e.g., Ma (1976a) and Rudnick (1978). There is an alternative way of calculation which begins with the equation of state $h = h(\langle\varphi\rangle)$, where the fluctuations $\delta\varphi$ are taken into account in, say, the one- or the two-loop approximation; for details, see Wallace (1976). Then $\Phi(t_i)$ and, hence, $\phi(t_i)$ are obtained by the integration of the equation of state written in the form $(\partial\Phi/\partial h) = -\langle\varphi\rangle$.

8.8.2 *Scaling relations*

The scaling laws for the second order derivatives (susceptibilities) of the thermodynamic potential discussed in Chapter 1 can be derived from the scale invariant form of $\Phi(\mu_i)$; see Eq. (8.166) and Eq. (8.171), which holds for $t \neq 0$. In order to obtain the specific heat in a fixed field h we must differentiate twice Eq. (8.171) with respect to T; the specific heat per unit volume near T_c is $C = -T_c\left[\partial^2\phi(t_i)/\partial T^2\right]$, or, using the relation $(\partial/\partial T) = T_c^{-1}(\partial/\partial t)$,

$$C = -\frac{1}{T_c}\frac{\partial^2\phi(t_i)}{\partial t^2}. \tag{8.172}$$

We shall focus the attention on those terms in Eq. (8.171) which may be the cause for the appearance of the thermodynamic quantities like the specific heat. In particular we shall discuss the effect of the relevant variables t and h. Since the generalization to more than two relevant variables $t_p \neq 0$ is straightforward we shall simplify the mathematical expression by dropping the second and the third terms in Eq. (8.171) and by setting $t_p \equiv 0$. Thus we obtain for the specific heat in a zero field ($h = 0$) the following expression:

$$C = C_\pm |t|^{-2+d/\lambda_t} + O\left[|t|^{y_c}; y_c > \left(\frac{d}{\lambda_t} - 2\right)\right], \tag{8.173}$$

where $O(|t|^{y_c})$ stands for the neglected terms. The scaling amplitudes $C_\pm$ are given by

$$C_\pm = \frac{d(d - \lambda_t)}{\lambda_t T_c}\phi_\pm \tag{8.174}$$

and $\phi_\pm = \phi(\pm 1, h = 0) \equiv \phi(\pm, 0)$. As in Chapter 1 we shall denote the specific heat scaling exponent by α. Therefore,

$$\alpha = 2 - \frac{d}{\lambda_t}. \tag{8.175}$$

In order to clarify the meaning of the exponent λ_t we shall consider the scaling relation (8.79a) for the correlation length given by

$$\xi(t_i) = b\xi(t', h', \ldots), \tag{8.176}$$

or, using Eq. (8.168) with $h = t_p = \cdots = 0$,

$$\xi(t) = \xi_\pm |t|^{-1/\lambda_t}; \tag{8.177}$$

the scaling amplitudes are $\xi_\pm = \xi(\pm 1, 0)$. The correlation length exponent is usually denoted by ν, therefore,

$$\nu = \frac{1}{\lambda_t}, \tag{8.178}$$

and from Eq. (8.175) we obtain the Josephson scaling relation $d\nu = (2 - \alpha)$; cf. Eq. (1.21).

Now it becomes clear how the scaling laws and the scaling relation between the critical exponents can be found by RG arguments. In this way the RG approach confirms the homogeneity (scaling) hypothesis discussed in Chapter 1. For the sake of completeness, we shall briefly discuss the remaining part of scaling laws and relations for the static critical phenomena.

The order parameter $\langle\varphi\rangle \sim (-t)^{\beta}$, $t < 0$ is calculated from the formula $\langle\varphi\rangle \sim [\partial\phi(t,h)/\partial h]_{h=0}$ and Eq. (8.169). As a result the exponent β is obtained:

$$\beta = d\nu - \Delta_h. \tag{8.179}$$

Remember that the susceptibility χ of the system for $h = 0$ is given by the second derivative of $\phi(t,h)$ with respect to h. Using the expression (8.171) for $\phi(t,h)$ we have

$$\chi \sim \left(\frac{\partial^2\phi}{\partial h^2}\right)_{h=0} \sim |t|^{2-\alpha-2\Delta_h}. \tag{8.180}$$

From $\Delta_h = \lambda_h\nu$ and the equation $\chi_T \equiv \chi \sim |t|^{-\gamma}$, $\gamma = \gamma'$, (see Table 1.1) we obtain the following relation

$$\gamma = \alpha - 2 + 2\nu\lambda_h, \tag{8.181}$$

where γ is the susceptibility exponent. From Eqs. (8.179) and (8.181), $d\nu = 2-\alpha$, and $\lambda_h\nu = \Delta_h$, we can derive the Rushbrooke relation $\alpha+\gamma+2\beta = 2$; see also Eq. (1.21).

Our discussion proceeds with the exponent λ_h of the scaling field h. In the strong-field region (Section 4.10.1), where $|t| \ll h$, we set $|t| = 0$ and consider $h \equiv |h| > 0$. Then the rescaling factor b in Eq. (8.166) can be chosen in the form

$$b = h^{-1/\lambda_h}, \tag{8.182}$$

so $\phi(t,h)$ becomes

$$\phi(0,h) = h^{d/\lambda_h}\phi_h(0,1) \tag{8.183}$$

where $\phi_h(0,1)$ is the scaling function ("scaling amplitude") for $t = 0$ and $h \neq 0$. When $t \neq 0$, the same scaling function will be

$$\phi_h^{(\pm)} = \phi_h\left(\pm h^{-\lambda_t/\lambda_h}\cdot t, 1\right). \tag{8.184}$$

Note that in the strong-field region the first argument ($\sim t$) is much less than unity in a close vicinity of T_c. Since the order parameter is given by $\langle\varphi\rangle \sim h^{1/\delta}$ it is easy to find the exponent δ by differentiating Eq. (8.183) with respect to h:

$$\delta = \frac{\lambda_h}{d - \lambda_h} \tag{8.185}$$

Therefore, $\lambda_h = \delta d/(1+\delta)$, and

$$\Delta_h = \frac{d\nu\delta}{1+\delta}. \tag{8.186}$$

Making a comparison of this equation with Eq. (8.179) we find $d\nu = \beta(1+\delta)$ and, using the Rushbrooke relation we obtain $\gamma = \beta(\delta - 1)$ — the Widom relation discussed in Chapter 1; see Eq. (1.21).

The Fisher exponent and the Fisher relation $\gamma = (2-\eta)\nu$, see Eqs. (1.21), can be derived by applying the R_b operation to the full correlation function $G(k) = \langle|\varphi_\alpha(k)|^2\rangle$:

$$\langle|\varphi_\alpha(k)|^2\rangle = b^{2-\eta}\langle|\varphi'_\alpha(k')|^2\rangle; \tag{8.187}$$

here the symbol $\langle\cdots\rangle$ stands for averaging over the total Hamiltonian $\mathcal{H}(\varphi)$. The RG relation (8.187) can be written in an equivalent way

$$G(t,k) = b^{2-\eta}G\left(b^{\lambda_t}t, bk\right); \qquad h = 0. \tag{8.188}$$

At $T = T_c$, we set $b = k^{-1}$, $k \equiv |k|$, and therefore $G(0,k) \sim k^{-2+\eta}$; this coincides precisely with the scaling form of $G(0,k)$ discussed in Chapter 1 (for the Fisher's definition of η; see Eq. (1.15). For $k = 0$ and $t \neq 0$ we have

$$G(t,0) \sim |t|^{-(2-\eta)\nu}, \tag{8.189}$$

and from $G(t,0) \sim \chi \sim |t|^{-\gamma}$ we obtain the Fisher's relation $\gamma = (2-\eta)\nu$.

Thus we have related the exponent y of the order parameter field $\varphi(k) = \{\varphi_\alpha(k)\}$ and the exponents λ_t and λ_h of the relevant fields $t_r \sim r \sim t$ and $t_h \sim h$ with the scaling exponent β of the order parameter and the scaling exponents $(\alpha, \gamma, \ldots)$ of the thermodynamic susceptibilities which are directly measured in experiments on the critical behaviour of real systems.

8.8.3 *Remarks on the crossover phenomena in the critical region*

The scaling form (8.166) of $\Phi(\mu)$ and the scaling form Eq. (8.188) of $G(\mu)$ are valid for any FP that may appear from $\mathcal{H}(\varphi)$. The different FPs describe different critical behaviour (Gaussian, nontrivial, tricritical, etc.). If there are another scaling exponents they should also obey scaling relations; for the definition of other exponents and a thorough treatment of the scaling corrections see Wegner (1976), Barber (1977), and Fisher (1983).

There are situation where the type of the critical behaviour of the system is altered. This is so when the system passes from one possible critical behaviour (one stable FP) either to another one (corresponding to another stable FP) or to a first-order phase transition (when no stable FP exists). These changes are united under the name (*crossover phenomena*) in the fluctuation region. For their proper treatment a classification of the possible FPs and their properties should be done. We have seen that the HFP describing an ordinary critical point (or ordinary second-order transition) is doubly unstable with respect to t and h. Since this instability is just what the general notion of the ordinary critical behaviour implies and no other (unnatural) instabilities occur we say that the HFP is "stable" (for $2 < d < 4$). If however a third relevant parameter appears, it becomes unstable and, hence, the modified (unstable) HFP has nothing to do with the standard critical behaviour. The causes for the possible instability are various, for example, the HFP becomes often unstable towards perturbations, like $\mu_i O_i$, that lead to a breaking of the initial symmetry of the Hamiltonian. FPs that describe the standard tricritical behaviour can be considered "stable" provided no more than three relevant scaling fields exist. In a similar way we may introduce the notion of a (relative) stability or instability of the different FPs that appear as a result of the RG analysis of simple and complex Hamiltonians.

The absence of stable FPs is usually interpreted as a signal that the usual (multi)critical behaviour is lacking. There can be three possibilities:

(i) FPs do not exist at all.
(ii) FPs exist but all of them are unstable in the domain of the μ-space of interest to the particular study (the physical domain).
(iii) One or more FPs exist but they are inaccessible because the values of the initial parameters are such that by R_b-operations applied to them, it is not possible to reach the FPs.

The above enumerated cases have one common feature: the correlation length ξ at the equilibrium transition point does not exhibit any divergence and, consequently, this can be considered as the occurrence of certain types of first-order phase transition (or, more generally non-continuous transitions). Sometimes the so-called fluctuation-driven first-order phase transition can take place. This is so when the transition is found to be of second order within the MF theory but taking into account the interacting fluctuations leads to the absence of infrared RG FPs (or, more generally, one of the three conditions enumerated above is satisfied). Then it is said that the system undergoes a crossover from the second-order to the (fluctuation-driven) first order transition due to the relevant fluctuation interaction. Very often this effect appears when there exist extreme anisotropy and/or a change of the symmetry due to the coupling of the order parameter to supplementary macroscopic modes (see also Chapter 9 where several examples are presented). The opposite phenomenon, i.e., the fluctuation-driven second order transition is also possible. Of course, these fluctuation changes of the order of the phase transition can be observed in the critical region. For details about the crossover phenomena and, in particular, about the fluctuation-driven change of the order of the phase transitions the reader is referred to Aharony (1976), Mukamel and Krinsky (1976), Bak, Krinsky and Mukamel (1976), Domany, Mukamel and Fisher (1977), Rudnick (1978), and Bruce (1980).

8.8.4 *Scaling and ϵ-expansion*

Now we can write the critical exponents $(\alpha, \beta, \gamma, \ldots)$ with an accuracy to the first order in ϵ. For the GFP these exponents take the "Gaussian" values, that is the values predicted by the Gaussian (Ornstein–Zernicke) approximation for the fluctuations (Table 8.2). The only Gaussian exponent which does not coincide with its MF (classical) value is α. It is obvious that the Gaussian and the MF approximations for the critical behaviour give one and the same results only for the borderline dimension $d_U = 4$. The MF exponents $\nu = 1/2$ and $\alpha = 0$ obey the Josephson relation $d\nu = (2 - \alpha)$ only for $d = 4$. Scaling relations, in which the dimensionality d enters explicitly are sometimes called *hyperscaling relations*; for the violation of the hyperscaling relations for dimensions $d > d_U$ and the connection of this problem with the effect of the so-called *dangerous scaling fields*; see Ma (1976a), and Fisher (1983).

As we know the Gaussian and MF exponents do not describe quantitatively correctly the critical behaviour in the critical region. The FP, which comes into being as a result of the fluctuation interaction in isotopic systems and, therefore, describes the fluctuation interaction in these systems is the HFP. Since the exponent $\eta = 0$ to the first order in ϵ in order to have the total set of six static critical exponents we need to calculate only one

Table 8.2 Exponents corresponding to the Gaussian and Heisenberg FPs.

exponent	Gaussian FP	Heisenberg FP
α	$2-\frac{2}{d}$	$\frac{(4-n)\epsilon}{2(n+8)}$
β	$\frac{1}{2}$	$\frac{1}{2}-\frac{3\epsilon}{2(n+8)}$
γ	1	$1+\frac{(n+2)\epsilon}{4(n+8)}$
δ	3	$3+\epsilon$
ν	$\frac{1}{2}$	$\frac{1}{2}+\frac{(n+2)\epsilon}{4(n+8)}$
η	0	$O(\epsilon^2)$

more exponent and to find the rest with the help of the scaling relations Eq. (1.21). For example, let us obtain ν. As long as the expansion parameter is ϵ and our calculation is valid for $\epsilon \ln b \sim \epsilon \ln(1/|t|) \ll 1$, we have to expand $\lambda_t = 1/\nu$ in ϵ. From Eq. (8.121a) we have

$$\nu = \frac{1}{2} + \frac{(n+2)}{4(n+8)}\epsilon; \tag{8.190}$$

it is the same as given in Table 8.2. When $d = 3$ and $n = 1$, $\nu = 7/12 > 1/2$. Clearly, the accuracy, to which the exponents are calculated to the first order in ϵ is not sufficient for a comparison with precise experiments; for example, the experiment predicts $\eta > 0$ rather than $\eta = 0$. This problem shows the necessity to investigate the RG equations to the order higher than first in ϵ.

8.9 Renormalization Group in the Two-loop Approximation

The basic RG treatment is carried out in the one-loop approximation. We should emphasize that the RG approach is applied only after we have obtained and compared with the experimental results from the MF analysis or from the phenomenological Landau expansion. If there is some discrepancy between the experiment and the MF theory one should try to reconcile the theory with the experiment using the RG. The RG within the one-loop approximation then yields all possible FPs and associated stable or unstable critical states together with their main characteristic quantities — the critical exponents. It can hardly be accepted that a new FP with coordinate(s) of order $O(\epsilon^2)$, that is a FP which does not exist within the ϵ^1-analysis, can be interpreted as something else than an artifact of the theory. There are situations however when the RG equations are to be investigated in higher than first order in $\epsilon = (d_U - d)$. For example, it often happens in the studies of complex Hamiltonians with competing interactions that two or more FPs are equally stable (or unstable) for some values of n and d, or one or more FPs exhibit a marginal stability. In order to resolve such problems one must consider the RG equations in the two-loop (ϵ^2-) approximation.

8.9.1 *Recursion relations to order ϵ^2*

Following the paper of Bruce, Droz and Aharony (1974) we shall present how the finite (non-differential) recursion relation for the φ^4-model are derived and analyzed. The parameters $\tilde{r}$, $\tilde{u}$, and $\tilde{c}$ obtained by the operation I_b performed in the two-loop approximation are the same as those given by Eq. (8.131). We assume that the quantities in the Hamiltonian are dimensionless so that $c = \Lambda = 1$.

The rescaling operation S_b will result in the following recursion relations for $c'(=1)$, r', and u':

$$1 = b^{-\eta}\left[1 - 32(n+2)u_0^2 \Delta B\right], \tag{8.191}$$

with

$$\Delta B = \left(\frac{1}{k'}\right)^2 \left[B(0,k') - B(0,0)\right] + O(r_0), \tag{8.192}$$

see Eqs. (8.31c) and (8.32), and

$$\begin{aligned} r' = b^{2-\eta}\,&[r_0 + 4(n+2)u_0 A_1(r_0) \\ &-16(n+2)^2 u_0^2 A_1(r_0) A_2(r_0) - 32(n+2) u_0^2 B(r_0)]\,, \end{aligned} \tag{8.193}$$

and

$$\begin{aligned} u' = b^{\epsilon - 2\eta}\,&\left[u_0 - 4(n+8)u_0^2 A_2(r_0)\right. \\ &+ 16(n^2+6n+20)u_0^3\left[A_2(r_0)\right]^2 + 64(5n+22)u_0^3 C(r_0) \\ &\left.+32(n+2)(n+8)u_0^3 A_1(r_0) A_3(r_0)\right]. \end{aligned} \tag{8.194}$$

In the above formulae,

$$A_m(r_0) = \int_{k_1}' \frac{1}{(r_0 + k^2)^m}, \qquad (m = 1, 2, \ldots), \tag{8.195}$$

$$B(r_0) = \int_{k_1}' \int_{k_2}' \int_{k_3}' \frac{\delta(k_1 + k_2 + k_3)}{(r_0 + k_1^2)(r_0 + k_2^2)(r_0 + k_3^2)}, \tag{8.196}$$

and

$$\begin{aligned} C(r_0) &= \int_{k_1}' \int_{k_2}' \int_{k_3}' \frac{\delta(k_1 + k_2 + k_3)}{(r_0 + k_1^2)(r_0 + k_2^2)(r_0 + k_3^2)^2} \\ &= -\frac{1}{3}\frac{\partial B(r_0)}{\partial r_0}. \end{aligned} \tag{8.197}$$

8.9.2 *Fixed points*

The GFP is again given by the coordinates $r_G = u_G = 0$ and the Gaussian exponents. The more interesting problem is how the coordinates of H look. We can write them in the form:

$$r_H = a_1\epsilon + a_2\epsilon^2, \tag{8.198a}$$

$$u_H = b_1\epsilon + b_2\epsilon^2. \tag{8.198b}$$

Since the coefficients a_1 and b_1 are known from Eqs. (8.113) and (8.114), our task is to obtain the coefficients a_2 and b_2. Before doing this, we shall calculate the exponent η with the help of Eq. (8.191); in which the integral difference ΔB can be found from Eq. (E.40). Bearing in mind that $b^{-\eta} \equiv 1 - \eta \ln b$ we obtain

$$\eta(u_0) = 8K_d^2(n+2)u_0^2 \tag{8.199a}$$

$$\eta = 8K_d^2(n+2)u^{*^2} + O(\epsilon^3). \tag{8.199b}$$

For H, we have to choose $u^* = u_H$. As our aim is to calculate η to order ϵ^2, the linear term in ϵ in the Eq. (8.198b) is sufficient and so we get

$$\eta = \frac{(n+2)}{2(n+8)^2}\epsilon^2. \tag{8.200}$$

The next step is to clarify two simple technical points. Exponents like $b^{f(\epsilon,n)}$ with the function $f(\epsilon, n)$ of the type

$$f(\epsilon, n) = A(n)\epsilon + B(n)\epsilon^2 \tag{8.201}$$

can be expanded in ϵ,

$$e^{f \ln b} = 1 + A\epsilon \ln b + \frac{1}{2}(A\epsilon)^2(\ln b)^2 + B\epsilon^2 \ln b + O[(\epsilon \ln b)^3], \tag{8.202}$$

$(\epsilon \ln b) \ll 1$. In addition to the condition $(\epsilon \ln b) \ll 1$ we have $b > 1$. The perturbation integrals will be calculated at the FPs where r^* is either $r_G = 0$ or $r_H \sim O(\epsilon)$. Within the present technique we shall expand the integrands of the perturbation integrals in ϵ. Then r_H must be much less than the minimal momentum of integration $k^2_{\min} \sim b^{-2}(\Lambda \equiv 1)$. Therefore a third condition on the rescaling factor b is $|r_H| < b^{-2}$. So b is subject to the conditions

$$1 \ll b^2 \ll \frac{1}{\epsilon} \sim \frac{1}{|r_H|}. \tag{8.203}$$

Now we shall calculate the coefficients a_2 and b_2 in Eq. (8.198a) and Eq. (8.198b). The equation for b_1 and b_2 is obtained by setting $u_0 = u' = u_H$ in Eq. (8.194) and expanding the factor $b^{\epsilon-2\eta}$ according to Eq. (8.202). The only integral in Eq. (8.194) which remains to be calculated to the first order in $\epsilon = (4 - d)$ is $A_2(r_0)$. Since within the approximation we are working with all terms of order $O(\epsilon^3)$ neglected, the integral $A_2(r_0)$ in the second term ($\sim u_0^2$) of Eq. (8.194) should be written as

$$A_2(r_H) = A_2(0) - 2A_3(0)r_H, \tag{8.204}$$

where we have used the identity $2A_3(r_0) = -[\partial A_2(r_0)/\partial r_0$; for $A_2(0)$ to ϵ^1-order and $A_3(0)$ to ϵ^0-order, see Eq. (E.27) and Eq. (E.28). The integrals participating in the terms proportional to $u^3 \sim \epsilon^3 + O(\epsilon^4)$ in Eq. (8.194) are taken to the zero order in ϵ and $r_0 \sim r_H = 0$. Thus we obtain that at the HFP Eq. (8.194) becomes

$$\begin{aligned}
&- \epsilon \ln b + \frac{1}{2}(\epsilon \ln b)^2 + 2\eta \ln b \\
&= -4(n+8)b_1\epsilon[A_2(0) - 2A_3(0)r_H] - 4(n+8)b_2\epsilon^2 A_2(0) \\
&+ 16(n^2 + 6n + 20)(\epsilon b_1)^2 A_2^2(0) + 64(5n + 22)(\epsilon b_1)^2 C(0) \\
&+ 32(n+2)(n+8)(\epsilon b_1)^2 A_1(0)A_3(0);
\end{aligned} \tag{8.205}$$

we have used Eq. (8.198b) for u_H. In the above equation the term $\sim \epsilon r_H$ is completely cancelled by the last term ($\sim \epsilon^2 A_1 A_3$).

From Eq. (8.107a), we have

$$r_H = -\frac{4(n+2)b_1 \epsilon A_1(0)}{1-b^{-2}} + O(\epsilon^2). \tag{8.206}$$

We shall put back this expression for r_H in Eq. (8.205) setting the denominator equal to unity in view of the condition $b^{-2} \ll 1$. Then it is readily seen that the above mentioned cancellation holds to an accuracy of order ϵ^2. Such a procedure of compensation is the ϵ-version of the renormalization within the standard perturbation theory (see, e.g., Section 7.10). In consequence, three types of terms remain in Eq. (8.205): $\sim \epsilon \ln b$, $\sim \epsilon^2 (\ln b)^2$, and $\sim \epsilon^2 \ln b$. The integral $A_2(0)$ to the zero order in ϵ is given by Eq. (8.112) and the integral $C(0)$ is

$$C(0) = \frac{1}{2} K_d^2 [\ln b + (\ln b)^2] + O(1); \tag{8.207}$$

cf. Eq. (E.36). In the $b_1\epsilon$-term of Eq. (8.205), $A_2(0)$ should be considered to the first order in ϵ; see Eq. (E.37). The results following from the terms proportional to $\epsilon \ln b$ confirm the formula (8.114) for u_H with $b_1 = 1/4(n+8)K_d$.

In order to write the respective equations in a more concise form we shall introduce the following notations:

$$\eta = \eta_0 \epsilon^2, \tag{8.208a}$$

and

$$c_i = b_i K_d; \qquad c_1 = \frac{1}{4(n+8)}. \tag{8.208b}$$

Then

$$\eta_0 + 2(n+8)c_2 = 16(5n+22)c_1^2 \tag{8.209}$$

and

$$c_1^{-2} = 16(n^2+6n+20) + 32(5n+22). \tag{8.210}$$

The expression (8.210) with c_1 from Eq. (8.208b) is an identity. It merely confirms the ϵ^1-result for u_H and the suitable choice of the present renormalization scheme. Identities like Eq. (8.210) can be used to check whether the calculations have been carried out in a proper way. From Eq. (8.209), we find c_2 and express it by the initial parameters b_i following Eq. (8.208a)–(8.208b). Putting back the obtained result in Eq. (8.198a) and Eq. (8.198b) we get

$$u_H = \frac{\epsilon}{4(n+8)K_d}\left[1 + \frac{(9n+42)}{(n+8)^2}\epsilon\right]. \tag{8.211}$$

In expressions like Eq. (8.211) K_d can be expanded in $\epsilon = (4-d)$, when $d_U \neq 4$, $\epsilon = (d_U - d)$, to become $K_d = K_{d_U} + O(\epsilon)$; see, e.g., Eq. (E.25); so one may represent u_H in another form (see, e.g., Barber, 1977). We should bear in mind that the $\epsilon \ln b$ contributions to the RG equations coming from the different terms, which contain the expanded form of K_d, cancel out each other. That is why, in order to avoid complicated calculations we may work with K_d (without any expansion). As it turns out the precise formula for the coordinates of FP, say, (r_H, u_H) is not very important, i.e., we should know them to the extent that gives an opportunity to determine the type of the considered FP and the correct values of the stability exponents λ_i.

The next task is to calculate r_H, i.e., the coefficient a_2 in Eq. (8.198a) and Eq. (8.198b): the coefficient $a_1 = -(n+2)/2(n+8)$, as given by the ϵ^1-analysis in Section 8.7; see Eq. (8.113). In fact, the value of r_H to the order ϵ^2 is not necessary for the ϵ^2-analysis of the critical behaviour, so we can skip the calculation of the coefficient a_2. The expression for r_H is obtained from Eq. (8.193) by setting $r' = r_0 = r_H$, neglecting the factor $b^{-\eta} \equiv 1 - \eta \ln b$, which yields corrections of order $O(\epsilon^3)$, and using that

$$A_1(r_H) = A_1(0) - A_2(0)r_H, \tag{8.212}$$

by analogy with Eq. (8.204). Furthermore, we drop all terms of order $O(\epsilon^3)$. The term $\sim u_0^2 A_1 A_2$ in Eq. (8.193) is completely cancelled by the term $\sim -u_H r_H A_1 A_2$; see also Eq. (8.106)) for r_H. Therefore, the equation for r_H reads

$$(1 - b^{-2})r_H = -4(n+2)u_H A_1(0) + 32(n+2)u_H^2 B(0), \tag{8.213}$$

where $A_1(0)$ should be calculated to the ϵ^1-order and $B(0)$ — to the zero order in ϵ; see Eq. (E.35). Substituting r_H and u_H from Eqs. (8.198a) and (8.198b) and neglected all terms of order ϵ^2 in the above equation, we can obtain r_H within the first-order ϵ-analysis. The calculation of the second order ϵ-correction, $a_2\epsilon^2$, to r_H is also straightforward.

8.9.3 *Exponents and stability properties*

We shall show in detail how the elements L_{ij} of the linearized RG matrix $\widehat{L}$, see Eq. (8.117), are calculated in the two-loop approximation.

We need the derivative $(\delta r'/\delta r_0)_{\mu H} = L_{11}$, which is found by a formal differentiation of Eq. (8.193). In the obtained expression all terms $\sim O(\epsilon^3)$ are dropped, in addition the terms like $u_H^2 A_1 A_2$ and $u_H r_H A_2$ are cancelled out in pairs; this is obvious from Eq. (8.204) and Eq. (8.206). The term $\sim u_H^2[\partial B(r_0)/\partial r_0]_{r_H=0}$ is represented by the integral $C(0)$; see Eq. (8.197) and Eq. (E.36). The result will be

$$\frac{\delta r'}{\delta r_0} = b^{2-\eta}\left[1 - 4(n+2)u_H A_2(0) + 16(n+2)^2 u_H^2 A_2^2(0) \right.$$
$$\left. +48(n+2)u_H^2 K_d^2 \ln b(1+\ln b)\right]. \tag{8.214}$$

The next step is to calculate the integral $A_2(0)$, see Eq. (E.27). In order to remain in the approximation we use, $A_2(0)$ should be taken in the different terms to a different order in *epsilon*: in $u_H A_2(0)$ — to the first order in ϵ, and in $u_H^2 A_2^2(0) \sim O(\epsilon^2)$ — to the zero order. The parameter u_H in Eq. (8.214) is replaced with its FP value, given either in terms of b_i by Eqs. (8.198a)–(8.198b), or, by the coefficients c_i; see Eq. (8.208a) and Eq. (8.208b). Again terms of order higher than ϵ^2 are neglected. So, the expression in the curly brackets of Eq. (8.214) becomes

$$b^{\lambda_t'} = 1 + \lambda_1\epsilon \ln b + \frac{1}{2}(\lambda_1\epsilon)^2(\ln b)^2 + \lambda_2\epsilon^2 \ln b, \tag{8.215}$$

where

$$\lambda_t' = \lambda_1\epsilon + \lambda_2\epsilon^2. \tag{8.216}$$

Here $\lambda_1 = -(n+2)/(n+8)$ according to the ϵ^1-analysis and

$$\lambda_2 = -\frac{6(n+2)(n+3)}{(n+8)^3}. \tag{8.217}$$

The final result for $(\delta r'/\delta r_0)_{\mu H}$ is

$$\left(\frac{\delta r'}{\delta r_0}\right)_{\mu H} = b^{\lambda_t} + O(\epsilon^3); \tag{8.218}$$

the explicit expression for $\lambda_t = 2 - \eta + \lambda_t'$ is

$$\lambda_t = 2 - \frac{(n+2)}{(n+8)}\epsilon - \frac{(n+2)(13n+44)}{2(n+8)^3}\epsilon^2. \tag{8.219}$$

We shall see below that this stability exponent is connected with the correlation length exponent ν through $\lambda_t = 1/\nu$, which, together with Eq. (8.219) yields

$$\nu = \frac{1}{2} + \frac{(n+2)}{4(n+8)}\,\epsilon + \frac{(n+2)(n^2+23n+60)}{8(n+8)^3}\,\epsilon^2 + O(\epsilon^3). \tag{8.220}$$

The quantity $L_{22} = (\delta u'/\delta u_0)$ at (r_H, u_H) is obtained from Eq. (8.194). Bearing in mind that

$$\frac{\partial}{\partial u_0} b^{-2\eta(u_0)} = -4\eta(u_0)(\ln b) b^{-2\eta(u_0)} \tag{8.221}$$

and using Eq. (8.204) for $A_2(r_H)$, it is not difficult to find that

$$\left(\frac{\delta u'}{\delta u_0}\right)_{\mu H} = L_{22}^{(1)} + L_{22}^{(2)}, \tag{8.222}$$

where

$$L_{22}^{(1)} = b^{\epsilon - 2\eta}\left[1 - 8(n+8)u_H A_2(0) + 48(n^2+6n+20)u_H^2 A_2^2(0)\right.$$
$$\left. +192(5n+22)u_H^2 C(0)\right], \tag{8.223}$$

and

$$L_{22}^{(2)} = 32(n+2)(n+8)u_H^2 A_1(0) A_3(0) + O(\epsilon^3). \tag{8.224}$$

We have explained above the way in which the calculations should be carried out. So we can render Eq. (8.223) in a more simple form:

$$L_{22}^{(1)} = b^{\epsilon - 2\eta}\left[1 - 2\epsilon \ln b + 2\epsilon^2 (\ln b)^2 + \frac{10n+44}{(n+8)^2}\epsilon^2 \ln b\right]; \tag{8.225}$$

here we have omitted the terms $\sim O(\epsilon^3)$. From this formula and Eq. (8.200),

$$L_{22}^{(1)} = b^{\lambda_u}, \tag{8.226}$$

where

$$\lambda_u = -\epsilon + \frac{(9n+42)}{(n+8)^2}\epsilon^2. \tag{8.227}$$

We shall see that $L_{22}^{(2)}$ can be expressed by L_{12} and L_{21} in the following way $L_{22}^{(2)} = L_{12}L_{21}b^{-2}$. This is not a general relation, rather it is a result from the subsequent calculations and the used approximation. Bearing that in mind, we shall proceed with the calculation of L_{12} and L_{21}. To the lowest order in ϵ they are given by

$$L_{12} = 4(n+2)b^2 A_1(0) + O(\epsilon), \tag{8.228}$$

and

$$L_{21} = 8(n+8)u_H^2 A_3(0) + O(\epsilon^3). \tag{8.229}$$

Combining all the above results we can write the matrix $\widehat{L}$ as

$$\widehat{L} = \begin{pmatrix} b^{\lambda_t} & L_{12} \\ L_{21} & b^{\lambda_u} + L_{22}^{(2)} \end{pmatrix}. \tag{8.230}$$

The eigenvalues λ of $\widehat{L}$ are obtained from the equation

$$\lambda^2 - (L_{11} + L_{22})\lambda + (L_{11}L_{22} - L_{12}L_{21}) = 0, \tag{8.231}$$

whose solutions $\lambda_\pm$ formally are given by the expressions

$$\lambda_+ = L_{11} + L', \tag{8.232a}$$

and

$$\lambda_- = L_{22} - L', \tag{8.232b}$$

where

$$L' = \frac{b^2 L_{22}^{(2)}}{L_{11} - L_{22}} \sim \epsilon^2. \tag{8.233}$$

The effect of the term L' is manifested in the appearance of $C_\pm\epsilon^2$ — amplitudes of the eigenvalues,

$$\lambda_+ = (1 + C_+\epsilon^2)b^{\lambda_t}, \tag{8.234a}$$

and

$$\lambda_- = (1 + C_-\epsilon^2)b^{\lambda_u}. \tag{8.234b}$$

The stability exponents $\tilde{\lambda}_\pm = \ln \lambda_\pm$, $\tilde{\lambda}_+ = \lambda_t$ and $\tilde{\lambda}_- = \lambda_u$, are in fact equal to λ_t and λ_u because $C_\pm\epsilon^2$ can be neglected as small. The exponent $\tilde{\lambda}_t \approx \lambda_t$ describes the scaling field $r \sim t$ and the exponent $\tilde{\lambda}_- \approx \lambda_u$ gives the variation δu of the interaction u near u_H. Therefore the exponent $(1/\lambda_t)$ is identified with the correlation length exponent ν.

One may try to find the (critical) dimension, below which λ_u becomes positive. Setting formally $\lambda_u = 0$ in Eq. (8.227) one obtains $\epsilon_c = (n + 8)^2/(9n + 42)$. Above the value ϵ_c, $\epsilon > \epsilon_c$, $\lambda_n(n, \epsilon) > 0$. This corresponds to dimensions $d < 4 - \epsilon_c$.

However such a result can be hardly accepted seriously because the ϵ^2-corrections to various quantities, in particular, to the stability exponents like λ_u can be considered reliable if only they are small compared to the ϵ^1-corrections. This fact, obviously, is the essence of the ϵ-expansion. For example, the value of λ_u is reliable for $\epsilon \ll \epsilon_c$.

The presentation of the RG equations in the two-loop (ϵ^2-) approximation and their analysis confirm the main features of the critical behaviour following from the first-order RG treatment of the φ^4-Hamiltonian. This means that the ϵ^2-corrections do not change in a drastic way the result from the one-loop approximation. In the same way one can apply the large-b version of RG for the study of more complex Hamiltonians. The newly obtained results concern the stability properties of the FPs. The method outlined above has been used for the first time by Aharony (1973) for systems with cubic anisotropy; see also Aharony (1976).

The reader who has traced out the above calculations should notice that the effect of the mass-renormalization diagrams of order $O(u_0^2)$ on the values of the critical exponents is totally compensated by the renormalization counter-terms for r_0, u_0 and the field φ which naturally arise in the Wilson–Fisher renormalization scheme. This fact can be used in the application of the RG analysis to complex models. From a practical point of view this means all mass-renormalization terms are to be ignored (diagrams for r and u with mass-renormalized internal G-lines see, e.g., Fig. 7.20b). So, when using the RG method in the two-loop approximation we can work with the diagrams, for r and u shown in Figs. 7.13(b, d), and Fig. 8.5(c). Of course, the total RG equations (with mass-renormalization diagrams included) can be derived and analyzed in the way shown in this section. In this case the natural procedure of compensation of the irrelevant (mass-renormalization) terms can be used to check whether the calculations are done correctly.

The limitations of the large-b RG method have been discussed in detail by Bruce, Droz and Aharony (1974). The extension of this method to the third order in ϵ is still a dubious question. The reason is not only because of the complexity of the calculations but also the irrelevant variable such as, for example, u_6 will affect the results. We should emphasize that the approximation of the large-b version of the RG recursion relations to the second order in ϵ is sufficient to investigate the scaling properties of a wide class of fluctuations. In most of the cases even the first-order RG recursion relations are sufficient. For a further information about the way, in which the large-b RG equations for complex systems are to be investigated and their reliability interpreted, the reader is referred to Lawrie, Millev and Uzunov (1987).

8.10 Other Theoretical Schemes and Methods of Calculation

Up to now we have been discussing the RG method in the Wilson–Fisher version of the recursion relations for a fluctuation field Hamiltonian in the momentum (k-) representation. This new RG method is in many aspects equivalent to the RG schemes developed previously in the quantum field theory (see also Chapter 1). It is remarkable that all results for the critical behaviour obtained within the new (Wilson–Fisher) RG scheme have been rederived within the older field-theoretical RG method based on the RG equations for the vertex functions or the Callan–Symanzik equations (Callan Jr., 1970; Symanzik, 1970); for a review of the RG see Brézin, Le Guillou and Zinn–Justin (1976), Di Castro and Jona–Lasinio (1976), Amit (1978). While the new RG scheme is naturally connected with the underlying ideas of the scaling and universality of the critical phenomena, the older field-theoretical version of RG seems to be more convenient for practical calculations to the higher order of the loop expansion.

8.10.1 *Direct calculation of the critical exponents*

The main achievement of the RG theory is that it makes possible to control and to interpret correctly the "dangerous" divergences of the perturbation series for the spatial dimensionalities $d \leq d_U$, i.e., for dimensions d which corresponds to the dimensionality of real systems. We have seen in Section 8.9 that the extension of the RG recursion method for calculations in the three-, four-, etc. loop approximations is quite difficult, if not impossible. The integration operation I_b generates a large number of effective interactions which cannot be ignored in these high orders of the theory. The critical exponents are then calculated within the ϵ-expansion to a high order in ϵ (ϵ^3, ϵ^4, etc.) by a direct investigation of the Dyson equation for the vertex function $\Gamma^{(2)}(k,r) = G^{-1}(k,r)$, the equation for $< \varphi >$, i.e., the equation of state and equations like that for the specific heat $C \sim < \varphi^2(x)\varphi^2(y) >$ and the four-point vertex $\Gamma^{(4)}(k_1,k_2,k_3)$. This direct method for the calculation of the critical exponents (Wilson, 1972) gives the opportunity to avoid doing the rescaling (S_b-) procedure (for a very instructive presentation of the method see Barber, 1977; the same results can also be rederived by the skeleton graph technique; see Tsuneto and Abrahams (1973), Abrahams and Tsuneto (1975); for a table of the critical exponents in powers of ϵ see, e.g., Ma (1976a). The high-order ϵ-calculations are of essential theoretical interest as they demonstrate the asymptotic nature of the ϵ-expansion; see Kazakov and Shirkov (1980), Vladimirov, Kazakov and Tarasov (1979), Chetyrkin and Tkachov (1982).

8.10.2 *Renormalization group in three dimensions*

Since the correct interpretation of the critical experiments requires to have predictions for the actual spatial dimension (most often $d = 3 < d_U$) of a real system it is absolutely necessary for the theory to describe in a proper way the three-dimensional systems. At this point we shall assume that $d_U = 4$ so that the results from the ϵ-expansion are to be compared with the experiments in ($d = 3$)-dimensional systems by an extrapolation from $\epsilon \ll 1$ to $\epsilon = 1$.

Let a quantity $A(\epsilon, n)$ be expanded in powers of ϵ:

$$A = A_0 + A_1\epsilon + A_2\epsilon^2 + \cdots , \tag{8.235}$$

where the coefficients depend on the symmetry index n; $A_l = A_l(n)$, $l \geq 0$. The expansion in powers of ϵ, like (8.235) for A will give reliable predictions provided the correction to order $(l+1)$, $A_{l+1}\epsilon^{l+1}$ is much less than that of order l, $A_l\epsilon^l$; for $l = 0$, i.e., $d = 4$, (A_0 represents the MF value of A). The conditions like $|A_1(n)\epsilon| = |A_2(n)\epsilon^2|$which present an equation for $\epsilon = \epsilon_c$ can certainly be used to determine the critical dimension $d_c = 4 - \epsilon_c$, below which the results from the ϵ-series are no longer valid. This point is widely discussed by Lawrie, Millev and Uzunov (1987).

The RG scheme itself does not require an ϵ-expansion ($\epsilon = d_U - d$). This expansion is intended only as a tool for obtaining explicit quantitative results concerning a wide class of fluctuation Hamiltonians. Such a widely used method should be however justified by both theoretical and experimental arguments. The most precise experiments are in an excellent agreement with the results from the ϵ-series to order ϵ^2 and their extrapolation to $\epsilon = 1$ (the higher orders in ϵ, $O(\epsilon^3)$, yield worse results because of the asymptotic nature of the expansion). Theoretically, the ϵ-method is justified by several general arguments and the comparison of the ϵ-results with those from the high-temperature expansions and the RG methods especially developed for 3-dimensional system; for the high-temperature expansion see Stanley (1971), Fisher (1983); for the RG in three dimensions see Baker Jr. *et al.* (1976), Baker Jr., Nickel and Meiron (1978), Le Guillou and Zinn–Justin (1980), Nickel (1981), Parisi (1980, 1988), Baker Jr. (1984). The divergence of the perturbation series in powers of the interaction constant u of the φ^4-model for 3-dimensional systems requires the use of nonperturbative methods. The perturbation terms $\sim u^l$ have numerical coefficients proportional to $l!$, in the FP, $u^* \sim O(\epsilon)$ with $\epsilon = 1$, and these terms are large. In this situation, the Borél resummation method and the Padé approximation are successfully applied to the study of the asymptotic critical behaviour $r \to 0$; see, e.g., Baker Jr. (1975), Parisi (1988), Joyce and Guttmann (1982), Zinn–Justin (1971), Hardy (1948). An interesting method for the evaluation of the high order ($l \gg 1$) perturbation terms $\sim u^l$ is presented by Lipatov (1977).

The results from the direct RG calculations for ($d = 3$)-dimensional systems and those from the ϵ-expansion to order ϵ^2 are in a good agreement, and this confirms the usefulness of the ϵ-series (Fisher, 1983; Baker Jr. and Kincaid, 1981; Baker Jr., 1984). An analytical method for RG treating of 3-dimensional systems has been developed by Ginzburg (1975) on the basis of the usual φ^4-model. The results from this method are in a good agreement with those from the ϵ-series; for an application of this method to dynamic critical phenomena (Hohenberg and Halperin, 1977) see Li (1983). Another RG approach in three dimensions has been presented by Newman and Riedel (1984).

A characteristic feature of the direct "three-dimensional" RG methods is their relative sophistication. That is why they cannot be easily applied to effective Hamiltonians with competing interactions. When this is possible the results are slightly different from those of the ϵ-expansion; see, e.g., Jug (1983, 1984)

8.10.3 $(1/n)$–*expansion*

One simplified method for the direct calculation of the critical exponents in three dimensions ($d = 3$) is the $(1/n)$–expansion (Ma, 1972; Fisher, Ma and Nickel , 1972; Abe, 1972, 1973; Suzuki, 1972; Ma, 1973a,b); for a review see Ma (1976a,b); Barber (1977). A notion of this method can be received by approximating the n-dependent coefficients in the two-loop RG equations (8.191)–(8.194) with the leading n-dependencies, for example, in the second term of Eq. (8.193) one should replace $(n+2) \approx n$; $n \gg 1$. Within this (large-n) approximation the RG equations will obey the condition of a double (ϵ- and $1/n$-) expansion and, hence, will be valid for both $\epsilon \ll 1$, and $(1/n) \ll 1$. Actually, the $(1/n)$-expansion is developed for the complete interval of dimensions d between two and four; $2 < d < 4$ (or for $d_L < d < d_U$ when $d_L \neq 2$ and $d_U \neq 4$).

The important $(1/n)$–calculations of the critical exponents are performed for the φ^4-Hamiltonian. The zero approximation, $(1/n)^0$, of the $(1/n)$-expansion for the φ^4-model is equivalent to the approximation of this Hamiltonian with the Berlin–Kac spherical model or, which is the same, to the Hartree approximation; see Sections 5.7.5 and 6.6.3, where the n-vector semi-classical lattice model with the spherical condition (6.133) is transformed to an effective fluctuation Hamiltonian with $u \sim 1/(n+2) \sim 1/n$ (for $n \gg 1$) (for the Hartree approximation, see Section 8.4). When making the $(1/n)$-expansion one must assume that the interaction constant u is proportional to $(1/n)$. There are no technical difficulties in the calculation of the critical exponents to the first order in $(1/n)$ and when some of the exponents, for example η, are calculated to order $(1/n)^2$. However considerable difficulties arise when one tries to find the higher-order $(1/n)$–corrections or to investigate complex Hamiltonians.

The RG recursion method can also be presented through the $(1/n)$–expansion (the so-called large-n RG recursion relations; see, e.g.,, Ma (1976a,b). This method gives explicit results to order $(1/n)^0$ — the Hartree limit; for details, see also Szepfalusy and Tél (1980), Vvedensky (1984).

In most of the cases the real systems are described by order parameters φ with symmetry index $n \sim 1 \div 10$. The extrapolation of the results for $n \gg 1$ to values of n like $1, 2, 3, \ldots$ cannot easily be controlled or evaluated. The qualitative agreements of the $(1/n)$–expansion results with the experiments in certain systems as well as the justification of the $(1/n)$-expansion as a method being intended to fill the "gap" between realistic φ^4-fluctuation Hamiltonian and its "spherical limit" $(n \to \infty)$ are out of doubt.

8.10.4 *Renormalization group in the x-space and references to other topics*

Another RG approach is developed for lattice ("spin-like") models. These type of RG transformations are performed in the x-space (x-space RG) and are conceptually close to the original Kadanoff idea of the "block-spin" transformations (Chapter 1).; see the review by Niemeijer and van Leeuwen (1976); for an introduction see Barber (1977), and Fisher (1983); for a review of recent results, simplified x-space RG transformations and applications to real systems see the review articles in the book edited by Burkhardt and van Leeuwen (1982). A simplified version of x-space RG is, for example, the *Migdal–Kadanoff transformation* (Migdal, 1975; Kadanoff, 1975, 1976).

We should mention also the $\epsilon = (d-2)$-expansion for dimensions $d \sim 2$ which is made on the basis of the σ-model — first introduced in the field theory; see Polyakov (1975), Brézin and Zinn–Justin (1976), Brézin, Zinn–Justin, and Le Guillou (1976); for instructive applications to particular systems see Nelson and Pelcovitz (1977), Sak (1977). The σ-model describes the ($k \sim 0$)-asymptotic behaviour of the Heisenberg model in the low-temperature limit; see, e.g., Stanley (1971).

Finally we shall mention the Fisher's finite-size scaling theory by which one can compare the critical behaviour of infinite systems with the behaviour of large but finite systems. A detailed description of this theory including RG studies and references to original papers is given by Barber (1983) and Diehl (1986), Suzuki (1986) has developed a new method, the so-called coherent-anomaly method (CAM), which is a combined theory of Kubo's response theory (Kubo, 1957, 1965) and Fisher's finite-size scaling theory; see also Suzuki (1988, 1991).

8.11 Interrelations with other theories

Starting from the Gibbs stability theory in the equilibrium thermodynamics (Chapter 2) we have traced out the development of the ideas for the description of stable phases and the phase transition points within the framework of thermodynamics and statistical mechanics .The basic concepts coming from Gibbs for understanding the phase transitions in the terms of the order parameter and the interpretation of the phase transition points as states in the phase diagram where the system changes its stability properties is the general framework, in which all knowledge about phase transitions is included.

We have illustrated in Chapter 4 the Landau expansion as a particular stability theory that is actually a compromise between a purely thermodynamic and a purely statistical theory. The Landau approach to phase transitions is based on the assumption that the saddle-point solution for the partition integral always exists and the mean (non fluctuating) order parameters $< \varphi >$ can be obtained from this solution. As this procedure cannot be practically accomplished for realistic microscopic Hamiltonians one is forced to use either general heuristic arguments to construct a quasi-macroscopic effective Hamiltonian of the system or to derive such macroscopic models with the help of the MF approximation for microscopic Hamiltonians (or to use more sophisticated coarse-graining procedures).

Thus, being well supported by the experimental observations one arrives at the stage in the development of the theory, where it becomes clear that the phase transitions and the critical phenomena are successfully described when the quasi-macroscopic (long-scale) fluctuations are taken into consideration. While basic approach to the solution of this problem is given by the standard statistical theory it becomes very difficult in a close vicinity of the stability (phase transition) points where the phases changes to each other and the harmonic approximation of free fluctuations is not valid. In such states and, in particular, in the (multi)critical points of stability where several types of continuous transitions may occur the stability of the system is ensured by the effect of the fluctuation interactions; the free fluctuations of the order parameter describe a marginal stability of the critical states. The theory becomes equivalent to a field theory with strong (fluctuation) interaction. The problems of the strong interactions in theoretical physics are always extremely difficult to solve and one of the approaches — the RG based on the ideas of scaling and universality successfully developed in the theory of critical phenomena presents one of the possible solutions.

We shall emphasize that the use of sophisticated theoretical methods as RG is especially required when the critical region is investigated. The MF theories and the Gaussian (or Ornstein–Zernicke) free field approximation is absolutely sufficient to provide a fairly well information about the order parameter fluctuations away from the (multi)critical point (outside the critical region). On the other side, the RG transformation and the RG scaling form of the thermodynamic and correlation quantities are a generalization of the MF theory and its lowest-order correction due to Ginzburg (1960) so as they are valid both in and outside the critical region. Of course, the scaling and crossover functions as well as the values of the critical exponents are gradually transformed to those from the standard Landau expansion when the system is driven away from the transition point. This "crossover" from MF behaviour far from the critical point to the regime of interacting fluctuations near the critical point is described by the full (scaling) thermodynamic functions (when they can be calculated explicitly) and by "effective" critical exponents.

The thermodynamic scaling functions have two limiting forms — the (asymptotic) scaling form near T_c and the simple MF form far from T_c. The effective critical exponents also depend essentially on the "distance" $t \sim (T - T_c$ from T_c. When t is large enough, they accept MF (classical) values and when $T \to T_c$, they acquire their asymptotic non-classical values like those obtained in this Chapter by the RG method within the ϵ-expansion. This theoretically established dependence of the main characteristics of critical behaviour on the distance from the critical point is very like the experimental evidence for the variation of the form of the scaling dependencies and the values of the critical exponents with the variation of parameters like t and h; for details about the theoretical description of effective exponents and scaling functions far from T_c; see, e.g., Dohm and Folk (1982), Bagnuls and Bervillier (1985, 1986a,b), and Folk and Moser (1987). There is an intermediate region of temperatures when the system is not far enough from T_c and so it cannot follow the simple MF scaling laws and critical exponents, at the same time the temperature is not sufficiently near T_c so as the non-classical critical behaviour to be observed. Then the behaviour of the system is described by the effective exponents which depend also on the temperature and this is confirmed by precise experiments (see Chapter 1).

Within the RG approach to the critical phenomena the thermodynamic and statistical stability theories gain an essentially new development. The problem of calculating the partition integral is replaced with the problem of investigating the invariant properties of the Hamiltonian with respect to a special (RG) transformation which consists of the removal of the "fast" (irrelevant) degrees of freedom $\varphi_\alpha(k; k > b^{-1}\Lambda)$ and the subsequent length-scale (S_b–) transformation. The elimination of the fast (short-range) degrees of freedom by integrating out from the partition integral can be performed by an infinitesimal step ($b = e^l \approx 1 + l, l \ll 1$). This procedure is intended to suggest the "theoretical spectroscopy" of the symmetry properties of the Hamiltonian rather than to ensure an actual integration in the partition integral. The results are obtained from the RG equations, the analysis of which presents the "spectroscopy" picture of the critical behaviour that results from *the structure* of the Hamiltonian, *the symme-*

try index n and *the spatial dimensionality* d. These are the features of the Hamiltonian that can be investigated by RG. In consequence, diverse real systems are systematized in universality classes as discussed in Section 7.11.3.

The fact that the universality classes will depend on the dimension d and the symmetry characteristics of the system like the structure of the Hamiltonian and the index n of the ordering is clear enough from the standard perturbation theory (Chapter 7). The main achievement of the RG theory is that it makes possible the calculation of the critical exponents and the scaling functions of the various fluctuation models and, therefore, the description of the critical properties within every universality class.

In some cases the RG equations describe first-order transitions rather than critical and multicritical behaviour. A special attention should be paid to the crossover phenomena from one to another universality class, i.e., from one to another type of critical behaviour or to the fluctuation-driven change of the order of the phase transition. All these phenomena are caused by the fluctuation interaction; they are described in the parameter (μ-) space of the Hamiltonian. The RG equation have their geometrical interpretation as RG flow lines (trajectories) in the μ-space and the possible types of phase transitions and critical phenomena described by a Hamiltonian depend on the properties of these RG trajectories.

The μ-space is mapped by a microscopic coarse-graining procedure to the phase diagram (T, h, Y) of the system and, at least in principle, the stability properties of the FPs, μ^*, in the μ-space can be related with the stability of the thermodynamic states (T, h, Y). The detailed investigation of the relation between the general thermodynamic stability theory and the stability analysis within the RG method is beyond the scope of this book; see also the instructive reviews by Ma (1973b), Fisher (1974), Ravndal (1976), Aharony (1976), Patashinskii and Pokrovskii (1977), Wallace and Zia (1978), and Bruce (1980).

Finally we shall mention that the RG ideas are not restricted to the field of phase transitions. It is now accepted that the ideas of scale invariance have been introduced for the first time in the theory of turbulent flows in the hydrodynamics; see, e.g.,, Landau and Lifshitz (1959). Since the appearance of the RG theory of phase transitions it has found a wide application in many problems of physics and related sciences; see, e.g., Wilson (1983), Anderson (1984a).

Chapter 9

Some Applications

9.1 Preliminary Remarks

This Chapter is intended as a brief review of the application of the phase transition theory to selected problems. For this purpose we shall use: the direct methods of the statistical calculation, in particular, when simple models are considered, the MF approximation within the framework of the Landau expansion, the lowest-order perturbation calculation together with the derivation of the Ginzburg criterion for complex systems, and the RG method. In the preceding Chapters these methods have been presented in detail and no principle difficulties arise in their application to selected simple models or to effective Hamiltonians describing complex systems. Sometimes, as it is, for example, for quantum and disordered systems the treatment of the phase transition properties requires the use of a modified technique of calculation. In such cases we shall either explain briefly the theoretical problems or the reader will be referred to the original articles. Some topics will be mentioned together with the results and the main sources of reference; so it will be a good exercise for the reader to try to re-derive these results with the help of the brief instructions presented in the text and the references cited in.

The selection of the subject matter in this Chapter follows two principles. Firstly, we have tried to avoid applications which have already been reviewed in details; see, e.g., the review articles mentioned in Chapter 8 and these in the series "Phase transitions and critical phenomena" (Domb and Green , 1972). Secondly, we shall choose those applications which are related to outstanding problems in condensed matter physics and statistical mechanics. The reference list contains the original papers and some of the most comprehensive review articles and books. As throughout the whole book, here the discussion is self-contained and the numerous references are not necessary for the first reading and understanding of the content.

In Sections 9.2–9.4 fundamental problems of the theory of quantum phase transitions are discussed; a quite more detailed review of the same topics is presented by Shopova and Uzunov (2003a). The quantum effects on critical phenomena included in this Chapter is mostly related to dilute Bose gases (Pitaevskii and Stringari, 2003; Pethick and Smith, 2008) and field theories of magnets, conventional and unconventional superconductors, and related topics (Leggett, 2006; Annett, 2004); for quantum phase transitions in strongly correlated electron systems, see Vojta (2003), Sachdev (1999), and Continentino (2001). Interesting applications of the phase transition theory to quantum phase transitions are presented also in the comprehensive review article by Pelissetto and Vicari (2002) and the books by Herbut (2007), and Schakel (2008).

In Sections 9.5 and 9.6 the basic types of disorder – quenched impurities and inhomogeneities (alias "random critical temperature") and random field disorder, are presented together with a generalization of the Ginzburg criterion for such systems. Anisotropy effects and several types of multicritical behaviour are considered in Section 9.6. These topics are important in the investigation of both classical and quantum phase transitions, where a variety of anisotropy effects and multicritical points occur.

Sections 9.7–9.8 are devoted to a detailed discussion of the gauge theory of phase transitions in superconductors (related topics in liquid crystals and quantum field theory are also mentioned). The magnetic fluctuations in superconductors obey the Coulomb gauge and this leads to the interesting phenomenon of a fluctuation driven change of the order of the phase transition. The phases and the phase diagram of unconventional ferromagnetic superconductors with a spin-triplet Cooper pairing of electrons are considered in Section 9.9. These systems exhibit a completely new universality class of critical behaviour (Section 9.9.5).

9.2 Ideal Bose Gas

The Bose–Einstein condensation (BEC) of the three-dimensional (3D) ideal Bose gas (IBG) is widely discussed in textbooks on statistical mechanics; see, e.g., Huang (1963), Pathria (1972), Landau and Lifshitz (1980). Although the bosons in the IBG do not interact, the existence of quantum correlations between them results in a special (Bose) statistics leading to the formation of the macroscopic Bose–Einstein condensate at low temperatures, $T < T_0$ — the condensation (or, quantum degeneracy) temperature. This phenomenon is very important for the understanding of the properties of real systems with inter-particle interactions (interacting real bosons or boson excitations); see, e.g., Kohn and Sherrington (1970), Comte and Noziéres (1982). The same phenomenon, surely, in a very modified form, lies in the basis of the treatment of the superfluid condensate, for example, in ^{4}He and ^{3}He or of the condensation of charged bosons in superconductors. A number of interacting systems can be approximated, of course, to a very low (zero) order by effective models of free bosons.

For the theory of phase transitions, the IBG is of considerable interest because: (i) it is an exactly solvable model, (ii) it describes the net effect of the quantum correlations of Bose type, and (iii) the thermodynamic behaviour of the IBG depends crucially on whether this system is at fixed density $\rho = N/V$, at fixed pressure P, or, at fixed temperature T.

In 1995, the theoretically predicted in 1924 BEC (Bose, 1924; Einstein, 1924) has been experimentally confirmed by experiments on ultra-cold gases of Rubidium and Sodium atoms (Anderson *et al.*, 1995; Davis *et al.*, 1995); for a detailed theory of BEC, see Pitaevskii and Stringari (2003). Since then the theoretical and experimental research on BEC and related topics is considerably enhanced.

9.2.1 *Thermodynamics*

The thermodynamics of the IBG is determined with the help of the grand canonical potential $\Omega = -PV$ and the mean number of particles N, whose mathematical expressions are well known from the books on statistical physics:

$$\Omega = T \sum_k \ln\left\{1 - z_0 e^{-\varepsilon_k/T}\right\}, \tag{9.1}$$

and

$$N = \sum_k \frac{1}{z_0^{-1} e^{\varepsilon_k/T} - 1}, \tag{9.2}$$

where $z_0 = e^{-r_0/T}$ with $r_0 = -\mu_0 \geq 0$ is the fugacity; $\mu_0 \leq 0$ is the chemical potential of the IBG; $k_B = 1$.

The free bosons have a quadratic energy spectrum $\varepsilon_k = \hbar^2 k^2/2m \equiv ck^2$, where m is the mass and k is the wave vector (below, k is often referred to as the momentum and units, in which $\hbar = 1$ will be often used). For models, which describe the interacting (fermionic and spin) systems only effectively through free (or interacting) boson excitations, the lower momentum energy spectrum is sometimes modified to become $\varepsilon_k = \hbar^2 k^\theta/2m$, where $0 < \theta \leq 2$. Then $\theta = 2$ stands for the short-range interactions and $0 < \theta < 2$ for the long-range interactions; see Section 7.11, where the spectrum $\varepsilon_k \sim k^\theta$ is derived for classical spin models and Fisher, Ma and Nickel (1972), and Sak (1973), where the RG study of these systems is discussed.

Eqs. (9.1) and (9.2) are investigated for d–dimensional IBGs in a number of papers; see, e.g., Cooper and Green (1968), Gunton and Buckingham (1968), Hauge (1969), Lacour–Gayet and Toulouse (1974). When the density $\rho = N/V$ is constant one has to solve Eq. (9.2) with respect to the chemical potential μ (or the fugacity $\hat{z}$) and to substitute the solution $\mu_0(\rho, T)$ in Eq. (9.1) in order to obtain the equation of state $P = P(\rho, T)$ of the IBG. At constant pressure, $P = (-\Omega/V) = const$, it is the Eq. (9.1) that is solved with respect to μ and the result $\mu(T; P)$ is substituted in Eq. (9.2) in order to obtain the equation of state (below, the index "0" of μ_0 and z_0 will be omitted).

These two cases, $\rho = const$ and $P = const$ represent quite different thermodynamics. As far as the critical region in vicinity of the condensation temperature T_0 is concerned the critical exponents for $P = const$ and $\rho = const$ differ from each other and obey the general renormalization scheme intended to describe thermodynamic systems with hidden variables (or internal constraints like $\rho = const$) (Fisher, 1968). In all cases one is faced with the mathematical problem to invert equations like (9.2) or (9.1) in order to express μ as a function of T, ρ (or P).

There are important limiting cases like $z = e^{\mu/T} \ll 1$ (classical limit) when Eqs. (9.1) and (9.2) yield the Clapeyron equation of state of the ideal classical (Boltzmann) gas with small quantum corrections ($\sim z^n, n = 1, 2, \ldots$). In such a situation it is not necessary to look for an exact expression for the equation of state. For our discussion of the critical state ($T \sim T_0$) another limiting case $r = |\mu| \sim 0$ or, which is the same, $z \sim 1$ will be of essential importance. It has been shown by Leonard (1968) with the help of the complex function analysis how Eqs. (9.1) and (9.2) can be inverted so as to be solved with respect to μ (for the same problem in the relativistic quantum gases; see Nieto (1970); relativistic Bose gases are discussed also by Landsberg and Dunning–Davies (1965), Haber and Weldon (1981), Singh and Pandita (1982), da Frota and Goulard Rosa, Jr. (1982), and da Frota, Silva and Goulard Rosa Jr. (1982).

Eqs. (9.1) and (9.2) can be easily studied in the thermodynamic limit ($N, V \to \infty$), where the pressure $P_0 = -\Omega_0/V$ of the particles with zero momentum ($k \equiv |k| = 0$) can be neglected as small compared to the total pressure P. Even if the number N_0 of the zero momentum particles is extremely high, i.e., $N_0 \sim N$, the pressure

$$P_0 = -\frac{T}{V}\ln(1-z) \tag{9.3}$$

will be $P_0 \sim (T/N)\ln N$ and, therefore, $P_0 \to 0$ when $N \to \infty$. The reader is advised to re-derive this result by using Eqs. (9.1)–(9.3) and the usual thermodynamic formulae; see also Pathria (1972). Replacing the summation in Eq. (9.1) and Eq. (9.2) by integration ($\sum_k \to V\int_k$) we obtain

$$\frac{\Omega}{V} = -TA(d,\theta)\lambda_T^{-d} g_{1+(d/\theta)}(z), \tag{9.4}$$

and

$$\frac{N}{V} = A(d,\theta)\lambda_T^{-d} g_{d/\theta}(z), \tag{9.5}$$

where

$$\lambda_T = \left(\frac{h^2}{2\pi m T}\right)^{1/\theta} \tag{9.6}$$

is the (de Broglie) thermal wave length;

$$g_\nu(z) = \frac{1}{\Gamma(\nu)}\int_0^\infty \frac{x^{\nu-1}dx}{z^{-1}e^x - 1} \tag{9.7}$$

is the Bose integral (Robinson, 1951; Isihara, 1971; Pathria, 1972), and

$$A(d,\theta) = \frac{2^{1-d+2d/\theta}\Gamma(d/\theta)}{\theta\pi^{(d/2)-(d/\theta)}\Gamma(d/2)} \tag{9.8}$$

is a constant; $A(d,2) = 1$.

Note that for $d > \theta$ one must divide the sum in the original expression (9.2) for N in two: the number

$$N_0 = \frac{1}{z^{-1}-1} \tag{9.9}$$

of the $(k = 0)$-particles and the remaining part $N - N_0 = N'$ for $k \neq 0$ and only after that the summation over k is replaced by an integration. When $d \leq \theta$ this procedure of removing N_0 from the total number N of particles would be wrong as the ($k = 0$–momentum particles are taken into account automatically in the integral for N. This statement can be easily verified if we recall that the sum (9.2) corresponds to the integral

$$N \sim \int_0^\infty dk.k^{d-1} n(k), \tag{9.10}$$

where

$$n(k) = \frac{1}{e^{(\varepsilon_k + r)/T} - 1} \tag{9.11}$$

is the Bose distribution and the momentum cutoff Λ is taken to be infinity ($\Lambda = \infty$) because of the specific properties of the integrand. From Eq. (9.10) and Eq. (9.11) one can obtain the Bose integral (9.7), where the integration variable $x \sim \varepsilon_k/T$ is the dimensionless energy ($\varepsilon_k \sim k^\theta$). It is then easy to see that the zero energy ($\varepsilon_k = k = 0$) particles will contribute to the integral (9.7) provided $d \leq \theta$ but when $d > \theta$ the "weight" $\varepsilon^{-1+d/\theta}$ of the particles with energies equal to ε will tend to zero as $\epsilon \sim k^\theta \to 0$.

The influence of the zero-momentum bosons turns out to be the physical reason for the well-known divergence of the function $g_{d/\theta}(z)$ in *low-dimensional Bose systems* ($d \leq \theta$). These properties of $g_{d/\theta}(z)$ are essential at extremely low temperatures ($T \sim 0$), where the role of the zero-momentum particles becomes very important despite the lack of any macroscopic condensation at $k = 0$ for any $T \neq 0$ in systems, where $\rho = const$.

Now we shall discuss briefly the critical behaviour of the IBG in two cases: (a) $T_0 > 0$, and (b) $T_0 = 0$.

9.2.2 *Bose–Einstein condensation at finite temperatures*

When $T_0 > 0$ it is usually said that the system exhibits BEC. The IBG at $\rho = const$ will have $T_0 > 0$ only if $d \geq \theta$. The main scaling behaviour is the same as that of the mean spherical (Berlin–Kac) model in the magnetism; the zero (Hartree) approximation of the large n φ^4-Hamiltonian; see Section 5.7.2 and Section 8.4. At $P = const$, the IBG has $T_0 > 0$ for any dimension $d > 0$. The asymptotic scaling features of this phase transition are described by the classic φ^4-Hamiltonian taken in the Gaussian approximation as suggested by Lacour–Gayet and Toulouse (1974), and Pfeuty and Toulouse (1975); see also the continuation of the n-vector models to the negative value $n = -2$ of the symmetry index n which gives results for the free energy and the two-point correlation function $G(k)$ equivalent to those following from the Gaussian approximation (Balian and Toulouse, 1973; Fisher, 1973). The above mentioned results can be derived from the general equations Eq. (9.4) and Eq. (9.5).

9.2.3 *Zero temperature condensation*

The low-dimensional ($d \leq \theta$) IBG at $\rho = const$ undergoes a special type of zero-temperature ($T_0 = 0$) BEC, which is described by "inverted spherical model exponents". The macroscopic condensate (BEC) abruptly appear at $T = 0$ but some onset of accumulation of bosons at very low energy levels at ultra-low temperature is also predicted (for details and a discussion of the universality of this zero-temperature critical behaviour; see Busiello, De Cesare and Uzunov (1985). Again the main thermodynamic results can be obtained from Eqs. (9.4) and (9.5). These results have been recently confirmed by alternative studies; see, e.g., Su and Chen (2010).

9.3 Nonideal Bose Gas

Here we shall point out several investigations of the nonideal Bose gas (NBG) by means of the RG method. The first application (Singh 1975–1978) of the Wilson–Fisher RG method to the NBG was done with the help of the operator formalism from the quantum statistical theory; see, e.g., Abrikosov, Gor'kov, and Dzyaloshinskii (1962), Fetter and Walecka (1971), and (Rickayzen, 1980).

9.3.1 *Notion of universality of the critical behaviour at finite critical temperatures*

The main results from the early RG studies of the NBG is the direct proof of the intuitively clear supposition that this system should belong to the universality class of the classical XY model in the magnetism. In fact, the creation and annihilation Bose operators $(a_k^\dagger, a_k)$ in the a^4-model of the NBG correspond to a complex order parameter field $\varphi = \varphi' + i\varphi''$ which at a quasi-macroscopic level of description should behave as the two-component, $\varphi = (\varphi_1, \varphi_2)$ order parameter of the classical φ^4-Hamiltonian.

Near the phase transition temperature (Λ-point) $T_c > 0$, the correlation length ξ of the usual (thermal) fluctuations tends to infinity whereas the de Broglie thermal wave length λ_T remains finite at any $T_c \neq 0$; see, e.g., Eq. (9.6) for λ_T of the IBG. Therefore, the effect of the finite-scale ($\sim \lambda_T$) quantum correlations will become negligible when $T_c(\neq 0)$ is approached and the asymptotic critical behaviour will be determined by the thermal fluctuation effects represented by their characteristic (correlation) length ξ as it is in the classical φ^4-model. These arguments have been confirmed by the direct RG calculations (Singh, 1975).

When we say that the NBG (or another quantum model) with a temperature $T_c \neq 0$ exhibits an universal critical behaviour or, shortly, universality, we have in mind that the quantum correlations described by the same model do not affect the (asymptotic) scaling properties of the system and, hence, the initially supposed quantum critical behaviour will belong to some of the universality classes known from the "sea" of the classical Hamiltonians.

Starting from the operator formalism of the quantum statistical mechanics one has to choose the way, in which the Bose operators a_k are to be rescaled ($a_k \to \eta(b)a'_{k'}$). There is a possibility to rescale these operators in such a way that after a large number of R_b-operations they become equivalent to commuting quantities. Alternatively one may introduce a rescaling of the boson mass m, keeping at the same time the invariance of the commutation rules; see Stella and Toigo (1976), and Baldo, Catara and Lombardo (1976). Both versions for performing RG have a reasonable physical interpretation of growing "blocks" of bosons of size $(ab)^d$, where a is the mean inter-particle distance, $a \sim \Lambda$ — the upper cutoff in the momenta, and b is the RG rescaling factor. As the "block" is enlarged the boson cluster of particles behaves classically and simultaneously the cluster mass $m_{\rm cl}$ grows; $m_{\rm cl} \sim m(ab)^d \sim mN_{\rm cl}$, where $N_{\rm cl}$ is the number of boson in a cluster ("block").

9.3.2 *Functional formulation*

The universality of the static critical behaviour of the NBG can also be discussed by means of the functional formulation of quantum statistical mechanics; see, e.g., Casher, Lurié and Revzen (1968), Leibler and Orland (1981), Kleinert (1978), Wiegel (1975, 1978), Popov (1976, 1987). In this approach the Hamiltonian of the NBG is a functional of a commuting (c–number) complex field $\psi(x,\tau)$, where $0 \leq \tau \leq 1/T$ is the imaginary "time" known from the theory of the temperature Green functions; see Rickayzen (1980), Popov (1976). In Fourier components,

$$\psi(x,\tau) = \left(\frac{T}{V}\right)^{1/2} \sum_q e^{ikx - i\omega_l \tau} \psi(q), \tag{9.12}$$

where $q = (\omega_l, k)$ is a $(d+1)$-dimensional vector, k is the wave vector and $\omega_l = 2\pi l T$ with $l = 0, \pm 1, \pm 2, \ldots$ is the (Bose) Matsubara frequency.

When the τ- dependence in $\psi(x,\tau)$ can be neglected or, which is the same the frequency ω_l in the Fourier amplitudes $\psi(\omega_l, k)$ is set identically zero ($\omega_l \equiv 0$), the Bose field $\psi(q)$ becomes, a classical two-component field $\psi(k) = \psi(0,k) = \psi' + i\psi''$ like the two-component order parameter $\varphi(k) = (\varphi_1, \varphi_2)$ of the classical systems. The ψ^4-Hamiltonian for the complex field $\psi(k)$ is equivalent to the φ^4-Hamiltonian of a two-component vector field $\varphi = [\varphi_1(k), \varphi_2(k)]$, i.e., to the XY model in the continuum limit.

If however the τ-dependence of the Bose field $\psi(x,\tau)$ in the (x,τ)-representation or the ω_l-dependence of $\psi(q)$ in the $q = (\omega_l, k)$-representation is preserved one can investigate the simultaneous effect of the quantum and classical fluctuation correlations and deduce the conditions, under which the quantum correlation are unessential for the critical behaviour. Thus the quantum effects are described by the presence of the frequency ω_l and ω_l-dependent terms in the Hamiltonian $\mathcal{H}[\psi(q)]$ of the system.

The functional formulation of the NBG is very convenient for RG studies. In this formalism the grand canonical potential becomes

$$\Omega = -T \ln \int D\psi e^{-\mathcal{H}(\psi)}, \tag{9.13}$$

where $\mathcal{H} \equiv (-\mathcal{H}/T)$ is the generalized "Hamiltonian" (in fact, $-\mathcal{H}(\psi) = S(\psi)$ is the "action" of the NBG (Popov, 1976). In Eq. (9.13), the functional integration

$$\int D\psi = \prod_{\omega_l} \prod_k d\psi^*(q) d\psi(q) \tag{9.14}$$

is over all degrees of freedom, namely, over the independent field amplitudes $\psi(q)$.

In order to make the comparison with the classical φ^4-Hamiltonian more comprehensive we shall assume that ψ is a $(n/2)$-component field, $\psi =$

$\{\psi_\alpha, \ldots, n/2\}$, where the field components ψ_α are complex functions of $q = (\omega_l, k)$; for the original NBG, $n = 2$ and ψ is a complex scalar field. The $q = (\omega_l, k)$-representation of $\mathcal{H}(\psi)$ for the NBG, i.e.,

$$\mathcal{H}(\psi) = -\sum_{\alpha,q} (i\omega_l - \varepsilon_k + \mu_0)\, |\psi_\alpha(k)|^2 + \frac{vT}{2V} \sum_{q_1 q_2 q_3;\alpha\beta} \psi^*_\alpha(q_1)\psi^*_\beta(q_2)\psi_\alpha(q_3)\psi_\beta(q_1 + q_2 - q_3) \tag{9.15}$$

is obtained by the Fourier transformation of $\mathcal{H}(\psi)$ in the (x, τ)-representation

$$\mathcal{H}(\psi) = -\int_0^{1/T} d\tau \int d^d x \left[\sum_\alpha \psi^*_\alpha(x,\tau)\frac{\partial}{\partial\tau}\psi_\alpha(x,\tau) - \frac{1}{2m}|\nabla\psi(x,\tau)|^2 + \mu_0|\psi(x,\tau)|^2 - \frac{v}{2}|\psi(x,\tau)|^4\right]. \tag{9.16}$$

In Eqs. (9.15) and (9.16), v is the interaction constant and μ_0 is the chemical potential of the IBG (i.e., NBG with $v = 0$). The natural upper cutoff Λ of the wave vectors, $0 < |k| < \Lambda$ is proportional to the mean inter-particle distance a; $\Lambda \sim 1/a$. We shall use the notation $r = -\mu > 0$ and

$$\xi(k) = \varepsilon_k + r. \tag{9.17}$$

For the model (9.15) of a $(n/2)$-component complex field one should substitute the field ψ in Eq. (9.14) by ψ_α; an additional multiplication is implied, i.e., a symbol $\prod_\alpha$ will appear in Eq. (9.14).

The perturbation treatment of the model (9.14)–(9.15) follows the same principles as explained in Chapter 7 and Chapter 8; see also Wiegel (1978), Popov (1976, 1987). Since $\psi(x, \tau)$ is a complex $(n/2)$-component vector we should use orienting arrows of the diagrammatic lines for the Green functions and the external legs $[\psi_\alpha(q)]$ of the diagrams within the RG. The bare ($v = 0$) Green function can be chosen in the form

$$G_0(q) \equiv -\langle|\psi_\alpha(q)|^2\rangle = \frac{1}{i\omega_l - \xi(k)}. \tag{9.18}$$

The diagrams will now be represented by G_0-lines which include $(d+1)$-dimensional (frequency-momentum) vectors $q = (\omega_l, k)$. In the continuum limit the summation over k is substituted by integration. Of course, a summation over the integral Matsubara frequencies ω_l of the diagrams should be performed for finite temperatures ($T > 0$); see Section E.4. In the limit $T \to 0$ one can apply the continuum limit (integration) over ω_l, too. This is made by the correspondence

$$T \sum_{\omega_l} = \int_{-\infty}^{\infty} \frac{d\omega}{2\pi}, \tag{9.19}$$

where ω is a continuous variables (sometimes it has a cutoff Λ_ω so that $-\Lambda_\omega < \omega < \Lambda_\omega$). The time ($t$-) dependent Green function and the real frequency ($\omega' \to -i\omega$) are related with the temperature Green function G_0 and the Matsubara frequency ω_l by several well-known rules; see Abrikosov, Gor'kov, and Dzyaloshinskii (1962), Fetter and Walecka (1971), and Rickayzen (1980).

The most important characteristic lengths of the NBG are included in Eq. (9.18) for G_0; so we can write it in the following way

$$G_0 = \frac{(2m/\hbar^2)}{\alpha_l \lambda_T^{-1} - k^2 - \xi^{-2}}, \tag{9.20}$$

where λ_T is the thermal wavelength (9.6) $\alpha_l = 8\pi^2 il$, and $\xi = (\hbar^2/2mr)^{1/2}$ is the correlation length of the IBG (here, $\hbar^2/2m$ corresponds to the parameter c in the classical φ^4-model). Another characteristic length of the NBG is the "scattering" length $a_{sc} = mv/4\pi\hbar^2$.

9.3.3 *Renormalization group results*

There are no difficulties in applying the RG recursion method to the model (9.15) in the one-loop (ϵ^1-) approximation; for details, see the papers by De Cesare (1978), and Busiello and De Cesare (1980), where the results correspond to a field version of the quantum XY model (Gerber and Beck, 1977) rather than to the NBG (the condition Eq. (9.28) for the ($\tau \to 0$)-type of summation over the Matsubara frequency of $G_0(q)$ is not fulfilled in these papers and, hence, the correct expression (9.28) for the respective self-energy contribution has not been achieved; see also the calculation in Section E.4.

The first order RG studies confirm the previously obtained results (Hertz, 1976) that the finite temperature critical behaviour ($T_c \neq 0$) of the NBG as well as of a number of systems modelled by Bose fields is universal, i.e., it falls into one of the universality classes (d, n) already known for classical fluctuation Hamiltonians (see Section 9.3.1).

The second important conclusion is concerned with the so-called *quantum* (or *zero-temperature;* $T_c = 0$) *critical behaviour.*[1] For some reason or other the critical temperature T_c in a number of real systems may tend to zero. The point $T_c = 0$, when it exists, is very special because at this point the thermal wave length λ_T is infinite. As $T \to T_c = 0$, $\lambda_T \to \infty$ and this has an effect on the usual critical behaviour. In this situation, the quantum correlations (effects) are essential for the asymptotic scaling behaviour near $T_c = 0$. A thorough RG study of a variety of effective quantum models (including the NBG) has been made by Hertz (1976) who has systematized earlier works (see Section 9.4) and has shown that in the "quantum" limit $(T \to 0)$ the systems exhibit the so-called classical-to-quantum dimensional crossover, i.e., as $T \to 0$ the upper critical dimension d_U corresponding to finite-temperature critical points T_c reduces to $d_U^0 = d_U - z$, where $z > 0$ is the dynamic critical exponent.

Bearing in mind these explanations and using the above mentioned papers the reader can easily derive the RG recursion relations for the model (9.15) to the first order in $\epsilon = d_U - d$. They can be written in the form (Uzunov, 1981a):

$$\omega_l' = b^{2-\eta}\omega_l, \tag{9.21}$$

$$m' = b^{\eta} m, \tag{9.22}$$

$$r' = b^{2-\eta}\left[r + \frac{1}{2}(n+2)\,uI_1(r)\right], \tag{9.23}$$

and

$$u' = b^{(4-d-2\eta)}\left\{u - \frac{1}{2}u^2\left[(n+6)I_2(r) + 2\tilde{I}_2(r)\right]\right\}, \tag{9.24}$$

where $u = (vT)$, $I_1(r)$, $I_2(r)$, and $\tilde{I}_2(r)$ are given by the expressions:

$$I_1(r) = \frac{1}{T}\int_k' n(k), \tag{9.25}$$

[1] The zero temperature critical behaviour may not be quantum in two cases: (i) when we neglect the quantum effects in the zero temperature limit $(T \to 0)$, and (ii) when, as it may happen in some systems, the classical (thermal) effects suppress the quantum correlations (fluctuations).

$$I_2(r) = \left(\frac{1}{2T}\right)^2 \int_k' \sinh^{-2}\left[\frac{\xi(k)}{2T}\right], \tag{9.26}$$

and

$$\tilde{I}_2(r) = \frac{1}{2T}\int_k' \frac{\coth\left[\xi(k)/2T\right]}{\xi(k)}. \tag{9.27}$$

The integrals (9.25)–(9.27) over k in the shell $b^{-1}\Lambda < k < \Lambda$ are obtained after the summation over the internal frequencies ω_l in the corresponding perturbation expressions (the diagrams in Fig. 9.1) has been done.

Note, that the summation over the frequencies ω_l of the internal lines formed by the legs of one and the same Hamiltonian (see Fig. 9.1a) is performed according to the rule:

$$\sum_{\omega_l} e^{i\omega_l \tau} G_0(q). \tag{9.28}$$

with ($\tau = +0$), which yields the correct density of states $n(k)$; see Section E.4. A further summation in Eq. (9.28) over k leads to the term $\sim uI_1$ in Eq. (9.23). The summations over ω_l in the order diagrammatic expressions (those in Fig. 9.1(b) and of higher than second order in n) can be accomplished in the way explained in Section E.4.

The relation (9.22) for the mass m is similar to that for c^{-1} in the classical φ^4-model. The requirement of the invariance of the mass m with respect to the R_b-operation leads to $\eta = 0$ which is a standard result in the one-loop approximation. The RG relation that has no analogue in the classical case is that for ω_l, see Eq. (9.21). Since $\omega_l = 2\pi l T$ we can write Eq. (9.21) by means of the temperature:

$$T' = b^{2-\eta}T \tag{9.29}$$

and, as we shall see, this relation is essential in the description of the critical behaviour in the quantum limit $\{T \sim T_c \to 0, T\xi(k) \gg 1\}$.

In the classical limit $\{T_c \neq 0, T\xi(k) \to 0\}$ the integral I_2 becomes equal to $\tilde{I}_2$; in addition the relation (9.24) naturally turns out an equation for the interaction parameter $u = vT$. Thus one has to analyze the RG equations for r and u which take the form of Eqs. (8.105a)–(8.105c) for the standard φ^4-Hamiltonian (universality). In the classical limit the frequency ω_l (or the temperature T) becomes a redundant variable; the reader should distinguish between $t \sim (T - T_c)$ or r, and T.

When the quantum limit is considered, we obtain

$$T' = b^2 T, \qquad r' = b^2 r, \tag{9.30}$$

$$v' = b^{2-d} v(1 - av), \tag{9.31}$$

where

$$a = \frac{1}{2} \int_k' \frac{1}{\xi(k)}, \tag{9.32}$$

and $\eta = 0$. For $T_c = 0$ and $T \to 0$, $I_1(r)$ and $I_2(r)$ from Eqs. (9.25) and (9.26) are equal to zero, i.e., the diagrams (a)–(e) in Fig. 9.1 do not give contributions to the RG equations. The only nonzero one-loop diagram for $T \to 0$ is the diagram (f) in Fig. 9.1.

As we see the properties of the perturbation terms in the zero temperature limit essentially depend not only on the topology of the corresponding diagrams but also on the orientation of the arrows of the internal G_0-lines. The reason for this simplification of the perturbation series for r' and v' is in the special dependence $G_0(\omega_l, 0) \sim (i\omega_l)^{-1}$ of the bare Green function G_0 of the NBG and the specific form of the Bose distribution $n(k)$.

As it seems at first glance the relevant parameters in the quantum limit are r and T and at the stable FP we have $T^* = r^* = 0$. Including the temperature T as an additional variable in the parameter space of the RG transformation has a simple physical reason. The parameter $r = -\mu > 0$ represents the deviation of the system from the critical point T_c; T_c is defined by $r(T_c) = 0$ whole the explicit dependence $r(T)$ is determined by external conditions; for example, when $\rho = const$ one may expect that $r \sim (T - T_c)^2$ and when $P = const$, $r \sim (T - T_c)$. For $T_c \neq 0$, the temperature T which enters in ω_l is a redundant parameter because it participates in the parameter space of the Hamiltonian through the product $u = vT$ only and therefore, it does not influence the critical thermodynamics.

When $T_c = 0$, the same temperature (or ω_l) can be accepted as the relevant parameter that describes the deviation of the system from the critical temperature ($T_c = 0$). In this case r can be considered as a redundant parameter. Note that within the standard model of the NBG the critical point is approached only through temperature variations. In some effective boson models (Hertz, 1976) r is an effective parameter which can depend on supplementary thermodynamic variables, say, on the variable Y, so that

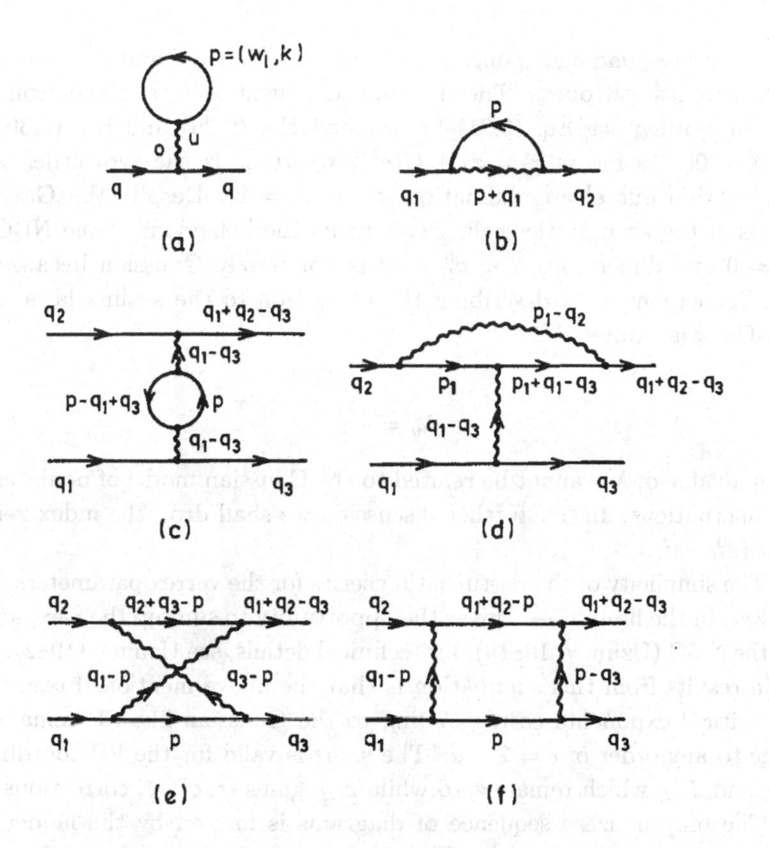

Fig. 9.1 Notation for the (ω_l, k)-dependence in the one-loop diagrams for the NBG; p, p_i-internal $(d+1)$-dimensional vectors, $p = (\omega_l, k)$, $-p = (-\omega_l, -k)$, q and q_i are the external (ω_l, k)-vectors.

the zero temperature critical point $r\,[Y_0, T_c(Y_0) = 0] = 0$ is approached by the variations of both T and Y.

The analysis of Eqs. (9.30)–(9.32) gives an essentially new (*Gaussian-like*) FP with coordinates

$$r_{\rm GL} = T_{\rm GL} = 0, \tag{9.33a}$$

$$v_{GL} = \frac{2\pi\hbar^2}{m}\epsilon_0, \tag{9.33b}$$

where

$$\epsilon_0 = d_U^0 - d > 0, \qquad d_U^0 = 2. \tag{9.34}$$

The corresponding exponents $\nu = 1/\lambda_r = 1/2$, $\eta = 0$ and $z = 2$ coincide with the classical ones. The dynamic exponent $\hat{z}$ is obtained from the rescaling of ω_l; see Eq. (9.21) for ω_l, and Eq. (9.29) and Eq. (9.30) for T; $\eta = 0$. As far as $z = z_0 + O(\epsilon^2)$, where z_0 is the zero order value of $\hat{z}$, within our ϵ^1-approximation, $z = z_0 = 2$. Despite the Gaussian values of the asymptotic scaling exponents the behaviour of the NBG for $T_c = 0$ and dimensions $d < d_U^0 = 2$ is not purely Gaussian because the stability exponent λ_v describing the correction to the scaling laws has a non-Gaussian value,

$$\lambda_v = -\epsilon_0. \tag{9.35}$$

Such a value of λ_v cannot be related to the Gaussian model of noninteracting fluctuations. In our further discussion we shall drop the index zero of $\epsilon_0 = (d_U^0 - d)$.

The simplicity of the perturbation series for the vertex parameters (r, v, $\hbar^2/2m$) in the limit $T \to 0$ gives the opportunity to sum up the ϵ-expansion for the NBG (Uzunov, 1981a); for technical details, see Uzunov (1982). The main results from this summation is that the above mentioned results for the critical exponents corresponding to the Gaussian-like FP remain the same to any order of $\epsilon = 2 - d$. The same is valid for the FP coordinates r_{GL} and T_{GL} which remain zero while v_{GL} gains $\epsilon^2, \epsilon^3, \dots$ corrections.

The only nonzero sequence of diagrams is formed by the ladder diagrams shown in Fig. 9.2(a). The diagrams with the same topology but another orientation of the arrows Figs. 9.2(b–c) give analytical expression that tends to zero as $T \to 0$. These ladder diagrams in Fig. 9.1(a) result in a geometrical progression for v', from which one obtains that the stability exponent $\lambda_u = -\epsilon$ does not acquire $O(\epsilon^2)$-corrections and the $O(\epsilon^2)$ corrections to v^*; ($\epsilon = 2 - d$). Thus the Gaussian-like FP of the NBG describes a non-universal quantum critical behaviour, i.e., a violation of the classical-to-quantum crossover (see Section 9.4).

9.3.4 *Notes*

The dependence of $r = -\mu$ on T or $\Delta T = (T - T_c)$ is determined by the thermodynamic conditions (constraints). For the IBG at $P = const$, $|\mu| \sim (T - T_c)$ and therefore, the model (9.15) will correspond to a usual critical behaviour as given by the classical φ^4-Hamiltonian ($T_c > 0$ for $P = const$ at any $d > 0$). The only difference is that at temperatures,

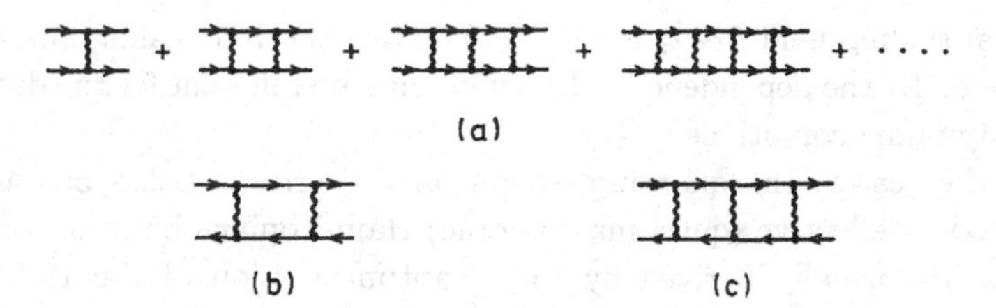

Fig. 9.2 (a) Ladder diagrams for $v'(T \to 0)$. (b) and (c) Ladder diagrams which do not give corrections to v' for $T \to 0$ because of the antiparallel orientation of the G_0-lines.

subject to the requirement

$$\xi \sim (T - T_c)^{-\nu} \lesssim \lambda_T \tag{9.36}$$

the quantum effects will be essential for the thermodynamics. Near T_c the inequality (9.36) is satisfied only if $T_c = 0$.

For the IBG at $\rho = const$, $|\mu| \sim (T - T_c)^{2/(d-2)}$ as it follows from Eqs. (9.4) and (9.5); see also Lacour–Gayet and Toulouse (1974), and Busiello, De Cesare and Uzunov (1985). The critical behaviour of the NBG at constant density for space dimensions $2 < d < 4$ has been discussed by Uzunov and Walasek (1985). The case $d \leq 2$ and the quantum limit $(T \to 0)$ are presented by Shopova and Uzunov (2003a). A large-n limit RG analysis has been performed by De Cesare (1982).

When the energy spectrum is $\varepsilon_k \sim k^\theta$, $0 < \theta \leq 2$, $d_L = \theta$, $d_U = 2\theta$, $d_U^0 = d_U(T \to 0) = \theta$, and the Gaussian-like FP is obtained from the $\epsilon = (\theta - d)$–expansion. The critical exponents for the NBG are $\eta = 2 - \theta$, $\nu = 1/\theta$, $z = \theta$ and $\lambda_v = (d - \theta)$.

The re-derivation of the above mentioned RG results would be a good exercise for the reader.

9.4 Quantum Critical Phenomena and Classical-to-Quantum Crossover

The reader who has already repeated the RG analysis from Section 9.3 (in the one-loop approximation at least) is certainly convinced that the quantum correlations between the bosons have an effect on the critical behaviour of the NBG in a way similar to that known from the IBG in $(d = 3)$ dimensions. The example of the NBG is in some sense special because the RG is applied to the chemical potential $\mu(= -r)$ and not to the usual scaling field $t \sim t_r \sim (T - T_c)$. Thus the RG scaling analysis is determined by the dependence $\mu(T)$ which may be different for the different thermodynamic conditions.

In other cases, the quantum effects on the critical behaviour are investigated by effective (quasi-macroscopic) Hamiltonians built out of Bose fields $\psi_\alpha(q)$, usually derived by the quantum version of the Hubbard–Stratonovich transformation; see, e.g., Stinchcombe (1973), Young (1975), and Lawrie (1978) for the transformation of the quantum Ising model in a transverse field, Hertz (1976) for microscopic models of fermion particles (itinerant ferro- and antiferromagnets and superconductors), Gerber and Beck (1977) for the quantum XY model in a transverse field, Baba, Nagai and Kawasaki (1979) for exciton systems, and Dacol (1980) for spin-1/2 planar Heisenberg ferromagnets.

The RG studies of all effective Hamiltonians of quantum (Bose) fields reveal the universality for critical points $T_c \neq 0$. In the quantum limit ($T \to 0$) for the zero-temperature critical points, the scaling behaviour is described by *the classical-to-quantum dimensional crossover* ($d_U \to d_U^0$); first discussed in the papers by Elliott and Wood (1971), Young (1975), Suzuki (1976), and Hertz (1976).

Following (Hertz, 1976) we shall write down the bare correlation function $G_0(q)$ that covers a number of effective ψ^4-models:

$$G_0^{-1}(q) = -\left[r_0 + ck^\theta + \frac{|\omega_l|^m}{k^{m'}}\right]. \tag{9.37}$$

In Eq. (9.37) $G_0(q)$ is defined by the first equality (9.18) and the parameters r_0, $c > 0$, as well as the exponents $\theta > 0$, $m > 0$ and m' are effective parameters which are usually derived from microscopic fermion and spin models (Hertz, 1976). Note, that in some papers, for example, in Hertz (1976), the correlation function is defined by $G_0(q) = \langle|\psi_\alpha(q)|^2\rangle$ which differs from the definition in Eg. (9.18) by a factor (−1). For the microscopic NBG, the Eq. (9.37) is not valid and the ω_l-dependent term in $G_0^{-1}(q)$ is $i\omega_l$; cf. Eq. (9.18). Note, that in some papers, for example, in Hertz (1976), the correlation function is defined by $G_0(q) = \langle|\psi_\alpha(q)|^2\rangle$ which differs from the definition in Eg. (9.18) by a factor (−1). Then the sign minus on the r.h.s. of Eq. (9.37) should be changed to plus.

The application of the simple S_b-operation to the ψ^4-Hamiltonian with the bare Green function of form (9.37) shows that

$$d_U^0 = \left(2 - \frac{1}{m}\right)\theta - \frac{m'}{m} = 2\theta - z_0, \tag{9.38a}$$

where

$$z_0 = \frac{\theta + m'}{m} \tag{9.38b}$$

is the dynamic critical exponent corresponding to the initial unrenormalized Hamiltonian (tree approximation); note, under renormalization within the loop approximation, the exponent z_0 as well as the values of the exponents θ, m, and m' in the renormalized (true) correlation function $G(q)$ acquire corrections. For example, the bare value z_0 of the dynamic exponent changes to a renormalized ("true") value $\hat{z}$. The correction $(z - z_0)$ is often in two-loop order but in particular cases this correction occurs as a result of one-loop self-energy contributions. For the same reason we may distinguish between the bare values of the exponents θ, m and m' and their renormalized counterparts.

In the quantum limit $(T \to 0)$, the asymptotic critical behaviour of a d-dimensional quantum system is equivalent to that of the corresponding $(d + z)$-dimensional classical system (where $\omega_l = 0$):

$$d \to d + z_0. \tag{9.39}$$

When a d-dimensional $|\psi(x, \tau)|^p$–theory $(p \geq 3)$, for example, a d-dimensional $|\psi(x, \tau)|^4$-theory $(p = 4)$ obeys the classical-to-quantum dimensional crossover (9.39), in the limit $T \to 0$, it describes a critical behaviour which is equivalent to that of the corresponding $(d+z_0)$-dimensional classical $|\psi(x)|^p$-theory; the last one is obtained by setting $\omega_l \equiv 0$ in the $[\psi(q)]^p$-Hamiltonian. In this case we say that the quantum critical phenomena obey a special form of universality (Hertz, 1976); see also Millev and Uzunov (1982). This form of universality is not a general rule and is not valid for all quantum systems. We have shown in Section 9.3 that the NBG does not enter into this scheme owing to the specific dependence of $G_0(q)$ on $i\omega_l$ which leads to unusual properties of the loop expansion for $T \to 0$ (Uzunov, 1981a).

At the end of this section, we shall present some of the numerous studies on quantum critical phenomena:

Basic problems — See the $(1/n)$-expansion performed by Abe (1974), Abe and Hikami (1974a,b), and Kondor and Szépfalusy (1974), the ϵ-expansion for the dynamic exponent $\hat{z}$ (Suzuki and Igarashi, 1974), the study of the classical-to-quantum crossover (Green, Sneddon, and Stinchcombe, 1979; Zannetti, 1980; Ruggiero and Zannetti, 1981, 1983).

Interacting Bose fluids — See Toyoda (1982), Hohenberg (1982), Creswick and Wiegel (1983), Walasek (1984), Olinto (1985, 1986), Weichman *et al.* (1986), and Fisher *et al.* (1989); for the critical dynamics of Bose fluids (including the λ-point in ^{4}He) see, e.g., (Suzuki, 1975), Ferrell and Bhattacharjee (1979), Dohm and Folk (1982) (and the references therein). Quantum Monte Carlo techniques have been used by Batrouni, Scalettar and Zimanyi (1990) to investigate 1D Bose fluids; recent studies of effects of disorder (see Section 9.5) in Bose fluids have been performed by Ma, Halperin and Lee (1986), Fisher and Fisher (1988), Weichman and Kim (1989), Schakel (1997); for experiments on disorder effects in ^{4}He see Kotsubo and Williams (1986), Chan *et al.* (1988), and Finotello *et al.* (1988). For dynamic effects in ^{3}He-^{4}He mixtures see (Dohm and Folk , 1983), and Moser and Folk (1990, 1991).

Magnetic models — See Goldhirsch (1979), Lewis (1979), Kopeć and Kozlowski (1983a), Lukierska–Walasek and Walasek (1981), Lukierska–Walasek (1983), Chubukov (1984), and Iro (1987).

Structural phase transitions — See Oppermann and Thomas (1975), Holz and Medeiros (1975), Beck, Schneider and Stoll (1975), Morf, Schneider and Stoll (1977), Millev and Uzunov (1982, 1983), and Uzunov (1983).

Problems of quantum phase transitions are widely discussed in the books by Continentino (2001), Herbut (2007), Sachdev (1999), and Schakel (2008) as well as in review articles by Vojta (2003), and Shopova and Uzunov (2003a).

9.5 Disorder Effects

The disorder which is always present in real systems may essentially influence the phase transition properties. There can be different sources of disorder. Here we shall briefly discuss the so-called *quenched disorder* caused by the nonequilibrium (irregular) distribution of impurities and defects in solid medium.

The main features of this type of disorder is that the characteristic relaxation time of the defects in the crystal lattice is much larger than the time, for which the experiment is performed. One example is the window glass — a thermodynamic nonequilibrium, i.e., quenched state of SiO_2, the crystal (equilibrium) form of which at room temperature is the quartz. Unlike the window glass which is an example of a *strong disorder* we shall assume that the disorder is weak, i.e., the medium is almost perfect. However, near the critical point the perfect (pure) system is unstable towards any perturbations and it may be expected that even *the weak disorder* will, under certain conditions, affect the critical behaviour. The way, in which the quenched disorder acts is thoroughly described in the book of Ma (1976a). Here we shall briefly acquaint the reader with two types of quenched disorder: (i) *random impurities* (or, for reasons, which become clear below, it is sometimes called disorder of type "random critical temperature"), and (ii) *random fields.*

9.5.1 *Random critical temperature*

Consider the Hamiltonian

$$\mathcal{H} = \frac{1}{2}\int d^dx \left\{ r(x)\varphi^2(x) + c(x)\left[\nabla\varphi(x)\right]^2 + 2u(x)\varphi^4(x) \right\}, \tag{9.40}$$

where

$$r(x) = r_0 + \varphi_r(x), \tag{9.41a}$$

$$c(x) = c + \varphi_c(x), \tag{9.41b}$$

and

$$u(x) = u + \varphi_u(x). \tag{9.41c}$$

In the above expressions, $\varphi_i(x)$, $i = (r, c, u)$ are random functions, which obey, for example, the Gaussian distribution

$$P(\varphi_i) = \exp\left\{ -\frac{1}{2}\sum_{ij}\frac{1}{\Delta_{ij}}\int d^dx\varphi_i(x)\varphi_j(x) \right\}, \tag{9.42}$$

$\Delta_{ij} = \Delta_{ji}$ are parameters describing the disorder ($\Delta_{ii} \equiv \Delta_i$).

Obviously the Hamiltonian (9.40)–(9.42) is a generalization of the usual φ^4-Hamiltonian (the corresponding pure system). The random functions (i.e., the random "potentials" taking part in $\mathcal{H}$) with the Gaussian distribution (9.42) which yields the Gaussian averages $\langle \cdots \rangle_R$ over the disorder, $\langle \varphi_i \rangle_R = 0$, and

$$\langle \varphi_i(x) \varphi_j(x') \rangle_R = \Delta_{ij} \delta(x - x'), \tag{9.43}$$

describe the disordered (impure) system.

The model (9.40)–(9.42) can be introduced by both phenomenological and microscopic arguments. The microscopic justification of the fluctuation Hamiltonian (9.40) is rather straightforward and we shall briefly discuss this point. If the exchange matrix $\widehat{J} = \{J_{ij}\}$ of a lattice model (see Sections 5.6 and 5.7) is not a regular function of the sites i and j, or in translationally invariant systems — of $|i - j|$, this means that the inter-site interaction J_{ij} will not be uniform but rather it will be expressed in a complicated way through the coordinates i and j of the sites in the lattice. Certainly, this can be caused by quenched impurities or another quenched defects (inhomogeneities). The exchange constants J_{ij} can be represented as a sum,

$$J_{ij} = J_{ij}^0 + \delta J_{ij} \tag{9.44}$$

of the regular $J_{ij}^0 = J^0(|i - j|)$ and an irregular (δJ_{ij}) parts (*random exchange model*).

The irregular part (δJ_{ij}) describing the quenched disorder is supposed unfamiliar or too complicated, so, instead of the function δJ_{ij} itself we shall introduce its random (Gaussian) distribution. Using the Hubbard–Stratonovich transformation (Section 6.6) of such a lattice model of disorder one can prove that the corresponding fluctuation field Hamiltonian will be of the form (9.40) where the random functions $\varphi_i(x)$ and their distribution parameters Δ_{ij} are connected in a rather complex way with the irregular part δJ_{ij} of the microscopic model and its random distribution parameters say, $\Delta_{ij}^{(J)}$.

One of the first studies of Hamiltonians like (9.40) is that of Larkin (1970), and Larkin and Ovchinnikov (1971) connected with the influence of the inhomogeneities on the thermodynamic properties of the superconductors. The simple dimensional analysis (the S_b-operation) or the lowest order perturbation calculations show that the random functions φ_c and φ_u are irrelevant near T_c, therefore, only one random function, $\varphi_r(x)$, is used in the consideration of the phase transition properties. In fact, Eq. (9.41a) for $r(x)$ with $r_0 = \alpha_0(T - T_c^0)$ can be written in the form

$$r(x) = \alpha_0 \left[T - T_c(x) \right]. \qquad (9.45)$$

In this equality,

$$T_c(x) = T_c^0 - \frac{1}{\alpha_0} \varphi_r(x) \qquad (9.46)$$

is the random (x-dependent) critical temperature. It is quite plausible that if the microscopic interaction (J_{ij}) responsible for the phase transition in a disordered medium varies with x, the critical temperature which is directly connected with the interaction strength will also vary from site to site in a random way. Thus the mean (averaged) over the impurities critical temperature $\langle T_c(x) \rangle_R$ will be equal to T_c^0 because $\langle \varphi_r(x) \rangle_R = 0$ according to the Gaussian distribution. This result is valid with respect to the bare critical temperature T_c^0. As far as the true critical temperature T_c is concerned it will depend on $\langle \varphi_r(x) \varphi_r(x') \rangle_R$; see also Section 9.5.5. Non-Gaussian distributions of J_{ij} and the random potentials φ_i can also be introduced.

In consequence, it is evident from the above discussion that the "random T_c"-φ^4-Hamiltonian will be

$$\mathcal{H} = \frac{1}{2} \int d^d x \left\{ r(x) \varphi^2(x) + c \left[\nabla \varphi(x) \right]^2 + 2u\varphi^4(x) \right\}; \qquad (9.47)$$

here $r(x)$ is given by Eq. (9.41a) and $\varphi_r(x)$ obeys the Gaussian distribution (9.43). The "random T_c" model. (9.47) has been widely investigated by the RG method (Grinstein, 1974; Lubensky, 1975; Khmel'nitskii, 1975; Grinstein and Luther , 1976); the same subject is reviewed by Ma (1976a), Grinstein (1985), Hertz (1985), and Lubensky (1979).

The "random T_c" term $\varphi_r \varphi^2$ in Eq. (9.47) causes the appearance of a new "random" FP (RFP) which is stable for $n > 4$. The asymptotic critical behaviour of random impurity systems with symmetry index $n < 4$ is given by the RFP rather than the usual HFP. This picture is consistent with the Harris (1974) criterion for the possibility of a sharp (i.e., usual) second order transition in systems with disorder (for a simple heuristic derivation of the Harris criterion see Grinstein (1985), and Hertz (1985). The "random T_c" isotropic model has been investigated by the RG method up to the two-loop approximation by (Jug, 1983, 1984) and the three-loop approximation by Mayer and Sokolov (1985). Another model of random impurities specially intended to describe impure magnets with the so-called "random axis" (*the random axis model*) has been analyzed by Mukamel and Grinstein (1982).

9.5.2 *Extended impurities and long-range random correlations*

Let the impurities be distributed in a regular way in some direction in the dD space of the system and randomly distributed in the other directions. For example, the random bonds in the 2D Ising model can be along one of the directions; in Fig. 9.3 this is the vertical direction. This is the McCoy-Wu model of bond disorder in 2D regular Ising lattices (McCoy and Wu, 1968, 1969) and describes the so-called *smeared* phase transition without the typical divergencies for a standard second order phase transition. Another remarkable feature of this model is that it, like the 2D Ising model with regular exchange, is exactly solvable (McCoy and Wu, 1968, 1969).

This picture of extended impurities has been generalized for linear or planar impurities in the dD space. Moreover, it has been suggested by Dorogovtsev (1980), Dorogovtsev(1981) and Boyanovsky and Cardy (1982a) a further generalization of the McCoy–Wu problem according to which the random bonds extend over ϵ_d dimensions, where $\epsilon_d < d$ need not be unity, two, or even an integer. The bonds along the remaining $(d-\epsilon_d)$ dimensions are completely correlated (as they are along the x-axis in Fig. 9.3). For such a "random T_c" φ^4-theory the upper critical dimension is $d = 4 + \epsilon_d$.

The ϵ-expansion about this dimension yields no stable FP for any $\epsilon_d > 0$; see Lubensky (1975), and Stolan, Pytte and Grinstein (1984). However, a double RG expansion in both $\epsilon = 4 - d$ and $\epsilon_d (\ll 1)$ gives a stable FP indicating a sharp (non-smeared) second order transition Dorogovtsev (1980, 1981).

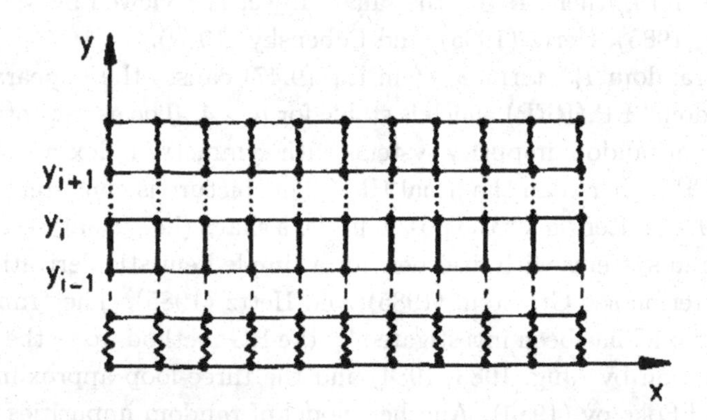

Fig. 9.3 An illustration of the McCoy–Wu model of disorder. The vertical bonds vary because of the disorder along the axis y; $J(y_{i+1} - y_i) \neq J(y_{j+1} - y_j)$ for $i \neq j$.

Therefore, in the case of *extended* (line, plane) impurities one should expect a sharp second order phase transition inside the critical region of the system provided the double (ϵ, ϵ_d)-expansion is well justified.

The (point) impurities which produce a distribution like (9.43) in the continuum limit are called short-range or even extremely short-range (δ-correlated) random impurities. When point impurities with long-range Gaussian correlation are investigated within the "random T_c" model one has to replace the relation (9.43) for φ_r by the more general expression

$$\langle \varphi_r(x)\varphi_r(x')\rangle_R = \frac{\Delta_r}{|x-x'|^a}, \qquad a > 0, \tag{9.48}$$

which in the k-space yields

$$\langle |\varphi_r(k)|^2\rangle_R = \Delta_0 + \Delta_1 k^{a-d}; \tag{9.49}$$

see also Section D.2. For $a = 0$, $\Delta_r = 0$; the random correlations (9.48)–(9.49) coincide with the δ-type correlations (9.43). The "δ–type" random impurities are sometimes called uncorrelated impurities or impurities correlated at a single ($x = x'$) point. The case $a > d$ corresponds to the so-called *short-range impurities.* It is obvious from Eq. (9.49) that the Δ_1-term vanishes in the infrared limit ($k \to 0$) and, therefore impurities will give only correlations to the main scaling behaviour of the corresponding system with δ-type random correlations. The effect of the *long-range random correlations* on the asymptotic critical behaviour is clearly manifested when $0 < a < d$. This effect has been investigated by Weinrib and Halperin (1983) by a double (ϵ, δ)-expansion, where $\epsilon = 4 - d$ and the new expansion parameter $\delta = \epsilon + (d - a)$ are also supposed small; $(d - a) \ll 1$; (for the calculation of the critical exponent η following this scheme; see Korutcheva and Uzunov (1984a). The main conclusion from the RG study of the long-range "random T_c" φ^4-model is that the RG equations have a new stable fixed point describing this type disorder and thus implying a sharp second order transition similar to that predicted by Dorogovtsev (1980).

The dynamic critical behaviour of the "random T_c" φ^4-model has been studied by Grinstein, Ma, and Mazenko (1977). The instability of the zero-temperature quantum critical behaviour in systems with quenched impurities has been established by Korutcheva and Uzunov (1984b), Busiello, De Cesare and Rabuffo (1984, 1985); see also Busiello and Uzunov (1987a), and the review by De Cesare (1987). The random T_c effects on the tricritical behaviour have been considered for the first time by Busiello, De Cesare and Uzunov (1984).

9.5.3 *Random field*

Another type of disorder is described by random field (RF) models. The RF φ^4-effective Hamiltonian will be

$$\mathcal{H}(\varphi) = -\int d^d x h(x)\varphi(x) + \mathcal{H}_4(\varphi), \tag{9.50}$$

where $\mathcal{H}_4(\varphi)$ is the usual φ^4-model of the system in a zero external field. The Hamiltonian (9.50) looks like that of a system in a nonuniform external field. The difference consists of the following: when $h(x) = \{h_\alpha(x)\}$ in Eq. (9.50) is an external nonuniform field, the equilibrium potential Φ and other thermodynamic quantities will be functionals of $h(x)$, while when $h(x)$ is a random field, the equilibrium potential Φ is obtained by a subsequent averaging over the (Gaussian) distribution of the random functions $h_\alpha(x)$:

$$\langle h_\alpha(x)h)_\beta(x')\rangle_R = \delta_{\alpha\beta} h_0^2 g_h(x, x'). \tag{9.51}$$

The dimensionless function g_h describes the shape of the Gaussian distribution and h_0^2 is a positive constant which gives the random-field strength (like Δ_r for the "random T_c" model). Again one may consider δ-type random correlations, $g_h = \delta(x - x')$, short-range RF correlations

$$g_h = \frac{1}{R^a}, \qquad R = |x - x'|, \tag{9.52}$$

with $a > d$, and long-range RF correlations $(0 < a < d)$ which describe the qualitatively different scaling behaviour. Eq. (9.51) with g_h from Eq. (9.52) will have the following asymptotic $(k \sim 0)$ form

$$\langle |h_\alpha(k)|^2\rangle_R \sim \begin{cases} k^{-(d-a)} & d \geq a, \\ 1 & d < a. \end{cases} \tag{9.53}$$

The RF effects on the critical behaviour of systems described by the φ^4-model has first been studied by Imry and Ma (1975), Grinstein (1976), Aharony, Imry and Ma (1976), and Young (1977); see also the review articles by Imry (1984), Parisi (1984), Grinstein (1985), and Hertz (1985).

The main result from the action of RFs on the critical behaviour is the so-called *RF dimensional crossover.* The upper critical dimension $d_U = 4$ of the pure φ^4-Hamiltonian changes to $d_U^{RF} = d_U + 2 + (d - a)$ or $d_U^{RF} = d_U + \theta + (d - a)$ when the energy spectrum of the fluctuations is $\varepsilon_k \sim k^\theta$ $(0 < \theta \leq 2)$. The asymptotic critical behaviour of a dD pure system will be then equivalent to that of the corresponding $[d + \theta + (d - a)]$-dimensional RF system.

For δ-type random field correlations, one has to use the second line on the *r.h.s.* of the relation (9.53); hence, $d_U^{RF} = d_U + \theta$. Then the RF dimensional crossover is given by $d \to (d - \theta)$, i.e., a dD system with δ-type RF correlations behaves like a pure system in $(d - \theta)$ dimensions. If the RF system is 3D and $\theta = 2$, the corresponding pure system will be described by the 1D φ^4-Hamiltonian and, therefore, no phase transition will occur for any $n \geq 1$. The phase transition is destroyed by the simultaneous action of the thermal fluctuations and the RF effects.

The RF lattice model

$$\mathcal{H} = -\frac{J}{2} \sum_{<ij>} \sigma_i \cdot \sigma_j - \sum_i h_i \cdot \sigma_i \tag{9.54}$$

and its field counterpart (9.50) in the continuum limit can be used to simulate the behaviour of a number of real systems, for example: (i) random exchange antiferromagnets in uniform magnetic field (Fishman and Aharony, 1979), (ii) charge density waves randomly pinned by impurities (Moncton *et al.*, 1976), and (iii) gel systems (Brochard and de Gennes, 1983).

The experimental observations, in particular, on random exchange antiferromagnets and the experimental verification of the RF crossover are discussed in detail by Grinstein (1985); see also Hagen *et al.* (1983), Belanger *et al.* (1983), Birgeneau *el al.* (1983), and King *et al* (1988). Non-perturbative studies of the RF problem can be found in the papers, for example, by Shapir (1985), Schwartz (1985), Schwartz and Soffer (1985). A comprehensive exposition of the RF effects is presented by Fisher (1985).

The simultaneous effect of the classical-to-quantum and the RF crossovers on the critical behaviour has been investigated by Aharony, Gefen and Shapir (1982), Busiello, De Cesare and Rabuffo (1983), and Uzunov, Korutcheva and Millev (1985a,b). The low-temperature behaviour of RF Ising systems has been investigated by Bray (1983). The RF Potts model is discussed by Blankschtein, Shapir and Aharony (1984) and the MF analysis of the RF Blume–Capel model has been performed by Kaufman and Kanner (1990).

9.5.4 *Technical remarks*

The results presented above can be derived from the standard perturbation technique and the ϵ-expansion within the RG approach. For this purpose one may treat the terms $\varphi_r \varphi^2$, $\varphi_c \varphi^2$, and $\varphi_u \varphi^4$, in (9.40), and $h\varphi$ in Eq. (9.50) as perturbations (see Ma, 1976a) or, equivalently, to use a modified perturbation theory based on the replica trick (for the replica trick, see Edwards and Anderson (1975), Imry (1984), Hertz (1985), Uzunov (1987)). The exercises in the next subsection are a good introduction to the lowest order perturbation theory of "random T_c" and RF models.

9.5.5 *Ginzburg criterion*

Now we shall derive the fluctuation shift of T_c and the Ginzburg criterion (see Section 7.5.2 and Section 7.11.1) from the perturbation expansion for the model (9.40) to the first order in Δ_r, Δ_c and u; the parameter Δ_u can be neglected. For this aim we have to calculate the Green function $G(k, r; \varphi_r, \varphi_c) = \langle|\varphi_\alpha(k)|\rangle$ from the Dyson equation (7.171) as a functional of the random functions $\varphi_r(k)$ and $\varphi_r(k)$ to the order $O(\varphi_i^2)$; $i \equiv r, c$. The averaging of the Dyson equation over the disorder, $\langle(\cdots)\rangle_R$, yields the equation for the averaged Green function

$$\bar{G}(k, r) = \langle G(k, r; \varphi_r, \varphi_c)\rangle_R.$$

The corresponding averaged self-energy function will be

$$\Sigma(k, r) = [\Delta_r - (n+2)u]\, I_1(r) - \Delta_c J_c(k, r), \tag{9.55}$$

where $I_1(r)$ is given by Eq. (7.162) and

$$J_c(k, r) = \int_p \frac{(k \cdot p)^2}{(cp^2 + r)}, \qquad 0 < |p| < \Lambda; \tag{9.56}$$

$r = \alpha_0(T - T_c)$ is the renormalized value of r_0. Using the above calculation, one obtains (Uzunov, 1986) the shift of the critical temperature

$$T_c = T_c^0 - \frac{K_d}{(d-2)} \left(\frac{\Lambda^{d-2}}{\alpha_0 c}\right) [(n+2)u - \Delta_r], \tag{9.57}$$

and the critical region is defined by the inequalities

$$r < \frac{\pi K_d}{2c^{d/2}\sin[\pi(d-2)/2]}\left[(n+2)u - \Delta_r\right]^{2/(4-d)}, \tag{9.58}$$

and

$$c^2 > \frac{K_d\Gamma\left(\frac{d-1}{2}\right)\Lambda^d}{2\pi^{1/2}d^2\Gamma(d/2)} f_d(\xi\Lambda)\Delta_c. \tag{9.59}$$

Here

$$f_d(\xi\Lambda) = 1 - \frac{d}{d-2}(\xi\lambda)^{-2} - \frac{\pi d}{2\sin(\pi d/2)}(\xi\Lambda)^{-d}, \tag{9.60}$$

and $\xi = (c/r)^{1/2}$ is the correlation length; near T_c, $(\xi\Lambda) \gg 1$ so that $f_d(\xi\Lambda) \sim 1$. In order to obtain the above results one must perform the integration in I_1 and J_c for an infinite cutoff ($\Lambda = \infty$) in the cases when this does not introduce divergences.

Another useful exercise is the derivation and the analysis of the Ginzburg criterion for the RF model (9.50); for details, see Kaufman and Kardar (1985), and Busiello and Uzunov (1987b).

9.6 Anisotropy, Coupled Order Parameters and Multicritical Behaviour

9.6.1 *Anisotropic systems*

The basic types of anisotropy that are considered in the theory of phase transitions are:

(1) Spatial anisotropy
(2) Quadratic anisotropy
(3) Cubic anisotropy.

If all (1)–(3) anisotropies are introduced in the effective field φ^4-Hamiltonian, it will take the form

$$\mathcal{H} = \frac{1}{2}\int d^d x\left\{\sum_{i=1}^{d} c_i(\nabla_i\varphi)^2 + \sum_{\alpha=1}^{n} r_\alpha\varphi_\alpha^2 + 2u\varphi^4 + 2v\sum_{\alpha=1}^{n}\varphi_\alpha^4\right\}. \tag{9.61}$$

In Eq. (9.61), $c_i \neq c_j$ $(i, j = 1, \ldots, d)$, $r_\alpha \neq r_\beta$, $\alpha, \beta = (1, \ldots, n)$ and $\nabla_i \equiv \partial/\partial x_i$. However one can consider particular cases, in which some of the parameters c_i or r_α coincide.

When $c_i = c_j \equiv c$, and $r_\alpha = r$, the fluctuation Hamiltonian (9.61) describes the cubic anisotropy of the crystal (iii) or, as it is sometimes called, quartic anisotropy; see the quartic Landau invariants (the v-terms) in Eq. (9.61). In Section 4.11.2 this anisotropy has been briefly discussed within the Landau expansion of a x-dependent order parameter. Of course, the crystal symmetry may be different from the simple cubic symmetry, and then one should use other quartic Landau invariants.

The pure spatial anisotropy (i) is described by $c_i \equiv c$ for any $v = 0$; see also Section 3.8.3. The pure quadratic anisotropy (ii) is given by $r_\alpha \equiv r$ and $v = 0$. For some systems one ought to consider the model (9.61) for $r_1 = \cdots = r_m = r^{(1)}$ and $r_{m+1} = \cdots = r_n = r^{(2)}$, $(1 \leq m < n)$. Then the quadratic anisotropy parameters in the n-dimensional space of the order parameter components φ_α are only two: $r^{(1)}$ and $r^{(2)}$. The same type of equalities may take place for the parameters c_i and this is one of the most frequently met examples of the spatial anisotropy. Then $c_1 = \cdots = c_m = c^{(1)}$ and $c_{m+1} = \cdots = c_d \equiv c^{(2)}$, $(1 \leq m < d)$, and choosing one of the axes $\hat{x}_1, \ldots, \hat{x}_m$ in the dD space, say, the axis $\hat{x}_1$, the anisotropy will be along the $(m+1), \ldots, d$ axes perpendicular to the axis $\hat{x}_1$. In the particular case of $m = 1$ and $v = 0$ we can denote $c^{(1)} = c_\parallel$ and $c_2 = \cdots = c_d = c_\perp$ so that one has the parameters $c_\parallel$ and $c_\perp$ of the well-known special case of spatial anisotropy, the planar anisotropy; see also Section 9.7.3; an example is presented in Section F on the basis of bct lattice.

The spatial anisotropy (i) stands for the anisotropy of the correlation length ξ. So it can be given by the "correlation length vector" $\xi = (\xi_1, \ldots, \xi_d)$, where $\xi_i = (c_i/r)^{1/2}$; $r_\alpha \equiv r$. One may introduce the effective "mass" tensor $\hat{m} = \{m_{ij}\}$ with $m_{ij} = \delta_{ij} m_i$ and eigenvalues $m_i = 1/c_i$ so that the k-representation of the gradient term in Eq. (9.61) becomes

$$\sum_{i,k} \frac{k_i^2}{2m_i}. \tag{9.62}$$

Near the critical point $(r \sim 0)$ all components ξ_i of the correlation length vector $\xi = \{\xi_i\}$ are very large $(c_i \neq 0)$. Far from T_c, the fluctuations of the order parameter along a spatial direction, say, j can be neglected provided $c_j \gg c_i$ for all $i \neq j$.

In contrast to the spatial anisotropy, the quadratic and cubic anisotropies refer to the $(n \geq 2)$–dimensional space of the order parameter $\varphi = \{\varphi_\alpha\}$; for a scalar field φ these anisotropies do not exist. As long as $c_i > 0$, the spatial anisotropy in Eq. (9.61) is related with the fluctuation effects; the mean field ordering $\varphi \equiv \bar{\varphi}$ for $r_\alpha = r$ and $v = 0$ will be x-independent. On the contrary, the anisotropies (ii) and (iii) can be described within the MF approximation of a uniform order parameter.

Another type of spatial anisotropy may arise from the different scale of the interactions along the different axes. For example, the interactions along the fist m axes, $\hat{x}_1, \dots, \hat{x}_m$, may be of short range type, and along the other directions of space — of long range type. Then the respective wave vector components k_i will enter in the Hamiltonian in different ways: $c^{(1)}(k_1^2 + \dots + k_m^2)$, and $c^{(2)}(k_{m+1}^\theta + \dots + k_d^\theta)$, where $0 < \theta < 2$ (Section 6.7).

Various types of anisotropy in the φ^4-term of the effective Hamiltonian can also be considered depending on the particular crystal symmetry of interest; see, e.g., Cowley (1980), Gufan (1982). Below we shall mentioned several results connected with the quadratic and cubic, and orthorhombic anisotropies.

9.6.2 *Mean-field analysis of anisotropic systems*

Neglecting the gradient terms in Eq. (9.61) we have

$$\phi = \frac{1}{2} \sum_{\alpha=1}^{n} r_\alpha \varphi_\alpha^2 + u\varphi^4 + v \sum_{\alpha=1}^{n} \varphi_\alpha^4, \tag{9.63}$$

where $\phi = \Phi/V$, and φ_α is the uniform order parameter. The n equations

$$\left(\frac{\partial \Phi}{\partial \varphi_\alpha}\right)_{\bar{\varphi}_\alpha} = 0, \tag{9.64}$$

which determine the equilibrium order parameter components $\bar{\varphi}_\alpha$ (denoted below by φ_α) are

$$(r_\alpha + 4u\varphi^2 + 4v\varphi_\alpha^2)\varphi_\alpha = 0. \tag{9.65}$$

The solutions of Eq. (9.65) are the possible equilibrium phases. But the latter should be investigated with respect to their stability. The sufficient, but not necessary, condition for the stability of the phases, i.e., the solution of Eqs. (9.65) is given by the requirement for the positive definiteness of the matrix (the inverse susceptibility tensor):

$$\hat{a} = \{a_{\alpha\beta}\} \equiv \left(\frac{\partial^2 \phi}{\partial\varphi_\alpha \partial\varphi_\beta}\right). \tag{9.66}$$

For the model (9.63) the elements of $\hat{a}$ are

$$a_{\alpha\beta} = (r_\alpha + 4u\varphi^2 + 12v\varphi_\alpha^2)\delta_{\alpha\beta} + 8u\varphi_\alpha\varphi_\beta, \tag{9.67}$$

As we see, the values of $a_{\alpha\beta}$ depend on the particular solution of Eq. (9.65), i.e., these values vary with the change of the phases. Therefore the stability conditions for the various phases will different for the different phases. IN order to perform the stability analysis we recall that from a pure mathematical point of view the sufficient condition of stability is equivalent to the requirement that all major minors of the matrix $\hat{a}$ are positive, i.e.,

$$a_{11} > 0, \qquad \begin{pmatrix} a_{11} \; a_{12} \\ a_{21} \; a_{22} \end{pmatrix} > 0, \ldots, \tag{9.68a}$$

or, generally, for the mth major minor,

$$\begin{pmatrix} a_{11} & \cdots & a_{1m} \\ a_{21} & \cdots & a_{2m} \\ \vdots & \ddots & \vdots \\ a_{m1} & \cdots & a_{mm} \end{pmatrix} > 0, \qquad 1 \leq m \leq n. \tag{9.68b}$$

When the Landau expansion is restricted to terms of quartic order in φ_α, like Eq. (9.63), but terms $\sim O(\varphi_\alpha^5)$ are not present the matrix $\hat{a}$ is a quadratic form of φ_α. So the requirements (9.68a)–(9.68b) are reduced to the conditions for the positive definiteness of this quadratic form (the inverse susceptibilities of the possible phases; cf. Eqs. (4.128) and $\Phi = F$ for $h = 0$).

For any solution $\bar{\varphi} = \{\bar{\varphi}_\alpha\}$ of Eqs. (9.65) the form (9.66) will be positively definite for different values of the parameters $(r, u, v\ )$ of the model. Since $r, u,$ and v are connected with the thermodynamic parameters $(T, Y, \ldots)$ the phases will have different regions of stability in the (T, Y) phase diagram of the system. It may happen that two or more phases are stable in the same domain in the (r, u, v)-space or, their stability domains have a subdomain of overlap. Then one has to compare the minima $\phi_i[\varphi^{(i)}]$ for this overlapping stable or metastable phases $\bar{\varphi}^{(i)}$, $(i = 1, \ldots)$, and to find the absolute minimum $\varphi^{(j)}$ (the stable phase): $\phi_j = \min\left\{\phi_i\left[\varphi^{(i)}\right]\right\}$. Thus one of the minima, $\varphi^{(j)}$ of ϕ will describe the stable phase and the other minima will describe the metastable phases (this occurs near the first-order transitions; see also Chapter 4).

When one (or more) of the inequalities (9.68a)–(9.68b) becomes equality one has to continue the analysis to the higher-order invariants ($\sim \varphi^6, \dots$) as the phase stability theory implies. These remarks are a supplement to the MF analysis within the Landau expansion discussed in Chapter 4.

After the above explanations the MF analysis of the expansion (9.63) is a relatively simple (although lengthy) mathematical calculus. It can be straightforwardly performed at least for $n = 2, 3$ and 4, but it seems more convenient at first to restrict ourselves to the case of the pure cubic anisotropy ($r_\alpha = r$).

9.6.3 *Cubic anisotropy in mean-field approximation*

It will be quite instructive for the reader to obtain the phases described by the model (9.63) with $r_\alpha \equiv r$ and, say $n = 3$, together with their stability properties. The possible phase are found from Eqs. (9.65) for $r_\alpha = r$ and $n = 3$. They are

$$(1) \qquad \varphi_1 = \varphi_2 = \varphi_3 = 0, \tag{9.69a}$$

$$(2) \qquad \varphi_1^2 = -\frac{r}{4(u+v)}, \quad \varphi_2 = \varphi_3 = 0, \tag{9.69b}$$

$$(3) \qquad \varphi_1^2 = \varphi_2^2 = -\frac{r}{4(2u+v)}, \quad \varphi_3 = 0, \tag{9.69c}$$

$$(4) \qquad \varphi_1^2 = \varphi_2^2 = \varphi_3^2 = -\frac{r}{4(3u+v)}. \tag{9.69d}$$

The symmetry of the model (9.63) is such that the states like $(\varphi_1)_\pm = \pm\left[-r/4(u+v)\right]^{1/2}$, see the phase (2), cannot be distinguished and, therefore, they belong to the same phase 2. The model (9.63) is invariant with respect to a change of the indices α, ($\alpha_1 \leftrightarrow \alpha_2$), so that the state

$$\varphi_1 = \varphi_2 = 0, \quad \varphi_3 = -\frac{r}{4(u+v)}, \tag{9.70}$$

which also exists as a solution of the Eqs. (9.65), belongs to the phase 2. A similar argument can be used for the phase 3 where Eqs. (9.65) have solutions ($\varphi_1 = 0$, $\varphi_2 = \varphi_3 \neq 0$, $\varphi_1 = \varphi_3 \neq 0$, $\varphi_2 = 0$).

When two (or more) solutions, like $(\varphi_1)_+$ and $(\varphi_1)_-$, given by Eq. (9.69b) and $(\varphi_3)_+$, and $(\varphi_3)_-$, given by Eq. (9.70) for the phase 2 as well as the above mentioned three types of solutions for the phase 3 correspond to one and the same value of the thermodynamic potential Φ (or ϕ) for the same values of the parameters (r, u, v) we say that these solutions belong to one and the same phase. Solutions like $(\varphi_1)_+$ and $(\varphi_1)_-$, given by Eq. (9.69b) and/or $(\varphi_3)_+$ and $(\varphi_3)_-$, given by and Eq. (9.70) are sometimes called "domains" or "configurations" of the same phase(2). They can be distinguished when an additional breaking of the symmetry of the model, say, an external force (field) is present. In the case of a local (x-dependent) breaking of the symmetry, say, by nonuniform external fields or inhomogeneities, the various "domains" can occur in the same sample.

The potential density ϕ corresponding to the phases 1–4 is $\phi_1 = 0$, and

$$\phi_2 = -\frac{r^2}{16(u+v)}, \tag{9.71a}$$

$$\phi_3 = -\frac{r^2}{8(2u+v)}, \tag{9.71b}$$

$$\phi_4 = -\frac{3r^2}{16(3u+v)}. \tag{9.71c}$$

Clearly, the considered case of $(n = 3)$–systems includes the $(n = 2)$–systems; one needs only to neglect the φ_3-component and the phase 4 as impossible in two-component systems. The phases can be symbolically written as vectors $(\varphi_1, \varphi_2, \varphi_3)$. For example, (para)phase 1 is $(0,0,0)$, phase 2 is $(\pm 1, 0, 0)$ or, equivalently $(0, \pm 1, 0)$ and $(0, 0, \pm 1)$ or, shortly, $(1, 0, 0)$, phase 3 is given by $(11, 0)$ and phase 4 is $(1, 1, 1)$.

Let us consider the stability properties of the phases 1–3 for $(n = 2)$–systems. Solutions 1–3 of Eqs. (9.65) for $n = 2$ and $r_\alpha = r$ describe minima of ϕ in the following domains of the parameter space (r, u, v):

$$\begin{aligned}
&(1) \quad r \geq 0, \\
&(2) \quad r < 0, \quad v < 0, \quad (u+v) > 0, \\
&(3) \quad r < 0, \quad v > 0, \quad (v + 2u) > 0.
\end{aligned} \tag{9.72}$$

Since no overlap of the domains 1–3 of the minima of the phases 1–3 occurs there is no need to compare their free energies. The disordered (para) phase is stable for any $r \geq 0$, u and v. The ordered phases 2 and 3 are stable for $r < 0$ but in different domains of the (u, v) plane; see Fig. 9.4(a).

This picture changes for $(n = 3)$–systems. In this case the phase $(1, 1, 0)$ has no domain of stability. The phase $(1, 1, 1)$ gives a minimum of ϕ for

$$(4) \qquad r < 0, \quad v > 0, \quad (3u + v) > 0. \tag{9.73}$$

Phase 2, i.e., $(1, 0, 0)$ is stable again under the condition (9.71). The stability domains of phases $(1, 0, 0)$ and $(1, 1, 1)$ are shown in Fig. 9.4(b). In both Fig. 9.4(a) and Fig. 9.4(b) OA and OB are borderlines of stability. For each point (u, v) above these lines the system will undergo a second order phase transition as r decreases from $r > 0$ to $r < 0$.

The type of ordered phases will depend on the value of the couple (u, v) as shown in Fig. 9.4 but the important point is that any of these phases will be stable for any $r < 0$. When the point (u, v) approaches the line AOB the corresponding ordered phase loses its stability. Below this line no stable ordered phases exist. This effect is completely analogous to the lack of an ordered phase in the isotropic ($v = 0$) φ^4-model for $u < 0$. As in the last case (Section 4.8) the analysis of the domains below the line AOB in Fig. 9.4 can be performed by including into the consideration the φ^6-invariants of the model (9.63). Crossing the line AOB from above one should observe a crossover from a second order phase transition to a first-order phase transition which passes through a tricritical point.

A further information about the mean field analysis of systems with cubic (and other) anisotropies is presented by Cowley (1980), Gufan (1982), Izyumov and Syromiatnikov (1984). For a MF analysis of a rather complex Landau model of axial and planar antiferromagnets see Plumer, Caillé and Hood (1989).

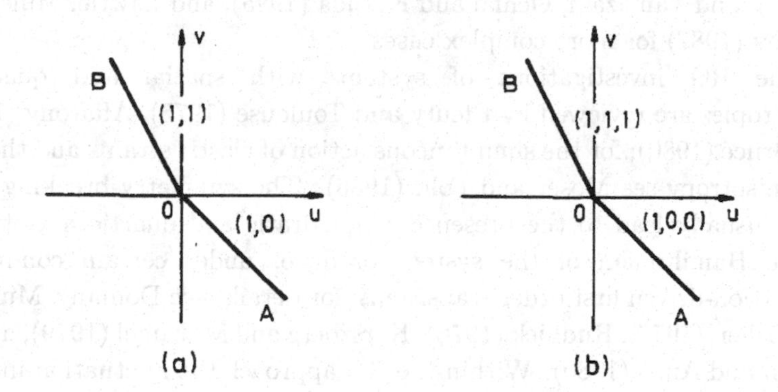

Fig. 9.4 (a) The stability domains of phases 2 and 3 for $(n = 2)$-systems. (b) The stability domains of the phases for $(n = 3)$-systems. OA is the line $u + v = 0$ and OB is the line $2u + v = 0$.

9.6.4 *Renormalization group studies of anisotropic systems*

The RG investigation of the cubic anisotropy effects has been performed by Aharony (1973) in the two-loop approximation. For the model (9.61) with $c_i \equiv c$ and $r_\alpha = r$ the cubic anisotropy (v) causes the appearance of a new (cubic) FP of the RG equations; for details, see Aharony (1973, 1976); for a three-loop calculation see Mayer and Sokolov (1988); for anisotropic antiferromagnets see Bak and Mukamel (1976); a detailed investigation of the relation between the symmetry properties of ($n = 4$)–component anisotropic systems and the stability properties of the RG FPs is given by Tolédano *et al.* (1985). The crossover to a fluctuation-driven first-order transition in "cubic systems" has been described within the RG method by Rudnick (1978). The simultaneous effect of the cubic anisotropy and disorder of type "random T_c" (see Section 9.5.1) has been investigated in the two-loop RG equations by Lawrie, Millev and Uzunov (1987); see also Yamazaki (1976, 1978), Yamazaki *et al.* (1986) and for experimental problems see Cox, Gibaud and Cowley (1988).

A characteristic feature of Hamiltonians like (9.61) with relatively large number of parameters is that they exhibit several FPs which describe second or first order transitions. For some values of the symmetry index n the RG equations may turn out degenerate and then one is forced to perform an expansion in $\epsilon^{1/2}$, $\epsilon = (4 - d)$ or, even in $\epsilon^{1/m}$, $m > 2$; see Khmel'nitskii (1975), Ma (1976a) and Shalaev (1977) for the case of impure Ising systems ($n = 1$), and Yamazaki, Ochiai and Fukuda (1985), and Lawrie, Millev and Uzunov (1987) for more complex cases.

The RG investigations of systems with spatial and quadratic anisotropies are reviewed by Pfeuty and Toulouse (1975), Aharony (1976), and Bruce (1980); for the simultaneous action of elastic strains and the spatial anisotropy see Moser and Folk (1986). The symmetry-breaking fields which usually lead to the presence of quadratic and quartic anisotropies in the Hamiltonian of the system produce, under certain conditions, fluctuation-driven first order transitions; for details, see Domany, Mukamel and Fisher (1977), Rudnick (1978), Kerszberg and Mukamel (1979), and Iacobson and Amit (1980). Within the RG approach the fluctuation-induced first order transition in anisotropic systems may occur either when the appropriate model does not possess a stable FP or when the stable FP is not physically accessible (Section 8.8.3); see also Brazovskii and Dzyaloshinskii (1975), Nattermann and Trimpler(1975), Nattermann (1976).

9.6.5 *Tricritical points and renormalization group*

The definition of tricritical points (TPs) in isotopic systems and their MF analysis have been discussed in Section 3.7.4 and Section 4.8, respectively. The general scaling analysis of TPs has been performed by Griffiths (1973). Within the fluctuation theory of phase transitions the TPs in isotopic systems are described by including the additional term

$$u_6 \int d^d x \varphi^6(x), \qquad u_6 > 0, \tag{9.74}$$

in the standard fluctuation φ^4-Hamiltonian. The TP — $(T_{\rm tr}, Y_{\rm tr})$ in the (T, Y) diagram of the system is then determined from the equations $r(T, Y) = u(T, Y) = 0$, where r and u are the renormalized parameters of the Hamiltonian.

When $r = u = 0$, the upper critical dimension of the systems with quadratic energy spectrum $\epsilon_k \sim k^2$ is $d_U = 3$ so the critical fluctuations for dimensions $d > 3$ are of Gaussian type; the system exhibits logarithmic singularities for $d = 3$ (and $r \to 0$), and for $d < 3$ the tricritical behaviour is non-Gaussian, i.e., with corrections coming from the interactions. According to this picture the critical exponents for $d \geq 3$ stick to their MF values (Riedel and Wegner, 1972; Wegner and Riedel, 1973); for a discussion of this point and the equation of state for TPs; see also Nelson and Rudnick (1975).

For dimensions $d < 3$ the critical exponents for TPs have also been calculated; see Stephen and McCauley Jr. (1973); Stephen, Abrahams and Straley (1975); Stephen (1980). The tricritical susceptibility near four dimensions has been investigated by Lawrie and Gedling (1982). Fluctuation-induced tricritical points are discussed by Blankschtein and Mukamel (1982), and Blankschtein and Aharony (1983). The effect of disorder of type "random T_c" (see Section 9.5.1) in the classical ($T_{\rm tr} \neq 0$) and quantum ($T_{\rm tr} \to 0$) limits has been discussed for the first time by Busiello, De Cesare and Uzunov (1984); see also Sokolov (1987).

A careful test of the tricritical scaling in liquid ^{3}He–^{4}He mixture has been performed by Riedel, Meyer and Behringer (1976); for experiments in metamagnets and other systems with TPs; see Salamon and Shang H.-T. (1980), the review by Knobler and Scott (1984), and the book edited by Pynn and Skjoltorp (1984). An exactly solvable model of structural phase transitions which exhibits TPs is discussed by Sarbach and Schneider (1975). TPs in ferromagnets with a single-ion anisotropy have been investigated by Sznajd (1984). The main theoretical results are presented in details by Lawrie and Sarbach (1984).

9.6.6 *Bicritical and tetracritical points*

These multicritical points occur in phase diagrams of complex systems where two (or more) order parameters of different nature (magnetic and superconducting or magnetic and structural) may be present and the phase transition lines cross each other as shown in Fig. 9.5; see also Landau and Lifshitz (1980).

We shall briefly discuss these points for isotropic systems described by the Hamiltonian:

$$\mathcal{H} = \frac{1}{2}\int d^d x\, [r_1\varphi_1^2 + r_2\varphi_2^2 + c_1(\nabla\varphi_1)^2 + c_2(\nabla\varphi_2)^2 + 2u_1\varphi_1^4 + 2u_2\varphi_2^4 + 4w\varphi_1^2\varphi_2^2]\,, \tag{9.75}$$

where $\varphi_1 = \{\varphi_{1\alpha}\}$ is a n_1-component field and $\varphi_2 = \{\varphi_{2\alpha}\}$ is a n_2-component field. The order parameters φ_1 and φ_2 describe two different phase transitions with bare critical temperatures $T_{c1}^0(Y)$ and $T_{c2}^0(Y)$. The functions $T_{c1}^0(Y)$ and $T_{c2}^0(Y)$ are represented in Fig. 9.5 by the lines ABA' (and ATA') and CBC' (and CTC'), respectively. In Eq. (9.75), $r_i = \alpha_{0i}(T - T_{ci}^0)$; $i = (1, 2)$. As the interaction parameters u_i and w are usually assumed to be smooth functions of T and Y so that one can use the approximation $u_i \approx u_i(T_{ci})$ and $w \approx$ const.

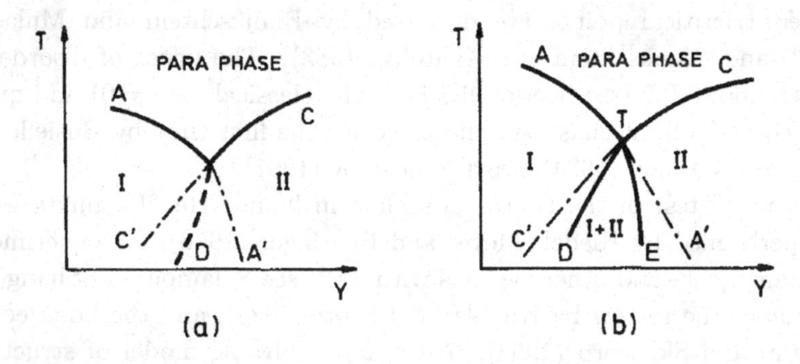

Fig. 9.5 (a) Bicritical (B) and (b) tetracritical (T) points. The continuous lines describe second order transitions. The line BD is a first order transition.

Near the multicritical points (B and T) in Fig. 9.5 both order parameters φ_1 and φ_2 can be different from zero. Their interaction is described by the parameter w.

We should mention that the model (9.75) is obtained by the Hubbard–Stratonovich transformation of the corresponding coupled lattice models (see Section 6.6.3). Besides, it seems useful to show the relationship between the models (9.75) and (9.61) when the latter does not contain spatial anisotropy, i.e., when $c_i = c = 1$ for any $i = 1, \ldots, d$. By a suitable choice of the length unit we may have $c = 1$. In the same way we may transform one of the c–parameters in (9.75) to unity, for example, $c_1 = 1$. The coefficient c_2 can be absorbed in the field φ_2, $\sqrt{c_2}\varphi_2 = \varphi_2'$ which leads to the transformations $u_2 = c_2^2 u_2'$ and $w = c_2 w'$ (after this simple procedure the superscript "prime" can be omitted). For $u_1 = u_2 = w \equiv u$ the sum of the three quartic terms in Eq. (9.75) takes the form $u\varphi^4$ of the third term on the r.h.s. of Eq. (9.61), provided the field φ is the $n = (n_1 + n_2)$–component field formed by the direct sum of the fields φ_1 and φ_2: $\varphi = (\varphi_1, \varphi_2) = (\varphi_{11}, \ldots, \varphi_{1n_1}, \varphi_{21}, \ldots, \varphi_{2n_2}) \rightarrow (\varphi_1, \ldots, \varphi_n)$. Now we see that the two models will be identical if in Eq. (9.61) we set $v = 0$, $r_1 = r_2 = \cdots = r_{n_1} = r^{(1)}$, $r_{n_1+1} = \cdots = \cdots = r_n = r^{(2)}$, and identify $r^{(1)}$ and $r^{(2)}$ with r_1 and r_2 in Eq. (9.65), respectively.

The mean field analysis can be performed using the scheme given in Section 9.6.2; see also Liu and Fisher (1973) and the detailed analysis by Imry (1975). The re-derivation of the mean field results, in particular, the stability conditions for the phases and the outline of the diagrams in Fig. 9.5 may be a useful exercise. Here we shall present the expressions for the order parameters of the ordered phases:

$$\begin{aligned}
&\text{I.} \quad && \varphi_1 = \pm\sqrt{\frac{-r_1}{4u_1}}\,, && \varphi_2 = 0, \\
&\text{II.} \quad && \varphi_1 = 0, && \varphi_2 = \pm\sqrt{\frac{-r_2}{4u_2}}\,, \\
&\text{III.} \quad && \varphi_1 = \pm\sqrt{\frac{r_1 u_2 - r_2 w}{4D}}\,, && \varphi_2 = \pm\sqrt{\frac{r_2 u_1 - r_1 w}{4D}}\,,
\end{aligned} \tag{9.76}$$

where $D = (w^2 - u_1 u_2)$. The corresponding potential densities ($\phi = \Phi/V$) are $\phi_i = -r_1^2/16u_1$, $\phi_{\rm II} = -r_2^2/16u_2$ and

$$\phi_{\rm III} = -\frac{1}{16D}\left[2r_1 r_2 w - r_1^2 u_2 - r_2^2 u_1\right]. \tag{9.77}$$

Phase III is also called *the coexistence* (or *mixed*) phase; in it phases I and II coexist.

If $w = 0$ (decoupled case), Fig. 9.5(a) and Fig. 9.5(b) will coincide. The lines BD and DTE will not exist but rather the domains of the phases will be given by the second order phase transition lines ABA' and CBC' (or, which is the same, ATA' and CTC'). Above the line ABC the paraphase ($\varphi_1 = \varphi_2 = 0$) will be stable in the domain ABC' — phase I will be stable, domain CBA' will be for the phase II and in the wing $C'BA'$ both I and II will be stable (coexistence).

When $w^2 < u_1 u_2$ the tetracritical point in Fig. 9.5(b) will appear. The phase transition lines below this point will be DT and ET rather than $C'T$ and $A'T$ (the reason is because of the coupling constant w). Phase I will be stable in the domain DTA, phase II will be stable in the domain ETC and the mixed phase III is stable below the line DTE.

As w^2 increases to $w^2 > u_1 u_2$ (strong coupling) the picture changes to that in Fig. 9.5(a). The line BD is the first-order equilibrium transition line between phases I and II (the mixed phase cannot occur). All these results are deduced from the mean field analysis of the model (9.75).

The RG study of the model (9.75) has been performed by Nelson, Kosterlitz and Fisher (1974), and Kosterlitz, Nelson and Fisher (1976). The model (9.75) exhibits two new (coupled $w^* \neq 0$) FPs: the *bicritical* FP which describes the bicritical behaviour and the *biconical* FP which describes the coupled ($w \neq 0$) tetracritical point; see also Bushev *et al.* (1980).

The Ginzburg criterion for bicritical and tetracritical points has been derived by Hornreich and Schuster (1979). It has been shown that an effect of the enhancement of the critical fluctuations near such multicritical points can take place in several systems; for a detailed discussion, see Tonchev and Uzunov (1981a). Bicritical points produced by quadratic ($\simeq \varphi^2$) symmetry-breaking terms are studied by Amit and Goldschmidt (1978). The influence of the vector potential of the magnetic field on the fluctuation properties near bicritical and tetracritical points in superconductors has been investigated by Tonchev and Uzunov (1981b). The specific heat in the Gaussian approximation is discussed by Uzunov (1981b). The RG analysis of impure systems with bicritical and tetracritical points have been studied by Laptev and Skryabin (1980), and Busiello and Uzunov (1989). Other fluctuation effects near bicritical and tetracritical points are described by Lyuksyutov, Pokrovskii and Khmel'nitskii (1975); see also Patashinskii and Pokrovskii (1977). Two-loop order RG study of the critical behaviour of bicritical and tetracritical points has also been performed (Folk, Holovatch, and Moser, 2008a). The critical dynamics has been studied in the framework of the two-loop RG equations, too; see Folk, Holovatch, and Moser (2008b, 2009).

9.6.7 *Lifshitz's and tricritical Lifshitz's points*

In isotropic systems, these multicritical points can be described by the Hamiltonian:

$$\mathcal{H} = \frac{1}{2}\int d^dx \left\{c_2(\nabla\varphi)^2 + c_4(\nabla^2\varphi)^2 + r\varphi^2 + u\varphi^4 + u_6\varphi^6\right\} \qquad (9.78)$$

The critical Lifshitz point (CLP) is defined by the equations:

$$c_2(T, Y) = r(T, Y) = 0, \qquad (9.79)$$

i.e., this point lies on the critical line $r = 0$, where $c_2 > 0$ decreases to $c_2 = 0$. Thus there should be a region in the (T, Y) diagram where c_2 changes sign from $c_2 > 0$ to $c_2 < 0$. For $c_2 < 0$ the system is unstable towards x-dependent fluctuations of φ and one must include into consideration the next gradient term allowed by the symmetry of the system, i.e., the term $c_4(\nabla\varphi)^4$, $c_4 > 0$; see also Section 6.6.2.

The MF analysis of the Hamiltonian (9.78) has been presented in details by Michelson (1977). The model (9.78) describes three phases: (i) the disordered phase ($\varphi = 0$) for $r > 0$, (ii) the usual ordered phase for $c_2 > 0$ and $r < 0$, and (iii) a new modulated (x-dependent) phase for $c_2 < 0$ and $r < 0$. The RG analysis has been done by Hornreich, Luban and Shtrikman (1975, 1977, 1978); see also the investigation of complex critical points performed by Chang, Tuthill and Stanley (1974), Tuthill, Nicoll and Stanley (1975) and Nicoll et al. (1977). For the CLP of the nematic-smectic A-smectic C transition in liquid crystals see Gorodetskii and Podneks (1989).

The coordinates of the tricritical Lifshitz point (TLP) in the (T, Y) diagram of the system are obtained from Eqs. (9.79) and

$$u(T, Y) = 0. \qquad (9.80)$$

So, at the TLP, u also changes sign from $u > 0$ to $u < 0$; $u_6 > 0$. The RG analysis of the TLP is presented by Dengler (1985); Aharony *et al.* (1985), and Tonchev and Uzunov (1985); for a discussion see also Uzunov (1987). There is a discrepancy between the RG results for the TLP presented in the paper by Aharony *et al.* (1985) and the paper by Tonchev and Uzunov (1985). On the basis of $O(\epsilon)$ RG study, the former concludes about the existence of stable FP which describes TLP. The second paper, based on $O(\epsilon)$ results and heuristic arguments predicts an instability of this type of quite complex criticality. A detailed recent RG investigation (Bervillier,

2004), based on an advanced RG procedure (Bagnuls and Bervillier, 2001), supports the result by Tonchev and Uzunov (1985) about the instability of the TL behaviour. For experiments on LPs and TLPs see the paper by Bindilatti, Becerra and Oliveira Jr. (1989) and the references therein.

9.7 Critical Properties of Superconductors and Liquid Crystals

Here we shall discuss some aspects of the theory of phase transitions in superconductors and liquid crystals (see also Sections 3.8.6 and 3.8.7), and related topics in quantum field theory. The superconductors can be classified as conventional and unconventional superconductors (see Sections 9.7.1 and 9.7.2) according to the symmetry of the electron pairing. We shall consider the entire picture of phases and phase transitions but the focus will be on the gauge effects in critical phenomena.

9.7.1 *Conventional superconductors*

From a phenomenological point of view the ordinary (or "conventional") superconductors are described at a (quasi)microscopic level by a complex *scalar* field $\psi(x)$ — the order parameter and the well-known Ginzburg-Landau (GL) free energy functional (Ginzburg and Landau, 1950):

$$F = \int d^d x \left[a|\psi|^2 + \frac{\hbar^2}{4m} \left| \left(\nabla - \frac{2ie}{\hbar c} \boldsymbol{A} \right) \psi \right|^2 + \frac{b}{2}|\psi|^4 + \frac{1}{8\pi} (\nabla \times \boldsymbol{A})^2 \right], \tag{9.81}$$

where $\boldsymbol{A}$ is the vector potential of the magnetic induction $\boldsymbol{B} = \nabla \times \boldsymbol{A}$[2]. So, the thermodynamic potential (9.81) is the generalized (nonequilibrium) Helmholtz free energy of the superconductor. Here we follow the notations introduced by Lifshitz and Pitaevskii (1980). In Eq. (9.81), $a = \alpha_0(T - T_{c0})$ with T_{c0} the (bare) critical temperature; the parameter b is supposed temperature independent; c is the velocity of light, $2m$ and $2e$ are the effective mass and the electric charge of the superconducting carriers.

[2] The vector product is defined for 3D systems only and this restricts the validity of the last term in Eq. (9.81); as well in Eq. (9.88) (Section 9.7.2). A more general expression of the magnetic energy $B^2/8\pi$ by the vector potential $\boldsymbol{A}$, which is valid for dD systems (spaces), is presented in Section 9.7.7.

Within this phenomenological theory the electric carriers can be considered as indeterminate classical objects (particles). As known from the microscopic theory of superconductivity, these carriers are the electron (fermion) Cooper pairs which appear below the phase transition temperature $T_{c0}(H)$ and are described by the field ψ (Lifshitz and Pitaevskii, 1980; Ketterson and Song, 1999). Thus m and e can be identified as the effective electron mass in the superconducting metal and the electron charge, respectively.

The electron Cooper pairs in usual (conventional) superconductors are in the simplest (s-wave) quantum state with opposite spins. The s-wave ψ-function is spherically symmetric and this is related with the fact that the macroscopic order parameter $\psi(x)$ in Eq. (9.81), which describes the Cooper pairs at a quasi-macroscopic classical level, is a complex scalar field. In unconventional superconductors (Section 9.7.2), where the Cooper pairs are not in the simplest (s-) quantum state and the Cooper pair is not spherically symmetric, the order parameter field $\psi(x)$ is a complex vector (Uzunov, 1991; Mineev and Samokhin, 1999).

Using the Legendre transformation

$$G = F - \int d^d x \boldsymbol{H} \cdot \boldsymbol{M}, \tag{9.82}$$

with $\boldsymbol{H}$ being the external magnetic field and $\boldsymbol{M} = (\boldsymbol{B} - \boldsymbol{H})/4\pi$ being the magnetization vector, we can obtain the generalized Gibbs potential

$$G(\psi, \boldsymbol{A}) = F(\psi, \boldsymbol{A}) - \frac{1}{4\pi} \int d^d x \left(\boldsymbol{H} \cdot \boldsymbol{B} - \frac{H^2}{2} \right). \tag{9.83}$$

In Eq. (9.83) the energy of the magnetic field is taken into account through the term $(-H^2/8\pi)$, which is added to both F and G. In the potential G the magnetic energy connected with the penetration of the magnetic field in the superconducting sample is also considered. As our discussion will be mainly restricted to the bulk properties of the superconductors we shall turn to the potential (9.81).

The functionals (9.81) and (9.83) correctly describe the spatial variations of the order parameter $\psi(x)$ and the possible superconducting phases ($\psi \neq 0$). It follows from the GL functional (9.81) that there can be two phases in a zero external field, where the theory reduces to the usual ψ^4-theory of the complex field ψ; the three A-dependent terms in Eq. (9.81) do not exist. These are the phases $|\psi| = 0$ for $a \geq 0$, and $|\psi| = \pm(|a|/b)^{1/2}$ for $r < 0$.

In a nonzero external magnetic field $\boldsymbol{H}$ the possible stable phases are:

$$\begin{aligned} &(\psi = 0, \boldsymbol{B} \neq 0)\,, \quad \text{[normal metal (i.e., “para phase”)]} \\ &(\psi \neq 0, \boldsymbol{B} = 0)\,, \quad \text{[Meissner phase]} \\ &(\psi \neq 0, \boldsymbol{B} \neq 0)\,, \quad \text{[mixed (vortex) phase]}. \end{aligned} \tag{9.84}$$

These phases are obtained from the GL equations $(\delta F/\delta\psi) = 0$ and $(\delta F/\delta \boldsymbol{A}) = 0$; for details see London (1950), Landau and Lifshitz (1960), Saint–James, Sarma and Thomas (1969), and Tilley and Tilley (1974); Ketterson and Song (1999); Poole, Farash and Creswick (2000).

Another important point is that the superconductors can be classified into two types according to their characteristic lengths — the Pippard coherence (or correlation) length

$$\xi(T) = \frac{\xi_0}{|t|^{1/2}}, \qquad \xi_0 = \left(\frac{\hbar^2}{4m\alpha_0 T_{c0}}\right)^{1/2}, \tag{9.85a}$$

and the London penetration depth

$$\lambda_L(T) = \frac{\lambda_0}{|t|^{1/2}}, \qquad \lambda_0 = \left(\frac{mc^2 b}{8\pi e^2 \alpha_0 T_{c0}}\right)^{1/2}, \tag{9.85b}$$

where ξ^0 and λ_0 are often referred to as “the zero temperature coherence and penetration lengths”, respectively. The coherence length ξ indicates the scale of spatial variations of the field $\psi(x)$ (in the phase transition theory we use the term “correlation length”) and the penetration length characterizes the spatial variations of the field $A(x)$. So, in the superconductors we have two correlation lengths. Sometimes, λ_0 is written in the form $\lambda_0 = \left(mc^2/8\pi e^2|\psi_0|^2\right)^{1/2}$, where $\psi_0^2 = |\psi_0|^2 = |a|/b$ is the square of the uniform superconducting order parameter in the bulk Meissner–Ochsenfeld phase. The type I-superconductors are those, for which the Ginzburg–Landau (GL) parameter

$$\kappa = \frac{\lambda_L(T)}{\xi(T)} = \frac{mc}{e\hbar}\left(\frac{b}{2\pi}\right)^{1/2} \tag{9.85c}$$

is less than $1/\sqrt{2}$, and the type II superconductors are those, for which $\kappa > 1/\sqrt{2}$. Relatively large effective mass of electrons and parameter b correspond to type II superconductors. In some theoretical papers the large “effective charge e” indicated for a small κ and type I superconductivity.

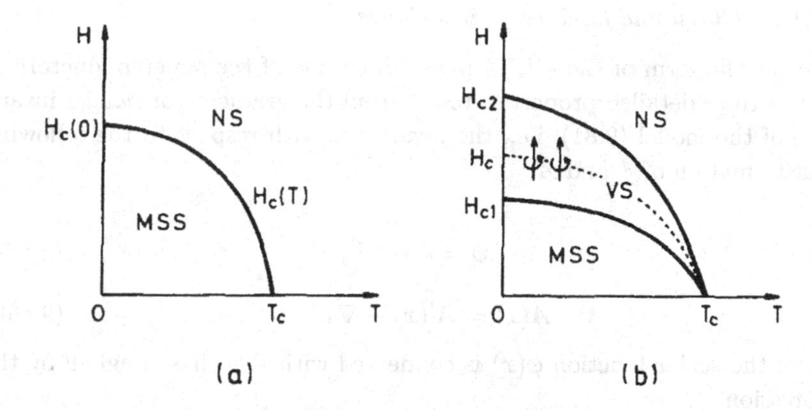

Fig. 9.6 (H, T) phase diagram of: (a) type I superconductors; the transition line is $H_c(T)$, (b) type II superconductors with two critical fields, $H_{c_1}(T)$ and $H_{c_2}(T)$ and the vortex (mixed) phase (VS) for $H_{c_1}(T) < H < H_{c_2}(T)$.

It seems important to emphasize, that the second relation (9.85c) has been obtained after the exact compensation of the inverse root dependencies of ξ and λ_L on t. This is a MF result, and may not be valid in a close vicinity of the critical point. In some regimes of strong fluctuations, the dependencies of ξ and λ_L on t could be more general: $\xi = \xi_0|t|^{-\nu}$ and $\lambda_L = \lambda_0|t|^{-\nu_A}$, where the correlation length exponents, ν and ν_A, may have different values near some particular critical points. Then κ will have a t-dependence of type $\kappa \sim |t|^{\Delta}$, $\Delta = (\nu - \nu_A)$. Now the type of the superconductor may be effectively changed for very small t: (a) from type II to type I, provided $\Delta > 0$, and (b) from type I to type II, provided $\Delta < 0$. If $\nu = \nu_A$ up to $t = 0$, the second equality (9.85c) will be always valid.

The two types of superconductors have quite different phase diagrams (see Fig. 9.6) and fluctuation properties (for the fluctuation effects see the review by Skocpol and Tinkham (1975). For example, the vortex phase, where the well-known Abrikosov lattice of magnetic fluxes coexists with the superconducting phase can appear only in superconductors of the second type; see Fig. 9.6(b).

9.7.1.1 *Global and local gauge invariance*

The specific form of the (H,T) phase diagrams of the superconductors as well as their detailed properties result from the gradient (or *gauge*) invariance of the model (9.81), i.e., the invariance with respect to the following transformation of ψ and A:

$$\psi = \psi' e^{i\theta(x)}, \tag{9.86a}$$

$$\boldsymbol{A}(x) = \boldsymbol{A}'(x) + \nabla\varphi(x), \tag{9.86b}$$

where the scalar function $\varphi(x)$ is connected with the phase angle θ by the expression:

$$\frac{2e}{\hbar c}\varphi(x) = \theta(x). \tag{9.87}$$

It is convenient to choose the Coulomb gauge $\nabla \cdot \boldsymbol{A} = 0$ for the vector potential $\boldsymbol{A}$; so the function φ in Eq. (9.86b) will satisfy the equation $\nabla^2\varphi = 0$ (provided $\nabla \cdot \boldsymbol{A}' = 0$).

The invariance of the GL functional (9.81) with respect to the gradient transformation (9.86a)–(9.87) is called *local gauge invariance.* This local gauge invariance is broken in the vortex (mixed) phase, see Eq. (9.84), where the supercondicting and normal states coexists owing to the formation of the (Abrikosov) vortex lattice (Abrikosov, 1957; Saint–James, Sarma and Thomas, 1969; Tilley and Tilley, 1974). Certainly, the vortex state in superconductors is possible because the phenomenon of superconductivity possesses the property of local gauge invariance. The vortex state always exists as a solution of the GL equations but only in type II superconductors, where $\kappa > 1/\sqrt{2}$, it occurs as a stable phase for certain T and H (Fig. 9.6). So the vortex phase is a remarkable example of breaking of the local gauge invariance.

In the particular case, when the phase $\theta(x)$ in the transformation Eq. (9.86a) is x-independent (uniform) $[\theta(x) = \theta]$, the second equality (9.86b) is trivial. Therefore, the GL functional is invariant also with respect to *global gauge transformation* of type $\psi(x) = \psi'(x)e^{i\theta}$ (any real number θ). The term "global" indicates that the transformation is the same for any spatial point x. The Meissner phase (uniform superconducting state) is then a phenomenon of breaking of the *global gauge invariance.*

These two properties of gauge invariance define the global and local $U(1)$ symmetries of the GL free energy. In both cases the gauge group is a one-dimensional Abelian continuous group, $U(1)$. As we shall see in Sections 9.7.5–9.7.7 and 9.8 these symmetries and, in particular, the local gauge symmetry, produce quite interesting fluctuation phenomena near the point of symmetry breaking, $T_{0c}(H)$, in particular, a fluctuation change of the order of the phase transition.

9.7.1.2 *Gauge symmetries in other condensed matter systems*

Another interesting gauge theory is that of the nematic-to-smectic A phase transition in liquid crystals (see also Section 3.8.7). According to the Kobayashi-McMillan-de Gennes theory (Kobayashi, 1970; McMillan, 1971; de Gennes, 1972), the smectic-A order is described by two order parameters: the nematic director vector $n(x)$ and the complex scalar $\psi(x)$ (the latter describes the plane arrangement of the long molecule mass centers).[3] When a description, analogous to that of superconductors, is introduced, as suggested by de Gennes (1972), the director vector is substituted with a gauge vector field, which is quite similar to the vector potential $A(x)$ in the GL functional. Apart from some specific features intended to take into account the liquid crystal anisotropy, the effective free energy of the smectic A looks very much like the GL free energy of superconductors (Halperin and Lubensky, 1974; Lubensky and Chen, 1978; de Gennes and Prost, 1993).

A gauge theory in condensed matter physics that has some (but not very close) similarity with the GL free energy of superconductors is the Chern–Simons–Ginzburg–Landau (CSGL) effective functional of the quantum Hall liquid state, in particular, in the description of phase transitions between the plateaus in the quantum Hall effect (Zhang, Hansson and Kivelson, 1989; Schakel, 1995, 2008; Wen, 1993; Pryadko and Zhang, 1994). Let us mention also the liquid metallic hydrogen, where the problem of the superconducting-to-superfluid phase transition (Babaev, Sudbø, and Aschroft, 2004) is also related to the topics discussed here. More examples can be found in the books by Herbut (2007) and Schakel (2008).

9.7.1.3 *Interrelationship with quantum field theory problems*

In elementary particle physics the gauge invariant theory similar to the GL theory of superconductivity is called Abelian Higgs Model (Higgs, 1964, 1966; Guralnik, Hagen, and Kibble, 1964; Englert and Brout, 1964; Kibble, 1967; Coleman and Weinberg, 1985; Guidry, 2004). The same global $U(1)$

[3]Henceforth we shall avoid the bold face for vectors quantities like $\boldsymbol{n}(\boldsymbol{x})$, $\boldsymbol{A}(\boldsymbol{x})$, $\boldsymbol{B}(\boldsymbol{x})$, etc.

and local $U(1)$ gauge symmetries are present there but the phenomenon of spontaneous symmetry breaking occurs only for imaginary mass of the Higgs (ψ–) field, which is the analogue of the field ψ, describing the Cooper pair in a superconductor (in the Higgs model the role of the parameter a in the GL theory is played by the square of the particle mass m: $a \sim m^2$).

In the absence of spontaneous symmetry breaking this model would describe an ordinary electrodynamics of charged scalars, but the situation becomes more interesting when the mentioned mass is imaginary and the breaking of symmetry occurs. Now the symmetry breaking phenomena receive other names and another physical interpretation.

The breaking of both global and local gauge symmetries ensures a mechanism for a transformation of the two initial scalar fields – analogues of the components ψ' and ψ'' of the complex field $\psi = \psi' + i\psi''$, and two massless photon fields – analogues of the two independent components A_j of the vector potential $A(x) = \{A_j(x); j = 1, 2, 3; \nabla.A(x) = 0\}$ in a superconductor, to four massive particle fields: the so-called *Higgs boson*, which is analogue of the spontaneous order $|\psi| > 0$ in a superconductor, and three massive vector field components, i.e., a massive three dimensional vector field. The mass of this new vector field is proportional to the electric charge $|q|$ and magnitude $|\psi|$ of the Higgs field and, as a matter of fact, this is exactly the way, in which the London penetration length in a superconductor depends on the electron charge $|e|$ and the modulus $|\psi|$ of the superconducting order parameter ψ.

Thus the spontaneous breaking of the local gauge symmetry leads to the formation of massive particles without spoiling the gauge invariance of the theory. This is called "the Higgs mechanism". It plays a fundamental role in the unified theory of electromagnetic, weak and strong interaction (Coleman and Weinberg, 1985; Guidry, 2004).

It is easy to see that there is something quite common between the phenomena of spontaneous breaking of the continuous symmetries in superconductivity theory and in quantum field theory. The superconducting phase $|\psi| > 0$ is the exact analogue of the Higgs boson in the Abelian-Higgs model, whereas the appearance of a massive vector field has its analogue in the finite London penetration length, mentioned above. Of course, there is no obstacles in interpreting the phenomena of spontaneous symmetry breaking in quantum field theory as phase transitions by taking the Higgs (ψ–) mode mass as tuning parameter. The phase transition will occur at zero mass of the Higgs boson.

A similar phenomenon of spontaneous breaking of both global $U(1)$ and local gauge symmetries is possible also within the scalar electrodynamics due to mass insertions from the radiation corrections (Coleman and Weinberg, 1973). The radiation corrections, in analogy with the magnetic fluctuations in a superconductor, generate an imaginary mass to the initially massless scalar field in this theory and the latter becomes very similar to the Abelian-Higgs model. Here the symmetry breaking leads to the appearance of massive scalar and vector fields describing neutral scalar meson and vector meson, respectively (Coleman and Weinberg, 1973). The radiation corrections to the Lagrangian of the massless scalar electrodynamics (Coleman and Weinberg, 1973) resemble very much, in particular in their mathematical form at $D = 4$, the magnetic fluctuation corrections to the GL free energy of 4D superconductors (Chen, Lubensky and Nelson, 1978). The interrelationships between the superconductivity theory and the gauge theories of elementary particles has been comprehensively discussed by (Linde, 1979). Note also the relation between the superconductivity GL functional and the CP^{N-1} confinement model (Hikami, 1979) and extensions to non-Abelian theories (Brézin *et al.*, 1979).

The gauge theories, mentioned so far, and their extensions have a wide application in the description of the Early Universe (Linde, 1979; Vilenkin and Shellard, 1994).

9.7.2 *Unconventional superconductors*

Within the same phenomenological approach we call unconventional superconductors those systems where the order parameter ψ in the GL functional is a complex vector $\psi = \{\psi_\alpha, \alpha = 1, \dots, n/2\}$, $n = 4, 6, \dots$. The vector nature of ψ gives the possibility several types of quartic anisotropies to exist (see Section 9.5). An example is the GL functional:

$$F(\psi, A) = \int d^dx \left\{ a|\psi|^2 + \frac{u}{2}|\psi|^4 + \frac{\bar{u}}{2}|\psi^2|^2 + \frac{v}{2}\sum_{\alpha=1}^{n/2} |\psi_\alpha|^4 \right.$$
$$\left. +\gamma\Big|[\nabla - iq_0 A(x)]\,\psi\Big|^2 + \frac{1}{8\pi}[\nabla \times A(x)]^2 \right\}, \tag{9.88}$$

where the parameters $\gamma = (\hbar^2/4m)$ and $q_0 = (2e/\hbar c)$ have been introduced for a further convenience.

The free energy (9.88) is discussed in the theory of heavy-electron and high-T_c superconductors; see, e.g., Volovik and Gor'kov (1985), Ueda and Rice (1985), Annett, Randeria and Renn (1988), Annett, Goldenfeld and Renn (1989), and the reviews by Annett (1990, 2004), Sigrist and Ueda (1991), Uzunov (1991), and the monograph by Mineev and Samokhin (1999). In Eq. (9.88), the v-term takes into account the crystal anisotropy of cubic type (to ensure a quite general example of crystal anisotropy) and the $\bar{u}$-term describes the (orthorhombic) symmetry of the unconventional (non-s state) Cooper pairs of fermions (electrons); for these symmetries, see Landau and Lifshitz (1980), Cowley (1980), and Mineev and Samokhin (1999). Here we assume a spin-triplet (p-wave) Cooper pairing, which defines the symmetry of the superconducting electron pairs. In this case, the field ψ is a three dimensional complex vector ($n/2 = 3$), which corresponds to a six-component real vector ($n = 6$). Note, that for some unconventional superconductors the gradient term has a quite more complex structute(Sigrist and Ueda, 1991; Mineev and Samokhin, 1999). In Eq. (9.88) we have used the so-called "one constant approximation", and this choice is entirely consistent with studies in which the structure of the vortex phase is not of interest.

In a nonzero external magnetic field, a vortex lattice is expected to occur as a solution of the GL equation for $\kappa > 1/\sqrt{2}$; but in this case the mean field analysis is rather complicated by the presence of the spatial, quadratic and quartic anisotropies together with the gauge field $A(x)$; see Burlakov (1985), Joynt (1988), Joynt *et al.* (1990), Tukuyasu, Hess and Sauls (1990), Palombo, Muzikar and Sauls (1990).

Here we briefly summarize the known results (Volovik and Gor'kov, 1985; Blagoeva *et al.*, 1990) for the uniform (Meissner) superconducting phases described by the free energy (9.88) when the magnetic induction $B(x)$ is zero. The possible phases can be classified by the structure of the complex vector order parameter $\psi = (\psi_1, \psi_2, \psi_2)$. We shall often use the moduli vector (ϕ_1, ϕ_2, ϕ_3) with magnitude $\phi = (\phi_1^2 + \phi_2^2 + \phi_3^2)^{1/2}$ and the phase angles θ_j; $\psi_j = |\psi_j|e^{\theta_j}$, $|\psi_j| \equiv \phi_j$, $j = 1, 2, 3$.

The analysis of the six equations of state $\partial F/\partial\phi_j = 0$ and $\partial F/\partial\theta_j = 0$ is performed in the way explained for anisotropic systems (Section 9.6). One has to solve the equations of state with respect to ϕ_j and θ_j (for the rotational invariance one of the three angles θ_j will be redundant), and then investigate the stability properties of the solutions with the help of the stability matrix, which is analogous to the matrix $\hat{a}$, given by Eq. (9.66). Another example of this type of calculation is given in Section 9.9.3 for the more complex case of unconventional ferromagnetic superconductors.

The normal phase (0,0,0) always exists. It is stable for $r \geq 0$, and corresponds to a free energy $f = 0$. Under certain conditions, six ordered phases occur for $r < 0$. The simplest ordered phase is of type $(\psi_1, 0, 0)$ with equivalent domains: $(0, \psi_2, 0)$ and $(0, 0, \psi_3)$. Multi-domain phases of more complex structure also occur, but we shall not always enumerate the possible domains. For example, the "two-dimensional" phases can be fully represented by domains of type $(\psi_1, \psi_2, 0)$ but there are also other two types of domains: $(\psi_1, 0, \psi_3)$ and $(0, \psi_2, \psi_3)$. As we consider the general case when the crystal anisotropy is present $(v \neq 0)$, this type of phases possesses the property $|\psi_i| = |\psi_j|$.

The two-dimensional phases are two and have different free energies. To clarify this point let us consider, for example, the phase $(\psi_1, \psi_2, 0)$. The two complex numbers, ψ_1 and ψ_2 can be represented either as two-component real vectors, or, equivalently, as rotating vectors in the complex plane. One can easily show that we shall have two phases: a collinear phase, when $(\theta_2 - \theta_1) = \pi k (k = 0, \pm 1, ...)$, i.e. when the vectors ψ_1 and ψ_2 are collinear, and another (non-collinear) phase when the same vectors are perpendicular to each other: $(\theta_2 - \theta_1) = \pi(k + 1/2)$. Having in mind that $|\phi_1| = |\phi_2| = \phi/\sqrt{2}$, the domain $(\psi_1, \psi_2, 0)$ of the collinear phase is given by $(\pm 1, 1, 0)\phi/\sqrt{2}$ whereas the same domain for the non-collinear phase will be $(\pm i, 1, 0)\phi/\sqrt{2}$. The two domains of these phases have similar representations.

In addition to the mentioned three ordered phases, three other ordered phases exist. For these phases all three components ψ_j have nonzero equilibrium values. Two of them have equal to one another moduli ϕ_j, i.e., $\phi_1 = \phi_2 = \phi_3$. The third phase is of the type $\phi_1 = \phi_2 \neq \phi_3$ and is unstable so it cannot occur in real systems. The two three-dimensional phases with equal moduli of the order parameter components have different phase angles and, hence, different structure. The difference between any couple of angles θ_j is given by $\pm\pi/3$ or $\pm 2\pi/3$. The characteristic vectors of this phase can be of the form $(e^{i\pi/3}, e^{-i\pi/3}, 1)\phi/\sqrt{3}$ and $(e^{2i\pi/3}, e^{-i2\pi/3}, 1)\phi/\sqrt{3}$. The second stable three dimensional phase is "real", i.e. the components ψ_j lie on the real axis; $(\theta_j - \theta_j) = \pi k$ for any couple of angles θ_j and the characteristic vectors are $(\pm 1, \pm 1, 1)\phi/\sqrt{3}$.

When the crystal anisotropy is not present $(v = 0)$ the picture changes. The increase of the level of degeneracy of the ordered states leads to an instability of some phases and to a lack of some non-collinear phases. Both two- and three-dimensional real phases, where $(\theta_j - \theta_j) = \pi k$, are no more constrained by the condition $\phi_i = \phi_j$ but rather have the freedom of a variation of the moduli ϕ_j under the condition $\phi^2 = -r > 0$. The two-dimensional non-collinear phase exists but has a marginal stability. All other noncollinear phases even in the presence of a crystal anisotropy ($v \neq 0$) either vanish or are unstable. This discussion demonstrates that the crystal anisotropy stabilizes the ordering along the main crystallographic directions, lowers the level of degeneracy of the ordered state related with the spontaneous breaking of the continuous symmetry and favors the appearance of non-collinear phases.

9.7.3 *Layered superconductors*

Another effect which can modify or drastically change the phase transition properties of both conventional and unconventional superconductors is the spatial anisotropy. Some superconductors, especially those with relatively high critical temperatures ($T_c > 40$K) exhibit a quasi two-dimensional superconductivity. The superconductor can be considered as one-dimensional chain, say, along the $\hat{z}$-axis of quasi two-dimensional superconducting ($x-y$) layers with relatively weak inter-layer interaction; see, e.g., Chu (1989), where the high-T_c superconductor Y–Ba–Cu–O is widely discussed.

These circumstances open the road for the research of two- and even one-dimensional superconductivity. As far as low-dimensional systems are considered it must be kept in mind that strong fluctuations may lower the phase transition temperatures in these systems and, in some cases to prevent the macroscopic ordering (see Sections 7.11.1–7.11.4); see also Ginzburg (1989). Despite the strong anisotropy it seems unreliable to approximate the superconductor as a set of independent layers and, therefore, a weak (and in some cases, strong) inter-layer interaction is to be taken into account.

On a phenomenological level the layered structure can be described by considering a one-dimensional $(x - y)$-layer chain along the $\hat{z}$-axis. If the superconducting order parameter ψ of a layer i is ψ_i, the inter-layer coupling can be included, at least in the first approximation, by a weak link (Josephson) term of the type $(\psi_i\psi_j^* + \psi_j\psi_i^*)$ or, in case of coupling only between the nearest neighbour layers, $(\psi_i\psi_{i+1}^* + \text{c.c.})$. Then one should take the GL functional F_i of a single layer i; see Eq. (9.8) or Eq. (9.88) with ψ replaced by ψ_i. The total GL functional of the layered structure will be the sum $\sum_i F_i$. On this line of studies, a phenomenological approach based on the Lawrence–Doniach model of layered superconductors can be developed (Lawrence and Doniach, 1971); see also Brandt (1990a,b), and Ebner and Stroud (1989).

Another way to treat the layer anisotropy is to accept the Ginzburg (1952) suggestion to replace the effective mass m^* in the GL functional with the effective mass tensor m^*_{ij}; see also Werthamer (1969) and Section 9.6. This leads to another generalization of the functionals Eq. (9.81) and (9.98) in which the mass m^* is substituted by the eigenvalues m^*_j of the tensor m^*_{ij}; i.e., the mass terms will be given by the following expression:

$$\sum_j^d \frac{\hbar^2}{2m^*_j} \int d^d x \left|(\nabla_j - iq_0 A_j)\psi\right|^2 . \qquad (9.89)$$

In three-dimensional ($d = 3$) superconductors the layered structure gives $m_x = m_y = m_{||}$ and $m_z = m_\perp$; $i = (x, y, z)$. The justification of this approach to the layered structures and its relationship with the phenomenological treatment based on the Lawrence–Doniach model are discussed by Bulaevskii, Ginzburg and Sobyanin (1988), and Brandt (1990b).

9.7.4 *Fluctuations of the superconducting order parameter*

Now we shall consider the fluctuation rate of the superconducting field ψ. The critical region of superconductors can be written as

$$|T - T_c| < \left(\frac{\tilde{b}}{\alpha_0^2 \xi_0^d}\right)^{2/(4-d)}, \quad 2 < d < 4, \qquad (9.90)$$

where the numerical factor has been omitted, $\tilde{b} = b$ for conventional superconductors and $\tilde{b} = (u + \bar{u} + v)$ for unconventional superconductors; cf. Eq. (7.105) and Eq. (7.235b). The correlation length in the low temperature ($T_c < 20 \div 25$K) superconductors is usually large enough so that the Ginzburg region is extremely small, for example, it is $\Delta T_G \lesssim 10^{-10} \div 10^{-14}$. This indicates that the critical fluctuations are weak and can be neglected. The same reason allows the fluctuation shift of the critical temperature $(T_{c0} - T_c)$ to be set equal to zero.

As the size of the Ginzburg region is a nonuniversal characteristic of the superconductor it depends on the microscopic interaction responsible for the superconductivity. The parameters b, r_0 and ξ_0 and T_c are certainly connected with the microscopic parameters of the system and, hence, the small size of the critical region is a characteristic of a particular superconducting class of materials rather than an unavoidable feature of the superconductivity as a whole.

Specific heat measurement near the critical point of several high-T_c superconductors ($T_c > 70$K) demonstrate a well-established departure from the MF jump of the specific heat; see Inderhees *et al.* (1988), Ginsberg *et al.* (1988), and Fossheim *et al.* (1988). These experiments demonstrate that there are strong fluctuations near the phase transition points of certain high-T_c superconductors. In contrast with the usual superconductors with $T < 25$K, where $\xi_0 \sim 10^3$ Å, in the above mentioned superconductors with strong fluctuations, $\xi_0 \sim 10$ Å.

In ordinary superconductors the critical (interacting) fluctuations act in a very narrow critical region ($\lesssim 10^{-10}$K) but the Gaussian fluctuations are not restricted close to T_c and extend far above T_c giving rise to the phenomenon of paraconductivity; see Skocpol and Tinkham (1975); Tinkham (2004). Because of the much stronger fluctuations this phenomenon should be observed in high-T_c superconductors as well; for a theoretical work; see Lobb (1987), Quader and Abrahams (1988), Shuster (1989), Fujita, Hikami and Brezin (1990); for experiments see Laegreid *et al* (1989) and the review by Annett (1990).

9.7.5 *Magnetic fluctuations in superconductors*

In a nonzero external field the phase transition in superconductors is of the first order and the reason is in the special gauge invariant structure at the GL expansion (9.1). In a zero magnetic field the phase transition is of second order. It occurs at the critical temperature $T_c(H = 0)$; see Fig. 9.6. When ψ is a complex scalar, this second order transition is described within the universality class of the XY model in the magnetism (see Section 5.7 and Section 3.8).

This seems to be true when the magnetic-field fluctuations are neglected. The (H, T) diagrams in Fig. 9.6 are drawn with respect to the mean (thermodynamic) magnetic field H. Let the mean magnetic field H be equal to zero. We shall discuss the properties near the critical temperature $T_c(0)$. For a zero H fluctuations δB of the magnetic induction can always exist and they are expressed by a fluctuation vector potential $\delta B(x) = \nabla \times \delta A(x)$; $\delta A(x)$ is denoted below by $A(x)$. Let us elucidate this point in more details.

In nonmagnetic superconductors, where the mean value $\langle M(x)\rangle = [M(x) - \delta M(x)]$ of magnetization vector $M(x)$ is equal to zero in the normal state in a zero external magnetic field ($H = 0$), the magnetic induction in presence of external magnetic field takes the form:

$$B(x) = H_0 + \delta H(x) + 4\pi\delta M(x), \tag{9.91}$$

where H_0 is the (uniform) regular part of the external magnetic field and $\delta H(x)$ is an irregular part of $H(x)$ created by uncontrollable laboratory effects. We neglect the irregular part $\delta H(x)$ and set $H_0 = 0$, then $B(x)$ contains only a fluctuation part $B(x) \equiv \delta B(x) = 4\pi\delta M(x)$ that describes the diamagnetic variations of $M(x)$ around the zero value $\langle M \rangle = 0$ due to fluctuations $\delta\psi(x)$ of the ordering field $\psi(x)$ above ($T > T_{c0}$) and below ($T < T_{c0}$) the normal-to-superconducting transition at T_{c0}. Note, that the non-fluctuation part $A_0(x) = [A(x) - \delta A(x)]$ corresponds to the regular part $B_0 = (H_0 + \langle M \rangle) = 0$ of $B(x)$ in nonmagnetic superconductors ($\langle M \rangle = 0$) in a zero external magnetic field ($H_0 = 0$). Then we can set $A_0(x) = 0$ and, hence, $\delta A(x) = A(x)$, so we have an entirely fluctuation vector potential $A(x)$, which interacts with the order parameter $\psi(x)$. This interaction can be of type $|\psi|^2 A$ and $|\psi|^2 A^2$ and generates all effects discussed in this Section.

It has been shown by Halperin, Lubensky and Ma (1974) that such vector-potential fluctuations together with the special gauge invariant structure of $F(\psi, A)$ will produce the so-called fluctuation-induced weakly first order transition at $T_c(0)$; *Halperin–Lubensky–Ma (HLM) effect.*

The HLM effect has been theoretically established by both MF-like and RG calculations (Halperin, Lubensky and Ma, 1974). When a perfect superconductor of the first type ($\kappa \ll 1$) is considered, the spatial variations of $\psi(x)$ are very weak in comparison with those of $A(x)$. The reason is in the inequality $\lambda_L \ll \xi$; ξ is the characteristic length of the x-variations of ψ and λ_L describes those of $A(x)$. Thus setting $\psi = \psi_0 = const$ in Eq. (9.81) one obtains the free energy $F(\psi_0, A)$, in which $A(x)$ is the only one fluctuating field. The vector potential $A(x)$ in the partition function

$$\mathcal{Z} \sim \int DA(x)\delta[\nabla \cdot A(x)] e^{-F(\psi_0, A)/T} \sim e^{-F_{\text{eff}}(\psi_0)/T} \tag{9.92}$$

can be integrated out and, hence, one obtains an effective model $F_{\text{eff}}(\psi_0)$, in which a third order term of the type $-a^2 e^2 |\psi_0|^3$ is present (the positive quantity a^2 depends on e^2 and the parameter γ). The field ψ has been taken in the simplest MF approximation of uniform order parameter and for this reason such calculations are often referred to as *MF-like approximation.* The reader can easily perform these calculations using the original paper by Halperin, Lubensky and Ma (1974); see also Section 9.8.6, Eq. (9.142) and Eq. (9.143).

The third order term, $-a^2|\psi_0|^3$, indicates the weakly-first-order transition in type I superconductors caused by the vector potential fluctuations. The transition is *weakly-first-order* because the parameter a^2 is relatively small and, consequently, the temperature interval (ΔT_1) about T_c where this special transition can be observed is very narrow. In fact, ΔT_1 is the ratio of the latent heat to the MF jump of the specific heat. It is given by

$$\Delta T_1 \sim (T_G - T_c)\kappa^{-6}, \qquad (d = 3). \tag{9.93}$$

The weakly first order phase transition can be therefore observed in systems where the size $\Delta T_G = |T_G - T_c|$ of the critical region and the inverse GL parameter κ^{-1} are sufficiently large. In the usual $(T_c \lesssim 25\text{K})$ type I superconductors (where $\kappa \ll 1/\sqrt{2}$), ΔT_1 is estimated (Halperin, Lubensky and Ma, 1974) to be of order 10^{-6}K so that the weakly first transition can hardly be observed (for a more detailed discussion of this problem, see Section 9.8).

According to Halperin, Lubensky and Ma (1974) the phase transition to the superconducting state at zero mean magnetic field $(H_c = 0)$ should always be of weakly first order. Then, as it seems, the experimental observation of such an effect can be made in extremely $(\kappa \ll 1/\sqrt{2})$ type I superconductors where the fluctuations $\delta\psi(x)$ of $\psi(x)$ are relatively weak with respect to the fluctuations $\delta A(x)$ of $A(x)$. The RG based theoretical arguments presented (Halperin, Lubensky and Ma, 1974) in favour of the weakly first order transition are valid for both type I and type II superconductors with the only difference that this special transition can easier be observed in type I materials.

However, detailed Monte Carlo calculations performed by Dasgupta and Halperin (1981) by a lattice variant of the GL functional demonstrate, that in the regime of extreme type II superconductivity ($\kappa \gg 1/\sqrt{2}$; i.e., "small charge q_0 regime"), the transition described by the GL model (9.81) reverts to second order. Another MC study (Bartholomew, 1983) concludes that at fixed effective charge $(|e^2| = 5)$ the HLM effect strongly depends on the GL parameter $\kappa = \lambda/\xi$: it is well established for $\kappa \ll 1$, becomes very weak for $\kappa \sim 1$, and vanishes for large κ. On the basis of this picture, a proposal is made (Bartholomew, 1983) that a tricritical point exists at some "tricritical" value of κ. As κ is related with the charge $|e|$, this proposal seems to be in conformity with other investigations.

Note, that the first prediction for a change of the phase transition order with the variation of κ and for the possibility of "tricritical point" to occur at some value of κ, has been made by (Kleinert, 1982) on the basis of duality arguments (Peshkin, 1978) and analytical calculations; see further development of this approach by (Kiometzis, Kleinert, and Schekel, 1994, 1995; Herbut, 1997; de Callan and Nogueira, 1999). This results are in conformity with the RG studies (Section 9.7.7).

In high-T_c superconductors, the zero temperature London penetration depth λ_0 is effectively of the order of $10^2 \div 10^3$ Å and the zero temperature coherence length is $\xi_0 \sim 10$ Å so that $\kappa \sim 10 \div 10^2$. The estimations of ΔT_G are in favour of the values $\Delta T_G \sim 1$K; for the above values of the parameters see Bardeen, Ginsberg and Salamon (1987), Gor'kov and Kopnin (1988). Therefore, in this class of superconductors $\Delta T_1 \lesssim 10^{-6}$ again and the HLM effect can hardly be experimentally detected. An experimental observation of this effect can be made for superconducting materials where $\Delta T_1 \gtrsim 10^{-3} \div 10^{-4}$ and, according to Dasgupta and Halperin (1981) result, $\kappa \lesssim 1 \div 10$. Recent studies predict a quite enhanced HLM effect in thin superconducting films (see Section 9.8).

9.7.6 *Weakly-first-order smectic A–nematic phase transition*

The HML effect of the fluctuation-induced weakly-first-order transition has been detected in a suitably chosen temperature interval on the line of the nematic-to-smectic A transition in liquid crystals (Anisimov, 1988; Anisimov *et al.*, 1987, 1990); note, that earlier experimental data indicated for second order rather than first order phase transition (Davidov *et al.*, 1979; Litster *et al.*, 1980). The argument that this is the same effect is based on the fact that the nematic–smectic A transition can be described within the de Gennes' liquid crystal model which is quite analogous to the GL functional (9.81) of superconductors; see Lubensky and Chen (1978), and also Toner (1982). The first theoretical prediction about the possibility of such a fluctuation-driven smectic A–nematic first order transition is given by Halperin and Lubensky (1974).

The density of the mass centers of the long molecules in the smectic-A phase of the liquid crystal is described by a complex (wave) function like the order parameter $\psi(x)$ in the superconductors. This periodic function represents the layered structure of the smectic A ordering; see Fig. 3.15. The director vector $n(x)$ of the long molecules, i.e., the nematic order parameter interacts with the smectic-A order parameter $\psi(x)$. After a suitable transformation of $n(x)$ and $\psi(x)$ the nematic order parameter can be substituted by a gauge field $A(x)$, which has the gauge properties of the vector potential of the magnetic field (Lubensky and Chen, 1978). Apart from the specific liquid crystal anisotropy the de Gennes model (Lubensky and Chen, 1978; Anisimov *et al.*, 1990) of the smectic A–nematic transition has the same gauge invariant form as the model (9.81) and this structure of the liquid crystal effective Hamiltonian leads to the theoretical prediction of the HLM effect.

The attractive features of searching for the experimental observation of the HLM effect at the nematic–smectic A transition is the existence of a nematic–smectic A tricritical point. Near this TP the fluctuations of the smectic order parameter $\psi(x)$ are relatively weak and can be neglected; $\psi(x) \approx$ *const*. Therefore, near the TP the original MF-like arguments will be valid; for details, see Anisimov (1988), and Anisimov *et al.* (1990). Triple and tricritical points in nematic-smectic-A-smectic-C systems and relationships with the present problem are discussed also by Huang and Lien (1981), Garland and Nounesis (1994), and Lelidis and Durand (1994).

The appearance of a third-order term ($\sim |\psi|^3$) in the effective free energies $F_{\text{eff}}(\psi_0)$ of the superconductor and the smectic A liquid crystal can be interpreted as a fluctuation ($-\delta A$) creation of an isolated Landau point (see Section 4.7.3). A point of this type is supposed to exist, under certain conditions on the line of another liquid crystal transition — the nematic-to-isotropic liquid transition; see, e.g., Vause and Sak (1978).

9.7.7 *Renormalization group investigations*

In order to perform RG study of the GL functional (9.81) we must write it in a form which is valid for dD superconductors. As the vector product $\nabla \times A(x)$ is defined only for 3D space, we will write the last term in Eq. (9.81) – the pure magnetic energy term, in a more general form

$$\frac{B^2}{8\pi} = \frac{1}{16\pi} \sum_{i,\, j=1}^{d} \left(\frac{\partial A_j}{\partial x_i} - \frac{\partial A_i}{\partial x_j} \right)^2 . \tag{9.94}$$

where the vector potential $A(x)$ obeys the Coulomb gauge $\nabla \cdot A(x) = 0$. For 3D superconductors the relation $B(x) = \nabla \times A(x)$ follows directly from the relation (9.93), which reads $B^2 = (\nabla \times A)^2$.

When $B(x) = B_0$ is uniform along the $\hat{z}$- axis, the Landau gauge $A_0(x) = B_0(-y/2, -x/2, 0)$ can be applied. This representation can be generalized for $(D > 2)$-dimensional systems, where the magnetic induction B_0 is a second rank tensor:

$$B_{0ij} = B_0(\delta_{i1}\delta_{j2} - \delta_{j2}\delta_{i1}). \tag{9.95}$$

If we use the vector notation $x = (x_1, x_2, r)$, where r is a $(D-2)$-dimensional vector, perpendicular to the plane (x_1, x_2), in the $3D$ case we have $r = (0, 0, z)$, and

$$B_j = \frac{1}{2}\epsilon_{jkl}B_{0kl} = B_0\delta_{j3} \, , \tag{9.96}$$

where ϵ_{jkl} is the antisymmetric Levi-Civita symbol; see, e.g., Lawrie (1983), and Shopova and Uzunov (2008). The Landau gauge and Eqs. (9.94)–(9.95) can be used for uniform $B = B_0$ when the spatially dependent $\delta B(x)$ fluctuations are neglected.

In the prevailing part of our consideration we will use the general Coulomb gauge of the field $A(x)$, which does not exclude spatial dependent magnetic fluctuations $\delta B(x)$. The more restrictive gauge, described by Eqs. (9.94)–(9.95), is important for some RG studies, where the magnetic induction is assumed to be uniform (Brézin, Nelson and Thiaville, 1985; Affleck, 1985; Brézin, Fijita, and Hikami, 1990; Mikitik, 1992; Radzihovsky, 1995, 1996) (Section 9.7.7.5). The approximation of uniform induction $B = (H + 4\pi M)$ can be justified for cases of relatively large and uniform external magnetic field H_0, when the magnetic fluctuations $\delta B(x) = 4\pi\delta M(x)$ can be ignored. Then one may consider the net effect of the uniform external magnetic field H_0 and the uniform fluctuation δB_0. Obviously, this scheme is not applicable in a close vicinity of the critical point $T_c(H = 0)$.

The RG studies of GL functional begin with the paper by Halperin, Lubensky and Ma (1974) where the RG equations within the one-loop (ϵ^1-) approximation are obtained and analyzed. For a more thorough investigation of the stability properties of the RG equations the scalar order parameter ψ in the usual GL functional is replaced by a $(n/2)$-component complex field: $\psi = \{\psi_\alpha, \alpha = 1, \ldots, n/2\}$. The RG equations for the model (9.81) can be derived by the method described in Section 8.7; see also Figs. 9.7(a, b). The reader may obtain these equations (Halperin, Lubensky and Ma, 1974) using the bare correlation function $G_0(k) = \langle|\psi_{\alpha(k)}|^2\rangle$ for the ψ_α-fields of the type $G_0^{-1}(k) = (r + \gamma k^2)$ and the bare correlation function of the A_i field in the Coulomb gauge of the type:

$$G_0^A(k, k') = \langle A_i(k)A_j(k')\rangle = G^A_{(0)ij}(k)\delta(k + k'), \tag{9.97}$$

where

$$G^A_{(0)ij} = \frac{4\pi}{k^2}\left(\delta_{ij} - \hat{k}_i\hat{k}_j\right); \tag{9.98}$$

where $\hat{k} = k/|k|$ is the unit vector, hence, $\hat{k}_j = k_j/|k|$. This form of the correlation function for the vector potential can be obtained by taking into account the Coulomb gauge $k.A(k) = 0$; see also Abrikosov, Gor'kov, and Dzyaloshinskii (1962), Lifshitz and Pitaevskii (1980).

Another important point in the derivation of the RG equations is related with the special structure of the model (9.81). The number of the vertex "parameters" is six and it is equal to the number of the possible vertices of type $|\psi|^2$, $|\nabla\psi|^2$, $|\psi|^4$, $(A\psi\nabla\psi^* + cc)$, $A^2|\psi|^2$, and the term given by Eq. (9.93). The RG equations will be written for the vertex "parameters" r, $\gamma = \hbar^2/4m$, u, γq_0, γq_0^2, and $1/8\pi$, where $q_0 = (2e/\hbar c)$ (in other units $1/8\pi$ is replaced by $1/8\pi\mu_0$, where μ_0 is the magnetic permeability of the vacuum).

As usual, the infrared FPs relevant to the phase transition problem are obtained from the requirement γ and $(1/8\pi)$, i.e., the factors in the front of the gradient terms to be invariant under the RG transformation. Performing the calculations in this way, we obtain three RG equations for γ, (γq_0), (γq_0^2) for two parameters γ and q_0. The RG calculations show that the RG transformation is consistent so that the three equations γ, (γq_0), and (γq_0^2) can be represented by two independent RG equations for γ, q_0. Another specific feature of the GL functional (9.81) is that the charge q_0 participates in the RG equations as a parameter of order $\epsilon^{1/2}$ and the actual charge-dependent parameter which is used in the analysis of the RG equations is the square of the charge $q_0^2 \sim \epsilon = (4 - d)$. Because of this feature of the model (9.81) the complete derivation of the RG equations for (γq_0), (γq_0^2) requires the investigation of the standard perturbation series up to the fourth order in (γq_0).

The last peculiarity of the GL functional (9.81) which will be mentioned here is that this model loses its gauge invariance, see Eqs. (9.86a) and (9.86b), when the Wilson-Fisher RG scheme is applied. The RG generates mass (k-independent) terms in the renormalized $\langle|A_j(k)|^2\rangle$ correlation function; see Eqs. (9.97)–(9.98) and the one-loop diagram in Fig. 9.7(c). Once such mass terms appear the gauge invariance of the renormalized model breaks down.

The inconsistency between the fundamental requirement for a gauge invariant form of the renormalized GL model and the Wilson–Fisher RG

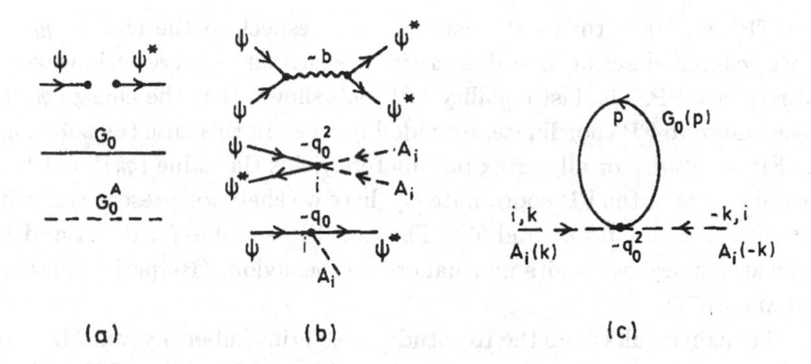

Fig. 9.7 (a) Diagrammatic symbols for G_0, G_0^A, ψ_α, ψ_α^* and A_i; (b) interaction vertices for the model (9.81); (c) the first-order mass diagram for G_0^A.

scheme is resolved by removing the non-zero symmetry-breaking diagrams such as that in Fig. 9.7(c). Thus, one obtains reliable conclusions from the RG analysis to the first order in $\epsilon = (4-d)$ but calculational difficulties appear in extending the RG investigation to the second order in ϵ; see also Chen, Lubensky and Nelson (1978) and the paper by Kölnberger and Folk (1990) discussed below. The application of dimensional regularization makes possible to give a solution of this problem (Kölnberger and Folk, 1990).

Here we present the RG relations in one-loop approximation (Halperin, Lubensky and Ma, 1974):

$$\begin{aligned}
a' &= e^{(2-\eta_\psi)l}\left[a + \frac{(n+2)b}{16\pi^2\gamma}\left(\frac{1-e^{-2l}}{2} - al\right) + \frac{3\gamma q_0^2}{4\pi}\left(1-e^{-2l}\right)\right], \\
b' &= e^{(\epsilon-2\eta_\psi)l}\left[b - \frac{(n+8)b^2}{16\pi^2\gamma^2}l - 12\left(\gamma q_0^2\right)^2 l\right], \\
\gamma' &= e^{-\eta_\psi l}\gamma\left(1 - \frac{3}{2\pi}q_0^2 l\right), \\
1 &= e^{-\eta_A l}\left(1 + \frac{n}{12\pi}q_0^2 l\right), \\
(q_0^2)' &= e^{(\epsilon-\eta_A)l}q_0^2.
\end{aligned} \tag{9.99}$$

It is seen from the last relation in (9.99) that the charge q_0 always has a FP value $q_0^* - 0$. This value corresponds to the usual critical behaviour, when the system is described by the pure $|\psi|^4$-model and magnetic fluctuations are not taken into account. The further analysis shows that the usual FPs

— GFP and HFP, turn out unstable with respect to the charge q_0 and, hence, cannot describe a stable critical behaviour. There is however an entirely new FP. The last equality in (9.99) shows, that the charge q_0 may have a nonzero FP coordinate, provided $\eta_A = \epsilon$. In this case the solution of the FP equations for all vertex parameters yields the value $(q_0^2)^* = 12\pi\epsilon/n$ for the square of the FP coordinate q_0^* (here we shall not present the values of the FP coordinates a^* and b^*). The new FP is stable for $d < 4$ and $n > 365.9$ and describes a quite unusual critical behaviour (Halperin, Lubensky and Ma, 1974).

The main results from the RG study (Halperin, Lubensky and Ma, 1974) is that the RG equations do not give a stable real FP, i.e., a FP with real coordinates μ^*, $\mu = (r, u, \gamma, q_0^2)$, for symmetry indices $n < 365.9$. For conventional superconductors $n = 2$ and for unconventional superconductors $n = 4, 6$ or higher, there are no physical arguments for the existence of real systems with, for example, symmetry indices of the ordering $n \gtrsim 300$ or 400.

The absence of a stable FP for $d < 4$ and $n < 365.9$ or, more precisely, the existence of a single unstable GFP implies that the considered transition is not the standard second order transition (see also Section 8.8.3). Bearing in mind the MF-like calculations which lead to a third-order ψ-term one can conclude that this fluctuation-driven transition is nothing else but a (weakly) first order transition. In contrast to the MF calculations at constant ψ, the RG calculations are valid for both first and second type superconductors.

The RG flows and the runaway of the RG trajectories near the unstable FPs have been investigated (Chen, Lubensky and Nelson, 1978) with the help of the Rudnick–Nelson method of integration of the differential RG equations (Rudnick and Nelson, 1976). A systematic study of the HLM effect has been performed by Lawrie (1982, 1983). It has been shown by Lawrie that there should be an effective value of the charge q_0, below which the phase transition at the critical point of superconductors is of the second order. This result is certainly consistent with the Monte Carlo studies by Dasgupta and Halperin (1981) and Bartholomew (1983), mentioned in Section 9.7.5; see also Eq. (9.85b), where $\lambda_L \sim 1/q_0$ and, hence the lowering of the effective charge (q_0) leads to an increase of the GL parameter κ. The one-loop analysis performed by (Lovesey, 1980) of 2D and 3D superconductors with annealed disorder confirms the HLM effect as well as the availability of second order phase transition in the Hartree limit $(n \to \infty)$.

The RG equations for the model (9.81) within the two-loop approximation have been derived and analyzed by Kölnberger and Folk (1990). The two-loop calculations reveal new stability properties of the RG equations but the results do not contradict to the initial HLM prediction of a weakly-first-order transition in conventional ($n = 2$) superconductors. Kölnberger and Folk (1990) have applied the field theoretical version of RG (see Chapter 1), where, owing to the dimensional regularization, diagrams such as that in Fig. 9.7(c) do not give contributions to the RG equations; see also Leibbrandt (1975), Collins (1984), and Schloms and Dohm (1989).

The two-loop RG studies have been further developed and reviewed by Folk and Holovatch (1996, 1999). These two-loop RG studies show that the account of higher orders of the loop expansion allows a lowering of the critical value $n_c = 365.9$, but the RG equations still have no real FP at $D = 3$ and $n = 2$ (conventional 3D superconductors), except for a treatment by Padè analysis. In $(2 + \epsilon)$ dimensions, the nonlinear σ model exhibits a second-order phase transition for all values of $n > 0$ (Lawrie, 1982). Owing to this result and more extensive studies (Freire and Litim, 2001; Bergerhoff *et al.*, 1996) of the dependence of the effective free energy on the GL parameter κ, one may suppose that the critical value n_c vanishes at some dimension $2 < d < 4$, that is, that a tricritical point "located" at some specific value of κ exists, where the superconducting phase transition changes its order; see also, a discussion by (Lawrie, 1997). RG studies (Bergerhoff *et al.*, 1996; Herbut and Tesenović, 1996) at fixed dimension d have also been carried out with the aim to determine the phase transition order; see, also, the comment by Lawrie (1997).

9.7.7.1 *Quantum effects and critical dynamics*

The dynamical criticality of the ψ-field has been investigated on the basis of a quantum ψ^4-model of superconductors, derived from the microscopic theory of superconductivity (Uzunov, 1980). The ψ field contains the imaginary time τ and after an analytical continuation to real time t allows for the calculation of the relaxation time and the dynamic critical exponent (see the discussion in Sections 9.3 and 9.4). The dynamic critical exponent z has been calculated to second order in $\epsilon = (4 - d)$. The result for z coincides with the respective result for the so-called model A of critical dynamics (lack of conservation laws) of the time dependent classical GL model (TDGL) (Uzunov, 1980); for the dynamics within the TDGL see Hohenberg and Halperin(1977).

In a generalized RG treatment, which includes the effects of the gauge field $A(x)$ on the critical dynamics of superconductors, the RG equations have been analyzed in one-loop approximation and the dynamic critical exponent has been obtained in the form $z = 2 + 18\epsilon/n$ (Bushev and Uzunov, 1980); latest RG studies (Lannert, Vishveshwara, and Fisher, 2004; Dudka, Folk, and Moser, 2007), performed in somewhat different theoretical framework, confirm this result. Besides it has been explicitly shown on the basis of a general GL functional, written in terms of Bose fields $\psi(x,\tau)$, that the quantum effects do not affect the behaviour in the asymptotic vicinity of the phase transition point and the RG results by Halperin, Lubensky and Ma (1974) remain valid for any $T > 0$ (Bushev and Uzunov, 1980). In quantum limit $T \to 0$ the picture may change and new instability may occur due to the classical-to-quantum dimensional crossover.

Another paper (Fisher and Grinstein, 1988) on a gauge model of type (9.81) is intended to the investigation of the quantum phase transition ($T_{c0} = 0$) in granular superconductors. The simultaneous effect of the local $U(1)$ symmetry, disorder and quantum fluctuations has also been considered in studies of disordered electronic systems at $T = 0$ (Fisher, Grinstein, and Girvin, 1990; Herbut, 1998).

Continentino (2001) discusses in details the quantum phase transition at zero temperature from superconducting state to the state of normal metal. In the limit $T \to 0$, a dimensional crossover of type $d \to (d+1)$ occurs due to the quantum fluctuations of the gauge field $A(\tau, x)$, where τ is the imaginary time (Section 9.3.2). The effective free energy of the 3D "quantum system" at $T = 0$ corresponds to the free energy of the respective 4D classical system. Thus the MF like analysis of 3D superconductor in the zero-temperature limit $T \to 0$ will give a fluctuation-driven first order phase transition, described by a logarithmic term of type $|\psi|^4\ln|\psi|$ rather than the $|\psi|^3$-term (Sections 9.7.5 and 9.8). Accordingly, in the same zero-temperature limit ($T_c = 0$) the 2D and q2D superconductors will be described by an effective free energy, containing a $|\psi|^3$-term as is in 3D superconductors with $T_c \neq 0$. The complete analysis of the quantum phase transitions ($T_c = 0$) of type II superconductors beyond the MF-like approximation (Section 9.8) requires the account of both thermal and quantum fluctuations of the order parameter field $\psi(\tau, x)$ with the help of RG; for example, such fluctuations have been systematically taken into account by Fisher and Grinstein (1988); Fisher, Grinstein, and Girvin (1990).

9.7.7.2 *Disorder effects on the critical behaviour of superconductors*

Disorder effects of type random impurities and random fields in superconductors have been also studied; see Boyanovsky and Cardy (1982b), Uzunov, Korutcheva and Millev (1983), Athorne and Lawrie (1985), and Busiello, De Cesare and Uzunov (1986). The RG analysis of the fluctuation effects near the critical field H_{c_2} in type II superconductors has been performed by Brézin, Nelson and Thiaville (1985).

The random field effect on the critical behaviour of conventional superconductors has been investigated in one-loop RG approximation by Busiello, De Cesare and Uzunov (1986). Both short-range and long-range random correlations (Section 9.5.3) have been used to show that the HLM effect still exists under the action of the random field. Disorder effects for unconventional superconductors are discussed in Section 9.7.7.3.

Here we will discuss mainly disorder described by quenched impurities and inhomogeneities with Gaussian distributions (Sections 9.5.1 and 9.5.2). The effects of annealed and quenched disorder in classical versions of Abelian-Higgs models, equivalent to the GL model (1) have been investigated by Hertz (1978) in the context of the theory of spin glasses in case of Dzyaloshinskii-Moriya interaction (Hertz, 1985). The specific feature of this approach is that the disorder is associated with the vector gauge field $A(x)$ and can be used to describe superconductors in random, uncorrelated, or, which is the same, short-range (δ-) correlated magnetic fields rather than usual ones. In high energy physics this approach makes contact with gauge fields, where a Higgs field is coupled to a random color field. It is natural to expect, as has been proven Lovesey (1980) and Hertz (1978), that the annealed gauge model will bring a fluctuation-driven first order phase transition at usual symmetry indices ($n \sim 2$) and a second order phase transition in the Hartree limit ($n \to \infty$). Lovesey (1980) has shown in a calculation in one-loop order that the phase transition in case of quenched impurities is most probably of second order, in contradiction to the conclusion for a lack of stable nontrivial FP of the RG equations within the same one-loop order (Hertz, 1978). The origin of this discrepancy is clear from the argument that the lack of stable FP may result without any change of the Hamiltonian structure, whereas the conclusion for the second order of phase transition in the paper by (Lovesey, 1980) has been made, perhaps, incorrectly, based on the argument of the $|\psi|^3$-term absence for $D = 3$ and the $|\psi|^2 \ln|\psi|$-term absence for $D = 2$ (see Section 9.8).

More usual case of quenched disorder of type random critical temperature in Abelian-Higgs models has also been considered (Boyanovsky and Cardy, 1982b; Uzunov, Korutcheva and Millev, 1983). This is the case as described in Section 9.5.1, when the parameter a has a random part $\varphi(x)$, i.e. a is substituted with $\tilde{a} = a + \varphi(x)$, where the Fourier amplitudes $\delta(k)$ of the *random* function $\varphi(x)$ obey, for example, the Gaussian distribution

$$\langle \varphi(k)\varphi(k')\rangle_R = \Delta(k)\delta_{-k,k'}. \tag{9.100}$$

Here the parameter $\Delta(k) \geq 0$ is a non-negative function for any $0 \leq |k| \leq \Lambda$ (henceforth, we set $\Lambda = 1$)(see Section 9.5.1). In case of the so-called "uncorrelated", or, (δ-) short range correlated quenched impurities [$\Delta(k) \equiv \Delta = const$], the system exhibits a spectacular competition between the impurities and gauge effects. We shall show this by the example of the RG equations derived by Uzunov, Korutcheva and Millev (1983). Using units, in which $k_B = T_{c0} = \Lambda = 1$ and the notations $q_0 = (2e/\hbar c)$ and $\gamma = \hbar^2/4m$, the RG equations can be written in the form

$$\begin{aligned}
a' &= e^{(2-\eta_\psi)l}\left[a + \frac{1}{8\pi^2\gamma}\left(\frac{n+2}{2}b - \Delta\right)\left(\frac{1-e^{-2l}}{2} - al\right)\right. \\
&\quad \left. + \frac{3\gamma q_0^2}{4\pi}\left(1 - e^{-2l}\right)\right], \\
b' &= e^{(\epsilon-2\eta_\psi)l}\left\{b + \frac{b}{8\pi^2\gamma^2}\left[6\Delta - \frac{1}{2}(n+8)b\right]l - 12\left(\gamma q_0^2\right)^2 l\right\}, \\
\Delta' &= e^{(\epsilon-2\eta_\psi)l}\left\{\Delta + \frac{\Delta}{8\pi^2\gamma^2}\left[4\Delta - (n+2)b\right]l\right\}, \\
\gamma' &= e^{-\eta_\psi l}\gamma\left[1 - \frac{3}{2\pi}q_0^2 l\right], \\
1 &= e^{-\eta_A l}\left(1 + \frac{n}{12\pi}q_0^2 l\right), \\
q_0' &= e^{(\epsilon-\eta_A)l/2}q_0.
\end{aligned} \tag{9.101}$$

When the disorder is neglected ($\Delta = 0$; pure superconductors) the Eqs. (9.101) coincide with those for pure superconductors and give the results discussed in Section 9.7.7; see Eq. (9.99). For impure superconductors ($\Delta > 0$), these equations describe a new stable FP of focal type (Uzunov, Korutcheva and Millev, 1983) for dimensions $D < 4$, which exists for symmetry indices $n > 1$ and has a physical meaning at dimensions $D > D_c(n) = 2(n+36)/(n+18)$ (Uzunov, Korutcheva and Millev,

1983). In the impure superconductor, the new focal FP occurs exactly in the domain of symmetry indices $n < 365.9$, where real FP does not exist for pure superconductors (Section 9.7.7).

Having in mind the asymptotic nature of the ϵ-expansions within the RG approach, one may conclude that this focal FP governs the critical behavior at the real dimension $d = 3$ in conventional superconductors ($n = 2$), although the direct substitution of $n = 2$ in the above expression for the "lower critical" dimension $d_c(n)$ yields $d_c(2) = 3.8$. This problem has been further investigated by Athorne and Lawrie (1985). Long-range quenched disorder within the same RG approach to the model (1) has been considered by Korutcheva and Millev (1984).

9.7.7.3 *Renormalization group study of unconventional superconductors*

The influence of orthorhombic and cubic anisotropy on the phase transition properties of the unconventional superconductors in a zero magnetic field described by the functional (9.88) with $A(x) \equiv 0$ has been studied within the RG method up to the two-loop approximation (Blagoeva *el al.*, 1990); for preceding one-loop RG results see Blagoeva *et al.* (1989). The anisotropy, especially the orthorhombic $\bar{u}$-term, representing the symmetry of the superconducting electron pairs, is the reason for the absence of stable FPs of the RG equations for $n > 4 - 2\epsilon$. When the orthorhombic ($\bar{u}$-) term in Eq. (9.88) is neglected a new (complex-cubic) FP becomes stable for symmetry indices $n > 4 - 2\epsilon$. The values of the critical exponents corresponding to this complex-cubic FP are very close to those of the exponents for the ordinary cubic FP (Aharony, 1973, 1976). Therefore, the unconventional superconductors with the orthorhombic anisotropy in a zero external magnetic field, and described by the GL functional (9.88) where the magnetic fluctuations are neglected ($A = 0$), will exhibit an anisotropy-driven first order transition into the critical region while the unconventional superconductors with cubic anisotropy ($\bar{u} = 0$) will exhibit the critical behaviour described by the cubic FP; see also Uzunov (1991).

When the magnetic fluctuations are taken into account ($A \neq 0$) in the model (9.88) the RG analysis (Millev and Uzunov, 1990) reveals two different sources of fluctuation-driven first order transitions at the critical point $T_c(0)$ of superconductors: the anisotropy and the magnetic fluctuations ($A \neq 0$). The effect of quenched impurities of type "random T_c" for the GL functional (9.88) have been investigated by Busiello and Uzunov (1990), and Busiello *et al.* (1991). The simultaneous effect of the spatial anisotropy, the thermal fluctuations and the quenched disorder may cause a drastic change in the properties of the Meissner and vortex phases in type II superconductors. The strong disorder can produce, under certain conditions, new Meissner glass and vortex glass superconducting phases; see, e.g., Fisher (1989), Fisher *et al.* (1989).

9.7.7.4 *Multi-criticality in superconductors*

Other critical properties related to the effect of the magnetic fluctuations in superconductors with a supplementary magnetic (or nonmagnetic) ordering have been studied by Grewe and Schuh (1980), Tonchev and Uzunov (1981b), Schuh and Grewe (1982), and Kopeć and Kozlowski (1983b). It has been shown that both bicritical and tetracritical critical phenomena are unstable with respect to the charge (q_0) and that the results for the weakly-first order phase transition are essentially changed for such multicritical points (Tonchev and Uzunov, 1981b). Another application of gauge models with the structure of the GL functional (9.81) has been made in the theory of spin glasses with Dzyaloshinski–Moriya interactions; see Hertz (1978), Dzyaloshinskii and Obukhov (1982), Ritala (1984).

9.7.7.5 *External field effect*

In real experiments the external magnetic field can hardly be completely eliminated. The study (Lawrie, 1983) of the HLM effect in the wider context of uniform external magnetic field H_0, shows that the weakly first phase transition, discussed so far can be obtained in Landau gauge from a renormalized theory of the equation of state in one loop order; see also Ref. (Lawrie, 1983). The phase transition at the second critical magnetic field H_{c2} has been studied by a $\epsilon = (6 - d)$ expansion within the one loop approximation with the conclusion that the magnetic fluctuation effects should produce a fluctuation induced first order phase transition (Brézin, Nelson and Thiaville, 1985). Further investigations of this problem have been performed in the Hartree limit ($n \to 0$) (Affleck, 1985; Radzihovsky, 1995), and in higher orders of the loop expansion (Brézin, Fijita, and Hikami, 1990; Mikitik, 1992); for comments, see Radzihovsky (1996), and Herbut and Tesenović (1996).

9.8 Magnetic Fluctuations in Thin Superconducting Films

9.8.1 *Preliminary notes*

Here we continue the discussion in Section 9.7.5 of magnetic fluctuations in superconductors. We shall use the MF-like approximation in which the field ψ is taken uniform and only spatially dependent magnetic fluctuations are taken into account. In this case the partition function (9.92) can be calculated without further approximations, and we shall consider the results for dD systems, in particular, for 2D, q2D, 3D and 4D superconductors; for more details, including applications to real superconductors, see papers by Folk, Shopova, and Uzunov (2001); Shopova, Todorov, and Uzunov (2003); Shopova and Todorov (2003a,b,c) and review articles by Shopova, Tsvetkov and Uzunov (2006); Shopova and Uzunov (2008).

In the present consideration we accept periodic boundary conditions for the superconductor surface. This means to ignore the surface energy including the additional energy due to the magnetic field penetration in the surface layer of thickness equal to the London penetration depth $\lambda_L(T)$. This approximation is adequate for superconductors of thickness $L_0 \gg \lambda(T) \gg a_0$, where a_0 is the lattice constant and $L_0 = \min\{L_i, i = 1, ..., D\}$; the volume of the superconductor is given by $V \equiv V_d = L_0 L_1 \dots L_{d-1}$. As we suppose the external magnetic field to be zero ($H_0 = 0$) or very small in real experiments, the requirement $L_0 \gg \lambda(T)$ can be ignored and we have the simple condition $L_0 \gg a_0$.

In this Section the wave vector will be denoted by $k = (k_0, k_1, \dots k_{d-1})$. In microscopic models of periodic structures the periodic boundary conditions confine the wave vector components $k_l = \{k_l = (2\pi n_l/L_l);\ l = 0, 1, ..., (d-1)\}$ in the first Brillouin zone $[-(\pi/a_0) \le k_l < (\pi/a_0)]$ and the expansion of their values beyond this zone can be made either by neglecting the periodicity of the crystal structure, or, on the basis of assumption that large wave vector moduli $|k|$ have a negligible contribution to the calculated quantities. The last argument is widely accepted in the phase transition theory, where the long-wavelength limit ($ka_0 \ll 1$) can be used (here, as well as at certain places below, $k = |k|$). In particular, this argument is valid in the continuum limit ($V_D/a_0^D \to \infty$). Therefore, for both crystal and non-periodic structures we can use the cutoff $\Lambda \sim (\pi/a_0)$ and afterwards extend this cutoff to infinity, provided the main contributions in the summations over the wave vector k come from the relatively small wave numbers ($k \ll \Lambda$). In fact, this is a quasi-macroscopic description based on the GL functional Eq. (9.81), so the microscopic phenomena are excluded from our consideration.

The GL free energy functional takes into account phenomena with characteristic lengths ξ_0 and λ_0 or larger (ξ and λ_L). In low-temperature superconductors ξ_0 and λ_0 are much bigger than the lattice constant a_0. Having in mind this argument we will assume that $\Lambda \ll (\pi/a_0)$. Whether the upper cutoff Λ for the wave vector modulus $|k|$ is chosen to be either $\Lambda \sim 1/\xi_0$ or $\Lambda \sim 1/\lambda_0$ is a problem that has to be solved by additional considerations. According to arguments presented by Folk, Shopova, and Uzunov (2001) we will often make the choice $\lambda \sim \xi_0^{-1}$ (this choice will be justified at a next stage of our discussion).

We will use the usual Fourier expansion

$$A_j(x) = \frac{1}{V_d^{1/2}} \sum_k A_j(k) e^{ikx} \tag{9.102}$$

and

$$\psi(x) = \frac{1}{V_d^{1/2}} \sum_k \psi(k) e^{ikx}, \tag{9.103}$$

where the Fourier amplitudes $A_j(k)$ obey the relations $A_j^*(k) = A_j(k)$ and $k \cdot A(k) = 0$. The Fourier amplitude $\psi(k)$ is not equal to $\psi^*(-k)$ because $\psi(x)$ is a complex function. For the same reason $\psi(0) \equiv \psi(k=0)$ is a complex number.

The effect of the superconducting fluctuations $\delta\psi(x)$ on the phase transition properties is very weak, as in usual superconductors, and is restricted in a negligibly small vicinity ($|t_0| \sim 10^{-12} \div 10^{-16}$) of temperature T_{c0}. This allow us to neglect the fluctuations $\delta\psi(x)$ and use only the mean value of ψ: $\psi \approx \langle\psi(x)\rangle$; Therefore, we apply the MF approximation with respect to the order parameter $\psi(x)$. Within this approximation we will take into account the magnetic fluctuations $\delta A(x)$ of $B_0 = 0$, i.e., $A(x) = \delta A(x)$ (see Section 9.7.5).

Furthermore, the fluctuations $A(x)$ can be integrated out from the partition function, defined by

$$Z_d(\psi) = \int DA\, e^{-F(\psi,A)/k_BT}, \tag{9.104}$$

where a label d has been added to Z to indicate that we work with dD systems, and we have restore the Boltzmann constant k_B. The integral symbol in Eq. (9.104) denotes

$$\int DA \equiv \int_{-\infty}^{\infty} \prod_{j=1}^{d} A_J(0) \prod_{k>0}^{|k|<\Lambda} d\Re A_j(k) d\Im A_J(k) \delta[k \cdot A(k)]. \quad (9.105)$$

Note, that only the independent field amplitudes are integration variables and this fact is taken into account in the expression (9.105) (see Section 7.3). The effective free energy $F_d(\psi)$ will be given $F_d = -k_B T \ln Z_d$.

The magnetic fluctuations will be completely taken into account, if only we are able to solve exactly the integral (9.104). The exact solution of this integral can be done for a uniform order parameter ψ. But it seems more instructive to present the same exact solution as a summation of a simple loop expansion and we will follow this way of calculation.

The uniform value of ψ is different from the MF value of $\psi_0 = \langle\psi\rangle$, because the uniform fluctuations of $\psi(x)$ always exist, so we should choose one of these two possibilities (Folk, Shopova, and Uzunov, 2001; Shopova and Todorov, 2003a). The problem of this choice arises after calculating the integral (9.104), at the next stage of consideration, when the effective free energy $F_d(\psi)$ is analyzed and the properties of the superconducting phase ($|\psi| > 0$) are investigated. The effective free energy is a particular case of the effective thermodynamic potential in the phase transition theory and we must treat the uniform ψ in the way prescribed in the field theory of phase transitions. It will become obvious from the next discussion that we will use a loop-like expansion, which can be exactly summed up to give a logarithmic dependence on $|\psi|$.

We begin our investigation by setting ψ uniform but at some stage we will also ignore the uniform fluctuation $\delta\psi$ and deal only with the equilibrium value ψ_0 of ψ. The equilibrium value will be calculated after taking into account magnetic fluctuations, so it will be different from the usual result $|\psi_0| = (|a|/b)^{1/2}$, where both magnetic and superconducting fluctuations are ignored. This simplest approximation for the equilibrium value of ψ is obtained from the GL free energy (9.81), provided $e = 0$ and the gradient term is neglected. Hereafter we will keep the symbol ψ_0 for the equilibrium order parameter in the more general case, when the magnetic fluctuations are not neglected, and will denote the same quantity for $e = 0$ by $\eta \equiv |\psi_0(e = 0)| = (|a|/b)^{1/2}$.

The above described approximation neglects the saddle point solutions of GL equations, where $\langle\psi(x)\rangle$ is x-dependent. Therefore, the vortex state that is stable in type II superconductors cannot be achieved. This is consistent with setting the external magnetic field to zero, so the vortex state cannot occur in any type superconductor. Our arguments can be easily verified with the help of GL equations Lifshitz and Pitaevskii(1980) for a zero external magnetic field; the only nonzero solution for ψ in this case is given by $\eta = (|a|/b)^{1/2}$, although the magnetic fluctuations $A(x) = \delta A(x)$ are properly considered.

9.8.2 General form of the effective free energy

When the order parameter ψ is uniform the functional (9.81) is reduced to

$$F(\psi, A) = F_0(\psi) + F_A(\psi) \tag{9.106}$$

with

$$F_0(\psi) = V_d(a|\psi|^2 + \frac{b}{2}|\psi|^4) \tag{9.107}$$

and

$$F_A(\psi) = \frac{1}{8\pi}\int d^dx \left\{\rho(\psi)A^2(x) + \frac{1}{2}\sum_{i,j=1}^{d}\left(\frac{\partial A_j}{\partial x_i} - \frac{\partial A_i}{\partial x_j}\right)^2\right\}. \tag{9.108}$$

Here $\rho = \rho_0|\psi|^2$ and $\rho_0 = (8\pi e^2/mc^2)$. It is convenient to calculate the partition function $Z(\psi)$ in the k-space, where Eq. (9.108) takes the form

$$F_A(\psi) = F_A(0) + \Delta F_A(\psi), \tag{9.109}$$

where

$$F_A(0) = \frac{1}{8\pi}\sum_{j,k} k^2 \left|A_j(\vec{k})\right|^2, \tag{9.110}$$

and

$$\Delta F_A(\psi) = \rho \sum_{j,k} |A_j(k)|^2 \tag{9.111}$$

(note, that we have used the gauge $k.A(k) = 0$).

Then the partition function becomes

$$Z(\psi) = e^{-F_0(\psi)/k_BT} Z_A(\psi), \tag{9.112}$$

where

$$Z_A(\psi) = \int DAe^{-F_A(\psi)/k_BT} \tag{9.113}$$

with $F_A(\psi)$ given by (9.109) and the functional integration defined by the rule (9.105). The effective free energy $F_D(\psi)$ will be

$$F_D(\psi) = F_0(\psi) + F_f(\psi), \tag{9.114}$$

where $F_0(\psi)$ is given by Eq. (9.107) and

$$F_f(\psi) = -k_BT \ln\left[\frac{Z(\psi)}{Z(0)}\right] \tag{9.115}$$

is the ψ-dependent fluctuation part of $F(\psi)$. In Eq. (9.114) the ψ-independent fluctuation energy $\{-k_BT \ln[\mathcal{Z}_A(0)]\}$ has been omitted. This energy should be ascribed to the normal state of the superconductor. Defining the statistical averages as

$$\langle(...)\rangle = \frac{\int DAe^{-F_A(0)/k_BT}(...)}{Z_A(0)}, \tag{9.116}$$

we may write Eq. (9.115) in the form

$$F_f(\psi) = -k_BT \ln\langle e^{-\Delta F_A(\psi)/k_BT}\rangle. \tag{9.117}$$

Equation (9.117) is a good starting point for the perturbation calculation of $F_f(\psi)$. We expand the exponent in Eq. (9.117) and also take into account the effect of the logarithm on the infinite series. In result we obtain

$$\mathcal{F}_f(\psi) = \sum_{l=1}^{\infty} \frac{(-1)^l}{l!(k_BT)^{l-1}} \langle \Delta F_A^l(\psi)\rangle_c\,, \tag{9.118}$$

where $\langle ...\rangle_c$ denotes connected averages. Now we have to calculate averages of type

$$\langle A_\alpha(k_1), A_\beta(k_2) \dots A_\gamma(k_n)\rangle_c. \tag{9.119}$$

We shall apply the Wick theorem and the correlation function of form given by Eqs. (9.97)–(9.98). In this Section we take into account the factor k_B and, hence, the r.h.s. of Eq. (9.98) should be multiplied by k_B.

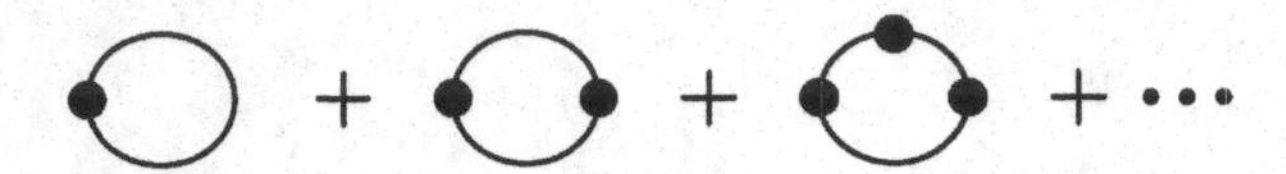

Fig. 9.8 Diagrammatic representation of the series (9.118); • represents the ρ-vertex in Eq. (9.108) and Eq. (9.111), and the solid lines represent bare correlation functions $\langle |A_j(k)|^2 \rangle$.

The calculation of the lowest order terms ($l = 1, 2, 3$) in Eq. (9.118) is straightforward. The perturbation terms are shown by diagrams in Fig. 9.8. The infinite series Eq. (9.118) can be exactly summed up and the result is the following logarithmic function

$$\mathcal{F}_f(\psi) = \frac{(d-1)}{2} k_B T \sum_k \ln \left[1 + \frac{\rho(\psi)}{k^2} \right]. \tag{9.120}$$

The same result for $\mathcal{F}_f(\psi)$ can be obtained by direct calculation of the Gaussian functional integral (9.104). This is done with the help of the integral representation of δ-function in (9.105). Thus one should firstly perform the integration over $A_j(x)$ and then an additional functional integration produced by the integral representation of the δ-function should be accomplished, as well.

Eqs. (9.114) and (9.120) give the effective free energy density

$$f_d(\psi) = \mathcal{F}_d(\psi)/V_d \tag{9.121}$$

in the form

$$f_d(\psi) = f_0(\psi) + \Delta f_d(\psi) , \tag{9.122}$$

where

$$f_0(\psi) = a|\psi|^2 + \frac{b}{2}|\psi|^4 \tag{9.123}$$

and

$$\Delta f_d(\psi) = \frac{(d-1)k_B T}{2V_d} \sum_k \ln \left(1 + \frac{\rho}{k^2} \right). \tag{9.124}$$

We should mention that the fluctuation contribution $\Delta f_d(\psi)$ to $f(\psi)$ transforms to convergent integral in the continuum limit

$$\frac{1}{V_d}\sum_{k_0,k_1,k_2} \rightarrow \int \frac{d^dk}{(2\pi)^d} = K_d \int_0^{\Lambda} dk.k^{d-1}. \tag{9.125}$$

(in the r.h.s. integral, $k = |k|$). But the terms in the expansion of the logarithm in (9.124) are power-type divergent with the exception of several low-order terms. Therefore, we will work with a finite sum of an infinite series with infinite terms. In our further calculations we will keep the cutoff Λ finite for all relevant terms in $\Delta f_d(\psi)$. This is the condition to obtain correct results.

9.8.3 *Dimensional crossover*

The dimensional (2D-3D) crossover has been considered by Rahola (2001), and Shopova and Todorov (2003b). Here we follow the paper by Shopova and Todorov (2003b), where the effective free energy density $f(\psi) \equiv f_3(\psi) = F_3(\psi)/V_3$ of a thin superconducting film of thickness L_0 and volume $V = V_3 = (L_0L_1L_2)$ is derived in a more general way, which allows the investigation of the dimensional crossover.

At this stage, we shall not specify the thickness L_0 of the film leaving it in the broad length interval (a_0, ∞), but we shall definitely assume that the dimensions L_1 and L_2 are macroscopic ($L_1 \gg a_0$, $L_2 \gg a_0$) and then the continuum limit along the axes $\hat{x}_1$ and $\hat{x}_2$ can be taken without a substantial approximation. To take the continuum limit we use the rule (9.125) with respect to the wave vector components k_1 and k_2 for the respective sums over k_1 and k_2 in Eq. (9.124)). The result for the effective free energy density is

$$f(\psi) = a|\psi|^2 + \frac{b}{2}|\psi|^4 + k_BTJ\left[\rho\left(\psi\right)\right], \tag{9.126}$$

where

$$J(\rho) = \int_0^{\Lambda} \frac{dq}{2\pi} qS\left(q,\rho\right) \tag{9.127}$$

where $q = |q|$, and

$$S(q,\rho) = \frac{1}{L_0} \sum_{k_0=-\Lambda_0}^{+\Lambda_0} \ln\left[1 + \frac{\rho(\psi)}{q^2 + k_0^2}\right], \tag{9.128}$$

and the 2D vector q is given by $q = (k_1, k_2)$; $q^2 = (k_1^2 + k_2^2)$.

In Eqs. (9.126)–(9.128), the integral $J(\rho)$ and the sum $S(q,\rho)$ over the wave vector $k = (k_0, q)$ are truncated by the upper cutoffs Λ and Λ_0. The finite cutoff Λ is introduced for the wave vector modulus $|q|$ and Λ_0 stands for the wave vector component k_0.

As our study is based on the quasi-macroscopic GL approach the second cutoff Λ_0 should be again related to ξ_0 rather than to the lattice constant a_0, i.e. $\Lambda_0 \sim (1/\xi_0)$, which means that phenomena at distances shorter than ξ_0 are excluded from our consideration. We will assume that the lowest possible value of Λ_0 is (π/ξ_0), as is for Λ, but we will keep in mind that both Λ_0 and Λ can be extended to infinity, provided the main contribution to the integral $J(\rho)$ and the sum $S(q,\rho)$ comes from the long wavelength limit ($q\xi_0 \ll 1$).

In a close vicinity of the phase transition point T_{c0} from normal ($\psi = 0$) to Meissner state ($|\psi| > 0$) the parameter $\rho \sim |\psi|^2$ is small and the main contribution to the free energy $f(\psi)$ will be given by the terms in S with small wave vectors $k \ll \Lambda$. This allows an approximate but reliable treatment of the 2D-3D crossover by expanding the summation over k_0 in (9.128) to infinity - $\Lambda_0 \sim \infty$. A variant of the theory when Λ_0 is kept finite ($\Lambda = \Lambda_0 = \pi/\xi_0$) can also be developed but the results are too complicated. Performing the summation and the integration in Eqs. (9.127)–(9.128) we obtain $J(\rho) = (\Lambda^2/2\pi L_0)I(\rho)$, where

$$I(\rho) = \int_0^1 dy \ln\left[\frac{\sinh\left(\frac{1}{2}L_0\Lambda\sqrt{\rho + y}\right)}{\sinh\left(\frac{1}{2}L_0\Lambda\sqrt{y}\right)}\right], \tag{9.129}$$

The integral (9.129) has a logarithmic divergence that corresponds to the infinite contribution of magnetic fluctuations to the free energy of normal phase ($T_{c0} > 0, \varphi = 0$). Such type of divergence is a common property of many phase transition models. In the present case, as is in other systems, this divergence is irrelevant, because the divergent term does not depend on the order parameter ψ and the free energy $f(\psi)$ is defined as the difference between the total free energies of the superconducting and normal phases: $f(\psi) = (f_S - f_N)$.

Introducing a dimensionless order parameter $\varphi = (\psi/\psi_0)$, where $\psi_0 = (\alpha_0 T_{c0}/b)^{1/2}$ is the value of ψ at $T = 0$, we obtain the free energy (9.126) in the form

$$f(\varphi) = \frac{H_{c0}^2}{8\pi}\left[2t_0\varphi^2 + \frac{b}{2}|\varphi|^4 + 2(1+t_0)CI(\mu\varphi^2)\right], \tag{9.130}$$

with $I(\mu\varphi^2)$, given by Eq. (9.129), $\mu = (1/\pi\kappa)^2$, $\Lambda = \pi/\xi_0$, and

$$C = \frac{2\pi^2 k_B T_{c0}}{L_0\xi_0^2 H_{c0}^2}. \tag{9.131}$$

From the equation of state $(\partial f/\partial\varphi = 0)$ we find two possible phases: $\varphi_{00} = 0$ and the superconducting phase $(\varphi_0 > 0)$, defined by the equation

$$t_0 + \varphi_0^2 + \frac{(1+t_0)CL_0\xi_0}{4\pi\lambda_0^2}K(\mu\varphi_0^2) = 0\,, \tag{9.132}$$

where

$$K(z) = \int_0^1 dy\, \frac{\coth\left(\frac{1}{2}\, L_0\Lambda\sqrt{y+z}\right)}{\sqrt{y+z}}. \tag{9.133}$$

The analysis of the stability condition $(\partial^2 f/\partial\varphi^2 \geq 0)$ shows that the normal phase is a minimum of $f(\varphi)$ for $t_0 \geq 0$, whereas the superconducting phase is a minimum of $f(\varphi)$, if

$$1 > \frac{1}{4}(1+t_0)CL_0\Lambda\mu^2\tilde{K}(\mu\varphi_0^2)\,, \tag{9.134}$$

where

$$\tilde{K}(z) = \int_0^1 \frac{dy}{y+z}\left[\frac{\coth\left(\frac{1}{2}L_0\Lambda\sqrt{y+z}\right)}{\sqrt{y+z}} + \frac{L_0\Lambda}{2\sinh^2\left(\frac{1}{2}L_0\Lambda\sqrt{y+z}\right)}\right]. \tag{9.135}$$

The entropy jump $\Delta s = (\Delta S/V) = [-df(\varphi_0)/dT]$ per unit volume at the equilibrium point of phase transition $T_c \neq T_{c0}$ is obtained in the form

$$\Delta s(T_c) = -\frac{H_{c0}^2\varphi_{c0}^2}{4\pi T_{c0}}\left[1 + \frac{CI(\varphi_{c0})}{\varphi_{c0}^2}\right], \tag{9.136}$$

where $\varphi_{c0} \equiv \varphi_0(T_c)$ is the jump of the dimensionless order parameter at T_c.

The second term in Δs can be neglected. In fact, taking into account the equation $f[\varphi_0(T_c)] = 0$ for the equilibrium phase transition point T_c we obtain that $|CI(\varphi_0)/\varphi_0^2|$ is approximately equal to $|t_{c0} + \varphi_{c0}^2/2|$, where φ_{c0}^2 and the dimensionless shift of the transition temperature $t_{c0} = t_0(T_c)$ are expected to be much smaller than unity. The latent heat $Q = T_c \Delta S(T_c)$ and the jump of the specific heat capacity at T_c, $\Delta C = T_c(\partial \Delta S/\partial T)$ can be easily calculated with the help of Eq. (9.136). For this purpose we need the function $\varphi_0(T)$, which cannot be obtained in analytical form from Eq. (9.132).

Eqs. (9.130) and (9.132) can be analyzed numerically. These relatively simple 2D-3D crossover formulae can be used in investigations of specific substances by variation of thickness L_0 of films from $L_0 \gg \Lambda^{-1}$ (3D system) to $a_0 < L_0 \ll \Lambda^{-1}$ (quasi-2D system) and even to a 2D system for $L_0 = a_0$. Then one can vary the effective dimension of the system $d_{\text{eff}}(L_0\Lambda)$ as a function of $L_0\Lambda_0$ from $D = 2$ to $D = 3$ (Craco et al, 1999). However, from a purely calculational point of view we have found more convenient to consider particular dimensions of interest separately and then to compare the results in order to demonstrate the relevant differences between the bulk (3D) and thin (q2D) film properties. This approach is applied in the remainder of this Section.

9.8.4 *Free energy for particular spatial dimensions*

Here we discuss in more details the calculation of the free energy for 2D, q2D and 3D superconductors of type I. For purely 2D superconductor, consisting of a single atomic layer (unrealistic case), we can use Eqs. (9.121)–(9.124) by setting $d = 2$, and calculate $\Delta f_2(\psi)$ with the help of rule (9.125):

$$\Delta f_2(\psi) = \frac{k_B T}{8\pi}\left[(\Lambda^2 + \rho_0|\psi|^2)\ln\left(1 + \frac{\rho_0|\psi|^2}{\Lambda^2}\right)\right.$$
$$\left. - \rho_0|\psi|^2 \ln\left(\frac{\rho_0|\psi|^2}{\Lambda^2}\right)\right]. \tag{9.137}$$

The first term of the above free energy can be expanded in powers of $|\psi|^2$:

$$\Delta f_2(\psi) = \frac{k_B T}{8\pi}\left[\rho_0|\psi|^2 + \rho_0|\psi|^2 \ln\left(\frac{\Lambda^2}{\rho_0|\psi|^2}\right) + \frac{\rho_0^2|\psi|^4}{2\Lambda^2}\right]. \tag{9.138}$$

Thus we obtain a result, given by (Lovesey, 1980). This case is of special interest because of the logarithmic term in the Landau expansion for $f(\psi)$, but it has no practical application for the lack of ordering in purely 2D superconductors (mono-atomic layers).

For q2D superconductors we assume that $(2\pi/\Lambda) > L_0 \gg a_0$, where L_0 is the thickness of the superconducting film, and the more precise choice of the upper cutoff $\Lambda \ll (1/a_0)$ for the wave numbers k_i is a matter of additional investigation (Folk, Shopova, and Uzunov, 2001). In order to justify this definition of quasi-2D system one can use the 2D–3D crossover description presented in Section 9.8.3. The summation over the wave number $k_0 = (2\pi n_0/L_0)$ in Eq. (9.128) cannot be substituted with an integration because $L_0 \ll L_j$ and the dimension L_0 does not obey the conditions, valid for L_j, $(j = 1, 2)$ (Uzunov and Suzuki, 1994, 1995; Craco et al, 1999).

Therefore, for such 3D system we must sum over k_0 and integrate over two other components (k_1 and k_2) of the wave vector k (Section 9.8.3). This gives an opportunity for a systematic description of the 2D–3D crossover as shown in Section 9.8.3. In the limiting case of very small thickness the 2D–3D crossover theory (Section 9.8.3) leads to a result, which is obtained more simply in an alternative way, namely, by ignoring all terms corresponding to $k_0 \neq 0$ in the sum in Eq. (9.124). This corresponds to the supposition that the q2D film thickness cannot exceed $2\pi\Lambda$.

Assuming this point of view, the real physical size of the q2D film thickness will depend on the choice of the cutoff Λ. It is certain at this stage that $\Lambda \geq \xi_0$, because the quasi-phenomenological GL theory does not account phenomena of size less than ξ_0. The upper cutoff Λ of wave numbers can be defined in a more precise way at next stages of consideration.

For a q2D film we have the expression

$$\Delta f(\psi) = \frac{2}{L_0} \Delta f_2(\psi) , \tag{9.139}$$

where $\Delta f_2(\psi)$ is given by Eq. (9.138).

For bulk (3D) superconductor we obtain

$$\Delta f_3(\psi) = \frac{k_B T}{2\pi} \left[\frac{\Lambda^3}{3} \ln\left(1 + \frac{\rho_0 |\psi|^2}{\Lambda^2}\right) + \frac{2}{3}\rho_0 |\psi|^2 \Lambda \right.$$
$$\left. - \frac{2}{3}\rho_0^{3/2} |\psi|^3 \arctan\left(\frac{\Lambda}{\sqrt{\rho_0 |\psi|^2}}\right)\right] . \tag{9.140}$$

The Landau expansion in powers of $|\psi|$ in this form of $f_3(\psi)$ confirms the results by Halperin, Lubensky and Ma (1974), and Chen, Lubensky and

Nelson (1978); moreover, it correctly gives the term of type $\rho_0^2|\psi|^4$, which has been considered small and neglected in these papers.

For 4D-systems $\Delta f_{\rm D}(\psi)$ becomes

$$\Delta f_4(\psi) = \frac{3k_BT}{64\pi^2}\left[\Lambda^2\rho_0|\psi|^2 + \Lambda^4 \ln\left(1+\frac{\rho_0|\psi|^2}{\Lambda^2}\right)\right.$$
$$\left. -\rho_0^2|\psi|^4 \ln\left(1+\frac{\Lambda^2}{\rho_0|\psi|^2}\right)\right]. \tag{9.141}$$

The above expression for $\Delta f_4(\psi)$ can be also expanded in powers of $|\psi|$ to show that it contains a term of the type $|\psi|^4 \ln\left(\sqrt{\rho_0}|\psi|/\Lambda\right)$, which produces a first order phase transition; this case is considered in the scalar electrodynamics (Coleman and Weinberg, 1973). In our further investigation we will focus our attention on 3D and q2D superconductors.

The free energy density $\Delta f_{\rm D}(\psi)$ can be expanded in powers of $|\psi|$ but the Landau expansion can be done only in an incomplete way for even spatial dimensions. Thus $f_2(\psi)$, $f_4(\psi)$, and $f(\psi)$ - the free energy density corresponding to the q2D films, contain logarithmic terms, which should be kept in their original form in the further treatment of the function $\Delta f_{\rm D}(\psi)$ in the Landau expansion.

The analysis can be performed in two ways: with and without Landau expansion of $\Delta f_d(\psi)$. These variants of the theory are called "exact" theory (ET) and "Landau" theory (LT), respectively (Shopova and Todorov, 2003a). It has been shown (Shopova and Todorov, 2003c) that these two ways of investigation give the same results in all cases except for q2D films with relatively small thicknesses ($L_0 \ll \xi_0$).

It seems important to establish the differences between these two variants of the theory because the HLM effect is very small and any incorrectness in the theoretical analysis may be a cause for an incorrect result. Following the same arguments one can investigate the effect of the factor T in $\Delta f_d(\psi)$ on the thermodynamics of q2D films. The factor T can be represented as $T = T_{c0}(1 + t_0)$ and one may expect that the usual approximation $T \approx T_{c0}$, which is well justified in the Landau theory of phase transitions (Lifshitz and Pitaevskii, 1980), may be applied (for a comprehensive discussion see Section 5.8.6). This way of approximation can be performed by neglecting terms in the thermodynamic quantities smaller than the leading ones. On the other hand practical calculations lead to the conclusion that this approximation cannot be made without a preliminary examination because for some q2D films it produces a substantial error of about 10% (Shopova and Todorov, 2003c). LT, in which the factor T is substituted by T_{c0} is referred to as "simplified Landau theory" - SLT. All three variants of the theory, ET, LT and SLT have been investigated by (Shopova and Todorov, 2003c); see also, Shopova, Tsvetkov and Uzunov (2006); Shopova and Uzunov (2008).

9.8.5 *Limitations of the theory*

The general result (9.121)–(9.124) for the effective free energy $f_d(\psi)$ has the same domain of validity (Lifshitz and Pitaevskii, 1980) as the GL free energy functional in a zero external magnetic field. When we neglect a sub-nano interval of temperatures near the phase transition point we can use Eq. (9.81), provided $|t_0| = |T - T_{c0}|/T_{c0} < 1$, or in the particular case of type I superconductors, $|t_0| < \kappa^2$ (Lifshitz and Pitaevskii, 1980). Note, that the latter inequality does not appear in the general GL approach. It comes as a condition for the consistency of our approach with the microscopic BCS theory for type I superconductors (Lifshitz and Pitaevskii, 1980).

Taking the continuum limit we have to assume that all dimensions of the body, including the thickness L_0, are much larger than the characteristic lengths ξ and λ. The exception of this rule is when we consider thin films. Especially for thin films of type I superconductors, where $((2\pi/\Lambda) > L_0 \gg a_0)$, we should have in mind that $\xi(T) > \lambda(T)$, so the inequalities $\xi > \lambda > \xi_0 > \lambda_0$ hold true in the domain of validity of the GL theory $|t_0| < \kappa^2 < 1$. In the paper by Folk, Shopova, and Uzunov (2001) a comprehensive choice of the cutoff Λ has been made, namely, $\Lambda = \xi_0$ (the problem for the choice of cutoff Λ is discussed also in Sections 9.8.6 and 9.8.7). The respective conditions for q2D films of type II superconductors are much weaker and are reduced to the usual requirements: $\kappa > 1/\sqrt{2}$, $|t_0| < 1$ and $(2\pi/\Lambda) > L_0 \gg a_0$.

If we perform a Landau expansion of $f_d(\psi)$ in powers of $|\psi|^2$, the condition $\rho \ll \Lambda^2$ should be satisfied. In order to evaluate this condition we substitute $|\psi|^2$ in $\rho = \rho_0|\psi|^2$ with $\eta^2 = |a|/b$, which corresponds to $e = 0$. As $\lambda^2(T) = 1/\rho$, the condition for the validity of the Landau expansion becomes $[\Lambda\lambda(T)]^2 \gg 1$, i. e., $(\Lambda\lambda_0)^2 \gg |t_0|$. Choosing the general form of $\Lambda_\tau = (\pi\tau/\xi_0)$, where τ describes the deviation of Λ_τ from $\Lambda_1 \equiv \Lambda = (\pi/\xi_0)$, we obtain $(\pi\tau\kappa)^2 \gg |t_0|$; $\kappa = (\lambda_0/\xi_0)$ is the GL parameter.

Thus we can conclude that in type II superconductors, where $\kappa = (\lambda_0/\xi_0) > 1/\sqrt{2}$, the condition $(\rho/\Lambda^2) \ll 1$ is satisfied very well for values of the cutoff in the interval between $\Lambda = (\pi/\xi_0)$ and $\Lambda = (\pi/\lambda_0)$, i.e., for $1 < \tau < (1/\kappa)$. For type I superconductors, where $\kappa < 1/\sqrt{2}$ the cutoff value $\Lambda \sim (1/\xi_0)$ leads to the BCS condition ($|t_0| < \kappa^2$) for the validity of the GL approach. Substantially larger cutoffs ($\Lambda \gg \pi/\xi_0$), for example, $\Lambda \sim (1/\lambda_0)$ for type I superconductors with $\kappa \ll 1$ lead to a contradiction between the BCS condition and the requirement $\rho \ll \Lambda^2$.

In our calculations we may use the parameter $\mu_\tau = (1/\pi\tau\kappa)^2$ and, in particular, $\mu \equiv \mu_1 = (1/\pi\kappa)^2$. In terms of μ the condition for the validity of $f_d(\psi)$ expansion becomes $\mu|t_0| \ll 1$, or, more generally, $\mu_\tau|t_0| \ll 1$. Choosing $\tau = 1/\pi$, we obtain the BCS criterion for the validity of the GL free energy for type I superconductors (Lifshitz and Pitaevskii, 1980). The choice $\tau = (\xi_0/\pi\lambda_0)$ corresponds to the cutoff $\Lambda_\tau = 1/\lambda_0$. As we will see in Section 9.8.6 the thermodynamics near the phase transition point in 3D systems has no substantial dependence on the value of the cutoff Λ_τ but it should be chosen in a way that is consistent with the MF-like approximation. The cutoff problem for q2D films has been investigated by Shopova and Todorov (2003c). It has been shown that the choice $\Lambda \sim \pi/\xi_0$ is consistent in the framework of GL theory.

Alternatively, the inequality $(\rho/\Lambda^2) \ll 1$ can be investigated with the help of the reduced order parameter φ defined by $\varphi = |\psi|/\eta_0$, where $\eta_0 \equiv \eta(T=0) = (\alpha_0 T_{c0}/b)^{1/2}$ is the so-called zero-temperature value of order parameter for the GL free energy $f_0(\psi)$, given by Eq. (9.123). The reduced order parameter φ will be equal to $|t_0|$ for $t_0 < 0$, if only the magnetic fluctuations are ignored, i.e., when $|\psi| = \eta$. Using the notation φ, we obtain the condition $(\rho/\Lambda^2) \ll 1$ in the form $\mu_\tau\varphi^2 \ll 1$. This condition seems to be more precise because it takes into account the effect of magnetic fluctuations on the order parameter ψ.

9.8.6 *Bulk superconductors*

9.8.6.1 *Thermodynamics*

The effective free energy $f_3(\psi)$ of bulk (3D-) superconductors is given by Eqs. (9.121)–(9.124) an Eq. (9.140). The analytical treatment of this free energy can be done by Landau expansion in small $(\sqrt{\rho_0}|\psi|/\Lambda)$. Up to order $|\psi|^6$ we obtain

$$f_3(\psi) \approx a_3|\psi|^2 + \frac{b_3}{2}|\psi|^4 - q_3|\psi|^3 + \frac{c_3}{2}|\psi|^6\,, \tag{9.142}$$

where

$$a_3 = a + \frac{k_B T \Lambda \rho_0}{2\pi^2} \,, \tag{9.143}$$

$$b_3 = b + \frac{k_B T \rho_0^2}{2\pi^2 \Lambda} \,, \tag{9.144}$$

$$q_3 = \frac{k_B T \rho_0^{3/2}}{6\pi} \,, \tag{9.145}$$

and

$$c_3 = -\frac{k_B T \rho_0^3}{6\pi^2 \Lambda^3} \,. \tag{9.146}$$

The cutoff Λ in Eqs. (9.143) - (9.146) is not specified and can be written in the form $\Lambda_\tau = (\pi\tau/\xi_0)$ as suggested in Section 9.8.5.

It can be shown by both analytical and numerical calculations (Shopova *et al.*, 2002) that $|\psi|^6$-term has no substantial effect on the thermodynamics, described by the free energy (9.142). That is why we ignore this term. The possible phases $|\psi_0|$ are found as a solution of the equation of state:

$$[\partial f(\psi)/\partial|\psi|]_{\psi_0} = 0 \,. \tag{9.147}$$

There always exists a normal phase $|\psi_0| = 0$, which is a minimum of $f_3(\psi)$ for $a_3 > 0$. The possible superconducting phases are given by

$$|\psi_0|_\pm = \frac{3q_3}{4b_3}\left(1 \pm \sqrt{1 - \frac{16a_3b_3}{9q_3^2}}\right) \geq 0. \tag{9.148}$$

Having in mind the existence and stability conditions of $|\psi_0|_\pm$–phases, we obtain that the $|\psi_0|_+$-phase exists for $(16a_3b_3) \leq 9q_3^2$ and this region of existence always corresponds to a minimum of $f_3(\psi)$. The $|\psi_0|_-$-phase exists for $0 < a_3 < 9q_3^2/16b_3$ and this region of existence always corresponds to a maximum of $f_3(\psi)$, i.e., this phase is absolutely unstable. For $a_3 = 0$, $|\psi_0|_- = 0$ and hence, it coincides with the normal phase. For $9q_3^2 = 16a_3b_3$ we have $|\psi_0|_+ = |\psi_0|_- = 3q_3/4b_3$ and $f_3(|\psi_0|_+) = f_3(|\psi_0|_-) = 27q_3^4/512b_3^3$. Furthermore $f_3(|\psi_0|_-) > 0$ for all allowed values of $|\psi_0|_- > 0$, whereas $f_3(|\psi_0|_+) < 0$ for $a_3 < q_3^2/2b_3$, and $f_3(|\psi_0|_+) > 0$ for $(q_3^2/2b_3) < a_3 < 9q_3^2/16b_3$. The equilibrium temperature $T_{\rm eq}$ of the first order phase transition is defined by the equation $f(|\psi_0|_+) = 0$, which gives the following result:

$$2b_3(T_{\rm eq})a_3(T_{\rm eq}) = q_3^2(T_{\rm eq}) \,. \tag{9.149}$$

These results are confirmed by numerical calculations of the effective free energy (9.142) (Shopova *et al.*, 2002).

The equilibrium entropy jump is $\Delta S = V\Delta s$ and $\Delta s = -(df_3(|\psi|)/dT)$ can be calculated with the help of Eq. (9.145) and the equation of state (9.147):

$$\Delta s = -|\psi_0|^2\Phi(|\psi_0|) , \tag{9.150}$$

where $\Phi(|\psi_0|)$ is the following function:

$$\Phi(y) = (\alpha_0 + \frac{k_B\Lambda\rho_0}{2\pi^2}) - \frac{\rho_0^{3/2}k_B}{6\pi}y + \left(\frac{k_B\rho_0^2}{4\pi^2\Lambda}\right)y^2 . \tag{9.151}$$

The specific heat capacity per unit volume $\Delta C = T(\partial\Delta s/\partial T)$ is obtained from Eq. (9.150):

$$\Delta C = -\left(\frac{T}{T_{c0}}\right)\left(\frac{\partial|\psi_0|^2}{\partial t_0}\right)\Phi(|\psi_0|) . \tag{9.152}$$

The quantities $\Delta s(T)$ and $\Delta C(T)$ can be evaluated at the equilibrium phase transition point T_{eq}, which is found from Eq. (9.149):

$$\frac{T_{\text{eq}}}{T_{c0}} \approx 1 - \frac{k_B\rho_0\Lambda}{2\pi^2\alpha_0} + \frac{\left(\rho_0^{3/2}k_B/6\pi\right)^2}{b + (\rho_0^2k_B/2\pi^2\Lambda)T_{c0}}\left(\frac{T_{c0}}{\alpha_0}\right) , \tag{9.153}$$

provided $|\Delta T_c| = |T_{c0} - T_{\text{eq}}| \ll T_{c0}$. Further we will see that the condition $|\Delta T_c| \ll T_{c0}$ is valid in real substances. The second term in r.h.s. of Eq. (9.153) is a typical negative fluctuation contribution, whereas the positive third term in r.h.s. of the same equality is typical for first-order transitions.

To obtain the jumps Δs and ΔC at T_{eq} we have to put the solution $|\psi_0|_+$, given by Eq. (9.148) in Eqs. (9.150)–(9.152). The result will be

$$\Delta s = -\frac{q_{3c}^2}{b_{3c}^2}\left[\alpha_0 + \frac{k_B\rho_0\Lambda}{2\pi^2} - \left(\frac{k_B\rho_0^{3/2}}{6\pi}\right)^2\frac{T_{\text{eq}}}{b_{3c}^2}\right] , \tag{9.154}$$

and

$$\Delta C = \frac{4\alpha_0}{b_{3c}}\left(\alpha_0T_{c0} - \frac{q_{3c}^2b}{b_{3c}^2}\right) , \tag{9.155}$$

where b_{3c} and q_{3c} are the parameters b_3 and q_3 at $T = T_{\rm eq}$. As $|\Delta T_c| = |T_{c0} - T_{\rm eq}| \ll T_{c0}$ we can set $T_{\rm eq} \approx T_{c0}$ in r.h.s. of Eqs. (9.154) and (9.155) and obtain $q_{3c} \equiv q_3(T = T_{\rm eq}) \approx q_3(T_{c0})$ and $b_{3c} \approx b_3(T_{c0})$. The latent heat $Q = -T_{\rm eq}\Delta s$ per unit volume of the first order phase transition at $T_{\rm eq}$ can be calculated from Eq. (9.154).

Now we shall discuss the ratio

$$(\Delta T)_{\rm eq} = \frac{Q}{\Delta C} \tag{9.156}$$

introduced (Halperin, Lubensky and Ma, 1974) as a measure of the temperature size of the HLM effect – a temperature interval, where the effect can be observed (in the paper of Halperin, Lubensky and Ma (1974) the same ratio is denoted by ΔT_1; see also Eq. (9.93) in Section 9.7.5). Keeping only the first terms in Eqs. (9.154)–(9.156), we obtain a result for $(\Delta T)_{\rm eq}$ which is four times smaller than that for ΔT_1 in the paper by Halperin, Lubensky and Ma (1974). To explain the difference let us mention that Eq. (9.155) gives the jump ΔC at the equilibrium phase transition point of the first order phase transition, described by $|\psi|^3$ term, while the specific heat jump, considered by Halperin, Lubensky and Ma (1974) is equal to the specific heat jump at the standard second order transition $\Delta\tilde{C} = (\alpha_0^2 T_{c0}/b)$, and is four times smaller. In fact, neglecting the second term in Eq. (9.155) and having in mind that $b_3 \equiv b$, see Eq. (9.158) below, we have that $\Delta C = 4\Delta\tilde{C}$.

Therefore, we obtain that $(\Delta T)_{\rm eq}$ is four times smaller than the respective value ΔT_1, given by Halperin, Lubensky and Ma (1974). This is valid in an approximation in which the second term in Eq. (9.155) is neglected (this approximation is justified in Section 9.8.6.2). The approximation of Δs and ΔC by the first (leading) terms in Eqs. (9.154) and (9.155) does not change essentially these quantities provided the cutoff Λ is comprehensively chosen (Section 9.8.6.2). Note that in the papers by Halperin, Lubensky and Ma (1974), and Chen, Lubensky and Nelson (1978) as well as in all preceding papers, the corrections to the leading terms in Δs and ΔC have not been taken into account.

9.8.6.2 *Results for Al*

In order to do the numerical estimates we represent the Landau parameters α_0 and b with the help of the zero-temperature coherence length ξ_0 and the zero-temperature critical magnetic field H_{c0}. The connection between them is given by formulae of the standard GL superconductivity theory (Lifshitz

Table 9.1 Values of T_{c0}, H_{c0}, ξ_0, κ, and $|\psi_0|$ for W, Al, and In (Madelung, 1990)

Substance	T_{c0} (K)	H_{c0} (Oe)	ξ_0 (μm)	κ	$\lvert\psi_0\rvert \times 10^{-11}$
W	0.015	1.15	37	0.001	0.69
Al	1.19	99.00	1.16	0.010	2.55
In	3.40	281.5	0.44	0.145	2.0

and Pitaevskii, 1980): $\xi_0^2 = (\hbar^2/4m\alpha_0 T_{c0})$ and $H_{c0}^2 = (4\pi\alpha_0^2 T_{c0}^2/b)$. The expression for the zero-temperature penetration depth $\lambda_0 = (\hbar c/2\sqrt{2}eH_{c0}\xi_0)$ is obtained from the above relation and $\lambda_0 = (b/\alpha_0 T_{c0}\rho_0)^{1/2}$. We will use the experimental values of T_{c0}, H_{c0} and ξ_0 for Al as given in Table 9.1. The experimental values for T_{c0}, H_{c0} and ξ_0 vary about 10-15%, depending on the method of measurement and the geometry of the samples (bulk material or films; see Madelung (1990)) but such deviations do not affect the results of our numerical investigations.

The evaluation of the parameters a_3 and b_3 for Al gives

$$a_3 = (\alpha_0 T_{c0})\left[t_0 + 0.972 \times 10^{-4}(1+t_0)\tau\right], \tag{9.157}$$

and

$$\frac{b_3}{b} = 1 + \frac{0.117}{\tau}. \tag{9.158}$$

Setting $\tau = 1$ corresponds to the cutoff $\Lambda_1 = (\pi/\xi_0)$. For $\tau = (1/\kappa)_{\text{Al}} = 10^2$, which corresponds to the much higher cutoff $\Lambda = (\pi/\lambda_0)$, we have $b_3 \approx b$, i.e., the ρ_0^2 -term in b_3, given by Eq. (9.144), can be neglected. However, as we can see from Eq. (9.158), for $\tau = 1$ the same ρ_0^2-correction in the parameter b_3 is of order $0.1b$ and cannot be automatically ignored in all calculations, in contrast to preceding suppositions (Halperin, Lubensky and Ma, 1974; Chen, Lubensky and Nelson, 1978). However, the more important fluctuation contribution in 3D superconductors comes from the $\tau-$term in Eq. (9.157) for the parameter a_3. This term is of order 10^{-4} for $\tau \sim 1$ and this is consistent with the condition $|t_0| < \kappa^2 \sim 10^{-4}$, but for $\tau \sim 10^2$, i.e., for $\Lambda \sim (\pi/\lambda_0) \sim 10^6\mu$m, the same $\tau-$ term is of order 10^2, which exceeds the temperature interval ($T_{c0} \pm 10^{-4}$) for the validity of BCS condition for Al.

These results demonstrate that for this theory to be consistent, we must choose the cutoff $\Lambda_\tau = (\pi\tau/\xi_0)$, where τ is not a large number ($\tau \to 1 \div 10$).

To be more specific, we set $\Lambda = \Lambda_1 = (\pi/\xi_0)$ as suggested in the paper by Folk, Shopova, and Uzunov (2001).

The temperature shift $t_{\text{eq}} = t_0(T_{\text{eq}})$ for bulk Al can be estimated with the help of Eq. (9.153). We obtain that this shift is negative and very small: $t_{\text{eq}} \sim -10^{-4}$. Note, that the second term in the r.h.s. of Eq. (9.153) is of order 10^{-4}, provided $\Lambda \sim (1/\xi_0)$ whereas the third term in the r.h.s. of the same equality is of order 10^{-5}. Once again the change of the cutoff Λ to values much higher than (π/ξ_0) will take the system outside the temperature interval, where the BCS condition for Al is valid. Let us note, that in the paper by Shopova *et al.* (2002) the parameter t corresponds to our present notation t_0. But the numerical calculation of the free energy function $f_3(\psi)$ the paper by Shopova *et al.* (2002) was made for the SLT variant of the theory and the shifted parameter $(t_0 + 0.972 \times 10^{-4})$ was incorrectly identified with t, and this has led to the wrong conclusion for its positiveness at the equilibrium phase transition point T_{eq}. As a matter of fact, the shifted parameter $(t_0 + 0.972 \times 10^{-4})$ is positive at T_{eq} but $t_{\text{eq}} \equiv t_0(T_{\text{eq}})$ is negative, as firstly noted by Shopova and Todorov (2003c).

Having in mind these remarks, when we evaluate Δs and ΔC for bulk Al we can use simplified versions of (9.154) and (9.155) which means to consider only the first terms in the r.h.s and to take $q_{3c} \approx q_3$ and $b_{3c} \approx b$ at T_{c0}. In this way we obtain

$$Q = -T_{c0}\Delta s = 0.8 \times 10^{-2} \left[\frac{\text{erg}}{\text{K . cm}^3}\right] , \tag{9.159}$$

and

$$\Delta C = 2.62 \times 10^3 \left[\frac{\text{erg}}{\text{cm}^3}\right] . \tag{9.160}$$

The results are consistent with an evaluation of ΔC for Al as a jump $(\Delta\tilde{C} = \alpha_0^2 T_{c0}/b)$ at the second order superconducting transition point (Halperin, Lubensky and Ma, 1974) that, as we have mentioned above, is four times smaller than the jump ΔC given by Eq. (9.160).

A complete numerical evaluation of the function $f_3(\psi)$ and the jump of the order parameter at T_{eq} for bulk Al was presented for the first time by Shopova *et al.* (2002). The results there confirm that the order parameter jump and the latent heat Q for bulk type I superconductors are very small and can hardly be observed in experiments.

We will finish the presentation of bulk Al with a discussion of the ratio (9.156). It can be also written in the form

$$(\Delta T)_{\text{eq}} = \frac{32\pi}{9}\left(\frac{k_B^2 T_{c0}^2}{b\alpha_0}\right)\left(\frac{e^2}{mc^2}\right)^3, \tag{9.161}$$

and it differs by a factor 1/4 from the respective result byHalperin, Lubensky and Ma (1974) for a reason which was explained in the discussion of Eq. (9.156). From Eq. (9.161) we have

$$(\Delta T)_{\text{eq}} = 6.7 \times 10^{-12}(T_c^3 H_{c0}^2 \xi_0^6), \tag{9.162}$$

and multiplying the number coefficient in the above expression by four we obtain the Eq. (10) in the paper by (Halperin, Lubensky and Ma, 1974).

9.8.7 *Thermodynamics of quasi-2D films*

Following receding research (Shopova and Todorov, 2003b,c) we will present the free energy density $f(\psi) = (F(\psi)/L_1 L_2)$ in the form

$$f(\varphi) = \frac{H_{c0}^2}{8\pi}\left[2t_0\varphi^2 + \varphi^4 + C(1+t_0)\Gamma(\mu\varphi^2)\right], \tag{9.163}$$

where

$$\Gamma(y) = (1+y)\ln(1+y) - y\ln y, \tag{9.164}$$

To obtain Eqs. (9.163)–(9.164) we have set $\Lambda = (\pi/\xi_0)$ and introduced the notation $\varphi = |\psi|/\eta_0$; η_0 is defined in Section 9.8.5. Some of the properties of free energy (9.163) were analyzed for Al films (Shopova and Todorov, 2003b) as well as for films of Tungsten (W), Indium (In), and Aluminium (Al) (Shopova and Todorov, 2003c). Here we will briefly discuss the main results.

The equilibrium order parameter $\varphi_0 > 0$, corresponding to the Meissner phase, can be easily obtained as a solution of the equation of state $\partial f(\varphi)/\partial\varphi = 0$, where $f(\varphi)$ is given by and Eq. (9.163). In explicit form this equation of state is

$$t_0 + \varphi_0^2 + \frac{C\mu(1+t_0)}{2}\ln\left(1 + \frac{1}{\mu\varphi_0^2}\right) = 0\,. \tag{9.165}$$

The logarithmic divergence in Eq. (9.165) has no chance to occur because φ_0 is always positive and does not tend to zero.

The largest terms in the entropy jump δs and the specific heat jump δC at the equilibrium first order phase transition point T_{eq} are given by

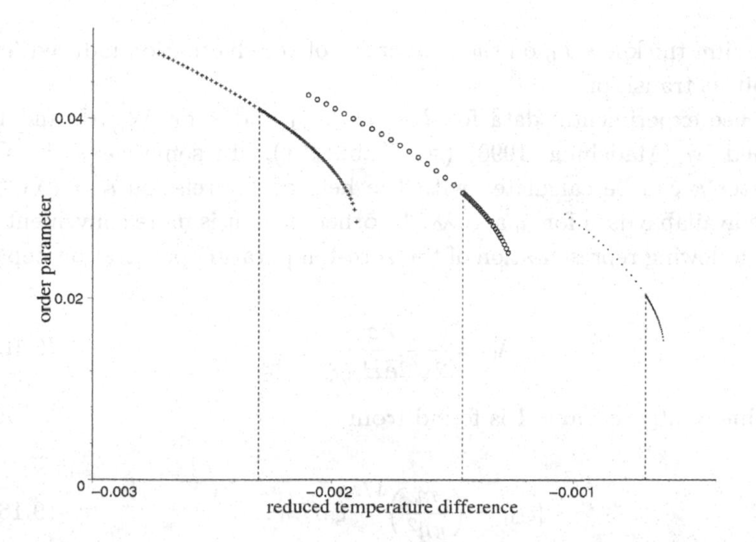

Fig. 9.9 Order parameter profile $\varphi(t_0)$ of Al films of different thicknesses: $L_0 = 0.05\ \mu$m (+-line), $L_0 = 0.1\ \mu$ m (∘), and $L_0 = 0.3\ \mu$m (·) (Shopova and Todorov, 2003c).

$$\Delta s = -\frac{H_{c0}^2}{4\pi T_{c0}}\varphi_{\text{eq}}^2, \tag{9.166}$$

where $\varphi_{\text{eq}} = \varphi_0(T = T_{\text{eq}})$, and

$$\Delta C = \frac{H_{c0}^2}{4\pi T_{c0}}. \tag{9.167}$$

The latent heat of the phase transition is given by $Q = -T_{\text{eq}}\Delta s$ and Eqs. (9.165)-(9.166). Since the temperatures T_{eq} and T_{c0} have very close values, the difference between the values of Q, Δs, and ΔC at T_{c0} and T_{eq}, respectively, can also be ignored, for example, $|\Delta C(T_{\text{eq}}) - \Delta C(T_{c0})|/\Delta C(T_{c0}) \ll 1$ and we can use either $\Delta C(T_{c0})$ or $\Delta C(T_{\text{eq}})$ (Shopova and Todorov, 2003b).

The equations (9.163) - (9.165) corresponding to q2D films are quite different from the respective equations for 3D superconductors but it is easily seen that the relatively large value of the order parameter jump φ^2 in thin films again corresponds to relatively small value of the GL parameter κ. That is why we consider element superconductors with small values of κ and study the effect of this parameter, the critical magnetic field H_{c0}

and the film thickness L_0 on the properties of the fluctuation-induced first order phase transition.

We use experimental data for T_{c0}, H_{c0}, ξ_0 and κ for W, Al, and In, published by (Madelung, 1990) (see Table 9.1). In some cases the GL parameter κ can be calculated with the help of the relation $\kappa = (\lambda_0/\xi_0)$ and the available data for ξ_0 and λ_0. In other cases it is more convenient to use the following representation of the zero-temperature penetration depth:

$$\lambda_0 = \frac{\hbar c}{2\sqrt{2}eH_{c0}\xi_0}\,. \tag{9.168}$$

The value of $|\psi_0|$ in Table 1 is found from

$$|\psi_0| = \left(\frac{m}{\pi\hbar^2}\right)^{1/2} \xi_0 H_{c0}\,. \tag{9.169}$$

The order parameter dependence on the reduced temperature difference t_0 is shown in Fig. 9.9 for Al films of different thicknesses. It is readily seen that the behavior of the function $\varphi_0(t_0)$ corresponds to a well established phase transition of first order. The vertical dashed lines in Fig. 9.9 indicate the respective values of $t_{\rm eq} = t_0(T_{\rm eq})$, at which the equilibrium phase transition occurs, as well as the equilibrium jump $\varphi_0(T_{\rm eq}) = \varphi_{\rm eq}$ for different thicknesses of the film. The parts of the $\varphi_0(t_0)$–curves, which extend up to $t_0 > t_{\rm eq}$, describe the metastable (overheated) Meissner states, which can appear under certain experimental circumstances (see in the parts of the curves on the r.h.s. of the dashed lines). The value of $\varphi_{\rm eq}$ and the metastable region decrease with the increase of the film thickness, which shows that the first order of the phase transition is better pronounced in thinner films and this confirms a conclusion of Shopova and Todorov (2003b).

These results are justified by the behavior of the free energy as a function of t_0. We used Eqs. (9.163)-(9.165) for the calculation of the equilibrium free energy $f[\varphi_0(t_0)]$. The free energy for Al films with different thicknesses is shown in Fig. 9.10. The equilibrium points $T_{\rm eq}$ of the phase transition correspond to the intersection of the $f(\varphi_0)$-curves with the t_0-axis. It is obvious from Fig. 9.10 that the temperature domain of overheated Meissner states decreases with the increase of the thickness L_0.

The shape of the equilibrium order parameter $\varphi_0(t_0)$ in a broad vicinity of the equilibrium phase transition for thin films ($L_0 = 0.05\mu$m) of W, Al, and In was found from Eq. (9.165). The result is shown in Fig. 9.11. The

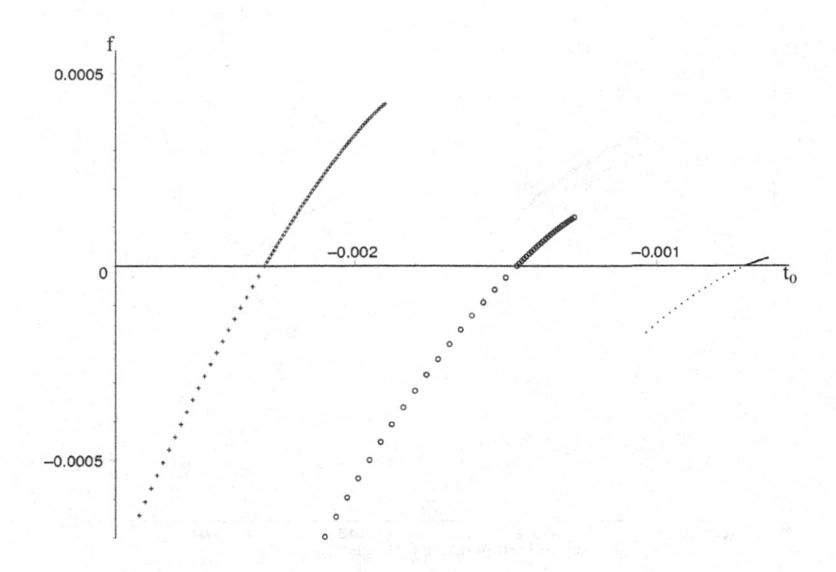

Fig. 9.10 The free energy $f(t_0)$ for Al films of thickness: $L_0 = 0.05\ \mu$m (+-line), $L_0 = 0.1\ \mu$m (∘), $L_0 = 0.3\ \mu$m (·) (Shopova and Todorov, 2003c).

vertical dashed lines in Fig. 9.11 again indicate the respective values of $t_{\text{eq}} = t_0(T_{\text{eq}})$, at which the equilibrium phase transition occurs as well as the equilibrium jump $\varphi_0(T_{\text{eq}}) = \varphi_{\text{eq}}$ in the different superconductors.

The order parameter jump at the phase transition point of In (the In curve is marked by points in Fig. 9.11) is relatively smaller than for W, and Al, where the GL parameter has much lower values. The same is valid for the metastability domains; see the parts of the curves in Fig. 9.11 on the left of the vertical dashed lines. As shown by Shopova and Todorov (2003a) the equilibrium jump of the reduced order parameter φ_{eq} of W has a slightly smaller value than that of Al, although the GL number κ for W has a ten times lower value, compared to κ of Al. Note, that in Fig. 9.11 we show the jump of φ_{eq}, but the important quantity is $|\psi|_{\text{eq}} = |\psi_0|\varphi_{\text{eq}}$. Using the data for $L_0 = 0.05\mu$m from Tables 1 and 2 we find for $|\psi|_{\text{eq}}$ the following values: 0.1×10^{11} for Al, 0.05×10^{11} for In, and 0.02×10^{11} for W.

This result shows that the value of the critical filed H_{c0} is also important and should be taken into account together with the smallness of GL number when the maximal values of the order parameter jump are looked for. Thus the value of the order parameter jump at the fluctuation-induced phase

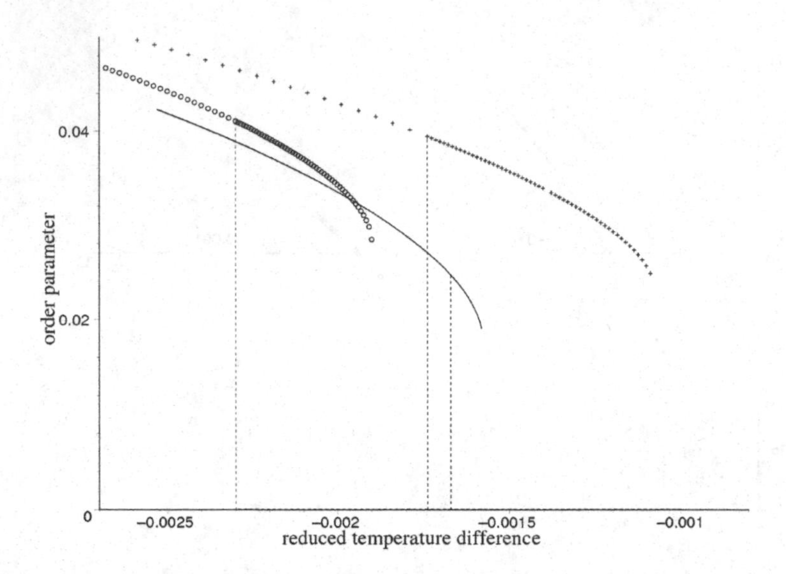

Fig. 9.11 Order parameter profile $\varphi(t_0)$ of films of thickness $L_0 = 0.05$ μm: W ("+"–line), Al (o), and In (·) (Shopova and Todorov, 2003c).

transition is maximal, provided small values of the GL parameter κ are combined with relatively large values of the critical field H_{c0}. In our case Al has the optimal values of these two parameters.

The importance of the zero-temperature critical magnetic field H_{c0} for the enhancement of the jumps of the certain thermodynamic quantities at the equilibrium phase transition point $T_{\rm eq}$ becomes obvious from Eqs. (9.166), (9.167) and (9.169). The equality (9.169) shows that the order parameter jump $|\psi|_{\rm eq} = |\psi_0|\varphi_{\rm eq}$ is large for large values of H_{c0} and ξ_0. This is consistent with the requirement for relatively small values of the GL parameter κ, as shown by Eq. (9.168).

Therefore, the unmeasurable ratio $Q/\Delta C$ discussed in Ref. (Halperin, Lubensky and Ma, 1974) does not depend on the value of the critical field H_{c0} but the quantities Q and ΔC themselves, as well as the order parameter jump $|\psi|_{\rm eq}$ depend essentially on H_{c0}. The values of the reduced order parameter jump $\varphi_{\rm eq}$ for films of Al, In and W of the same thickness have the same order of magnitude while the respective order parameter jump $|\psi|_{\rm eq}$ is one order of magnitude higher for Al than for W, as shown above.

The effect of the critical magnetic field H_{c0} on the latent heat Q is, however, much stronger(Shopova and Todorov, 2003a) and, for this reason the latent heat Q in W films is very small and can be neglected while in Al and In films it reaches values, which could be measured in suitable experiments. This is so because the latent heat is proportional to the difference $[H_{c0}^2/8\pi \sim b|\psi_0|^4]$ between the energies of the ground state (superconducting phase at $T = 0$) and the normal state. It is consistent with the fact that the fluctuation contribution to the free energy, i.e., the $C-$term in the r.h.s. of Eq. (9.163) is generated by the term of type $|\psi|^2 \int d^D x A^2(x)$ in the GL free energy. At $T = 0$ this free energy term is also proportional to the mentioned difference between the free energies of the superconducting ground state and the normal state.

This analysis shows that the MF studies of the HLM effect have a well defined domain of validity for both 3D and q2D superconductors. Our conclusion is that the MF theory of the magnetic fluctuations in superconductors and, in particular, the MF prediction for the fluctuation driven weakly first order phase transition in a zero external magnetic field in bulk superconductors (Halperin, Lubensky and Ma, 1974) and q2D superconducting films (Folk, Shopova, and Uzunov, 2001; Shopova and Todorov, 2003b,c) is reliable and can be tested by experiments. While the HLM effect in bulk systems is inobservantly small, in q2D superconductors this effect is much stronger and may be observed with available experimental techniques.

This consideration of q2D superconductors is highly nontrivial in view of the relevance of the effective Landau parameters dependence on the thickness of the films, L_0. We have justified this dependence by simple heuristic arguments and by a reliable consideration of the 2D–3D crossover. In contrast to initial expectations (Halperin, Lubensky and Ma, 1974) that films made of superconductors with extremely small GL parameter κ such as Al and, in particular, W will be the best candidates for an experimental search of the HLM effect, the more careful analysis (Shopova and Todorov, 2003c) definitely gives somewhat different answer.

The Al films still remain a good candidate for transport experiments, through which the jump of the order parameter at the phase transition point could be measured, but surprisingly the W films turn out inconvenient for the same purpose, due to their very low critical field H_{c0}. The requirement of relatively high values of the critical magnetic field H_{c0} for the clearly manifested first-order phase transition has been established. Although In has ten times higher GL number κ than Al, for the relatively high critical field, the In films can be used on equal footing with the Al films in experiments intended to prove the order parameter jump.

As shown in this Section experiments, designed to search the HLM effect, can be performed by both type I and type II superconductors. In experiments the external magnetic field cannot be completely eliminated. Then vortex states may occur for $H > 0$ below $T_c = T_c(H) \leq T_{c0}$ in type II superconducting films and this will obscure the HLM effect. Note, that in both type I and type II superconductors the external magnetic field H generates additional entropy jump at the phase transition point $T_c(H)$ and this effect can hardly be separated from the entropy jump (9.166) caused by the magnetic fluctuations in the close vicinity of T_{c0}. Therefore, in experiments intended for searching of HLM effect, the external magnetic field should be minimized as much as possible. For q2D superconductors, where the HLM effect is relatively strong and the latent heat can exceed several ergs, one should ensure external fields less than 10 Oe, or, in more reliable experiments, less than 1 Oe (Shopova and Todorov, 2003c).

The results for Al, In, and W shows that the metastability domains (of overheating and overcooling) are much larger than the Ginzburg critical region. This result justifies the reliability of the MF treatment. As one may see from the figures the metastability temperature interval is relatively larger for smaller values of the GL number κ, but in order to ensure large values of some measurable thermodynamic quantities as, for example, the latent heat Q and the specific heat capacity C we must choose a material with large critical field H_c.

The experimental verification of the order parameter jump can be made by transport experiments. As we see from the figures, this quantity has maximal values for W, where the parameter κ is very small. Having in mind also the large metastability regions in this material, one may conclude that W is a good candidate for a testing the HLM effect by transport experiments (measurements of the superconducting currents), provided some specific disadvantages of this material (quite low T_{c0}, etc.) from the experimental point of view do not contradict to this conclusion.

Another effect, which is relevant to the present discussion, is the known variation of the GL parameter κ with the variation of the thickness L_0 (Huebener, 2001). The parameter $\kappa < 1/\sqrt{2}$ of a type I bulk superconductor may change up to values corresponding to a type II superconductor with the decrease of the thickness L_0 below 10^{-7}m.

9.9 Phase Transitions in Unconventional Ferromagnetic Superconductors

In the beginning of this Century (2000–2003), the coexistence of itinerant ferromagnetism and unconventional superconductivity has been discovered in several inter-metallic compounds. It is natural to suppose that new unconventional ferromagnetic superconductors with the same property of coexistence of ferromagnetism and superconductivity will be found soon. Up to date experimental and theoretical studies show that the unconventional ferromagnetic superconductors exhibit interesting phase transitions, including quantum phase transitions, multi-critical points, and a new type of critical behaviour – topics, which will be discussed in this Section as a good example of application of the phase transition theory.

9.9.1 *Coexistence of unconventional superconductivity and itinerant ferromagnetism*

The remarkable coexistence of itinerant ferromagnetism and unconventional (spin-triplet) superconductivity at low temperatures ($T < 1$ K) was discovered experimentally in the intermetallic compounds UGe_2 (Saxena *et al.*, 2000; Tateiwa *et al.*, 2001; Harada *et al.*, 2007), $ZrZn_2$ (Pfleiderer *et al.*, 2001), and URhGe (Aoki *et al.*, 2001). Other metallic compounds, such as UCoGe (Huy *et al.*, 2007, 2008) and UIr (Akazawa *et al.*, 2005; Kobayashi *et al.*, 2006), were also found to be spin-triplet ferromagnetic superconductors. In $ZrZn_2$, URhGe, and UCoGe, the mixed phase of coexistence of ferromagnetism and unconventional superconductivity (labeled the FS phase) occurs over a wide range of pressure (i.e., from ambient pressure $P \sim 1$ bar up to 10 kbar). By contrast, in other compounds (e.g., UGe_2 and UIr) this FS phase is found only in the high-pressure part ($P \sim 10$ kbar) of the $T - P$ phase diagram.

Another feature of the above compounds is that the FS phase occurs only in the ferromagnetic phase domain of the $T - P$ diagram. Specifically, at equilibrium and a given pressure P, the temperature $T_F(P)$ of the normal-to-ferromagnetic phase (or N-FM) transition is never lower than the temperature $T_{FS}(P)$ of the ferromagnetic-to-FS phase (or FM-FS) transition. This is consistent with the point of view that the superconductivity in these compounds is triggered by the spontaneous magnetization $M(x)$, by analogy with the well-known triggering of the superfluid phase A_1 in ^{3}He at mK temperatures by the external magnetic field $H(x)$. This helium-analogy has been used in some theoretical studies (see, e.g., (Machida and Ohmi, 2001; Walker and Samokhin, 2002; Shopova and Uzunov, 2003b, 2005a)), where Ginzburg-Landau (GL) free energy terms to describe the FS phase were derived by symmetry arguments.

For the spin-triplet ferromagnetic superconductors the trigger mechanism was recently examined in detail (Shopova and Uzunov, 2005a). The main system properties are affected by a term in the GL expansion of the form $M|\psi|^2$, which represents the interaction of $M = \{M_j; j = 1, 2, 3\}$ with the complex superconducting vector field $\boldsymbol{\psi} = \{\psi_j\}$. Specifically, this term triggers $\boldsymbol{\psi} \neq 0$ for certain T and P values. An analogous trigger mechanism is familiar in the context of improper ferroelectrics (Cowley, 1980).

A crucial consideration here is the nonzero magnetic moment of the spin-triplet Cooper pairs of the electrons. While the spin-singlet (s-wave) Cooper pairs in conventional superconductors have net spin zero and are quite sensitive to the magnitude of the magnetic induction $B(x)$, the spin-triplet pairs are known to be robust with respect to relatively large $B(x)$. The phenomena of spin-triplet superconductivity and itinerant ferromagnetism are both due to the same electron bands of the compounds: the f-band electrons in uranium-based compounds and the d-band electrons in $ZrZn_2$. However, the microscopic band theory of magnetism and superconductivity in non-Fermi liquids of strongly interacting heavy electrons is either too complex or insufficiently developed to describe the complicated behavior in itinerant ferromagnetic compounds. Consequently, several authors (Machida and Ohmi, 2001; Walker and Samokhin, 2002; Shopova and Uzunov, 2003b, 2005a, 2009; Linder and Sudbø, 2007; Linder *et al.*, 2008; Cottam, Shopova and Uzunov, 2008) have explored a phenomenological description within self-consistent MF theory, and here we will discuss this approach; for reviews on this problem, see Shopova and Uzunov (2005b, 2006).

Let us emphasize that here we shall present a general thermodynamic analysis in which the surface and bulk phases are treated on the same footing. For this reason, the results do not contradict to experiments (Yelland *et al.*, 2005) showing a lack of bulk superconductivity in $ZrZn_2$ but the occurrence of a surface FS phase at surfaces with higher Zr content than that in $ZrZn_2$.The generalized GL free energy which will be introduced in Section 9.9.2 has been derived from the microscopic theory by Dahl and Sudbø (2007).

9.9.2 *Free energy of ferromagnetic superconductors with spin-triplet electron pairing*

The order parameter (ψ field) in the theory of spin-triplet (p-wave) superconductivity is a three component complex vector: $\psi = (\psi_1, \psi_2, \psi_3)$. Neglecting some anisotropy in the gradient terms (Uzunov, 1991; Mineev and Samokhin, 1999; Shopova and Uzunov, 2005a) the GL free energy is given by Eq. (9.88), where the symmetry index n should be taken equal to six ($n = 6$). For our convenience, here we shall change the notations of the parameters in Eq. (9.88) in the following way: $a \to a_s = \alpha_s(T - T_s)$, $u \to b_s$, $\bar{u} \to u_s$, and $v \to v_s$. To the free energy Eq. (9.88), now denoted by $F_s(\psi, A)$, we should add the ferromagnetic free energy

$$F_{\mathrm{F}}(M) = \int_V d^d x \left(c_f \sum_{j=1}^{3} |\nabla M_j|^2 + a_f M^2 + \frac{b_f}{2} M^4 \right), \qquad (9.170)$$

where $b_f > 0$, and $a_f(T) = \alpha_f(T - T_f)$ is given by the parameter $\alpha_f > 0$ and the critical temperature T_f of the generic ferromagnetic phase transition.

The interaction between the ferromagnetic and superconducting subsystems (i.e., between the fields ψ and M) has been deduced by rigorous symmetry arguments (Machida and Ohmi, 2001; Walker and Samokhin, 2002). It is given by two terms:

$$F_{\mathrm{I}}(\psi, M) = \int_V d^d x \left[i\gamma_0 M \cdot (\psi \times \psi^*) + \delta M^2 |\psi|^2 \right], \qquad (9.171)$$

where $\gamma_0 \sim J_{ex}$ (with J_{ex} the ferromagnetic exchange constant), and δ is the second interaction parameter, which describes the symmetry ($\pm M$) conserving interaction ($\psi^2 M^2$). In general, the parameter δ for ferromagnetic superconductors may take both positive and negative values but the magnitude of the negative values is restricted by requirements for the stability of the model.

The values of the material parameters (T_s, T_f, α_s, α_f, b_s, u_s, v_s, b_f, γ_0 and δ) depend on the choice of the particular substance and on thermodynamic parameters as temperature T and pressure P. The γ_0-term in the above expression is the most substantial for the description of experimentally found ferromagnetic superconductors and the $\delta M^2|\psi|^2$–term makes the model more realistic in the strong coupling limit as it gives the opportunity to enlarge the phase diagram including both positive and negative values of the parameter a_s. In this way the domain of the stable ferromagnetic order is extended down to zero temperatures for a wide range of values of material parameters and the pressure P — a situation that corresponds to the experiments in ferromagnetic superconductors. Now we sum these three free energies to define the total free energy $F = (F_s + F_f + F_I)$, and the respective total energy density $f(\psi, M) = F(\psi, M)/V$.

We shall investigate the uniform phases, where the vector fields ψ and $M(x)$ do not depend on the space vector x; the possibility for the appearance of Meissner phase in ferromagnetic superconductors was discussed by Shopova and Uzunov (2005b) and Linder and Sudbø (2007). It should be stressed that the investigation of the uniform phases is an obligatory stage in the understanding of the properties of unconventional ferromagnetic superconductors irrespectively on whether these phases appear in particular substances at $T > 0$.

The Meissner phase occurs when the magnetization M is enough low and the vortex phase is energetically unfavourable. To elucidate this argument, let us mention, that here the magnetization M affects on the x-dependence of the equilibrium order parameter $\langle\psi(x)\rangle$, in a way, which is similar to that of the external magnetic field H in nonmagnetic superconductors (see, e.g., Fig. 9.6). The magnetization vary with the variation of the thermodynamic parameters $(T, P, ...)$ and the composition of the superconductor and, therefore, at some T and P the uniform Meissner phase may appear, depending on the particular superconductor.

In case of a strong easy axis type of magnetic anisotropy, as is in UGe_2 (Saxena *et al.*, 2000), the overall complexity of MF analysis of the free energy $f(\psi, M)$ can be avoided by performing an "Ising-like" description: $M = (0, 0, \mathcal{M})$, where $\mathcal{M} = \pm|M|$ is the magnetization along the $\hat{z}$-axis. Because of the equivalence of the "up" and "down" physical states $(\pm M)$ the thermodynamic analysis can be performed within the "gauge" $\mathcal{M} \geq 0$. When the magnetic order has a continuous symmetry we can take advantage of the symmetry of the total free energy $f(\psi, M)$ and avoid the consideration of equivalent thermodynamic states that occur as a result of the respective symmetry breaking at the phase transition point but have no effect on thermodynamics of the system. In the isotropic system one may again choose a gauge, in which the magnetization vector has the same direction as $\hat{z}$-axis $(|M| = M_z = \mathcal{M})$ and this will not influence the generality of thermodynamic analysis. Here we shall prefer the alternative description within which the ferromagnetic state may appear through two equivalent *up* and *down* domains with magnetizations $\mathcal{M}$ and $(-\mathcal{M})$, respectively.

At a certain stage of our MF analysis of the uniform phases and the possible phase transitions between such phases in a zero external magnetic field ($H = 0$), we shall neglect the crystal anisotropy. The anisotropy causes small changes (of shape and size) of the domains of stability of the possible phases and this effect can be ignored in most of the considerations (Shopova and Uzunov, 2005a,b, 2006)

It is convenient to introduce the notation

$$b = (b_s + u_s + v_s) \tag{9.172}$$

and redefine the order parameters and the other quantities in the following way:[4]

$$\begin{gathered} \varphi_j = b^{1/4}\psi_j = \phi_j e^{i\theta_j}, \quad M = b_f^{1/4}\mathcal{M}, \\ r = \frac{a_s}{\sqrt{b}}, \quad t = \frac{a_f}{\sqrt{b_f}}, \quad w = \frac{u_s}{b}, \quad v = \frac{v_s}{b}, \\ \gamma = \frac{\gamma_0}{b^{1/2}b_f^{1/4}}, \quad \gamma_1 = \frac{\delta}{(bb_f)^{1/2}}, \end{gathered} \tag{9.173}$$

where $\phi_j \equiv |\psi_j|$; $j = 1, 2, 3$ (cf. equivalent representation of the field ψ in Section 9.7.2). Having in mind that the order parameters ψ and M are considered uniform [the gradient terms $|\nabla M_j|^2$ and $|(\nabla - iq_0 A)\psi|^2$ are ignored], and using Eqs. (9.172) and (9.173), we can write the total free energy density $f(\psi, M)$ in the form

$$\begin{aligned} f(\psi, M) = {} & r\phi^2 + \frac{1}{2}\phi^4 + 2\gamma\phi_1\phi_2 M \sin(\theta_2 - \theta_1) \\ & + \gamma_1\phi^2 M^2 + tM^2 + \frac{1}{2}M^4 \\ & - 2w[\phi_1^2\phi_2^2 \sin^2(\theta_2 - \theta_1) + \phi_1^2\phi_3^2 \sin^2(\theta_1 - \theta_3) \\ & + \phi_2^2\phi_3^2 \sin^2(\theta_2 - \theta_3)] \\ & - v[\phi_1^2\phi_2^2 + \phi_1^2\phi_3^2 + \phi_2^2\phi_3^2], \end{aligned} \tag{9.174}$$

where $\phi^2 = (\phi_1^2 + \phi_2^2 + \phi_3^2)$.

9.9.3 *Phases*

The equilibrium phases are obtained from the equations of state

[4] The parameter γ introduced in Eq. (9.173) should not be confused with the notation $\gamma = \hbar^2/4m$ used only in Section 9.7.

$$\frac{\partial f(\mu_0)}{\partial \mu_\alpha} = 0, \tag{9.175}$$

where the series of symbols μ can be defined as, for example, $\mu = \{\mu_\alpha\} = (M, \phi_1, ..., \phi_3, \theta_1, ..., \theta_3)$; μ_0 denotes an equilibrium phase. The stability matrix $\tilde{F}$ of the phases μ_0 is defined by

$$\hat{F}(\mu_0) = \{F_{\alpha\beta}(\mu_0)\} = \frac{\partial^2 f(\mu_0)}{\partial \mu_\alpha \partial \mu_\beta}. \tag{9.176}$$

The possible (stable, metastable and unstable) phases are given in Table 9.2 together with the respective existence and stability conditions. The normal or disordered phase — "N-phase" in Table 9.2, always exists (for all temperatures $T \geq 0$) and is stable for $t > 0$, $r > 0$. The superconducting phase denoted by SC1 is unstable. The same is valid for the phase of coexistence of ferromagnetism and superconductivity denoted in Table 9.2 by CO2. The N–phase, the ferromagnetic phase (FM), the superconducting phases (SC1–3) and two of the phases of coexistence (CO1–3) are generic phases because they appear also in the decoupled case ($\gamma \equiv 0$) (see also Section 9.7.2). When the $M\phi_1\phi_2$–coupling is not present, the phases SC1–3 are identical and represented by the order parameter ϕ with components ϕ_j that participate on equal footing. The asterisk attached to the stability condition of the second superconductivity phase (SC2) indicates that our analysis is insufficient to determine whether this phase corresponds to a minimum of the free energy.

It can be shown that the phase SC2, two other purely superconducting phases and the coexistence phase CO1, have no chance to become stable for $\gamma \neq 0$. This is so, because the phase of coexistence of superconductivity and ferromagnetism (FS in Table 9.2) that does not occur for $\gamma = 0$ is stable and has a lower free energy in their domain of stability. A second domain ($M < 0$) of the FS phase is denoted in Table 9.2 by FS*. Here we shall describe only the first domain FS. The domain FS* is considered in the same way.

For $\gamma = 0$, these results can directly be taken from the theory of bicritical and tetracritical points (Section 9.6.6), or, which is almost the same, from the theory of ferromagnetic superconductivity in ternary compounds (Vonsovsky, Izyumov and Kurmaev, 1982; Maple and Fisher, 1982).

For $r > 0$, i.e., for $T > T_s$ Table 9.2 shows that three phases exist: the N–phase, FM and FS. As this case is most consistent with the experi-

Table 9.2 Phases and their existence and stability properties [$\theta = (\theta_2 - \theta_1)$, $k = 0, \pm 1, ...$].

Phase	order parameter	existence	stability domain
N	$\phi_j = M = 0$	always	$t > 0, r > 0$
FM	$\phi_j = 0,\ M^2 = -t$	$t < 0$	$r > 0,\ r > r_e(t)$
SC1	$\phi_1 = M = 0, \phi^2 = -r$	$r < 0$	unstable
SC2	$\phi^2 = -r,\ \theta = \pi k,\ M = 0$	$r < 0$	$(t > 0)^*$
SC3	$\phi_1 = \phi_2 = M = 0,\ \phi_3^2 = -r$	$r < 0$	$r < 0, t > 0$
CO1	$\phi_1 = \phi_2 = 0,$ $\phi_3^2 = -r,\ M^2 = -t$	$r < 0,$ $t < 0$	$r < 0$ $t < 0$
CO2	$\phi_1 = 0, \phi^2 = -r$ $\theta = \theta_2 = \pi k,\ M^2 = -t$	$r < 0$ $t < 0$	unstable
FS	$2\phi_1^2 = 2\phi_2^2 = \phi^2 = -r + \gamma M$ $\phi_3 = 0,\ \theta = 2\pi(k - 1/4)$ $\gamma r = (\gamma^2 - 2t)M - 2M^3$	$\gamma M > r$	$3M^2 > (-t + \gamma^2/2)$ $M > 0$
FS*	$2\phi_1^2 = 2\phi_2^2 = \phi^2 = -(r + \gamma M),$ $\phi_3 = 0,\ \theta = 2\pi(k + 1/4)$ $\gamma r = (2t - \gamma^2)M + 2M^3$	$-\gamma M > r$	$3M^2 > (-t + \gamma^2/2)$ $M < 0$

ments on the inter-metallic compounds mentioned in Section 9.9.1, we shall consider namely this variant of the theory.

For $r > 0$, i.e., for $T > T_s$ Table 9.2 shows that three phases exist: the N–phase, FM and FS. As this case, together with the assumption $T_f \gg T_s$, is most consistent with the experiments on the inter-metallic compounds mentioned in Section 9.9.1, we shall consider namely this variant of the theory. Two of these three phases are quite simple: the N-phase with existence and stability domains ($T > \max\{T_f, T_s\}$), shown in Table 9.2, and the FM phase with the existence condition $t < 0$, as shown in Table 9.2, and a stability domain defined by the inequality $r_e^{(1)} \leq r$. Here

$$r_e^{(1)} = \gamma_1 t + \gamma\sqrt{-t}, \tag{9.177}$$

The third phase – FS, is described by the following equations (Table 9.2):

$$\phi_1 = \phi_2 = \frac{\phi}{\sqrt{2}}, \quad \phi_3 = 0, \tag{9.178}$$

$$\phi^2 = (\pm\gamma M - r - \gamma_1 M^2), \tag{9.179}$$

$$(1 - \gamma_1^2)M^3 \pm \frac{3}{2}\gamma\gamma_1 M^2 + \left(t - \frac{\gamma^2}{2} - \gamma_1 r\right) M \pm \frac{\gamma r}{2} = 0, \tag{9.180}$$

and

$$(\theta_2 - \theta_1) = \mp\frac{\pi}{2} + 2\pi k, \tag{9.181}$$

($k = 0, \pm 1, ...$). The upper sign in Eqs. (9.179)–(9.181) corresponds to the FS domain where $\sin(\theta_2 - \theta_1) = -1$ and the lower sign corresponds to the FS* domain with $\sin(\theta_2 - \theta_1) = 1$.

The analysis of the stability matrix (9.176) for these phase domains shows that FS is stable for $M > 0$ and FS* is stable for $M < 0$. As these domains belong to the same phase, namely, have the same free energy and are thermodynamically equivalent, we shall consider one of them, for example, FS.

In order to outline a theoretical phase diagram in the plane (t, r) one should take in mind the stability conditions, available in Table 9.2, which define the domains of stability of the three phases. The stability conditions for FS seem more complex, and using the data presented in Table 9.2, we can present them in the following form:

$$\begin{aligned}
&2\gamma M - r - \gamma_1 M^2 \geq 0, \\
&\gamma M \geq 0, \\
&3(1-\gamma_1^2)M^2 + 3\gamma\gamma_1 M + t - \gamma_1 r - \gamma^2/2 \geq 0.
\end{aligned} \tag{9.182}$$

it is convenient to treat the stability conditions (9.182) together with the existence condition $\phi^2 \geq 0$, where ϕ is given by Eq. (9.179).

On the basis of the existence and stability analysis (Shopova and Uzunov, 2005a) we draw in Fig. 9.12 the (t, r) phase diagram. For this aim we have chosen $\gamma = 1.2$, $\gamma_1 = 0.8$ and, as mentioned above, we set the symmetry parameter w equal to zero. The domains of the phases and the phase transition lines are clearly indicated in Fig. 9.12. The phase transition between the N-phase and FS phase is of first order (solid line) and goes along the equilibrium line AC in the t-interval: ($t_A = \gamma^2/2$, $t_C = 0$).

N, FM, and FS phases coexist at the triple point C with coordinates $t = 0$, and $r_C^{(1)} = \gamma^2/4(1+\gamma_1)$. On the left of the point C, for $t_B^{(1)} = -\gamma^2/4(1+\gamma_1)^2 < t < 0$ the first order phase transition line is given by the equation

$$r_{eq}^{(1)*}(t) = \frac{\gamma^2}{4(1+\gamma_1)} - t. \tag{9.183}$$

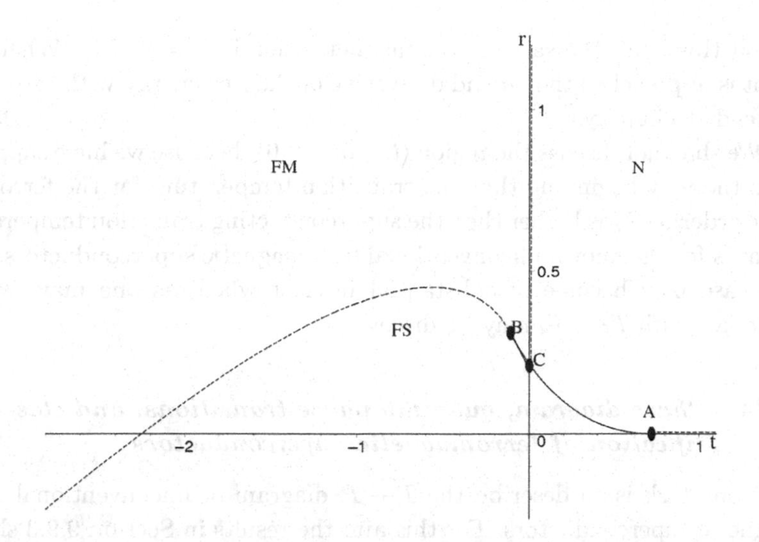

Fig. 9.12 The phase diagram in the (t, r) plane for $\gamma = 1.2$, $\gamma_1 = 0.8$ and $w = 0$. The parameters of the theory (r, t, γ, γ_1, $w, \dots$) are defined by Eq. (9.173). The domains of stability of the phases N, FM and FS are indicated. A and B are tricritical points of phase transitions separating the dashed lines (on the left of point B and on the right of point A) of second order phase transitions from the solid line ABC of first order phase transitions. The FS phase is stable in the whole domain of the (t, r) below the solid and dashed lines. The vertical dashed line coinciding with the r-axis above the triple point C indicates the N-FM phase transition of second order.

This function is illustrated by the line BC in Fig. 9.12 that terminates at the tricritical point B with coordinates $t_B^{(1)}$ (given above), and

$$r_B^{(1)} = \frac{\gamma^2(2+\gamma_1)}{4(1+\gamma_1)^2}, \tag{9.184}$$

which is obtained by substituting t in Eq. (9.183) with $t_B^{(1)}$.

To the left of the tricritical point B the second order phase transition curve is given by the relation (9.177). Here the magnetization is $M = \sqrt{-t}$ and the superconducting order parameter is equal to zero ($\phi = 0$). This line intersects the t-axis at $t_{S2} = -(\gamma/\gamma_1)^2$ and is well defined also for $r < 0$. The function $r_e^{(1)}(t)$ has a maximum at the point

$$(t_m, r_m) = \left(-\frac{\gamma^2}{4\gamma_1^2}, \frac{\gamma^2}{4\gamma_1}\right) \tag{9.185}$$

and at this point the value of the magnetization is $M = \gamma/2\gamma_1$. When this point is approached the second derivative of the free energy with respect to M tends to infinity.

We shall not discuss the region ($t > 0, r < 0$), because we have supposed from the very beginning that the transition temperature for the ferromagnetic ordering T_f is higher then the superconducting transition temperature T_s, as is for the known unconventional ferromagnetic superconductors. But this case may become of substantial interest when, as one may expect, materials with $T_f < T_s$ may be discovered.

9.9.4 *Phase diagram, quantum phase transitions, and classification of ferromagnetic superconductors*

Now our task is to describe the $T - P$ diagram of unconventional ferromagnetic superconductors. For this aim the results in Section 9.9.3 should be connected with the experimental data. We have to choose the best normalization of the theory so as to minimize the theory parameters and to assume an appropriate dependence of the theory parameters on the pressure P. This problem has been solved by Cottam, Shopova and Uzunov (2008); for more details, see also Shopova and Uzunov (2009).

A convenient dimensionless free energy can now be defined by $\tilde{f} = f/(b_f M_0^4)$, where $M_0 = [\alpha_f T_{f0}^n/b_f]^{1/2} > 0$ is the value of M corresponding to the pure magnetic subsystem ($\psi \equiv 0$) at $T = P = 0$ and $T_{f0} = T_f(0)$. We use the notation $b_s = b$, and recall that $u_s = v_s = 0$. On scaling the order parameters as $m = M/M_0$ and $\varphi = \psi/[(b_f/b)^{1/4}M_0]$ we obtain

$$\tilde{f} = r\phi^2 + \frac{\phi^4}{2} + tm^2 + \frac{m^4}{2} + 2\gamma m\phi_1\phi_2\sin\theta + \gamma_1 m^2\phi^2, \tag{9.186}$$

where $\phi_j = |\varphi_j|$, $\phi = |\varphi|$, and θ is the phase angle between the complex φ_2 and φ_1. The dimensionless constants are

$$t = \tilde{T}^n - \tilde{T}_f^n(P), \quad r = \kappa(\tilde{T} - \tilde{T}_s), \tag{9.187}$$

where $n = 1$ corresponds to the usual case of MF theory, and $n = 2$ stands for a possible description within the framework of the spin-fluctuation theory (Yamada, 1993),

$$\kappa = \frac{\alpha_s b_f^{1/2}}{\alpha_f b^{1/2} T_{f0}^{n-1}}, \quad \gamma = \frac{\gamma_0}{(\alpha_f b)^{1/2} T_{f0}^{n/2}}, \quad \gamma_1 = \frac{\delta}{(bb_f)^{1/2}}. \tag{9.188}$$

The reduced temperatures are

$$\tilde{T} = \frac{T}{T_{f0}}, \quad \tilde{T}_f(P) = \frac{T_f(P)}{T_{f0}}, \quad \tilde{T}_s(P) = \frac{T_s(P)}{T_{f0}}. \tag{9.189}$$

The analysis involves making simple assumptions for the P dependence of the t, r, γ, and γ_1 parameters in Eq. (9.186). Specifically, we assume that only T_f has a significant P dependence, described by $\tilde{T}_f(P) = (1 - \tilde{P})^{1/n}$, where $\tilde{P} = P/P_0$ and P_0 is a characteristic pressure deduced later. In $ZrZn_2$ and UGe_2 the P_0 values are very close to the critical pressure P_c at which both the ferromagnetic and superconducting orders vanish, but in other systems this is not necessarily the case. As we will discuss, the nonlinearity ($n = 2$) of $T_f(P)$ in $ZrZn_2$ and UGe_2 is relevant at relatively high P, at which the N-FM transition temperature $T_F(P)$ may not coincide with $T_f(P)$.

The simplified model in Eq. (9.186) is capable of describing the main thermodynamic properties of spin-triplet ferromagnetic superconductors. There are three stable phases:

(i) the normal (N) phase, given by $\phi = m = 0$,

(ii) the pure ferromagnetic (FM) phase, given by $m = (-t)^{1/2} > 0$, $\phi = 0$, and

(iii) the FS phase, given by $\phi_1^2 = \phi_2^2 = (\gamma m - r - \gamma_1 m^2)/2$, $\phi_3 = 0$, where $\sin\theta = -1$ and m satisfies

$$(1 - \gamma_1^2)m^3 + \frac{3}{2}\gamma\gamma_1 m^2 + \left(t - \frac{\gamma^2}{2} - \gamma_1 r\right) m + \frac{\gamma r}{2} = 0. \tag{9.190}$$

We note that FS is a two-domain phase (Shopova and Uzunov, 2003b, 2005a). Although Eq. (9.190) is complicated, some analytical results follow, e.g., we find that the second order phase transition line $\tilde{T}_{FS}(P)$ separating the FM and FS phases is the solution of

$$\tilde{T}_{FS}(P) = \tilde{T}_s + \frac{\gamma_1}{\kappa} t(T_{FS}) + \frac{\gamma}{\kappa} \left[-t\left(T_{FS}\right)\right]^{1/2}. \tag{9.191}$$

Under certain conditions, the $T_{FS}(P)$ curve has a maximum at

$$\tilde{T}_m = \tilde{T}_s + \frac{\gamma^2}{4\kappa\gamma_1} \tag{9.192}$$

with pressure P_m found by solving $t(T_m, P_m) = -(\gamma^2/4\gamma_1^2)$. Examples will be given later, but generally this curve extends from ambient P up to the tri-critical point labeled B, with coordinates (P_B, T_B), where the FM-FS phase transition occurs at a straight line of first order transition up to a critical-end point C. The lines of all three phase transitions (N-FM, N-FS, and FM-FS) terminate at C. For $P > P_C$ the FM-FS phase transition occurs on a rather flat, smooth line of equilibrium transition of first order up to a second tricritical point A with $P_A \sim P_0$ and $T_A \sim 0$. Finally, the third transition line terminating at C describes the second order phase transition N-FM. The temperatures at the three multi-critical points correspond to $\tilde{T}_A = \tilde{T}_s$, $\tilde{T}_B = \tilde{T}_s + \gamma^2(2+\gamma_1)/4\kappa(1+\gamma_1)^2$, and $\tilde{T}_C = \tilde{T}_s + \gamma^2/4\kappa(1+\gamma_1)$, while the P values can be deduced from the previous equations. These results are valid whenever $T_f(P) > T_s(P)$, which excludes any pure superconducting phase ($\psi \neq 0, m = 0$) in accord with the available experimental data. Note that, for any set of material parameters, $T_A < T_C < T_B < T_m$ and $P_m < P_B < P_C$.

A calculation of the $T-P$ diagram from Eq. (9.186) for any material requires some knowledge of P_0, T_{f0}, T_s, κ γ, and γ_1. The temperature T_{f0} can be obtained directly from the experimental phase diagrams. The model pressure P_0 is either identical to or very close to the critical pressure P_c at which the N-FM phase transition line terminates at $T \sim 0$. The characteristic temperature T_s of the generic superconducting transition is not available from the experiments and thus has to be estimated using general consistency arguments. For $T_f(P) > T_s(P)$ we must have $T_s(P) = 0$ at $P \geq P_c$, where $T_f(P) \leq 0$. For $0 \leq P \leq P_0$, $T_s < T_C$ and therefore for cases where T_C is too small to be observed experimentally, T_s can be ignored. For systems where T_C is measurable this argument does not hold. This is likely to happen for $T_s > 0$ (for $T_s < 0$, T_C is very small). However, in such cases, pure superconducting phase should be observable. To date there are no experimental results reported for such a feature in $ZrZn_2$ or UGe_2, and thus we can put $T_s = 0$. We remark that negative values of T_s are possible, and they describe a phase diagram topology in which the FM-FS transition line terminates at $T = 0$ for $P < P_c$. This might be of relevance for other compounds, e.g., URhGe.

Typically, additional features of the experimental phase diagram must be utilized. For example, in $ZrZn_2$ these are the observed values of $T_{FS}(0)$ and the slope $\rho_0 \equiv [\partial T_{FS}(P)/\partial P]_0$ at $P = 0$. For UGe_2 one may use T_m, P_m, and P_{0c}, where the last quantity denotes the other solution (below P_c) of $T_{FS}(P) = 0$. The ratios γ/κ and γ_1/κ can be deduced using Eq. (9.191) and the expressions for T_m, P_m, and ρ_0, while κ is chosen by requiring a suitable value of T_C.

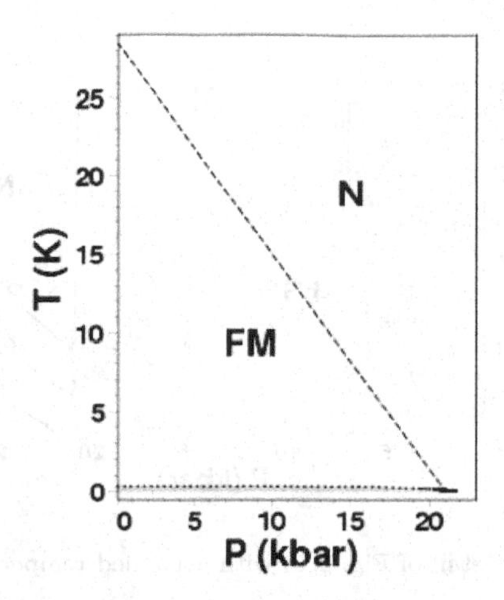

Fig. 9.13 $T - P$ diagram of $ZrZn_2$ calculated taking $T_s = 0$, $\rho_0 = 0.003$ K/kbar, $T_{f0} = 28.5$ K, $P_0 = 21$ kbar, $\kappa = 10$, $\gamma/\kappa = 2\gamma_1/\kappa \approx 0.2$. The low-$T$ domain of the FS phase is seen more clearly in the following figure.

Experiments (Pfleiderer *et al.*, 2001) for $ZrZn_2$ indicate $T_{f0} = 28.5$ K, $T_{FS}(0) = 0.29$ K, $P_0 \sim P_c = 21$ kbar, and $T_F(P) \sim T_f(P)$ is almost a straight line, so $n = 1$ describes the P-dependence. The slope for $T_{FS}(P)$ at $P = 0$ is harder to estimate; its magnitude should not exceed $T_{f0}/P_c \approx 0.014$ on the basis of a straight-line assumption, implying $-0.014 < \rho \leq 0$. However, this ignores the effect of a maximum, although it is unclear experimentally in $ZrZn_2$, at (T_m, P_m). If such a maximum were at $P = 0$ we would have $\rho_0 = 0$, whereas a maximum with $T_m \sim T_{FS}(0)$ and $P_m \ll P_0$ provides us with an estimated range $0 \leq \rho_0 < 0.005$. The choice $\rho_0 = 0$ gives $\gamma/\kappa \approx 0.02$ and $\gamma_1/\kappa \approx 0.01$, but similar values hold for any $|\rho_0| \leq 0.003$. The multi-critical points A and C cannot be distinguished experimentally. Since the experimental accuracy (Pfleiderer *et al.*, 2001) is less than ~ 25 mK in the high-P domain ($P \sim 20 - 21$ kbar), we suppose that $T_C \sim 10$ mK, which corresponds to $\kappa \sim 10$. We employed these parameters to calculate the $T - P$ diagram using $\rho_0 = 0$ and 0.003. The differences obtained in these two cases are negligible, with both phase diagrams being in excellent agreement with experiment.

The latter value is used in Fig. 9.13, which gives $P_A \sim P_c = 21.10$ kbar, $P_B = 20.68$ kbar, $P_C = 20.99$ kbar, $T_A = T_F(P_c) = T_{FS}(P_c) = 0$ K, $T_B = 0.077$ K, $T_C = 0.038$ K, and $T_{FS}(0) = 0.285$ K. The low-T region is seen in more detail in Fig. 9.14, where the A, B, C points are shown

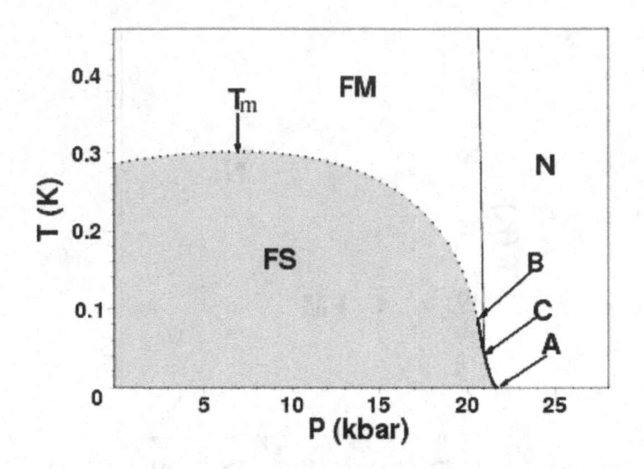

Fig. 9.14 Detail of Fig. 9.13 with expanded temperature scale.

and the order of the FM-FS phase transition changes from second to first order around the critical end-point C. The $T_{FS}(P)$ curve has a maximum at $P_m = 6.915$ kbar and $T_m = 0.301$ K. These results account well for the main features of the experimental behavior (Pfleiderer *et al.*, 2001), including the claimed change in the order of the FM-FS phase transition at relatively high P. Within the present model the N-FM transition is of second order up to $P_C \sim P_c$. Moreover, if the experiments are reliable in their indication of a first order N-FM transition at much lower P values, the theory can accommodate this by a change of sign of b_f, leading to a new tricritical point located at a distinct $P_{tr} < P_C$ on the N-FM transition line. Since $T_C > 0$ a direct N-FS phase transition of first order is predicted in accord with conclusions from de Haas–van Alphen experiments (Kimura *et al.*, 2004) and some theoretical studies (Uhlaz, Pfleiderer and Hayden, 2004). Such a transition may not occur in other cases where $T_C = 0$. In SFT ($n = 2$) the diagram topology remains the same but points B and C are slightly shifted to higher P (typically by about 1 bar).

The experimental data (Saxena *et al.*, 2000; Tateiwa *et al.*, 2001; Harada *et al.*, 2007) for UGe_2 indicate $T_{f0} = 52$ K, $P_c = 1.6$ GPa ($\equiv$ 16 kbar), $T_m = 0.75$ K, $P_m \approx 1.15$ GPa, and $P_{0c} \approx 1.05$ GPa. Using again the variant $n = 1$ for $T_f(P)$ and the above values for T_m and P_{0c} we obtain $\gamma/\kappa \approx 0.098$ and $\gamma_1/\kappa \approx 0.168$. The temperature $T_C \sim 0.1$ K corresponds to $\kappa \sim 5$. Using these, together with $T_s = 0$, leads to the $T - P$ diagram in Fig. 9.15, showing only the low-T region of interest. We obtain $T_A = 0$ K, $T_B = 0.481$ K, $T_C = 0.301$ K, $P_A = 1.72$ GPa, $P_B = 1.56$ GPa, and $P_C = 1.59$ GPa. There is agreement with the main experimental findings, although

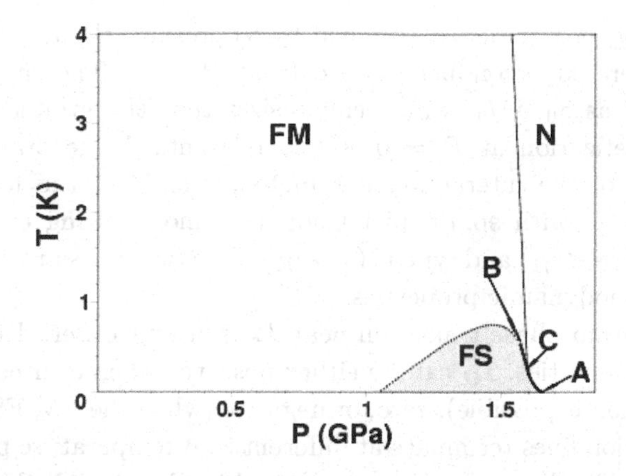

Fig. 9.15 Low-T part of the $T-P$ diagram of UGe_2 calculated taking $T_s = 0$, $T_{f0} = 52$ K, $P_0 = 1.6$ GPa, $T_m = 0.75$ K, $P_{0c} = 1.05$ GPa ($\kappa = 5$, $\gamma/\kappa = 0.098$, and $\gamma_1/\kappa = 0.168$).

P_m corresponding to the maximum (found at ~ 1.44 GPa in Fig. 9.15) is about 0.3 GPa higher than suggested experimentally. If the experimental plots are accurate in this respect, this difference may be attributable to the so-called (T_x) meta-magnetic phase transition in UGe_2, which is related to an abrupt change of the magnetization in the vicinity of (T_m, P_m). Thus, one may suppose that the meta-magnetic effects, which are outside the scope of our current model, significantly affect the shape of the $T_{FS}(P)$ curve by lowering P_m (along with P_B and P_C). It is possible to achieve a lower P_m value (while leaving T_m unchanged), but this has the undesirable effect of modifying P_{c0} to a value that disagrees with experiment. In SFT ($n = 2$) the multi-critical points are located at slightly higher P (by about 0.01 GPa), as for $ZrZn_2$.

The estimates for UGe_2 imply $\gamma_1\kappa \approx 1.9$, so the condition for $T_{FS}(P)$ to have a maximum found from Eq. Eq. (9.191) is satisfied. As we discussed for $ZrZn_2$, the location of this maximum can be hard to fix accurately in experiments. However, P_{c0} can be more easily distinguished, as in the UGe_2 case. Then we have a well-established quantum (zero-temperature) phase transition of second order, i.e., a quantum critical point (Shopova and Uzunov, 2003a). From Eq. (9.191) the existence of this type of solution in systems with $T_s = 0$ (as UGe_2) occurs for $\gamma < \gamma_1$. Such systems (which we label as U-type) are essentially different from those such as $ZrZn_2$ where $\gamma_1 < \gamma$ and hence $T_{FS}(0) > 0$. In this latter case (Zr-type compounds) a maximum $T_m > 0$ may sometimes occur, as discussed earlier. We note that the ratio γ/γ_1 reflects a balance effect between the two $\psi - M$ interactions.

When the trigger interaction (typified by γ) prevails, the Zr-type behavior is found where superconductivity exists at $P = 0$. The same ratio can be expressed as $\gamma_0/\delta M_0$, which emphasizes that the ground state value of the magnetization at $P = 0$ is also relevant. In general, depending on the ratio of the interaction parameters γ and γ_1, the ferromagnetic superconductors with spin-triplet Cooper fermion pairing can be of two types: type I ($\gamma < \gamma_1$) and type II ($\gamma > \gamma_1$). The two types are distinguished in their thermodynamic properties.

The quantum phase transition near P_c is of first order. Depending on the system properties, T_C can be either positive (when a direct N-FS first order transition is possible), zero, or negative (when the FM-FS and N-FM phase transition lines terminate at different zero-temperature phase transition points). The last two cases correspond to $T_s < 0$. All these cases are possible in Zr- and U-type compounds. The zero temperature transition at P_{c0} is found to be a quantum critical point, whereas the zero-temperature phase transition at P_c is of first order. As noted, the latter splits into two first order phase transitions. This classical picture may be changed through quantum fluctuations (Shopova and Uzunov, 2003a). An investigation (Uzunov, 2006, 2007) of the quantum fluctuations and the quantum dimensional crossover by RG methods revealed a fluctuation change in the order of this transition to a continuous phase transition belonging to an entirely new class of universality (Section 9.9.5). However, this option exists only for magnetically isotropic order (Heisenberg symmetry) and is unlikely to apply to spin-triplet ferromagnetic superconductors with Ising or XY symmetries.

Even in its simplified form, this theory has been shown to be capable of accounting for a wide variety of experimental behavior. A natural extension to the theory is to add a M^6 term which provides a formalism to investigate possible metamagnetic phase transitions (Huxley, Sheikin and Braithwaite, 2000) and extend some first order phase transition lines. Another modification of this theory, with regard to applications to other compounds, is to include a P dependence for some of the other GL parameters.

9.9.5 *Peculiar renormalization group analysis and unusual universality class of critical behaviour*

Here we present some results for the fluctuation effects on the phase transition properties of unconventional ferromagnetic superconductors. The MF analysis shows several interesting phase transitions and multicritical points.

It seems interesting to understand their fluctuation properties. For this aim we perform a RG investigation (Uzunov, 2006, 2008).

The dimensional analysis (Section 8.7.5) shows that all but one interaction terms in the total free energy have upper (critical) dimension $d_U = 4$. Only the first term in Eq. (9.171) corresponds to $d_U = 6$. This term generates fluctuations which strongly affect on the critical behaviour at any spatial dimensionality $d \leq 6$, whereas the fluctuation interactions generated by the other interaction terms become irrelevant for $d \geq 4$. Therefore, performing RG analysis near $d_U = 6$ with an expansion parameter $\epsilon = (6 - d)$, we may take into account only the interaction term of type $M\psi^2$ and ignore all other interaction terms as unessential participants in the fluctuation phenomena.

The relevant part of the fluctuation free energy (fluctuation Hamiltonian) can be written in the form

$$\mathcal{H} = \sum_k \left[\left(r + k^2\right) |\psi(k)|^2 + \frac{1}{2}\left(t + k^2\right) |M(k)|^2 \right] + \frac{ig}{\sqrt{V}} \sum_{k_1,k_2} M(k_1) \cdot \left[\psi(k_2) \times \psi^*(k_1 + k_2)\right] \tag{9.193}$$

where $g \equiv \gamma_0 \geq 0$, $V \sim L^d$ is the volume of the dD system, the length unit is chosen so that the wave vector k is confined below unity $(0 \leq |k| \leq 1)$, $g \geq 0$ is a coupling constant, describing the effect the scalar product of the m-dimensional real vector $M(k) = \{M_j(k), j = 1, \ldots, m\}$ and the vector product $(\psi \times \psi^*)$ of the $n/2$-complex vector $\psi(k) = \{\psi_\alpha(k), \alpha = 1, \ldots, n/2\}$ and its complex conjugate field $\psi^*(k)$ for $m = n/2 = 3$.

The fluctuation Hamiltonian describes fluctuation phenomena near phase transition points where both parameters r and t take values near to zero, i.e., when the generic critical temperatures T_s and T_f are very near to each other. When the phase transition lines of the ferromagnetic and superconducting phase transitions are not enough close to each other the correlation lengths the fields $M(k)$ and $\psi(k)$ are not simultaneously divergent and the interactions between these fields do not contribute to the critical fluctuations.

One may consider several cases: (i) uniaxial magnetic symmetry, $M = (0, 0, M_3)$, (ii) tetragonal crystal symmetry when $\psi = (\psi_1, \psi_2, 0)$, (iii) XY magnetic order $(M_1, M_2, 0)$, and (iv) the general case of cubic crystal symmetry and isotropic magnetic order ($m = 3$) when all components of the 3D vectors $M(x)$ and ψ may have nonzero equilibrium and fluctuation components. The latter case is of major interest to real systems where fluctuations of all components of the fields are possible despite the presence of spatial crystal and magnetic anisotropy that nullifies some of the equilibrium field components. In one-loop approximation, the RG analysis reveals different pictures for anisotropic (i)-(iii) and isotropic (iv) systems.

As usual, a Gaussian ("trivial") FP ($g^* = 0$) exists for all $d > 0$ and, as usual, this FP is stable for $d > 6$ where the fluctuations are irrelevant. In the reminder of this letter the attention will be focussed on spatial dimensions $d < 6$, where the critical behavior is usually governed by nontrivial FPs ($g^* \neq 0$). In the cases (i)-(iii) only negative, i.e., "unphysical" FP values of g^2 have been obtained for $d < 6$. For example, in the case (i) the RG relation for g takes the form

$$g' = b^{3-d/2-\eta} g \left(1 + g^2 K_d \ln b\right), \tag{9.194}$$

where g' is the renormalized value of g,

$$\eta = \left(\frac{K_{d-1}}{8}\right) g^2 \tag{9.195}$$

is the anomalous dimension (Fisher's exponent) of the field M_3. Using Eq. (9.194) one obtains the FP coordinate

$$(g^2)^* = -96\pi^3\epsilon. \tag{9.196}$$

For $d < 6$ this FP is unphysical and does not describe any critical behavior. For $d > 6$ the same FP is physical but unstable towards the parameter g as one may see from the positive value $y_g = -11\epsilon/2 > 0$ of the respective stability exponent y_g defined by

$$\delta g' = b^{y_g} \delta g. \tag{9.197}$$

Therefore, a change of the order of the phase transition from second order in MF approximation to a fluctuation-driven first order transition when the fluctuation g–interaction is taken into account, takes place. This conclusion is supported by general concepts of RG theory and by the particular property of these systems to exhibit first order phase transitions in MF approximation for broad variations of T and P (Sections 9.9.3 and 9.9.4).

In the case (iv) of isotropic systems the RG equation for g is degenerate and the ϵ-expansion breaks down. A similar situation is known from

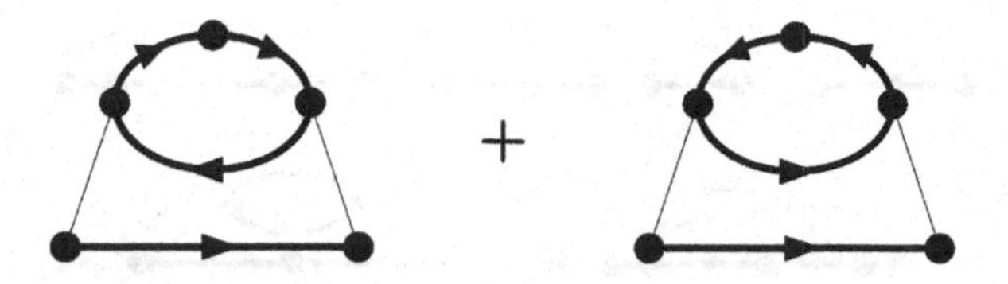

Fig. 9.16 A sum of g^5–diagrams equal to zero. The thick and thin lines correspond to correlation functions $\langle|\psi_\alpha|^2\rangle$ and $\langle|M_j|^2\rangle$, respectively; vertices (•) represent g–term in Eq. (9.193).

the theory of disordered systems (Lubensky, 1975; Grinstein, 1976; Lawrie, Millev and Uzunov, 1987) but here the physical mechanism and details of description are different. Namely for this degeneration one should consider the RG equations up to the two-loop order.

The derivation of the two-loop terms in the RG equations is quite non-trivial because of the special symmetry properties of the interaction g-term in Eq. (9.193). For example, some diagrams with opposite arrows of internal lines, as the couple shown in Fig. 9.16, have opposite signs and compensate each other. The terms bringing contributions to the g–vertex are shown diagrammatically in Fig. 9.17. The RG analysis is carried out by a completely new $\epsilon^{1/4}$-expansion for the FP values and $\epsilon^{1/2}$-expansion for the critical exponents; again $\epsilon = (6-d)$.

The RG equations are quite lengthy and here only the equation for g is presented. It has the form

$$g' = b^{(\epsilon-2\eta_\psi-\eta_M)/2} g\left[1 + Ag^2 + 3(2B+C)g^4\right], \tag{9.198}$$

where

$$A = \frac{K_d}{2}\left[2\ln b + \epsilon(\ln b)^2 + (1-b^2)(2r+t)\right], \tag{9.199a}$$

$$B = \frac{K_{d-1}K_d}{192}\left[9(b^2-1) - 11\ln b - 6\,(\ln b)^2\right], \tag{9.199b}$$

and

$$C = \frac{3K_{d-1}K_d}{64}\left[\ln b + 2\,(\ln b)^2\right], \tag{9.199c}$$

η_M and η_ψ are the anomalous dimensions of the fields $M(k)$ and $\psi(k)$, respectively. The one-loop approximation gives correct results to order

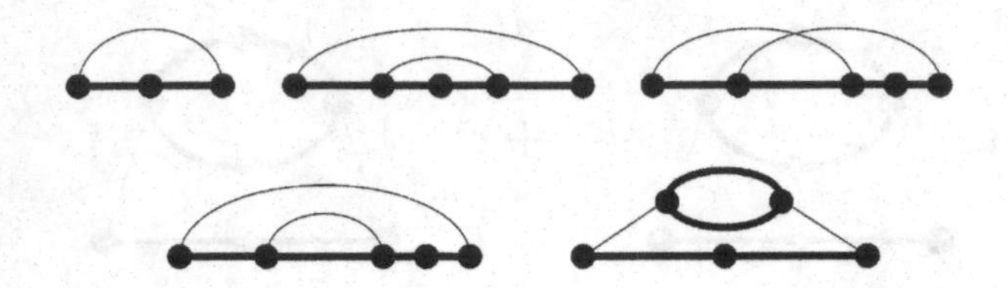

Fig. 9.17 Diagrams for g' of third and fifth order in g. The arrows of the thick lines have been omitted.

$\epsilon^{1/2}$ and the two-loop approximation brings such results up to order ϵ. In Eq. (9.199a), r and t are small expansion quantities with equal FP values

$$t^* = r^* = K_d g^2, \tag{9.200}$$

Using the condition for invariance of the two k^2-terms in Eq. (9.193) one obtains $\eta_M = \eta_\psi \equiv \eta$, where

$$\eta = \frac{K_{d-1}}{8} g^2 \left(1 - \frac{13}{96} K_{d-1} g^2\right). \tag{9.201}$$

Equation (9.198) yields a new FP

$$g^* = 8\left(3\pi^3\right)^{1/2} \left(\frac{2\epsilon}{13}\right)^{1/4}, \tag{9.202}$$

which corresponds to the critical exponent

$$\eta = 2\sqrt{\frac{2\epsilon}{13}} - \frac{2\epsilon}{3} \tag{9.203}$$

(for $d = 3$, $\eta \approx -0.64$).

The eigenvalue problem for the RG stability matrix

$$\hat{\mathcal{M}} = \left(\frac{\partial \mu_i}{\partial \mu_j}\right)_{\mu^*}, \qquad (\mu_1, \mu_2, \mu_3) = (r, t, g) \tag{9.204}$$

can be solved by the expansion of the matrix elements up to order $\epsilon^{3/2}$. When the eigenvalues $\lambda_j = A_j(b)b^{y_j}$ of $\hat{\mathcal{M}}$ are calculated dangerous large terms of type b^2 and $b^2(\ln b)$, $(b \gg 1)$, in the off-diagonal elements of the matrix $\hat{\mathcal{M}}$ ensure the compensation of redundant large terms of the same type in the diagonal elements $\hat{\mathcal{M}}_{ii}$. This compensation is crucial for the validity of scaling for this type of critical behavior. Such a problem does not appear in standard cases of RG analysis. As in the usual ϕ^4-theory the amplitudes A_j depend on the scaling factor b:

$$A_1 = A_2 = 1 + \frac{27}{13} b^2 \epsilon, \tag{9.205a}$$

and

$$A_3 = 1 - \frac{81}{52} \epsilon (\ln b)^2. \tag{9.205b}$$

The critical exponents $y_t = y_1$, $y_r = y_2$ and $y_g = y_3$ are b–invariant:

$$y_r = 2 + 10\sqrt{\frac{2\epsilon}{13}} + \frac{197}{39}\epsilon, \tag{9.206}$$

$$y_t = y_r - 18\sqrt{\frac{2\epsilon}{13}}, \tag{9.207}$$

and $y_g = -\epsilon > 0$ for $d < 6$. The correlation length critical exponents $\nu_\psi = 1/y_r$ and $\nu_M = 1/y_t$ corresponding to the fields ψ and M are

$$\nu_\psi = \frac{1}{2} - \frac{5}{2}\sqrt{\frac{2\epsilon}{13}} + \frac{103}{156}\epsilon, \tag{9.208}$$

and

$$\nu_M = \frac{1}{2} + 2\sqrt{\frac{2\epsilon}{13}} - \frac{5\epsilon}{156}, \tag{9.209}$$

respectively.

These exponents describe a quite particular multi-critical behavior which differs from the numerous examples known so far. For $d = 3$, $\nu_\psi = 0.78$ which is somewhat above the usual value $\nu \sim 0.6 \div 0.7$ near a standard phase transition of second order, but $\nu_M = 1.76$ at the same dimension ($d = 3$) is unusually large. The fact that the Fisher's exponent η is negative for $d = 3$ does not create troubles because such cases are known in complex systems, for example, in conventional superconductors (Halperin, Lubensky and Ma, 1974).

The present ϵ-expansion is valid under the conditions $\epsilon^{1/2} b < 1$, $\epsilon^{1/2}(\ln b) \ll 1$ provided $b > 1$. These conditions are stronger than those corresponding to the usual ϕ^4-theory. This means that the present expansion in non-integer powers of ϵ has a more restricted domain of validity than the standard ϵ-expansion. Using the known relation $\gamma = (2 - \eta)\nu$, the susceptibility exponents for $d = 3$ take the values $\gamma_\psi = 2.00$ and $\gamma_M = 4.65$. These values exceed even those corresponding to the Hartree approximation ($\gamma = 2\nu = 2$ for $d = 3$) and can be easily distinguished in experiments.

9.9.5.1 *Quantum criticality*

The critical behavior discussed so far may occur in a close vicinity of finite temperature multi-critical points ($T_c = T_f > 0$) in systems possessing the symmetry of the model (9.193). In certain systems, this multi-critical points may occur at $T = 0$. In the quantum limit ($T \to 0$), or, more generally, in the low-temperature limit [$T \ll \mu; \mu \equiv (t, r); k_B = 1$] the thermal wavelengths of the fields $M(x)$ and $\psi(x)$ exceed the inter-particle interaction radius and the quantum correlations fluctuations become essential for the critical behavior (Hertz, 1976; Shopova and Uzunov, 2003a). The quantum effects can be considered by RG analysis of a comprehensively generalized version of the model (9.193), namely, the action $\mathcal{S}$ of the referent quantum system. The generalized action is constructed with the help of the substitution

$$(-\mathcal{H}/T) \to \mathcal{S}[M(q), \psi(q)]. \tag{9.210}$$

Now the description is given in terms of the (Bose) quantum fields $M(q)$ and $\psi(q)$ which depend on the $(d+1)$-dimensional vector $q = (\omega_l, k)$; $\omega_l = 2\pi l T$ is the Matsubara frequency ($\hbar = 1; l = 0, \pm 1, \ldots$). The k-sums in Eq. (9.193) should be substituted by respective q-sums and the inverse bare correlation functions $(r + k^2)$ and $(t + k^2)$ in Eq. (9.193) contain additional ω_l–dependent terms, for example(Hertz, 1976; Shopova and Uzunov, 2003a)

$$\langle |\psi_\alpha(q)|^2 \rangle^{-1} = |\omega_l| + k^2 + r. \tag{9.211}$$

The bare correlation function $\langle |M_j(q)| \rangle^2$ contains a term of type $|\omega_l|/k^\theta$, where $\theta = 1$ and $\theta = 2$ for clean and dirty itinerant ferromagnets, respectively (Hertz, 1976). The quantum dynamics of the field ψ is described by the bare value $z = 2$ of the dynamical critical exponent $z = z_\psi$ whereas the quantum dynamics of the magnetization corresponds to $z_M = 3$ (for $\theta = 1$), or, to $z_M = 4$ (for $\theta = 2$). This means that the classical-to-quantum dimensional crossover at $T \to 0$ is given by $d \to (d+2)$ and, hence, the system exhibits a simple MF behavior for $d \geq 4$. Just below the upper quantum critical dimension $d_U^{(0)} = 4$ the relevant quantum effects at $T = 0$ are represented by the field ψ whereas the quantum (ω_l–) fluctuations of the magnetization are relevant for $d < 3$ (clean systems), or, for even for $d < 2$ (dirty limit) (Hertz, 1976). This picture is confirmed by the analysis of singularities of the relevant perturbation integrals. Therefore, the quantum fluctuations of the field ψ have a dominating role below spatial dimensions $d < 4$.

Taking into account the quantum fluctuations of the field ψ and completely neglecting the ω_l–dependence of the magnetization M, $\epsilon_0 = (4-d)$–analysis of the generalized action $\mathcal{S}$ has been performed within the one-loop approximation (order ϵ_0^1). In the classical limit ($r/T \ll 1$) one re-derives the results already reported above together with an essentially new result, namely, the value of the dynamical exponent

$$z_\psi = 2 - \sqrt{\frac{2\epsilon}{13}} \tag{9.212}$$

which describes the quantum dynamics of the field ψ. In the quantum limit ($r/T \gg 1, T \to 0$) the static phase transition properties are affected by the quantum fluctuations, in particular, in isotropic systems ($n/2 = m = 3$). For this case, the one-loop RG equations corresponding to $T = 0$ are not degenerate and give definite results. The RG equation for g,

$$g' = b^{\epsilon_0/2} g \left(1 + \frac{g^2}{24\pi^3} \ln b\right), \tag{9.213}$$

yields two FPs: (a) a Gaussian FP ($g^* = 0$), which is unstable for $d < 4$, and (b) a FP $(g^2)^* = -12\pi^3\epsilon_0$ which is unphysical [$(g^2)^* < 0$] for $d < 4$ and unstable for $d \geq 4$. Thus the new stable critical behavior corresponding to $T > 0$ and $d < 6$ disappears in the quantum limit $T \to 0$.

At the absolute zero and any dimension $d > 0$ the P–driven phase transition is of first order. This can be explained as a mere result of the limit $T \to 0$. The only role of the quantum effects is the creation of the new un-physical FP (b). In fact, the referent classical system described by $\mathcal{H}$ from Eq. (9.193) also looses its stable FP (9.202) in the zero-temperature (*classical*) limit $T \to 0$ but does not generate any new FP because of the lack of g^3–term in the equation for g'; see Eq. (9.213). At $T = 0$ the classical system has a purely MF behavior (Shopova and Uzunov, 2003a) which is characterized by a Gaussian FP ($g^* = 0$) and is unstable towards T–perturbations for $0 < d < 6$. This is a usual classical zero temperature behavior where the quantum correlations are ignored. For the standard ϕ^4– theory this picture holds for $d < 4$. One may suppose that the quantum fluctuations of the field ψ are not enough to ensure a stable quantum multi-critical behavior at $T_c = T_F = 0$ and that the lack of such behavior is in result of neglecting the quantum fluctuations of M. One may try to take into account these quantum fluctuations by the special approaches from the theory of disordered systems, where additional expansion parameters are used to ensure the marginality of the fluctuating modes at the same borderline dimension d_U (Shopova and Uzunov, 2003a). It may be conjectured that the techniques known from the theory of disordered systems with extended impurities cannot be straightforwardly applied to the present problem and, perhaps, a completely new supposition should be introduced.

9.9.5.2 *Final remarks*

An important final remark is addressed to the diamagnetic effects. The latter have not been included in our RG investigation. As we know from Sections 9.7.5–9.7.7 the coupling of the ψ–field and the vector potential $\boldsymbol{A}$ of the magnetic induction $\boldsymbol{B}$ leads to the weakly first order phase transition of Halperin–Lubensky–Ma (HLM effect). This effect will appear within a more complex RG analysis carried out for the fields ψ and $\boldsymbol{A}$. Then one may conjecture that the critical behaviour in isotropic ferromagnetic superconductors will be suppressed by the diamagnetic effects. The HLM effect is very weak and may appear in type II superconductors only under certain conditions. Thus one may suppose that the critical behavior established in Section 9.9.5 effectively comprises the behaviour of real isotropic ferromagnetic superconductors up to a satisfactory level of accuracy.

In conclusion, the present results may be of use in interpretations of recent experiments (Pfleiderer and Huxley, 2002) in UGe_2, where the magnetic order is uniaxial (Ising symmetry) and the experimental data, in accord with the present consideration, indicate that the phase transition from the phase of coexistence to the disordered phase is of first order. Systems with isotropic magnetic order are needed for an experimental test of the new multi-critical behavior.

Appendix A

Useful Mathematical Formulae

A.1 Homogeneous Functions and Euler Theorem

The function $f(x_1, \ldots, x_n)$ is homogeneous of order p with respect to the variables $x_1, \ldots, x_n$ if

$$f(\lambda x_1, \ldots, \lambda x_n) = \lambda^p f(x_1, \ldots, x_n) . \tag{A.1}$$

If f is a continuously differentiable homogeneous function,

$$\sum_{i=1}^{n} x_i \frac{\partial f}{\partial x_i} = pf(x_1, \ldots, x_n) . \tag{A.2}$$

The equality Eq. (A.2) which is obtained by differentiating Eq. (A.1) with respect to λ is called *Euler theorem* for homogeneous functions.

A.2 Useful Formulae for the Gamma Function

$$\Gamma(z) = \int_0^\infty x^{z-1} e^{-x} dx$$

$$\Gamma(z+1) = z\Gamma(z) \tag{A.3}$$

$$\Gamma(z)\Gamma(1-z) = \frac{\pi}{\sin(\pi z)} \qquad (\hat{z}\text{-noninteger}) \tag{A.4}$$

$$\Gamma\left(\frac{1}{2}\right) = \sqrt{\pi}\,, \qquad \Gamma\left(\frac{3}{2}\right) = \frac{\sqrt{\pi}}{2}. \tag{A.5}$$

$$\Gamma(n) = (n-1)! \qquad (\text{integer } n > 0) \tag{A.6}$$

$$\Gamma(n+\frac{1}{2}) = 1\cdot 3\cdot 5\cdots(2n-3)(2n-1)\sqrt{\pi}/2^n \equiv \frac{(2n-1)!!\sqrt{\pi}}{2^n} \tag{A.7}$$

$$\Gamma(2z) = (2\pi)^{-1/2}2^{2z-\frac{1}{2}}\Gamma(z)\Gamma(z+\frac{1}{2}) \tag{A.8}$$

$$\ln\Gamma(1+z) = -C_E z + \sum_{n=2}^{\infty}\frac{(-1)^n}{n}\left[\zeta(n)-1\right]z^n \qquad (|z|<2). \tag{A.9}$$

In (A.9), $\zeta(s) = \sum_{k=1}^{\infty} k^{-s}$, for $\Re s > 1$ is the Riemann (zeta) function and $C_E = 0.57722\ldots$ is the Euler constant.

A.3 Logarithm of Matrices

The logarithms of Eq. (B.6) and Eq. (B.11) can be taken with the help of the identities $\det(\hat{K}^{1/2}) = (\det\hat{K})^{1/2}$ and

$$\ln\det\hat{K} = \mathrm{Tr}(\ln\hat{K}). \tag{A.10}$$

If $\hat{K} = \hat{I} + \hat{L}$, where $\hat{I} = \{\delta_{\alpha\beta}\}$ is the unit matrix,

$$\ln\hat{K} = \hat{L} - \frac{1}{2}\hat{L}^2 + \frac{1}{3}\hat{L}^3 - \ldots \tag{A.11}$$

This algorithm can be applied also to infinite-rowed matrices $\hat{K}$, which appear in the functional integrals; see, e.g., Section C.5.

Appendix B

Gaussian Integrals and Transformations

B.1 Gaussian Integrals

$$\int_0^\infty x^l e^{-(rx)^m} dx = \frac{\Gamma\left(\frac{l+1}{m}\right)}{mr^{l+1}}, \qquad (r, l+1, m > 0); \qquad \text{(B.1)}$$

$\Gamma(z)$ is the gamma function; see Section B.1.

$$\int_{-\infty}^\infty x^l e^{-x^{2k}} dx = \begin{cases} \frac{1}{k}\Gamma\left(\frac{l+1}{2k}\right), & \text{even } l, \\ 0, & \text{odd } l, \end{cases} \qquad (l+1>0). \qquad \text{(B.2)}$$

$$\int_{-\infty}^\infty x^{2k\nu-1} e^{-ax^{2k}-bx^{4k}} dx = \frac{\Gamma(\nu)e^{a^2/8b}}{k(2b)^{\nu/2}} D_{-\nu}(a/\sqrt{2b}). \qquad \text{(B.3)}$$

In (B.3), $D_{-\nu}$ is the parabolic cylinder function (see, e.g., Abramowitz and Stegun, 1965).

For $m = 2$, Eq. (B.1) represents a *standard Gaussian integral.* For $m > 0$, we may refer to this equality as representing *a Gaussian-like* integral. Changing the integration variable ($x \to y^{2/m}$) one can easily obtain the usual Gaussian integral with the exponent of type $e^{-r^m y^2}$. When $l = mp$, where $p = 1, 2, \ldots$, one may obtain the values of the integral (B.1) by a p-multiple differentiation of the value $\Gamma(1/m)/mr$ of the same integral for $l = 0$ with respect to the parameter r^m: $[\Gamma(1/m)/m][\partial^p r^{-1}/\partial(r^m)^p]$; note, that $\partial/\partial r^m = (\partial/\partial r)(\partial r/\partial r^m) = (1/mr^{m-1})(\partial/\partial r)$.

The integrals (B.1)–(B.3) are used in the calculation of partition integrals (for example, see Sections 6.3 and 7.3) and in the Feynman method of calculation of perturbation integrals (Binney *et al.*, 1992; Ivanchenko ans Lisyanskii, 1995; Zinn–Justin, 1996; Kleinert, 2001).

B.2 Multiple Gaussian Integrals

Let $\varphi = \{\varphi_\alpha\}$ and $h = \{h_\alpha\}$ be real n-component vectors ($\alpha = 1, \ldots, n$) whose scalar product is defined by $\varphi h = \sum_\alpha \varphi_\alpha h_\alpha$. If

$$\varphi \hat{K} \varphi' = \sum_{\alpha\alpha'} \varphi_\alpha K_{\alpha\alpha'} \varphi'_{\alpha'} \tag{B.4}$$

is a positively definite bilinear form and $D\varphi = d\varphi_1 \cdots \varphi_n$, the Gaussian integral in the n-dimensional space of the vector components φ_α is defined by

$$J(K,h) = \int D\varphi\, e^{-\frac{1}{2}\varphi\hat{K}\varphi + h\varphi}. \tag{B.5}$$

The solution is

$$J(K,h) = \det\left[\left(\frac{\hat{K}}{2\pi}\right)^{-\frac{1}{2}}\right] e^{\frac{1}{2}h\hat{K}^{-1}h}, \tag{B.6}$$

where $\hat{K}^{-1}$ is the matrix inverse to $\hat{K}$.

Note 1: The solution Eq. (B.6) of Eq. (B.5) is obtained by the shift transformation

$$\varphi_\alpha = \varphi'_\alpha + A_\alpha \tag{B.7}$$

of φ_α to the new variables φ'_α. Substituting Eq. (B.7) in the integrand exponent Eq. (B.5) and requiring that all coefficients of the terms which are linear in φ'_α are zero, one obtains the shift variables A_α

$$A_\alpha = \hat{K}^{-1}.h \equiv \sum_\beta \hat{K}^{-1}_{\alpha\beta} h_\beta, \tag{B.8}$$

and

$$J(\hat{K},h) = e^{\frac{1}{2}h\hat{K}^{-1}h} J(K,0). \tag{B.9}$$

The integral $J(K,0)$ is calculated by a change of the variables, which corresponds to the rotation

$$\varphi' = \hat{K}^{-1/2} \cdot \varphi''. \tag{B.10}$$

The Jacobian $(D\varphi'/D\varphi'')$ of this transformation is $\det(\hat{K}^{-1/2})$ and the integral over φ'' is $J(1,0) = (2\pi)^{n/2}$; see Eq. (B.1) with $l = 0$, $m = 2$, and $r = 1$. Finally, we have to use the identity $\det(z\hat{K}) = z^n \det(\hat{K})$, where z is a number.

Note 2: Eq. (B.6) is valid for any non-degenerate symmetric matrix $\hat{K}$.

Note 3: For complex vectors φ and h, where $\varphi_\alpha = \varphi'_\alpha + i\varphi''_\alpha$ and $h_\alpha = h'_\alpha + ih''_\alpha$, the Gaussian integral is

$$\begin{aligned} J(\hat{K}, h) &= \int D\varphi e^{-\varphi^* K\varphi + h\varphi^* + h^*\varphi} \\ &= \det\left[\frac{K}{\pi}\right]^{-1} \cdot e^{h^* \hat{K}^{-1}.h}, \end{aligned} \tag{B.11}$$

where

$$\int D\varphi \equiv \int d\varphi' d\varphi'' \equiv \prod_\alpha \int d(\Re\varphi_\alpha) d(\Im\varphi_\alpha) \tag{B.12}$$

and φ^*, h^* are the complex conjugates of φ and h. The solution Eq. (B.11) is obtained in the way explained in Note 1.

Note 4: If $\hat{K}$ is diagonal matrix, $K_{\alpha\beta} = \delta_{\alpha\beta} K_\alpha$,

$$J(K, h) = \left[\prod_\alpha \left(\frac{2\pi}{K_\alpha}\right)^{1/2}\right] \exp\left\{\frac{1}{2}\sum_\alpha \left(\frac{h_\alpha^2}{K_\alpha}\right)\right\}. \tag{B.13}$$

Appendix C

Some Functionals

C.1 Definition

A functional is a real function defined on a vector space of functions. The vector space could be a real n-dimensional space, or, a complex vector space.

Our most common example is the Hamiltonian in the field theory, e.g., the Hamiltonian Eq. (7.1) defined on a real n-dimensional vector space of functions — the components of the vector $\varphi(x) = \{\varphi_\alpha(x), \alpha = 1, \ldots, n\}$, and Eqs. (9.15) and (9.16), where the Hamiltonian is defined on the vector space of the complex functions $\psi_\alpha(x)$ and $\psi_\alpha(q)$, respectively.

C.2 Functional Power Series

Let $A(\varphi)$ be a functional of $\varphi(q) \equiv \varphi_\alpha(x)$; $q \equiv (\alpha, x)$, $x = (x_1, \ldots, x_d) \equiv \{x_i\} \in V$, $\alpha = 1, \ldots, n$. It can be represented by the series

$$A(\varphi) = \sum_{l=0}^{\infty} \frac{1}{l!} \int d^d q_1 \ldots d^d q_l \, a_l(q_1, \ldots, q_l) \varphi(q_1) \ldots \varphi(q_l), \tag{C.1}$$

where we have used the short notation

$$\int d^d q_\nu = \sum_{\alpha_\nu} \int d^d x_\nu, \qquad \nu = 1, \ldots, l. \tag{C.2}$$

(Here the integration over the continuous variables (x_i) and the summation over the discrete variables (α_i) is implied.)

Examples:

(a) *Linear functional*

$$A(\varphi) = \int d^d q a_1(q)\varphi(q), \tag{C.3}$$

(b) *Bilinear functional*

$$A(\varphi) = \int d^d q_1 d^d q_2 \; a_2(q_1, q_2)\varphi(q_1)\varphi(q_2), \tag{C.4}$$

(c) *The Taylor series* for $A(\varphi)$ at $\bar{\varphi} = 0$ is given by Eq. (C.1) with

$$a_l = \left[\frac{\delta^{(l)} A(\varphi)}{\delta\varphi(q_1)\ldots\varphi(q_l)}\right]_{\bar{\varphi}=0} \tag{C.5}$$

(see also Section C.3).

(d) *The Taylor series at* $\bar{\varphi} \neq 0$ is also given by Eq. (C.1) but now the functional derivatives Eq. (C.5) are taken at $\bar{\varphi} \neq 0$ and $\varphi(q_i)$ in Eq. (C.1) are replaced by $\delta\varphi(q_i) = \varphi(q_i) - \bar{\varphi}(q_i)$.

(e) *Gaussian series*

$$\begin{aligned} A(\varphi) &= \sum_{l=0}^{\infty} \frac{1}{l!} \left[\int d^d q_1 d^d q_2 \; \varphi(q_1) K(q_1, q_2) \varphi(q_2)\right]^l \\ &\equiv \exp\left[\int d^d q_1 d^d q_2 \; \varphi(q_1) K(q_1, q_2) \varphi(q_2)\right], \end{aligned} \tag{C.6}$$

where $K(q_1, q_2)$ is an infinite-rowed matrix with respect to the continuous variables x_1, x_2.

(f) *Exponential series*

$$\begin{aligned} A(\varphi) &= \sum_{l=0}^{\infty} \frac{1}{l!} \left[\int d^d q \; a(q)\varphi(q)\right]^l \\ &= \exp\left[\int d^d q \; a(q)\varphi(q)\right]. \end{aligned} \tag{C.7}$$

Useful notations:

$$\varphi K \varphi = \int d^d q_1 d^d q_2 \; \varphi(q_1) K(q_1, q_2) \varphi(q_2), \tag{C.8}$$

$$a\varphi = \int d^d q \; a(q)\varphi(q). \tag{C.9}$$

C.3 Functional Differentiation

$a(q_0)$ will be the functional derivative of $A(\varphi)$ with respect to the variation of the function $\varphi(x)$ at x_0 if the first order variation of $A(\varphi)$ is

$$\delta A(\varphi) = A(\varphi + \delta\varphi) - A(\varphi) = \int a(q_0)\delta\varphi(q_0)dq_0, \tag{C.10}$$

where $a(q_0)$ is the functional derivative,

$$a(q_0) = \frac{\delta A}{\delta\varphi(q_0)}, \tag{C.11}$$

at fixed discrete variable α in $q \equiv (\alpha, x)$. The functional derivatives of order $l > 1$, see Eq. (C.5), are defined as functional coefficients of the variations $\delta^{(l)}A(\varphi)$:

$$\delta^{(l)}A(\varphi) = \int d^d q_1 \ldots d^d q_l \left[\frac{\delta^{(l)}A}{\delta\varphi(q_1)\ldots\delta\varphi(q_l)}\right]_{\bar{\varphi}} \delta\varphi(q_1)\ldots\delta\varphi(q_l). \tag{C.12}$$

The functional derivative as a result of continuum limit:

Let $A[\varphi]$ is a "functional" of the lattice field $\varphi = \{\varphi_i\}$. A small variation of this lattice functional is given by

$$\delta A[\varphi] = A[\varphi + \delta\varphi] - A[\varphi] \approx \sum_i \left(\frac{\delta A}{\delta\varphi_i}\right)\delta\varphi_i. \tag{C.13}$$

In the continuum limit, the r.h.s. of Eq. (C.13) will be

$$\lim_{\substack{V\to\infty \\ N\to\infty}} \frac{1}{v}\sum_i v_i \left\{\frac{\delta A[\varphi]}{\delta\varphi(i)}\right\}\delta\varphi(i) = \frac{N}{V}\int d^d x \left\{\frac{\delta A[\varphi]}{\delta\varphi(x)}\right\}\delta\varphi(x), \tag{C.14}$$

where $v_i \sim v = (V/N)$ is the size of the cell i around site i and the density limit $\varphi(i) = (\varphi_i/v_i) \to \varphi(x)$ has been used (see Section 6.5.1). The derivative in the curly brackets in the integral on the r.h.s. of Eq. (C.14) is the functional derivative corresponding to the functional $A[\varphi]$ in the continuum limit.

In order to establish the correspondence between the derivatives of the lattice and continuum versions of the functional $A[\varphi]$ we take the continuum limit of the lattice derivative

$$\frac{\delta A[\varphi]}{\delta\varphi_i} = \frac{\delta A[\varphi]}{v_i\delta\varphi(i)} \longrightarrow \frac{1}{v}\frac{\delta A[\varphi]}{\delta\varphi(x)}.$$

Now we apply the operation $\delta/\delta\varphi(y)$ to the r.h.s. of Eq. (C.14) and keep only the term which does not vanish at $\delta\varphi(x) = 0$. The calculation is straightforward:

$$\frac{1}{v}\int d^d x \left\{\frac{\delta A[\varphi]}{\delta\varphi(x)}\right\}\frac{\delta\varphi(x)}{\delta\varphi(y)} = \frac{1}{v}\int d^d x \left\{\frac{\delta A[\varphi]}{\delta\varphi(x)}\right\}\delta(x-y) = \frac{1}{v}\frac{\delta A[\varphi]}{\delta\varphi(y)}.$$

This result indicates the following correspondence

$$\frac{\delta A[\varphi]}{\delta\varphi_i} \Longleftrightarrow \frac{\delta A[\varphi]}{\delta\varphi(x)}.$$

The result can be straightforwardly written for n variables $\{\varphi_i^\alpha\}$. In the continuum limit $A[\varphi]$ is a functional of the field $\varphi(q) = \varphi(\alpha, x)$.

C.4 Functional Integration

The functional integration is defined by

$$A(\varphi) = \int d\varphi(q) a[\varphi(q)] = \lim_{\substack{v_i\to 0\\ N\to\infty}} \int \prod_\alpha \prod_i d\varphi_i^\alpha a(\{\varphi_i^\alpha\}), \tag{C.15}$$

provided the continuum limit ($v_i \to 0, N \to \infty$) for the multiple integral over φ_i^α exists.

C.5 Functional Gaussian Integral

If the components φ_α and h_α of the vectors $\varphi = \{\varphi_\alpha\}$ and $h = \{h_\alpha\}$ are functions of the continuous variable x, $x \in V$, we can write $\varphi(q) \equiv \varphi_\alpha(x)$, and $h(q) \equiv h_\alpha(x)$; $q = (\alpha, x)$. Using the notations Eq. (C.2) and Eqs. (C.8)–(C.9), the functional Gaussian integral will be

$$\begin{aligned} J(K,h) &= \int D\varphi(q) \exp\left[-\frac{1}{2}\int d^d q_1 d^d q_2 \varphi(q_1)\hat{K}(q_1,q_2)\varphi(q_2) \right. \\ &\qquad \left. + \int d^d q h(q)\varphi(q)\right] \\ &\equiv \int D\varphi(q) \exp\left[-\frac{1}{2}\varphi\hat{K}\varphi + h\varphi\right], \end{aligned} \tag{C.16}$$

where

$$\int D\varphi(q) \equiv \prod_{x\in V} \prod_{\alpha=1}^{n} \int d\varphi_\alpha(x); \tag{C.17}$$

cf. Eq. (B.5). The solution is given by Eq. (B.6), where $\hat{K}$ is the infinite-rowed (with respect to x) symmetric matrix $\hat{K}(q_1, q_2) \equiv \hat{K}_{\alpha_1\alpha_2}(x_1, x_2)$.

For complex vectors φ and h one can use Eq. (B.11) and Eq. (B.12), where $\hat{K} = \hat{K}(q_1, q_2)$, and

$$\int D\varphi = \prod_{x\in V} \prod_{\alpha=1}^{n} \int d\Re\varphi_\alpha(x) d\Im\varphi_\alpha(x). \tag{C.18}$$

Appendix D

D-dimensional Integration and Some Fourier Transformations

In the main text we often avoid the bold-face notations ($\boldsymbol{x}$ and $\boldsymbol{k}$) for the x-space and k-space vectors. Here and in some next appendix chapters we use the usual bold face for the vectors.

D.1 D-dimensional Angular Integration

D.1.1 *Angular integration*

Here we recall some rules of angular integration over space domains of spherical shape (this can often be made as a good approximation to large spaces of another form, for example, cubs).

The integration operation in the infinite space ($V \sim L^d \to \infty$) of the d-dimensioned vectors $\boldsymbol{x}$ is given by

$$\int d\boldsymbol{x} \equiv \int d^d x = \int_0^\infty dx . x^{d-1} \int d\Omega, \tag{D.1}$$

where the angular part of the integration is

$$\int d\Omega = \int_0^\pi d\theta_1 \sin^{d-2}\theta_1 \int_0^\pi d\theta_2 \sin^{d-3}\theta_2 \\ \ldots \int_0^\pi d\theta_{d-2} \sin\theta_{d-2} \int_0^{2\pi} d\varphi \tag{D.2}$$

is an integration over the $(d-1)$ angles in spherical coordinates $x = |\boldsymbol{x}|$, θ_i; $i = 1, 2, \ldots (d-1)$, $\theta_{d-1} \equiv \varphi$. With the help of Eq. (D.1) and Eq. (D.2), any function $f(\boldsymbol{x}) \equiv f(x, \theta_i)$ is integrated by the rule

$$\int d\boldsymbol{x}\, f(\boldsymbol{x}) \equiv \int d^d x\, f(\boldsymbol{x}) = \int_0^\infty dx\, x^{d-1} \int d\Omega\, f(x, \theta_i), \tag{D.3}$$

The volume $V_d(R)$ of the dD sphere of radius R will be

$$V_d(R) = \int_0^R dx.x^{d-1} \int d\Omega.1 = \frac{R^d}{d}\Omega_d, \tag{D.4}$$

where the area Ω_d of the unit ($R = 1$) dD sphere is calculated with the help of Eq. (D.2) and the integral

$$\int_0^\pi d\alpha \sin^p \alpha = \frac{\sqrt{\pi}\,\Gamma\left(\frac{p+1}{2}\right)}{\Gamma\left(\frac{p}{2}+1\right)}, \qquad (p > -1). \tag{D.5}$$

As a result

$$\Omega_d = \int d\Omega.1 = \frac{2\pi^{d/2}}{\Gamma(d/2)}. \tag{D.6}$$

The area $S_d(R) = \partial V_d(R)/\partial R$ of the dD sphere can be expressed by the area of the corresponding unit sphere through the relation $S_d(R) = R^{d-1}\Omega_d$. For example, when $d = 3$, $V_3(R) = 4\pi R^3/3$, and $S_3(d) = 4\pi R^2$. For $d = 1$, $S_1(R) = \Omega_1 = 2$, and $V_1(R) = 2R$. The "area" of the "1D sphere" does not depend on R. The "volume" $V_1(R) = 2R$ gives the length of the straight line between the points, $-R$ and R, situated on the real axis and so $S_1 = 2$ represents the "count" of its ends.

If $f(\boldsymbol{x})$ depends on x, $f = f(|x|)$,

$$\int d^d x f(x) = \Omega_d \int_0^\infty f(x) x^{d-1} dx. \tag{D.7}$$

For functions like $f(x, \theta_1)$,

$$\int d^d x f(x, \theta_1) = \Omega_{d-1} \int_0^\infty dx.x^{d-1} \int_0^\pi d\theta_1 \sin^{d-2}\theta_1 f(x, \theta_1), \tag{D.8}$$

for functions of type $f(x, \theta_1, \theta_2)$,

$$\begin{aligned} &\int d^d x f(x, \theta_1 \theta_2) \\ &= \Omega_{d-2} \int_0^\infty dx x^{d-1} \int_0^\pi \int_0^\pi d\theta_1 d\theta_2 \sin^{d-2}\theta_1 \sin^{d-3}\theta_2\, f(x, \theta_1, \theta_2), \end{aligned} \tag{D.9}$$

etc.

When calculating integrals of functions $f(k)$ in the momentum space of the vectors $\boldsymbol{k}$, i.e.,

$$\int_k \equiv \int d\boldsymbol{k} \equiv \int \frac{d^d k}{(2\pi)^d}, \tag{D.10}$$

we have to replace Ω_d with

$$K_d = \frac{\Omega_d}{(2\pi)^d} = \frac{2^{1-d}\pi^{-d/2}}{\Gamma(d/2)}. \tag{D.11}$$

The integrals of the functions $f(k)$, $k \equiv |\boldsymbol{k}|$, and $f(k, \theta_1, \ldots, \theta_l)$ for any $1 \leq l \leq (d-2)$ become

$$\int_k f(k) \equiv \int \frac{d^d k}{(2\pi)^d} f(k) = K_d \int_0^\infty dk.k^{d-1} f(k), \tag{D.12}$$

and

$$\begin{aligned}\int_k f(k, \theta_1, \ldots, \theta_l) = \frac{K_{d-l}}{(2\pi)^l} \int_0^\infty dk k^{d-1} \int_0^\pi \ldots \int_0^\pi d\theta_1 \ldots d\theta_l \\ \times \sin^{d-2}\theta_1 \sin^{d-3}\theta_2 \ldots \sin^{d-2-(l-1)}\theta_l \, f(k, \theta_1, \ldots, \theta_l).\end{aligned} \tag{D.13}$$

D.1.2 *Relationship of Cartesian and spherical coordinates in d-dimensional space*

Usually, $f(\boldsymbol{x})$ or $f(\boldsymbol{k})$ are represented by the Cartesian components $x_i = x\hat{x}_i$ of $\boldsymbol{x} = x(\hat{x}_1, \ldots, \hat{x}_d) \equiv x\hat{x}$, $|\hat{x}| = 1$; or in the momentum (k) space, $\boldsymbol{k} = k\,\hat{k}$, $\hat{k} = (\hat{k}_1, \ldots, \hat{k}_d)$. The relations between the spherical and the Cartesian components of $\boldsymbol{x}$ (or $\boldsymbol{k}$) are:

$$
\begin{aligned}
\hat{x}_1 &= \prod_{\nu=1}^{d-1} \sin\theta_\nu, \\
\hat{x}_2 &= \left(\prod_{\nu=1}^{d-2} \sin\theta_\nu\right) \cos\theta_{d-1}, \\
\hat{x}_3 &= \left(\prod_{\nu=1}^{d-3} \sin\theta_\nu\right) \cos\theta_{d-2}, \\
&\ \vdots \\
\hat{x}_p &= \left(\prod_{\nu=1}^{d-p} \sin\theta_\nu\right) \cos\theta_{d-p+1}, \\
&\ \vdots \\
\hat{x}_{d-2} &= \sin\theta_1 \sin\theta_2 \cos\theta_3, \\
\hat{x}_{d-1} &= \sin\theta_1 \cos\theta_2, \\
\hat{x}_d &= \cos\theta_1,
\end{aligned}
\tag{D.14}
$$

where $d > 1$, $1 \leq p \leq d$, $\theta_d \equiv 0$, $\theta_{d-1} \equiv \varphi$, d and p are integers, and by convention we set $\prod_{\nu=1}^{0} \sin\theta_\nu = 1$ and $\theta_d = 0$. The angles defined by Eq. (D.14) participate in the angular integration defined by Eq. (D.2).

In the 2D space, using the condition $1 \leq p \leq d$ we may take into account only the first two relations Eq. (D.14) and consider the other relations as redundant. From these relations we obtain the known result $\hat{x}_1 = \sin\theta_1$ and $\hat{x}_2 = \cos\theta_1$. To obtain this result we have used the convention $(\prod_\nu^0 \dots) = 0$ (see above). Now, following the most common notations we may identify x_2 with the coordinate x and x_1 – with y in the (x, y) Cartesian coordinate system. In these notations, we have $\hat{x} = \cos\theta$, and $\hat{y} = \sin\theta$; $(\theta \equiv \theta_1)$. The same result can be obtained from the last two relations in Eq. (D.14), again on the basis of the condition $1 \leq p \leq d$ (!). Now we have to use the convention $\theta_d = \theta_2 = 0$.

For the 3D space, we repeat the procedure outlined for the 2D case. We may again use either the first three, or, alternatively, the last three relations in Eq. (D.14). Using the same change in the notations ($\hat{x}_2 \to \hat{x}$ and $\hat{x}_1 \to \hat{y}$, $\theta_1 = \theta$) as well as $\hat{x}_3 = \hat{z}$ and $\theta_2 = \varphi$, we obtain

$$\begin{aligned}\hat{x} &= \sin\theta\cos\varphi \\ \hat{y} &= \sin\theta\sin\varphi \\ \hat{z} &= \cos\theta.\end{aligned} \tag{D.15}$$

The integral

$$I_{kl\ldots}^{\alpha\beta\ldots} = \int d^d x \left[f_0(x)\hat{x}_k^\alpha \hat{x}_l^\beta \ldots \right], \tag{D.16}$$

where $\hat{x}_k^\alpha \equiv (\hat{x}_k)^\alpha$, can be calculated using (D.2), (D.5),(D.14), and

$$\int_0^\pi d\alpha \sin^p\alpha \begin{Bmatrix}\sin m\alpha \\ \cos m\alpha\end{Bmatrix} = \frac{2^{-p}\pi\Gamma(p+1)}{\Gamma\left(\frac{p+m}{2}+1\right)\Gamma\left(\frac{p-m}{2}+1\right)} \begin{Bmatrix}\sin(\pi m/2) \\ \cos(\pi m/2)\end{Bmatrix},$$

$$[p+1, p+m+2, p-m+2 > 0], \tag{D.17}$$

or,

$$\int_0^{\pi/2} d\alpha\, \sin^p\alpha\cos^q\alpha = \frac{\Gamma\left(\frac{p+1}{2}\right)\Gamma\left(\frac{q+1}{2}\right)}{2\Gamma\left(\frac{p+q}{2}+1\right)}, \qquad (p, q > -1). \tag{D.18}$$

Using (A.8) one may show that the solutions of the integrals (D.5), (D.17) and (D.18) are in a total conformity one another for $m = q = 0$.

If one or more exponents $\alpha, \beta, \ldots$ in Eq. (D.16) are odd numbers,

$$I_{kl\ldots}^{2n+1,\beta\ldots} = 0. \tag{D.19}$$

When all exponents $\alpha, \beta \ldots$ are even numbers, the integral (D.16) is different from zero. It can be calculated with the help of (D.1)–(D.2) and (D.14). For example, choosing $f = f_0(x)\hat{x}_k^2$ we obtain that $I_k^{(2)}$ does not depend on k:

$$I^{(2)} = \frac{\Omega_d}{d}\int_0^\infty f_0(x)x^{d-1}dx \tag{D.20}$$

[see also Aharony and Fisher (1973)].

D.2 Some Fourier Transformations

The identity

$$\int d\boldsymbol{R} e^{i\boldsymbol{k}\cdot\boldsymbol{R}} = \Omega_d \Gamma(d/2) 2^{\frac{d}{2}-1} \int_0^\infty dR R^{d-1} J_{\frac{d}{2}-1}(kR)(kR)^{1-\frac{d}{2}}, \quad \text{(D.21)}$$

where $J_\nu(y)$ is the Bessel function of the first kind and $R \equiv |\boldsymbol{R}|$, $k \equiv |\boldsymbol{k}|$, can be used for the calculation of perturbation integrals and Fourier transforms (Bateman, 1953). In this section we present some useful Fourier transforms of functions describing short- and long-range potentials and correlations; for a more detailed discussion, see Uzunov, Korutcheva and Millev (1985b).

The Fourier transform of the function $f(R)$ is given by

$$\tilde{f}(k) = \frac{(2\pi)^{d/2}}{k^{d/2-1}} \int_0^\infty dR.R^{d/2} J_{\frac{d}{2}-1}(kR) f(R). \quad \text{(D.22)}$$

The inverse transformation will be

$$f(R) = \frac{1}{(2\pi)^{d/2} R^{(d/2)-1}} \int_0^\infty dk.k^{d/2} J_{\frac{d}{2}-1}(kR) \tilde{f}(k). \quad \text{(D.23)}$$

Sometimes the functions $f(R)$ and $\tilde{f}(k)$ are not absolutely integrable. Then the standard trick of the regularization is used:

$$\varphi'(y) = \varphi(y) e^{-\delta y}, \qquad \varphi \equiv (f, \tilde{f}), \qquad \delta > 0. \quad \text{(D.24)}$$

After performing the Fourier transformation of φ' one takes the limit $\delta \to +0$ (the so-called *regularized transformation*).

For the function

$$f(R) = \frac{1}{R^{d+\theta}} \quad \text{(D.25)}$$

the Fourier transformation will be

$$\tilde{f}(k) = \frac{\pi^{d/2} \Gamma\left(-\frac{\theta}{2}\right)}{2^\theta \Gamma\left(\frac{d+\theta}{2}\right)} k^\theta, \qquad -d < \theta < 0. \quad \text{(D.26)}$$

It is found by a direct calculation of the integral Eq. (D.22) for values in the interval $-(d+1)/2 < \theta < 0$. The same result is obtained after applying the regularization Eq. (D.24) but now θ takes values in the interval $-d < \theta < 0$.

A cutoff $R > 1/\Lambda$ must be used when $\theta \geq 0$. The integration over R is splitted in two integrals: the first one between the limits $1/\Lambda$ and 1, and the second one in the limits 1 and ∞. The result for $\tilde{f}(k)$, see also Uzunov, Korutcheva and Millev (1985b), can be represented in the form

$$\tilde{f}(k) = (2\pi)^{d/2} \left\{ \frac{\Gamma(-\theta/2)\, k^{\theta}}{2^{(d/2)+\theta}\Gamma\left(\frac{d+\theta}{2}\right)} \right. \\ \left. - 2^{1-d/2} \sum_{l=0}^{\infty} \frac{(-1)^l}{4^l (2l-\theta)\, l!\, \Gamma\left(l+\frac{d}{2}\right)} \frac{k^{2l}}{\Lambda^{2l-\theta}} \right\} \tag{D.27}$$

for $\theta \neq 2m (m = 0, 1, \ldots)$. Taking the limit $\theta \to +0$, and $\theta \to +2$ in Eq. (D.27) we obtain

$$\tilde{f}_{\theta=0}(k) = \frac{\pi^{d/2}}{\Gamma(d/2)} \left\{ 2 \ln\left(\frac{\Lambda}{k}\right) \right. \\ \left. - \Gamma\left(\frac{d}{2}\right) \sum_{l=1}^{\infty} \frac{(-1)^l}{4^l\, l\, \Gamma\left(l+\frac{d}{2}\right)} \left(\frac{k}{\Lambda}\right)^{2l} \right\}, \tag{D.28}$$

and

$$\tilde{f}_{\theta=2}(k) = \frac{\pi^{d/2}}{\Gamma(d/2)} \left\{ \Lambda^2 - \frac{k^2}{d} \ln\left(\frac{\Lambda}{k}\right) \right. \\ \left. - \Gamma\left(\frac{d}{2}\right) \sum_{l=2}^{\infty} \frac{(-1)^l}{4^l (l-1)\, l!\, \Gamma\left(l+\frac{d}{2}\right)} \frac{k^{2l}}{\Lambda^{2l-2}} \right\}. \tag{D.29}$$

The Fourier transform of the function

$$g(R) = \frac{e^{-\lambda R}}{R^{d+\theta}}, \qquad \theta < 0, \quad \lambda > 0, \tag{D.30}$$

is

$$\tilde{g}(k) = \frac{2\pi^{d/2}\Gamma(-\theta)}{\Gamma(-\theta/2)\Gamma[(d+\theta-1)/2]}\left(\lambda^2+k^2\right)^{\theta/2}$$
$$\times \sum_{l=0}^{\infty} \frac{(-1)^l \Gamma\left(l-\dfrac{\theta}{2}\right)\Gamma[l+(d+\theta-1)/2]}{l!\,\Gamma\left(l+\dfrac{d}{2}\right)} \cdot \frac{k^{2l}}{\left(\lambda^2+k^2\right)^l}. \tag{D.31}$$

This formula can be expanded in power series when $(k/\Lambda) \ll 1$. For $\lambda = 0$ (and $\theta < 0$) it gives Eq. (D.26). For $\theta > 0$ the exponential factor λ in Eq. (D.30) will be unessential when k is small and $\tilde{g}(k)$ will follow the behaviour of $\tilde{f}(k)$; cf. Eq. (D.27), and see Uzunov, Korutcheva and Millev (1985b), where these and other transformations are discussed.

Appendix E

Perturbation Integrals and Sums

E.1 Simple Perturbation Integrals

Consider the integral

$$I = \int_0^\Lambda \frac{k^{d-1}dk}{(r_0 + ck^2)^2}. \tag{E.1}$$

This integral appears in the expression for the fluctuation specific hear, see Eq. (7.57), as well as in the perturbation series; see, e.g., Eq.(7.166). For $\Lambda \to \infty$ (or in situations when the infinite upper limit is a good approximation) it can be calculated with the help of the formula

$$\int_0^\infty \frac{x^{m-1}}{(a+bx)^{m+n}} = \frac{\Gamma(m)\Gamma(n)}{a^n b^m \Gamma(m+n)}, \qquad [a, b, m, n > 0]. \tag{E.2}$$

Changing the variables $(c/r_0)k^2 = y$ and replacing the new cutoff $(\Lambda^2 c/r_0)$ with infinity when $r_0 \to 0$ in Eq. (E.1), we obtain an integral like Eq. (E.2). Replacing the cutoff with infinity means, to neglect corrections of the type $O(t^a)$, $a > 0$ and $t \sim r_0$ to the main dependence (7.58). This calculation is valid for dimensions $d < 4$. For $d \geq 4$, the upper cutoff in Eq. (E.1), $(\Lambda^2 c/r_0)$ of y, cannot be changed to infinity; for $(\Lambda^2 c/r_0) \to \infty$ and $d \geq 4$ the integral becomes infinite (see also Eq. (E.3) and Eq. (E.4) which give the logarithmic divergences of C_{0f} for $d = 4$ and the variation of C_{0f} for $d = 5$):

$$I_{(d=4)} \sim \ln(\zeta_0\Lambda) + \text{const} + O(t), \tag{E.3}$$

where $\xi_0 = (c/r_0)^{1/2}$ is the correlation length, and

$$I_{d=5} \sim \text{const} + O(t^{1/2}). \tag{E.4}$$

One can calculate in the similar way (finite cutoff) the potentials (7.53a) and (7.52); provided the amplitudes $h_\alpha(k)$ are known. These free energy parts are divergent for infinite cutoff ($\Lambda = \infty$). This calculation yields a number of temperature dependent (and temperature independent) terms.

E.2 Treatment of Perturbation Integrals

The simplest lowest-order perturbation integrals, see Eq. (7.166) for $l = 1, 2$, are straightforwardly calculated (see Section E.1). In other cases, one needs more general methods of calculation; one of them is based on the identity Eq. (D.21); see also Tuthill, Nicoll and Stanley (1975).

In order to illustrate this method we shall consider the diagrams in Fig. 7.9. These diagrams give the integral in the expression (7.148). The Fourier transformations of the G_0-functions in the continuum limit are

$$G_0(\boldsymbol{R}) = \int_k \frac{d^dk}{(2\pi)^d} G_0(\boldsymbol{k}) e^{-i\,\boldsymbol{k}.\boldsymbol{R}} \tag{E.5a}$$

and

$$G_0(\boldsymbol{k}) = \int d\boldsymbol{R}\, G_0(\boldsymbol{k}) e^{i\,\boldsymbol{k}.\boldsymbol{R}}, \tag{E.5b}$$

where $G_0(\boldsymbol{k}) \equiv G_0(k)$; $k = |\boldsymbol{k}|$.

The substitution of the Fourier transformation Eq. (E.5a) of the G_0-functions in the expression (7.148) yields

$$\int \frac{d^k}{(2\pi)^d} G_0^2(\boldsymbol{k}) J(r_0, \boldsymbol{k}) e^{-i\,\boldsymbol{k}.(\boldsymbol{x}-\boldsymbol{y})}, \tag{E.6}$$

where $J(r_0, \boldsymbol{k})$ is the self-energy integral (7.167); see also Eq. (7.190c). The integrand $G_0^2(\boldsymbol{k})$ is not present in the Dyson equation (7.171) for $G^{-1}(\mathbf{x}_1' - \boldsymbol{x}_2')$ in the $\boldsymbol{x}$-representation. In fact, $G^{-1}(\mathbf{x}_1' - \boldsymbol{x}_2')$ is given by

$$G^{-1}_{\gamma\delta}(\boldsymbol{x}_1' - \boldsymbol{x}_2') = \sum_{\alpha\beta} \int d\boldsymbol{x}_1 d\boldsymbol{x}_2 G^{-1}_{\gamma\alpha}(\boldsymbol{x}_1' - \boldsymbol{x}_1) \times G^{-1}_{\alpha\beta}(\boldsymbol{x}_1 - \boldsymbol{x}_2) G^{-1}_{\beta\delta}(\boldsymbol{x}_2 - \boldsymbol{x}_1'). \tag{E.7}$$

The corresponding self-energy integral in the $\boldsymbol{x}$-representation will be

$$\begin{aligned}\int d\boldsymbol{x}d\boldsymbol{y}d\boldsymbol{x}_1d\boldsymbol{x}_2G_0^{-1}(x_1'-\boldsymbol{x}_1)G_0^{-1}(\boldsymbol{x}_2'-\boldsymbol{x}_2)\\ \times G_0(\boldsymbol{x}_1-\boldsymbol{x})G_0(\boldsymbol{y}-\boldsymbol{x}_2)G_0^3(\boldsymbol{x}-\boldsymbol{y}).\end{aligned} \tag{E.8}$$

The Fourier transformation of the G_0-functions in Eq. (E.8) gives

$$\int \frac{d^dk}{(2\pi)^d} J(r_0,\boldsymbol{k})e^{-i\,\boldsymbol{k}.(\boldsymbol{x}-\boldsymbol{y})} \tag{E.9}$$

or, regarding the Dyson equation (7.171) for G^{-1} in the $\boldsymbol{k}$-representation the integral should be $J(r_0,\boldsymbol{k})$; cf. Eq. (7.190c).

Clearly, the problem is the calculation of integrals in the k-representation like $J(r_0,\boldsymbol{k})$ and integrals like Eq. (E.6), Eq. (E.5a) and Eq. (E.5b) in the x-representation. With the φ^{2l}–theories ($l \geq 2$) one has to consider diagrams as those for the self-energy function shown in Fig. E.1, and those for the vertex functions $\Gamma^{(2l)}$ shown in Fig. E.2; $\Gamma^{(2l)}$ are defined in Section 8.2. The corresponding perturbation integrals can be written in the form

$$I_m(\boldsymbol{k},r_0) = \int d^dR\,G_0^m(\boldsymbol{R})\,e^{i\,\boldsymbol{k}.\boldsymbol{R}} \tag{E.10}$$

or, using Eqs. (E.5a) for $G_0(\boldsymbol{R})$,

$$I_m(\boldsymbol{k},r_0) = \int d^dR\,I_1^m(R,r_0)\,e^{i\,\boldsymbol{k}.\boldsymbol{R}} \tag{E.11}$$

and

$$I_1(R,r_0) = \int \frac{d^dp}{(2\pi)^d}\,G_0(\boldsymbol{p})\,e^{-i\boldsymbol{p}.\boldsymbol{R}}. \tag{E.12}$$

The representation (E.11)–(E.12) of $I_m(\boldsymbol{k},r_0)$ is trivial because $I_1(\boldsymbol{R},r_0)$ is exactly the function $G_0(R)$ ($R = |\boldsymbol{R}|$); see Eq. (7.31). If $G_0(p) = G_0(\boldsymbol{p})$ in Eq. (E.12) has the form (7.26) with $c = 1$, $I_1(\boldsymbol{R},r_0)$ will be given by Eq. (7.33) or, in another form $I_1(\boldsymbol{R},r_0) = I_1(R,r_0)$, where

$$I_1(R,r_0) = \frac{1}{(2\pi)^{d/2}}\left(\frac{\sqrt{r_0}}{R}\right)^{\frac{d}{2}-1} K_{\frac{d}{2}-1}(R\sqrt{r_0}), \qquad d<5, \tag{E.13}$$

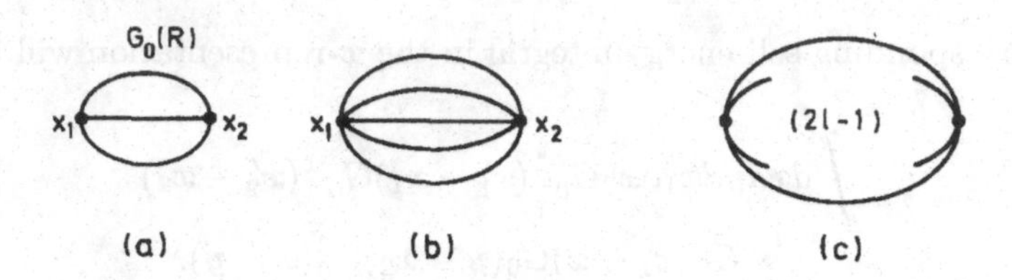

Fig. E.1 Momentum-dependent self-energy diagrams of order u_{2l}^2: (a) φ^4-theory, (b) φ^6-theory, (c) φ^{2l}-theory, $l \geq 2$; the number of internal G_0-lines is $(2l-1)$.

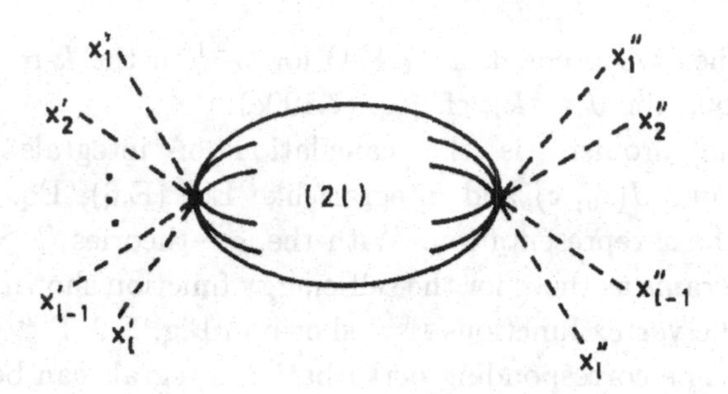

Fig. E.2 Diagrammatic contribution to the vertex $\Gamma^{(2l)}$; $l > 1$. The number of the internal $G_0(x_1 - x_2)$-lines is $2l$.

$R = |\boldsymbol{R}|$. From Eq. (D.21) and Eq. (E.11) we have $I_m(\boldsymbol{k}, r_0) = I_m(k, r_0)$, where

$$I_m(k,r_0) = (2\pi)^d K_d \Gamma(d/2) \times \int_0^\infty dR\, R^{d-1} \left(\frac{kR}{2}\right)^{1-\frac{d}{2}} J_{\frac{d}{2}-1}(kR) I_1^m(R, r_0) \tag{E.14}$$

and using Eq. (E.13),

$$I_m(k,r_0) = \frac{2^{1-d}\pi^{-d/2}}{(2\pi)^{\frac{md}{2}-d}} r_0^{\frac{m}{2}(\frac{d}{2}-1)} \times \int_0^\infty dR.R^{d-1+m(1-\frac{d}{2})} \left(\frac{kR}{2}\right)^{1-\frac{d}{2}} J_{\frac{d}{2}-1}(kR) K^m_{\frac{d}{2}-1}(R\sqrt{r_0}). \tag{E.15}$$

For important values of d and m, the integral Eq. (E.15) is divergent. Then one has to introduce a cutoff, for example, the short-distance cutoff

$R > \Lambda_R^{-1}$ which correspond to the upper cutoff Λ over the momenta ($k < \Lambda$); see, e.g., Tuthill, Nicoll, and Stanley (1975). This way of calculation can be used for more complex forms of $G_0(k)$, for example, $G_0(k) = 1/(ck^\theta + r)$, $\theta > 0$.

Another general method due to R.Feynman is thoroughly presented in the book of Amit (1978). The method is used for the momentum integrals over functions of the type $(a_1^{\alpha_1}, a_2^{\alpha_2} \dots a_n^{\alpha_n})^{-1}$ where a_i are functions of the integral momenta of the perturbation diagrams. The simplest example is the integral (7.244), often denoted by $\text{II}(k, r)$ or $\Pi(\boldsymbol{k}, r)$,

$$\Pi(\boldsymbol{k}, r) = \int \frac{d^d p}{(2\pi)^d} \frac{1}{(r + cp^2)\,[r + c(\boldsymbol{p} + \boldsymbol{k})^2]}. \tag{E.16}$$

The calculation is based on the identity

$$\frac{1}{a_1 a_2} = \int_0^1 \frac{d\alpha}{[\alpha a_1 + (1 - \alpha) a_2]^2}. \tag{E.17}$$

Denoting $[r + c(\boldsymbol{k}+\boldsymbol{p})^2] = a_1$ and $(r_0 + ck^2) = a_2$ and integrating Eq. (E.17) over $\boldsymbol{k}$ one obtains $\Pi(\boldsymbol{k}, r)$ in the form

$$\Pi(\boldsymbol{k}, r) = \frac{1}{c^2} \int_0^1 d\alpha \int_p \frac{1}{\left(\frac{r}{c} + p^2 + \alpha k^2 + 2\alpha \boldsymbol{k}.\boldsymbol{p}\right)^2}. \tag{E.18}$$

It is not difficult to see that the denominator in Eq. (E.18) can be written as

$$\left[\frac{r}{c} + (\boldsymbol{p} + \alpha \boldsymbol{k})^2 + \alpha(1 - \alpha)k^2\right]^2$$

and the integration wave vector $\boldsymbol{p}$ can be changed to $\boldsymbol{p}' = \boldsymbol{p} + \alpha \boldsymbol{k}$. Therefore, we obtain

$$\Pi(\boldsymbol{k}, r) = \Pi(k, r) = \frac{1}{c^2} \int_0^1 d\alpha \int \frac{d^d p'}{(2\pi)^2} \frac{1}{[rc + p'^2 + \alpha(1 - \alpha)k^2]^2}. \tag{E.19}$$

Thus the angular integration (over the angle θ between the vectors k and p; $k.p = kp\cos\theta$) in Eq. (E.18) has been avoided, and instead of it a supplementary integration over the (Feynman) parameter α has been introduced. This method is used when such a change in the integration is convenient. The integral $\Pi(k, r)$ is represented by the hypergeometric function; see, e.g., Ma (1973a, 1976a,b).

E.3 ϵ-Expansion of Perturbation Integrals

Let us consider the integrals $A_m(r)$ given by Eqs. (8.106). Within the RG recursion method (Chapter 8) these integrals are calculated for the FP value r^* of r ($r_H \sim O(\epsilon), r_G = 0$). Using the relation $-mA_{m+1}(r) = (\partial A_m(r)/\partial r)$ we can write

$$A_1(r^*) = A_1(0) - A_2(0)r^* + A_3(0)r^{*2} - A_4(0)r^{*3} + \dots \tag{E.20}$$

and

$$A_m(r^*) = A_m(0) - mA_{m+1}(0)r^* + \dots \tag{E.21}$$

Next, we have to calculate

$$A_m(0) = K_d \int_{b^{-1}\Lambda}^{\Lambda} \frac{k^{d-1}dk}{k^{2m}} = \begin{cases} \dfrac{K_d}{d-2m}\Lambda^{d-2m}\left(1 - b^{2m-d}\right), & d \neq 2m \\ K_d \ln b, & d = 2m, \end{cases} \tag{E.22}$$

as a power series in $\epsilon = 4 - d$. To do this one has to use the identity

$$a^{\epsilon} = e^{\epsilon \ln a} = 1 - \epsilon \ln a + \cdots$$

and the expansion of $\ln \Gamma(1 + z)$ for small $\hat{z}$; see Eq. (A.9).

For example,

$$A_1(0) = \frac{K_{4-\epsilon}}{2-\epsilon}\Lambda^{2-\epsilon}\left(1 - b^{-2+\epsilon}\right), \tag{E.23}$$

or, neglecting terms of order $O(\epsilon^2)$ and $O(\epsilon b^{-2} \ln b)$,

$$A_1(0) = \frac{\Lambda^2}{16\pi^2}\left(1 - b^{-2}\right)\left\{1 - \epsilon\left[\ln \Lambda - 1 + \frac{C_E}{2} - \frac{1}{2}\ln 4\pi\right]\right\}, \tag{E.24}$$

where we have used that

$$K_{4-\epsilon} = K_4\left\{1 + \frac{\epsilon}{2}\left[1 - \frac{C_E}{2} + \ln 4\pi\right]\right\} + O(\epsilon^2); \tag{E.25}$$

C_E is the Euler constant; see Eq. (A.9).

The integral $A_1(r^*)$ to the first order in ϵ (and r^*) is obtained from Eq. (E.20), Eq. (E.22) for $m = 2$ and Eq. (E.24):

$$A_1(r^*) = \frac{\Lambda^2}{16\pi^2}\left[\left(1 - b^{-2}\right) - \epsilon \ln \frac{\Lambda e^{-1+C_E/2}}{\sqrt{4\pi}}\right] - (K_4 \ln b)\, r^*. \tag{E.26}$$

Other simple integrals are

$$\begin{aligned} A_2(0) &= K_{4-\epsilon} \int_{b^{-2}\Lambda}^{\Lambda} dk\, k^{d-5} \\ &= K_{4-\epsilon} \int_{b^{-1}\Lambda}^{\Lambda} \frac{dk}{k}\left[1 - \epsilon \ln k + O\left(\epsilon^2\right)\right] \\ &= K_{4-\epsilon} (\ln b) \left\{1 - \epsilon 2 \ln\left(\Lambda^2/b\right)\right\}, \end{aligned} \tag{E.27}$$

and

$$A_3(0) = K_d \int_{b^{-1}\Lambda}^{\Lambda} k^{d-7} dk; \tag{E.28}$$

cf. Eq. (8.195). For $d = 4$,

$$A_3(0) = \frac{K_4}{2\Lambda^2}\left(b^2 - 1\right) + O(\epsilon). \tag{E.29}$$

The expansion in powers of $r^* \sim O(\epsilon)$ can be used for integrals like $A_2(r, q_1 b^{-1})$ as well; see, e.g.,, Eq. (8.28) with $c = 1$ and Eq. (8.78). Thus one obtains

$$A_2\left(r, q_1 b^{-1}\right) = A_2\left(0, q_1 b^{-1}\right) + A_2'\left(q_1 b^{-1}\right) r^* + \cdots, \tag{E.30}$$

where

$$A_2'\left(q_1 b^{-1}\right) = -\frac{1}{c^3} \int_k' \frac{1}{k^2 \left(\boldsymbol{k} + \boldsymbol{q}_1 b^{-1}\right)^2}\left[\frac{1}{k^2} + \frac{1}{\left(\boldsymbol{k} + \boldsymbol{q}_1 b^{-1}\right)^2}\right]. \tag{E.31}$$

This expansion is possible because $b^{-1}\Lambda < k = |\boldsymbol{k}| < \Lambda$, $b^{-1}\Lambda < |\boldsymbol{k} + \boldsymbol{q}_1 b^{-1}| < \Lambda$ and $r^* \sim \epsilon \ll (b^{-1}\Lambda)^2$; see the condition (8.203).

It has been shown in Section 8.6 that the first order coefficiont in the expansion of the integral $A_2(0, q_1 b^{-1})$ in powers of $(q_1 b^{-1})^2$ is equal to zero provided $d = 4$. It can be shown with the help of the identity

$$\int_0^\pi \frac{d\theta \sin^2\theta}{a+b\cos\theta} = \frac{\pi}{a+\sqrt{a^2+b^2}}, \qquad [a, (a^2-b^2)>0], \tag{E.32}$$

that $A_2(0, q_1b^{-1})$ for $d=4$ does not depend on (q_1b^{-1}). If $a = p_1^2+p_2^2$ and $b = 2p_1p_2\cos\theta$, it follows from Eq. (E.32) that

$$\begin{aligned}\int_0^\pi \frac{d\theta \sin^2\theta}{(p_1^2+p_2^2+2p_1p_2\cos\theta)} &= \frac{\pi}{p_1^2+p_2^2+|p_1^2-p_2^2|}\\ &= \begin{cases}\dfrac{\pi}{2p_1^2}, & p_1^2>p_2^2,\\ \dfrac{\pi}{2p_2^2}, & p_2^2>p_1^2,\end{cases}\\ &\equiv \frac{\pi}{2}\min\left(\frac{1}{p_1^2}, \frac{1}{p_2^2}\right).\end{aligned} \tag{E.33}$$

The relation Eq. (E.33) is useful for the calculation of angular integrals. Substituting $p_1 = k$ and $p_2 = (q_1b^{-1})$ in the integral over θ in Eq. (8.82) for $d=4$, and bearing in mind $1>k>b^{-1}>q_1b^{-1}$, we obtain $\pi/2k^2$, that is, the integral (8.82) does not depend on the external momentum q_1.

The integral

$$B(0) = \int_{p_1}^{\prime}\int_{p_2}^{\prime} \frac{1}{p_1^2p_2^2(\boldsymbol{p}_1+\boldsymbol{p}_2)^2}, \qquad (c=\Lambda=1), \tag{E.34}$$

cf. Eq. (8.196), is calculated for $d=4(\epsilon=0)$ with the help of Eq. (E.33):

$$\begin{aligned}B(0) &= \frac{1}{4}K_3K_4\int_{b^{-1}}^1 dp_1.p_1\int_{b^{-1}}^1 dp_2.p_2\min\left(\frac{1}{p_1^2}, \frac{1}{p_2^2}\right)\\ &= K_4^2\int_{b^{-1}}^1 dp_1.p_1\left[\frac{1}{p_1^2}\int_{b^{-1}}^{p_1} p_2.dp_2 + \int_{p_1}^1 d\ln p_2\right]\\ &= \frac{K_4^2}{2} - K_4^2b^{-2}\ln b + O(\epsilon);\end{aligned} \tag{E.35}$$

$K_3 = 4K_4$.

In the same way, one can obtain $C(0)$; see Eq. (8.197):

$$C(0) = \int_{p_1}^{\prime}\int_{p_2}^{\prime}\frac{1}{p_1^4p_2^2(\boldsymbol{p}_1+\boldsymbol{p}_2)^2} = \frac{K_4^2}{2}\left[\ln b + (\ln b)^2 + O(1)\right]. \tag{E.36}$$

Finally we shall calculate the integral difference

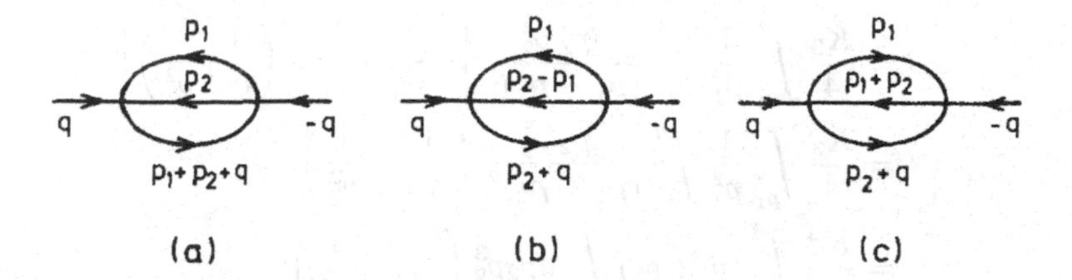

Fig. E.3 Relation between the notations for the internal momenta and the momentum cutoff.

$$\Delta B(0) = \int_{p_1}' \int_{p_2}' \frac{1}{p_1^2 p_2^2} \left[\frac{1}{(\boldsymbol{p}_1 + \boldsymbol{p}_2 + \boldsymbol{q})^2} - \frac{1}{(\boldsymbol{p}_1 + \boldsymbol{p}_2)^2} \right]; \tag{E.37}$$

see Eqs. (8.31c) and (8.32) which takes part in the expression (8.191) for the exponent η; $r = 0$. According to Fig. E.3(a), the conditions on the momenta p_i, $i = (1, 2)$ and q in Eq. (E.37) should be: $0 < q = |\boldsymbol{q}| < b^{-1}$, $b^{-1} < p_i = |\boldsymbol{p}_i|$, $(i = 1, 2)$, $|\boldsymbol{p}_1 + \boldsymbol{p}_2 + \boldsymbol{q}| < 1$; $\Lambda = 1$. Although the momentum moduli satisfy $p_i > b^{-1}$ and $\boldsymbol{q}$ can be chosen small enough ($q \ll b^{-1}$), the denominator $(\boldsymbol{p}_1 + \boldsymbol{p}_2 + \boldsymbol{q})^{-2}$ in Eq. (D.28) cannot be straightforwardly expanded in powers of q because the sum $(\boldsymbol{p}_1 + \boldsymbol{p}_2)$ does not necessarily obey the restriction $|(\boldsymbol{p}_1 + \boldsymbol{p}_2| > b^{-1}$. But under the conditions $p_i > b^{-1}$, $|\boldsymbol{p}_1 + \boldsymbol{p}_2|$ can take small values including $0 \leq |\boldsymbol{p}_1 + \boldsymbol{p}_2| \leq q$. It is then convenient to introduce the shift $\boldsymbol{p}_2' = \boldsymbol{p}_1 + \boldsymbol{p}_2$ and to substitute $\boldsymbol{p}_2 = \boldsymbol{p}_2' - \boldsymbol{p}_1$ in Eq. (E.37). This shift corresponds to the notations in Fig. E.3(b), where the *prime* of $\boldsymbol{p}_2'$ is omitted. Further, one can change the sign of $\boldsymbol{p}_1$, $\boldsymbol{p}_1 \to -\boldsymbol{p}_1$ which corresponds to an opposite direction of the $\boldsymbol{p}_1$-line in Fig. E.3(c) (obviously this procedure does not affect the integral). In accordance with Fig. E.3(c), Eq. (E.37) takes the form

$$\Delta B(0) = \int_{\boldsymbol{p}_1}' \frac{1}{p_1^2} \int_{\boldsymbol{p}_2} \frac{1}{(\boldsymbol{p}_1 + \boldsymbol{p}_2)^2} \left[\frac{1}{(\boldsymbol{p}_2 + \boldsymbol{q})^2} - \frac{1}{p_2^2} \right], \tag{E.38}$$

where

$$b^{-1} < p_1, \qquad |\boldsymbol{p}_1 + \boldsymbol{p}_2|, \qquad |\boldsymbol{p}_2 + \boldsymbol{q}| < 1. \tag{E.39}$$

Now the momentum $\boldsymbol{p}_2$ obeys the inequalities Eq. (E.39) and the general condition $0 < p_2 < 1$. The integral Eq. (E.38) is calculated by the following scheme:

$$\begin{aligned}
\Delta B &= \frac{K_3}{4}\int_{p_1}^{\prime}\frac{1}{p_1^2}\int_0^1\frac{dp_2\,p_2^3}{(\boldsymbol{p}_1+\boldsymbol{p}_2)^2}\left[-\frac{1}{p_2^2}+\min\left(\frac{1}{q^2},\frac{1}{p_2^2}\right)\right]\\
&= \frac{K_3}{4}\int_{p_1}^{\prime}\frac{1}{p_1^2}\int_0^q\frac{dp_2\,p_2^3}{(\boldsymbol{p}_1+\boldsymbol{p}_2)^2}\left(\frac{1}{q^2}-\frac{1}{p_2^2}\right)\\
&= \frac{K_3^2}{16}\int_{b^{-1}}^1 d(\ln p_1)\int_0^q dp_2p_2^3\left(\frac{1}{q^2}-\frac{1}{p_2^2}\right)\\
&= -\frac{K_4^2}{4}q^2\ln b+O(\epsilon).
\end{aligned}\tag{E.40}$$

The calculation of the integrals in the other cases, non-quadratic momentum dependence of $G_0(k,r)$, for example, $G_0(k,0)\sim k^\theta$, or another borderline dimension $d_U\neq 4$ arising in the investigation of various multicritical phenomena, is similar to that presented here.

E.4 Summation over Matsubara Frequencies

The summation over the (Bose)Matsubara frequencies $w_l = 2\pi lT$, $l = 0,\pm 1,\dots$ (see Section 9.3) is done with the help of formulae like

$$\sum_{l=1}^{\infty}\frac{2x}{(\pi l)^2+x^2}=\coth x-\frac{1}{x},\tag{E.41}$$

and

$$\sum_{l=1}^{\infty}\frac{(\pi l)\sin lx}{(\pi l)^2+b^2}=\frac{1}{2}\frac{\sinh\left(1-\frac{x}{\pi}\right)b}{\sinh b}.\tag{E.42}$$

Let us calculate, for example, the sum

$$\lim_{\tau\to+0}\sum_{\omega_l}G_0(q)e^{i\omega_l\tau}=\lim_{\tau\to+0}\sum_{\omega_l}\frac{e^{i\omega_l\tau}}{i\omega_l+a},\tag{E.43}$$

where $a=-\xi(k)$; see Eqs. (9.18) and (9.27). The sum Eq. (E.43) for a finite τ can be easily represented in the form

$$\frac{1}{a}+\sum_{\omega_l>0}\frac{2a\cos\omega_l\tau}{\omega_l^2+a^2}+\sum_{\omega_l>0}\frac{2\omega_l\sin\omega_l\tau}{\omega_l^2+a^2}\tag{E.44}$$

The limit $\tau \to 0$ in the second term of Eq. (E.44) can be straightforwardly taken whereas the third term requires a more special consideration as $\tau \to 0$; for $\omega_l \gg 1$ and $\tau \ll 1$, $\omega_l \tau$ does not tend to zero. The third term in Eq. (E.44) can be rewritten in another form. Using Eq. (E.42) we obtain

$$\sum_{\omega_l > 0} \frac{2\omega_l \sin \omega_l \tau}{\omega_l^2 + a^2} = \frac{\sinh\left[(1 - 2\tau T)\, a/2T\right]}{2T \sinh(a/2T)} \tag{E.45}$$

which tends to $(1/2T)$ for $\tau \to 0$. In the limit $\tau \to 0$ the second term in Eq. (E.44) becomes

$$\frac{1}{2T} \coth\left(\frac{a}{2T}\right) - \frac{1}{a}; \tag{E.46}$$

we have used Eq. (E.41). Thus

$$\lim_{\tau \to +0} \sum_{\omega_l} G_0(q) e^{i\omega_l \tau} = \frac{e^{a/T}}{T\left(e^{a/T} - 1\right)}. \tag{E.47}$$

We have considered one nontrivial example, where the limit $\tau \to 0$ should be taken. In general, no difficulties arise when one takes the sum over the internal frequencies ω_l of the various perturbation integrals for quantum (Bose) models. The important point in such calculations is to represent the sum considered by sums of the form of Eq. (E.41).

Appendix F

Fourier Amplitudes of Lattice Exchange Interactions

The exchange parameters $J_{ij} \equiv J(|x_i - x_j|)$ in Eq. (5.110) can be expanded in Fourier series:

$$J(\Delta R_{ij}) = \sum_k J(k) \exp(ik \cdot \Delta R_{ij}) \tag{F.1}$$

where $\Delta R_{ij} = (x_i - x_j)$ is the dD dimensional vector difference corresponding to two given interacting sites (i and j), and k is the dD wave vector: $k = (k_1, \ldots, k_d)$. The inverse transformation will be

$$J(k) = \sum_{\Delta R_{ij}} J(|\Delta R_{ij}|) \exp(-ik \cdot \Delta R_{ij}) \tag{F.2}$$

For nn interactions, $J(\Delta R_{ij})$ will take the value $J_{\text{ex}} \neq 0$ only when the difference ΔR_{ij} corresponds to two nn sites i and j), and will be zero for other ΔR_{ij}. For a simple cubic lattice (sc) with a lattice constant a_0, the terms in Eq. (F.2) with nonzero contribution correspond to the following nn ΔR_{ij} vectors: $(\pm a_0, 0, \ldots, 0)$, $(0, \pm a_0, 0, \ldots, 0)$, $\ldots$, $(0, \ldots, 0, \pm a_0)$. We shall denote these vectors by ΔR_{ij}^{nn}. Now we can perform the summation in Eq. (F.2):

$$\begin{aligned} J(k) &= J_{\text{ex}} \sum_{\Delta R_{ij}^{nn}} \exp(-ik \cdot \Delta R_{ij}^{nn}) \\ &= J_{\text{ex}} \left(e^{-ik_1 a_0} + e^{ik_1 a_0}\right) + \cdots + \left(e^{-ik_d a_0} + e^{ik_d a_0}\right) \\ &= 2J_{\text{ex}} \sum_{\alpha=1}^{d} \cos(k_\alpha a_0) . \end{aligned} \tag{F.3}$$

For $k = 0$, $J(0) \equiv J_0 = 2dJ_{\text{ex}} = zJ_{\text{ex}}$, where $z = 2d$ is the number of *nn*. In the long-wavelength limit, $|k|a_0 \ll 0$, $\cos(k_\alpha a_0) \approx 1 - (k_\alpha a_0)^2/2$, Eq. (E.3) yields

$$J(k) = J_0 - \frac{J_0 a^2}{2d} k^2. \tag{F.4}$$

The same calculation can be performed for other lattices. For example, in case of body-centered tetragonal lattice (bct, (a_0, a_0, c_0) in 3D space with lattice constant a_0 along the axes $\hat{x}$ and $\hat{y}$, and lattice constant c_0 along the $\hat{z}$ axis, or, with the structure $(a_0, a_0, \ldots, a_0, c_0)$ of a dD (hyper-parallelepiped), we easily check that the ΔR_{ij}^{nn} vectors, attached to any site i, will be given by $(\pm a_0/2, \pm a_0/2, \ldots, \pm a_0/2, \pm c_0/2)$. Now the number of these vector differences is equal 2^d). Performing the summation in Eq. (E.2), we obtain

$$\begin{aligned}
J(k) &= J_{\text{ex}} \sum_{\Delta R_{ij}^{nn}} \exp(-ik \cdot \Delta R_{ij}^{nn}) \\
&= J_{\text{ex}} \sum_{n_1=\pm 1} \cdots \sum_{n_d=\pm 1} \exp\left[\tfrac{1}{2}\left(-ik_1 a_0 n_1 - \cdots - ik_{d-1} a_0 n_{d-1} - ik_n c_0 n_d\right)\right] \\
&= J_{\text{ex}} \sum_{n_1=\pm 1} \left(e^{-ik_1 a_0 n_1/2}\right) \cdots \\
&\quad \cdots \sum_{n_{d-1}=\pm 1} \left(e^{-ik_{d-1} a_0 n_{d-1}/2}\right) \sum_{n_d=\pm 1} \left(e^{-ik_d c_0 n_d/2}\right) \\
&= 2^d J_{\text{ex}} \cos\left(\frac{k_1 a_0}{2}\right) \cdots \cos\left(\frac{k_{d-1} a_0}{2}\right) \cos\left(\frac{k_d c_0}{2}\right),
\end{aligned} \tag{F.5}$$

which generalizes the result by Flax and Raich (1969). For $k = 0$ we have $J(0) \equiv J_0 = zJ_{\text{ex}}$ with $z = 2^d$, and in the long wavelength limit, Eq. (F.5) reads

$$J(k) = J_0 \left\{1 - \frac{1}{8}\left[(ka_0)^2 + k_d^2\left(c_0^2 - a_0^2\right)\right]\right\}. \tag{F.6}$$

For a body-centered cubic lattice (bcc), $a_0 = c_0$, the spatial anisotropy does not exists, and all coefficients ($\sim a_0^2$) in the front of the wave vector components are equal one another.

In both cases considered here, the coefficients in the front of k^2 are proportional to the square of the lattice constant a_0, which means that the latter is proportional to the zero-temperature correlation length ξ_{00} in models where these long wavelength limiting forms of $J_(k)$ are used. This is a typical feature of all effective theories developed in this way. Finally, we note that in the example of bct, the considered exchange interactions will be the actual nn interactions (the shortest-distance interactions), provided $(2a_0^2 + c_0^2)^{1/2} < 2a_0$, i.e., for $c_0 < \sqrt{2}a_0$.

Bibliography

Abe, R. (1972). *Prog. Theor. Phys.* **48**, 1414.
Abe, R. (1973). *Prog. Theor. Phys.* **49**, 1877.
Abe, R. (1974). *Prog. Theor. Phys.* **52**, 1135.
Abe, R. and Hikami, S. (1974a). *Prog. Theor. Phys.* **52**, 1453.
Abe, R. and Hikami, S. (1974b). *Phys. Lett. A* **47**, 341.
Abrahams, E. and Tsuneto, T. (1975). *Phys. Rev.* **B11**, 4498.
Abramowitz, M. and Stegun, I. A. (eds.) (1965). *Handbook of Mathematical Functions* (Dover, New York).
Abrikosov, A. A. (1957). *Zh. Eksp. Teor. Fiz.* **32**, 1442 [English translation: (1957). *Sov. Phys. JETP* **5**, 1174].
Abrikosov, A. A. (1965). *Physics* **2**, 5.
Abrikosov, A. A., Gor'kov, L. P. and Dzyaloshinskii, I. E. (1962). *Methods of Quantum Field Theory in Statistical Physics* (Fizmatgiz, Moscow) in Russian; English Translation (Prentice–Hall, Englewood Cliffs, New Jersey, 1963).
Aczel, J. (1966). *Lectures on Functional Equations and their Applications* (Academic Press, New York).
Aczel, J. (1969). *On Applications and Theory of Functional Equations* (Academic Press, New York).
Affleck, I. (1985). *Nucl. Phys. B* **257**[FS14], 451.
Aharony, A. (1973). *Phys. Rev. B* **8**, 4270.
Aharony, A. (1976). in C. Domb and M. S. Green (eds.), *Phase Transitions and Critical Phenomena*, Vol. 6 (Academic, London) p. 357.
Aharony, A. (1983). in F. J. W. Hahne (ed.), *Critical Phenomena* (Springer, Berlin) p. 209.
Aharony, A., Domany, E., Hornreich, R. M, Schneider, T. and Zannetti, M. (1985). *Phys. Rev. B* **32**, 3358.
Aharony, A. and Fisher, M. E. (1973). *Phys. Rev. B* **8**, 3323.
Aharony, A., Gefen, Y., and Shapir, Y.(1982). *J. Phys. C* **15**, 673.
Aharony, A., Imry, Y. and Ma, S. K. (1976). *Phys. Rev. Lett.* **37**, 1364.
Ahlers, G. (1970). in K. H. Bennemann and J. B. Ketterson (eds.), *The Physics of Liquid and Solid Helium* (Wiley, New York) Chapter 2, p. 1;

Ahlers, G. (1980). *Rev. Mod. Phys.* **52**, 489.
Athorne, C. and Lawrie, I. D. (1985). *Nucl. Phys. B*[FS14] **257**, 577.
Akazawa, T., Hidaka, H., Kotegawa, H., Kobayashi, T. C., Fujiwara, T., Yamamoto, E., Haga, Y., Settai, R. and Onuki, Y. (2005). *Physica B* **359-361**, 1138.
Amit, D. J. (1974). *J. Phys. C* **7**, 3369.
Amit, D. J. (1976). *J. Phys. A* **9**, 1441.
Amit, D. J. (1978). *Field Theory, the Renormalization Group and Critical Phenomena* (McGraw Hill, New York).
Amit, D. J. and Goldschmidt Y. Y. (1978).*Ann. Phys.* **114**, 356.
Anderson, M. M., Enster, J. R., Matthews, M. R., Wieman, C. E., and Cornell E. A. (1995). *Science* **269**, 198.
Anderson, P. W. (1984a). *Basic Notions of Condensed Matter Physics* (Benjamin, London).
Anderson, P. W. (1984b). *Phys. Rev. B* **30**, 1549.
Anisimov, M. A. (1987). *Critical Phenomena in Liquids and Liquid Crystals* (Nauka, Moscow) in Russian.
Anisimov, M. A. (1988). *Mol. Cryst. Liq. Cryst. A* **162**, 1.
Anisimov, M. A., Cladis, P. E., Gorodetskii, E. E., Huse, D. A., Podneks, V. E., Taratuta, V. G., van Saarloos W. and Voronov V. P. (1990). *Phys. Rev. A* **41**, 6749.
Anisimov, M. A., Voronov, V. P., Gorodetzkii, E. E., Podneks, V. E. and Kholmurodov F. (1987). *Pis'ma Zh. Eksp. Teor. Fiz.* **45**, 336. [(1987) *Sov. Phys. JETP Letts.* **45**, 425.]
Annett, J. F. (1990). *Adv. Phys.* **39**, 83.
Annett, J. F. (2004). *Superconductivity, Superfluids, and Condensates* (Oxford University Press, Oxford).
Annett, J. F., Goldenfeld, N. and Renn S. R. (1989). *Physica C* **162–164**, 377.
Annett, J. F., Randeria, M. and Renn S. R. (1988). *Phys. Rev. B* **38**, 4600.
D. Aoki, D., Huxley, A., Ressouche, E., Braithwaite, D., Flouquet, J., Brison, J-P., Lhotel, E. and Paulsen C. (2001). *Nature* **413**, 613.
Athorne, C. and Lawrie I, D. (1985). *Nucl. Phys. B* **257** [FS14], 577.
Baba, Y., Nagai, T. and Kawasaki, K. (1979). *J. Low Temp. Phys.* **36**, 1.
Babaev, E., Sudbø, A. and Aschroft, N. W. (2004). *Nature* **431**, 666.
Bagnuls, C. and Bervillier, C. (1985). *Phys. Lett. A* **107**, 299.
Bagnuls, C. and Bervillier, C. (2001). *Phys. Rep. C* **348**, 91.
Bagnuls, C. and Bervillier, C. (1986). *Phys. Rev. B* **34**, 299.
Bagnuls, C. and Bervillier, C. (1986b). *Phys. Lett. A* **115**, 84.
Bak, P., Krinsky, S. and Mukamel, D. (1976). *Phys. Rev. Lett.* **36**, 52.
Bak, P. and Mukamel, D. (1976). *Phys. Rev. B* **13**, 5086.
Baker, G. A. (1962). *Phys. Rev.* **126**, 2071.
Baker, G. A., Jr. (1975). *Essentials of Padé Approximations* (Academic, New York).
Baker, G. A., Jr. (1975). in C. Domb and J. L. Lebowitz (eds.), *Phase Transitions and Critical Phenomena*, Vol. 9 (Academic, London) p. 233.
Baker, G. A., Jr. and Kincaid, J. M. (1981). *J. Stat. Phys.* **24**, 469.

Baker, G. A., Jr., Nickel, B. G., Green M. S. and Meiron, D. I. (1976). *Phys. Rev. Lett.* **36**, 1351.
Baker, G. A., Jr., Nickel, B. G. and Meiron D. I. (1978). *Phys. Rev. B* **17**, 1365.
Baldo, M., Catara, E. and Lombardo, U. (1976). *Lett. Nuovo Cimento* **15**, 214.
Balian, R. and Toulouse, G. (1973). *Phys. Rev. Lett.* **30**, 544.
Barbara, B., Gignoux, D. and Vettier, C. (1988). *Lectures on Modern Magnetism* (Science Press & Springer, Beijing & Berlin).
Barber, M. N. (1977). *Phys. Reps. C* **29**, 1.
Barber M. N., (1983). in C. Domb and J. L. Lebowitz (eds.), *Phase Transitions and Critical Phenomena*, Vol 8 (Academic, London) p. 145.
Bardeen, J., Cooper, L. N. and Scrieffer, J. R. (1957). *Phys. Rev.* 108, 1175.
Bardeen, J., Ginsberg, D. M. and Salamon, M. B. (1987). in S. A. Wolf and V. Z. Krèsin (eds.), *Novel Superconductivity* (Plenum, New York) p. 333.
Bartholomew, J. (1983). *Phys. Rev. B* **28**, 5378.
Bateman, H. (1953). in A. Erdelyi (ed.), *High Transcendental Functions* (McGraw–Hill, New York).
Batrouni, G. G., Scalettar, R. T. and Zimanyi, G. T. (1990). *Phys. Rev. Lett.* **65**, 1765.
Baxter, R. J. (1982). *Exactly Solved Models in Statistical Mechanics* (Academic, London).
Beck, H., Schneider, T. and Stoll, E. (1975). *Phys. Rev. B* **12**, 5198.
Belanger, D. P., King, A. R., Jaccarino, V. and Cardy, J. L. (1983). *Phys. Rev. B* **28**, 2522.
Belanger, D. P. and Yoshizawa, H. (1987) *Phys. Rev. B* **35**, 4823.
Bergerhoff, B., Freire, F., Litim, D. F., Lola, S. and Wetterich, C. (1996). *Phys. Rev. B* **53**, 5437.
Berlin, T. H. and Kac M. (1952). *Phys. Rev.* **86**, 821.
Bervillier, C. (2004). *Phys. Lett. A* **331**, 110.
Bindilatti, V., Becerra, C. C. and Oliveira, N. F. Jr. (1989). *Phys. Rev. B* **40**, 9412.
Binney J. J., Dowrick, N. J., Fisher, A. J. and Newman, M. E. J. (1992). *The Theory of Critical Phenomena* (Oxford University Press, Oxford).
Birgeneau, R. J., Cowley, R. A., Shirane, G. and Yoshizawa, H. (1983). *Phys. Rev. B* **27**, 6747.
Birgeneau, R. J., Shirane, G., Blume, H. and Koehler, W. C. (1974). *Phys. Rev. Lett.* **33**, 1098.
Bjorken, I. D. and Drell, S. D. (1965). *Relativistic Quantum Fields* (McGraw–Hill, New York).
Blagoeva, E. J., Busiello, G., De Cesare, L., Millev, Y. T., Rabuffo, I. and Uzunov, D. I. (1989). *Phys. Rev. B* **40**, 7357; Err., ibid., **41**, 4785.
Blagoeva, E. J., Busiello, G., De Cesare, L., Millev, Y. T., Rabuffo, I. and Uzunov, D. I. (1990). *Phys. Rev. B* **42**, 6124.
Blankschtein, D. and Aharony, A. (1983). *Phys. Rev. B* **28**, 386.
Blankschtein, D. and Mukamel, D. (1982). *Phys. Rev.* **B25**, 6939.
Blankschtein, D., Shapir, Y. and Aharony, A. (1984). *Phys. Rev. B* **29**, 1263.
Blinc, R. and Ẑekš, B. (1974). *Soft Modes in Ferroelectrics and Antiferroelectrics* (North Holland, Amsterdam).

Blume, M. (1966). *Phys. Rev.* **141**, 517.
Blume, M., Corliss, L. M., Hastings, J. M. and Schiller. E. (1974). *Phys. Rev. Lett.* **32**, 544.
Blume, M., Emery, V. J. and Griffiths, R. B. (1971). *Phys. Rev. A* **4**, 1071.
Blume, M. and Watson, R. E. (1967). *J. Appl. Phys.* **38**, 991.
Bogoliubov, N. N. (1947). *J. Phys. USSR* **11**, 23. Reprinted by D. Pines *The Many Body Problem* (Benjamin, New York, 1962).
Bogoliubov, N. N. and Shirkov, D. V. (1955) *Dokladi Acad. Nauk.* (USSR) **103**, 203.
Bogoliubov, N. N. and Shirkov, D. V. (1959)*Introduction to the Theory of Quantized Fields* (Interscience, New York); 3rd Edition (Wiley, New York, 1980).
Bose, N. S. (1924). *Z. Phys.* **26**, 178.
Boyanovsky, D. and Cardy, J. L. (1982a). *Phys. Rev. B* **26**, 154.
Boyanovsky, D. and Cardy, J. L. (1982a). *Phys. Rev. B* **25**, 7058.
Brandt E. H. (1990a). *Physica B* **165–166**, 1129.
Brandt E. H. (1990b). *Int. J. Mod. Phys. B* **5**, 751.
Bray, A. (1983). *J. Phys. C* **16**, 5875.
Brazovskii S. A. and Dzyaloshinskii I. E. (1975). *Pis'ma Zh. Eksp. Teor. Fiz.* **21**, 360. [(1975) *JETP Lett.* **21**, 164].
Brézin, E., Fijita, A. and Hikami, S. (1990). *Phys. Rev. Lett.* **65**, 1949.
Brézin, E., Itzykson, C., Zinn-Justin, and Zuber, J-B. (1979). *Phys. Lett. B* **82**, 442.
Brézin, E., Le Guillou J. C. and Zinn–Justin J. (1976). in C. Domb and M. S. Green (eds.), *Phase Transitions and Critical Phenomena* (Academic, London) p. 127.
Brézin, E., Nelson, D. R. and Thiaville, A. (1985). *Phys. Rev. B* **31**, 7134.
Brézin, E. and Zinn–Justin, J. (1976). *Phys. Rev. Lett.* **36**, 691; *Phys. Rev. B* **14**, 3110.
Brézin, E., Zinn–Justin, J. and Le Guillou, J. C. (1976). *Phys. Rev. D* **14**, 2615.
Brochard, F. and de Gennes, P. C. (1983) *J. Physique Lett.* **44**, L785.
Brout, R. H. (1965) *Phase Transitions* (Benjamin, New York).
Brout, R. H. (1976). in J. Brey and R. B. Jones (eds.), *Critical Phenomena* (Lecture Notes in Physics, Vol 54) (Springer, Berlin).
Bruce, A. D. (1980). *Adv. Phys.* **29**, 111.
Bruce, A. D., Droz, M. and Aharony, A. (1974). *J. Phys. C* **7**, 3673.
Brush, S. G. (1967). *Rev. Mod. Phys.* **39**, 883.
Brush, S. G. (1983). *Statistical Physics and Atomic Theory of Matter* (Princeton University Press, Princeton).
Buckingham M. J. and Fairbank W. M. (1961). in C. J. Gorter (ed.), *Progress in Low Temperature Physics*, Vol 3 (North Holland, Amsterdam) p. 80.
Bulaevskii, L. N., Ginzburg, V. L. and Sobyanin, A. A. (1988). *Zh. Eksp. Teor. Fiz.* **94**, 355 [(1988)*Sov. Phys. JETP* **68**, 1494]; *Physica C* **152**, 378.
Burkhardt, J. W. and van Leeuwen, J. M. J. (eds.) (1982). *Real–Space Renormalization* (Springer, Berlin).
Burlakov, L. I. (1985). *Zh. Eksp. Teor. Fiz.* **89**, 1382. [*Sov. Phys. JETP* **62**, 800.]

Bushev, M. K., Mahaldiani, N. V., Tonchev, N. S. and Uzunov,D. I. (1980). *Compt. Rend. Acad. Bulg. Sci.* **33**, 897.
Bushev, M. K. and Uzunov D. I. (1980). *Phys. Lett A* **76**, 306; Er, ibid., **78** 491.
Busiello, G. and De Cesare, L. (1980). *Nuovo Cimento B* **59**, 327.
Busiello, G., De Cesare, L., Millev, Y. T., Rabuffo, I. and Uzunov, D. I. (1991). *Phys. Rev. B* **43**, 1150.
Busiello, G., De Cesare, L. and Rabuffo, I. (1983). *Phys. Rev. B* **28**, 6463; *Physica A* **117**, 445.
Busiello, G., De Cesare, L. and Rabuffo, I. (1984). *Phys. Rev. B* **29**, 4189.
Busiello, G., De Cesare, L. and Rabuffo, I. (1985). *J. Phys. A* **18**, L749.
Busiello, G., De Cesare, L. and Uzunov, D, I. (1984). *J. Phys. A* **17**, L441.
Busiello, G., De Cesare, L. and Uzunov, D, I. (1985). *Physica A* **132**, 199.
Busiello, G., De Cesare, L. and Uzunov, D, I. (1986). *Phys. Rev. B* **34**, 4932.
Busiello G. and Uzunov D. I., (1987a) in G. Busiello, L. De Cesare, F. Mancini and M. Marinaro (eds.), *Advances on Phase Transitions and Disorder Phenomena* (World Scientific, Singapore) p. 130.
Busiello G. and Uzunov D. I. (1987b). *Physica Status Solidi (b)* **139**, L107.
Busiello G. and Uzunov D.I. (1989). *Phys. Rev. B* **40**, 7321.
Busiello G. and Uzunov D. I. (1990). *Phys. Rev. B* **42**, 1018.
Callan, C. G. Jr. (1970). *Phys. Rev. D* **2**, 1541.
Callen, H. B. (1960). *Thermodynamics* (Wiley, New York).
Capel, H. W. (1966). *Physica* **32**, 966.
Capel, H. W. (1996). *Scaling and renormalization Group in Statistical Physics* (Cambridge University Press, Cambridge).
Carneiro, C. E. I., Henriques, V. B. and Salinas, S. R. (1987). *J. Phys. A* **20**, 189.
Casher, A., Lurié, D. and Revzen, M. (1968). *J. Math. Phys.* **9**, 1312.
Chan, M. H. W, Blum, K. I, Murphy, S. Q., Wong, G. K. S. and Reppy J. D. (1988). *Phys. Rev. Lett.* **61**, 1950.
Chang, T. S., Hankey, A. and Stanley, H. E. (1973). *Phys. Rev. B* **8**, 346.
Chang, T. S., Tuthill, G. F. and Stanley H. E., (1974). *Phys. Rev. B* **9**, 4882.
Chen, J-H., Lubensky T. C. and Nelson D. R. (1978). *Phys. Rev. B* **17**, 4275.
Chetyrkin, K. G. and Tkachov F. V. (1982). *Phys. Lett. B* **114**, 340.
Chu, C. W. (1989). *Phys. Scripta T* **27**, 11.
Chubukov, A. V. (1984). *Theor. Math. Phys.* **60**, 145.
Cochran, W. (1971). in E. J. Samuelson, E. Anderson and J. Feder (eds.) *Structural Phase Transitions and Soft Modes* (Universitetforlaget, Oslo) p. 1.
Coleman, S. (1985). *Aspects of Symmetry* (Cambridge Univ. Press, Cambridge).
Coleman, S. and Weinberg, E. (1973). *Phys. Rev. D* **7**, 1888.
Collins, J. (1984). *Renormalization* (Cambridge Univ. Press, Cambridge).
Comte, C. and Noziéres, P. (1982). *J. Physique* **43**, p. 1069, p. 1083.
Continentino, M. A. (2001). *Quantum Scaling in many-Body Systems* (Worls Scientific, Singapore).
Cooper, L. N. (1957). *Phys. Rev.* **104**, 1189.
Cooper, M. J. and Green, M. S. (1968). *Phys. Rev.* **176**, 302.
Cottam, M. G., Shopova, D. I. and Uzunov, D. I. (2008). *Phys. Lett. A* **373**, 152.
Cowley, R. A. (1980). *Adv. Phys.* **29**, 1.

Cox, U. J., Gibaud, A. and Cowley, R. A. (1988). *Phys. Rev. Lett.* **61**, 982.
Craco, L., De Cesare, L., Rabuffo, I., Takov, I. P. and Uzunov, D. I. (1999). *Physica A* **270**, 486.
Creswick, R. J. and Wiegel, F. W. (1983). *Phys. Rev. A* **28**, 1579.
Curie, P. (1895). *Ann. Chem. Phys.* **7**(5), 289; and Oeuvres, Paris (1908).
Dacol, D. K. (1980). *J. Low Temp. Phys.* **41**, 349.
da Frota, H. O. and Goulard Rosa, Jr. S. (1982). *J. Phys. A* **15**, 2221.
da Frota, H. O., Silva, M. S. and Goulard Rosa, Jr. S. (1984). *J. Phys. C* **17**, 1669.
Dahl, E. K. and Sudbø. A. (2007). *Phys. Rev. B* **75**, 144504.
Dasgupta, C. and Halperin B. I. (1981). *Phys. Rev. Lett.* **47**, 1556.
Davidov, D., Safinia, C. R., Kaplan, M., Dana, S. S., Schaetzing, R., Birgeneau, R. J. and Litster, J. D. (1979). *Phys. Rev. B* **19**, 1657.
Davis, K. B., Mewes, M.-O., Andrews, M. R., van Driten, N. J., Dufree, D. S., Kurn, D. M. and Ketterle, W. (1995). *Phys. Rev. Lett.* **75**, 3969.
de Boer, J. (1974). *Physica A* **73**, 1.
de Callan, C. and Nogueira, F. S. (1999). *Phys. rev. B* **60**, 4255.
De Cesare, L. (1978). *Lett. Nuovo Cimento* **22**, 325, 632.
De Cesare, L. (1982). *Nuovo Cimento* **D1**, 283.
De Cesare, L. (1987). *Rev. Solid State Sci.* (World Scientific) **3**, 1.
de Gennes, P. G. (1972). *Phys. Lett. A* **38**, 339; *Solid Satte Commun.* **10**, 753.
de Gennes, P. G. (1973). *Mol. Cryst. Liq. Cryst.* **21**, 49.
de Gennes, P. G. (1974). *The Physics of Liquid Crystals* (Clarendon Press, Oxford).
de Gennes, P. G. (1979). *Scaling Concepts in Polymer Physics* (Cornell Univ. Press, Ithaca & London).
de Gennes, P. G. and Prost, J. (1993). *The Physics of Liquid Crystals*, 2nd ed. (Clarendon Press, Oxford).
de Groot, S. R. and Mazur, P. (1969). *Nonequilibrium Thermodynamics* (North–Holland, Amsterdam).
de Jongh, L. J. and Miedema, A. R. (1974). *Adv. Phys.* **23**, 1.
Dengler, R. (1985). *Phys. Lett. A* **108**, 269.
Deonarine, S. and Birman, J. L. (1983). *Phys. Rev. B* **27**, 4261.
De Pasquale, F., Di Castro, C. and Jona–Lasinio, G. (1971). in M. S. Green (ed.), *Critical Phenomena* [Proceedings of the Int. School in Physics "E. Fermi," Course LI] (Academic, New York) p. 123.
Devonshire, A. F. (1949). *Philos. Mag.* **40**, 1040.
Devonshire, A. F. (1949). *Philos. Mag.* **42**, 1065.
Devonshire, A. F. (1954). *Adv. Phys.* **3**, 85.
Di Castro, C. and Jona–Lasinio, G. (1976). *Phase Transitions and Critical Phenomena* (Academic, London) p. 507.
Diehl, C. (1986). in C. Domb and J. L. Lebowitz, *Phase Transitions and Critical Phenomena*, Vol 10 (Academic, London) p. 75.
Diehl H. W. (1980). *Phys. Lett. A* **75**, 375.
Dohm, V. and Folk, R. (1982). *Physica B* **109–110**, 1549.
Dohm, V. and Folk, R. (1983). *Phys. Rev. B* **28**, 1332.

Domany, E,, Mukamel, D., and Fisher M. E. (1977). *Phys. Rev. B* **15**, 5432.
Domb, C. and Green, M. S. (eds.) (1972). *Phase Transitions and Critical Phenomena*, Vols. 1–7 (Academic Press, London) [Vols. 8–16··· of the same series is edited by C Domb and J L Lebowitz].
Domb, C. and Hunter, D. L. (1965). *Proc. Phys. Soc.* **86**, 1147.
Dorogovtsev, S. N. (1980). *Phys. Lett. A* **76**, 169.
Dorogovtsev, S. N. (1981). *Zh. Eksp. Teor. Fiz.* **80**, 2053. [(1981) *Sov. Phys. JETP* **53**, 1070.]
Dudka, M., Folk, R. and Moser, G. (2007). *Cond. Mat. Phys.* **10**, 189.
Dvorak, V. (1974). *Ferroelectrics* **7**, 1.
Dzyaloshinskii, I. E. (1964). *Zh. Eksp. Teor. Fiz.* **46**, 1352; **47**, 992. [*Sov. Phys. JETP* **19**, 960.]
Dzyaloshinskii, I. E. and Obukhov, S. P. (1982). *Zh. Eksp. Teor. Fiz.* **83**, 813.
Ebner, C. and Stroud, D. (1989). *Phys. Rev. B* **39**, 789.
Edwards, S. F. and Anderson, P. W. (1975). *J. Phys. F* **5**, 965.
Ehrenfest, P. (1933). *Commun. Kamerling Onnes Laboratory, Leiden, Suppl.* **75b**; *Proc. Acad. of Sci. Amsterdam* **36**, 153.
Einstein, A. (1924). *Sitzber. Kgl. Preuss. Acad. Wiss., Phys.–math* p. 261; (1925), ibid., p. 3.
Elliott, R. J. and Wood, C. (1971). *J. Phys. C* **4**, 2359.
Englert, F. and Brout, R. (1964). *Phys. Rev. Lett.* **13**, 1554.
Essam, J. W. and Fisher, M. E. (1963). *J. Chem. Phys.* **38**, 802.
Euch, G., Knops, H. J. and Verboven, E. J. (1970). *J. Math. Phys.* **11**, 1655.
Fairbank, W. M. (1963). in Careri C. (ed.) *Proceedings of the Int. School of Physics "Enrico Fermi"* Course XXI (Academic, New York) p. 293.
Fairbank, W. M., Buckingham, M. J. and Kellers C. F. (1958). in Dillinger J. R. *Proceedings of the Fifth Int. Conference on Low Temperature Physics* (University of Wisconsin, Madison) p. 50.
Fatuzzo, E. and Merz, W. J. (1967). *Ferroelectricity* (North–Holland, Amsterdam).
Ferrell, R. A. and Bhattacharjee, J. K. (1979). in J. L. Birman, H. J. Cummins and K. K. Rebane (eds.) *Light Scattering in Solids* (Plenum, New York); (1979) *J. Low Temp. Phys.* **36**, 165.
Fetter, A. L. and Walecka, J. D. (1971). (McGraw–Hill, New York); (1999), (2003) – next editions by Dover Publ., New York.
Feynman, R. P. and Hibbs, A. R. (1965). *Quantum Mechanics and Path Integrals* (McGraw–Hill, New York).
Finotello, D., Gillis, K. A., Wong, A. and Chan, M. H. W. (1988). *Phys. Rev. Lett.* **61**, 1954.
Fisher, D. S. (1985). *Phys. Rev. B* **31**, 7233.
Fisher, D. S. and Fisher, M. P. A. (1988). *Phys. Rev. Lett.* **61**, 1847.
Fisher, D. S., Fisher, M. P. A. and Huse, D. A. (1991). *Phys. Rev. B* **43**, 130.
Fisher, M. E. (1962). *Physica* **28**, 172.
Fisher, M. E. (1964a). *Archs. Ration. Mech. Analysis* **17**, 377.
Fisher, M. E. (1964b). *J. Math. Phys.* **5**, 944.
Fisher, M. E. (1964c). *Phys. Rev. A* **136**, 1599.

Fisher, M. E. (1965). in W. E. Brittin (ed.), *Lectures in Theoretical Physics* Vol. VIIC (Colorado Univ. Press, Boulder) p. 1.
Fisher, M. E. (1967). *Rep. Prog. Phys.* **30**, 615.
Fisher, M. E. (1968). *Phys. Rev.* **176**, 257.
Fisher, M. E. (1971). in M. S. Green (ed.), *Critical Phenomena*[Proceedings of the International School of Physics "E. Fermi", Course LI] (Academic, New York) p. 1.
Fisher, M. E. (1973). *Phys. Rev. Lett.* **30**, 679.
Fisher, M. E. (1974). *Rev. Mod. Phys.* **46**, 597.
Fisher, M. E. (1983). in F. J. W. Hahne (ed.), *Critical Phenomena* (Springer, Berlin) p. 1.
Fisher, M. E. (1984). in R. Pynn and A. Skjeltorp (eds.), *Multicritical Phenomena*, ed. R Pynn and A Skjeltorp (Plenum, New York) p. 1.
Fisher, M. E., Ma, S. K. and Nickel, B. G. (1972). *Phys. Rev. Lett.* **29**, 917.
Fisher, M. P. A. (1989). *Phys. Rev. Lett.* **62**, 1415.
Fisher, M. P. A. and Grinstein, G. (1988). *Phys. Rev. Lett.* **60**, 2008.
Fisher, M. P. A. and Grinstein, G. and Girvin, S. M. (1990). *Phys. Rev. Lett.* **64**, 587.
Fisher, M. P. A., Weichman, P. B., Grinstein, G., and Fisher, D. S. (1989).*Phys. Rev.* **B40**, 546.
Fishman, S. and Aharony, A. (1979). *J. Phys. C* **12**, L729.
Flax, L. and Raich, J. C. (1969). *Phys. Rev.* **185**, 797.
Folk, R. and Holovatch, Yu. (1996). *J. Phys. A: Math. Gen.* **29**, 3409.
Folk, R. and Holovatch, Yu. (1999). in D. V. Shopova and D. I. Uzunov (eds.) *Correlations, Coherence, and Order* (Klewer Academic/Plenum Publishers, New York–London).
Folk, R. and Moser, G. (1987). *Phys. Lett. A* **120**, 39.
Folk, R., Holovatch, Yu. and Moser, G. (2008a). *Phys. Rev. E* **78**, 041124.
Folk, R., Holovatch, Yu. and Moser, G. (2008a). *Phys. Rev. E* **78**, 041125.
Folk, R., Holovatch, Yu. and Moser, G. (2009). *Phys. Rev. E* **78**, 031109.
Folk, R., Shopova, D. V. and Uzunov, D. I. (2001). *Phys. Lett. A* **281**, 197.
Fossheim, K., Nes, O. M., Loegreid, T., Darlington, C. N. W., O'Conner, D. A. and Gough, C. E. (1988). *Int. J. Mod. Phys. B* **1**, 1171.
Fowler, R. H. (1938). *Proc. Camb. Philos. Soc.* **34**, 382.
Fowler, M. and Zawadowsky, A. (1971). *Solid State Commun.* **9**, 471.
Fradkin, E. S. (1955). *Zh. Eksp. Teor. Fiz.* **28**, 750.
Freire, F. and Litim, D. F. (2001). *Phys. Rev. D* **64**, 045014.
Fujita, A., Hikami, S. and Brezin, E. (1990). *Physica B* **165–166**, 1125.
Galam, S. (1985). *Phys. Rev. B* **31**, 1554.
Galam, S. and Birman, J. L. (1983). *J. Phys. C* **16**, L1145; *Phys. Rev. Lett.* **51**, 1066.
Garland, C. W. and Nounesis, G. (1994). *Phys. Rev. E* **49**, 2964.
Gell–Mann, M. and Low, F. (1954). *Phys. Rev.* **95**, 1300.
Gerber, P. R. and Beck, H. (1977). *J. Phys. C* **10**, 4013.
Getzlaff, M. (2008). *Fundamentals of Magnetism* (Springer–Verlag, Berlin–Heidelberg).

Gibbs, J. W. (1876). *Trans. Connect. Acad.* **3**, 108; reprinted in *The Collected Works*, see Gibbs (1948).
Gibbs, J. W. (1878). *Trans. Connect. Acad.* **3**, 343; *Abstract*, published in *Amer. J. Sci* **16**, 441; reprinted in *The Collected Works*, see Gibbs (1948).
Gibbs, J. W. (1948). *The Collected Works of J Willard Gibbs*, in two volumes (Longmans, Green and Co., New York); previous editions in 1908 and 1928.
Ginsberg, D. M., Inderhees, S. E., Salamon, M. B., Goldenfeld, N., Rice, J. P, and Pazol, B. G. (1988). *Physica C* **153–155**, 1082.
Ginzburg, S. L. (1975). *Zh. Eksp. Teor. Fiz.* **68**, 274.
Ginzburg, V. L. (1945). *Zh. Eksp. Teor. Fiz.* **15**, 739.
Ginzburg, V. L. (1949). *Usp. Fiz. Nauk (Sov. Phys. Uspekhi)* **38**, 490 (in Russian).
Ginzburg, V. L. (1952). *Zh. Eksp. Teor. Fiz.* **23**, 236.
Ginzburg, V. L. (1960). *Fiz. Tverd. Tela* **2**, 2031. [(1960) *Sov. Phys. Solid State* **21**, 1824.]
Ginzburg, V. L. (1989). *Phys. Scripta T* **27**, 76.
Ginzburg, V. L. and Landau, L. D. (1950). *Zh. Eksp. Teor. Fiz.* **20**, 1064; reprinted in Ref. Landau (1967).
Ginzburg, V. L. and Sobyanin, A. A. (1976). *Usp. Fiz. Nauk* **120**, 153 [(1976) *Sov. Phys. Uspekhi* **19**, 773].
Ginzburg, V. L. and Sobyanin, A. A. (1988). *Usp. Fiz. Nauk* **154**, 545 [(1988) *Sov. Phys. Uspekhi* **31**, 289].
Glansdorf, P. and Prigogine, I. (1971). *Thermodynamic Theory of Structure, Stability and Fluctuations* (Wiley, London).
G0ldenfeld (1992). *Lectures on Phase Transitions and the Renormalization Group* (Addison Wesley, New York).
Goldhirsch, I. (1979). *J. Phys. C* **12**, 5345.
Goldstone, J. (1961). *Nuovo Cimento* **19**, 159.
Gor'kov, L. P. and Kopnin, N. B. (1988). *Usp. Fiz. Nauk* **156**, 117. [(1988) *Sov. Phys. Uspekhi* **31**, 289.]
Gorodetskii, E. E. and Podneks, V. E. (1989). *Phys. Lett. A* **136**, 233.
Goshen, S., Mukamel D. and Shtrikman, S. (1974). *Int. J. Magnetism.* **6**, 221.
Grewe, N. and Schuh, B. (1980). *Phys. Rev. B* **22**, 3183.
Green, M. B., Sneddon, L., and Stinchcombe, R. B. (1979). *J. Phys. A* **12**, L189.
Griffiths, R. B. (1964). *J. Math. Phys.* **5**, 1215.
Griffiths, R. B. (1965a). *J. Math. Phys.* **6**, 1447.
Griffiths, R. B. (1965b). *J. Chem. Phys.* **43**, 1958.
Griffiths, R. B.(1965c). *Phys. Rev. Lett.* **14**, 623.
Griffiths, R. B. (1973). *Phys. Rev. B* **1**, 545.
Griffiths, R. B. (1974). *Physica A* **73**, 174.
Griffiths, R. B. and Wheeler, J. C. (1970). *Phys. Rev.* **A2**, 1047.
Grinstein, G. (1974). Ph. D. Thesis, Harvard University, USA.
Grinstein, G. (1976). *Phys. Rev. Lett.* **37**, 944.
Grinstein, G. (1985). in E. G. D. Cohen (ed.) *Fundamental Problems in Statistical Mechanics VI* (North–Holland, Amsterdam) p. 147.
Grinstein, G. and Luther, A. (1976). *Phys. Rev. B* **13**, 1329.

Grinstein, G., Ma, S. K. and Mazenko, G. (1977). *Phys. Rev. B* **15**, 258.
Gufan, Yu. M. (1982). *Structural Phase Transitions* (Nauka, Moscow), in Russian.
Gufan, Yu. M. and Larin, E. S. (1978). *Sov. Phys. Doklady* **23**, 754.
Gufan, Yu. M. and Larin, E. S. (1979). *Izvestia Acad. Nauk USSR* **43**, 1567.
Guggenheim, E. A. (1939). *Proc. Roy Soc. A* (London) **169**, 134.
Guggenheim, E. A. (1945). *J. Chem. Phys.* **13**, 253.
Guggenheim, E. A. (1967). *Thermodynamics* (North–Holland, Amsterdam).
Guidry, M. (2004). *Gauge Theories* (Wiley-VCH Verlag, Weinheim).
Gunton, J. D. and Buckingham, M. J. (1968). *Phys. Rev.* **166**, 152.
Guralnik, G. S., Hagen, C. R. and Kibble, T. W. B. (1964). *Phys. Rev. Lett.* **13**, 585.
Haber, H. E. and Weldon, H. A. (1981). *Phys. Rev. Lett.* **46**, 1497.
Hagen, M., Cowley, R. A., Satija, S. K., Yoshizawa, H., Shirane, G., Birgeneau, R. J. and Guggenheim, H.J. (1983). *Phys. Rev. B* **28**, 2602.
Halperin, B. I. and Hohenberg, P. C. (1967). *Phys. Rev. Lett.* **19**, 700.
Halperin, B. I. and Hohenberg, P. C. (1969). *Phys. Rev.* **177**, 952.
Halperin, B. I. and Lubensky, T. C. (1974). *Solid State Commun.* **14**, 994.
Halperin, B. I., Lubensky, T. C. and Ma, S. K. (1974). *Phys. Rev. Lett.* **32**, 292.
Hankey, A., Chang, T. S. and Stanley, H. E. (1973). *Phys. Rev. B* **8**, 1178.
Harada, A., Kawasaki, S., Mukuda, H., Kitaoka, Y., Haga, Y., Yamamoto, E., Onuki, Y., Itoh, K. M., Haller, E. E. and Harima, H. (2007). *Phys. Rev. B* **75**, 14502 (R).
Hardy, G. H. (1948). *Divergent Series* (Oxford Univ Press, Oxford).
Harris, A. B. (1974). *J. Phys. C* **7**, 1671.
Hauge, E. H. (1969). *Physica Norvegica* **4**, 19.
Heller, P. C. (1967). *Rep. Prog. Phys.* **30**, 731.
Herbut, I. F. (1997). *J. Phys. A: Math. Gen.* **30**, 423.
Herbut, I. F. (1998). *Phys. Rev. B* **57**, 13739.
Herbut, I. F. (2007). *A Modern Approach to Critical Phenomena* (Cambridge University Press, Cambridge).
Herbut, I. F. and Tesanović, Z. (1996). *Phys. Rev. Lett.* **76**, 4588.
Hertz, J. A. (1976). *Phys. Rev. B* **14**, 1165.
Hertz, J. A. (1978). *Phys. Rev. B* **18**, 4875.
Hertz, J. A. (1985). *Phys. Scripta T* **10**, 1.
Higgs, P. W. (1964). *Phys. Lett.* **12**, 132; *Phys. Rev. Lett.* **13**, 508.
Higgs, P. W. (1966). *Phys. Rev.* **145**, 1156.
Hikami, S. (1966). *Progr. Theor. Phys.* **62**, 226.
Hill, T. L. (1956). *Statistical Mechanics* (McGraw–Hill, New York).
Hohenberg, P. C. (1967). *Phys. Rev.* **158**, 383.
Hohenberg, P. C. (1982). *Physica B* **109&110**, 1436.
Hohenberg, P. C. and Halperin, B. I. (1977). *Rev. Mod. Phys.* **49**, 435.
Holz, A. and Medeiros, J. T. N. (1975). *J. Phys. A* **8**, 1115.
Hornreich, R. M., Luban, M. and Shtrikman, S. (1975). *Phys. Rev. Lett.* **35**, 1678.
Hornreich, R. M., Luban, M. and Shtrikman, S. (1977). *Physica A* **86**, 465.
Hornreich, R. M., Luban, M. and Shtrikman, S. (1978). *J. Magn. Magn. Materials* **7**, 121.

Hornreich, R. M. and Schuster, H. G. (1979). *Phys. Lett. A* **70**, 143.
Huang, K. (1963). *Statistical Mechanics* (Wiley, New York); (1987) a next edition (Wiley, New York).
Huang, C. C. and Lien, S. C. (1981). *Phys. Rev. Lett.* **47**, 1917.
Hubbard, J. (1958). *Phys. Rev. Lett.* **3**, 77.
Hubbard, J. (1979). *Phys. Rev. B* **20**, 4784.
Huebener, R. P. (2001). *Magnetic Flux Structures in Superconductors* (Springer Verlag, Berlin).
Huxley, A., Sheikin, I. and Braithwaite, D. (2000). *Physica B* **284-288** 1277.
Huxley, A., Sheikin, I., Ressouche, E., Kernavanois, N., Braithwaite, D., Calemczuk, R. and Flouquet J. (2001). *Phys. Rev. B* **63** 144519-1.
Huy, N. T., Gasparini, A., de Nijs, D., Huang, Y. K., Klaasse, J. C. P., Gortenmulder, T., Visser, A., Hamann, A., Görlach, T. and v. Löhneysen, H. (2007). *Phys. Rev. Lett.* **99**, 067006.
Huy, N. T., de Nijs, D., Huang, Y. K., Visser, A. (2008). *Phys. Rev. Lett.* **100**, 077001.
Iacobson, H. H. and Amit, D. J. (1980). *Ann. Phys.* **133**, 57.
Imry, Y. (1975). *J. Phys. C* **8**, 567.
Imry, Y. (1984). *J. Stat. Phys.* **34**, 849.
Imry, Y. and Ma, S. K. (1975). *Phys. Rev. Lett.* **35**, 1399.
Inderhees, S. E., Salamon, M. B., Goldenfeld, N., Rice, J. P., Pazol, B. G., Ginsberg, M. D., Lin, J. Z. and Crabtree, G. W. (1988). *Phys. Rev. Lett.* **60**, 1178.
Iro, H. (1987). *Z. Physik* **68**, 485.
Ishibashi, Y. and Dvorak, V. (1978). *J. Phys. Soc. Jpn.* **44**, 941.
Isihara, A. (1971). *Statistical Physics* (Academic, New York).
Ising, E. (1925). *Z. Physik* **31**, 253.
Ivanchenko, Yu. M. and Lisyanskii, A. A., Yu. A. (1995). *Physics of Critical Fluctuations* (Springer, New York).
Izyumov, Yu. A. and Skryabin, Yu. N. (1987). *Statistical Mechanics of Magnetically Ordered Systems* (Nauka, Moscow), in Russian.
Izyumov, Yu. A. and Syromiatnikov, N. V. (1984). *Phase Transitions and Symmetry of Crystals* (Nauka, Moscow), in Russian.
Josephson, B. D. (1967). *Proc. Phys. Soc.* **92**, 269. 276.
Joyce, G. S. (1966). *Phys. Rev.* **146**, 349.
Joyce, G. S. (1972). in C. Domb and M. S. Green *Phase Transitions and Critical Phenomena* Vol. 2 (Academic, London) p. 375.
Joyce, G. S. and Guttmann, A. J. (1982). in P. R. Graves–Morris *Padé Approximants and Their Applications* (Academic, London).
Joynt, R. (1988). *Supercond. Sci. Technology* **1**, 210.
Joynt, R. Mineev, V. P., Volovik, G. E. and Zhitomirskii, M. E. (1990). *Phys. Rev. B* **42**, 2014.
Jug, G. (1983). *Phys. Rev. B* **27**, 609; 4518.
Jug, G. (1984). in R. Pynn and A. Skjeltorp *Multicritical Phenomena* (Plenum, New York) p. 329.
Kac, M. (1968). in M. Chrétien, E. P. Gross, and S. Deser, *Statistical Physics,*

Phase Transitions and Superfluidity, Vol. 1 (Gordon & Breach, New York) p. 241.
Kac, M. and Thompson, C. J. (1971). *Physica Norvegica* **5**, 163.
Kac, M., Uhlenbeck, G. E. and Hemmer, P. C. (1963). *J. Math. Phys.* **4**, 216.
Kac, M., Uhlenbeck, G. E. and Hemmer, P. C. (1964). *J. Math. Phys.* **5**, 60.
Kadanoff, L. P. (1966). *Physics* **2**, 263.
Kadanoff, L. P. (1975). *Phys. Rev. Lett.* **34**, 1005.
Kadanoff, L. P. (1976). *Ann. Phys.* **100**, 359.
Kaufman, M. and Kanner M. (1990). *Phys. Rev. B* **42**, 2378.
Kaufman, M. and Kardar, M. (1985). *Phys. Rev. B* **31**, 2913.
Kazakov, D. I. and Shirkov, D. V. (1980). *Preprint* P2-80-462 JINR–Dubna (in Russian); English Translation: *Fortch. der Physik* **28**, 465.
Kerszberg, M. and Mukamel, D. (1979). *Phys. Rev. Lett.* **43**, 293.
Ketterson, J. B. and Song, S. N. (1999). *Supersonductivity* (Cambridge University Press, Cambridge).
Khachaturyan, A. G. (1983). *Theory of Structural Transformations in Solids* (Wiley, New York).
Khalatnikov, I. M. (1965). *Introduction to the Theory of Superfluidity* (Benjamin, New York).
Khmel'nitskii, D. E. (1975). *Zh. Eksp. Teor. Fiz.* **68**, 1960. [(1976) *Sov. Phys. JETP* **41**, 981.]
Kibble, T. W. B. (1967). *Phys. Rev.* **155**, 1554.
Kimura, N., Endo, M., Issiki, T., Minagawa, S., Ochiai, A., Aoki, H., Terashima, T., Uji, S., Matsumoto, T. and G. G. Lonzarich, G. G. (2004). *Phys. Rev. Lett.* **92**, 197002.
King, A. R., Ferreira, I. B., Jaccarino, V. and Belanger, D. P. (1988). *Phys. Rev. B* **37**, 219.
Kiometzis, M., Kleinert, H. and Schakel, A. M. J. (1994). *Phys. Rev. Lett.* **73**, 1975.
Kiometzis, M., Kleinert, H. and Schakel, A. M. J. (1995). *Forschr. Phys.* **43**, 697.
Kirkwood, J. G. (1935). *J. Chem. Phys.* **3**, 300.
Kittel, C. (1963). *Quantum Theorry of Solids* (Wiley, London).
Kittel, C. (1971). *Introduction to Solid State Physics* (Wiley, London).
Kleinert, H. (1978). *Fortschr. der Physik* **26**, 565.
Kleinert, H. (1982). *Al Nouvo Cimento* **35**, 405.
Kleinert, H. and Schulte–Frohlinde, V. (2001). *Critical Properties of the ϕ^4-Theories* (WWorld Scientific, Singapore).
Knobler, C. M. and Scott, R. L. (1984). in C. Domb and J. L. Lebowitz (eds.), *Phase Transitions and Critical Phenomena*, Vol. 9 (Academic, London) p. 164.
Kobayashi, K. K. (1970). *Phys. lett. A* **31**, 125; *J. Phys. Soc. Jpn.* **29**, 101.
Kobayashi, T. C., Fukushima, S., Hidaka, H., Kotegawa, H., Akazawa, T., Yamamoto, E., Haga, Y., Settai, R., and Y. Onuki, Y. (2006). *Physica B* **378-361**, 378.
Kociński, J. and Wojtczak. L. (1978). *Critical Scattering Theory* (Polish Sci. Publishers & Elsevier Sci. Publ. Company, Warszawa & Amsterdam).

Kohn, W. and Sherrington, D. (1970). *Rev. Mod. Phys.* **42**, 1.
Kohnstamm, Ph. (1926). in H. Geiger and K. Scheel (eds.), *Handbuch der Physik*, Vol 10 (Springer, Berlin) p. 223.
Kolnberger, S. and Folk, R. (1990). *Phys. Rev. B* **41**, 4083.
Kondor, I. and Szépfalusy, P. (1974). *Phys. Lett. A* **47**, 393.
Kopeć, T. K. and Kozlowski, G. (1983a). *Phys. Lett. A* **95**, 104.
Kopeć, T. K. and Kozlowski, G. (1983a). *J. Phys. F* **13**, L127.
Korutcheva, E. R. and Millev. (1984). *J. Phys. A: Math. Gen.* **17**, L511.
Korutcheva, E. R. and Uzunov, D. I. (1984a). *Phys. Status Solidi (b)* **126**, K19.
Korutcheva, E. R. and Uzunov, D. I. (1984b). *Phys. Lett. A* **106**, 175.
Kosterlitz, J. M., Nelson, D. R., and Fisher, M. E. (1976). *Phys. Rev. B* **13**, 412.
Kotsubo, V, and Williams, G. A. (1986). *Phys. Rev. B* **33**, 6106.
Kreuzer, H. J. (1981). *Nonequilibrium Thermodynamics and its Statistical Foundations* (Clarendon, Oxford).
Kubo, R. (1957). *J. Phys. Soc. Jap.* **12**, 570.
Kubo, R. (1965). *Statistical Mechanics* (North–Holland, Amsterdam).
Lacour–Gayet, P. and Toulouse, G. (1974). *J. Physique* **35**, 426.
Laegreid, T., Tuset, P., Nes, O–M., Slaski, M. and Fossheim, K. (1989). *Physica C* **162–164**, 490.
Landau, L. D. (1937a). *Zh. Eksp. Teor. Fiz.* **7**, 19. [*Phys. Z. Sowjet* **11**, 26.]; reprinted in Ref. Landau (1967);
Landau, L. D. (1937b). *Zh. Eksp. Teor. Fiz.* **11**, 26; reprinted in Ref. Landau (1967).
Landau, L. D. (1941). *J. Phys. USSR* **5**, 71; reprinted in refs. Landau (1967) and Khalatnikov (1965).
Landau, L. D. (1947). *J. Phys. USSR* **11**, 91; reprinted in refs. Landau (1967) and Khalatnikov (1965).
Landau, L. D. (1949). *Phys. Rev.* **75**, 884.
Landau, L. D. (1967). in D. Ter Haar (ed.), *Collected Papers by L D Landau* (Pergamon, New York).
Landau, L. D., Abrikosov, A. A. and Khalatnikov, I. M. (1954). *Dokladi Acad. Nauk (USSR)* **95**, 773; and p. 1177; *ibid* **96**, 261.
Landau, L. D. and Lifshitz, E. M. (1958). *Quantum Mechanics* (Pergamon, London).
Landau, L. D. and Lifshitz, E. M. (1959). *Fluid Mechanics* (Pergamon, London).
Landau, L. D. and Lifshitz, E. M. (1960). *Electrodynamics of Continuous Media* (Pergamon, Oxford).
Landau, L. D. and Lifshitz, E. M. (1980). *Statistical Physics*, Part I (Pergamon, London) (revised edition by L. P. Pitaevskii); see also editions in 1958 and 1959 (Pergamon, Oxford).
Landau, L. D. and Pomeranchuk, I. (1955). *Dokladi Acad. Nauk (USSR)* **102**, 489.
Landsberg, P. T. and Dunning–Davies, J. (1965). *Phys. Rev. A* **138**, 1049.
Lang, J. C. and Widom, B. (1975). *Physica A* **81**, 190.
Lange, R. (1965). *Phys. Rev. Lett.* **14**, 3.
Lange, R. (1966). *Phys. Rev.* **146**, 301.

Lannert, C., Vishveshwara S. and Fisher, M. P. A. (2004). *Phys. Rev. Lett.*. **92**, 097004.
Laptev, V. M. and Skryabin, Yu. N. (1980). *Fiz. Tverd. Tela* **22**, 2949. [(1980) *Sov. Phys. Solid State* **22**, 1722.]
Larkin, A, I,. (1970). *Zh. Eksp. Teor. Fiz.* **58** 1466. [(1970) *Sov. Phys. JETP* **31**, 784.]
Larkin, A. I. and Khmel'nitskii, D. E. (1969). *Zh. Eksp. Teor. Fiz.* **56**, 2087. [(1969) *Sov. Phys. JETP* **56**, 2087.]
Larkin, A .I., Mel'nikov, V. I. and Khmel'nitskii, D. E. (1971). *Zh. Eksp. Teor. Fiz.* **60**, 846. [(1971) *Sov. Phys. JETP* **33**, 458.]
Larkin, A. I. and Ovchinnikov, Yu. N. (1971). *Zh. Eksp. Teor. Fiz.* **61**, 1221. [(1972) *Sov. Phys. JETP* **34**, 651.]
Larkin, A. I. and Pikin, S. A. (1969). *Zh. Eksp. Teor. Fiz.* **56**, 1664. [*Sov. Phys. JETP* **29**, 891.]
Lawrence, W. E. and Doniach, S. (1971). in E. Kanda (ed.), *Proc. 12th Int. Conference on Low Temperature Physics (LT–12)* (Academic, Kyoto) p. 361.
Lawrie, I. D. (1978). *J. Phys. C* **11**, 1123.
Lawrie, I. D. (1982). *Nucl. Phys. B* **200** [FS4], 1; *J. Phys. C* **15**, L879.
Lawrie, I. D. (1983). *J. Phys. C* **16**, 3513; 3527.
Lawrie, I. D. (1997). *Phys. Rev. Lett.* **78**, 979.
Lawrie, I. D. and Athorne, C. (1983). *J. Phys. A: Math. Gen.* **16**, L587.
Lawrie, I. D. and Gedling, F. W. (1982). *J. Phys. A* **15**, 1705.
Lawrie, I. D., Millev Y, T., and Uzunov, D. I. (1987). *J. Phys. A* **20**, 1599; Erratum **20**, 6159.
Lawrie, I. D. and Sarbach S. (1984). in C. Domb and J. L. Lebowitz (eds.), *Phase Transitions and Critical Phenomena*. Vol. 9 (Academic, London) p. 1.
Lebowitz, J. L. (1974). *Physica A* **73**, 48.
Lebowitz, J. L. and Penrose, O. (1966). *J. Math. Phys.* **7**, 98.
Leggett, A. J. (1975). *Rev. Mod. Phys.* **47**, 331.
Leggett, A. J. (2006). *Quantum Fluids: Bose Condensation and Cooper Pairing in Cinensed matter Systems* (Oxford University Press, Oxford).
Le Guillon, J. C. and Zinn–Justin, J. (1980). *Phys. Rev. B* 21, 3976.
Leibbrandt, G. (1975). *Rev. Mod. Phys.* **47**, 849.
Leibler, S. and Orland, H. (1981). *Ann. Phys.* **132**, 277.
Lelidis, I. and Durand, G. (1994). *Phys. Rev. Lett.* **73**, 672.
Leonard A. (1968). *Phys. Rev.* **175**, 221.
Levanyuk, A. P. (1959). *Zh. Eksp. Teor. Fiz.* **36**, 810.
Levelt–Sengers J. M. H. (1974). *Physica A* **73**, 73.
Lewis, A. L. (1979). *Phys. Rev. Lett.* **42**, 907.
Li M. S. (1983). *Zh. Eksp. Teor. Fiz.* **84**, 985.
Lifshitz, E. M. (1941). *Zh. Eksp. Teor. Fiz.* **11**, 255; 269. [English Translation: (1942) *J. of Phys. (USSR)* **6**, 61, 251.]
Lifshitz, E. M. (1944). *Zh. Eksp. Teor. Fiz.* **14**, 353.
Lifshitz, E. M. and Pitaevskii, L. P. (1979). *Physical Kinetics* (Nauka, Moscow)(in Russian) [Vol. X of the Landau–Lifshitz Course in Theoretical Physics.]

Lifshitz, E. M. and Pitaevskii, L. P. (1980). *Statistical Physics* Part II (Pergamon, London) [Vol IX of the Landau–Lifshitz Course in Theoretical Physics.]

Linde, A. D. (1979). *Rep. Progr. Phys.* **42**, 389.

Linder, J. and Sodbø, A. (2007). *Phys. Rev. B* **76**, 054511.

Linder, J., Sperstad, I. B., Nevidomskyy, A. H., Cuoco, M. and Sodbø, A. (2008). *Phys. Rev. B* **77**, 184511.

Lipa, J. A. and Chui, J. C. P. (1983). *Phys. Rev. Lett.* **51**, 2291.

Lipatov, L. N. (1977). *Zh. Eksp. Teor. Fiz.* **72**, 411.

Litster, J. D., Birgeneau, R. J., Kaplan, M. and Safinia, C. R. (1980). in T. Riste (ed.) *Order in Strongly Fluctuating Condensed Matter Systems* (Plenum Press, New York).

Liu, K. S. and Fisher, M. E. (1973). *J. Low Temp. Phys.* **10**, 655.

Lobb, C. J. (1987). *Phys. Rev. B* **36**, 3930.

London, F. (1950). *Superfluids (Vol. I): Macroscopic Theory of Superconductivity* (Wiley, New York).

London, F. (1954). *Superfluids (Vol. II): Macroscopic Theory of Superfluid Helium* (Dover, New York).

Lovesey, S. W. (1980). *Z. Physik B – Cond. Matter* **40**, 117.

Luban, M. (1976). in C. Domb and M. S. Green (eds.), *Phase Transitions and Critical Phenomena*, Vol. 5a(Academic, London) p. 35.

Lubensky, T. C. (1975). *Phys. Rev. B* **11**, 3573.

Lubensky, T. C. (1979). in R. Ballian, R. Maynard and G. Toulouse (eds.), *III–Condensed Matter*[Proceedings of "Les Houches" School, Session XXXI] (North–Holland, Amsterdam) p. 407.

Lubensky, T. C. and Chen, J.–H. (1978). *Phys. Rev. B* **17**, 366.

Lukierska–Walasek, K. (1983). *Phys. Lett. A* **95**, 377.

Lukierska–Walasek, K. and Walasek, K. (1981). *Phys. Lett. A* **81**, 527.

Lyuksyutov, I. F., Pokrovskii, V. L. and Khmel'nitskii, D. E. (1975). *Zh. Eksp. Teor. Fiz.* **69**, 1817. [*Sov. Phys. JETP* **42** 923.]

Ma, S. K. (1972). *Phys. Rev. Lett.* **29**, 1311.

Ma, S. K. (1973a). *Phys. Rev. A* **7**, 2172.

Ma, S. K. (1973b). *Rev. Mod. Phys.* **45**, 589.

Ma, S. K. (1976a). *Modern Theory of Critical Phenomena* (Benjamin, London).

Ma, S. K. (1976b). in C. Domb and M. S. Green (eds.), *Phase Transitions and Critical Phenomena*, Vol. 6 (Academic, London) p. 249;

Ma, S. K. (1985). *Statistical Mechanics* (World Scientific, Singapore).

Ma, M., Halperin, B. I. and Lee, P. A. (1986). *Phys. Rev. B* **34**, 3136.

Machida, K. and Ohmi, T. (2001). *Phys. Rev. Lett.* **86**, 850.

Madelung, O. (ed.) (1990). *Numerical Data and Functional Relationships in Science and Technology*, New Series, 21, *Superconductors* (Springer, Berlin).

Maple, M. B. and Fisher, F. (eds.) (1982). *Superconductivity in Ternary Compounds*, Parts I and II (Springer Verlag, Berlin).

Mattis, D. C. (1965). *The Theory of Magnetism* (Harper Row, New York).

Mattis, D. C. (1985). *The Theory of Magnetism II* (Springer, Berlin).

Mayer, I. O. and Sokolov, A. I. (1985). *J. Appl. Phys.* **24**, (Suppl. 24-2) 185.

Mayer, I. O. and Sokolov, A. I. (1988). *Ferroelectrics Lett.* **9**, 95.

McCoy, M. and Wu, T. T. (1968). *Phys. Rev. Lett.* **21**, 549; *Phys. Rev.* **176**, 631.
McCoy, M. and Wu, T. T. (1969). *Phys. Rev.* **188**, 982; 1014.
McMillan, W. L. (1971). *Phys. Rev. A* **4**, 1238; (1972), ibid., **6**, 936.
Mermin, N. D. (1968). *Phys. Rev.* **176**, 250.
Mermin, N. D. and Wagner, H. (1966). *Phys. Rev. Lett.* **17**, 1133.
Michelson, A. (1977). *Phys. Rev. B* **16**, 577; 585.
Micnas, R. (1979). *Physica A* **98**, 403.
Migdal, A. A. (1968). *Zh. Eksp. Teor. Fiz.* **55**, 1964. [*Sov. Phys. JETP* **28**, 1036.]
Migdal, A. A. (1975). *Zh. Eksp. Teor. Fiz.* **69**, 810; 1457. [(1976) *Sov. Phys. JETP* **42**, 413; 743.]
Mikitik, G. P. (1992). *Zh. Eksp. Teor. Fiz.* **101**, 1042. [(1976) *Sov. Phys. JETP* **74**, 558.]
Millev, Y. T. and Uzunov, D. I. (1982). *Communication* E17–82–22 (JINR, Dubna).
Millev, Y. T. and Uzunov, D. I. (1983). *J. Phys. C* **16**, 4107.
Millev, Y. T. and Uzunov, D. I. (1990). *Phys. Lett. A* **145**, 287.
Mitag, L. and Stephen, M. J. (1974). *J. Phys. A* **7**, L109.
Mineev, V. P. and Samokhin, K. V. (1999). *Introduction to Unconventional Superconductivity* (Gordon and Breach, Amsterdam).
Moldover, M. R., Sengers, J. V., Gamman, R. W. and Hocken, R. J. (1979). *Rev. Mod. Phys.* **51**, 79.
Moncton, D. E., Di Salvo, F. J., Axe, J. D., Sham, L. J. and Patton, B. R. (1976). *Phys. Rev. B* **14**, 3432.
Morf, R., Schneider, T. and Stoll, E. (1977). *Phys. Rev. B*, **16**, 462.
Moriya, T. (1979). *J. Magn. Magn. Materials* **14**, 1.
Moser, G. and Folk, R. (1986). *Solid State Commun* **57**, 707.
Moser, G. and Folk, R. (1990). *Physica B* **165 – 166**, 559.
Moser, G. and Folk, R. (1991). *Phys. Rev. B* **44**, 819.
Mróz, B., Kiefte, H., Clouter, M. J. and Tuszynski, A. (1971). *Phys. Rev. B* **43**, 641.
Mukamel, D. and Grinstein, G. (1982). *Phys. Rev. B* **B25**, 381.
Mukamel, D. and Krinski, S. (1976). *Phys. Rev. B* **13**, 5065.
Münster, A. (1969). *Statistical Thermodynamics* (Springer, Berlin).
Münster, A. (1970). *Thermodynamics* (Wiley, London).
Nattermann, T. (1976). *J. Phys. C* **9**, 3337.
Nattermann, T. and Trimpler, S. (1975). *J. Phys. A* **8**, 2000.
Néel, L. (1932). *Ann. de Physique* **17**, 64.
Nelson, D. R. Kosterlitz, J. M. and Fisher M. E. (1974). *Phys. Rev. Lett.* **33**, 817.
Nelson, D. R. and Pelcovitz, R. A. (1977). *Phys. Rev. B* **16**, 2191.
Nelson, D. R. and Rudnick, J. (1975). *Phys. Rev. Lett.* **35**, 178.
Newman, K. E. and Riedel, E. K. (1984). *Phys. Rev. B* **30**, 6615.
Nickel, B. (1981). *Physica A* **106**, 48.
Nicoll, J. F., Tuthill, G. F., Chang, T. S. and Stanley, H, E. (1977). *Physica B* **86–88**, 618.
Niemeijer, Th. and van Leeuwen, J. M. J. (1976). in C. Domb and M. S. Green (eds.), *Phase Transitions and Critical Phenomena*, Vol. 6 (Academic, London) p. 425.

Nieto, M. M. (1970). *J. Math. Phys.* **11**, 1346.
Olinto, A. C. (1985). *Phys. Rev. B* **31**, 4279.
Olinto, A. C. (1986). *Phys. Rev. B* **33**, 1849.
Olsson, P. and Teitel, S. (1998). *Phys. Rev. Lett.* **80**, 1964.
Onsager, L. (1944). *Phys. Rev.* **65**, 117.
Oppermann, R. and Thomas, H. (1975). *Z. Physik B* **22**, 387.
Ornstein,, L. S. and Zernicke, F. (1914). *Proc. Acad. Sci Amsterdam* **17**, 793.
Osheroff, D. D., Richardson, R. C. and Lee, D. M. (1972). *Phys. Rev. Lett.* **28**, 885.
Palombo, M., Muzikar, P. and Sauls, J. A. (1990). *Phys. Rev. B* **42**, 2681.
Pathria, R. K. (1972). *Statistical Mechanics* (Pergamon, Oxford).
Parisi, G. (1980). *J. Stat. Phys.* **23**, 49.
Parisi, G. (1984). in J.–B Zuber and R. Stora (eds.) *Recent Advances in Field Theory and Statistical Mechanics* [Les Houches School, Session XXXIX] (Elsevier Amsterdam) p. 475.
Parisi, G. (1988). *Statistical Field Theory* (Addison–Westlley, Redwood City).
Patashinskii, A. Z. and Pokrovskii, V. L. (1964). *Zh. Eksp. Teor. Fiz.* **46**, 994 [*Sov. Phys. JETP* **19**, 677].
Patashinskii, A. Z. and Pokrovskii, V. L. (1966). *Zh. Eksp. Teor. Fiz.* **50**, 439 [*Sov. Phys. JETP* **23**, 292].
Patashinskii, A. Z. and Pokrovskii, V. L. (1977). *Usp. Fiz. Nauk* **121**, 55 [*Sov. Phys.-Uspkhi* **20**, 31].
Patashinskii, A. Z. and Pokrovskii, V. L. (1979). *Fluctuation Theory of Phase Transitions* (Pergamon, Oxford).
Pearce, P. A. and Thompson, C. J. (1977). *J. Stat. Phys.* **17**, 189.
Peierls, R. E. (1934). *Helv. Phys. Acta* **7**, Suppl. 2, 81.
Peierls, R. E. (1935). *Ann. Inst. Henri Poincaré* **5**, 177.
Peierls, R. E. (1936). *Proc. Camb. Philos. Soc.* **32**, 471.
Peierls, R. E. (1936). *Surprises in Physics* (Princeton Univ. Press, Princeton).
Pelcovits, R. A. and Nelson, D. R. (1976). *Phys. Lett. A* **57**, 23.
Palissetto, A. and Vicari, E. (2002). *Phys. Rep. C* **368**, 549.
Penrose, O. and Lebowtz, J. L. (1971). *J. Stat. Phys.* **3**, 211.
Percus, J. K. and Yévick, G. J. (1958). *Phys. Rev.* **110**, 1.
Peshkin, M. E. (1978). *Ann. Phys.*(N.Y.) **113**, 122.
Pethick, J. and Smith, H. (2000). *Bose-Einstein Condensation in Dilute Gases* (Cambridge University Press, Cambridge).
Pfeuty, P. and Toulouse, G. (1975). *Introduction to the Renormalization Group and Critical Phenomena* (Wiley, Chichester).
Pfleiderer, C. and Huxley, A. D. (2002). *Phys. Rev. Lett.* **89**, 147005.
Pfleiderer, C., Uhlatz, M., Hayden, S. M., Vollmer, R., v. Löhneysen, H., Berhoeft, N. R. and Lonzarich G. G. (2001). *Nature* **412**, 58.
Pines, D. (1990). *Physica B* **163**, 78.
Pippard A. B. (1957). *The Elements of Classical Thermodynamics* (Cambridge Univ. Press, Cambridge).
Pitaevskii, L. P. (1959). *Zh. Eksp. Teor. Fiz.* **37**, 1784 [(1960) *Sov. Phys. JETP* **10**, 1267].

Pitaevskii, L. P. ans Stringari, S. (2003). *Bose Einstein Condensation* (Oxford University Press, Oxford).
Plumer, M. L., Caillé, A. and Hood. K. (1989). *Phys. Rev. B* **39**, 4489.
Poole, C. K., Farash, H. A. and Creswick, R. J. (2000). *Handbook of Superconductivity* (Academic Press, New York).
Polyakov, A. M. (1968). *Zh. Eksp. Teor. Fiz.* **55**, 1026 [*Sov. Phys. JETP* **28**, 533].
Polyakov, A. M. (1975). *Phys. Lett. B* **59**, 79.
Pomeranchuk, I. (1950). *Zh. Eksp. Teor. Fiz.* **20**, 1919.
Popov, V. N. (1976). *Functional Integrals in Quantum Field Theory and Statistical Physics* (Atomisdat, Moscow) (in Russian); English Translation in 1983 (Reidel).
Popov, V. N. (1987). *Functional Integrals and Collective Excitations* (Cambridge Univ. Press, Cambridge).
Prigogine, I. and Defay, R. (1954). *Chemical Thermodynamics* (Longmans Green, London).
Potts, R. B. (1952). *Proc. Camb. Philos. Soc.* **48**, 106.
Pryadko, L. and Zhang, S. C. (1994). *Phys. Rev. Lett.* **73**, 3282.
Pynn, R. and Skjeltorp, A. (eds.). (1984). *Multicritical Phenomena* (Pleuum, New York).
Quader, K, F. and Abrahams, E. (1988). *Phys. Rev. B* **38**, 11977.
Radzihovsky, L. (1995). *Phys. Rev. Lett.* **74**, 4722.
Radzihovsky, L. (1996). *Phys. Rev. Lett.* **76**, 4451.
Rahola, J. C. (2001). *J. Phys. Studies* **5**, 304.
Ravndal, F. (1976). *Scaling and Renormalization Groups* (The Niels Bohr Institute and NORDITA, Copenhagen).
Reichl, L. E. (1980). *A Modern Course in Statistical Physics* (Edward Arnold Publ., Texas).
Rickayzen, G. (1980). *Green's Functions and Condensed Matter* (Academic, London).
Riedel, E. K., Meyer, H. and Behringer, R. P. (1976). *J. Low Temp. Phys.* **22**, 369.
Riedel, E. K. and Wegner F. J. (1972). *Phys. Rev. Lett.* **29**, 349.
Ritala, R. K. (1984). *J. Phys. C* **17**, 449; see also Ref. Pynn R. and Skjeltorp A. (eds.) (1984) p. 405.
Robinson, J. E. (1951). *Phys. Rev.* **83**, 678.
Roulet, B,, Gavoret, J. and Nozières, P. (1969). *Phys. Rev.* **178**, 1072.
Rowlinson, J. S. (1959). *Liquids and Liquid Mixtures* (Butterworth, London).
Rubinowicz, A. (1968). *Quantum Mechanics* (Elsevier Publishers & Polish Sci. Publishers, Amsterdam & Warszawa).
Rudnick, J. (1975). *Phys. Rev. B* **16**, 363.
Rudnick, J. (1978). *Phys. Rev. B* **18**, 1406.
Rudnick, J. (1976). *Phys. Rev. B* **13**, 2208.
Ruelle, D. (1963a). *Helv. Phys. Acta* **36**, 183, 789.
Ruelle, D. (1963b). *Ann. Phys* **25**, 109.
Ruelle, D. (1969). *Statistical Mechanics: Rigorous Results* (Benjamin, New York).

Ruggiero, P. and Zannetti, M. (1981). *Phys. Rev. Lett.* **47**, 1231.
Ruggiero, P. and Zannetti, M. (1983). *Phys. Rev.* **27**, 3001.
Runnels, L. K. (1972). in C. Domb and M. S. Green (eds.), *Phase Transitions and Critical Phenomena*, Vol. 2 (Academic, London) p. 305.
Rushbrooke, G. S. (1938). *Proc. Roy. Soc. A (London)* **166**, 296.
Rushbrooke, G. S. (1963). *J. Chem. Phys.* **39**, 842.
Saam, W. F. (1970). *Phys. Rev. A* **2**, 1401.
Sachdev, S. (1999). *Quantum Critical Phenomena* (Canbridge University Press, Cambridge).
Saint–James, D. Sarma, G. and Thomas, E. J. (1969). *Type II Superconductivity* (Pergamon, London).
Sak, J. (1973). *Phys. Rev.* **8**, 281.
Sak, J. (1977). *Phys. Rev. B* **15**, 4344.
Salamon, M. B. and Shang, H.–T. (1980). *Phys. Rev. Lett.* **44**, 879.
Salsburg, Z. W. (1971). in R. Balescu *et al.* (eds.), *Lectures in Statistical Physics* (Springer, Berlin) p. 20.
Sarbach, S. and Schneider, T. (1975). *Z. Physik B* **20**, 399.
Saxena, S. S., Agarwal, P., Ahilan, K., Grosche, F. M., Haselwimmer, R. K. W., Steiner, M. J., Pugh, E., Walker, I. R., Julian, S. R., Monthoux, P., Lonzarich, G. G., Huxley, A., Sheikin, I., Braithwaite, D. and Flouquet, J. (2000). *Nature* **406**, 587.
Schakel, A. M. J. (1995). *Nucl. Phys. B* **453**[FS], 705.
Schakel, A. M. J. (1997). *Phys. Lett. A* **224**, 287.
Schakel, A. M. J. (2008). *Boulevard of Broken Symmetries: Effective Field Theories of Condensed Matter* (World Scientific, Singapore).
Schloms, R. and Dohm, V. (1989). *Nucl. Phys. B* **44**, 189.
Schuh, B. and Grewe, N. (1982). *Z. Phys B* **46**, 149.
Schwartz, M. (1985). *Phys. Lett. A* **107**, 199.
Schwartz, M. and Soffer, A. (1985). *Phys. Rev. Lett.* **55**, 2499.
Scott, J. F. (1974). *Rev. Mod. Phys.* **46**, 83.
Shalaev, B. N. (1977). *Zh. Eksp. Teor. Fiz.* **73**, 2301 [*Sov. Phys. JETP* **46**, 1204].
Shapir, Y. (1985). *Phys. Rev. Lett.* **54**, 154.
Shirane, G. (1974). *Rev. Mod. Phys.* **46**, 437.
Shopova, D. V. and Todorov, T. P. (2003a). *J. Phys. Condensed Matter* **15**, 5783.
Shopova, D. V. and Todorov, T. P. (2003b). *Phys. Lett. A* **314**, 250.
Shopova, D. V. and Todorov, T. P. (2003c). *J. Phys. Studies* **7**, 330.
Shopova, D. V., Todorov, T. P., Tsvetkov, T. E. and Uzunov, D. I. (2002). *Mod. Phys. Lett. B* **16**, 829.
Shopova, D. V., Tsvetkov, T. E. and Uzunov, D. I. (2006). *J. Phys. Studies* **10**, 330.
Shopova, D. V., Todorov, T. P. and Uzunov, D. I. (2003). *Mod. Phys. Lett. B* **17**, 141.
Shopova, D. V. and Uzunov, D. I. (2003). *Phys. Rep. C* **379**, 1.
Shopova, D. V. and Uzunov, D. I. (2003). *Phys. Lett. A* **313**, 139.
Shopova, D. V. and Uzunov, D. I. (2005a). *Phys. Rev. B*, **72**, 024531.

Shopova, D. V. and Uzunov, D. I. (2005b). *Bulg. J. Phys.*, **32**, 81.
Shopova, D. V. and Uzunov, D. I. (2006). in Murrey, V. N. (ed.), *Progress in Ferromagnetism Research* (Nova Science Publishers, New York), pp. 223.
Shopova, D. V. and Uzunov, D. I. (2008). in Oliver Chang (ed.), *Progress in Superconductivity Research* (Nova Science Publishers, New York), pp. 13-53.
Shopova, D. V. and Uzunov, D. I. (2009). *Phys. Rev. B*, **79**, 064501.
Shuster, G. V. (1989). *Pis'ma Zh. Eksp. Teor. Fiz.* **50**, 93 [*JETP Letts.* **50**, 107].
Sigrist, M. and Ueda, K. (1991). *Rev. Mod. Phys.* **63**, 239.
Singh, K. K. (1975). *Phys. Lett. A* **51**, 27; *Phys. Rev. B* **12**, 2819.
Singh, K. K. (1976). *Phys. Lett. A* **57**, 309.
Singh, K. K. (1977). *Prog. Theor. Phys.* **58**, 1045.
Singh, K. K. (1978). *Phys. Rev. B* **13**, 3192.
Singh, S. and Pandita, P. N. (1982). *Phys. Lett. A* **92**, 65.
Singsaas, A. and Ahlers, G. (1984). *Phys. Rev. B* **30**, 5103.
Skocpol, W. J. and Tinkham, M. (1975). *Rep. Prog. Phys.* **38**, 1049.
Slonczewski, J. C. and Thomas, H. (1970). *Phys. Rev. B* **1**, 3599.
Smart, J. S. (1966). *Effective Field Theories in Magnetism* (Saunders, Philadelphia).
Sokolov, A. I. (1987). *Fiz. Tverd. Tela* **29**, 2787.
Stanley, H. E. (1968). *Phys. Rev.* **176**, 718.
Stanley, H. E. (1969). *J. Phys. Soc. Japan* **265**, 102.
Stanley, H. E. (1971). *Introduction to Phase Transitions and Critical Phenomena* (Clarendon, Oxford).
Stanley, H. E. (1973). (ed.), *Cooperative Phenomena near Phase Transitions, a bibliography with selected readings* (MIT, Cambridge,Massachusetts).
Stanley H. E., Chang, T. S., Habrus, R. and Li, L. L. (1976). in K A. M.üller and A. Rigamonti (eds.), *Local Properties of Phase Transitions* [Proceedings of the Int. School of Physics "E. Fermi", Course LIX] (North–Holland, Amsterdam).
Stella, A. L. and Toigo, F. (1976). *Nuovo Cimento B* **35**, 207.
Stephen, M. J. (1980). *J. Phys. C* **13**, L83.
Stephen, M. J., Abrahams, E. and Straley, J. P. (1975). *Phys. Rev. B* **12** 256.
Stephen, M. J. and McCauley, J. L. Jr. (1973). *Phys. Lett. A* **44**, 89.
Stinchcombe, R. B. (1973). *J. Phys. C* **6**, 2459.
Stolan, B,, Pytte, E. and Grinstein, G. (1984). *Phys. Rev. B* **30**, 1506.
Stratonovich, R. L. (1957). *Dokladi Acad. Nauk. (USSR)* **115**, 1097 [(1958) *Sov. Phys. Doklady* **2**, 416].
Stueckelberg, E. C. G. and Petermann, A. (1953). *Helv. Phys. Acta* **26**, 499.
Su, G. and Chen, J. (2010). *Eur. J. Phys.* **31**, 143.
Suzuki, M. (1968). *J. Math. Phys.* **9**, 2064.
Suzuki, M. (1972). *Phys. Lett. A* **42**, 5.
Suzuki, M. (1975). *Prog. Theor. Phys.* **53**, 97.
Suzuki, M. (1976). *Prog. Theor. Phys.* **56**, 1007, 1454.
Suzuki, M. (1986). *J. Phys. Soc. Jpn.* **55**, 4205.
Suzuki, M. (1988). *J. Phys. Soc. Jpn.* **57**, 2310.

Suzuki, M. (1991). in M. Suzuki and R. Kubo *Evolutionary Trends in Physical Sciences* [Springer Proceedings in Physics] Vol. 57 (Springer, Berlin) p. 141.
Suzuki, M. and Igarashi, G. (1974). *Phys. Lett. A* **47**, 361.
Symanzik, K. (1970). *Comun. Math. Phys* **18**, 227.
Szepfalusy, P. and Tél, T. (1980). *Z. Phys. B* **36**, 343.
Sznajd, J. (1984). *J. of Magn. and Magn. Materials* **42**, 269.
Tarski, J. (1968). in A. O. Barut and W. E. Brittin (eds.), *Quantum Theory and Statistical Physics* [Lectures in Theoretical Physics, Vol. X-A] (Gordon & Breach, New York) p. 433.
Tateiwa, N., Kobayashi, T. C., Hanazono,K., Amaya, A., Haga, Y., Settai, R. and Onuki Y. (2001). *J. Phys. Condensed Matter*, **13**, L17.
Temperley, H. N. V. (1972). in C. Domb and M. S. Green (eds.), *Phase Transitions and Critical Phenomena* (Academic, London) p. 227.
Temperley, H. N. V. and Travena, D. H. (1978). *Liquids and Their Properties* (Wiley, Chichester).
Taylor, A. F. and Coy, D. C. (1980). *Introduction to Functional Analysis* (Wiley, New York).
Tilley, D. R. and Tilley, J. (1974). *Superfluidity and Superconductivity* (Van Nostrand Reinhold, New York).
Tinkham, M. (2004). *Introduction to Superconductivity* (Dover Publications, New York)(Third edition); (1996) Second edition (McGraw–Hill, New York).
Tisza, L. (1951). in R. Smoluchovski, J. E. Mayer and W. A. Weyl (eds.), *Phase Transitions in Solids* (Wiley, New York) p. 1; reprinted in Tisza L. (1966). *Generalized Thermodynamics* (MIT, Cambridge, Massachusetts) p. 194.
Tisza, L. (1961). *Ann. Phys.* **13**, p. 1; reprinted in Tisza, L. (1966). *Generalized Thermodynamics* (MIT, Cambridge, Massachusetts) p. 102.
Tolédano, J. C., Michel, L., Tolédano, P. and Brezin, E. (1985). *Phys. Rev. B* **31**, 7171.
Tolédano, J. C. and Tolédano, P. (1987). *The Landau Theory of Phase Transitions* (World Scientific, Singapore).
Tonchev, N. S. and Uzunov, D. I. (1981). *J. Phys. A* **14**, L103.
Tonchev, N. S. and Uzunov, D. I. (1981). *J. Phys.A* **14**, 521.
Tonchev, N. S. and Uzunov, D. I. (1985). *Physica A* **134**, 265.
Toner, J. (1982). *Phys. Rev. B* **26**, 462.
Toyoda, T. (1982). *Ann. Phys.* **141**, 154.
Tsuneto, T. and Abrahams, E. (1973). *Phys. Rev. Lett.* **30**, 217.
Trikey, S,, Kirk, W. and Adams F. (1972). *Rev. Mod. Phys.* **44**, 668.
Tukuyasu, T. A., Hess, D. W. and Sauls, J. A. (1990). *Phys. Rev. B* **41**, 8891.
Tuszyński, J. A., Clouter, M. J. and Kiefte, H. (1986). *Phys. Rev. B* **33**, 3423.
Tuthill, G. F., Nicoll, J. F. and Stanley, H. E. (1975) *Phys. Rev. B* **11**, 4579.
Ueda, K. and Rice, T. M. (1985). *Phys. Rev. B* **31**, 7114.
Uhlaz, M., Pfleiderer, C. and Hayden, S. M. (2004). *Phys. Rev. Lett.* **93**, 256404.
Uhlenbeck, G. E. and Ford, G. W. (1963). *Lectures in Statistical Mechanics* (American Math. Society, Providence).
Uzunov, D. I. (1980). *Phys. Lett. A* **78**, 395.
Uzunov, D. I. (1981a). *Phys. Lett. A* **87**, 11.

Uzunov, D. I. (1981b). *Phys. Lett. A* **84**, 489.
Uzunov, D. I. (1982). *Communication* E17–82–21 (JINR, Dubna).
Uzunov, D. I. (1983). *Phys. Status Solidi (b)* **120**, K39.
Uzunov, D. I. (1986). *Compt. Rend. Acad. Bulg. Sci.* **39**, 85.
Uzunov, D. I. (1987). in G. Busiello, L. De Cesare, F. Mancini and M. Marinaro (eds.), *Advances on Phase Transitions and Disorder Phenomena* (World Scientific, Singapore) p. 256.
Uzunov, D. I. (1991). in E. R. Caianiello (ed.), *Advances in Theoretical Physics* (World Scientific) p. 96.
Uzunov, D. I. (1996). in D. I. Uzunov (ed.), *Lectures on Cooperative Phenomena in Condensed Matter* (Heron Press, Sofia).
Uzunov, D. I. (2006). *Phys. Rev. B* **74**, 134514.
Uzunov, D. I. (2007). *Eur. Phys. Lett.* **77**, 20008.
Uzunov, D. I. (2008). *Phys. Rev. E* **78**, 041122.
Uzunov, D. I., Korutcheva, E. R. and Millev, Y. T. (1983). *J. Phys. A* **16**, 247; see, also, (1984). *J. Phys. A* **17**, 247.
Uzunov, D. I., Korutcheva, E. R. and Millev, Y. T. (1985a). *Physica A* **129**, 535.
Uzunov, D. I., Korutcheva, E. R. and Millev, Y. T. (1985). *Phys. Status Solidi (b)* **130**, 243.
Uzunov, D. I. and Suzuki M. (1994). *Physica A* **204**, 702.
Uzunov, D. I. and Suzuki M. (1994). *Physica A* **216**, 489.
Uzunov, D. I. and Walasek, K. (1985). *Phys. Lett. A* **107**, 207; Err., ibid, **110**, 482.
Vaks, V. G. and Larkin, A. I. (1965). *Zh. Eksp. Teor. Fiz.* **45**, 975 [(1966) *Sov. Phys. JETP* **22**, 678].
Vaks, V. G., Larkin, A. I. and Pikin, S. A. (1966). *Zh. Eksp. Teor. Fiz.* **51**, 361 [(1967) *Sov. Phys. JETP* **24**, 240].
Van der Waals, J. D. (1873). Ph. D. Thesis, (Leiden University).
van Hove L. (1949). *Physica* **15**, 951.
Vassilev, A. N. (1976). *Functional Methods in Quantum Field Theory and Statistics* (St. Peterburg University, St. Peterburg).
Vause, C. and Sak, J. (1978). *Phys. Rev. B* **18**, 1455.
Vilenkin, A. and Shellard, E. P. S. (1994). *Cosmic Strings and Other Topological Deffects* (Cambridge University Press, Cambridge).
Vladimirov, A. A., Kazakov, D.I. and Tarasov, O. V. (1979). *Zh. Eksp. Teor. Fiz.* **77**, 1035 [*Sov. Phys. JETP* **50**, 521].
Vojta, M. (2003). *Re. Progr. Phys.* **66**, 2069.
Vollhardt, D. and Wölffe, P. (1990). *The Superfluid Phases in Helium 3* (Taylor and Francis, London).
Volovik, G. E. and Gor'kov, L. P. (1985). *Zh. Eksp. Teor. Fiz.* **88**, 1412 [*Sov. Phys. JETP* **61**, 843].
Volterra. (1959). *Theory of Functionals and of Integral and Integrodifferential Equations* (Dover, New York).
Vonsovsky, S. V., Izyumov, Yu. A. and Kurmaev, E. Z. (1982). *Superconductivity of Transition metals* (Springer Verlag, Berlin).

Voronel, A. V. (1976). in C. Domb and M. S. Green (eds.), *Phase Transitions and Critical Phenomena,* Vol. 5b (Academic, New York) p. 343.
Vvedinsky D. D. (1984). *J. Phys. A* **17**, 709.
Wagner, H. (1966). *Z. Physik* **195**, 273.
Walasek K. (1984). *Phys. Lett. A* **101**, 343.
Wallace, D. C. (1972). *Thermodynamics of Crystals* (Wiley, New York).
Walker M. B. and Samokhin K. V. (2002). *Phys. Rev. Lett.* **88**, 207001.
Wallace, D. J. (1976). in C. Domb and M. S. Green (eds.), *Phase Transitions and Critical Phenomena,* Vol. 6 (Academic, London) p. 294.
Wallace, D. J. and Zia, R. K. P. (1978). *Rep. Prog. Phys.* **41**, 1.
Wegner, F. J. (1972). *Phys. Rev. B* **5**, 4529.
Wegner, F. J. (1974). *J. Phys. C* **7**, 2098.
Wegner, F. J. (1976). in C. Domb and M. S. Green (eds.), *Phase Transitions and Critical Phenomena,* Vol. 6 (Academic, London) p. 8.
Wegner, F. J. and Houghton, A. (1973). *Phys. Rev. A* **8**, 401.
Wegner, F. J. and Riedel, E. K. (1973). *Phys. Rev. B* **7**, 248.
Weichman, P. B. and Kim, K. (1989). *Phys. Rev. B* **40**, 813.
Weichman, P. B., Rasolt, M., Fisher, M.E. and Stephen, M. J. (1986). *Phys. Rev. B* **33**, 4632.
Weinrib, A. and Halperin, B. I. (1983). *Phys. Rev. B* **27**, 413.
Weiss, P. J. (1907). *J. Physique* **6**, 661.
Weiss, X. G. and Wu, Y. S. (1993). *Phys. Rev. Lett.* **70**, 1501.
Werthamer, N. R. (1969). in R. D. Parks (ed.), *Superconductivity,* Vol. 1 (Marcel Dekker, New York) p. 321.
Western, A. B., Baker, A. G., Bucon, C. R. and Schmidt, V. H. (1978). *Phys. Rev. B* **17**, 4461.
White, R. M. (1970). *Quantum Theory of Magnetism* (McGraw–Hill, New York); see also the third, completely revised edition (Springer–Verlag, Berlin–Heidelberg, 2008).
Widom, K. G. (1964). *J. Chem. Phys.* **41**, 1633.
Widom, K. G. (1965). *J. Chem. Phys.* **43**, 3892. 3898.
Wiegel, F. W. (1975). *Phys. Reps. C* **16**, 58.
Wiegel, F. W. (1978). in G. J. Papadopoulos and J. T. Devreese (eds.), *Path Integrals* [NATO Series **B34**] (Plenum, New York) p. 419.
Wilks, J. and Betts, D. S. (1987). *An Introduction to Liquid Helium* (Clarendon, Oxford).
Wilson, K. G. (1971). *Phys. Rev. B* **4**, 3174, 3184.
Wilson, K. G. (1972). *Phys. Rev. Lett.* **28**, 548.
Wilson, K. G. (1973). *Rev. Mod. Phys.* **55**, 583.
Wilson, K. G. and Fisher, M. E. (1972). *Phys. Rev. Lett.* **28**, 248.
Wilson, K. G. and Kogut, J. (1974). *Phys. Reps. C* **12**, 75.
Witschel, W. (1980). *J. Chem. Phys.* **50**, 265.
Witschel, W. (1981). *Z. Naturforsch. A* **36**, 481.
Yamada, Y. (1993). *Phys. Rev. B* **47**, 11211.
Yamazaki, H. (1976). *Prog. Theor. Phys.* **56**, 1733.
Yamazaki, Y. (1978). *Can. J. Phys.* **56**, 139.

Yamazaki, Y., Holtz, A., Ochiai, M. and Fukuda, Y. (1986). *Phys. Rev. B* **33**, 3960, 3474.
Yamazaki, Y., Ochiai, M. and Fukuda, Y. (1985). *J. Phys. C* **18**, 4987.
Yang, C. N. and Lee, T. D. (1952). *Phys. Rev.* **87**, 404, 410.
Yelland, E. A., Hayden, S. M., Yates, S. J. C., Pfleiderer, C., Uhlarz, M., Vollmer, R., v Löhneysen, H., Bernhoeft, N. R., Smith, R. P., Saxena, S. S., N. Kimura, N. (2005). *Phys. Rev. B* **72**, 214523.
Yeomans, J. M. (1992). *Statistical Mechanics of Phase Transitions* (Oxford University Press, Oxford).
Yethiraj, A., and Bechhoefer, J. (2000). *Phys. Rev. Lett.* **84**, 3642.
Young, A. P. (1975). *J. Phys. C* **8**, L309.
Young, A. P. (1977). *J. Phys. C* **10**, L257.
Zhang, S. C., Hansson T. H. and Kivelson, S. (1989). *Phys. Rev. Lett.* **63**, 903.
Zannetti, M. (1980). *Phys. Rev. B* **22**, 5267.
Zia, R. K. P. and Wallace, D. J. (1975). *J. Phys. A* **8**, 1495.
Ziman, J. M. (1979). *Models of Disorder* (Cambrodge Univ. Press, Cambridge).
Zimm, B. H. (1951). *J. Chem. Phys.* **19**, 1019.
Zinn-Justin, J. (1971). *Phys. Rep. C* **1**, 55.
Zinn-Justin, J. (1996). *Quantum Field Theory and Critical Phenomena* (Oxford University Press, Oxford).

Index

g-factor, 215
k-space, 615
x-space, 615

Abelian group, 531
action, 493
  least, 36
additivity, 31–33, 65
alloy, 95
analysis
  dimensional, 451
angular momentum, 214, 217
  orbital, 214, 221
  spin, 215
  total, 219, 220, 228
anisotropy, 166, 179, 513
  crystal, 110
  crystal field, 214
  cubic, 179, 513
  extreme, 219
  extreme uniaxial, 111
  hyper-cubic, 179
  quadratic, 513
  spatial, 513
  uniaxial, 214
antiferroelectrics, 116
antiferromagnet, 68, 111
  Ising, 217
  isotropic, 112
  random exchange, 511
  uniaxial, 115
approximation
  classical, 186, 222
  Gaussian, 263, 266, 268, 290, 340
  Hartree, 423, 424
  Hooke, 88, 110
  large–n, 480
  mean field (MF), 192, 230, 232
  mean-field-like, 539, 548
  one-constant, 215, 217
  one-loop, 411, 444
  Ornstein–Zernicke, 12
  saddle point, 192
  semi-classical, 220
  two-loop, 425
asymptote
  infrared, 415
  ultraviolet, 21
asymptotic
  behaviour, 11
  critical exponent, 11
  law, 11
  limit, 9
  power law, 11, 18
availability, 53
average, 187, 316
  connected, 269, 348
  connected Gaussian, 269
  disconnected, 348
  Gaussian, 264, 265, 270, 353
  irreducible, 333
  statistical, 187

binodal line, 71, 72

biophysics, 8
Bohr magneton, 215
Boltzmann constant, 3, 187
Bose gas, 118
  ideal (IBG), 486
  nonideal (NBG), 491
Bose integral, 489, 490
Bose liquid, 118
Bose operator, 492
  annahilation, 491
  creation, 491
boundary condition, 317
  periodic, 279
Brillouin zone, 302, 317

Cartesian component, 617
chemistry, 8
classification
  Brout, 85
  Ehrenfest, 83
  Pippard, 85
coarse-graining, 23, 198, 205
coefficient
  magnetic stiffness, 109
  stiffness, 88, 89
  thermal expansion, 82
coexistence, 523, 579
  curve, 73, 74
  domain, 73
  isotherm, 104
  line, 69, 100
  surface, 96, 99
coherence length, 528
combinatorial factor, 357, 362, 363, 365
commutation relation, 216
compound
  inter-metallic, 579
compressibility
  isothermal, 43, 82, 83
concentration, 94
condensation, 249
  Bose-Einstein, 4, 119, 486
condition
  boundary, 189, 196, 197, 318
  extremum, 54
  material, 37
  mechanical, 37
  thermal, 37
connodal, 72, 75
constant
  fine structure, 215
  Curie, 110
  exchange, 216
  Lamé, 90
constraint, 36, 500
  spherical, 227
continuous transitions, 7
Cooper pairs, 4, 119, 527
correlation, 206
  function, 11, 316
    dynamic, 20
    static, 20
  inter-particle, 1
  length, 20, 323, 462
  length exponent, 462
  long-range, 620
  quantum, 4, 23, 118, 487, 492, 496
  random, 509
  short-range, 620
  spatial, 206
Coulomb gauge, 221
criterion
  Ginzburg, 274, 343, 389, 512
  Harris, 507
  Lifshitz, 301
  of stability, 52
critical
  (borderline) dimension (dimensionality), 22
  exponent, 19
  Lifshitz point, 525
  amplitude, 9
  coexistence, 100
  dimension, 394, 496
  dynamics, 547
  end point, 101, 163
  equilibria, 47, 48
  experiment, 8
  exponent, 9, 11, 23
  fluctuation, 12
  Hamiltonian, 437

hypersurface, 100
instability, 48
line, 2, 99
opalescence, 7, 324
phase, 47
phenomena, 78
point, 2, 8, 78, 79
gas-liquid, 7
region, 8, 343, 345, 390, 537
scattering, 12
stability, 48
state, 2, 19, 47
temperature, 78, 383
critical behaviour, 2
large–n, 422
nontrivial, 447
quantum, 496
zero temperature, 496
critical exponent
dynamical, 496, 503, 547
critical temperature, 389
crossover, 465
classical-to-quantum, 496, 502, 548
dimensional, 323, 559
exponent, 178, 179, 458
phenomena, 174
random field, 511
cumulant, 212, 266
Gaussian, 266, 270
irreducible, 269
Curie constant, 110, 240
Curie law, 240
Curie point, 2
Curie–Weiss law, 109
cutoff, 388, 389, 399

defect, 504
degree of freedom, 3, 35, 186
diagram
disconnected, 350
Feynman, 349
Hartree, 418
irreducible, 361, 380
loop, 367
mass-renormalization, 387
one-particle reducible, 349
reducible, 379
two-loop, 413
two-particle irreducible, 349
two-particle reducible, 349
weight factor, 365
diamagnetic term, 221
dimension
anomaluos, 12
borderline, 323, 452
borderline (critical), 394
spatial, 186, 393
upper borderline, 395
director vector, 531
discontinuity, 3, 10
disorder, 218
quenched, 504
strong, 505
weak, 505
distribution
Gaussian, 505
grand canonical, 193
Guassian, 550
statistical, 186
dynamic
critical behaviour, 13
critical exponent, 13
critical phenomena, 13

Early Universe, 533
econo-physics, 8
effect
disorder, 549
fluctuation, 7, 8
Halperin–Lubensky–Ma, 539, 552
Meissner–Ochsenfeld, 119
paramagnetic, 108, 233
quantum, 4, 23, 548
quantum Hall, 531
strain, 180
Ehrenfest equations, 84
electric polarization
spontaneous, 116
energy
available, 53
internal, 49, 187
mean, 205

ensemble
  canonical, 190, 198
  grand canonical, 193
  isothermal-isobaric, 192
  microcanonical, 190
enthalpy, 53, 205
entropy, 27, 33, 146, 188, 189
  extrema, 39
  function, 27, 34
  inclease, 25, 34
  increase, 53, 64
  magnetic, 75
  maximum, 46
  maximum principle, 47
  munimum, 46
  representation, 50
equation
  Clapeyron–Clausius, 74, 103
  Dyson, 375
  Euler, 29, 52
  Euler–Lagrange, 318, 319
  fundamental, 27, 49
  Landau–Khalatnikov, 13
  of state, 17, 38, 246
    equilibrium, 38
  Van der Waals, 249
equilibrium, 35, 46
  metastable, 46
  neutral, 46
  stable, 46
Euler constant, 604
Euler theorem, 603
expansion
  $1/n$, 424, 480
  asymptotic, 268, 310
  cumulant, 212, 269
  epsilon (ϵ–), 476, 478
  Fourier, 554
  Landau, 166, 180
  loop, 367, 411, 426, 477
  perturbation, 349, 376
  weak coupling, 272

Fermi surface, 119
ferroelectrics, 116
  improper, 117
ferromagnet, 68
  Ising, 217, 222
  uniaxial, 111
ferromagnetism
  itinerant, 579
  weak, 115
Feynman diagram, 349
field, 275
  applied electric, 116
  Bose, 493
  canonical scaling, 457
  critical, 446
  discrete (lattice), 275
  external, 11
    conjugate, 167
  external magnetic, 30, 106
  Higgs, 532, 549
  homogeneous, 190
  lattice, 220
  linear scaling, 446
  longitudinal, 223
  magnetic, 1, 2, 29, 221
  massless scalar, 533
  molecular, 7, 235
    internal, 229
  random, 505, 549
  random color, 549
  relevant scaling, 447
  scaling, 456, 459, 466
    irrelevant, 458
    marginal, 458
    relevant, 458
  staggered, 112
  thermodynamic, 30, 71, 103
  transverse, 223
Fisher exponent, 12
fixed point, 436
  biconical, 524
  bicritical, 524
  complex-cubic, 551
  Gaussian (GFP), 438
  Gaussian-like, 499–501
  Hamiltonian, 437
  Heisenberg (HFP), 438
  nontrivial, 438
  random, 507

flow line, 429
fluctuation, 190, 191, 230, 259
  anomalous, 323
  correlation, 7, 324
  critical, 323, 324
  Gaussian, 8, 262, 264
    uniform, 262
  Gaussian-like, 263
  Goldstone, 343
  interacting, 8, 263, 314
  interaction, 8, 346
  magnetic, 551, 553, 555
  massless, 343
  non-Gaussian, 263, 264
  noncritical, 323
  noninteracting, 22
  pseudo-Gaussian, 263
  quantum, 548
  spatially uniform, 191
  spontaneous, 65
  superconducting, 555
  uniform, 263
free energy, 53
  Helmholtz, 51, 53, 188
function
  Bessel, 274, 322
  Brillouin, 232, 233, 245, 294
  characteristic, 53
  connected Green, 212
  correlation, 206, 207, 209, 332
  crossover, 176
  density distribution, 91
  Dirac, (δ–), 188
  distribution, 187, 193, 257
  gamma, 265, 266
  Gaussian partition, 264
  Gibbs, 51, 52
  Green, 212, 320, 334, 493
    l–point, 212
  Helmholtz, 50
  homogeneous, 17, 28, 603
  irreducible correlation, 208, 209, 213, 270, 316
  irreducible Green, 334
  Langevin, 236
  Massieu, 55
  partition, 187, 258, 259, 314
  random, 505
  reducible correlation, 333
  rescaling, 426
  Riemann, 604
  scaling, 463
  self-energy, 375
  stationary, 275
  total Green, 332
  vertex, 407, 408
functional
  bilinear, 610
  Chern–Simons–Ginzburg–Landau, 531
  Gaussian integral, 612
  generating, 207, 208
  integral, 289, 316
  series, 276
  Taylor series, 290

gauge
  Coulomb, 221, 530, 542
  Landau, 542
gauge group, 530
gauge invariance, 530
  local, 530
gauge symmetry, 531, 532
gauge theory, 533
Gaussian integral, 191, 201, 263, 317
  multiple, 291
Gibbs rule, 60, 61
  second, 100
Ginzburg number, 344
Ginzburg region, 537
global gauge, 530
  invariance, 530
gyromagnetic ratio, 215

Hamiltonian, 186, 187
  (of) Landau–Ginzburg–Wilson, 315
  BEG, 227
  counter-term, 387
  effective, 204, 205, 240, 241, 259, 294, 315, 404, 427
  flow line, 429
  fluctuation, 240, 346

generalized, 493
spin, 215
Higgs boson, 532
Higgs mechanism, 532
homogeneity, 17
hypothesis, 17–20
property, 18, 33

imaginary time, 492
impurity, 504
(long-range) random, 509
extended (line, plane), 509
random, 505, 549
random, δ–type, 509
short-range, 509
inhomogeneity, 506
integral
Gaussian, 266, 605
Gaussian-like, 263, 605
multiple Gaussian, 265
non-Gaussian, 263
partition, 263, 290
interaction
competeing, 112
constant, 314
correction, 347
dipole-dipole, 220
exchange, 215, 240
fluctuation, 263, 324
inter-layer, 536
long-range, 309
magnetic, 220
nearest-neighbour (nn), 218
self–, 218, 229
short-range, 299, 309
spin, 214
invariance
length-scale, 20
translational, 195, 218, 219, 240
isotherm, 75

Jacobian, 295

Kadanoff block picture, 23
Kadanoff transformation, 23
Kronecker symbol, 193, 279

Lagrange multiplier, 59, 60
Landau invariants, 180
latent heat, 74
lattice
d–dimensional (dD), 218
constant, 218, 281
field, 220, 275
hyper-cubic, 218, 318
model, 220
tetragonal
body-centered, 219, 514, 636
limit
asymptotic, 192
classical, 228
continuum, 212, 240, 282, 283, 285, 286, 289, 553
Hargtee, 549
infrared, 391
large–b, 415
large–n, 421
long-wavelength, 300, 553
quantum, 496, 498
semi-classical, 220, 228
small momentum, 300
thermodynamic, 3, 4, 283, 288, 489
zero-temperature, 548
liquid crystal, 121
anisotropy, 531
local gauge, 530
invariance, 530
logarithmic correction, 22

magnetic induction, 119
magnetic moment, 106, 194, 214, 215, 219
effective, 110
magnetic susceptibility, 243
magnetization, 29, 30, 68
alternating, 112
instantaneous, 194
nonequilibrium, 217
per particle, 106
saturated, 233
spontaneous, 2
staggered, 68, 228
sub(-magnetization), 112

mass operator, 375
Matsubara frequency, 493, 495
Maxwell relations, 82
Maxwell rule, 77
mean field (MF), 7
mean value, 187, 315
metamagnet, 115
metastability, 46
  branch, 75, 78
method
  of generating functional, 212
  saddle point, 194
mixture, 94
  binary, 95
mode
  Goldstone, 166, 344
  Goldstone (boson), 342
model
  n–component (vector), 226, 227
  (Berlin–Kac) spherical, 227, 422
  Abelian–Higgs, 531, 532, 550
  Blome–Capel, 227
  free field, 317
  Gaussian, 317, 319
  Heisenberg, 222
  Higgs, 532
  Ising, 214, 222
  Lawrence–Doniach, 536, 537
  non-Gaussian, 267
  quantum Ising, 502
  quantum XY, 502
  random T_c, 507
  random axis, 507
  random exchange, 506
  random field, 511
  rotator, 304
  sigma (σ–), 481
  time dependent, 547
  transverse XY, 223
  XXZ, 223
  XY, 223
momentum space, 118
multicritical
  phenomena, 102
  point, 102

Neél temperature, 112
nearest-neighbour (nn), 215
nematics, 121
normal mode
  vibrational, 116

operator
  composite, 213
  redundant, 459
  scaling, 456, 459
orbital momentum
  quantum number, 219
order
  local, 91
  macroscopic, 186
  short-range, 91
order parameter, 4, 67, 71, 103, 186, 205
  n–component, 258
  n-vector, 314
  equilibrium, 7, 53
  field, 13
  nonequilibrium, 186, 204
  secondary, 117
  space, 165, 262

paramagnet, 1
parameter
  bare, 374
  effective expansion, 391
  expensive, 27
  Ginzburg–Landau, 528
  intensive, 30
  irrelevant, 451, 459
  macroscopic, 33
  marginal, 451
  microscopic, 31
  redundant, 459
  relevant, 451
  renormalized, 383, 387, 406
  space, 426, 429
partition function, 192
  canonical, 187
penetration depth, 528
perturbation
  first order, 348

integral, 605
second order, 348
term, 347
phase, 1
disordered, 5
distorted, 116
ferromagnetic, 2, 106
gas, 5
high temperature, 70
higher-symmetry, 116
liquid, 5
low temperature, 70
Meissner, 119, 528
metastable, 46, 157
mixed, 119, 523
nematic, 121
ordered, 3, 165, 166
paramagnetic, 106
separation, 95
smectic, 121
stable, 46
superconducting, 1
transiton point, 7
vortex, 120, 528
phase transition
equilibrium, 7
fluctuation induced, 552
fluctuation-driven, 549
line, 69
nematic-to-smectic *A*, 531
of first order, 83
of second order, 2, 83
quantum, 504, 548
smeared, 508
structural, 88, 115
superconducting-to-superfluid, 531
vapour-liquid, 224
weakly first order, 540
phonon, 119
point
"lambda"–, (λ–), 118
bicritical, 524
critical-end, 101, 163
isolated critical, 153
multicritical, 160, 522
saddle (stationary), 192
stationary, 39, 275
tetracritical, 524
tricritical, 101, 155, 160, 521
Lifshitz, 525
triple, 86
polarization
electric, 116
sublattice, 116
potential
chemical, 30, 37
exchange, 218
Gibbs, 53, 59, 103, 131, 192
nonequilibrium, 204
grand canonical, 52
intermolecular, 250
Landau, 139
Lennard–Jones, 224
long-range, 620
nonequilibrium, 131, 204, 237
pairwise, 91
saddle point, 193, 200
short-range, 620
thermodynamic, 53, 314
pressure, 29
principle
of entropy increase, 25
of least action, 36
variational, 35
process
irreversible, 34
natural, 34
reversible, 34, 49
spontaneous, 36
unnatural, 35
propagator, 21
fluctuation, 22
particle, 349

quantum
Hall liquid, 531
quasiparticle, 4
quenched impurity, 549

radiation correction, 533
random
axis, 507

critical temperature, 507, 550
fixed point, 507
impurity system, 507
random field, 511
correlation, 511
long-range, 510
short-range, 510
recursion
approximate formula, 23
formula, 23
regularization, 620
dimensional, 547
relation
Fisher, 20
fundamental, 32, 53
Gibbs–Duhem, 39
hyperscaling, 466
Josephson, 20, 462
recursion, 404, 496
renormalization group, 427
recursive, 278
renormalization group, 407
Rushbrooke, 20, 463
scaling, 462
thermodynamic, 26
Widom, 20
renormalization, 382, 417
group, 23, 382, 404, 427
mass–, 382
transformation, 23
renormalization group
approach, 19
differential, 453
linearized, 431, 444
trajectory, 429, 447
representation
diagrammatic, 349
energy, 28
entropy, 27, 38
irreducible, 117
rescaling factor, 400, 425
reversibility, 34

saddle point, 194
scaling, 17
amplitude, 462, 463
correction, 19
equation of state, 19
factor, 20
hypothesis, 17, 123
law, 11, 19
phenomenological, 19
relation, 20
tricritical, 521
separability, 200
shift exponent, 178
singularity
logarithmic, 3, 10
nonanalytic, 3
power, 3
weak, 10
socio-physics, 8
solute, 94
solution, 94
solvent, 94
space
d–dimensional, 318
n–dimensional, 606, 609
complex vector, 609
momentum (k–), 279, 400
reciprocal (k–), 279, 286
vector, 609
specific heat, 43, 146, 189
capacity, 43
jump, 147
spherical component, 617
spherical coordinate, 615
spin
angular momentum, 219
localized, 214
model, 220
moment
effective, 240
quantum number, 215, 219
variable, 220
spinodal line, 76
stability
condition, 516
marginal, 48
matrix, 584
state
corresponding, 252, 253

distorted, 117
equilibrium, 25, 34, 36
ferrimagnetic, 115
homogeneous, 38
inhomogeneous, 38
metamagnetic, 115
metastable, 1, 2, 46
nonequilibrium, 34, 36
single-phase, 40, 67
spin-flop, 115
stable, 1, 2
superconducting, 4
unstable, 46, 47
subgroup, 116
sublattice, 111
superconductivity, 119
(quasi-) two-dimensional, 536
one-dimensional, 536
type I, 528
unconventional, 579
superconductor, 68, 119, 526
conventional, 526, 527
ferromagnetic, 579
heavy-electron, 120, 534
high–T_c, 534, 536
spin-triplet
ferromagnetic, 580
type II, 528
unconventional, 120, 533, 534, 551
ferromagnetic, 579
supercooling, 76
superfluid, 68
charged, 119
unconventional, 120
superfluidity, 118
superheating, 76
surface effect, 318
susceptibility, 105, 136, 169, 191, 210
isothermal, 11, 105
lognitudinal (parallel), 173
magnetic, 109
tensor, 515
thermodynamic, 3
transverse (perpendicular), 173
symmetry
Z_2, $Z_{(2J+1)}$, 235
argument, 7
continuous, 165, 236
discrete, 236
global
continuous, 342
index, 393, 484
operation, 116
orthorombic, 534
rotational, 5
spontaneous breaking, 236
up-down, 234, 235
symmetry breaking, 5, 165, 167
spontaneous, 165, 236, 532
transition, 315
system
d-dimensional, 197
binary, 95
compressible, 86
disordered, 218, 219
impure, 506
exciton, 502
fluid, 68
gas-liquid, 68
isolated, 34
nano-size, 186
separable, 200
solid, 68

tensor
strain, 87
thermal expansion
coefficient, 43
thermal wave length, 489, 496
trace, 187
transformation
differential, 453
Fourier, 218, 279, 321, 324, 620
Gaussian, 291
Hubbard–Stratonovich, 291, 502
integral, 205
Kadanoff block, 400
Legendre, 51, 52, 103, 527
Migdal–Kadanoff, 481
regularized, 620
renormalization-group, 411, 427
scale, 426

shift, 325, 606
symmetry, 5
transition
adiabatic, 104
continuous, 2, 85
displacive, 116
ferroelectric, 116
fluctuation-driven
first order, 551
fluctuation-induced, 539
gas-liquid, 5, 86, 250
gas-solid, 86
liquid-solid, 86
nematic-to-smectic A, 121, 541
order-disorder, 116
paramagnetic-to-ferromagnetic, 2, 5
point, 85
superfluid, 1
symmetry breaking, 165
symmetry conserving, 166
weakly first order, 154
weakly-first-order, 540

universality, 85
class, 252, 253, 486, 491

variable
block, 21
continuous, 204
discrete, 193
extensive, 4, 25
relevant, 447
vector
n–component, 606
director, 121
magnetization, 166
vector potential, 221, 526
fluctuation, 539
vortex line, 119

Wick theorem, 335

Zeeman energy, 215, 216, 235
Zeeman term, 216, 221, 235